Handheld Computing for Mobile Commerce: Applications, Concepts and Technologies

Wen-Chen Hu
University of North Dakota, USA

Yanjun Zuo
University of North Dakota, USA

Information Science REFERENCE
INFORMATION SCIENCE REFERENCE
Hershey · New York

Director of Editorial Content: Kristin Klinger
Director of Book Publications: Julia Mosemann
Acquisitions Editor: Mike Killian
Development Editor: Christine Bufton
Publishing Assistant: Kurt Smith
Typesetter: Deanna Zombro
Quality control: Jamie Snavely
Cover Design: Lisa Tosheff
Printed at: Yurchak Printing Inc.

Published in the United States of America by
Information Science Reference (an imprint of IGI Global)
701 E. Chocolate Avenue
Hershey PA 17033
Tel: 717-533-8845
Fax: 717-533-8661
E-mail: cust@igi-global.com
Web site: http://www.igi-global.com/reference

Library of Congress Cataloging-in-Publication Data

Handheld computing for mobile commerce : applications, concepts and technologies / Wen-Chen Hu and Yanjun Zuo, editors.
p. cm.
Includes bibliographical references and index.
Summary: "This book looks at theory, design, implementation, analysis, and application of handheld computing under four themes: handheld computing for mobile commerce, handheld computing research and technologies, wireless networks and handheld/mobile security, and handheld images and videos"--
Provided by publisher.
ISBN 978-1-61520-761-9 (hardcover) -- ISBN 978-1-61520-762-6 (ebook) 1. Mobile commerce. 2. Electronic commerce--Technological innovations. 3. Business enterprises--Computer networks. 4. Mobile computing. I. Hu, Wen Chen, 1960- II. Zuo, Yanjun, 1969-
HF5548.34.H364 2010
658.8'72--dc22

2009035778

British Cataloguing in Publication Data
A Cataloguing in Publication record for this book is available from the British Library.

Table of Contents

Detailed Table of Contents

It focuses on the improvement of rendering performance by reducing the impacts of these problems with UbiWave, an end-to-end framework to enable real time mobile access to high resolution graphics using wavelets. The framework tackles the issues including simplification, transmission, and resource efficient rendering of graphics content on mobile device based on wavelets by utilizing (i) a Perceptual Error Metric (PoI) for automatically computing the best resolution of graphics content for a given mobile display to eliminate guesswork and save resources, (ii) Unequal Error Protection (UEP) to improve the resilience to wireless errors, (iii) an Energy-efficient Adaptive Real-time Rendering (EARR) heuristic to balance energy consumption, rendering speed and image quality, and (iv) an energy-efficient streaming technique. The results facilitate a new class of mobile graphics application which can gracefully adapt the lowest acceptable rendering resolution to the wireless network conditions and the availability of resources and battery energy on mobile device adaptively.

The authors define a theoretical model for autonomic and altruistic computational entities, and they use it to build a framework for peer-to-peer service-oriented infrastructures, focusing on three key aspects: overlay scheme, dynamic service composition and self-configuration of peers. Based on this framework, JXTA-SOAP Mobile Edition is a software component that completes the Sun MicroSystem's JXTA platform, supporting peer-to-peer sharing of Web Services.

It is about programming mobile handheld devices with a scripting language called AmbientTalk. This language has been designed with the goal of easily prototyping applications that run on mobile devices interacting via a wireless network. Programming such applications traditionally involves interacting with low-level APIs in order to perform basic tasks like service discovery and communicating with remote services. The authors introduce the AmbientTalk scripting language, its implementation on top of the Java Micro edition platform (J2ME) and finally introduce Urbiflock, a pervasive social application for handheld devices developed entirely in AmbientTalk.

This chapter introduces the differences of interrupt handling in many different aspects to measure these differences effect on mobile applications performance and throughput. The major contributions to this chapter are first to introduce the interrupt handling mechanism in mobile system with through elaboration on the types of interrupt handling that a Mobile OS may use. Then a deep analysis for both interrupt handling mechanisms used by the Symbian and RT-Linux OS is presented. A comprehensive conclusion is explained about the major differences in all aspects between Symbian and RT Linux mobile OS.

The authors present two systems for mobile phone users in order to provide comfortable Web browsing experience. One system provide various presentation functions for Web browsing so that users can select appropriate one based on their browsing situations. The other system provides functions to navigate users within a Web page so that they can reach information of interest without getting lost in the page. This chapter introduces designs of these systems and introduces results of user experiments, through which the authors show that the browser can reduce users' burden on mobile Web by enabling to select appropriate presentation functions adapted to their situations and by navigating them on a large Web page with the entertaining interface.

Traditional Web pages are mainly designed for desktop or notebook computers. They usually do not suit the devices well because the pages, especially the large files, can not be properly, speedily displayed on the microbrowsers due to the limitations of mobile handheld devices: (i) small screen size, (ii) narrow network bandwidth, (iii) low memory capacity, and (iv) limited computing power and resources. Therefore, loading and visualizing large documents on handheld devices become an arduous task. Various methods are created for browsing the mobile Web efficiently and effectively. This chapter investigates some of the methods: (i) page segmentation, which is used to segment Web pages, (ii) component ranking, which is used to rank page components after segmentation, and (iii) other ad hoc methods, such as text summarization, transcoding, and Web usage mining. Though each method employs a different strategy, their goals are the same: conveying the meaning of Web pages by using minimum space. The major problem of the current methods is that it is not easy to find the clear-cut components in a Web page.

Section 3
Wireless Networks and Handheld/Mobile Security

Wireless networks are an essential component of a mobile-commerce system and handheld security is the must for the success of mobile commerce. This section including five chapters gives related issues of wireless networks, handheld security, and location-based services.

In this chapter the authors present how to achieve location privacy during LBS without a centralized and trusted middleware. First, they review the recent progress on location positioning technologies. Second, they investigate how to perform location cloaking without users exposing their accurate locations to a trusted third party. They decompose the problem into two subproblems: proximity minimum k-clustering and secure bounding. Third, the authors study how to perform nearest neighbor query with guaranteed privacy. A framework called 2PASS is proposed that allows the client to control what objects to request in order to minimize their number while not compromising location privacy of the user. The core component of 2PASS is a lightweight WAG-tree index from which the client can compute out the objects to request from the server.

It discusses survivability issues related to RFID systems. For mission-critical systems empowered by the RFID technology, any interruption of essential services, even for a short period of time, is not acceptable. Hence, survivability must be provided to ensure that the critical services can be continuously delivered, despite of malicious attacks and system failures. This chapter studies and survey survivability enhancing techniques in face of the special challenges that limited computational capacities, high mobility, and sensitive nature of RFID devices pose.

Mobile and handheld devices are becoming an integral part of people's work, life and entertainment. These lightweight pocket-sized devices offer great mobility, acceptable computation power and friendly user interfaces. As people are making business transactions and managing their online bank accounts via

It discusses mobile vision on movement. In the early days of photography, camera movement is a nuisance that can blur a picture. Once movement becomes measurable by micro-mechanical means, the effects can be compensated by optical, mechanical or digital technology to enhance picture quality. Alternatively movement can be quantified by processing image streams. This opens up for new functionality upon convergence of the camera and the mobile phone, for instance by 'actively extending the hand' for remote control and interactive signage.

This chapter provides a detailed overview of DVC by explaining the underlying principles and results from information theory and introduces a number of application scenarios. It also discusses the most important practical architectures that are currently available. One of these architectures is analyzed step-by-step to provide further details of the functional building blocks, including an analysis of the coding performance compared to traditional coding schemes. Next to this, it is demonstrated that the computational complexity in a video coding scheme can be shifted dynamically from the encoder to the decoder and vice versa by combining conventional and distributed video coding techniques. Lastly, this chapter discusses some currently important research topics of which it is expected that they can further enhance the performance of DVC, i.e., side information generation, virtual channel noise estimation, and new coding modes.

An up-to-date critical survey of fast mode decision techniques for the H.264/AVC standard is provided in this chapter. The motivation for this chapter is twofold: Firstly to provide an up-to-data review of the existing techniques and secondly to offer some insights into the studies of fast mode decision techniques.

Foreword

Mobile handheld devices such as smartphones have become extremely popular and are now an integral part of our daily activities. People carry them everywhere and expect to be able to access a wide range of handheld applications whenever they wish. A major part of the applications is related to mobile commerce, which is defined as the exchange or buying and selling of commodities, services, or information on the Internet through the use of mobile handheld devices. Mobile commerce includes various mobile applications such as location-based services, mobile advertisements, mobile entertainments, mobile inventory and tracking, mobile payments and banking, just to name a few. For about a decade, mobile commerce has become the hottest new trend in business transactions.

- The future of mobile commerce is bright, as shown by the following predictions:
    - Even with the economic downturn in 2008, the smartphone sales were still strong. In the fourth quarter of 2008, worldwide sales of smartphones reached 38.1 million units, an increase of 3.7 percent compared to the fourth quarter of 2007 (Megna, 2009).
    - The sales of mobile content and services will reach to $150 billion by 2011 according to FierceMarkets, Inc. (2007). Among them:
    - SMS (short message service) and related messaging applications will generate $93 billion globally, accounting for more than half of projected mobile data revenues, multimedia services including music, video games, TV and adult content will reach to about $38 billion, and user-generated content such as social networking service will grow to a $13 billion market.
- Informa Telecoms & Media (Mobile Marketing Magazine, 2009) has the following forecasts:
    - In 2013, almost 300 billion transactions, worth more than US $860 billion, will be conducted using a smartphone. It is a twelve-fold increase in gross global transaction values in just five years.
    - By 2013, over 445 million mobile subscribers will use their smartphones to purchase physical goods and services regularly.
    - By 2013, there will be 977 million users of mobile banking services worldwide, a dramatic increase from approximately 67 million at the end of 2008.
- 204 million mobile users will adopt mobile payments, which generate almost $22 billion of transactions, by 2011 according to Glenbrook Partners, LLC (2008).

Although people perform mobile-commerce transactions all the time, most mobile users have no idea how they work because mobile applications involve such a wide variety of disciplines and technologies and new technologies are being created every day. For example, the handheld technologies include energy saving, handheld data management, handheld HCI (human computer interface), handheld peripherals, mobile operating systems, Web content adaptation, and wireless networks. Researchers working on innovative mobile-commerce applications must therefore be familiar with new ideas and concepts from many fields. For example, many of the popular mobile applications offered by the iPhone App Store are location-based and involve activities such as finding the nearest gas station or a specific type of ethnic restaurant. This kind of application does not rely solely on traditional computing approaches but also requires the use of handheld computing techniques such as GPS (global positioning system) tracking and map services.

To my surprise and knowledge, there is no journal or magazine dedicated to smartphone research currently. (The inaugural issue of International Journal of Handheld Computing Research, edited by the one of the editors of this book, will be published in the beginning of 2010—from the book editors.) Two magazines, *Handheld Computing* and *Smartphone & Pocket PC*, are out of print now because of lack of subscriptions. By the way, these two magazines were not really related to handheld research. Introduction of smartphones and PDAs and their applications is the magazines' major mission. Some smartphone books are available in the bookstores now, but most of them are related to specific devices such as iPhone or BlackBerry and they are application/development-oriented instead of research-oriented. With the extreme popularity of cell phones and smartphones, I believe there is a knowledge gap of handheld computing for mobile commerce needed to be filled. The book *Handheld Computing for Mobile Commerce: Applications, Concepts and Technologies* is a long awaited book for readers interested in handheld computing and mobile commerce. It covers a broad range of handheld topics for mobile commerce, both in depth and breadth. It is a must-read book for IT personnel and students who want to keep up with the fast-evolving IT.

Wenchang Fang, Professor and Dean
College of Business
National Taipei University
Taipei, Taiwan

Wenchang Fang *received his PhD from the Northwestern University, USA in 1994. He is currently a professor and the dean of the College of Business at the National Taipei University, Taiwan. He is the Editor-in-Chief of two journals: Electronic Commerce Studies and Contemporary Management Research. His current research interests include inventory management, electronic commerce, information management, and artificial intelligence.*

REFERENCES

Fierce Markets, Inc. (2007). *Forecast: Mobile Content and Services $150B by 2011*. Retrieved March 14, 2009, from http://www.fiercemobilecontent.com/story/forecast-mobile-content-and-services-150b-by-2011/2007-02-02

Glenbrook Partners, LLC. (2008). *Forecast: $22 Billion in Mobile Payments by 2011.* Retrieved July 21, 2009, from http://www.paymentsnews.com/2008/01/forecast-22-bil.html

Megna, M. (2009). *Smartphone Sales: 2009 Forecast Calls for Pain.* Retrieved May 02, 2009, from http://www.internetnews.com/stats/article.php/3810441/Smartphone+Sales+2009+Forecast+Calls+for+Pain.htm

Mobile Marketing Magazine. (2009). *Informa Bullish about Mobile Banking.* Retrieved June 17, 2009, from http://www.mobilemarketingmagazine.co.uk/2009/02/informa-bullish-about-mobile-banking.html

Preface

This book, *Handheld Computing for Mobile Commerce: Applications, Concepts and Technologies* collects high-quality research papers and industrial and practice articles in the areas of handheld computing for mobile commerce from academics and industrialists. It includes research and development results of lasting significance in the theory, design, implementation, analysis, and application of handheld computing. Twenty-two excellent articles from 71 world-renowned scholars and IT professionals are included in this book, which covers four themes: (i) handheld computing for mobile commerce, (ii) handheld computing research and technologies, (iii) wireless networks and handheld/mobile security, and (iv) handheld images and videos.

INTRODUCTION

With the advent of the World Wide Web, electronic commerce has revolutionized traditional commerce, boosting sales and facilitating exchanges of merchandise and information. The emergence of wireless and mobile networks has made possible the introduction of electronic commerce to a new application and research area: mobile commerce. In just a few years, mobile commerce has emerged from nowhere to become the hottest new trend in business transactions. The success of mobile commerce relies on the widespread adoption by consumers of more advanced handheld devices such as smartphones, which include some data-processing capability and thus permit vital activities such as mobile Internet browsing and location-based services. Table 1 gives the numbers of units of mobile phones, PCs and servers, and handheld devices shipped in the years from 2002 to 2008 based on reports from market researchers (BNET, 2004; Canalys, 2007; CNET, 2003, 2006a, & 2006b; Gartner, 2005a, 2005b, 2005c, 2006, 2007, 2008a, 2008b, & 2009; GsmServer, 2004; IDC, 2008). The table reveals that smartphones enjoyed the highest rate of increase compared to the sales of mobile phones and PCs and servers and that by 2008 the number of PDAs sold had dwindled to almost nothing. It is expected that smartphones will overtake the number of PCs shipped in the very near future. Handheld computing research is thus becoming a critical area as mobile users ask for more and more functions from their smartphones.

Mobile commerce prevails and mobile phones have become ubiquitous in today's society. However, mobile users are no longer satisfied with simple phones, but instead expect ever more powerful functions to be available from their mobile devices. Advanced phones, known as smartphones, allow mobile users to perform a wide variety of advanced handheld functions such as browsing the mobile Internet or finding a nearby theater showing a specific movie. The design and development of these new, improved handheld functions require the help of handheld computing research. A timely book covering handheld computing and mobile commerce is therefore needed.

Table 1. Mobile phones, PCs and servers, and handheld devices shipped from 2002 to 2008

	Mobile Phones	PCs and Servers	Smartphones	PDAs (without phone capabilities)
Number of Units Shipped in 2002 (Million)	432	148	—	12.1
Number of Units Shipped in 2003 (Million)	520	169	—	11.5
Number of Units Shipped in 2004 (Million)	713	189	—	12.5
Number of Units Shipped in 2005 (Million)	991	209	—	14.9
Number of Units Shipped in 2006 (Million)	991	239	64	17.7
Number of Units Shipped in 2007 (Million)	1153	271	122	—
Number of Units Shipped in 2008 (Million)	1220	302	139	—

AIM OF THE BOOK AND TARGET AUDIENCE

Mobile commerce is a trend of electronic commerce. Mobile handheld devices and computing are used to realize and assist mobile commerce. The handheld industry has applied handheld computing for many years. However, handheld devices and computing are diverse and there does not exist a formal approach to mobile commerce implementation. Our book is one of the first few books which systematically covers mobile handheld devices and computing and provides various approaches to mobile commerce implementation. It will help IT students, researchers, and professionals to better understand handheld devices and concepts and therefore produce more useful, effective handheld applications and products. Various handheld topics are covered in this book. Some of them are:

- Client-side mobile-commerce computing, applications, and programming
- Context/location-based services, computing, and applications
- Energy saving for handheld devices
- Handheld devices, architecture, and systems
- Handheld specifications, standards, guidelines, software, and tools
- Java ME systems, computing, applications, and programming
- Mobile advertising and sales
- Mobile and wireless networks
- Mobile commerce applications and systems
- Mobile Web 2.0 and plus
- Mobile Web and Internet
- Mobile/handheld algorithms and methodologies
- Mobile/handheld human computer interface and user interface design and implementation
- Mobile/handheld images and videos
- Mobile/handheld operating systems and platforms
- Mobile/handheld programming languages and environments
- Mobile/handheld security
- Web content adaptation for handheld devices

The target audience of this book will be composed of students, IT professionals, and researchers working in the fields of handheld computing and mobile commerce. It especially benefits the IT personnel of corporations because companies are gradually setting up the mobile versions of their electronic commerce systems. This book will help IT workers smoothly build mobile commerce systems based on their traditional IT knowledge. It could be used for a textbook of an advanced computer science (or related disciplines) course and could be a reference book for IT professionals and students. Since this book covers the handheld computing for mobile commerce systematically, it is also for people desiring to learn the topics on their own. The benefits of this book include:

- Fill the gap of lack of handheld-computing books.
- Help IT students and professionals master the handheld technology.
- Provide a textbook for a course of handheld computing, mobile commerce, or mobile computing.
- Can be used as a reference book for IT workers and students.

ORGANIZATION OF THE BOOK

Mobile commerce and handheld computing include such a wide variety of subjects and technologies that it is almost impossible for a single book to adequately cover all the subjects involved. This book therefore focuses on introducing the major topics concerning mobile commerce and handheld computing and provides extensive references for readers interested in discovering more information. It is divided into the following four sections, with a total of twenty-two chapters:

- *Handheld computing for mobile commerce*, which discusses how handheld computing supports mobile commerce,
- *Handheld computing research and technologies*, which covers major handheld technologies, methodologies, algorithms, and programming,
- *Wireless networks and handheld/mobile security*, which gives related issues of wireless networks and handheld security, and
- *Handheld images and videos*, which covers images and videos used by mobile commerce.

Section 1: Handheld Computing for Mobile Commerce

Handheld computing is the use of handheld devices like smart cellular phones to perform wireless, mobile, handheld operations such as browsing the mobile Web and finding the nearest gas stations. Mobile commerce is the most important application of handheld computing. This section discusses some handheld-computing methods for mobile commerce.

- *Chapter 1. A User Context-Aware Advertising Framework for the Mobile Web*, which elaborates over context-aware advertising on mobile web, discusses the benefits and challenges of adapting user contexts to the mobile advertising process, and classifies user contexts into three categories according to their characteristics and usage. The authors present a novel user context-aware advertising framework for mobile web that integrates the user contexts into the process of generating, selecting, matching, and presenting advertisements customized to mobile web pages.
- *Chapter 2. Plugging into the Online Database and Playing Secure mobile Commerce*, which discusses cloud computing, which is capable of appearing ubiquitously with mobile devices and intends to outstretch its various applications by the devices. The next generation of mobile devices will use wireless broadband access and human-computer interaction technologies which support cloud services and interface designs respectively advances to allow remote plug-and-play with web 2.0

applications that is suitable for mobile commerce in which this chapter emphasizes. Besides, for sustainable development of a mobile commerce solution, workable but not securable is absolutely not enough. Therefore, a secure information retrieval and reveal protocol for mobile commerce based on modified RSA digital signature is also proposed and demonstrated.

- *Chapter 3. Quality Evaluation of B2C M-Commerce Using the ISO9126 Quality Standard*, in which a new method has been introduced which measures the value of relevance for each m-commerce system attribute. The theoretical framework for this metric is also presented. The validity of the presented measures should further examine with different user groups in alternative evaluation cases and it is included in future work. It should be mentioned that the values presented are not strictly defined as numerical results but present the correlation among m-commerce systems attributes and external quality characteristics.
- *Chapter 4. A Picture and a Thousand Words: Visual Scaffolding for Mobile Communication in the Developing World*, which introduces Picture Talk, a software application that the authors designed for use in environments with low literacy, limited Internet connectivity, and little familiarity with information services. Because basic mobile phones are the most common devices used by BoP populations, the authors have implemented Picture Talk on mobile phones. The authors are now investigating ways of providing access to some Picture Talk features on less expensive mobile phones using voice and text messaging. The limitations of using these devices to access rich structured content by users with limited literacy skills exposes human-computer interaction challenges that are keys to enabling broad access to information by people in BoP populations.
- *Chapter 5. Web Applications on the Move: Opening Up New Opportunities for Mobile Developers*, which shows that there are a number of activities on the way to extend the Mobile Web platform towards a "hybrid" platform, which can compete with platforms for locally installed "fat" applications. The authors present a prototype of a hybrid platform, the FOKUS Mobile Widget Runtime and sample applications to demonstrate how these future hybrid applications may look like.
- *Chapter 6. A J2ME Mobile Application for Normal and Abnormal ECG Rhythm Analysis*, which presents a novel, but low cost and relatively equitable ECG signal analysis and alert system for telecardiology. This system fully harnesses the computational power of a plain mobile phone to perform real-time data mining tasks. The evaluation results not only prove it is a feasible approach but also show its potential for future practical applications.
- *Chapter 7. Factors Facing Mobile Commerce Deployment in United Kingdom*, which discusses the challenges facing mobile commerce deployment in United Kingdom. Although the number of mobile phone users is increasing and the technology is available for successful implementation of m-commerce, only a small number of users utilize m-commerce services. At the same time, mobile phones are becoming smarter, and the most of latest phones are capable of connecting to the Internet. The chapter looks at the background of m-commerce as well as the technological development of mobile phone to the current stage. Also, technical and non technical issues which hinder the adoption of m-commerce are discussed and solutions and recommendations given.

Section 2: Handheld Computing Research and Technologies

Handheld computing involves different disciplines such as wireless networks and mobile platforms and various technologies like Java and C/C++ handheld programmig. This section gives some of the major handheld technologies including energy saving, mobile platforms, handheld programming, and Web content adaptation.

- *Chapter 8. UbiWave: An Novel Energy-Efficient End-to-End Solution for Mobile 3D Graphics*, which presents UbiWave, an end-to-end framework using wavelets to transmit and render graphics content at various resolutions on mobile devices. Ubiwave improves the performance of mobile graphics applications by balancing energy consumption, rendering speed and image quality. Ubiwave includes four parts: (i) a perceptual error metric to guide the scaling of mobile graphics scenes to the lowest LoD at which users do not perceive distortion due to simplification (called the PoI); (ii) a novel Forward Error Correction (FEC) scheme based on the principles of Unequal Error Protection (UEP); (iii) an Energy-efficient Adaptive Real-time Rendering (EARR) heuristic to balance energy consumption, rendering speed and image quality and (iv) an energy-efficient 3D streaming technique. By combining PoI, UEP, EARR and our streaming technique, the rendering speed and image quality of mobile graphics applications in wireless networks can be maximized, while minimizing energy consumption.
- *Chapter 9. Peer-to-Peer Service Sharing on Mobile Platforms*, which introduces the *Networked Service-oriented Autonomic Machine (NSAM)*, which is a theoretical model of a hardware/software entity that is programmed to be altruistic in sharing its resources. The focus is on NSAMs whose hardware resources can be classified as mobile devices, offering and consuming services. In this context, the author present a framework for peer-to-peer service sharing, based on three key aspects: overlay scheme, dynamic service composition and self-configuration of peers. This framework is suitable to characterize many existing platforms and to define new ones.
- *Chapter 10. Scripting Mobile Devices with AmbientTalk*, which describes AmbientTalk, a distributed object-oriented scripting language specifically designed to deal with the hardware characteristics inherent to mobile ad hoc networks. What makes AmbienTalk a suitable scripting language for the implementation of mobile computing applications are its event-driven application model, its automatic buffering of messages to deal with intermittent connectivity and its built-in peer-to-peer service discovery abstractions to discover nearby applications.
- *Chapter 11. Interrupt Handling in Symbian and Linux Mobile Operating Systems*, which introduces a survey on differences among interrupts in the Linux and Symbian Mobile operating systems; we concluded that both interrupt mechanisms are similar in some ways and different in another, especially in organizational. In Symbian OS the pending interrupts are handled in a FIFO order but in the RT-Linux they are handled in a prioritized order.
- *Chapter 12. Web Page Adaptation and Presentation for Mobile Phones*, which presents two systems for mobile phone users in order to provide comfortable Web browsing experience. One system provide various presentation functions for Web browsing so that users can select appropriate one based on their browsing situations. The other system provides functions to navigate users within a Web page so that they can reach information of interest without getting lost in the page. This chapter introduces designs of these systems and introduces results of user experiments, through which the authors show that the browser can reduce users' burden on mobile Web by enabling to select appropriate presentation functions adapted to their situations and by navigating them on a large Web page with the entertaining interface.
- *Chapter 13. Technologies and Systems for Web Content Adaptation*, which investigates some of the Web content adaptation methods: (i) *page segmentation*, which is used to segment Web pages, (ii) *component ranking*, which is used to rank page components after segmentation, and (iii) *other ad hoc methods*, such as text summarization, transcoding, and Web usage mining. Though each method employs a different strategy, their goals are the same: *conveying the meaning of Web pages by using minimum space.* The major problem of the current methods is that it is not easy to find the clear-cut components in a Web page. Other related issues such as mobile handheld devices and microbrowsers will also be discussed in this chapter.

Section 3: Wireless Networks and Handheld/Mobile Security

Wireless networks are an essential component of a mobile-commerce system and handheld security is mandatory for the success of mobile commerc. Related issues of LBS privacy, RFID system survivability, mobile Internet connectivity, handheld security, and wireless networks are discussed in this section.

- *Chapter 14. Positioning and Privacy in Location-Based Services*, in which the authors present how to achieve location privacy during LBS without a centralized and trusted middleware. First, they review the recent progress on location positioning technologies. Second, they investigate how to perform location cloaking without users exposing their accurate locations to a trusted third party. They decompose the problem into two sub-problems: proximity minimum k-clustering and secure bounding. Third, the authors study how to perform nearest neighbor query with guaranteed privacy. A framework called 2PASS is proposed that allows the client to control what objects to request in order to minimize their number while not compromising location privacy of the user. The core component of 2PASS is a lightweight WAG-tree index from which the client can compute out the objects to request from the server.
- *Chapter 15. Survivability in RFID Systems*, which discusses survivability enhancing techniques for RFID systems. Survivability is a relatively new research area. RFID survivability requires innovative techniques to address the limitations of low-cost RFID tags, highly mobile devices, and challenging environment in which an RFID system operates. This chapter summaries the potential survivability enhancing techniques in the literature and provides references for researchers and system developers to develop technologies towards resilient, secure, and survivable RFID systems.
- *Chapter 16. Mobile and Handheld Security*, which discusses the security issues and possible solutions of mobile security in three layers: mobile hardware, mobile operating system and mobile applications. In order to provide high level security and privacy good for business and daily life, it is essential to strengthen security in all three layers. Robust and reliable security is built on hardware that is initially designed and then implemented with security in mind. Mobile operating systems are expected to have better capability designed and management, while mobile applications need to be standardized and built with reliable quality. Mobile users need to gradually realize the importance of security and privacy on mobile systems and start to learn to utilize secure applications and secure features in the mobile OS to protect their mobile devices.
- *Chapter 17. Design and Performance Evaluation of a Proactive Micro Mobility Protocol for Mobile Networks*, which introduces the Proactive Micro Mobility (PMM) Protocol for the optimization of network load. A novel approach is proposed to design and analyze IP micro-mobility protocols. The cellular Micro Mobility Protocol provides passive connectivity in an intra domain. The PMM Protocol optimizes miss-routed packet loss in Cellular IP under handoff conditions and during time delay. A comparison is made between the PMM Protocol and the Cellular IP showing that they offer equivalent performance in terms of higher bit rates and optimum value. A mathematical analysis shows that the PMM Protocol performs better than the Cellular IP at 1 MHz clock speed and 128 kbps down link bit rate. The simulation shows that a short route updating time is required in order to guarantee accuracy in mobile unit tracking. The optimal rate of packet loss in the PMM Protocol in a Cellular IP are analyzes route update time. The results show that no miss-routed packets are found during handoff.
- *Chapter 18. A Comparative Review of Handheld Devices Internet Connectivity Revenue Models to Support Mobile Learning*, which provides a survey of mobile broadband revenue models deployed by mobile network operators in the UK, USA and Canada. The survey of exiting revenue models

highlights the technology adoption trends for handheld devices by consumers and identifies the future impact of these trends on the network operators and content providers with respect to educational content. The chapter focuses on innovations in consumer propositions that can support the Mobile Learning phenomenon. The study reveals that the various operators aim to differentiate their consumer propositions by branding, technology devices, and flexible pricing structures. From the results of the study it is clear that the current continuous convergence of multimedia applications, information services, digital networks, and devices will likely lead to an increase in adoption of mobile learning systems in the UK, Canada and the USA especially as the price per bandwidth drops and new innovative connectivity options are deployed such as built in mobile broadband processor in laptops and consumer devices.

Section 4: Handheld Images and Videos

Images and videos play an important role of mobile commerc. This section discusses critical issues of delivering images and videos to mobile handheld devics. It includes four chapters on vision movement (Spaanenburg and Malki), video coding (Lambert, et al.), fast mode decision techniques (Lambert, el al.), and video streaming (Lee and Smith).

- *Chapter 19. Mobile Vision on Movement*, which discusses mobile vision on movement. In the early days of photography, camera movement is a nuisance that can blur a picture. Once movement becomes measurable by micro-mechanical means, the effects can be compensated by optical, mechanical or digital technology to enhance picture quality. Alternatively movement can be quantified by processing image streams. This opens up for new functionality upon convergence of the camera and the mobile phone, for instance by "actively extending the hand" for remote control and interactive signage.
- *Chapter 20. Distributed Video Coding for Video Communication on Mobile Devices and Sensors*, which addresses the concept of distributed video coding which is currently emerging as a new video coding paradigm allowing the construction of ultra-low complex video encoder at the expense of a more complex decoder. The theoretical foundations of DVC were discussed briefly after which an overview was given of existing DVC solutions and architectures. One of these architectures was used as reference for a more in-depth discussion of the functional building blocks of a DVC system. As computational complexity plays an important role in the context of DVC, the latter DVC system was extended with a number of coding modes allowing to dynamically shift the complexity between encoder and decoder, facilitating the requirements of emerging video communication applications. Finally, they provided an outlook to some future research directions for which it is believed that advances in these domains will contribute to the overall coding performance of DVC systems.
- *Chapter 21. Fast Mode Decision in H.264/AVC*, which provides an up-to-date critical survey of fast mode decision techniques for the H.264/AVC standard. The motivation for this chapter is twofold: Firstly to provide an up-to-data review of the existing techniques and secondly to offer some insights into the studies of fast mode decision techniques.
- *Chapter 22. Mobile Video Streaming*, which introduces essential technical components for constructing mobile video streaming systems. They include the latest development on broadband wireless technology and video-capable mobile handheld devices. As many modern technologies are often

driven by consumer demand, user experience and expectation are discussed from the perspective of mobile video streaming. At the end, several cutting-edge research and development breakthroughs are presented as they may change the future of mobile video streaming systems.

Wen-Chen Hu and Yanjun Zuo
August 15, 2009

REFERENCES

BNET. (2004). *Gartner Says Worldwide PDA Industry Suffers 5 Percent Shipment Decline in 2003—Top Stories*. Retrieved April 02, 2009, from http://findarticles.com/p/articles/mi_m0NZB/is_2_6/ai_113888610/

Canalys. (2007). *64 Million Smart Phones Shipped Worldwide in 2006*. Retrieved March 12, 2009, from http://www.canalys.com/pr/2007/r2007024.htm

CNET. (2003). *Gartner Ups Estimate for 2003 PC Shipments*. Retrieved May 12, 2009, from http://news.cnet.com/Gartner-ups-estimate-for-2003-PC-shipments/2100-1003_3-5104019.html

CNET. (2006a). *PC Market Surged in 2005, Will Settle in 2006*. Retrieved May 12, 2009, from http://news.cnet.com/PC-market-surged-in-2005%2C-will-settle-in-2006/2100-1003_3-6028454.html?tag=mncol

CNET. (2006b). *Mobile Phone Sales Pass 800 Million*. Retrieved May 12, 2009, from http://news.cnet.com/Mobile-phone-sales-pass-800-million/2100-1039_3-6037984.html

Gartner. (2005a). *Gartner Says Worldwide PDA Shipments Grew 7 Percent While Revenue Increased 17 Percent in 2004*. Retrieved January 12, 2009, from http://www.gartner.com/it/page.jsp?id=492106

Gartner. (2005b). *Gartner Says Strong Mobile Sales Lift Worldwide PC Shipments to 12 Percent Growth in 2004*. Retrieved February 09, 2009, from http://www.gartner.com/it/page.jsp?id=492098

Gartner. (2005c). *Gartner Says Mobile Phone Sales Will Exceed One Billion in 2009*. Retrieved February 09, 2009, from http://www.gartner.com/press_releases/asset_132473_11.html

Gartner. (2006). *Gartner Says Worldwide PDA Shipments Reach Record Level in 2005*. Retrieved January 30, 2009, from http://www.gartner.com/it/page.jsp?id=492242

Gartner. (2007). *Gartner Says Worldwide PDA Shipments Top 17.7 Million in 2006*. Retrieved March 19, 2009, from http://www.gartner.com/it/page.jsp?id=500898

Gartner. (2008a). *Gartner Says Worldwide PC Market Grew 13 Percent in 2007*. Retrieved March 09, 2009, from http://www.gartner.com/it/page.jsp?id=584210

Gartner. (2008b). *Gartner Says Worldwide Mobile Phone Sales Increased 16 Per Cent in 2007*. Retrieved March 25, 2009, from http://www.gartner.com/it/page.jsp?id=612207

Gartner. (2009). *Gartner Says Worldwide Smartphone Sales Reached Its Lowest Growth Rate with 3.7 Per Cent Increase in Fourth Quarter of 2008*. Retrieved March 18, 2009, from http://www.gartner.com/it/page.jsp?id=910112

GsmServer. (2004). *Mobile Phone Sales in 2003*. Retrieved January 12, 2009, from http://gsmserver.com/articles/sales2003.php

IDC. (2008). *Handheld Devices Sink 53.2% During Fourth Quarter But Protracted Decline Appears to Be Slowing, Says IDC*. Retrieved April 08, 2009, from http://www.idc.com/getdoc.jsp?containerId=prUS21083408

Acknowledgment

Cell phones became popular more than ten years ago, but the popularity of smartphones just started a few years ago. The editors believe a book of handheld computing for mobile commerce is needed. This book project took exactly one year to finish. From August 14, 2008 of responding to the publisher's request to August 15, 2009 of submitting the final book. It is a large and hard, but also enjoyable, memorable, and rewarding work. The editors spent a great deal of time of communicating with (potential) authors via numerous emails and organizing and managing this book. The successful accomplishment of this book is a credit to many people. It consists of 22 chapters of more than 200,000 words, which are contributed by a total of 71 authors. The editors thank authors for their quality work and great effort of revising their work based on the reviewers' comments. The reviewers who provided such helpful feedback and detailed comments are particularly appreciated. Special thanks go to the staff at IGI Global, especially to Christine Bufton, Mehdi Khosrow-Pour, and Jan Travers. Finally, the biggest thanks go to our family members for their love and support throughout this project.

Wen-Chen Hu and Yanjun Zuo

Section 1
Handheld Computing for Mobile Commerce

Chapter 1
A User Context-Aware Advertising Framework for the Mobile Web

Nan Jing
University of Southern California, USA

Yong Yao
University of Southern California, USA

Yanbo Ru
University of Southern California, USA

ABSTRACT

Context-aware advertising is one of the most critical components in the Internet ecosystem today because most WWW publisher's revenue highly depends on the relevance of the displayed advertisement to the context of the user interaction. Existing research works in context-aware advertising mainly focus on analyzing either the content of the web page (in which it is also called contextual advertising), or the keywords of the user search. However, we have identified the limitations of these works when being extended into mobile web, which has become a major platform for users to access Internet with thanks to the new lightweight web technologies and the development of mobile devices. These mobile devices are equipped with networking capabilities and sensors that provide versatile contexts including physical environment, user internal and social community. These contexts, which are far beyond just page content and search keywords, should be well organized and utilized for online advertising to gain better user experience and reaction. In this chapter, we point out the aforementioned limitations of the existing works in context-aware advertising when being applied for mobile platforms. We also discuss the characteristics of the contexts that are available on mobile devices and clearly describe the challenges of utilizing these contexts to optimize the advertisement on mobile platforms. We then present a context-aware advertising framework that collects and integrates the user contexts to select, generate, and present advertising content. The purpose of this framework is to provide the mobile users with targeted and purposeful advertisement. Finally, we discuss the implementation aspects and one specific application of this framework and outline our future plans.

DOI: 10.4018/978-1-61520-761-9.ch001

INTRODUCTION AND MOTIVATION

Online advertising constitutes a large portion in the financial ecosystem of web sites nowadays, including search engines, commercials, blogs, news, reviews etc. Driven by recent Internet revolution and the tremendous increases in online traffic, a huge growth in spending on online advertising is seen in last few years. eMarketer (2007) reports a total Internet advertising spending of nearly 20 billion US dollars just in 2007. This number supports the World Wide Web (WWW) to be amongst the top 3 advertisement medium, along with TV and print media. In these online advertisements, contextual advertising is a main category that we have identified in providing the advertising content matching the keywords of the user searches or the content of the web pages where the advertising content will be placed. The main players in this domain are major search engines and yellow pages on WWW. How to optimize the advertising content in this method is always an important research topic with the dual goals of increasing revenue of both publisher and advertising business.

An optimized context-aware advertising web should only provide ads that very match with the content of the Web pages, which therefore provides the users with information to their interests and allow advertisers to reach their potential customers in a non-intrusive way (Chartterjee & Hoffman & Novak, 2003, Wang & Zhang & Eredita 2002). In order to find the matching ads, two issues have to be carefully addressed: first is to identify and organize the applicable contexts in a user activity. Second, matching and ranking ads must be based on the identified and organized contexts.

Meanwhile, mobile computing technologies have profoundly transformed the way how people communicate and receive information from various media including WWW. With mobile devices becoming more powerful and affordable, the user base has expanded from the early business elites to ordinary people. By the end of 2007, there are about 3 billion cellular phone subscribers, which is more than twice the number of PC users worldwide. Furthermore, the cellular phone coverage is estimated to reach the 90% of the world's 6 billion population by the year of 2010. The statistics clearly indicates that mobile phones are already the most pervasive information technology platform. In this regard, mobile information access is gaining widespread prominence with improving connection speed and access technologies leading to richer content explosion and user experience. The addition of mobility has opened up new prospects as devices are expected to be with users at all time providing reliable information on user intentions and contexts. The next generation of mobile applications would be adaptive in that they leverage mobility with context awareness in order to provide more customized information and, at meantime, more targeted advertisement. Thus, it is becoming imperative that context awareness be seen as one critical norm in developing advertising framework on mobile platforms.

Recent mobile computing research is investigating how to collect and analyze contexts of user activities in mobile environment (Bardram, 2004, Couder & Kermarrec, 1999, Pascoe, 1998, Wennlund, 2003). Because of the lack of heterogeneous context structures amongst different applications in this domain, the existing research works, however, have not identified and organized sufficient context resources from mobile user activities. In addition, even provided a large amount of contextual information, the existing works we have identified still cannot utilize this information to match and select advertising content. Considering addressing these challenges in mobile platforms that has limited processing capacities, a new framework is needed to provide well designed and illustrated solutions to these challenges.

Therefore, this chapter has described a user context-aware and processing framework applicable on mobile platforms. This framework defines context structure suitable for users' activities in mobile environment. This framework also provides approaches to select advertising content that matches with identified and organized contexts in the context structure. This chapter also presents the architecture design and application examples of a prototype system, called Skyhelper, which is implemented using the framework and the approach developed in this work.

Context Awareness for Mobile Web

Definition of Context and Context Awareness

In general context means situational information. One of its popular definitions (Dey, 2001) is "any information that can be used to characterize the situation of an entity. An entity is a person, place or object that is considered relevant to the interaction between a user and an application, including the user and applications themselves". In the studies we have reviewed relevant to using context information, there are mainly two ways in which context is used in software applications. First, applications can optimize their outputs according to the contexts. Major search engines using the keywords and web page content to provide more targeted advertising content fall in this category. Second, the context information can be used to create new types of applications, such as location based applications. In these studies, context is often separated into physical context representing the environment of the activity and logical context representing more abstract information about the stakeholder and the application. Physical context properties are at a very low level of abstraction and are continuously updated to take into account the fact that the state of the stakeholder and the application continuously changes, such as spatial and temporal information. Logical context information is needed to enrich the semantics of physical context information (e.g., stakeholder's preferences) thus making it meaningful for high-level purposes (e.g., stakeholder's visits to certain locations) (Kappel et al., 2002). Theoretically any information available in the course of an interaction can be used as context information, such as time of the interaction, user identity, application status. In our research, the focus is the context information that is useful and critical to determine the context-aware advertising content on mobile web.

In fact, context awareness is not a new topic. It has been pioneered by Mark Weiser around fifteen years ago who then focused on the context-aware computing area under the vision of ubiquitous computing (a.k.a. pervasive computing or ambient intelligence). Ubiquitous computing is a method devised to make distributed computing available by multiple computers throughout the physical environment and make them transparent to the stakeholders (Weiser 1991, 1994). Context awareness as a scientific term was first introduced by Schilit (Schilit & Adams & Want, 1994, Schilit & Theimer, 1994) in ubiquitous computing. In his research, context is divided into three categories: computing context, user context, and physical context. By these categories, Schmidt (1999) further defined context as knowledge of the user's and IT device's state, including surroundings, situation and locations. Other researchers have thoughts of dividing context into different categories. Besides what we have discussed earlier about physical and logical contexts, Prekop and Burnett (2003) proposed the external and internal context, where the external refers to context that can be measured by hardware, i.e., location, light, or air pressure, whereas the internal is mostly specified by the users or captured by monitoring user interactions, i.e., the users' goals, tasks, and social interactions. In general, common points of context classification are generating the information on the conditions and surroundings of the users, monitoring the user activities continuously, and providing them with needed information in real time. In addition, a context-aware model should be designed for recognizing contexts containing users' needs accurately. Technologies required for context awareness and processing include context extraction, context construction, database management for persisting contexts, and information generation and selection based on relevant contexts (Pascoe, 1998).

We acknowledge that context has no standard definition, since every school of study can give their understanding about context to a valid purpose. However, in a particular area such as mobile web platform, the target of using context is to better serve users by providing needed information to these users on mobile devices. Classifying of context should embody mobile-user-centric essence and, particularly, in our research, it should be directly helpful for us to generate and select more targeted and purposeful advertising content for the users.

In our study, we also recognize that in mobile and ubiquitous computing, the notion of context is often equated simply with key words and contents in PC web or just location information in mobile web. Actually the mobile context is more complex than that. Mobile application usage can vary continuously because of changing circumstances and differing user needs. To fit into these circumstances and satisfying these needs, manufacturers and developers have built numerous devices, databases, and communities to model these circumstances and capture the needs in order to better serve the users. The information that they have modeled and captured, which is usually open to the public, is very helpful and cannot be ignored in generating and selecting context-aware advertising content.

Characteristics of Contexts in Our Study

Based on the understanding and recognition we have discussed in the previous section, the contexts in our research, in a high level view, should be divided into three categories: physical, user internal, and social. Physical context represents the environment of the user, such as time, locations, devices, etc. User internal context refers to the information that can be constructed by the user herself, such as user objectives, preferences, and activity history. Social context relates to the user's social communities, i.e. the information that can be co-constructed by the user's social connections. Imagine the preferences and activity history of the user's friends for this purpose. Table 1 gives more details of these three categories of contexts. With these contexts, various heuristics and rules can be defined for different situations and purposes. Examples are like such: location information helps the mobile web site provide the user with the ads of the exact business nearby. The time of a user request should match with business hours. In a warm day (temperature), cold drinks may better attract users than hot coffee, unless the user has an important business meeting in one hour (user calendar). The user who walks in a big mall may favor an advertising coupon from Macy's inside (nearby business) than a restaurant which is two miles way. On a multimedia-powered phone the user will be likely amazed at a deliberately-designed multi-media ads that may look annoying elsewhere, such as to a user who uses a basic phone and looks at the 'not available' warning.

Table 1. Mobile context categories

Context Category	Context Details
Physical	Location, time, temperature, weather, traffic, building, nearby business, etc.
User Internal	User profile, user calendar, user contacts, device, device resources (manufacturer, model, touchscreen, resolution, keyboard, portrait/landscape, memory, stylus, multimedia, Bluetooth, etc.), provided services, history of service uses, service recommendation, service failure, user's physical condition, etc.
Social	Nearby friends, friends' recommendations, friends' history of service uses, friends' locations, etc.

When a user cannot make up her mind about which restaurant, the ads from one with the recommendations of her friends will surely help her out. These example heuristics can go on and on, refined by observing and analyzing user practices, while undoubtedly well utilizing these contexts is critical in the success of providing more targeted and purposeful advertising content.

Summary

To better utilize contexts for mobile web, a context-aware software framework has to be designed to generate and select advertising content based on these contexts. To illustrate such as a framework developed in our work, the rest of this chapter structures as follows: Section 3 reviews a few school of study which are relevant to this work. The new context-aware advertising framework for mobile web is proposed in Section 4. Section 5 discusses the implementation aspects of this framework. Finally section 6 concludes this chapter and outlines the open issues that are to be addressed to extend and improve this framework.

Literature Review

Online Context-Aware Advertising

As an emerging research topic, online advertising has very few publications, even less for context-aware advertising. Wang et al in their work (Wang, Zhang, Choi & Eredita, 2002) stated that the advertising contents must be relevant to the user's interest to match with the user's experience and promote the chances of later interactions. Ribeiro-Neto et. al. (2005) worked on a ground-breaking report from the information retrieval perspective in which they examined a number of strategies to match pages to ads based on search keywords. More recently, the fast-growing popularity of sponsored search in online advertising, such as major search engines, has motivated more researchers from multiple disciplines, such as information retrieval, query optimization, and database management, to study various topics. Dean and Ghemawat (2004) presented their approach of extracting keywords from web pages to match with advertising contents. Andrei Broder et al. (2007) proposed a framework for matching ads using a large taxonomy including both semantic and syntactic feature. Ribeiro-Neto et al. (2005) tried to use additional pages using a Bayesian model to overcome the difference between the vocabularies of Web pages and ads. Yih et al (2006) presented an original approach for context-aware advertising in reducing it to the problem of sponsored search advertising by extracting phrases from the page and matching them with the bid phrase of the ads. They used various features to determine the importance of page phrases for advertising purposes. Another school of study tries to estimate the click through rate of ads using data analysis tools such as clustering analysis for keyword matching and classification (Regelson & Fain, 2006). In this work, the ads are clustered by their bid phrases. The click through rate is averaged over each cluster.

In a summary of all the reviewed works, they have provided valuable references and solid grounds for building the frameworks and approaches to match online context information with the advertising content. However, most of them have only associated online context with either the content of web pages or the keywords of user searches and therefore, even with solid-grounded matching approaches, their works cannot be extend to the context-aware advertising challenges on mobile platforms. A new framework is needed to utilize the contexts on mobile platforms to generate and select targeted and purposeful advertising content for mobile users.

Context Awareness

Recently researchers have paid long due attention to context acquisition and utilization in various mobile platforms. Khedr et al (2005) apply agent-

Figure 1. Context-aware advertising framework for the mobile web

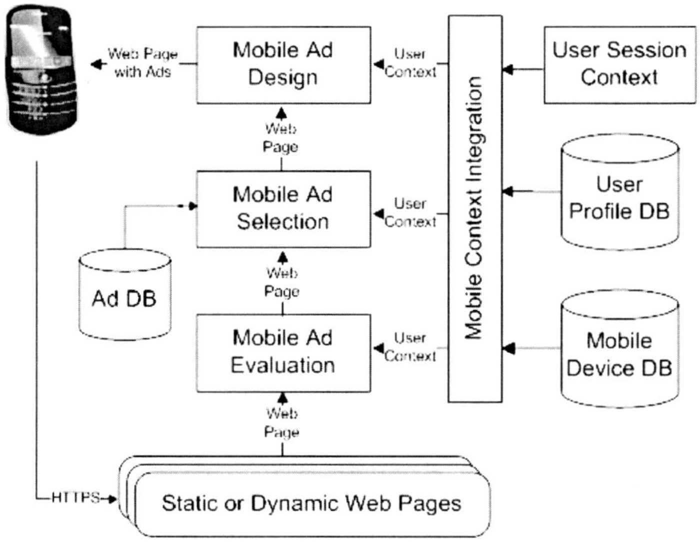

based approaches for building mobile context-aware platform using the network-level context. Biegel and Cahill (2004) described a framework of utilizing environmental observance for context aware application development in ubiquitous computing. Gu et al. (2004) described context models using ontology in mobile intelligent environments.

Major mobile organizations such as Open Mobile Alliance (OMA), W3C and IETF (Internet Engineering Task Force) have worked on standardization that has greatly influenced the research on mobile platforms. However, there is still lack of effective ways to utilize contexts for delivering more targeted and purposeful content. W3C's Cascading Style Sheet (Bos et. al., 2009) media queries determine a specific style sheet based on the type of media that is accessing the web page, such as PC, PDA, etc. Another standard, Synchronized Multimedia Integration Language (SMIL) (Bulterman et al., 2005) also supports checking the characteristics of the system whose dynamics are governed by the runtime mobile environment. The User Agent Profile (UAProf) (WAP User Agent Specification, 1999) by OMA is commonly used by mobile researchers and developers to identify device characteristics using a pre-defined vocabulary over RDF. WURFL (Passini & Trassati, 2009) is another popular resource description mechanism used by the mobile platform. One limitation of this mechanism is in that it needs the developers to constantly solicit information from client devices and update the database that holds all the resource information. More importantly, there is no well-established approach or procedure to apply the device resource information described by WURFL or such mechanism for optimizing the information provided to the users, not to mention effectively utilizing this information to generate and select advertising content on mobile web.

Context-Aware Advertising Framework

Figure 1 shows the high-level view of our context-aware advertising framework for the mobile web.

As shown in the figure, the user sends a mobile web request to the web server from the mobile device. After receiving the user's request, the web server first constructs a static or dynamic web page. The web page is not returned to the user immediately but passed to the *Mobile Ad Evaluation* component to check for potential advertising opportunity. The component takes the content and the context of the web page into account to decide whether it is appropriate to add advertisements to the page. The component also determines the type and the optimal number of advertisements to be added to the page. If it is a good practice to insert advertisements to the current page, then the web page, as well as the type and the number of potential advertisements, is passed to the next component, the *Mobile Ad Selection*, to select and rank advertisements from the database. The advertisements are selected according to relevancy, quality, and user context. Next, the *Mobile Ad Design* component chooses the format, resolution, page position, and other presentation details of the selected advertisements and inserts them into the web page. The web page is customized by the device context. Finally, the ad-extended web page is delivered to and displayed on the user's mobile device.

The mobile advertising framework has a dual goal of improving mobile advertising relevance without sacrificing the user's overall experience while browsing the mobile web. The resources taken by downloading and showing mobile advertisements must be carefully evaluated for mobile devices which have very limited resources compared to Desktop computers. The framework accomplishes this goal by considering the user context in each step. Explicitly, the three Mobile Ad components interact with the *Mobile Context Integration* component to acquire user context information to decide whether or not to enable advertisements, which advertisements to add, and how to present the advertisements, etc. The *Mobile Context Integration* component serves as a central point of the user context and provides a unique interface to access it.

In the following, we first describe the *Mobile Context Integration* component and then discuss the *Mobile Ad Evaluation*, the *Mobile Ad Selection*, and the *Mobile Ad Design* components in details.

User Context Integration

Section 2 describes the characteristics and classification of user context for mobile advertising. In fact, the *Mobile Context Integration* component is a repository that combines such contexts including physical contexts (such as the mobile device) and user contexts (such as the user profile, user session, and the content of the currently viewed web page). It is our future work to extend the framework to integrate social context. The mobile device context includes capabilities of the user's mobile device, which differs greatly from device to device. To acquire the mobile device context, the web server first extracts a signature from the request header, which is unique to the brand and the model of the mobile device. The server then uses the signature as a key to retrieve the complete device context, including screen resolution, supported input method, browser type, and other capabilities from the mobile device database.

The user profile context may include basic user information, such as address, email, and phone-number, and the user's behavior or preference, such as favorite restaurant types. Since mobile devices are very personal, the user profile context can be constructed directly from a user's inputs on web pages, or inferred indirectly by tracking the browsing history of requests initiated from the same device. Such information is stored in the user profile database. The server can detect the user identity by checking the login authority, cookie values, or a previously assigned special link to the user. Once the user identity is determined, the web server can retrieve the user profile context from the database.

For a new web request session, a user session context object is constructed to keep the context information of the current session. The session context object is updated and maintained continuously during the session. The session context is similar to the user profile context but more accurate and relevant to the current web page, and will be written back to the user profile database when the session is over. The session context may include the user's current location. For example, a user specifies his/her location before searching for nearby restaurants. The user's location is then added to the session context object reused by the following-up searching requests. In certain scenarios, it is even possible to infer environment and physical context of the user from the context of the web pages being browsed by the user in the current session.

Mobile Advertising Evaluation

The circumstances under which mobile webs are browsed are generally very different than those for Internet webs, and less comfortable. A user may be on his/her way to the airport to pick up some friend and trying to find out the arrival time of the flight. The amount of attention that the user can give to the mobile web also varies, as other elements in the environment may compete for the user's attention (Sidnal & Manvi, 2006). Mobile devices have limited screen size and capabilities. Thus, it is a general best practice to keep the size of a mobile web page small with a simple lay-out. Irrelevant advertisements can be intrusive to the user, so it is crucial to understand the context of the mobile user: why, where and when the user is accessing the mobile web, the content of the web page, and the mobile device capabilities, before adding advertisements to the web page. The user's web browsing and ad-click history also shows the user's attitude towards the advertisements on mobile web pages. Mobile web sites are usually structured to have multiple levels of navigational pages to balance between having too many links on a page and asking the user to follow too many links to reach what the user is looking for. The user may return to the same navigational page frequently while browsing the mobile web site. Advertisements can be added to the page only at the first time when the user opens the page. In another example, if the user is searching for information on the mobile device, each mobile web page usually shows a small number of results in order to reduce the page size and not forcing the user to scroll the page a lot. The component can decide to display advertisements only on the first result page.

The Mobile Advertising Evaluation component also settles down the type and the number of advertisements proper to the web page. The most popular online advertising types include sponsored search advertising and contextual advertising (Andrei & Marcus & Vanja & Lance, 2007). Depending on the content of the web page, either sponsored search advertisements or contextual advertisements is more appropriate. If the user is browsing a category of restaurants or searching for the closest gas station, a Sponsored Search advertisement is more relevant to the user context. Similarly, if the user is reading blogs on his mobile device, then a contextual advertisement selected according to the content of the blog is more likely to be relevant to the user.

The capabilities of mobile devices vary significantly. The new generation of mobile devices is usually equipped with a big touch screen, and thus can show more advertisements on the same page without interrupting the user. This is compared to old devices with a smaller screen, and the user can only scroll the page by repeatedly pressing navigational keys. In particular, it is acceptable to show several advertisements on a mobile phone with a screen size of 480*320, but it would take almost half of the screen displaying the same number of advertisements on an old phone with a resolution of only 160*128. The device context is a key factor to consider how many advertisements can be added to the web page.

Mobile Advertising Selection

Online advertisements are generally implemented as a quality-based bidding scheme. For instance,

Google and Yahoo! search marketing rank sponsored search advertisement by the bid price on matching key words plus a quality score evaluated by the advertisement's click-through-rate, keyword relevancy with landing page, and site quality (Bernard & Tracy, 2008). For mobile advertising, the matching process can be extended by considering the user context.

As described in section 2, the user's location is usually available and most relevant to mobile advertising. Research shows that in most economic transactions, the location of the buying and the selling parties are relevant (Sidnal & Manvi, 2006). For example, the advertisement of a nearby pizza restaurant is probably more attractive than a discount issued by a restaurant twenty miles away. User profile context can be explored to select the most relevant advertisements to the user by matching the user context to candidate advertisements. One example is that the server can use the user's favorite restaurant type from the user profile context to select restaurant advertisements.

Mobile Advertising Design

Mobile advertisements can be displayed in alternative formats on a mobile device as simple text links, colorful images, or animated images. The size of a mobile web page is much smaller compared to an Internet web page in order to reduce the download time and to fit the page to the small screen of mobile devices. An image advertisement is more eye-catching than a text link, but also takes longer to download and occupies a bigger part of the screen. An image/animated image advertisement can be more intrusive to some mobile users than a simple text link advertisement.

Mobile Web Banner Ad is a popular type of advertisements on mobile web pages, which composes a still or animated image and optional text Taglines. The aspect ratio and the size of the banner image need to be adjusted to the user's mobile device. If the users are unfamiliar with image banners on mobile web sites, many don't realize the image banners can be navigated to and clicked on, and a Text Tagline can be added to generate a higher click rates (Mobile Market Association, 2008). If the integrated User Context suggests that the user is familiar to image banners, then the Text Tagline can be removed to improve user browsing experience.

Exemplary Application and Case Study

In order to provide an exemplary application and conduct appropriate case studies to validate our approach, we have implemented a prototype system, namely Skyhelper, based on the framework and approaches described in previous sections. Skyhelper is a mobile web site that allows users to search for information about theatre locations, movie show time, gas price, restaurants and menus from their mobile devices. The web site returns search results with appropriate advertisements accurately selected based on the users' search criteria and contexts including their location and profiles. The site consists of three layers as shown in Figure 2.

Client: The client can be any browser-equipped mobile devices, from the out-of-date cell phones to the state-of-the-art high end PDAs such as Blackberry, iPhone and Google phone.

Web Server: We use a Tomcat web server as the container for search and advertising services. The web server consists of three modules: 1) Information Retrieval Module accepts HTTP requests from the client, searches from the database, constructs the results in terms of web pages, and sends these web pages to the Advertising Module for further processing. 2) Advertising Module accepts search results from the Information Retrieval Module, adds advertisements if applicable and returns the final web pages to the client. 3) Database Access Module works as the interface between the Information Retrieval and Advertising modules and the database server. The

Figure 2. Architecture of skyhelper prototype system

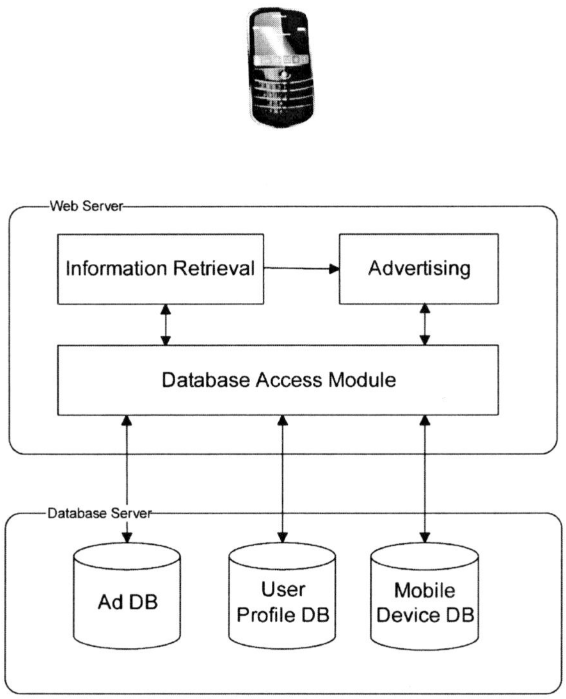

advantages of having a Database Access Module are to separate the functionality of the web server and the database server, and to balance the workload between these modules. So Information Retrieval and Advertising modules can focus on search and advertisement processing and do not need to worry about the implementation details of the database server.

Database Server: We use MySQL 5.0 to store user profiles, mobile device, and advertisement information. Current implementation hosts all databases on one server. In future, we suggest distributing the databases onto multiple servers running different DBMS systems to achieve better scalability and short response time.

The user context-aware advertising framework we have presented in this chapter focuses on supporting the design and implementation of the Advertising Module. In the rest of this section, we will discuss the details of this module.

Detecting the Capability of the User's Mobile Device

The capability of the user's mobile device differs greatly from device to device. While many high-end mobile devices have featured full function web browsers, browsing the Web on most mid- and low-end mobile devices has not become as convenient as expected. Mobile devices are quite restrictive on the format and length of the received content. There can be some information loss or malfunction if the web page is presented in some mode that the mobile device does not support. For example, Javascript, AJAX and Google map can provide an excellent use experience for the newest iPhone 3G users, but they may not work well on an out-of-date cell phone. To produce web pages that adapt to all kinds of mobile browsers, we must first detect the capability of the user's mobile device.

We built a Mobile Device Database which contains the information about the capabilities and features of more than ten thousand mobile devices. Capability information were collected from the WURFL project, which is an open source project that stores the information of many mobile devices and provides functionalities to use its information to identify a specific device (WURFL, 2009). Once the web server receives a request from a client, a signature is extracted from the request header, which is unique to the brand and the model of the mobile device. The server then uses the signature as a key to retrieve the complete device context, including screen resolution, supported input method, browser type, and other capabilities from the mobile device database. This device capability information will be used for better selection and presentation of the advertisements.

Advertisement Evaluation

The Advertisement Evaluation component decides if it is applicable to add some advertisement on a webpage. It also determines the number of ad-

Figure 3. Browsing restaurants on iPhone

Figure 4. Browsing restaurants on Nokia N70

vertisements to be added. In the example shown in Figure 3 and 4, we display tree advertisements on iPhone but only one advertisement on Nokia N70, by taking device context into account, since iPhone has a big screen of 480 * 320 pixels while the Nokia N70 has a much smaller screen resolution of 172 *208.

We implement the advertisement evaluation component as a C4.5 decision tree. The tree was built using Weka 3 data mining software (WEKA 2009). Weka contains a Java implementation of the C4.5 algorithm and a collection of visualization tools for data analysis and predictive modeling, together with graphical user interfaces for easy access to this functionality. We extended the functionality of Weka to output the decision tree as a Java class, which can be easily integrate into our Java based advertising module.

Advertisement Selection

Advertisements in the database are classified into categories. Each category is associated with fifteen keywords. User queries are classified to categories, by the textual similarity and semantic-based matching between the queries and the keywords associated with these categories. Only the advertisements belongs to the matched categories will be selected to display on the web pages.

Advertisements in the categories are ranked according to a score calculated using a set of heuristic rules. Top ranked advertisements are considered most relevant to user context and more interested to the user. To ensure freshness and diversity of the advertisement and create a better use experience, we keep an advertising log for each user session. No advertisement is allowed to be displayed on more than five pages in the whole session or on a sequence of more than three continuous pages.

Advertisement Presentation

Having the advertisements evaluated and selected, the final step is to present them in an appreciate format and layout. Typical mid- and low-end cell phones display less than twenty lines of text on the screen. High-end mobile services, such as Blackberry and iPhone, have bigger screens with higher resolution, but it is still aesthetically

Figure 5. Clicks of users who use context-aware advertising framework or not

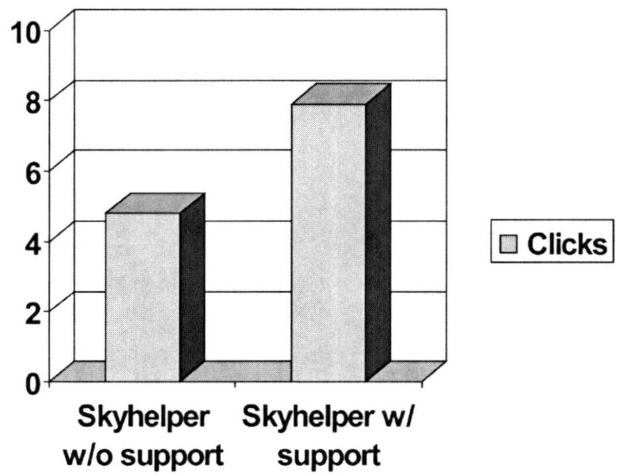

unpleasant to directly browse web pages originally designed for a desktop computer.

To adapt the advertisements to various mobile devices, we keep four different versions for each advertisement – a text string and three images in different sizes and resolutions. Some advertisements (about 15%) also have an animated banner version. Mobile devices are classified to low, medium and high levels based on their capabilities and features, based on which the appropriate version of the advertisement is chosen and displayed.

In addition to the device capability, user profile context are also used to adapt the advertisement presentation. For example, for a senior user, the font size of the text will be automatically enlarged for easy reading.

Case Study and Proof of Concept

In order to provide a proof of concept for our framework, we have evaluated the quality of the search results returned from the mobile site Skyhelper, which uses our user context-aware advertising system. Two versions of the site have been tested in our case studies, one with the support of our user context-aware advertising system and the other without. First, we configured a few emulators for the devices with different capabilities (screen resolution, Javascript support, and GPS enabled, etc) and various profiles from a small group of users we have gathered for this study. Second, we tested a set of queries provided by the users for common information such as show times and restaurants on both versions of the mobile site. Third, we provided the users with the results of both tests and let the users determine whether they will click the top (ten) links in both sets of the results which helps us make a comparison. After finishing these works, the comparison between the user interactions in both sets of the results gives us a clearer idea of the quality improvement in the search results with the support of our framework.

The user feedback data is shown in Fig. 5, where the average user clicks for the first set of the search results is compared with the average clicks for the second set. According to the data we have obtained, more user clicks (3 more clicks out of 10) have been observed for the second set of the search results, i.e. on Skyhelper with the support of our framework. In addition, most test users think that the results returned from the site built using our framework more suited to their need than the one without. And the user context-aware framework improves the mobile web search effectiveness and efficiency. Therefore, based on this preliminary analysis and with certain limitations caused by the nature of case studies (e.g. limited user profiles and case selections), it can still be clearly seen that this framework fulfills its objectives well.

CONCLUSION AND FUTURE WORK

In this chapter, we elaborate over context-aware advertising on mobile web. We discuss the benefits and challenges of adapting user contexts to the mobile advertising process, and classify user contexts into three categories according to their characteristics and usage. We present a novel user context-aware advertising framework for mobile web that integrates the user contexts into the process of generating, selecting, matching, and presenting advertisements customized to mobile web pages. We also show a prototype of the context-aware advertising framework as a part of the Skyhelper mobile online search application. In this work, we focus mainly on user internal contexts and some physical contexts that can already be acquired by the Skyhelper application. It is our next step to extend the framework to explore other types of user contexts, including social contexts and more types of physical contexts. We will also concentrate on mining context information and developing more intelligent context-aware advertisement matching and selection algorithms.

REFERENCES

Bardram, J. E. (2004, March 14 - 17). Applications of context-aware computing in hospital work: examples and design principles. In *Proceedings of the 2004 ACM symposium on Applied computing*, Nicosia, Cyprus.

Bernard, J. J., & Tracy, M. (2008). Sponsored search: an overview of the concept, history, and technology. *International Journal of Electronic Business*, *6*(2), 114–131. doi:10.1504/IJEB.2008.018068

Biegel, G., & Cahill, V. (2004). A framework for developing mobile, context-aware applications.In *Proc, Second IEEE Annual Conference on Pervasive Computing and Communications*, PERCOM, 2004

Bos, B., Celik, T., Hickson, I., & Håkon, W. L. (2009). Cascading Style Sheets (CSS 2.1). *W3C working note*. Retrieved, from http://www.w3.org/TR/CSS21/, 2009

Broder, A., Fontoura, M., Josifovski, V., & Riedel, L. (2007). A semantic approach to contextual advertising, In *SIGIR '07: Proceedings of the 30th annual international ACM SIGIR conference on Research and development in information retrieval* (pp. 559-566). New York: ACM.

Bulterman, D. (2005). Synchronized Multimedia Integration Language (SMIL2.1). *W3C recommendation*. Retrieved December 2005, fromhttp://www.w3.org/TR/2005/REC-SMIL2-20051213/

Chatterjee, P., Hoffman, D. L., & Novak, T. P. (2006). Modeling the clickstream: Implications for web-based advertising efforts . *Marketing Science*, *22*(4), 520–541. doi:10.1287/mksc.22.4.520.24906

Couder, P., & Kermarrec, A. M. (1999). *Improving Level of Service of Mobile User Using Context-Awareness*. Paper presented at the 18th IEEE Symposium on Reliable Distributed System, Lausanne, Switzerland.

Dean, J., & Ghemawat, S. (2004). Mapreduce: simplified data processing on large clusters. In *Sixth Symposium on Operating System Design and Implementation*, pages 137–150.

Dey, A. (2001, February). Understanding and using context. *Personal and Ubiquitous Computing*, *5*(1), 4–7. doi:10.1007/s007790170019

Fain, D. C., & Pedersen, J. O. (2006). Sponsored search: A brief history. *Bulletin of the American Society for Information Science and Technology*, *32*(2), 12–13. doi:10.1002/bult.1720320206

Gu, T., et al. (2004). An ontology-based context model in intelligent environments. In *Proc, Communication Networks and Distributed Systems Modeling and Simulation Conf., Soc, for modeling and simulation intl's, 2004.*

Kappel, G., Retschitzegger, W., Kimmerstorfer, E., Pröll, B., Schwinger, W., & Hofer, T. (2002, June 10). *Towards a Generic Customisation Model for Ubiquitous Web Applications.* 2nd International, Workshop on Web Oriented Software Technology, IWWOST'2002; Málaga, Spain.

Khedr, M., & Karmouch, A. (2005). ACAI: Agent-based contextaware infrastructure for spontaneous applications. *Journal of Network and Computer Applications*, 19–44. doi:10.1016/j.jnca.2004.04.002

Mobile Marketing Association. (2008). (n.d.). Mobile Advertising Guidelines. [fromhttp://mma-global.com/mobileadvertising.pdf]. *Retreived.*

Neto, B., Cristo, M., Golgher, P., & deMoura, E. (2005). Impedance coupling in content-targeted advertising. In*Proc. SIGIR, 2005.* eMarketer (2007). *eMarketer*Retrieved (n.d.)., from http://www.emarketer.com/Article.aspx?id=1004635.

Pascoe, J. (1998).Adding generic contextual capabilities to wearable computers. In *Proceedings of 2nd International Symposium on Wearable Computers*(pp. 92-99)

Passini, L., & Trassati, A. (2009). *Wireless Universal Resource File (WURFL)*. Retrieved 2009, from http://wurfl.sourceforge.net/

Prekop, P., & Burnett, M. (2003). Activities, context and ubiquitous computing. *Special Issue on Ubiquitous Computing Computer Communications*, *26*(11), 1168–1176.

Regelson, M., & Fain, D. (2006). Predicting click-through rate using keyword clusters. *In Proc. of the Second Workshop on Sponsored Search Auctions*, 2006.

Ribeiro-Neto, B., Cristo, M., Golgher, P. B., & de Moura, E. S. (2005). Impedance coupling in content-targeted advertising. In *SIGIR '05: Proc. of the 28th annual intl. ACM SIGIR conf.*,pages 496–503, New York: ACM.

Schilit, B., Adams, N., & Want, R. (1994, December). Context aware computing applications. In *Proceedings of IEEE Workshop on Mobile Computing Systems and Applications* (pp85-90). Santa Cruz, CA

Schilit, B., & Theimer, M. (1994). Disseminating Active Map Information to Mobile Hosts. *IEEE Network*, *8*(5), 22–32. doi:10.1109/65.313011

Schmidt, A., Aidoo, K. A., Takaluoma, A., Tuomela, U., Laerhoven, K. V., & de Velde, W. V. (1999, September). Advanced interaction in context. In *Proceedings of First International Symposium on Handheld and Ubiquitous Computing* (pp.89-101), Karlsruhe, Germany.

Sidnal, N. S., & Manvi, S. S. (2006). Context aware mobile commerce using agent technology. In*Ad Hoc and Ubiquitous Computing, 2006. ISAUHC '06. International Symposium,* (pp. 163-168).

User Agent Specification, W. A. P. (1999). Received (n.d.)., from http://www.wapforum.org/what/technical.htm, 1999

Wang, C., Zhang, P., Choi, R., & Eredita, M. (2002). Understanding consumers attitude toward advertising. In *Eighth Americas conf. on Information System* (pages 1143–1148)

Weiser, M. (1991, September). The computer for the 21st century. *Scientific American*, 94–104.

Weiser, M. (1993, July). Some computer science issues in ubiquitous computing. *Communications of the ACM, 36*(7), 75–84. doi:10.1145/159544.159617

Wennlund, A. (April 2003). *Context-aware Wearable Device for Reconfigurable Application Networks*, Department of Microelectronics and Information Technology(IMIT) WURFL (2008) Retrieved April, 2003, from http://wurfl.sourceforge.net/

Yih, W., Goodman, J., & Carvalho, V. R. (2006). Finding advertising keywords on web pages.In *WWW '06: Proc. of the 15th intl. conf. on World Wide Web* (pages 213–222), New York: ACM.

ADDITIONAL READING

Chen, G. L., & Kotz, D. (2000, November). *A survey of contextaware mobile computing research* (Technical Report TR2000-381). New Hampshire, USA: Dartmouth College Computer Science Department

Dey, A. K. (2001, February). Understanding and using context. *Personal and Ubiquitous Computing, 5*(1), 4–7. doi:10.1007/s007790170019

Engelmore, I. R., & Morgan, T. (1988). *Blackboard Systems*. Reading, MA: Addison-WesleyMitchell T. (1997), McGraw-Hill, 1997.

PearlJ. (1988). Probabilistic Reasoning in Intelligent Systems. San Mateo, CA: Morgan Kaufmann.

Prekop, P., & Burnett, M. (2003). Context and ubiquitous computing. *Special Issue on Ubiquitous Computing Computer Communications, 26*(11), 1168–1176.

Schilit, B., Adams, N., & Want, R. (1994, December). Contextaware computing applications. In *Proceedings of IEEE Workshop on Mobile Computing Systems and Applications*, Santa Cruz, CA, (pp85-90).

Schwinger, W., Grun, Ch., Proll, B., Retschitzegger, W., & Schauerhuber, A. (July 2005). *Context-Awareness in Mobile Tourism Guide – A Comprehensive Survey* (Technical Report). Johannes Kepler University Linz, Austria: IFS/TK.

Want, R. (1995, December). An Overview of the Parctab Ubiquitous Computing Eiroment. *IEEE Personal Communications, 2*(6), 28–43. doi:10.1109/98.475986

Weiser, M. (1993, July). Some computer science issues in ubiquitous computing. *Communications of the ACM, 36*(7), 75–84. doi:10.1145/159544.159617

Chapter 2

Plugging into the Online Database and Playing Secure Mobile Commerce

I-Horng Jeng
Chinese Culture University, Taiwan

ABSTRACT

Mobile commerce is one of emerging inter-discipline technology which integrates the network protocol, multimodal sensation, storage management, and other research areas. It intends to make paperless applications for both convenience and ecology on the mobile devices -- including those used for ticketing, coupons, loyalty rewards, payments, etc. By the innate limitations of the physical properties, mobile device -- particularly the handheld mobile device -- must make their best tradeoffs among the available hardware resources to reach their dedicated specifications. However, one of the recent progresses in the new technologies of the Internet, cloud computing, is capable of appearing ubiquitously with mobile devices and intends to outstretch its various applications by the devices. The next generation of mobile devices will use wireless broadband access and human-computer interaction technologies which support cloud services and interface designs respectively advances to allow remote plug-and-play with web 2.0 applications that is suitable for mobile commerce in which this chapter emphasizes. Besides, for sustainable development of a mobile commerce solution, workable but not securable is absolutely not enough. Therefore, a secure information retrieval and reveal protocol for mobile commerce based on modified RSA digital signature is also proposed and demonstrated.

INTRODUCTION

Mobile commerce is the ability to conduct e-commerce which consists of services over electronic systems such as the Internet or other networks by using mobile devices. It intends to make paperless applications for both convenience and ecology -- including those used for ticketing, coupons, loyalty rewards, payments, etc. -- through all kinds of mobile technologies and to make them pervasive and ubiquitous. By the innate limitations of the physical properties, mobile device -- particularly

DOI: 10.4018/978-1-61520-761-9.ch002

the handheld mobile device–must make their best tradeoffs among the available hardware resources to reach the specifications whatever they are general-purpose or special as mobile commerce. Not only the four main sections: the arithmetic and logic unit (ALU), the control unit, the memory, and the input and output devices must be included, but also demanded the overcritical uncluttered, minimalist interfaces to supply the consumer markets. The input and output devices embedded in the handheld device are essentially the typical **multimodal sensations** of haptic, auditory, and visual, such as keypad, microphone, camera for the input sensations, and display, speaker, battery vibration for the output sensations, respectively. Thus, at least four possible combinations of I/O pairs have been yielded (Jeng, Chang & Wang, 2008):

- Voice input, text/image output, which depends on speech recognition (a digital camera is an example of image output)
- Touch input, text/image output, which depends on handwriting recognition for textual data
- Voice input, sound output, which typically involves a digital recorder
- Touch input, sound output, which typically involves an MP3 or other music file player

The sensations described above applying perceptions of human are expected as touch or near-field ranges of sensations, however, the remote sensations (over a distance of 10 meters) such as ZigBee, Bluetooth, and WiFi playing an important role in the sensor networking for the two-way background information exchange, as shown in Fig. 1, benefit the cloud services via **wireless broadband access** (WBA). In other words, the near-field sensing of human-computer **interaction** (HCI) technologies use lots of short-distance sensors to deal with human-computer interactions, and the far-field sensing used by WBA technology maybe applies a few long-distance sensors to accelerate data communications in the viewpoint of multimodal sensation. One of the recent progresses in the new technologies of the Internet is capable to appear ubiquitously with mobile devices and intends to outstretch its various applications by the devices, which is called cloud computing (or cloud services) and explained accurately by a quote from (Hewitt, 2008, p. 96): "In the **cloud computing** paradigm, information is permanently stored in servers on the Internet and cached temporarily on clients that include ... handhelds". Cloud computing and mobile devices complement each other for mobile devices usually are lacking of enough storage and computing power but which usually solvable by cloud computing, and mobile devices could be an effective outstretch for cloud services. Also, there exist at least four possible data access forms for the storage:

- Local in-device data, in which the applications such as contact or calendar are the typical
- Local out-device data, in which the medias such as the SD/MMC cards are the most typical for the backup data
- Remote synchronous data, in which the audiovisual streaming maybe the most popular peer-to-peer application
- Remote asynchronous data, in which the digital data stored somewhere on the Internet can be retrieved anytime once connected

It seems that size is the fate of mobile handheld devices and weight is their destiny, but cloud computing appears as a contemporary solution for them to breakthrough their limits. The next generation of mobile devices will use WBA and HCI technologies which support cloud services and interface designs to allow remote plug-and-play with **web 2.0** applications (Jaokar & Fish, 2006) that is suitable for mobile commerce in which this chapter emphasizes. For a demonstration of mobile commerce discussed in this

Figure 1. The handheld device interfaced via multimodal I/O sensations and connected to cloud services

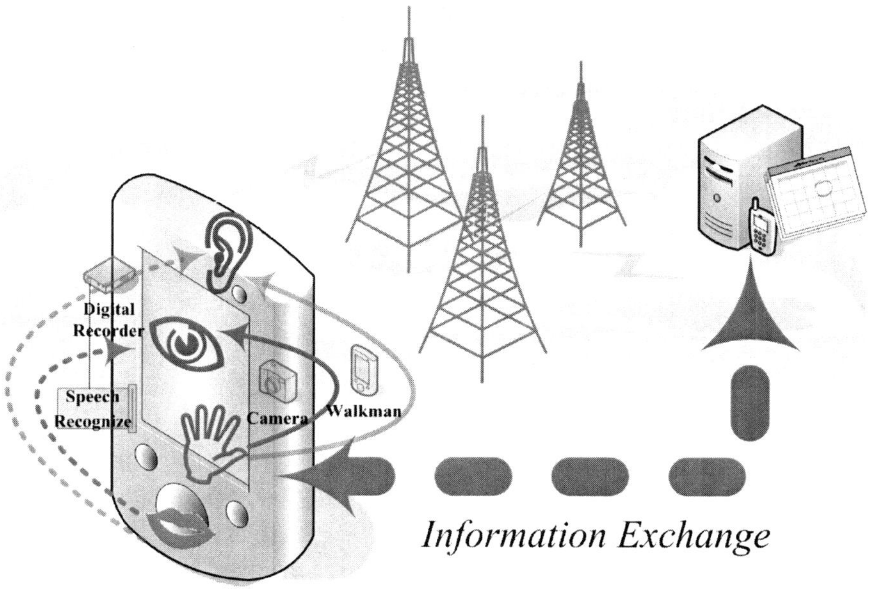

chapter, a case study of how the HCI and WBA technologies are combined for usage is revealed by a complete flow of manipulating mobile tickets from user login, entry preview, entry creation, till the barcode scanning, and finally the secure issues are furthermore presented.

BACKGROUND

Basically, the mobile commerce solutions make use of the mobile technologies such as the data communication and sensation technologies for information interaction, transmission, and storage as illustrated in Fig. 2a. Figure 2a is depicted in the viewpoint of mobile device to deal with commercial information via basic six-steps execution model named as send, transmit, receive, write, read, and display (abbreviated as STRWRD). For the interfaces of interaction, transmission, and storage gone through by the information, the **interaction** interface is basic while the other two are optional found in most mobile devices. In our demonstration, the same terminologies, issuers, clients and cashiers, called three parties in (Aigner, Dominikus, & Feldhofer, 2007) are adopted for necessary processing and handling of online **e-coupon** (or **e-ticket**) access, and we found that among different mobile commerce solutions, the **interaction** interface is indispensable between the clients and the issuers or cashiers, however, the transmission or storage maybe used or not.

One possible mobile commerce solution is enabled by near-field communication (NFC), which is a short-range wireless communication technology that enables the exchange of data between devices with a bandwidth of almost 2 MHz. In NFC-enabled solution, there exist the client, who wants to get a particular e-coupon for product or service, the issuer, who generates and hands over the desired e-coupon, and the cashier, who has a function that the **e-coupon** can be cashed in and has to proof the validity of the e-coupon (Aigner, Dominikus, & Feldhofer, 2007). NFC-enabled client devices can connect at once (less than 0.1 seconds) with the **issuer**, transmit or receive data at the same time with the **cashier** (refer to the step 1 and 6 in Fig. 2b), and also stored

Figure 2. Basic six-steps execution model (abbreviated as STRWRD) run in three interfaces for information interaction, transmission, and storage: (a) Base model with STRWRD, (b) NFC model with SWRD, (c) SMS model with variant STRWRD, and (d) Gosport model with STRD

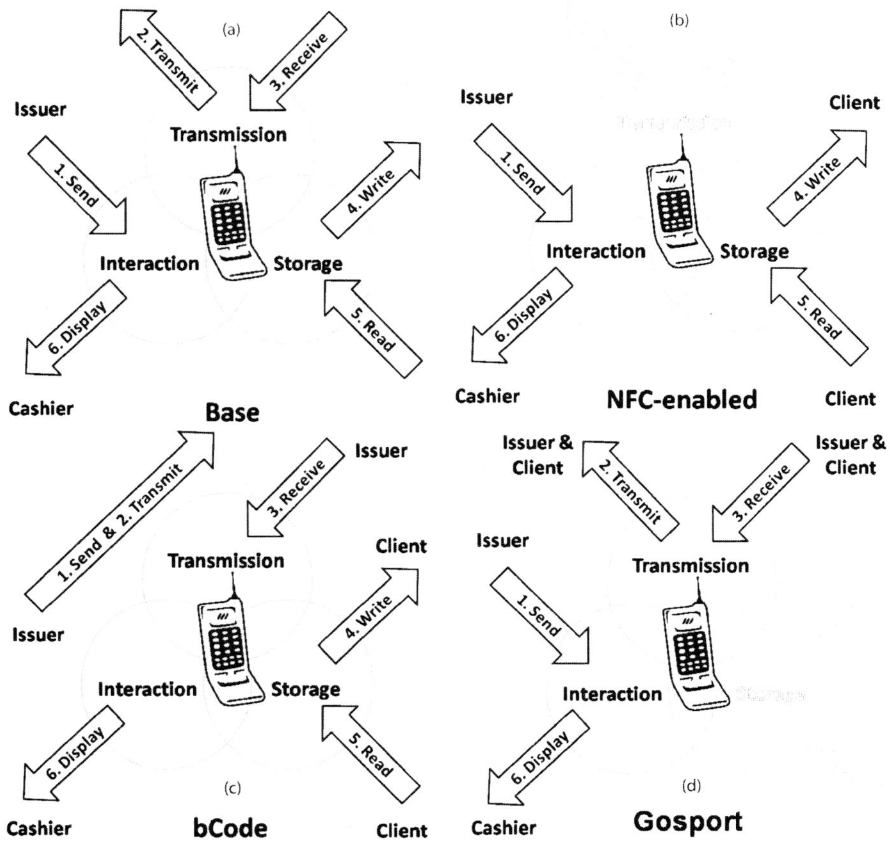

to and retrieved from local storage (as depicted in the step 4 and 5). The difference between Fig. 2a and 2b is the usage of transmission interface is unnecessary and thus the steps 3 and 4 can be disabled for NFC, so it is enough to use only four-steps execution model of send(connect), write, read, and display (abbreviated as SWRD). NFC's short-range broadband access and confirmable technology make it suitable for mobile ticketing applications, but it has other limitations such as insecure communications and a possible lack of local device memory for ticket storage.

Another possible mobile commerce solution is bCODE (http://www.bcode.com), one of the well-known mobile commerce technologies in which the narrowband short message service (SMS) has been used to send and transmit ticket by the issuer (refer to the step 1 and 2 in Fig. 2c) and finally the cashier's optical character reader (OCR) used to read ticket electronically from the screen of client's mobile device (step 6 in Fig. 2c). Every time the bCODE ticket is written to and read from client's SMS inbox storage (step 4 and 5) of a mobile device, after successfully received (step 3) and being displayed on the screen (step 6) if used. In comparison with Fig. 2a, Fig. 2c shows a different way of message sending and transmitting not to the **interaction** interface directly, but through the transmission interface instead. The main limitations of SMS used for mobile commerce is its protocol supports up to 160 character messages and makes connectionless communications in one direction from source to destination without checking receipt. bCODE could support more than 99 percent of existing devices, but simultaneously inherit SMS's limitations. As a result, messages might either fail to appear exactly as sent (for data loss) or fail to be saved as expected (for local storage shortages in the SMS inbox).

Jeng, Chang and Wang (2008) have proposed an alternative mobile commerce solution code name as Gosport 1.0 in which the **issuer** can send the commercial information to the client either via the mobile device as shown in steps from 1 to 3 of Fig. 2d, or via the non-mobile platform like the way used by bCode as shown in steps from 1 to 3 of Fig. 2c. It can be done for Gosport 1.0 all because of the **WBA** and cloud computing technologies in which the transmission and storage interfaces are combined and thus may accelerate the transition to a paperless society than a narrowband solution for more reliable transmission, wider message length, and larger storage capacitance. At final (step 6), Gosport reveals message by one-/two-dimension barcode image on screen being scanned with barcode reader efficiently and securely (e.g. http://qrcode.kaywa.com). However, Gosport 1.0 somehow needs an upgrade for security protected by the Google Calendar Service only in reliance on the authentication of under the Transport Layer Security (TLS) Protocol or its predecessor, Secure Sockets Layer (SSL).

For sustainable development of a mobile commerce solution, workable but not securable is absolutely not enough. That is the reason why the recent researches for NFC and SMS technologies pay more attention to the security issues such as Aigner, Dominikus, and Feldhofer (2007) proposed a system of virtual coupons using NFC technology based on the protocol according to the published standard of ISO/IEC 9798-2: 1999, and Toorani and Shirazi (2008) introduced a new secure application layer protocol, called SSMS, a secure SMS messaging protocol for the m-payment systems intends to efficiently embed the desired security attributes in the SMS messages. Aigner et al. (2007) focuses on integrity issues of the protocol for the possible attacks on generation, copying, manipulation, and multiple cash-in, while Toorani et al. (2008) are interested in the confidentiality, integrity, authentication, and non-repudiation issues. Similarly, the Gosport 1.0 project proposed by Jeng et al. (2008) maybe has some risks of security vulnerabilities such as the integrity issue, therefore, Jeng, Lee, Wang, and Cheng (2009) have proposed a protocol addressing on the secure application of information retrieval and reveal based on the upgraded project Gosport 2.0 and published at May 2009.

For information sent through an insecure channel, a properly implemented **digital signature** lets the receiver believe the information was sent by the claimed sender. As similar as RSA asymmetric cryptography (Rivest, Shamir, & Adleman, 1978), digital signature (Goldwasser, Micali, & Rivest, 1988) also uses a pair of public-private keys but in an opposite encryption/decryption directions, e.g. RSA uses public key first private key second as shown in Fig. 3a, but digital signature uses first private last public oppositely, as shown in Fig. 3b. For illustrating traditional digital signature scheme in Fig. 3b, firstly sender selects a private key via a key generation algorithm to sign a message with the key to produce a signature, then receiver uses the corresponding public key to decrypt the signature and make verification for them. RSA is widely used in e-commerce protocols and digital signature which satisfies the secure attributes of such as authentication, integrity, and non-repudiation is suitable for plenty of electronic systems.

Jeng, Lee, Wang, and Cheng (2009) propose a modified digital signature scheme which combines the attributes of RSA and digital signature to provide an alternative verification scheme especially for mobile commerce applications. It makes a secure Internet information protocol for both retrieval and reveal based on a modified RSA digital signature to offer a cost-down paperless mobile-ticket scheme and targets to impact human

Figure 3. Two cryptosystems encrypted with opposite public-key direction: (a) RSA workflow, and (b) Digital signature scheme

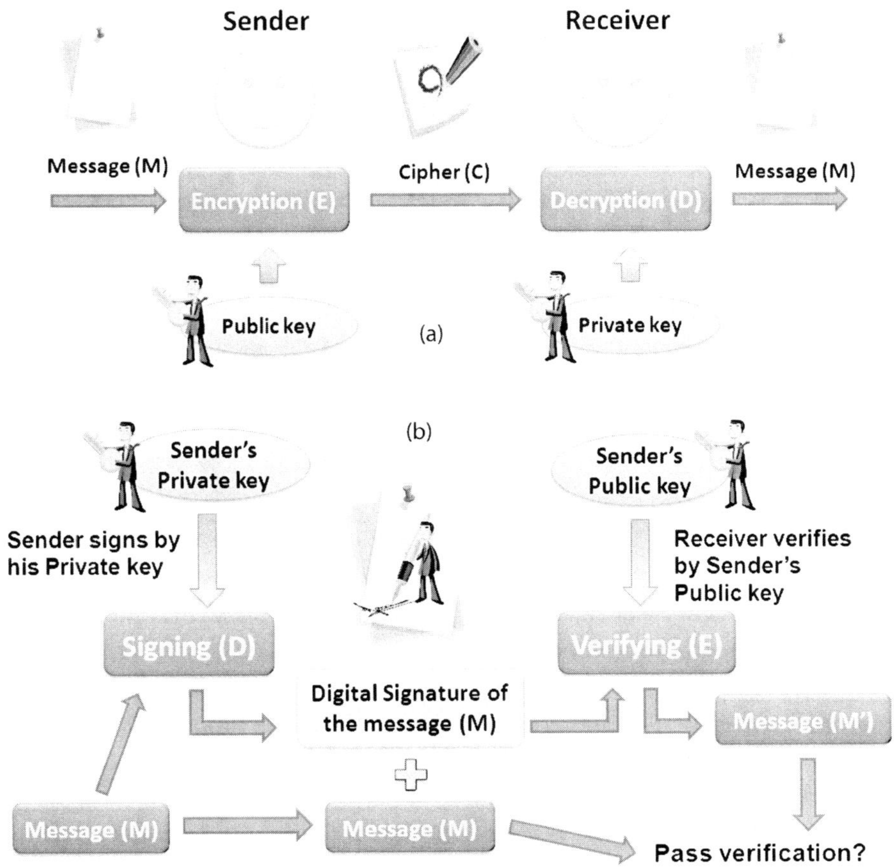

societies by a commerce infrastructure via WBA. In this chapter, the topics of the project Gosport 2.0 including 1.0 will be presented one by one in order to demonstrate how the scheme is feasible and applicable for mobile commerce including those used for ticketing, **coupons**, loyalty rewards, and advertisement.

OPEN SDK ACCELERATES MOBILE COMMERCE

The open SDK (Software Development Kit) is good for developers for the comprehensive APIs (Application Programming Interface) and the easy-to-use toolkits let developer experience rapid productivity gains. In October 2007, Apple's Steve Jobs announced that the company would make an iPhone SDK available to third-party developers in February 2008. Some weeks later, the Android platform (http://developer.android.com) announced with the founding of the Open Handset Alliance (OHA; http://www.openhandsetalliance.com) is a joint project and Google is one member of OHA. The iPhone SDK only supports one development environment (Mac OS X), which is generally believed that much less convenient than Android (Windows, Mac OS X, and Linux). However, both application frameworks support haptic HCI and Internet accessibility for networking.

Amazon Web Services maybe are the first-launched online databases, and Google Services are another typical examples but free of charge with limited web service functionality. Amazon Web Services launched in July 2002, offering online services billed on usage by client-side applications or other Web sites. Google provides a variety of APIs for web and desktop programmers alike, including Google data APIs, which let programmers create applications to read and write data from Google services for free with condition. Although most Google services didn't launch at the same time, they do offer methods to create HTTP requests and process HTTP responses as well as Amazon does.

It is important that regardless of platform choice, web services that are integrated with WBA technology can reach mobile users through well-built applications and thus become mobile services. That is why we select Google Calendar as our base service (because its five Ws format—when, where, what, who, and why—is suitable for all events, its extended properties are universal, and it's free) and Android as the development platform (because of its simple SDK installation and familiar programming language support). Exemplifying the mobile **web 2.0** paradigm, our mobile commerce application pushes information up into the Google Calendar database rather than merely bringing information down to the user.

CLOUD COMPUTING ENABLES MOBILE COMMERCE

Responding to the above discussion, cloud computing not only complements each other with mobile devices, but also starts to enable the applications on the mobile devices such as for mobile commerce. As a case study, the project Gosport is a start to the mobile commerce application being involved in the interaction between the cloud service of Google Calendar and the open Android platform. The name "Gosport" is chosen especially in accordance with the significance of "Google Passport," based on the Google Service, and intends to use the four suits of patterns following poker to be four symbols of mobile services as mobile coupons, loyalty rewards, advertisements, and tickets.

Cloud computing can accelerate to reduce the storage capacitance, benefit the multimedia contents, and increase the information security especially for mobile commerce. These items are presented and demonstrated one by one in the order of "diamond," "club," "heart," and "spade" modes as defined in the Gosport services.

Cloud Computing May Reduce the Storage Capacitance

Figure 4 shows some snapshots from using Gosport 1.0 to demonstrate how we can combine Android and Google Calendar together into mobile commerce application. Corresponding to the six-step execution model (STRWRD) depicted in Fig. 2, the steps 1 and 2 executed by the **issuer** should go through snapshots: ①-②-④-⑤, then in step 3 the client receiving the coupon may go through ①-②-③, and if the **cashier** need to scan the barcodes of the client for verification, which can go forwards into step 6 via ①-②-⑥ by skipping the redundant storage interface of the client (i.e., the steps 4 and 5). Two steps (WR) are reduced and only four-step model (STRD) is needed.

The six-snapshots represent a flow of **coupon** creation, selection, presentation, and verification. By pairing Google Calendar and Android, we can accomplish the following operations:

- The user (issuer or client) signs in to the Google Calendar account. With built-in HTTP client libraries, Android allows secured authentication via remote calling directly to the Google Calendar APIs (http://code.google.com/apis/calendar). WBA accelerates the connection-oriented authentication process with a 182-byte secure token, Google Calendar API returns to Gosport after the user signing in to the SSL-encrypted Google account. Before

Figure 4. Gosport mobile ticketing application 1.0 is designed to enable a quick and paperless of commercial flow. The six snapshots in the Gosport mobile ticketing process illustrate some significant aspect of developing mobile commerce apps such as authentication, content creation, content presentation, and content verification. (1) Signing, (2) Selecting, (3) Listing, (4) Creating, (5) Forwarding, and (6) Scanning

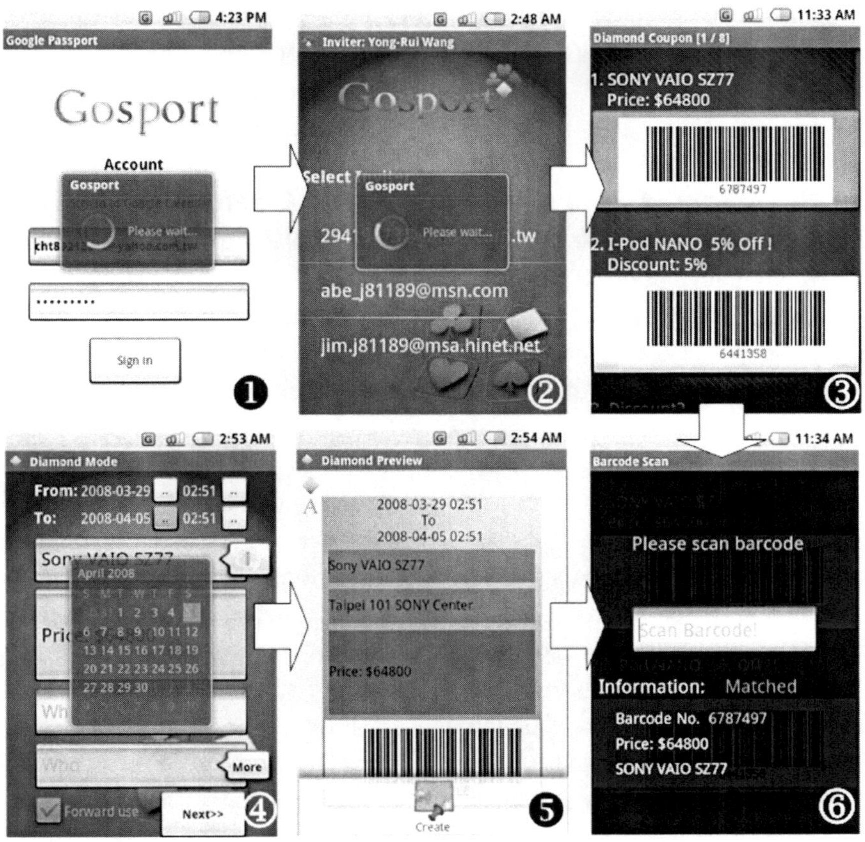

successful login, the snapshot 1 in Fig. 4 displays the progress circle bar which is a visual effect offered by Android API.

- The user makes a selection from one of the issuers who create coupons, and lists the coupons ready for matching with the query. Google Calendar follows the iCalendar standard (RFC 2445; http://www.ietf.org/rfc/rfc2445.txt) for calendar data exchange, which makes it easy to share, express in five-W form, and design for custom coupon formats. With query parameters or a specified date range, a Gosport user can retrieve arrays of coupons on demand without worrying about the size of local storage or the risk of data loss. After successful login, the emails of all issuers and the barcode coupons for the selected **issuer** (p.s. issuer is a Google Calendar user who invites other users in his/her calendar schedule) will be listed at snapshots 2 and 3, respectively.
- The **issuer** sends coupon in Gosport by creating the Calendar entries with fingertip pressing on the touch screen. The human fingertip isn't as fine as a pen point, so Android enlarges the GUI components for easy hand-eye coordination. Snapshots

4 and 5 depict the display for coupon transmission: Issuer creates the calendar entry by pressing the button "Next" and the menu item "Create". Google Calendar provides properties (arbitrary name-value pairs) that Gosport can use to store application-specific information, such as coupon type, as needed.

- The **cahier** scans the barcodes of the client's coupons on Gosport repeatedly for matches, as illustrated in snapshot 6, without other cash-in devices because the barcode scanning UI being built in Gosport. Most challenging part of narrowband small-display mobile ticketing is the risk of loss of data and loss of continuity for scanning barcode, but Gosport uses WBA technology to solve this and save time. The barcode scanning scheme downloads coupons from Google Calendar on demand, scans the coupons repeatedly on the Android display as listed in snapshot 3, and quickly matches barcode for each pair which maybe means a serial number for identification. This scheme connects the barcode reader and remote database, and thus reduces the cost of any cash-in device to just a pair of mobile devices embedded with Gosport.

All the visual components used such as the textfields, progress circle bar, lists, datetime dialog, buttons with or without icon images, etc. come from the packages of Android SDK based on J2SE. All the contents for creating, sending, storing, retrieving a coupon are made via the Google GData APIs based on the HTTP standard operations such as POST/PUT/GET fed with the XML described information needed to be parsed.

In Fig. 4, the only presented one-dimension (1D) **barcode** coupon is classified into "diamond" mode of anonymous coupon service for only the same serial number is encoded for scan. The 1,100 alphanumeric 2D barcode, larger than 150 word's 1D barcode, can be adopted to reveal information containing individual identification such as each email account for loyalty reward service which is classified as "club" mode in Gosport 1.0. Worthy to be mentioned, the barcode generations are also kinds of "cloud service" from the third-part websites, except that only the barcode specification such as code type, format, size, etc. is requested as arguments and no website storage is required.

Cloud Computing May Benefit the Multimedia Contents

The barcode images used in the Gosport solution and downloaded from third parties, as well as the other multimedia contents widely distributed in the cloud services, need WBA technologies to accomplish the transmission efficiently. As a result, the multimedia contents applied on mobile commerce can be benefited from the cloud computing technologies which are enabled by the WBA, i.e., the advertizing service of Gosport expressed in the "heart" mode can make posters, greeting cards, marquees, and so on.

Figure 5 show some screenshots for demonstrations: Firstly Fig. 5a is a typical heart-mode ad card introduces the information for sale and reveals in a visual marquee effect. Then, if interested in more information, the client can press two buttons on the right-hand side to explorer the website or locate the position on the map as illustrated in Fig. 5b and 5c respectively, and these all thanks to the WBA for efficiency and convenience.

Cloud Computing May Increase the Information Security

As introduced, the mobile commerce framework we proposed bases on one cloud service called Google Calendar and utilizes its entry-sharing mechanism by creating (sending + transmitting) and receiving messages (tickets) through HTTP request commands of POST and GET. These two-way POST and GET commands are the

Figure 5. Multimedia contents applied on Gosport project and demonstrated by (a) the "heart" mode of advertisement service, (b) the website exploring, and (c) the Google Map illustration

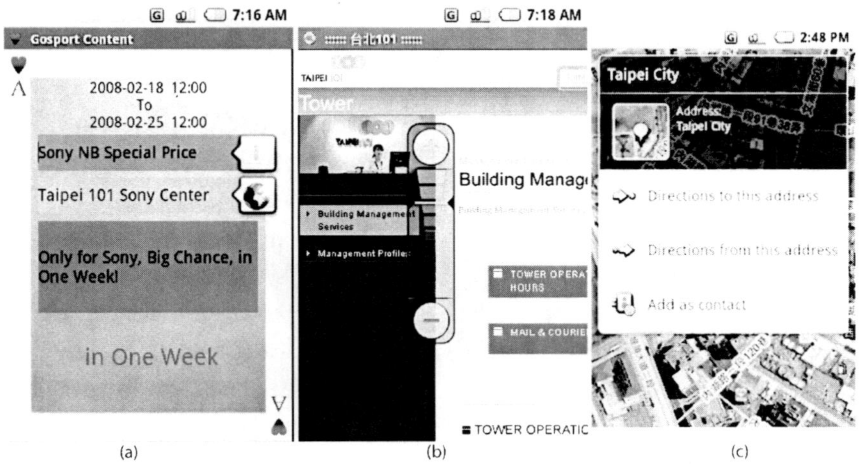

core operations of web 2.0 on which the cloud services are based. The interlaced requests made by the POST, PUT, and GET commands weave the secure nets for Gosport 2.0 tickets.

A Modified RSA Signature Scheme Built on Web 2.0 Infrastructure

As depicted in steps 1 and 2 of Fig. 6, the ordinary sending, transmitting, and receiving events occur as before just as in Gosport 1.0. The difference is there existing one additional operation called "join in" for the proposed secure protocol as illustrated in step 3 of Fig. 6, in which each client takes the **issuer's** public key to encrypt the information individually customized by each client. At the final step for check, as step 4 in Fig. 6, there occurs also an information display step same as step 6 in Fig. 2, but more complicated for the issuer (assuming also the cashier) uses the non-sharing

Figure 6. Secure information retrieval and reveal protocol proposed for mobile commerce based on modified RSA digital signature

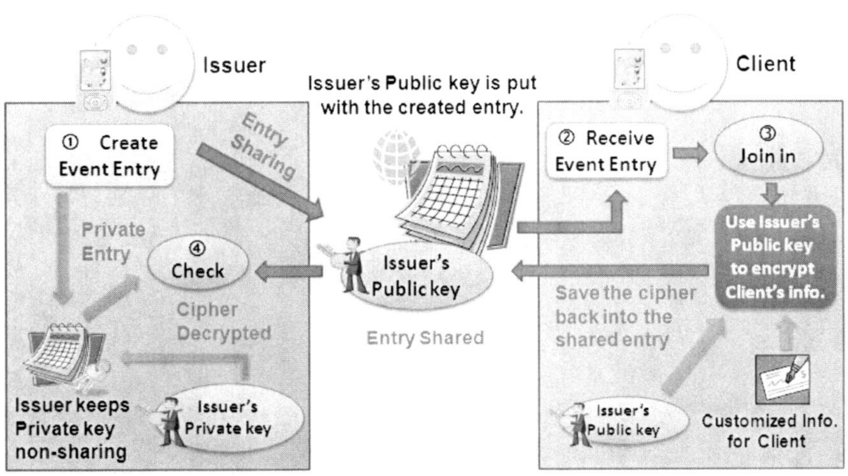

private key to decrypt the information scanned by the **barcode** reader to verify if the message (ticket) is accurate in the right time, right place, and especially for the right person.

The Workflow for the Modified RSA Signature Scheme Proposed

In Fig. 7, after successful login by the issuer as shown in snapshot 1 and using the same email account illustrated in snapshot 2, a spade mode for RSA-signed digital ticket can be created and sent (refer to the snapshot 3), and then waits to be verified by the cashier (as snapshot 6).

Each client of the ticket may decide to accept the invitation by pressing the "join in" button or not (snapshot 4). If the client accepts, there appears a formal invitation ticket with hypertext containing a two-dimension barcode on it which is encoded with some customized information for each client invited individually (shown in snapshot 5).

Tutorial for Exploring Cloud Computing Service

Cloud computing technologies usually become some embodiments of services in front of people such as web email service, blog service, remote image storage service, and so on. The related technologies seem follow the trends from standalone computing, to client/server model computing, and now the cloud computing. No matter what grid technologies for the background framework, the foreground services intend to bring the users

Figure 7. Six screen captures numbered at the lower right corner for "spade" mode demonstration

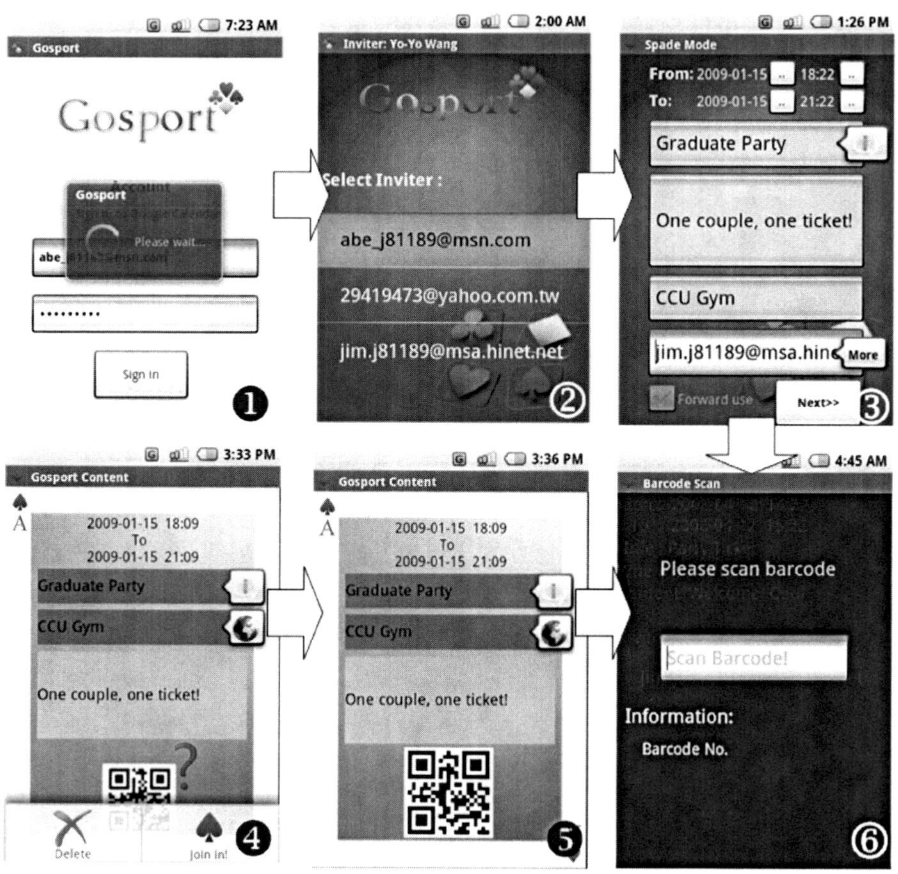

Figure 8. The screen shot of the web page for creating a Google Service account

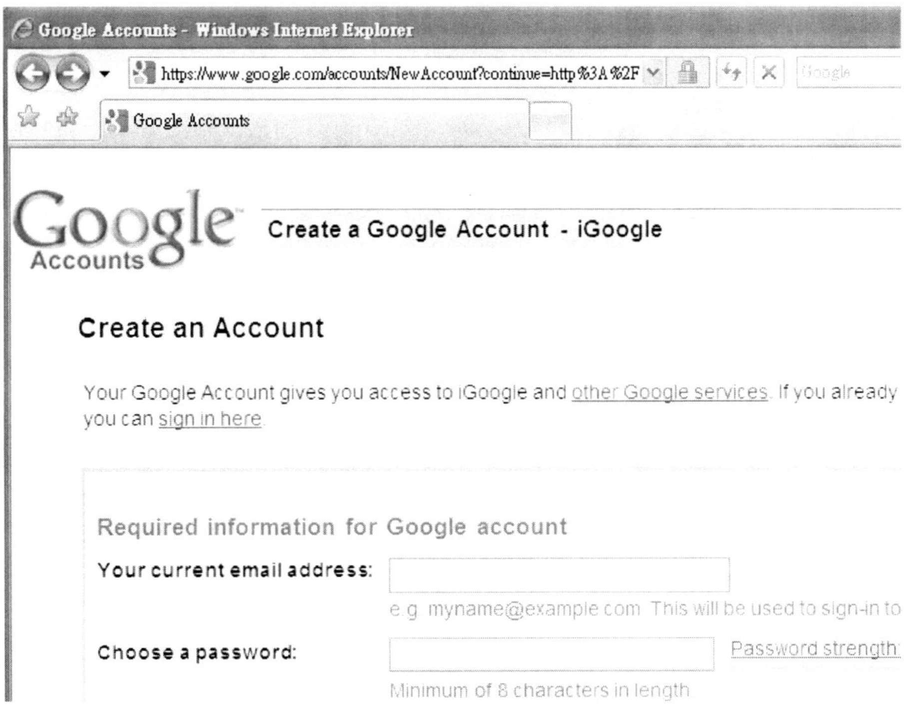

to a bank-like two-way information life style: save your information anytime, anywhere and withdraw them on your demand efficiently and securely. That is the key reasons why the web 2.0 concepts can be realized by the cloud computing. Here we will guide you a trip of Google Service to experience a different kind of computing way: The travel package of Google Calendar.

Start the Trip for Google Service from a Successful Authentication

First of all, you need an account as a passport for the trip. The account is composed of a legal email address and a password with minimum of 8 characters in length. If you haven't prepared your account yet, it is convenient to create one by starting from the URL of (http://www.google.com/ig?hl=en) and following the links of such as "Sign in" and then "Create an account now" to switch into the "Google Accounts" page been snapshot partially as shown in Fig. 8 for your reference.

After your successful login, you can go around the service there and enjoy yourself: News, Books, Groups, or something about multimedia services of Images, Photos, and YouTube. These two-way web services allow you to upload and download information for storing, sharing, processing, etc. announced as **Web 2.0**.

Go on the Trip for Further Adventure by Using Programming Language

Before we go for another trip of making the two-way service by using programming language, except for the username and password, other two stuffs need to be prepared: Google gdata client-side source code (Java client library is what we choose among other supported choices such as for .NET, PHP, Python, Objective-C, and JavaScript by corresponding to http://code.google.com/intl/en/apis/gdata/clientlibs.html) and its shell-based tool by using Ant, according to Ant's original author, James Duncan Davidson, the name is an acronym

Figure 9. The screen shot of the web page for downloading the tools for programming

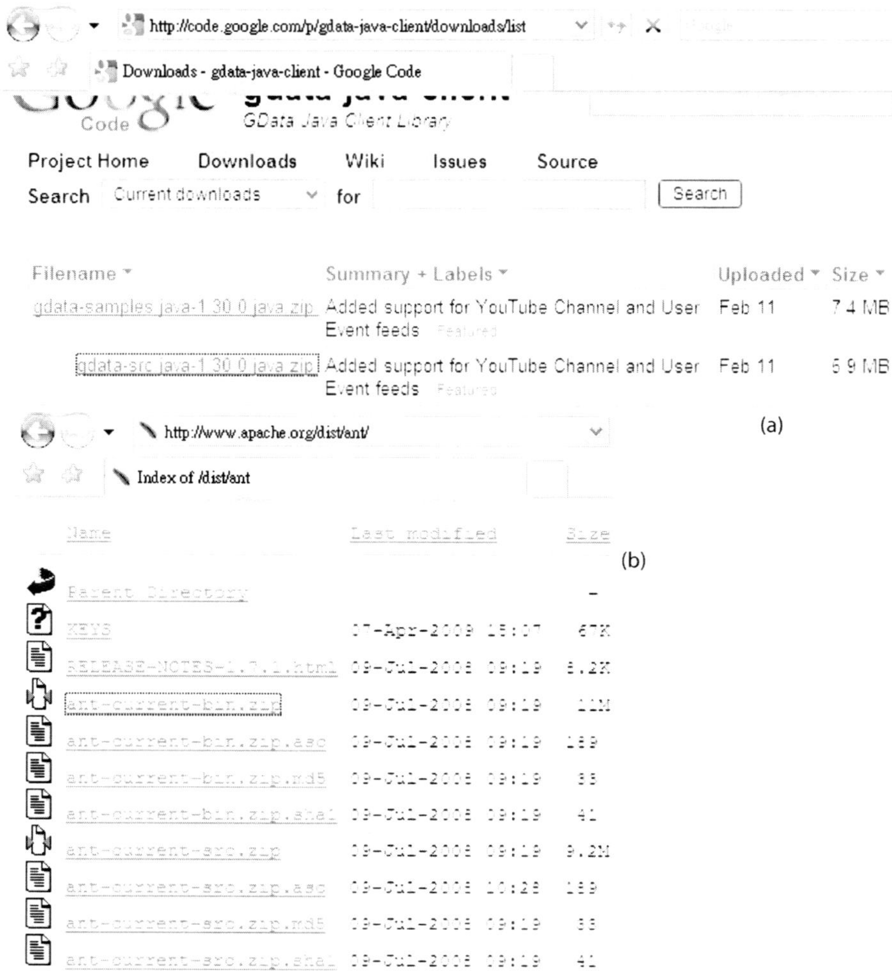

for "Another Neat Tool," which is a Java-based build tool (http://ant.apache.org).

Figure 9 (a) and (b) are the two screen shots of free downloading pages for these tools. The gdata-samples.java-1.30.0.java.zip contains the sample source for some but not all Google services such as blogger, book, calendar, photos, you-tube, etc., and ant-current-bin.zip contains the key batch file ant.bat which can be executed with proper parameters. After unzipping the Java sources in directory D:\, for example, and its build tool somewhere you can access, then we intend to build the execution by the following command:

```
D:\gdata-samples.java-
1.30.0.java\gdata\java>ant -f
build-samples.xml sample.calen-
dar.run
```

However, you may not be successful to execute this command if you have not prepared your username and password by finding the file of "build.properties" on the path of "gdata-samples.java-1.30.0.java\gdata\java\build-samples" and fill in these important two attributes of "sample.credentials.username" and "sample.credentials.password" accordingly.

Finish the Trip for Demonstration by Creating Event via Programming

Before we finish this tutorial by demonstrating how we can modify the Java program and recompile it, we can examine the xml file dedicated to the calendar sample by following the path below:

```
D:\gdata-samples.java-
1.30.0.java\gdata\java\build-
samples\calendar.xml
```

We can control the length of the demonstration by using the comment symbol-pair, "<!--" and "-->," to temporarily cancel the target runs and shorten the results accordingly:

```
<target name="sample.calendar.
run" depends="sample.calendar.
build"
description="Runs the Calendar
samples">
<antcall target="sample.calen-
dar.eventfeed.run"/>
<!--antcall target="sample.cal-
endar.calendarfeed.run"/-->
<!--antcall target="sample.cal-
endar.aclfeed.run"/-->
</target>
```

For the sample username jim.j81189@msa.hinet.net we used, the results of the final part are listed on the console something like

```
...
[java]
[java] Full text query
[java] Events matching Tennis:
[java]
[java]
[java] Events from 2007-01-05 to
2007-01-07:
[java]
[java]
[java] Successfully created
event Tennis with Mike
[java] Successfully created
quick add event Tennis with John
Aprilpm-pm
[java] Successfully created web
content event World Cup
[java] Successfully created re-
curring event Tennis with Dan
[java] Event's new title is "Im-
portant meeting".
[java] Set a 15 minute EMAIL re-
minder for the event.
[java] Successfully deleted all
events via batch request.
BUILD SUCCESSFUL
Total time: 14 seconds
D:\gdata-samples.java-
1.30.0.java\gdata\java>
```

Finally, we can make a little bit modification by temporarily remove one Java statement to show the new execution result after rebuilding the project. The statement we choose to remove appears in the file "EventFeedDemo.java" at line about 528 as:

```
deleteEvents(myService, event-
sToDelete);
```

and becomes as

```
//deleteEvents(myService, event-
sToDelete);
```

after being marked by "//." After rerun, the first three event created by the sample program such as single event of "Tennis with Mike," quick add event of "Tennis with John," and the web content event for World Cup really appear on the calendar of jim.j81189@msa.hinet.net as shown in Fig. 10 (a), (b) and (c), respectively. By the way, the title for the single event of "Tennis with Mike" appears as "Important meeting" not because of

Figure 10. Three calendar event screen clippings for the demonstration without deletions

an error, but a successful operation for updating title instead.

CONCLUSION

This chapter introduces a mobile commerce project Gosport based on an open mobile platform of Android and a cloud service of Google Calendar, compares this project with two well-known related works by the issues of execution steps, interfaces, security, and proposes a secure web 2.0 protocol for the information retrieval and reveal by a modified RSA digital signature scheme. The Google Service and Android platform we choose to make the mobile commerce project based on are the popular and free to access and might be an evidence for a proper application and technology for the handheld computing for mobile commerce. Besides, the tutorial for exploring cloud computing service may bring the readers into an adventure from being a service user to a program designer.

Finally, what the case study demonstrated not only is a feasible workflow, but also an applicable mobile commerce prototype hopes to make contribution to a more environmentally friendly commercial society.

REFERENCES

Aigner, M., Dominikus, S., & Feldhofer, M. (2007). A System of Secure Virtual Coupons Using NFC Technology. In*Proceedings ofthe5th Ann. IEEE Int'l Conf. Pervasive Computing and Communications Workshops (PerComW 07)*, (pp. 362-366). IEEE CS Press.

Goldwasser, S., Micali, S., & Rivest, R. (1988). A digital signature scheme secure against adaptive chosen-message attacks . *SIAM Journal on Computing*, *17*(2), 281–308. doi:10.1137/0217017

Haselsteiner, E. & Breitfuß, K. (2006). Security in near field communication (NFC), *Printed handout of Workshop on RFID Security, 6*. Amsterdam: Philips Semiconductors.

Hewitt, C. (2008). ORGs for Scalable, Robust, Privacy-Friendly Client Cloud Computing . *Internet Computing*, *12*(5), 96–99. doi:10.1109/MIC.2008.107

Jaokar, A., & Fish, T. (2006). *Mobile Web 2.0 -- The innovator guide to developing and marketing next generation wireless/mobile applications*, London: Futuretext. Retrieved March 22, 2009, from http://mobileweb20.futuretext.com

Jeng, I. H., Chang, A. Y., & Wang, Y. R. (2008). Plug into the online database and play Mobile Web 2.0. *IT Professional*, *10*(5), 34–38. doi:10.1109/MITP.2008.107

Jeng, I. H., Lee, C. J., Wang, Y. R., & Cheng, C. K. (2009). Secure Information Retrieval and Reveal for Mobile Apparatus Based on 2D Barcode Digital Signature. In*Proc. 13rd Ann. IEEE Int'l Symposium on Consumer Electronics (ISCE 09)*, (pp. 683-686). IEEE Press.

Jeng, I. H., & Wang, Y. R. (2008). *Gosport Video*. Retrieved March 30, 2009, from http://faculty.pccu.edu.tw/~zyh2/gosport/Gosport.AVI.

Rivest, R., Shamir, A., & Adleman, L. (1978). A Method For Obtaining Digital Signatures and Public-Key Cryptosystems . *Communications of the ACM*, *21*(2), 120–126. doi:10.1145/359340.359342

Toorani, M., & Shirazi, A. A. B. (2008). SSMS - A Secure SMS Messaging Protocol for the M-Payment Systems, In *Proceedings of the 13th IEEE Symposium on Computers and Communications (ISCC '08)*, (pp. 700-705)., Marrakesh, Morocco; IEEE ComSoc.

Chapter 3
Quality Evaluation of B2C M-Commerce Using the ISO9126 Quality Standard

John Garofalakis
University of Patras, Greece

Antonia Stefani
University of Patras, Greece

Vassilios Stefanis
University of Patras, Greece

ABSTRACT

Business to Consumer M-commerce applications, are data-intensive, user-driven, and have increasing needs for accessibility, efficiency, adaptivity, portability and competitiveness. However, their design process still lacks a systematic quality control method. In this chapter we explore m-commerce quality attributes using the external quality characteristics of the ISO9126 software quality standard. Our goal is to provide a quality map of a B2C m-commerce system so as to facilitate more accurate and in detail quality evaluation. The result is a new evaluation framework based on decomposition of m-commerce services to three distinct user-software interaction patterns and mapping to ISO9126 quality characteristics.

INTRODUCTION

A significant advance in the on-line business arena is the advent of mobile services, which are becoming a reality for enterprises and users alike. New technologies in mobile networking and mobile device hardware primarily and mobile software secondarily have permitted the realization of the vision of a mobile web. Or at least they promise to realize it; The first steps have already been made with commercially successful mobile services flourishing and promises for even more impressive attempts are on the way. There is an enthusiasm in business, academia and users for mobile services, and this enthusiasm is the impetus for not only the research of the novel but for the adaptation of the old (Bouwman et al., 2008).

DOI: 10.4018/978-1-61520-761-9.ch003

E-commerce, in the form of Business to Consumer transactions is one of the primary business successes of the WWW. It is only natural that enterprises sought to increase their market share by moving to the mobile Web as well. Mobile-commerce (m-commerce) systems are being developed at an increasing rate in recent years. As a business process, m-commerce can be viewed as particular type of e-commerce (Coursaris, 2002) and refers to transaction with monetary value that is conducted via a mobile network. When users conduct m-commerce such as e-banking or purchase products, they do not need to use a personal computer system. Indeed, they can simply use some mobile handheld devices such as Personal Digital Assistants (PDA) and mobile phones to conduct various e-commerce activities. In the past, these mobile devices or technologies were regarded as a kind of luxury for individuals. However, this situation has changed. Technology has driven the growth of the mobile services industry thus creating a new opportunity for the growth of m-commerce (Ngai, 2007; Huang et al., 2007). Location-based services are also attracting the attention of the business world (Junglas, 2007).

Focusing on B2C services (Business to Consumer services), this uniqueness is both a blessing and a curse. Being user-intensive, it is absolutely imperative that the software satisfies mobile user needs; mobile commerce user needs are, in many perspectives different than Internet-based e-commerce user needs mainly because the access medium is different. Thus, the quality of the software itself, that is the satisfaction of implied and non-implied user needs, is of primary importance. To date, most research efforts focus on Quality of Service which deals mostly with low-level network attributes (Ghinea & Angelides, 2004).

The research on the quality of B2C m-commerce systems is a new and challenging task; especially the quality of mobile commerce systems as it is perceived by the end-user is only now becoming a research issue. However providers of mobile services and mobile hardware have always paid attention to ergonomics and usability. Google's Android platform is an approach that aims to attract the novice user and actually increase the total target group of advanced mobile services by creating new users (Android, 2009). Usability is not the only dimension of software quality. According to ISO standards, there are many dimensions to software quality that need to be satisfied. A user perspective, rather than a developer perspective, of quality is important (Hong et al., 2008).

The quality of software is a principle concern to end-users and developers as well. It is increasingly difficult to evaluate diverse software such as m-commerce. The later provides a wealth of different services, different in the sense that different technologies and user-service interaction patterns are used. By identifying these differences in the level of basic services it becomes easier to apply different evaluation methods that are suitable for each case. Such a method would permit a detailed quality evaluation with an increased practical impact. After all, different software artefacts should be evaluated with methods focusing on their uniqueness. Having these in mind, one of the main questions posed is how to identify these differences and how to cluster the services according to them. Another problem that has to be dealt with is that a formal evaluation method should be used in order to provide a concise solution. It is with the above observation that this chapter examines the quality attributes of m-commerce systems adopting the ISO9126 software quality standard (ISO, 2001). ISO9126 is a general standard for software quality that is user-driven. Because of its generality, it can be applied to any kind of software. In order for it to be practical however it must be seen in the light of a specific application domain. Adopting and adapting ISO 9126 for specific domains is not new and not foreign to the standard itself (Losavio, 2004; Cote, 2005). A usual approach is to enhance the hierarchical and (by design) open scheme to include more attributes suitable for a domain (Stefani & Xenos, 2008).

Building on ideas initially presented in (Garofalakis et al., 2007), this chapter explores B2C m-commerce quality attributes using the external quality characteristics of ISO9126 of Functionality, Usability, Efficiency and Reliability. A new evaluation framework is proposed based on decomposition of m-commerce services to three distinct user-software interaction patterns and mapping to ISO9126 quality characteristics. The contribution of this work is the m-commerce specificity of the proposed technique, a technique that is flexible and extendable.

Background

Software Quality Evaluation

E-commerce as an application area is based on several primary research areas in computer science such as software engineering (Marca & Perdue, 2000; Saunders et al, 2006), data management, data communication and networks (Bhatti et al, 2000), computer security, Human Computer Interaction (HCI) and other disciplines. Some of them are closely related such as operations research and mathematics for cryptography (Bhimani, 1996; Lacoste, 2000) while others are more remote such as law, business, psychology and social behavior, marketing, and communication (Proctor et al, 2003; Li & Zhang, 2004; Kulkarni et al, 2005).

Research on B2C e-commerce system quality can usually be categorized into technology-centered or consumer-centered based on the main focus of each paper. The technological approach focuses on the quality of the infrastructure such as telecommunications, information technology, internet infrastructure (Zwass, 1996; Bilgoli, 2002; Elfriede & Rashka, 2001) and the services provided such customer self-service, customer relation management and business intelligence support (Papazoglou, 2001). The consumer-centered views may take a usability approach (Nielsen et al., 2001; Holzinger, 2005; Kelli & Vidgen, 2005) or a consumer behavior approach (Chen, 2005).

Historically, the success of computer applications has been based on the one hand on increased ease, decreased cost and new possibilities and on the other on convincing people, the potential 'users', of these attributes. When the people to be convinced are a class of professionals, then they either adopt the application as part of their professional tool set (as is the usual case with engineers) or they go out of business (as it happened with newspaper typesetters). There is a significant difference with applications such as e-learning, e-government and e-commerce whose success depends on convincing 'the public'.

Somehow 'proving' that shopping on-line is easier, costs less and gives new possibilities is not sufficient for B2C e-commerce to become the predominant mode of shopping. This is why the technology-centered approach to B2C e-commerce quality gives us only part of the story. Studies such as (Hong & Lerch, 2002; Holzinger, 2005) which focus on the quality of the infrastructure e.g. quality attributes and the services provided e.g. customer self-service, customer relationship management and business intelligence support try to explain and predict acceptance of B2C systems based on their technical aspects.

The consumer-centered usability approach on B2C system quality is based on premises about human behavior related to shopping. Several approaches such as (Olsina et al, 2000; Nielsen et al, 2001; Chen, 2005) conclude by giving usability guidelines as to how such a system will ensure efficient, effective, even enjoyable shopping, without frustrating unfruitful searches, doubtful transactions or impertinent queries on the part of the system.

Finally, the consumer belief centered approach uses trust as its key notion (Holsapple & Sasidharan, 2005; Moores, 2005). Trust is defined broadly, as the customer's willingness to spend time, money and hand over personal information to a B2C e-Commerce system. It includes numerous factors such as brand reputation, the reputation of the firm offering the on-line services, and differences between individuals in their general propensity to

Figure 1. ISO9126 software quality standard: Part 1

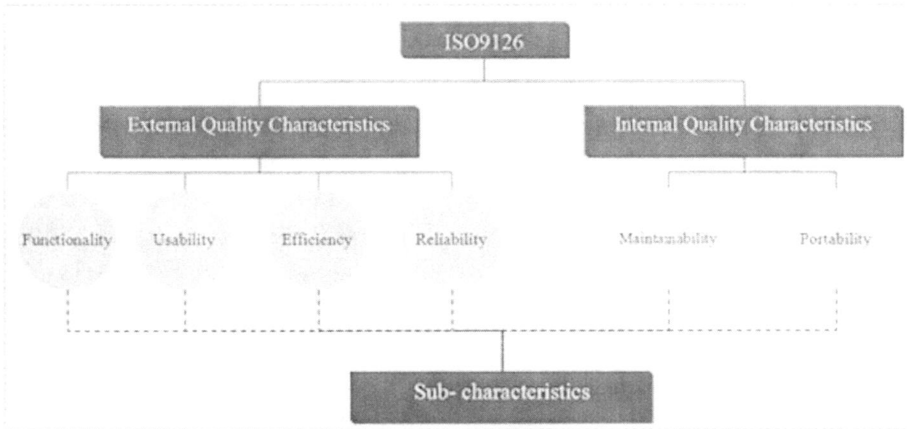

trust; interface properties such as graphic design and layout, content organization and usability; informational content including information the e-commerce system provides about products and services, privacy policies (Moores, 2005) and practices and trustworthy security; relationship management, including post-purchase communication and customer service; and fair pricing. Trust has to do mainly with issues outside the 'e' part: actual delivery and quality of goods and support of shopping decisions.

The ISO9126 Quality Standard

ISO attacks the problem of defining software quality by decomposing it to several sub-problems and by questioning about what are the different behavioral patterns of software as it interacts with the hardware, the users or other systems. As a result, many different standards were defined creating a lot of confusion. After a significant effort to reduce the numbers of standards, the ISO9126 standard release 2004 was defined and is considered to date the main software quality standard of ISO (ISO, 2004). It includes guidelines of how the software should behave internally and externally in order to be of good quality; it provides tangible tools called metrics as practical measures of quality. The standard, by definition does not provide guidelines on how to build quality software but guidelines on the characteristics of good quality software. For this it has received some criticism about its practicality, especially compared to relevant W3C initiatives. We consider them complementary as they have different goals.

According to ISO9126 quality is defined as a set of features and characteristics of a product or service that bear on its ability to satisfy stated or implied needs. In order to provide a developer view of software, besides the end user's view and guidelines for overall assessment of quality (Cote et al., 2005) the latest revision of four-part ISO9126 software quality standard has been proposed.

ISO9126: Part 1 defines the quality model for software products (figure 1). The other three parts discuss the metrics that are used to evaluate the quality characteristics defined in Part 1 which are internal metrics, external metrics, and quality in use metrics. The quality model is subdivided into two parts: the quality model for internal quality characteristics and external quality characteristics, and the quality model for quality in use. A quality characteristic defines a property of the software product that enables the user to describe and appraise some product quality aspect. A characteristic can be detailed into multiple quality sub- characteristics.

External quality characteristics are observed when software products are used, that is, they are measured and appraised when the products are tested, resulting in a dynamic view of the software. Evaluation of internal quality characteristics is accomplished by verifying the software project and source code, resulting in a static view. The quality model for internal and external characteristics categorizes quality attributes into six characteristics: Functionality, Usability, Efficiency, Reliability, Maintainability, and Portability. Each of these characteristics is subdivided into quality sub-characteristics. These quality characteristics can be used as goals to be reached in development, selection and acquisition of components and also as factors in predicting properties of component-based applications.

The external quality characteristics of ISO9126 quality model may be used as basis for m-commerce quality evaluation but further analysis and mapping of its characteristics is required. The main issue is how m-commerce system's quality can be analyzed using this standard. In this work, we use the following external quality characteristics of ISO 9126 to evaluate m-commerce systems: Functionality, Usability, Efficiency and Reliability. Each of the above mentioned characteristics provide the quality framework (actually the baseline) on which an m-commerce system may be built, taking into account end-users requirements. The external quality characteristics of ISO9126 are defined as follows:

- **Functionality** refers to a set of functions and specified properties that satisfy stated or implied needs. The meaning of Functionality is to provide integrative and interactive functions in order to ensure end-user convenience. Especially for m-commerce systems Functionality refers to the existence of these functions and services that support end user's interaction via the mobile system.
- **Usability** is defined as a set of attributes that bear on the effort needed for the use of a product or service, based on the individual assessment of such use by a stated or implied set of users. Usability is an important quality characteristic as all functions of an m-commerce system are usually developed in a way that seeks to facilitate the end-user by simplifying end-user's actions; this fact can however affect negatively the system in certain cases.
- **Efficiency** is a complex concept that entails both conceptual challenges as well as implementation difficulties. Efficiency is defined as the capability of the system to provide appropriate performance, relative to the amount of resources used, under stated conditions. It refers to a state where system functions are both usable and successful, i.e. they achieve their aim, the reason for their existence. One of the main criteria of efficiency of an m-commerce system is the quality relating to time and resource behavior.
- **Reliability** is the quality characteristic that refers to a set of attributes that bear on the capability of software to maintain its performance level under stated conditions for a stated period of time. Especially for m-commerce systems reliability refers to systems tolerance on end users actions.

Is the WWW Mobile-Ready?

The World Wide Web is not mobile-ready. Many Web pages are laid out for presentation on desktop size displays exploiting capabilities of desktop browsing software (Burigat et al., 2008). Accessing such a Web page on a mobile device often results in a poor experience. The main factor resulting in this negativity is page size and layout. Because of the limited screen size and the limited amount of material that is visible to the user, context and overview are often lost. A page may require considerable (vertical) scrolling to be visible, especially if the top of the page contains many images and/or navigation links. Layout patterns such as dense text and chunks of hyperlinks are also discouraging user from continuing their on-line experience. A few of the

Figure 2. A classic B2C site with dense hyperlink structure as seen by a mobile browser

parameters that affect mobile browsing in general include page layout, input devices used, network speed and device ergonomics with respect to software handling.

A psychological rule for successful browsing is to facilitate, as soon as possible, the creation of a mental picture of the site a user chooses to visit. This is a seamless process for most web sites. This is not however the case when a mobile device is used. Disorientation, difficulty to decode the structure of a web page, that is no immediate feedback as to whether information needs are fulfilled may result to increased drop-out rate (the user leaves the web site with a high probability of not visiting it again). Consistency is becoming a vital factor to success. Dense text, numerous hyperlinks, large images, lengthy forms and tables are negatively affecting the browsing experience. Figure 2 displays a web page with dense text. It is obvious that so much information cannot be read in the small screen of a mobile phone even if the user zooms in.

Mobile device input is often difficult and certainly very different from a desktop computer equipped with a keyboard. Mobile devices often have only a very limited keypad, with small keys, and sometimes with no pointing device. Latest releases include track balls and touch screens, an advance that significantly facilitates user input. Lengthy URLs and those that contain a lot of punctuation are particularly difficult to type correctly. Because of the limitations of screen and input, forms are hard to fill in as well. This is because the navigation between fields may not occur in the expected order and because of the difficulty in typing into the fields. While many modern devices provide back buttons, others do not, and in some cases, where back functionality exists, users may not know how to invoke it. This means that it is often very hard to recover from browsing errors.

Mobile networks can be slow compared to fixed data connections and still have a measurably higher latency. This can lead to long retrieval times, especially for lengthy content and for content that requires a lot of navigation between pages. Mobile data transfer costs money. The fact that mobile devices frequently support only limited types of content means that a user may follow a link and retrieve information that is unusable on their device. Even if the content type can be interpreted by their device there is often an issue with the experience not being satisfactory - for example, larger images may only be viewable in small pieces and require considerable scrolling. Web pages may contain content that the user has not specifically requested for - especially advertising-related images or large images. In the mobile world this data contributes to poor usability and may add considerably to the cost of the retrieval. Cost is an issue if the user is charged by the kb.

Mobile users typically have different interests compared to users of fixed or desktop devices. They are likely to have more immediate and goal-directed intentions than desktop Web users. Their intentions are often to find out specific pieces of information that are relevant to their context. An example of such a goal-directed application might be a user requiring specific information about schedules for a journey he/she is currently undertaking. Mobile users are typically less interested in lengthy documents or in browsing lengthy pages. The ergonomics of the device are frequently unsuitable for reading lengthy documents, and users will often only access such information from mobile devices only when more convenient access is not available.

Developers of commercial Web sites should have in mind that different commercial models are often at work when the Web is accessed from mobile devices as compared with desktop devices. For example, some mechanisms that are commonly used for presentation of advertising material (such as pop-ups and large banners) do not work well on small devices.

As noted above, the restrictions imposed by the keyboard and the screen typically require a different approach to page design than for desktop devices. Various other limitations may apply and these have an impact on the usability of the Web from a mobile device. Mobile browsers usually do not support scripting or plug-ins, which means that the range of content that they support is limited. In many cases, the user has no choice of browser and upgrading is not possible. Some activities associated with rendering Web pages are computationally intensive - for example re-flowing pages, laying out tables, processing unnecessarily long and complex style sheets and handling invalid markup. Mobile devices typically have quite limited processing power which means that page rendering may take a noticeable time to complete. As well as introducing a noticeable delay, such processing uses more power as does communication with the server. Many devices have limited memory available for pages and images, and exceeding their memory limitations results in incomplete display and can cause other problems.

The above mentioned limitations apply in e-commerce sites as well. In fact, e-commerce users are much more demanding than a regular Internet user with a general interest in information browsing. Frequent on-line buyers, having used to high quality e-commerce services in the WWW, (a level of quality which was reached after some years of maturing both technologically and ergonomically), are more demanding (and less forgiving) from m-commerce sites. Penalty for low quality (probably) affects both the device used and the site visited; and the penalty for poor services is the slow death of on-line commerce: a shrinking number of visits and a resulting reduced income.

Mobile Web Best Practices

The limitations presented briefly in the previous section were noticed early on, and significant efforts, especially by the W3C were initiated in order to overcome them. The W3C mobile web best practices were born as a result (W3C, 2007; W3C 2008). They can be considered as the first step towards increasing the usability and partially the efficiency of web sites when accessed from a mobile browser. Their advantage is that they are practical however they do not embrace quality as a whole; at least quality as it is addressed by ISO.

Mobile web best practices and mobile ok basic tests are the result of two different working groups of W3C Mobile Web Initiative (MWI). The Mobile Web Initiative is led by worldwide key players in the mobile production chain, including authoring tool vendors, content providers, adaptation providers, handset manufacturers, browser vendors and mobile operators. There are nineteen MWI Sponsors: Ericsson, France Telecom, HP, Nokia, NTT DoCoMo, TIM Italia, Vodafone Group Services Limited, Afilias, Bango, Jataayu Software, Mobileaware Ltd., Opera Software, Segala, Sevenval AG, Rulespace and Volantis Systems Ltd.

The mobile web best practices document specifies best practices for delivering Web content

Table 1. Default delivery context

Characteristic	Value
Usable Screen Width	*120 pixels, minimum*
Markup Language Support	*XHTML Basic 1.1 delivered with content type application/ xhtml+xml*
Character Encoding	*UTF-8*
Image Format Support	*JPEG and GIF 89a*
Maximum Total Page Weight	*20 kilobytes*
Colors	*256 Colors, minimum*
Style Sheet Support	*CSS Level 1. In addition, CSS Level 2@media rule together with the handheld and all media types*
HTTP	*HTTP ver1.0 or more recent HTTP ver1.1*
Script	*No support for client side scripting*

to mobile devices. The principal objective is to improve the user experience of the Web when accessed from such devices. The recommendations refer to delivered content and not to the processes by which it is created, nor to the devices or user agents to which it is delivered. In other words, mobile web best practices refer to how the web content should be presented to the end user, independently to his/her device or to the adaptation mechanisms the network may use (e.g. content adaptation proxies). There is no proposition yet specifically from m-commerce.

The sixty best practice statements are grouped in five categories a) Overall Behavior: General principles that underlie delivery to mobile devices b) Navigation and Links: Because of the limitations in display and of input mechanisms, the possible absence of a pointing device and other constraints of mobile devices, care should be exercised in defining the structure and the navigation model of a Web site c) Page Layout and Content: This category refers to the user's perception of the delivered content. It concentrates on design, the language used in its text and the spatial relationship between constituent components. It does not address the technical aspects of how the delivered content is constructed d) Page Definition and e) User Input: This section contains statements relating to user input. This is typically more restrictive on mobile devices than on desktop computers and often a lot more restrictive.

In order to allow content providers to share a consistent view of a default mobile experience, the W3C has defined the Default Delivery Context, a simple and largely hypothetical mobile user agent. This allows providers to create appropriate experiences in the absence of adaptation and provides a baseline experience where adaptation is used. The Default Delivery Context (DDC) has been determined by the W3C as being the minimum delivery context specification necessary for a reasonable experience of the Web. It is recognized that devices that do not meet this specification can provide a reasonable experience of other non-Web services. The Default Delivery Context is presented in Table 1.

It must be noted that many devices exceed the capabilities defined by the DDC. Content providers are encouraged not to diminish the user experience on those devices by developing only to the DDC specification, and are encouraged to adapt their content, where appropriate, to exploit the capabilities of the device used.

Decomposing Quality Attributes

Modeling a B2C M-Commerce System

The overall idea of modeling a B2C m-commerce system is that software artefacts that exhibit different behavior when invoked require a different evaluation approach. One cannot usually evaluate with the same method different thinks and expect to get precise measurements. Thus, by recognizing the different service categories that need to be handled differently by the user, either as a process or as a user-software interface and then by grouping the provided functions to these categories we create distinct function evaluation clusters. By mapping the functions to ISO9126 external sub-characteristics we provide a focus for the evaluation. There is a strength for each relation between a function and the sub-characteristics. We consider this strength to be mostly user-perceived and so we have contacted a survey to record it. Binding these two steps together we answer the question of how to evaluate which function.

In order to model the interaction among the end user and the m-commerce system we consider four different interaction patterns: Presentation, Navigation, Purchasing and Location-based. Presentation describes how a product or service is presented to the end user. For example, a book may be presented using an image-snapshot of its content and an electronic device by a 3D animation. Navigation describes the various mechanisms provided to the end user for accessing information and services of the m-commerce system. Site structure, menus, shortcuts and all those means that facilitate the browsing process are included here. Purchasing refers to the facilities provided for the commercial transaction per se. These interaction patterns are usually applied through a browser, just as in web e-commerce. Mobile device however take into account user location. Either push or pull m-commerce services are available. In the first category, the user's location triggers software proximity switches and adds or offers from nearby points of sale may appear. The user may choose to enable or disable such services or even make a list of the products or providers of interest. Information may come in the form of an SMS or a comment-like banner on a map. Pull services are invoked by the user usually through a query mechanism. Location-based pull services provide m-commerce information based on a proximity or a geographic area query. For example, queries such as "show me the electronic retails shops that sell iPhone near my current position" or "which electronic retail shops are offering discounts in an area of about 1km around my location" are common. This type of service is not the classic B2C commerce that we are used to, but it definitely requires an on-line presence since information, either pull or push must be available to mobile users. This is a type of m-marketing or m-recommender mechanism similar to the classic recommender mechanisms of a B2C system. It is however location-based driven. The most frequent medium to access such information are maps either provided through a browser or through a map service.

Applying the above steps to m-commerce requires an adjustment to the attributes that the system presents because of its wireless communication character. In the following paragraphs we present the functions (we call them attributes because they include both services and systems characteristics) of mobile systems that constitute end user purchasing process.

The aim of this chapter is not to describe all existing B2C m-commerce attributes or fully present their use but rather to offer a quality evaluation of these attributes and to present a quality framework for m-commerce systems. The patterns are discussed in the following sub-sections while their categorization and mapping is presented later. One could say that an attribute is mapped to all external quality sub-characteristics of ISO9126 so why should there be a need for mapping? The hypothesis is correct however not all relations of the type attribute-quality sub-characteristic are of the same importance to the quality evaluation process. In fact, some of the relations are stronger than others. On the other hand none is so weak to be considered negligible. We call the strength of the relation, weight. We consider the weight to be mostly user-dependable. This means that the quality performance of an attribute is actually evaluated by the user and thus the user contributes to the forming of a strong or a weak relation

Figure 3. Presentation of a book using image, text and hyperlinks. A limited snapshot of the contents is also available

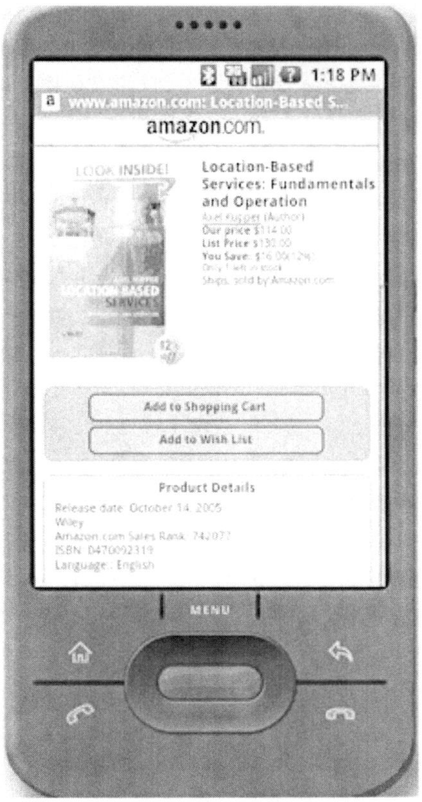

between the attribute and the expectations he/she has of the software. Expectations are closely linked to needs and thus to quality. It is difficult to measure the exact weight of a relation even when expert evaluators participate in a survey to determine them. Although exact measurements are not feasible within the scope of this research, a crude measure of the weight for each relation is calculated through a user survey.

The Presentation Pattern

Presentation is supported basically by text and images because mobile devices present limitation such as screen size and resolution, number of supported colors, computation power, memory size, rate of data transfer and energy required for proper functionality. Color usage is also important. Using colors obviously gives a pleasant and friendly interface, but a too colored screen confuses. All the pages of the m-commerce system must have the same colors so the user can feel that he/she is navigating in the same environment. By removing background images, background colors and text colors we increase the readability of the content. The use of images in Internet applications is common. Nevertheless, using images in mobile web applications significantly increases download and response time and thus, usage cost.

Presentation issues are also related with thematic consistency and the default delivery context which intends to provide an acceptable mobile environment for any end user from different mobile devices. The clarity of the text presented with meaningful, short and simple words and the presentation of the central meaning at the first page of each mobile device contributes attributes that an m-commerce system should provide to the end user for an accessible mobile environment. Additionally providing a descriptive title for the page allows easy identification of the content and by keeping the title short reduces page weight, and bear in mind that it may be truncated.

The Navigation Pattern

The navigability of an m-commerce system is a critical factor for its success. Navigation is an important design element, allowing users to acquire more of the information they are seeking and making that information easier to find.

Navigation issues support m-commerce systems quality by taking into account the quality of components such as indexes, navigation bars, site maps and quick links. The availability of these components facilitates access of information and services and enables users to locate efficiently the information they need, while avoiding usability bottlenecks. Additionally, navigation concerns the facilities for accessing information and the connectivity of the above systems.

Figure 4. Limited use of image links and a handful of only the most important shortcuts in this Amazon web page

Navigation refers at attributes that support the navigability of the m-commerce systems. These refer to navigation bars, which according to W3C Mobile Web Best Practices 1.0, should be placed on the top of the page. Any other secondary navigational element may be placed at the bottom of the page if really needed. It is important the users should be able to see page content once the page has loaded without scrolling. M-commerce systems, as e-commerce systems provide simple metaphors such as shopping cart where the end user can insert the products that intend to buy. Mobile devices present limitations on text inputting so an m-commerce system will be enabled by attributes such as access keys (keyboard short cuts), by providing defaults at any function that the user should select an action and also by avoiding free text with minimum text inputting. The navigability of the mobile system is also supported from search services which are also related with device capabilities and context presentation as well. Search with simple text inputting in an AND/OR operator format enables the user to find the information needed without navigating to several mobile pages. Search attributes can reduce the cost of mobile browsing and prevent navigability difficulties.

Additionally, because of the limitations in display and of input mechanisms, the possible absence of a pointing device and other constraints of mobile devices, care should be exercised in defining the structure and the navigation model of a Web site. Especially the use of links should be limited aiming to provide a balance between having a large number of navigation links on a page and the need to navigate multiple links to reach content.

The Purchasing Pattern

Purchasing refers to all B2C m-commerce systems attributes that strongly support their commercial character of web systems (figure 5). In particular, it refers to attributes that support the interaction with the m-commerce system. These attributes are also related to the navigability of the system but they are categorized differently because of their significant contribution to the purchasing process.

Purchasing process success is also related to the stability of the process via the m-commerce system and issues like error tolerance and error recovery at this crucial procedure. M-commerce systems success and trustworthiness is based on the system's tolerance on the above issues.

Authentication and personalization attributes support an m-commerce system where the end user can provide private information (i.e. Credit Card Number).

Figure 5. Two steps of the purchasing process: shipping details and credit cards information entry

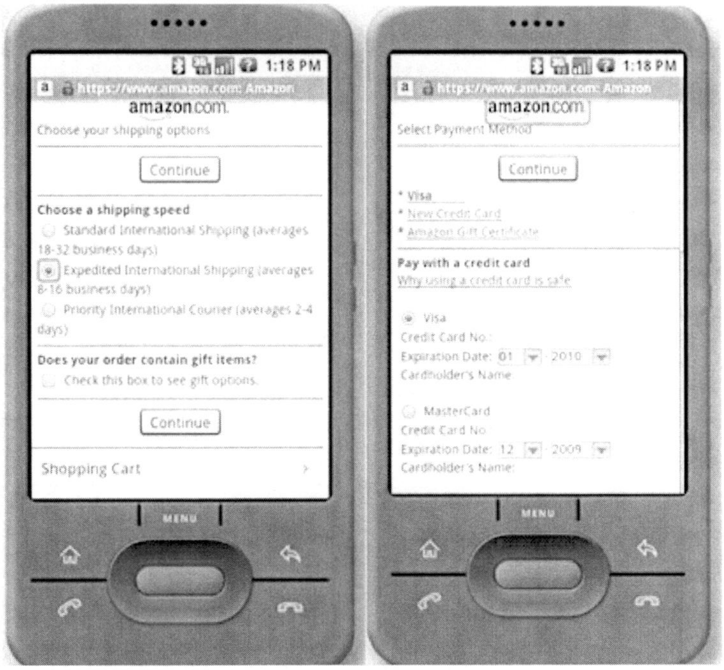

The Location-Based Pattern

Localization services can enable the presentation of the products and service because the m-commerce system can recommend the best selection based on end user's positioning (figure 6). Additionally notification services provide great advantage to m-commerce systems because they can also be combined with localized information. Alternative payment methods support either a complete transaction via the m-commerce system or otherwise combined with localized information can allow the mobile user to conclude a transaction to the closest sales point.

The main functions that make use of basic context information (e.g. the current location of the user) could be categorized as follows:

- **View:** they are generally four available view: map, traffic map, satellite and enriched map. The plain map depicts the roads and blocks of a city without any other information of interest to the user. Additional

Figure 6. Location of bookstores near the user's location. Most of them exist on-line as well

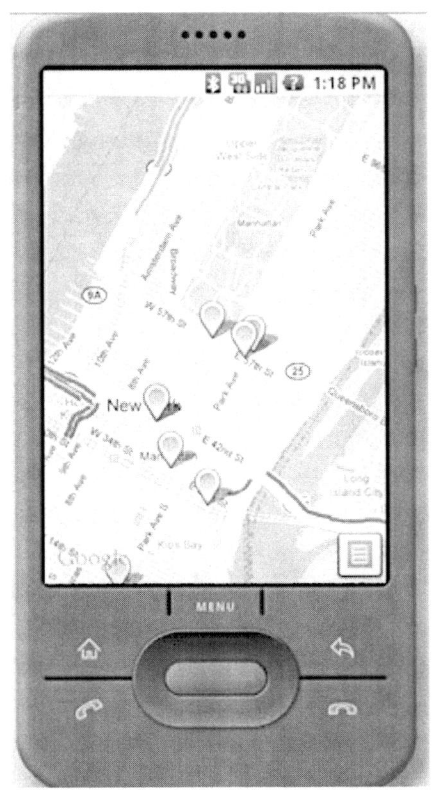

information is depicted only when the user performs a query. Traffic maps are an extension, provide traffic information. They are usefully mainly in Business to Business service like fleet management or goods monitoring. Satellite maps provide mainly terrain information and are useful for special applications that make use of geospatial services. An enhanced map contains points of interest (POI). The type of the POI is defined by the user. In m-commerce, it may include sale-points, hotels, bookstores etc.

- **Navigation functions:** In LBS the user browses information which is located in a map. Just as in the case of a browser there are standards and special functions that facilitate navigation. They include options for free moving over the map, zoom, Search for POIs in the map, directions (from starting point to finishing point), history, back and forth buttons, browsing over POIs, listing of POIs information. These functions belong to the Navigation facet of the system.
- **Context-Awareness:** attributes that make explicit use of the positioning mechanism include calculation of current location and appearance on the map, triggered messages, location-based billing (mobile vouchers), direction (from or to current location), local information services and POIs near current location.

Although the above-mentioned attributes do not directly constitute m-commerce functions they are often used as supportive functions. For example, the user query "show me in which stores near my location I can pick-up the book I bought on-line?" involves the pinpointing of the user's current location (localization) and a search for specific POIs in a region of interest.

Location-based services are either pull or push. Pull services are activated by the user (e.g. a query) and push services by the service provider (e.g. sales' offers near current location). In m-commerce notification services are the most commonly used. Location information can be mixed with time-dependent information especially as a support to mobile ticketing services.

New devices that make use of a wealth of sensors will be able to support more supporting functions that pull data depending not only on the location but on other parameters as well (e.g. orientation, speed etc.) (Wright, 2009).

A Survey for Defining Mapping Strengths

Methodology

In this experiment, three expert quality evaluators were selected in a heuristic evaluation method (Nielsen, 1990). Heuristic evaluation is done by looking at an interface and trying to come up with an opinion about what is good and bad about the interface. Ideally people would conduct such evaluations according to certain rules, such as those listed in typical guidelines documents. The evaluators for this method are IT experts with experience in quality evaluation and mobile systems as well. The nature of the presented evaluation method demands the use of expert evaluators because of its technical character.

There was a two step evaluation process. Firstly the evaluators were asked to proceed a complete purchase using a mobile phone and two different emulators from their PC. For the evaluation process we have used the Nokia N 70 mobile phone. The N70 has a screen with resolution 176 x 208 pixels and supports 262.144 colors. The phone can also connect to 3G networks for high rate data transfers using the Opera Mobile 8.51 browser. In order to avoid operability issues for the Nokia N70, help about the functionalities of the device was provided during the evaluation process. The emulators were Google's Android and OpenWave.

The three evaluators have browsed in three popular m-commerce systems according to Google Search in order to have a recent m-commerce experience. Each evaluator was asked to assess specific m-commerce attributes and evaluates each one by assigning one value of relevance (rij). Relevance defines the correlation among the m-commerce systems attribute i (presented in table 2) and software quality characteristic j ordered as they presented in the chapter (i.e. j=1 for Functionality, j=2 for Usability, j=3 for Efficiency and j=4 for Reliability) using a five-grade Liker-type scale. The evaluator may select from the Liker-type scale assigning one different value for each quality characteristic.

$$r_{i,j}\begin{cases}1, no\\2, weak\\3, strong\ \ corellation\\4, very\ \ strong\\5, critical\end{cases}$$

This provides a qualitative representation of m-commerce systems quality and especially gives emphasis on external quality characteristics.

Results

Quality evaluation of m-commerce systems attributes provides a quantitative representation of e-commerce systems' quality. The following table provides the evaluation results for the m-commerce attributes presented in the previous section. Especially presents the values of function relevance (r) for each attribute. These values are the average values of all evaluators approximated in monad.

Based on the evaluation results, quality of B2C m-commerce systems can be modeled in external quality characteristics and attributes. Providing a value for each attribute an ordered list for each external quality characteristic is provided. These values provide a first impression of end users preferences and perquisites about m-commerce systems' attributes.

The categorization of these attributes provides important feedback for m-commerce systems' assessment which is in an initial stage. By evaluating the attributes that an m-commerce system provides to the end user we also offer an end user perception of quality. End user's experience is a critical determinate of success in mobile web applications. If end users, who are also the customers, cannot find what they are searching for, they will not buy it; a site that buries key information impairs business decision making. Poorly designed interfaces increase user errors, which can be costly. A user-centered evaluation approach supports all the tasks users need to accomplish using different m-commerce systems' attributes. The above evaluation process provides measurement results which can be also be defined as metrics for a quantitative representation of m-commerce systems' quality.

In order to evaluate m-commerce systems features a new metric that summarizes the relevance of each attribute is introduced. This metric is called Mobile Attributes Weight (MAW) and it provides an evaluation weight with respect to the four quality characteristics. It is calculated by the following formula:

$$MAW = normalized\sum_{i=1}^{4} r_{ij} \in [0,1]$$

Where r_{ij} is the relevance for every listed m-commerce system attribute. The value for *MAW* provides a numerical value for every m-commerce system attribute and an ordered list about end user preference based on external quality characteristics. The values for *MAW* need to be further specified, probably with experience testing in future work and the use of different end users' groups.

MAW actually represents attributes importance for the end user and can be used at the development phase in order to define end user preferences.

Table 2. Mapping of attributes to quality characteristics per pattern and weights of the relations

M-commerce attributes	*Quality characteristics*			
	F	**U**	**E**	**R**
Presentation				
Product's description	3	5	3	3
Still images	3	5	3	1
Use of Text	4	5	3	2
Use of Colors	3	5	4	2
Use of Graphics	4	5	3	2
Clarity	3	5	4	2
Content Theme	3	5	4	1
Text inputting	4	5	4	1
Thematic consistency	2	5	4	2
Provide defaults	3	5	4	2
Navigation				
	F	**U**	**E**	**R**
Navigation mechanism	4	4	4	3
Uploading Time	3	3	5	4
Access keys	4	5	4	2
Use of Links	4	4	3	3
Help	5	5	3	3
Feedback	3	5	4	3
Undo functions	5	3	3	5
User oriented hierarchy	2	5	4	3
Redirection	5	3	4	3
navigation bar	5	5	3	1
Scrolling	3	5	4	2
Search response time	2	4	5	4
Search results processing	3	4	5	3
Purchasing				
	F	**U**	**E**	**R**
Shopping cart –Metaphor	4	5	4	2
Security mechanism	3	2	4	5
Pricing Mechanism	3	4	3	3
Alt. payment methods	4	4	3	4
Authentication	5	2	3	5
Personalization	4	5	4	2
Trans. recourses behavior	3	3	5	4
Error recovery	3	3	3	5
Errors tolerance	4	3	4	4
Stability	4	3	3	5
	F	**U**	**E**	**R**

Table continued following page

Table 2. continued

M-commerce attributes	*Quality characteristics*			
	F	U	E	R
Location-based				
	F	U	E	R
Mobile ticketing	5	5	2	3
Mobile vouchers	4	4	3	2
P2P information service	4	4	2	3
Localization	4	5	3	1
Notification service	3	5	3	4

The evaluation process provides also interesting results about the quality characteristics. In an up and down processing of values r_{ij} the WF=0,24, WU=0,30, WE=0,26, WR=0,20 values have been defined as the normalized average values for each quality characteristic. From these values arises that m-commerce end users gives great emphasis to Usability and Efficiency issues and less on Functionality and Reliability. These values differ from e-commerce systems where Usability and Functionality have equally great importance (Stefani & Xenos, 2008). In e-commerce systems the end users expects different and usable functions/services, but in m-commerce systems the end user desires the basic functions with increased efficiency as far as time and resource behavior are concerned.

CONCLUSION

In this chapter, we presented a quality evaluation for selected attributes of m-commerce systems and particularly B2C m-commerce systems. This evaluation provides an extendable framework useful for mobile system developers. We believe that this is a step towards more effective measurement of m-commerce systems' quality. We acknowledge that our attributes does not include a complete set and may not cover every aspect of m-commerce systems. The above evaluation results provide an initial research for m-commerce systems' quality.

In this chapter a new method has been introduced which measures the value of relevance for each m-commerce system attribute. The theoretical framework for this metric is also presented. The validity of the presented measures should further examine with different user groups in alternative evaluation cases and it is included in future work. It should be mentioned that the values presented are not strictly defined as numerical results but present the correlation among m-commerce systems attributes and external quality characteristics.

Practical application of the evaluation is always an issue. That is, providing tangible information to developers on how to design and develop quality m-commerce applications. A valuable tool to address this need are metrics, the bottom level of the ISO916 model. Metrics are measures of quality. While quality attributes provide a somewhat generic view of quality and for this reason they have attracted criticism for their practicality, metrics provide more information to the mobile application developer/designer. W3C mobileOK tests use such metrics for evaluating the appropriateness of web content for presentation through mobile devices. For example, the existence of long vertical scroll bars in a web site deteriorates its representation in a mobile phone where the screen is of a limited size. This metrics has two values, yes for a need for vertical scrolling and no otherwise. Although this is a somewhat rough approach to quality (i.e. there is no information on how much the vertical scrolling is, if it is existent), it provides an insight on what developers and designers expect from quality evaluation techniques: tangible information upon which design decisions can be relied. It must be noted that metrics do not make the use of ISO characteristics obsolete. They are actually the fine-grained level of the ISO9126 quality pyramid. ISO has recognized the usefulness and has included several metrics for software evaluation in the latest release of ISO9126. However, these metrics are too general to be applied in m-commerce in terms of practical impact. There is a need to produce a new set of mobile-specific web metrics, perhaps beginning with the existing corpus of web metrics and fine-tuning or alter were necessary.

There is a wealth of works that present, analyze or evaluate the use of web metrics, the majority focusing on web usability. There are no specific e-commerce metrics that could be considered the parental link to m-commerce metrics. Usability is of course an issue. But m-commerce quality is much more than that: it includes the process itself, the functionality, reliability and all the external characteristics defined in ISO9126.

Location-based services pose a new challenge. For example proximity post-sales services (e.g. special offers to clients approaching a sales point) could prove vital to a business engaged in m-commerce. M-commerce functions of the Purchasing facet may be mixed with time and location information services. So where does one start to present useful metrics for m-commerce. Using the patterns as a starting point and the existing corpus of web metrics as a basis, a categorization is possible.

Location-based will turn into context-aware in the near future. Sensors such as magnetic compasses and accelerators are already standard equipment in new mobile/smartphones. New challenges will arise when merging of personal and context data will be made available for processing by the AI-capable mobile devices of the near feature.

M-commerce is an intriguing research area with high dynamicity. New software and hardware create the opportunities for a large future user base. Increased user diversity and the provision of advanced functions to novice users requires software of high quality. Building such software is difficult and the fine-tuning of existing quality evaluation methods would help towards easing the burden of designers and programmers. User driven standards such as ISO9126, when suitably enhanced, are able complement practical initiatives as the ones of W3C. Although practicality will always remain an issue, insights on how to offer quality mobile services is feasible. The work presented in this chapter is a step towards this direction.

REFERENCES

W3C (2007). *Mobile Web Best Practices 1.0, W3C Proposed Recommendation*, 2007 Retrieved February 22, 2009, from http://www.w3.org/TR/mobile-bp/

W3C (2008). *Mobile Web Application Best Practices Working Draft*. 22 December 2008. Retrieved March 1, 2009, from http://www.w3.org/TR/2008/WD-mwabp-20081222/

Android(2009). *Android*. Retrieved March 26, 2009, from http://www.android.com

Bhatti, N., Bouch, A., & Kuchinsky, A. (2000). Integrating user-perceived quality into Web server design. *Computer Networks*, *33*, 1–16. doi:10.1016/S1389-1286(00)00087-6

Bhimani, A. (1996). Securing The Commercial Internet. *Communications of the ACM*, *39*(6), 29–35. doi:10.1145/228503.228509

BidgoliH. (2002). Electronic Commerce Principles and Practice. London: Academic Press.

BouwmanH.De VosH.HaakerT. (Eds.). (2008). Mobile Service Innovation and Business Models. New York: Springer. 10.1007/978-3-540-79238-3

Burigat, S., Chittaro, L., & Gabrielli, S. (2008). Navigation techniques for small-screen devices: An evaluation on maps and web pages. *International Journal of Human-Computer Studies*, *66*(2), 78–97. doi:10.1016/j.ijhcs.2007.08.006

Chen, R. (2005). *Modeling of User Acceptance of Customer E-Commerce Website*. Paper presented at WISE 2005, 454-462.

Clarke, I. (2001). Emerging value propositions for m-commerce. *The Journal of Business Strategy*, *18*(2), 133–148.

Cote, M., Suryn, W., Laporte, C., & Martin, R. (2005). The Evolution Path for Industrial Software Quality Evaluation Methods Applying ISO/IEC 9126:2001 Quality Model: Example of MITRE's SQAE Method. *Software Quality Journal*, *13*(1), 17–30. doi:10.1007/s11219-004-5259-6

Coursaris, C., & Hassanein, K. (2002). Understanding m-commerce. *Quarterly Journal of Electronic Commerce*, *3*(3), 247–271.

ElfriedeD.RashkaJ. (2001). Quality Web Systems, Performance, Security, and Usability. Reading, MA: Addison Wesley.

Garofalakis, J., Stefani, A., Stefanis, V., & Xenos, M. (2007). *Quality attributes of consumer-based m-commerce systems*. Paper presented at the 2007 ICETE-Business Conference, 130-136.

Ghinea, G., & Angelides, M. C. (2004). A User Perspective of Quality of Service in m-Commerce. *Multimedia Tools and Applications*, *22*(2), 187–206. doi:10.1023/B:MTAP.0000011934.59111.b5

Holsapple, C., & Sasidharan, S. (2005). *The dynamics of trust in B2C e-commerce: a research model and agenda*. Paper presented at ISeB, 377-403.

Holzinger, A. (2005). Usability Engineering Methods for Software Developers. *Communications of the ACM*, *48*(1), 71–74. doi:10.1145/1039539.1039541

Hong, S., Thong, J. Y., Moon, J., & Tam, K. (2008). Understanding the behavior of mobile data services consumers. *Information Systems Frontiers*, *10*(4), 431–445. doi:10.1007/s10796-008-9096-1

Hong, S. J., & Lerch, F. J. (2002). A Laboratory study of Customers' preferences and purchasing behavior with regards to software components. *The Data Base for Advances in Information Systems*, *33*(3), 23–37.

HuangW. W.WangY.DayJ. (2007). Global Mobile Commerce: Strategies, Implementation and Case Studies. Hershey, PA: Idea Group Reference.

ISO/IEC 9126 (2004). *Software Product Evaluation –Quality Characteristics and Guidelines for the User*. Geneva, Switzerland: International Organization for Standardization.

Junglas, I. (2007). On the usefulness and ease of use of location-based services: insights into the information system innovator's dilemma. *International Journal of Mobile Communications*, *5*(4), 389–408. doi:10.1504/IJMC.2007.012787

Kelli, B., & Vidgen, R. (2005). *A quality framework for web site quality: user satisfaction and quality assurance*. Paper presented at the WWW 2005, 930-931.

Kulkarni, N. Kumar, S. Mani, K. & Padmanabhuni, S. (2005). *Web Services: E-Commerce Partner Integration.*, IT-Pro, 23-29.

Kwon, O. B., & Sadeh, N. (2004). Applying case-based reasoning and multi-agent intelligent system to context-aware comparative shopping. *Decision Support Systems*, *37*(2), 199–213.

Lacoste, G., et al. (Eds.). (2000). *The Commerce Layer: A Framework for Commercial Transactions*. LNCS 1854, pp. 121–153.

Li, Q., & Zhang, X. (2004). *Three Dimensional Model: An Analyzing Sketch for E-commerce Theories and Applications*. Paper presented at the Sixth International Conference on Electronic Commerce, 207-212.

Losavio, F., Chirinos, L., Matteo, A., Levy, N., & Ramdane, A. (2004). ISO quality standards for measuring architectures. *Journal of Systems and Software*, *72*, 209–223. doi:10.1016/S0164-1212(03)00114-6

Marca, D., & Perdue, B. (2000).*A Software Engineering Approach and Tool Set for Developing Internet Applications*. Paper presented at ICSE 2000, Limerick, Ireland, 738-741.

Moores, T. (2005). Do customers understand the role pf privacy in E-commerce. *Communications of the ACM*, *48*(3), 86–91. doi:10.1145/1047671.1047674

Ngai, E. W. T., & Gunasekaran, A. (2007). A review for mobile commerce research and applications. *Decision Support Systems*, *43*, 3–15. doi:10.1016/j.dss.2005.05.003

Nielsen, J., & Molich, R. (1990). *Heuristic Evaluation of Users Interfaces*. Paper presented at CHI90, 249–256.

NielsenJ.MolishR.SnyderC.FarreliS. (2001). E – Commerce User Experience. Boston: Nielsen Norman Group.

Olsina, L. Lafuente, G. & Rossi, G. (2000). *E-commerce Site Evaluation: a Case Study*. Paper presented at EC-Web 2000, 239-252.

Papazoglou, M. (2001). Agent oriented support in e-business technology. *Communications of the ACM*, *44*(4), 71–77. doi:10.1145/367211.367268

Proctor, R., Vu, K., Najjar, L., Vaughan, M., & Salvendy, G. (2003). Content Preparation and Management for E-Commerce Web Sites. *Communications of the ACM*, *46*(12), 289–299. doi:10.1145/953460.953513

Saunders, S., Ross, M., Staples, G., & Wellington, S. (2006). The software quality challenges of service oriented architectures in e-commerce. *Software Quality Journal*, *14*, 65–75. doi:10.1007/s11219-006-6002-2

Stefani, A., & Xenos, M. (2008). E-commerce system quality assessment using a model based on ISO 9126 and Belief Networks. *Software Quality Control*, *16*(1), 107–129.

Wright, A. (2009). Get Smart. *Communications of the ACM*, *52*(1), 15–16. doi:10.1145/1435417.1435423

Zwass, V. (1996). Electronic Commerce: Structures and Issues. *International Journal of Electronic Commerce*, *1*(1), 3–13.

Chapter 4
A Picture and a Thousand Words:
Visual Scaffolding for Mobile Communication in the Developing World

Robert Farrell
IBM T J Watson Research Center, USA

Catalina Danis
IBM T J Watson Research Center, USA

Thomas Erickson
IBM T J Watson Research Center, USA

Jason Ellis
IBM T J Watson Research Center, USA

Jim Christensen
IBM T J Watson Research Center, USA

Mark Bailey
IBM T J Watson Research Center, USA

Wendy A. Kellogg
IBM T J Watson Research Center, USA

ABSTRACT

Mobile communication is a key enabler for economic, social and political change in developing regions of the world. Today's internet-enabled multimedia and touch-screen mobile smartphones could become the future platform for delivering information and communication technology (ICT) to these regions. We describe Picture Talk, a smartphone application framework designed to facilitate local information sharing in regions with sparse Internet connectivity, low literacy rates and having users with little prior experience with information technology. We argue that engaging citizens in developing regions in information creation and information sharing leverages peoples' existing social networks to facilitate transmission of critical information, exchange of ideas, and distributed problem solving. All of which can promote economic development.

DOI: 10.4018/978-1-61520-761-9.ch004

INTRODUCTION

We are interested in designing applications that enable people at the base of the economic pyramid (BoP) to create, share, and discuss information as is commonly done on the World-Wide Web today, but through mobile technologies. The BoP includes over one billion people with little access to computer technology living on less than $1US per day in some of the least developed countries in sub-Saharan Africa, the Indian Sub-continent, and parts of Asia and South/Central America. As others have recognized (Prahalad, 2004; Kumar et al, 2008), enabling connections among a wide spectrum of people can lead to the empowerment of the disenfranchised and enable people at the BoP to express their entrepreneurial tendencies. This could result, for example, in the creation of broader markets for local goods and services. The global reach of mobileM communication networks offers, for the first time, a broad platform for delivering applications and software services in BoP regions.

We have three long-term goals for the mobile applications we build. First, we want applications we develop and deploy to be usable by even the most disadvantaged users. Second, we want to enable these users to document local needs, problems, and issues by creating, storing, and sharing digital artifacts (e.g., maps, photos, graphics, radio news reports, music, games, TV segments, informal news). Third, we want to enable these users to engage in conversation about these digital artifacts to offer solutions, share perspectives, or to engage in social exchanges.

Our initial implementation toward these goals is Picture Talk, a social computing application framework that enhances persistent conversations with visual scaffolding. Picture Talk's social computing features support social behavior and social connections between users (Danis et al., 2009) through mobile phone conversations. Its persistent conversation feature allows users to engage in spoken discussion asynchronously. Visual scaffolding provides structure for these asynchronous voice-based communications, enabling parallel access rather than requiring serial access as is done in voice-only messaging systems. Participants in Picture Talk conversations can engage in topics of shared interest using multiple access channels: telephone (voice-only), web browser or mobile smartphone (w/data connection), and mobile phones with Multimedia Messaging Service (MMS).

This chapter first discusses some of the obstacles that BoP communities face in trying to access information technology, then introduces the Picture Talk application framework design and an implementation, and then discusses some of the particular challenges of the BoP environment for application developers.

BACKGROUND

In this section we provide background on some of the obstacles that BoP populations currently face in becoming part of the global community with access to information technology.

In the economically developed world, access to information technology has been largely through Internet-connected computers. An important benefit of access to the Internet has been the potential for contact with the worldwide community of users. The Usenet network, one of the earliest online discussion venues (created in 1979), supported threaded discussion on a wide variety of topics among participants distributed worldwide. Online communities became very popular in the 1980s and 1990s. For example, the WELL ("Whole Earth 'Lectronic Link") was a hybrid face-to-face and online group that served participants in the Bay area of San Francisco, California (Rheingold, 1993). Members of the WELL engaged in discussions of topics of common interest and the forum also served as a means of self-expression. Similar applications could be deployed to BoP communities to enable discussions on topics of local interest, provide a voice for individuals who would otherwise have no forum for their ideas, and enable solutions to communal problems through information exchange.

The rapid uptake of mobile phones in developing regions has yielded examples that demonstrate the feasibility of giving individuals at the BoP a voice and aggregating their contribution to provide value to a broader audience. For example, Ushahidi, meaning "testimony" in Swahili, is a platform for crowdsourcing crisis information. Ushahidi allows anyone to transmit geo-coded data via Short Message Service (SMS), email or web and visualize it on a map or timeline. Time-sensitive information from the public is aggregated and distributed widely (Ushahidi.com, 2009). Another example, the AfriGadget site (AfriGadget, 2009) aggregates reports "showcasing African ingenuity" that are provided through emails by individuals throughout Africa. We would like to include both citizen journalism and reader comments in our designs.

While these examples illustrate that people in economically developing regions are beginning to participate in the production and consumption of information, particularly as it is enabled by mobile telephones, they also illustrate some of the obstacles to their widespread use in developing regions. Three obstacles are germane to our arguments. First, despite initiatives such as One Laptop Per Child (OLPC, 2009), computing technology remains out of reach of the large majority of the BoP population. Lack of reliable networks to access the Internet further limits the ability of people in these regions to access information even where capable devices are available. BoP users are necessarily very cost conscious, driving a need for a low cost platform comprised of both a mobile wireless infrastructure and low-cost mobile access devices. Second, low literacy rates prevent significant portions of the BoP from using the Internet's predominantly textual interaction mode. Third, despite skills and experience in the social use of mobile phones, many BoP users may have little familiarity with or motivation to use the device to access information services, preferring face-to-face interaction. In this section we examine each of these obstacles in more detail.

Technology Landscape

The statistics for even basic access to electricity in developing regions are alarming. According to the Open Society Initiative for Southern Africa (OSISA, 2009), approximately 90% of Africa's one billion people have no regular access to electricity. Where power to homes is not available, people often travel to a centrally located solar-powered, wind-powered, or coin-operated charging station to maintain use of their mobile phone, though inventors are working on providing ways of generating electricity from personal movement (AfriGadget, 2009).

Global statistics for computer usage demonstrate huge differences between developed and developing countries. For example, the highest rates of access to the Internet in 2007 were in Sweden (82%), the US (81%) South Korea (81%), and other developed countries, whereas the lowest access rates included Tanzania (6%), Kenya (12%), and Uganda (11%). Similarly, computer ownership is lowest in Uganda and Tanzania, both at 2% (The Pew Research Center for the People and the Press, 2007).

The picture changes radically when considering mobile phones rather than computers. A recent survey by the International Telecommunications Union found that while only one quarter of the earth's population of 6.7 billion uses the Internet, nearly two thirds of the population uses mobile phones (ITU, 2009). Wireless phone use is exploding in the developing world: Sixty-eight percent of mobile phone subscribers worldwide are outside of North America and Europe. In Africa, mobile subscribers have jumped from 10 million to 400 million in the last five years (2003-2008) and the growth is still accelerating (ITU, 2009). The rate of mobile phone ownership in the Ivory Coast,

Mali, Nigeria and South Africa is over 60%, higher than in Canada (The Pew Global Attitudes Project, 2007). In 2009, India become the second largest wireless phone subscriber base in the world, after China (EE Times, 2008). Mobile phone use in economically developing regions crosses the barriers of gender, age, and education (Samuel et. al., 2005).

The exponential growth of mobile phone networks in BoP markets is fueled by the need for communication in environments where there are few alternatives. The lack of traditional wired infrastructure creates an opportunity: much of the developing world is a "green field" where new computer and communications technologies can be deployed without being hampered by existing business models, infrastructures, or user expectations. For example, in many parts of Africa, wireless networks have leapfrogged the public switched telephone network in terms of installed base. In 2007, the African continent had 280 million total telephone subscribers, but 260 million of these were mobile cellular subscribers. Building a new wireless network is faster, easier, more reliable, and less expensive than putting in a whole new wired infrastructure.

Despite the growth of wireless networks, few developing countries yet have data communications channels sufficient to provide rural populations with access to the public Internet. In 2008, only 7% of India (Internet World Stats, 2008) and 5% of Africa (Appfrica, 2008) had access to the Internet.

The capabilities of mobile phones are also increasing rapidly. In the early 1990s, few phone users would have been aware of mobile text messaging, but by 2008, almost 3.5 trillion SMS messages were sent worldwide (Portio Research, 2009). The first deployments of camera phones occurred in 2001 and by 2004, 370 million mobile phones with digital cameras were sold (InfoTrend/CAP Ventures, 2004). In the late 1980s and early 1990s, cell phones were used for voice communication only and users typed on a numeric keypad. Today's smartphones have high resolution touch screen displays, miniature keyboards, and other flexible input methods. Worldwide smartphone sales increased 12.7 percent in the first quarter of 2009 (Gartner Group, 2009) and sales are anticipated to grow at more than a 30% compound annual growth rate over the next five years. Today more smartphones are sold globally than laptops (INSTAT, 2007).

Literacy

Literacy is typically defined as the ability to read and write, however there is an inherent lack of precision that results from the methods of assessment and thus official figures often over-estimate functional literacy. For example, the commonly cited statistics, such as those compiled by UNESCO (2009), are based on census and other self-report methods which are fundamentally inexact. Also, the definition of literacy can vary from 'the ability to write a simple sentence' to 'being able to freely communicate ideas in literate society.' Individuals as young as fifteen who may have been counted as literate because they were attending primary or secondary school may, because of lack of language use, be functionally illiterate as adults (Seshagiri, Sagar & Joshi, 2007). According to UNESCO (2009), two-thirds of the world's 785 million illiterate adults are found in only eight countries (India, China, Bangladesh, Pakistan, Nigeria, Ethiopia, Indonesia, and Egypt). Low literacy rates are concentrated in South and West Asia, sub-Saharan Africa, and the Arab states (CIA, 2009), with percentages averaging in the 60s, though some countries like Mali and Niger report rates for 15 to 20 year olds of less than 30%. Men typically have higher rates of literacy than women in traditional societies (UNESCO, 2009).

While there are no generally accepted statistics on how much of the Internet is available in different languages, it is generally accepted that the dominant language on the Internet is English, making much of the Internet linguistically inaccessible to the large majority of the BoP (EnglishEnglish.com, 2009). The large number of languages spoken in BoP countries is intertwined with literacy and access to written information. While countries such as India have two official languages (Hindi and English), there are an additional 22 "scheduled" languages, and approximately 400 other languages in use by significant numbers of the population (Ethnologue, 2006). Thus individuals who may be literate in their native language may nevertheless be functionally illiterate if information is available only in one of the official languages (Plauché and Nallasamy, 2007). A report by UNESCO indicates that economically developed countries may be marginalizing speakers of hundreds of local languages (UNESCO, 2008).

Designers of applications geared towards illiterate users have focused on non-text modalities in order to design more generally accessible applications in the countries with low rates of literacy. For example, speech is a widely used modality, even in kiosks (Morris, 2000). However, limitations on the generality of speech recognition technology in multi-lingual environments (Plauché and Nallasamy, 2007) demands the use of other modalities for a broad set of functions needed to complement spoken language interfaces. For example, Joshi, Welankar, Kanitkar and Sheikh (2008) developed and tested a phonebook they call Rangoli aimed at low literacy populations. Rather than entering phone numbers based on alphabetical order, users are able to use a combination of color, icon and spatial location. Similarly, Froehlich and colleagues (2009) proposed applying digital storytelling (for example, video or sequences of still photographs accompanied with spoken annotations) as ways of enabling low literacy individuals to participate in information creation and sharing.) As noted above, even for literate users in the population, the large number of languages in BoP countries makes it unlikely that the user's preferred local language will be used in the user interface. Thus the use of pictures to augment spoken language may allow many more people to have meaningful access to information.

Social and Cultural Context

Technologies deployed in developing regions must be sensitive to the social and cultural contexts in which they operate. To provide one example, in the southeastern Indian state of Kerala, fishermen now use mobile phones to get market price information before deciding where to sell their fish (Abraham, 2007). About 40% report an increase in income and 50% report fewer losses due to unsold or spoiled fish when they start calling for prices. Interestingly, however, few of these fishermen consistently go to markets with the highest prices; instead many choose ports where their "commission agent" has a presence. Because commission agents invest in the fisherman's business (e.g., financing the purchase of a fishing vessel), the fisherman feels a social obligation to bow to the agent's wishes, even when doing so may prevent him from maximizing his income.

Several field reports illustrate specific ways in which trust in one's social network and distrust in official sources of information influence the use of computing technology. For example, farmers in the southern state of Tamil Nadu use web-connected kiosks (telecenters) fielded by a local sugar factory to ask only "simple" (i.e., low-stakes) questions of a purported agricultural expert who is not known to them, saving "high-stakes" questions for successful farmers with whom they have some pre-existing relationship (Srinivasan, 2007). Gopakumar (2006) explains that local people play a critical intermediary role in the success of telecenters. For example, living in the same village led target users of the Akshaya telecenter to develop trust in the entrepreneurs and intermediaries who ran the centers. By extension, they also developed trust in the abstract systems of medicine and government that were the ultimate sources of the information.

To summarize, these studies demonstrate the power that access to information can have in improving people's lives, but also how the impact of information is gated by social factors like trust, accountability, and social and institutional pressures.

The question we address in the remainder of the paper is: how can we address the impact of the factors we have discussed – constrained technology landscape, low literacy rates and a traditional social and cultural context – when designing systems appropriate for the billions of potential users at the base of the economic pyramid? We start by desescribing Picture Talk, a mobile social computing application framework we have developed.

A MOBILE SOCIAL COMPUTING APPLICATION FRAMEWORK

Picture Talk is a software application framework intended to support a wide range of social interactions that can be accomplished through asynchronous communication, including conversations with remote participants, question and answer exchanges, and peer production of localized content. This section begins by laying out the rationale that underlies Picture Talk by describing the scenarios and design sketches that marked the beginning of the design process. After presenting the initial vision, it goes on to describe a working prototype.

Rationale and Design Sketches

Because of the large numbers of local languages and widespread written illiteracy, speech seems like an obvious choice for supporting mediated interaction in many areas of the world. However, when speech is transposed into digital settings, many things change and a number of well-known problems arise. In the type of application we were envisioning, conversations would be asynchronous, carried out between people in different places speaking at different times. This means that Picture Talk conversations would lack some characteristics that are important for establishing and maintaining common ground (Clark & Brennan, 1991) -- "the knowledge that the participants have in common, and they are aware that they have it in common" (Olson & Olson, 2000, pp. 157). For example, it would mean that speakers would not be able to see one another, or share visual cues like glances, gestures and shrugs that in collocated speech enable interlocutors to control the conversation's flow, easily refer to objects, and verify that they are being understood (e.g., Yankelovich et al., 2004). It also potentially means that many more people can engage in a conversation, something that could be valuable but which also could exacerbate these problems.

The concept of Picture Talk arose out of consideration of these problems, and how they might be addressed in the context of a mobile phone-based communication system. The crux of the solution was to augment speech with three types of visual component: comment proxies, pictorial contexts and visual controls. Comment proxies are visual representations of digital speech that depict various types of meta-information, such as the identity of the speaker, the length of the comment, and the relationship of the comment to other comments (e.g., a reply); they also provide direct access to the comment they represent, thus mitigating the difficulty of navigating voice posts. Pictorial contexts are diagrams or photographs that provide a background for a particular conversation; pictorial contexts serve both to represent the conversation as a whole, and allow comment proxies to take on additional meaning by virtue of their location with respect to the pictorial background. Finally, visual controls are a variety of visual user interface components for controlling the system, for example, a message play button.

Figure 1 shows three early design sketches of Picture Talk developed in the context of a scenario set in rural India. (By 'design sketches'

Figure 1. Three design sketches of the Picture Talk concept for applications set in rural India: (a) Rice Talk, for farmers to discuss problems with their Rice plants; (b) Health Talk, for villagers to discuss health problems; and (c) TinkerTalk, for people in a region to indicate that they need the services of a traveling tinker. Background images of rice plant © Ivan Kopylov | Dreamstime.com, of human body © Dannyphoto80 | Dreamstime.com, and of map © Robert Adrian Hillman | Dreamstime.com. Used with permission

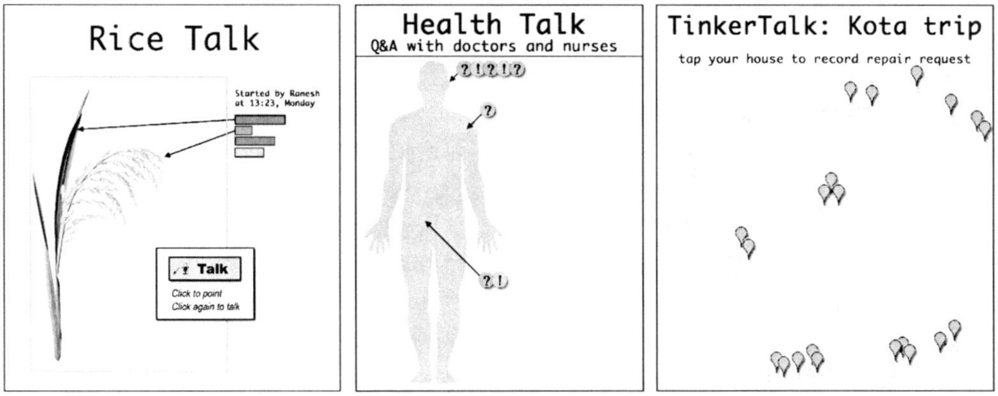

we mean provisional concepts that are intended as conversation starters with stakeholders, rather than as depictions of well-considered solutions.) The first sketch, Rice Talk, envisions an asynchronous conversation among farmers about pests and diseases affecting their rice plants. It consists of (1) a white 'card' showing a diagram of a rice plant (the pictorial context); (2) a series of colored bars (the comment proxies) that represent spoken comments, showing their durations, which of them have been made by the same speaker, and the part of the plant to which the comments refer; and (3) a floating 'talk' button (the visual control). The second sketch shows a health-oriented conversation with red and blue circles (the comment proxies) superimposed over a diagram of the human body (the pictorial context), the circles' positions indicating what aspect of the body or health they refer to and how they are related to other comments. The third sketch shows a conversation between a traveling tinker (i. e., a mender of pots) and potential customers, the pictorial context being a map of the region, and comment proxies (the red balloons) indicating where the speaker is located.

Besides communicating the basic idea behind Picture Talk – using pictures, and simple visual representations of voice comments to provide scaffolding for asynchronous speech-based communication – the sketches serve other purposes. First of all, they illustrate the flexibility of the basic concepts. The pictorial contexts, and similarly the comment proxies, can represent a large range of topics, and even when depicted as simple geometric shapes, they can represent a considerable array of meta-information. Perhaps more importantly, the sketches are useful in raising a number of questions both within the design team, and with other audiences. How do the pictures get into the system? What sort of meta-information should comment proxies depict? Do different conversations benefit from the display of different comment meta-information? What sort of visual representations will be understandable by the envisioned user populations? How do users find their ways to particular conversations? As the aim of this chapter is not to trace the trajectory of the design, it will not detail its evolution, but will instead move on to describe the user experience of the resulting working prototype.

Prototype Implementation

Our implementation of Picture Talk consists of a client application running on the Android ™ G1 ™ mobile phone and a centralized data server running an application-specific Web service in the Ruby on Rails™ (RoR, 2009) Web application server environment.

When users launch the client application on their mobile phone, their phone number is used to retrieve their user profile from the Web service. If this is the first time the user has accessed the service, they are prompted to record their name and take a picture of themselves. The user is then presented with a menu that has four options: take a picture, view the gallery of the pictures taken by other users, view the profiles of other users, or update one's own profile.

Users can start a discussion by simply taking a picture and tapping anywhere on the photo. The system stores the picture in the gallery of shared pictures and records various metadata (e.g., who started the discussion, the time and date). Additional metadata could be stored, such as the location where the picture was taken, using the built-in Global Positioning System (GPS) receiver on the G1 phone. Subsequently, users can join an ongoing discussion by finding the picture in the gallery and tapping on it, and being lead to the discussion screen.

The discussion screen (see Figure 2) has four elements: the context (a picture), comment proxies (graphics on the upper right depicting spoken comments about the picture), participant icons (a horizontal scrolling gallery of photos), and visual controls (buttons beneath the pictures of the participants to control audio recording and playback). In the spoken comments area, each graphic represents a single comment from a user. We are exploring various techniques for associating the speaker's photo with her comment.

We designed the audio controls to allow the user to compose and review a recording before posting it to the discussion for others to hear. A bar graphic is drawn under the audio buttons to reflect the length of the recording. While recording audio, the user can tap on the picture to point out something of interest in the picture, for example, the diseased part of a rice plant. The visual annotation will then be associated with the comment. When a user posts a comment, the bar graphic is posted to the discussion area to the right of the picture. The bar graphic provides a visual "residue" of the comment recording (Hollan, Hutchins, & Kirsh, 2000) for subsequent users. The length of the bar reflects the length of the recording. Pressing anywhere on the bar graphic starts playing the recorded audio and displays any corresponding visual annotation on the picture. The same set of controls is used for both recording and playback,

Figure 2. A picture talk discussion

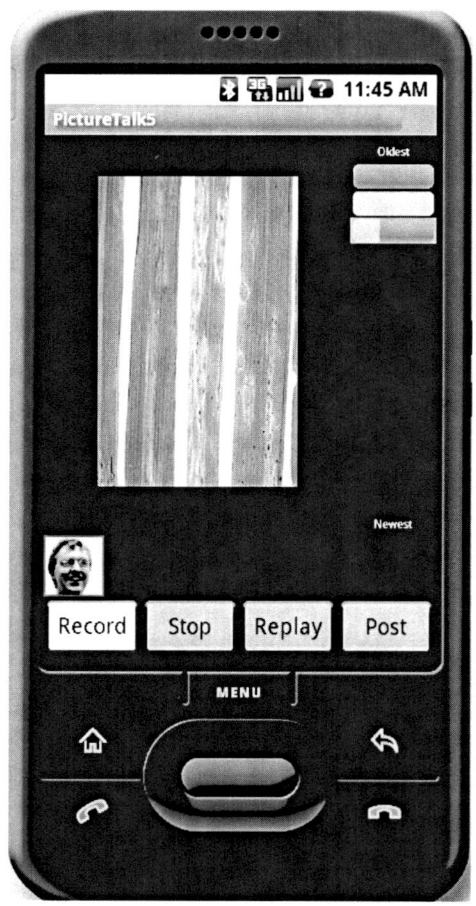

Figure 3. A visual menu of a picture talk user's social network

much like a music player. Users can pause the playback or replay the audio from the beginning. The bar graphics are listed chronologically from top to bottom in a scrolling window with the most recent always visible.

Posting a comment stores the audio, and any visual annotations, with the discussion so that subsequent users accessing the picture can access the comment's audio and visual elements. It also stores metadata (who made the comment, the date and time of their comment), posts the author's photo to the scrolling gallery of discussion participants, notifies other users in the discussion that there is a new comment, and makes the respondent's profile accessible to other discussion participants. The respondent's profile can help participants determine their trust in the information provided by the respondent. For example, a respondent may be a friend who is instantly recognizable from their photo or may be someone not known to the discussants but nonetheless reputable. The person starting the discussion is able to invite additional discussants. Individuals may block their ability to receive these notifications.

The photos of each user in the discussion are posted below the picture, in a scrolling picture gallery. Touching a user's photo leads to their user profile. The user profile has their photo, contact information (telephone number) and a scrolling gallery of the pictures anchoring discussions they have started. Touching a picture leads to a discussion screen with the given picture as the context. In this way, users can quickly find and engage in discussions started by other participants. This could be useful, for example, if a user has come across a farmer who has posted useful information about rice fungi and wants to see what other advice the farmer may have provided on other topics.

Picture Talk provides the option to view photos of one's co-discussants (see Figure 3). Touching a co-discussant's photo leads to the user profile. As users engage in discussions on a topic, their network of co-discussants grows. Photos of co-discussants can cue memory for relevant discussion contexts and serve as a visual index to organize the pictures anchoring discussions,

Picture Talk is architected as a client-server application (see Figure 4). A Java application, running on the Android Linux-based operating system, is launched from the Android phone and accesses the Web Picture Talk data server which is a server machine with a Web service running on Ruby on Rails. The Picture Talk data server provides a persistent data model for the application's objects (discussions, pictures, comments, audio clips, people, etc.). To minimize the data exchanged between clients and server (and hence conserve wireless bandwidth), the server assigns version numbers to the data objects so that both client and server know when data object updates

Figure 4. Picture talk architecture

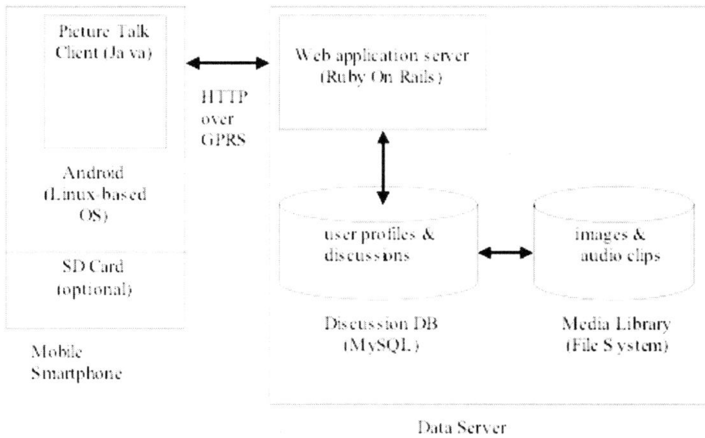

are needed to synchronize the data model. The server uses Rails' active record support to store and access the data objects in a MySQL® database. Pictures and voice recordings are stored in files.

The Picture Talk client is installed as a third-party application and runs on Google's Android open source operating system (OS). Communication with the Ruby on Rails server happens over General Packet Radio Service (GPRS), a package-oriented data service with increasing penetration into the developing world. Several wireless carriers offer compatible phones for the Android platform. The G1 ™ phone has suitable hardware for running Picture Talk client: a 3.2-inch touch-screen display, wireless networking, a microphone, built-in speakers, a camera, and gigabytes of external storage. A number of smartphones provide similar functionality, but Picture Talk takes advantage of the Android OS's capability of accessing the phone's hardware, including detecting the presence of wireless network services, recording and playback of audio, controlling the built-in camera and storing pictures on the phone and in external storage.

When the Android client has access to a wireless network, it sends pictures captured with the phone's camera and audio captured with the phone's microphone to the Picture Talk server and automatically updates the currently displayed discussion. When the phone is disconnected, new pictures from the camera are stored on its Secure Digital (SD) card, when available, or on the phone's local storage and users can still start discussions, make audio postings, listen to previously accessed audio postings, and update personal information. When disconnected, data objects are stored and retrieved from a database local to the phone using Android's SQLite software library.

Kiosk and Voice-Only Access

Given the current technology trajectory in developing nations, we expect to see increased adoption of smartphones in developing nations in the next three to five years. But in order to get early feedback on our designs, we are interested in deploying Picture Talk as widely as possible in the near term as well. Thus, we have developed a voice-only version of Picture Talk in order to make the application accessible to users of lower end phones. We have also created a web version, suitable for kiosk or telecenter use. The voice-only client allows people using basic mobile phones, commonly found in BoP environments, to listen to and record discussion comments, and even exchange pictures with the Picture Talk data server via MMS, if available. An additional server-side component, built using the open source Asterisk® Public Branch Exchange (PBX) telephony toolkit, provides voice and telephone keypad Interactive

Voice Response (IVR) interfaces for low-end mobile phones, and in turn uses the persistent data server (described above) to access discussion objects. Both of these clients access the same data as the Android client, but display that data in a suitable way for the platform at hand. For example, the rice plant anchoring the discussion in Figure 1 is sent using MMS. Subsequently, when another user wants to participate in the discussion, the Picture Talk server first sends the picture to their phone in another MMS message. Users listen to voice comments over a normal voice channel. We had to develop additional server-side functions to transform the audio and image objects into formats usable by and optimized for both wireless phones and desktop computers.

While the user experience for voice-only clients is necessarily more restrictive than with the smartphone or web browser clients, having the voice-only option makes Picture Talk discussions potentially available to a broad range of BoP users. Further research is needed to enable voice-only clients to more effectively find and navigate relevant content, share information, and tap into discussion databases that have heretofore been usable only from data-capable devices in the hands of literate users.

FUTURE RESEARCH DIRECTIONS

Many of Picture Talk's features represent general capabilities that could be applied in a variety of mobile applications. In this section, we look at several such features and discuss some additional challenges in developing applications for BoP markets and future research directions to address these challenges.

Identity

Like many social software applications, Picture Talk helps users share their appearance, contact information, and so on, to each other. However, in many developing regions of the world, it is common for mobile phones to be shared amongst members of a family or even an entire village. Adeya (2005) describes one African couple that shared one mobile phone: the wife used the phone during the day for business and the husband at night for personal calls. To address this issue, some handset manufacturers have added support for multiple address books on one phone. In some cases the very notion of ownership may be quite different from the idea of "personal property" common in developed nations. Mobile social computing applications in many BoP contexts will need to allow users to identify themselves to the system explicitly and in innovative ways (e.g., by selecting their picture or identifying a vocal sample).

Participation, Inclusion, and Viral Spread

Picture Talk promotes participation and inclusion in three ways. First, anyone who registers with the service can start and manage a Picture Talk discussion. Second, any registered user can discover and engage in discussions started by other users. Finally, as stated previously, Picture Talk provides the ability to invite others to join in a Picture Talk discussion. This ability to notify and invite people to participate, even people who are not currently registered users of the system, supports the possibility of "viral" growth of the Picture Talk user population. In developing regions where the idea of using technology to access information beyond one's social network may unfamiliar, viral spread can be a key bridging mechanism. If a friend recommends a Picture Talk discussion to a potential new user, they may be more likely to engage, find something of value, and become a "consumer" of information than if they needed to find the information themselves. And becoming a consumer can in turn lead to producing information – in the case of Picture Talk, starting a discussion oneself.

Blended Synchrony

Picture Talk implements a concept we call 'blended synchrony' (Erickson et al., 2006), meaning that the same application supports (near) synchronous and asynchronous interaction among participants. Picture Talk discussions persist over time, with remarks separated by seconds, minutes, days, or even months. Some discussions will feel quite immediate and rapid-fire, whereas others may be slower paced, or might be more like announcements than a true conversation. It just depends on the pattern of participation. Blended synchrony is useful in environments where communication needs to be close to real-time in some cases but can be asynchronous in others. Cultural and societal as well as pragmatic factors may come into play in deciding when and how to communicate with Picture Talk or any ICT application (Hudson, Christensen, Kellogg, & Erickson, 2002).

Navigational Affordances

As a conversational system, Picture Talk's audio postings could quickly grow to an unmanageable size as many users access the system. The problems with navigating a large amount of voice content are well known (Muller & Daniels, 1990). Time-varying multimedia do not offer the same navigational affordances as visual interfaces (Muller, Farrell, Cebulka & Smith, 1992). In Picture Talk, we have mitigated this problem by anchoring aural information to metadata that is made explicit through photos and graphics. For example, the author is depicted by the author's photo and the duration of the recording is shown using a graphic. Ultimately we would like users to be able to easily switch between visual or voice menus organized by authors, topics, time periods, locations, photos, tags (or other kinds of descriptive labels), and so on. A more complete solution will no doubt ultimately be needed.

Information Sharing

A number of researchers are looking at how to enable people in the developing world to share information using mobile technologies. For example, Steele and Tisselli (2006) describe three systems that enable BoP users to share information for mutual benefit. In the first, citizens documented cases of inaccessible spaces (e.g., a truck blocking a pedestrian walkway). In another case, messengers using motorcycles documented travel hazards. In a third case, nomadic Pygmies in the Congo Basin were provided with portable GPS-enabled PDAs with an iconic interface. They walked to various places and labeled trees and forest areas as food supplies, burial grounds, and so on, to prevent deforestation. In these cases, a map was used as the central visual device. Picture Talk could be extended to provide special support for maps or other types of special-purpose graphics, as we explored in the Tinker Talk design sketch.

Synchronization and Offline Use

Several applications in India and parts of Africa have been designed for mobile users without Internet access who periodically travel to areas with Internet access. For example, Prahalad (2004) reports on how ITC, one of India's largest private companies, developed a community of e-farmers with direct access to global prices, weather forecasts, farming techniques, et cetera, through centrally located Internet kiosks. Our Picture Talk implementation synchronizes content when an Internet connection is available. This provides a more flexible solution than a kiosk, where a single computer must be shared and users are unable to produce and consume information offline. However, more research is needed to understand when and where users might require connectivity and how this impacts the user experience of asynchronous conversation.

CONCLUSION

We are at an exciting point in the history of mobile computing. For the first time, the billions of people in some of the world's poorest countries have the promise of participating in the information revolution through mobile computing and communications devices. If successful, this could bring about positive social, political, and economic change in regions struggling with illiteracy, disease, poverty, natural disasters, oppression, and other challenges. Enabling ordinary citizens to become both producers and consumers of information could facilitate viral spread of critical information during crises, encourage broad exchange of ideas, connect experts with those needing help, strengthen social networks, and enable people at the base of the economic pyramid to become full participants in society and world economic markets.

We introduced Picture Talk, a software application we designed for use in environments with low literacy rates, limited Internet connectivity, and little familiarity with information services. Because basic mobile phones are the most common devices used by BoP populations, we have implemented Picture Talk on mobile phones. We are now investigating ways of providing access to some Picture Talk features on less expensive mobile phones using just voice and text messaging. The limitations of using these devices to access rich structured content by users with limited literacy skills exposes human-computer interaction challenges that are key to enabling broad access to information by people in BoP populations.

ACKNOWLEDGMENT

We thank Ketki Dhaneshal for sharing ethnographic studies of Indian villages and Nitendra Rajput, Arun Kumar, Amit Nanavati and all of the members of the IBM India Research Lab's Spoken Web team for helpful information about mobile phone use in India. We also thank John Ponzo for help with mobile phone platforms and Gail Hepworth and Steve Koeblen for helping us get started on projects in Africa.

REFERENCES

Abraham, R. (2007). Mobile phones and economic development: Evidence from the fishing industry in India. *MIT Press Journal*, *4*(1), 5–17.

Adeya, C. N. (2005). *Wireless technologies and development in Africa*. Unpublished report. Retrieved June 15,2009, Fromhttp://arnic.info/workshop05/Adeya_WirelessDev_Sep05.pdf

AfriGadget. (2009). *Harnessing Personal Movement for Power in Rural Africa.* Retrieved June 15, 2009, fromhttp://www.afrigadget.com/2009/02/12/harnessing-personal-movement-for-power-in-rural-africa/

Appfrica (2008). *The current state of Internet penetration in Africa.* Retrieved June 15, 2009, from http://appfrica.net/blog/archives/248

Central Intelligence Agency. (2009). *The World Factbook*. Retrieved June 15, 2009, from https://www.cia.gov/library/publications/the-world-factbook/fields/2103.html.

Clark, H. H., & Brennan, S. E. (1991). Grounding in Communication. In L. Resnick, J. Levine & S. Teasley (Eds.), *Perspectives on Socially Shared Cognition* (127-149). Hyattsville, MD: American Psychological Association.

Danis, C. Bailey, M., Christensen, J., Ellis, J., Erickson, T., Farrell, R., & Kellogg, W. A. (2009) Social Computing Applications for the Next Billion Users. In *Designing Future Mobile Software for Underserved Users Workshop* at CSCW 2008.

EnglishEnglish.com. (2003). *What percentage of the internet is in English?* Retrieved June 15, 2009, fromhttp://www.englishenglish.com/english_facts_8.htm

Erickson, T., Kellogg, W. A., Laff, M., Sussman, J., Wolf, T. V., Halverson, C. A., & Edwards, D. (2006). A persistent chat space for work groups: the design, evaluation and deployment of loops. In *Proceedings of the 6th Conference on Designing Interactive Systems 06* (pp. 331-340) New York: ACM Press.

Ethnologue (2006). *Ethnologue, Languages of the World.* Retrieved June 15, 2009, from http://www.ethnologue.com

Frohlich, D. M., Rachovides, D., Riga, K., Bhat, R., Frank, M., Edirisinghe, E., et al. (2009). StoryBank: mobile digital storytelling in a development context. In *Proceedings of CHI 2009* (1761-1770), New York: ACM.

Gartner Group. (2009). *Gartner Says Worldwide Mobile Phone Sales Declined 8.6 Per Cent and Smartphones Grew 12.7 Per Cent in First Quarter of 2009.* Press released dated May 20, 2009. Retrieved June 15, 2009, from http://www.gartner.com/it/page.jsp?id=985912

Gopakumar, K. (2006). E-governance services through telecentres: Role of human intermediary and issues of trust. *Information Technologies and Development*, *4*(1), 19–35.

HerringS. C.ScheidtL. A.KouperI.WrightE. (2006). A longitudinal content analysis of weblogs: 2003-2004. In TremayneM. (Ed.), Blogging, Citizenship, and the Future of Media (pp. 3–20). London: Routledge.

Hollan, J., Hutchins, E., & Kirsh, D. (2000). Distributed cognition: toward a new foundation for human-computer interaction research. *ACM Transactions on Computer-Human Interaction*, *7*(2), 174–196. doi:10.1145/353485.353487

Hudson, J. H., Christensen, J., Kellogg, W. A., & Erickson, T. (2002). I'd be overwhelmed, but it's just one more thing to do. In *Proceedings of the SIGCHI conference on Human Factors in Computing Systems*. (pp. 97-104) New York: ACM Press.

Infotrend/CAP Ventures. (2004). W*orldwide Camera Phone and Photo Messaging Forecast: 2004-2009.* London: Kluwer. Retrieved June 15, 2009, from http://store.infotrendsresearch.com/PhotoGallery.asp?ProductCode=MobileImagingStudy10106

INSTAT. (2007). *Size and Growth of Smartphone Market Will Exceed Laptop Market for Next Five Years.* Retrieved June 15, 2009, from http://www.instat.com/press.asp?ID=2148&sku=IN0703823WH.

Internet World Stats. (2008). *Internet usage in Asia.* Retrieved June 15, 2009, fromhttp://www.internetworldstats.com/stats3.htm

ITU. (2009). *New ITU ICT Development Index compares 154 countries.* (press release dated March 2, 2009).http://www.itu.int/newsroom/press_releases/2009/07.html

Joshi, A., Welankar, N., Bl, N., Kanitkar, K., & Sheikh, R. (2008, September 2-5). Rangoli: A Visual Phonebook for Low-literate Users. *MobileHCI 2008*, Amsterdam, The Netherlands.

Kumar, A., Rajput, N., Agarwal, S., Chakraborty, D., & Nanavati, A. A. (2008). Organizing the unorganized: Employing IT to empower the under-privileged. In *Proceedings of the International World-Wide Web Conference* (pp. 935-944), Beijing, China: ACM Press.

Lampe, C., Ellison, N., & Steinfield, C. (2006). A face(book) in the crowd: social searching vs. social browsing. In *Proceedings of the Conference on Computer-supported Cooperative Work* (pp. 167-170) New York: ACM Press.

MorrisT. (2009). Multimedia Systems. New York: Springer.

Muller, M., & Daniels, J. (1990). Toward a definition of voice documents. In Conference on Supporting Group Work. In *Proceedings of the ACM SIGOIS and IEEE CS TC-OA Conference on Office Information Systems* (pp. 174 – 183), New York: ACM Press.

MullerM. J.FarrellR.CebulkaK. D.SmithJ. G. (1992). Issues in the usability of time-varying multimedia. In BlattnerM. M.DannenbergR. B. (Eds.), Multimedia interface design (pp. 7–38). New York: ACM Press.

OLPC. (2009). *One Laptop Per Child.* Retrieved June 15, 2009, from http://laptop.org

Olson, G. M., & Olson, J. S. (2000). Distance matters. *Human-Computer Interaction, 15*(2), 139–178. doi:10.1207/S15327051HCI1523_4

OSISA. (2009). *Electricity for Africa?* Retrieved June 15, 2009, from http://www.osisa.org/node/4164

Plauché, M., & Nallasamy, U. (2008). Speech interfaces for equitable access to information technology. *Information Technologies and International Development, The MIT Press, 4*(1), 69–86. doi:10.1162/itid.2007.4.1.69

Portio Research (2009). *Mobile Messaging Futures 2009-2013.* Retrieved June 15, 2009, fromhttp://wwww.portioresearch.com/MMF09-13.html

PrahaladC. K. (2004). The Fortune at the Bottom of the Pyramid: Eradicating Poverty through Profits Pearson Education Inc. Upper Saddle River, NJ: Wharton School Publishing.

RheingoldH. (1993). The virtual community: Homesteading on the electronic frontier. Reading, MA: Addison Wes.

RoR. (2009). *Ruby on Rails.* Retrieved June 15, 2009, from http://rubyonrails.org

Samuel, J. (2005). Mobile communications in South Africa, Tanzania and Egypt: Results from Community and Business Surveys. *Africa: The Impact of Mobile Phones . The Vodafone Policy Paper Series, 2*(March), 44–52.

Seshagiri, S., Sagar, A., & Joshi, D. (2007). Connecting the 'Bottom of the Pyramid' – An Exploratory Case Study of India's Rural Communication Environment. *WWW 2007, May 8-12, 2007*, Alberta, Canada.

Srinivasan, J. (2007). The role of trustworthiness in information service usage: The case of Parry information kiosks in Tamil Nadu, India. In*Proceedings of the International Conference on Information and Communication Technologies for Development (ICTD),* (pp. 345-352) New York: ACM Press.

Steels, L., & Tisselli, E. (2008), Social Tagging in Community Memories. In *Proceedings of the 2008 AAAI Spring Symposium: Social Information Processing. Stanford University, ed.*, Menlo Park, CA: AAAI Press.

The Pew Research Center for the People and the Press. (2007). *The Pew Global Attitudes Project.* Retrieved October 4, 2007, from http://www.pewglobal.org

Times, E. E. (2008). *India's wireless network base will soon be the world's second largest.* Article by K.C. Krishnadas on March 24, 2008. Retrieved June 15, 2009, from http://www.eetimes.com/news/latest/showArticle.jhtml?articleID=206905386

UNESCO. (2008). *UNESCO WebWorld News | Point of View.* Retrieved June 15, 2009, from http://www.unesco.org/webworld/points_of_views/tawfik_2.shtml

UNESCO. (2009). *UNESCO Institute for Statistics.* Retrieved June 15, 2009, fromhttp://www.uls.unesco.org

Ushahidi.com. (2009). *Crowdsourcing Crisis Information (FOSS).* Retrieved June 15, 2009, from http://www.ushahidi.com/

Yankelovich, W. W., Roberts, P., Wessler, M., Kaplan, J., & Provino, J. (2004). Meeting Central: Making distributed meetings more effective. In *Proceedings of the Conference on Computer-supported Cooperative Work (pp.* 419-428) New York: ACM Press.

ENDNOTES

[1] Android is a trademark of Google, Inc.

[2] G1 is a trademark of T-Mobile USA, Inc.

[3] Asterisk is a registered trademark of Digium, Inc.

[4] MySQL is a registered trademark of MySQL AB in the United States, the European Union and other countries.

Chapter 5
Web Applications on the Move: Opening Up New Opportunities for Mobile Developers

Anna Kress
Fraunhofer Institute for Open Communication Systems (FOKUS), Germany

David Linner
Fraunhofer Institute for Open Communication Systems (FOKUS), Germany

Stephan Steglich
Fraunhofer Institute for Open Communication Systems (FOKUS), Germany

ABSTRACT

As a new platform for mobile applications the "Mobile Web" has recently gained importance. However, the Web as an application platform presents a number of limits to the application developer when compared to other application platforms, e.g. limited access to the local functionality of the mobile device. Those limits can be addressed through so-called "hybrid" application platforms which combine the best from the worlds of Web applications and locally installed applications. We believe that such hybrid applications will gain a significant market share in the nearby future. In this chapter we reflect the current state of those hybrid application platforms and analyze their advantages: After deriving general requirements for future mobile application platforms, we discuss the promises and limits of the Mobile Web platform and describe recent activities of public bodies addressing the discussed limits through "hybrid" extensions. Finally, we discuss the FOKUS Mobile Widget Runtime as a prototype for a hybrid application platform, and propose future research directions in this field.

INTRODUCTION

The market for mobile end-user applications is rapidly growing while still being shaped both in terms of new business models and underlying technologies. The interconnected questions that hereby arise are: What new kind of mobile applications will emerge in the future, as mobile devices are qualitatively different from stationary devices, and what engineering approaches and development and

DOI: 10.4018/978-1-61520-761-9.ch005

execution platforms are appropriate to enable that new kind of applications and to promote further innovation in the field.

As a new platform for mobile applications the "Mobile Web" has recently gained importance, in particular through the market appearance of a new generation of richly equipped smartphone devices. Driven by their Web access friendly hardware and software, faster mobile data networks and falling online costs, the amount of people using a mobile phone for daily Web access is constantly growing. Here, the appeal of the mobile Web is not only fuelled through a better mobile browsing experience of static Web content, but also significantly through an increasing range of user appealing innovative Web applications.

The term "mobile Web application" is often used broadly for any type of application that connects to content or services on the Web, no matter how the application is programmed, deployed and accessed, and therefore including "fat clients", that is, local, installed applications, and "thin clients", that is, applications based on the Web browser, where the bigger part of the application logic (and therefore the bigger part of the footprint) resides in the network.

Both types of applications have their specific strengths and weaknesses:

On the one side, using the Web browser provides the application developer with a ubiquitous client and opens up the possibilities of "zero install" and "zero config" as applications are accessed through simple URL browsing. Also, the usage of established Web technologies with a low learning curve (like HTML or AJAX) lowers the entry barrier for application developers both in the technological and the cost-related sense. Besides, the Web serves as a low-cost distribution platform.

On the other side, locally installed applications are usually better integrated into the target platform, allowing for example access to the data stored on the device or its sensory equipment, and allowing a more efficient execution. Equivalent possibilities for browser-based applications are still in their infancy, because the classical Web browser is a much more restricted environment.

However, the borders between browser-based and local applications are getting blurred already: On Personal Computers, browser-based applications are already provided with a set of "local" features like offline storage and offline execution (e.g. through Google's Gears), and vice versa local applications are being connected to the Web (e.g. Adobe's AIR). In the mobile field, the needs of browser-based applications to access mobile device functionalities have also been already recognized, e.g. through so-called Web widgets, that is, small, packaged Web applications running on specialized Web runtimes which are similar to a Web browser, but are more application-centric, and are better integrated into the underlying platform [Figure 1].

The result of those developments are "hybrid" application platforms profiting from both worlds.

The objective of this chapter is to reflect the current state of hybrid application platforms for mobile devices and show future market and research directions.

The outline of the chapter is the following: First, we discuss the qualitative difference of mobile devices and its influence on upcoming innovative applications and derive general requirements for a mobile applications platform. In the next step we discuss the advantages of the Web based approach, and analyze the promises and limits of the "Mobile Web platform" when compared to other application platforms. We then describe recent activities of public bodies addressing the limits of the Web as an application platform, namely, activities pursued by the W3C, the OpenAjax Alliance and OMTP BONDI. At last, we present and discuss in detail the FOKUS Mobile Widget Runtime as a prototype for a hybrid applications platform in the light of the requirements derived in the first chapter and two sample applications. Finally, we discuss future research directions.

Figure 1. Examples for extended platform access for web-based applications

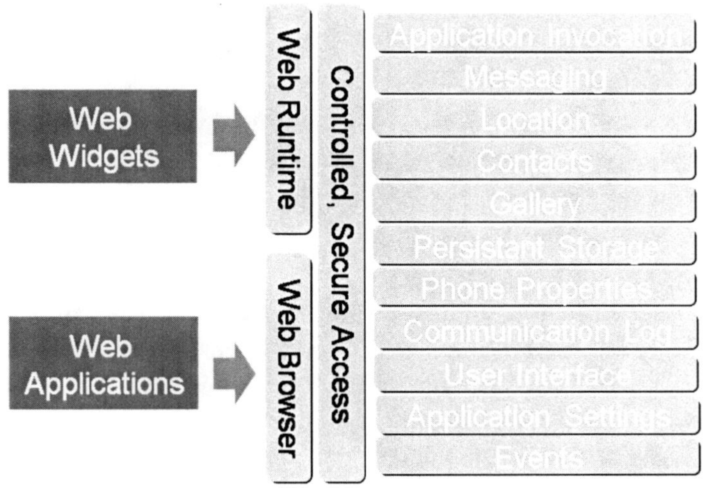

GENERAL REQUIREMENTS FOR FUTURE MOBILE APPLICATION PLATFORMS

Forthcoming mobile applications will make use of the rich on-device equipment, mash it up in a smart way with content and services on the Web, and interact with the user through novel interfaces. Here, the approach of simply "miniaturizing" existing Desktop applications – that is, scaling them down to resource-restricted devices – would fail to consider the potential for innovation resulting from how mobile devices differ from stationary devices.

Imagine for example a clever integration of your mobile address book, your and your contacts' physical location as determined through the on-device GPS sensor, and the huge amount of information and services on the Web: The result is a richer, more dynamic view of your personal relationships. It can for example additionally include a real-time map visualization of your contacts, direct links from your address book to more information about a particular contact on the Web like e.g. personal blogs or social network membership, real-time notifications about updated content like status updates or new shared photo albums, and automatic updates of your address book if contact data, like a phone number, was changed.

However, the new mobile usage scenarios will have to be supported in an appropriate way by the underlying application platform, that is, the layer on which the applications are running. While in this chapter we will derive general requirements for such future mobile applications platforms, we will discuss one application platform which we believe will gain a significant market share in the nearby future – namely the Mobile Web platform – in the next chapter.

The Qualitative Difference of Mobile Devices and Connected Challenges

Compared to stationary devices, mobile devices differ through additional capabilities, but also through additional restrictions. We therefore at first discuss those differences and connected challenges, and in the next step derive requirements for their appropriate incorporation into a mobile applications platform.

The three most distinguishing qualities of a mobile device are: its *ubiquity*, its *personalization* and its *rich hardware*.

Ubiquity: A mobile device accompanies its owner. It is this quality that determines that "location matters": A person's current physical location provides valuable input to context-sensitive applications like e.g. locators for nearby places or events that match the device owner's interests.

Personalization: The mobile device is a highly personal device storing and producing all kinds of information about the device owner. This information can be reused on application level for a personalized, context-sensitive and therefore value-added user experience. Here, especially the personal address book provides a hook to the social group of the owner and serves as a natural connection point to the world of Web content and services, like social networks, content sharing platforms and online communication tools ranging from instant messaging to blogs.

Rich hardware: Mobile devices are increasingly equipped with rich hardware including a broad range of sensors and input and output technologies (e.g. camera, GPS, acceleration sensors, touch screens, RFID chips, projectors etc.). While the sensors can provide valuable user context to applications, rich input and output technologies allow for multimodal interaction. In the latter case, for example voice- or gesture-based input can be used as a valuable alternative to the small on-device keyboard. Additionally, short range communication technologies like Near Field Communication (NFC) or Bluetooth can be used to access information or services on devices in the direct surroundings, and therefore support the mobile user when he is interacting with the nearby environment. Short range communication technologies are also suitable for peer to peer interaction without a centralized instance, like for example peer to peer exchange of contact information.

When taken together and connected to the huge amount of information and services on the Web, the ubiquity aspect, the personal information stored and produced on a mobile device, and its rich hardware equipment enable a new generation of applications which merge the virtual and the physical world.

However, a number of challenges remain:

Resource restrictions: Mobile devices are resource-restricted starting from reduced processing power to smaller screens and keyboards. Though resource-consuming operations can be sourced out into the network, mobile applications still have to deal with intermittent network disconnections and bandwidth limitations. Also, energy consumption of mobile devices still constitutes a big challenge. The gap between richly equipped and therefore power hungry devices on the one side, and the charge capacity of batteries on the other side is not yet resolved.

Security challenges: The massive usage of personal information also poses new severe security challenges, especially threats to privacy, when personal information flows are not transparently controlled and can for example be abused to track a person's behavior in a malicious way.

Heterogeneous mobile platforms: A special challenge for mobile application developers is that mobile devices differ strongly in available hardware and software, like sensors, the operating system, available libraries or browser plug-ins. This is also called the "mobile fragmentation problem", which results in the inability to "write once and run anywhere". As a result, applications have in practice to be rewritten for different platforms. Fragmentation therefore increases the required effort in the software life cycle, drives up costs, increases time-to-market and therefore imposes barriers to entry into the mobile application market, especially for smaller players.

Derived Requirements for Future Mobile Applications Platforms

From this outlined potential for innovation and connected challenges, paired with practical insights, which we gained through development and user trials of different prototypical mobile applications we derive the following requirements which should be accounted for by a mobile applications platform:

Integration of on-device content and content and services in the network: The platform should support both access to local content and services (like the mobile address book) and content and services on the network. Here, the network can serve as an execution and storage backbone: Though mobile devices are equipped with local storage and a processing unit, they cannot compete with the powerful resources available in the network. Also, the support of inter-connection of on-device content and content and services on the Web (so called mash-ups) will allow for more powerful, value-added applications.

Integration of rich hardware: The platform should provide applications access to the rich hardware of mobile devices. For example access to the GPS sensor allows the realization of innovative location based services.

Support for mobile user interfaces and mobile user interaction: Presentation and handling of applications should consider restrictions like small screens and keyboard, and allow usage even when the user is moving. Here, multimodal interaction through integration of rich input and output technologies can be used as a valuable alternative. Also, usage of applications "on the move" often implies that up-to-date information does matter; for example time critical application events in the network (like location-sensitive frequently updated information) should be pushed to the user immediately.

Connection awareness: Distributed applications should remain useable (to the extent allowed by the application logic) even when the network connection is interrupted, either through network failure or on purpose. That means that offline data storage and offline execution should be supported.

Cost awareness: Users should be aware of the costs caused by mobile applications, which holds for costs in terms of money as well as for costs in terms of physical resources like power consumption or data traffic.

Security: The platform should account for security threats, and here especially threats to a user's privacy. Personal information flows should be transparent to the user.

Addressing the fragmentation problem: The fragmentation problem is a complex issue, and it is questionable, if it can be solved completely, because differences in devices are sometimes intentional (Rajapakse, 2008). For example, users may have different preferences on the device size and, implicitly, on its hardware equipment, which determines the size of the device. However, some differences may be not intentional, like installed or missing libraries and available application programmer interfaces (APIs). Here, standardization efforts can help. How the application platform itself may ease the problem is still a valid research issue. For example information about the device which can be accessed by the application at runtime may allow appropriate dynamic adaptation of the application.

Figure 2. The mobile web platform

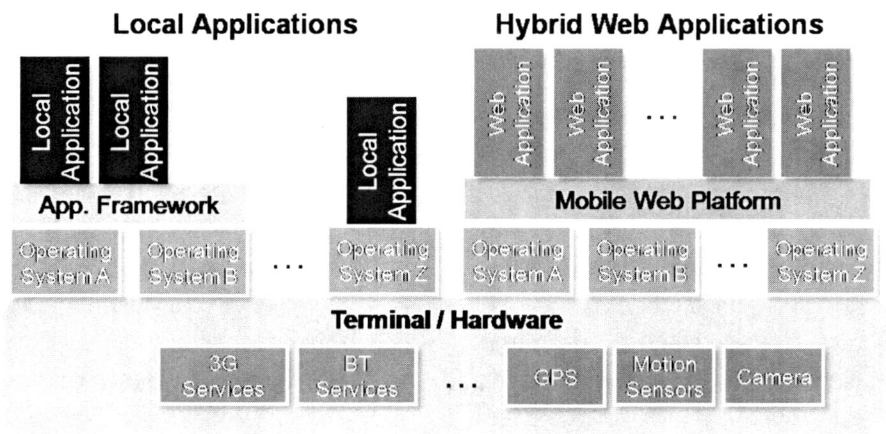

THE MOBILE WEB PLATFORM

As a new platform for mobile applications the "mobile Web" has recently gained importance. Here, the term "mobile Web application" is often used broadly for any type of application that connects to content or services on the Web, no matter how the application is programmed, deployed and accessed, and therefore including "fat clients", that is, local, installed applications, and "thin clients", that is, applications based on the Web browser, where the bigger part of the application logic (and therefore the bigger part of the footprint) resides in the network and can be accessed by browsing to its URL.

Recently, a third type of mobile Web applications running on top of so-called Web runtimes is gaining ground: Web runtimes use similar technologies as Web browsers, but are more application-centric and better integrated into the underlying hardware and software of the device. Considering their footprint and functionality, Web runtimes may be placed between local and Web-browser based clients, because they allow both local execution of applications and remote execution in the network. Additionally, Web runtimes usually address simple lightweight applications (often referred to as Web widgets).

This third type of Web applications is what we call "hybrid" applications, that is, applications that utilize the "Mobile Web platform", but can also execute locally and utilize local functionality of the mobile device. We believe that such "hybrid" applications – either running on top of Web runtimes or on extended Web browsers – will gain a significant market share in the nearby future.

In this chapter we discuss the reasons that lead us to our conclusion: We first describe our view of what the term "Mobile Web platform" means in this context, and analyze the promises and limits of the mobile Web when compared to other application platforms. We then describe recent activities of public bodies driving the evolution of the mobile Web to address these limits – namely initiatives pursued by the World Wide Web Consortium (W3C), the OpenAjax Alliance and OMTP BONDI. These initiatives show that the interest in "hybrid" application platforms is on the rise. In the next chapter we then present our own research in this area, namely a prototype of a hybrid application platform.

What Is the "Mobile Web Platform"?

The term "Mobile Web platform", which we understand as an abbreviation for "the Web as a platform for mobile applications" is not easy to nail down, because the Web platform is decentralized, open to extensions, and under perpetual heavy evolution. Though the foundations of the Web – and therefore also the Web platform – are laid by the World Wide Web Consortium (W3C) (like the HTTP protocol and HTML, the Hypertext Markup Language), in practice the Web platform is driven by different independent actors. Here, a prominent example from the recent years is the unforeseen rise of Ajax (Asynchronous JavaScript and XML). This is also the biggest advantage of the Web as a platform – it is open for innovation by its very nature.

None the less, we define some of the cornerstones of the Mobile Web platform as following:

In general, an application platform is the layer on top of which applications are running. From one point of view it can be described as a set of application programmer interfaces (APIs) and conceptual models, which define how the programmer is supported or restricted by the platform. From another point of view an application platform can be described as a set of hardware and software components, which define what has to be present (or has to be additionally installed) on the end user device or in the network (in case of a distributed application).

According to those views we define the Mobile Web platform as a collection of Web-related, non-proprietary, that is, *open* (though not necessarily formally standardized) protocols and languages (most notably HTTP, HTML and JavaScript); established architectural concepts for distributed Web applications (e.g. REST or Web Services); APIs and technical components allowing deployment and access to content and services located on the Web (the most important being the Web server and the Web browser or, alternatively, the Web runtime).

Additionally, advanced Web applications make heavy use of Ajax (Asynchronous JavaScript and XML), which is a combination of several of the technologies and components described above, and which evolved into an integral part of the Web platform. Ajax allows to build rich Web applications with advanced user interfaces, e.g. "drag&drop" of objects inside a Web browser, and a user interaction behavior similar to that of Desktop applications: Instead of reloading the whole interface as response to a user interaction as is the case with classical Web pages, Ajax asynchronously refreshes only the relevant parts of the interface. This is achieved through small amounts of data which are exchanged with the network in the background.

The requirements for Ajax (and therefore parts of what we regard as the Web platform) include support for a structured document format like HTML, where the document constitutes the user interface; JavaScript and document manipulation functions like DOM (Document Object Model) to control the behavior of the user interface; and the XMLHttpRequest object (XHR) for asynchronous retrieving of data from the network.

In some cases due to resource-restrictions of mobile devices, subsets or adoptions of the described technologies and components may be more appropriate (and exist in practice) for the Mobile Web platform as opposed to the "Desktop" Web platform (e.g. XHTML Basic). An important point here is what is actually supported by the end device. This is a difficult issue due to the "fragmentation problem" of the mobile market, which we discuss later on in this chapter.

Also, though a number of popular proprietary formats and technologies exist (like e.g. Adobe Flash) that are utilized for Web applications, we do not include them into our definition of the "Mobile Web platform" because they contradict the "openness" approach of the Web platform, which is in our view a crucial driver for innovation.

Besides the technological components described here, the "Mobile Web platform" is equally defined by its conceptual promises and limits. Those promises and limits are described in the next two subchapters.

Promises of the Mobile Web Platform

The "Desktop" Web already demonstrates the potential of the Web as an application platform by providing numerous examples of user-appealing interactive Web applications which are accessed through the Web browser. Though in practice some problems arise, which we will discuss later, this approach holds also the following promises for mobile Web applications:

Cross-platform portability: Ideally, browser-based applications are cross-platform portable, because they can be accessed from any of the various Web browsers, no matter e.g. on top of which operating system the browser is running. Here, the Web browser provides the application developer with a ubiquitous client, because one or the other browser variant is usually pre-installed on contemporary mobile phones.

No necessity for manual installation, configuration or updates: The Web browser opens up the possibilities of "zero install", "zero config" and automated application updates as applications are accessed through simple URL browsing, and can be configured and updated in the network of the application provider. This is especially attractive on mobile devices where the user does not wish to tamper with the device and application settings due to its small screen and keyboard.

Attractive user interfaces trough mobile Ajax: Mobile Ajax is the extension of Ajax to mobile devices. The advantages of mobile Ajax are the same as those of its Desktop version: a richer user experience without having to use proprietary technologies or need for additional software components beyond the browser; less data/bandwidth being consumed because only relevant data is refreshed; using open standard Web technologies developers are already familiar with, which means, a lower learning curve and a faster time to market. A further advantage of Ajax is its dependency on a set of technologies that come built-in with any contemporary Web browser. Ajax applications also reside in the network, and therefore result in a thin client.

The Web as an execution and storage backbone: The mobile device is a resource-restricted device. Here, the Web can be used as an execution or storage backbone, where resource-consuming operations can take place in the network. Also, when storing data in the network, users may become device independent, as they may access the data from other devices as well. Additionally, smart usage of third party content and services on the Web (so called mash-ups) can lead to more powerful, value-added applications.

Lower technology- and costs-related entry barrier: The Web as a platform lowers the entry barrier for application developers both in the technological and the cost-related sense, when compared to other technologies used in the mobile field: The usage of established Web technologies with a low learning curve (like HTML or JavaScript) allows tapping into the creative potential of an already existing big community of yet "not-mobile" Web developers. Also, the Web community already originated numerous solutions and algorithms for numerous of specialized and standard problems, which are shared as open source projects. Consequently, developers of applications for the Web can assemble their solutions by utilizing present code and components from third parties and significantly safe time and thus resources. Besides, the Web also serves developers as a low-cost distribution platform.

Limits of the Mobile Web Platform

The limits of the Mobile Web platform are the following:

> **Dependency on a network connection:** Web-based applications may turn useless when disconnected from the network, though here local storage and execution may help, at least to a certain degree allowed by the application logic.
> **Sandbox model of the classical Web browser:** The Web browser follows the sandbox security model, which isolates the content loaded into the browser from the underlying system, and therefore does e.g. usually not allow loaded Web applications to access sensors or other services running on the device. However, the sandbox model is currently being questioned, either through Web runtimes integrated into the device, or through direct Web browser extensions as e.g. proposed by OMTP BONDI initiative (described below).

However, device-level extensions of the Web platform have to deal with the fragmentation problem: An extension has to cover not only a significant share of the various browsers, but also of operation systems. Also, the "closed garden" model of many mobile OS vendors makes the extension of the Web platform impossible, if not technically, then legally.

> **Portability and fragmentation problem:** Compared to fat local clients, thin or lightweight (that is, Web browser or runtime) clients share a number of difficulties concerning portability, though in some aspects, thin or lightweight clients have a slight advantage:

The application platform of fat clients is either directly the operating system (OS) or an additional runtime component which is to a certain degree OS independent, but with hooks into OS functionality like access to the file system. Prominent examples for operating systems in the mobile field are Symbian OS, Windows Mobile or more recently Android; a prominent example for a fat client runtime component is Java Micro Edition (J2ME).

OS based clients are integrated into a particular OS, and therefore have the advantage not to depend on additional components that have to be installed on the device. But it can be a time- and cost-consuming task to port them to devices using other operating systems, which usually have different APIs and conceptual programming models.

On the contrary, a client based on a runtime environment (RE) like J2ME does not face the problem of portability in the same degree as long as the RE is provided for different operating systems and encapsulates their differences through an abstraction layer. A disadvantage of the RE is that is has to be additionally installed on the device and is usually not really a light-weight component.

Though here the Web browser has the advantage that it is in principle a ubiquitous client, in practice subtle problems arise. For example, Ajax is still not ubiquitous on mobile devices. If Ajax is available, developers cannot be sure if full Ajax or only a subset is supported. The problem of specific extensions was already discussed above.

> **Web programming model limits:** The capabilities of application engineering in Web browsers are limited: JavaScript, the Ajax scripting language, does not include advanced programming concepts like multi-threading; also the execution of an interpreted scripting language is less efficient than that of a compiled language.

The presentation model based on HTML is appropriate for text-based content, but not adjusted to graphics on which rich Web applications rely.

Also audio and video rendering are currently not natively supported. Though native support was introduced in HTML5, this standard is not implemented yet by the major browser vendors.

The classical browser lacks in general support for common application platform concepts (as e.g. found in the Java Virtual Machine or in operating systems), because it was originally designed as a viewer application for Web content, not as an execution environment for applications. The lacking concepts are e.g. application life cycle management, management and isolation of multiple concurrent applications and their allocated resources, or inter-application communication channels.

Evolution of the Mobile Web Platform Through Public Bodies

World Wide Web Consortium (W3C)

The World Wide Web Consortium (W3C) is an international consortium. Its mission is to "lead the World Wide Web to its full potential by developing protocols and guidelines that ensure long-term growth for the Web" (W3C, 2009). The W3C addresses "Web interoperability" by publishing open standards for Web languages and protocols to avoid Web fragmentation.

In this context especially the following activities of the W3C are important: the evolution of HTML to HTML5, the W3C Mobile Web initiative (W3C, 2009) the W3C Web Applications (WebApps) Working Group (W3C, 2008) and its Widget Working Draft documents (W3C, 2009):

The HTML5 specification (W3C, 2009) released last year introduced a set of new features which are relevant in this context, e.g. specification of APIs which allow Web applications running in browsers to store data in local databases. W3C's Mobile Web Initiative is currently focusing on developing best practices for mobile Web sites and Web applications, device information needed for content adaptation and test suites for mobile browsers. The W3C WebApps Working Group is documenting existing APIs for Web applications and developing new APIs for richer Web applications. The group is also working on Web widgets specifications. Here, a number of working drafts has already been released. The W3C widget draft documents also influence the OMTP BONDI initiative as W3C is a member of BONDI.

OpenAjax Alliance / Mobile Ajax

The OpenAjax Alliance is an organization of companies, open source projects and other bodies dedicated to the adoption and evolution of open and interoperable Ajax-based Web technologies (OpenAjax Alliance, 2009). Its members include e.g. industrial leader companies like Vodafone, Sony-Ericsson, Microsoft, Opera, Google and Oracle, but also standardization bodies like the W3C. Though the OpenAjax Alliance itself does not intend to be a formal standardization body, its members do engage in standards-related activities, e.g. in the context of the Mobile Web Initiative (MWI) of the W3C. They are also providing reviews and feedback on activities of the OMTP BONDI standardization initiative, which is described in the next subchapter.

One of the committees of the OpenAjax Alliance is the Mobile Task Force which focuses on Mobile Ajax activities. Here, the declared goal of the Alliance is not to create technology subsets or profiles, but to use the same standard HTML and JavaScript technologies as those used for the Desktop Web. However, the Alliance admits that Ajax is still an emerging technology for mobile phones, and though Mobile Ajax support is growing, the market will stay fragmented in the future. That is, developers cannot be sure that mobile browsers will fully support their Ajax application, especially, when advanced features are used. As a remedy, the Alliance in partnership with the W3C Mobile Web Initiative is proposing a device descriptions repository listing the Ajax capability of mobile devices, so that server-side adaptation tools can be used to deliver appropriate content to different devices.

Indicatively, future working directions of the OpenAjax Alliance show that its members are supporting the trend towards a "hybrid" Mobile Web application platform. The Alliance identifies as future key Ajax features: offline support for Ajax applications; access to device services like location, contact lists or the phone dialer through additional JavaScript APIs; and support for Ajax beyond the classical browser, that is, inside of Web runtime engines for locally installed Ajax applications.

OMTP BONDI

Some of the members of the OpenAjax Alliance are also members of OMTP BONDI (Open Mobile Terminal Platform; BONDI because "like the Australian beach, OMTP wants mobile customers to have the greatest surfing experience whilst making the experience as safe as possible").

OMTP BONDI is an operator driven initiative with the aims to standardize key APIs to sensitive functions on the mobile device, and to protect the user from malicious applications abusing those APIs through user controlled security policies (OMTP BONDI, 2009). Currently, full members of OMTP are AT&T, Hutchison 3G, Orange, Telecom Italia, Telefónica, Telenor, T-Mobile and Vodafone. Additionally, OMTP has the support of Ericsson and Nokia and further participants from all parts of the mobile industry, including hardware and operating systems providers and application software developers.

Besides the specification of the APIs and security policies, BONDI develops an open source reference implementation. A first version of the reference implementation for Windows Mobile including a set of sample mobile widgets is available at the BONDI website.

BONDI addresses both applications running in a browser and installable mobile widgets, that is, small packaged Web applications. For mobile widgets BONDI utilizes the W3C Widgets specifications, and works closely with the W3C Web Applications Working Group. The BONDI reference implementation and its sample widgets use Ajax; however, both W3C Widgets and OMTP BONDI specifications are language and technology independent, so that other alternatives like SVG instead of HTML may be used.

CONCLUSION

As stated in the previous chapter, we believe that forthcoming mobile applications will make use of the rich on-device equipment, mash it up in a smart way with content and services on the Web, and interact with the user through rich novel interfaces. Here, the approach of the "Web platform" as opposed to the approach of fat clients is promising for a number of reasons which we described, though Web platform based applications are still in their infancy.

It remains to be seen if and when the described standardization activities will change the situation. Standardized APIs open the mobile device up for 3rd party application providers, and therefore decrease the control of main actors in the mobile field like operators or device manufacturers over what applications are installed on the device by end users. As those actors increasingly play the role of application providers themselves, e.g. through branded (and strictly controlled) application stores, the open model may lead to conflicting interests. Additionally, operators are security-conscious as they fear that end customers may blame them for any security problems resulting from access to sensitive device APIs. Also, the evolution of the Web is a slow process as the standardization of HTML5 has proved again. So far it is realistic to assume that open platforms will co-exist with closed, non-interoperable application platforms for some time in the future.

Also, domain-specific extensions of the Web Runtime environment again raise the platform fragmentation problem. In practice the situation will stay difficult as the fragmentation problem is a complex problem. The application can then be either developed according to the lowest common denominator of the targeted range of mobile devices, or utilize a feature that is only available on a small range of devices.

HYBRID PLATFORM CASE STUDY: FOKUS MOBILE WIDGET RUNTIME

In this chapter we describe the FOKUS Mobile Widget Runtime, our prototype of a hybrid platform for mobile Web applications. Furthermore, two example applications are presented: a Geo-Caching Widget and a Car-Sharing Widget.

Overview

The interest for the Web as platform for mobile applications comes from the shift from document browsing to execution of interactive, networked applications. The application model for those applications was introduced and established more than a decade ago. The base of this model is called DOM scripting: When the Web server transmits a hypertext document, the document contains not only HTML code, but additionally interpretable code. The embedded code then defines the behavior of the document on user actions in the Web browser, e.g. mouse clicks or pressed keys.

The syntax for the embedded code is specified as ECMA262, better known as JavaScript. DOM scripting enables a Web developer to assign JavaScript functions to input actions of the user. Additionally, JavaScript code can modify the DOM of the document it is embedded with. These modification capabilities range from changing colors and fonts to adding or removing complete document elements. Thus, the code embedded within a Web page is able to listen to user inputs and can completely control the graphical output as far as DOM allows. For this reason, the current generation of Web browsers is more an application runtime environment than a crawler for static documents. Extensions to the JavaScript environment like the Ajax XMLHttpRequest object strengthen this new role (Kesteren, 2008).

Recently, besides Web applications Web Widgets are gaining ground. The most important innovation of Web Widgets is the pre-fetching of content. While in regular Web applications documents (e.g. HTML, images, style files, SVG, JavaScript files) are transmitted from server to client when needed, the request of a Widgets results in the delivery of all documents that belong to this Web application, regardless if needed at that moment or not. This model enables Web applications based on a Widget scheme to operate as usual even if the origin Web server is physically not available. In reverse, a Widget is a Web application that contacts its origin server exceptionally. From a conceptual point of view, Web Widgets and Web applications that can access client storage (e.g. as enabled by Google Gears) are equally powerful as e.g. Google Mail proves.

Our work was aiming at a runtime environment for rich, location-based applications on mobile phones, based on a Web application model. As communication with the origin server on mobile devices can be occasionally interrupted, we decided to use a widget-based model. The resulting widget user agent, the FOKUS Mobile Widget Runtime (MWR), realizes a rendering and execution environment for mobile Widgets as well as tools to manage Widgets on the mobile phone. As introduced above, a mobile Widget is a compressed package of Web documents, containing, for example, text, images, graphics or interpretable scripts. The package is deployed to the mobile phone through a Web download by the user. Each package contains at least enough information to provide meaningful presentation and usability to

Figure 3. FOKUS MWR architecture overview

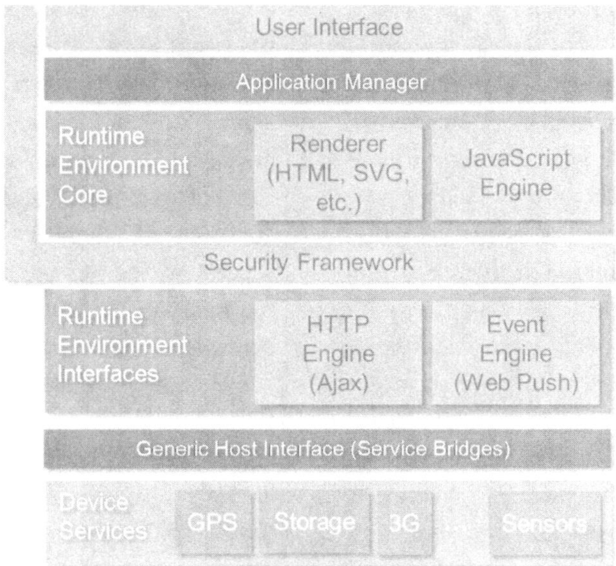

the user. Further documents that are required based on runtime decisions of the user can be requested on demand from the network.

An evaluation version of the FOKUS MWR with a restricted set of features, sample applications and a developer guide can be downloaded free of charge at myLab, our laboratory for research of technologies for Web and Web 2.0 Web site:

http://mylab.fokus.fraunhofer.de/platform/mobilewidgetruntime/overview

Architecture

The architecture of the FOKUS MWR comprises the seven major parts depicted in Figure 3: Runtime Environment Core, Runtime Environment Interfaces, Generic Host Interface, Device Services, Application Manager, User Interface, and Security Framework. The Runtime Environment Core contains components required for the operation of the basic Web application model explained above. The Runtime Environment Interfaces describe the means of communication through the network provided by MWR to Widgets. The Generic Host Interface extends the Widget application model and realizes a dynamic bridge to the services and capabilities uniquely provided by the host system the MWR is executed on. The User Interface provides access to the Application Manager, which enables the user to download, start and pause Widgets, as well as to the Security Framework. The Security Framework controls the access of Widgets to the network interface and services of the host system based on user-defined rules.

Runtime Environment Core

The core architecture of the MWR follows the REST architectural style (Fielding, 2000): On demand of a client the server responds with representations of the requested resources. To address temporal unavailability of the server, for example in case of an interrupted data connection, representations for all resources constituting an application are pre-fetched from the server. The types of representations supported by the Runtime Environment depend on the configuration in terms of available interpreters and renderers. For a minimum application model a renderer for textual and graphical representations and an interpreter for a script programming language are required.

MWR is intended as a platform for mobile applications based on Web technologies, not a common Web browser. Accordingly, the capability for rich graphical presentations is more important than the presentation of long hypertext documents. The basic markup language for presentations in MWR is Scalable Vector Graphics (SVG). Compared to Hypertext Markup Language (HTML) SVG has several advantages with regard to the targeted mobile environment. SVG supports by nature the presentation of 2D graphics (including color gradients, transparency and object transformations) and thus reduces the need for raster images. The reduced number of raster images has a direct impact on the amount of data that needs to be transferred from the server to the client and thus on the costs. However, form elements as known from HTML, e.g. text input fields, radio button, or checkboxes need to be implemented by the application developer.

In addition to the SVG renderer the Runtime Environment comprises an interpreter for JavaScript code. The application model enabled by SVG and JavaScript supports the application developer to control the presentation by manipulating the SVG immanent Document Object Model (DOM) and to observe keypad/mouse inputs by the user. In addition to these basic functions of the Runtime Environment Core, MWR implements access to several value-added functions and functions of the host platform.

Runtime Environment Interfaces

The basic application model of MWR Widgets supports communication with the Widget origin server based on the Ajax concept. However, Ajax interactions are client-initiated and based on polling. In mobile community applications the overall state of the application may change with any action of a user on the Widget. To keep the application responsive and information presented to the user up to date, MWR would have to poll for updates continuously. However, continuous polling affects the volume of data transferred between server and client, the processing load at the client, and the time of application availability. For this purpose the Ajax API is complemented with a Resource Event API. The Resource Event API enables developers to subscribe Widgets at runtime to resource updates. The Resource Event API is based on a publish/subscribe protocol. Only when a Widget receives the update notification, it may in the next step decide to request the current state of the resource.

The practical integration of both HTTP and the event protocol is illustrated in Figure 4: POST and GET denote common HTTP methods for interactions with a resource. SUBSCRIBE, UNSUBSCRIBE and NOTIFY are the methods of the event protocol. The names *uri*, d_1 and d_2 are constants, whereof *uri* denotes a resource by its Uniform Resource Identifier (URI), d_1 data that is supposed to change the state of this resource, and d_2 the resulting representation of this resource. If the resource addressed by *uri* is static, d_1 and d_2 can also be equal. A client subscribes for the change of a resource, e.g. initiated by requests of other clients through one of the methods PUT, POST, or DELETE. If such a change occurs the server notifies the subscriber via a push channel about the URI of the affected resource. The subscriber may then decide to obtain the latest state by another GET request. If no further notification on resource changes is required, for example if the application is terminated at the client, the server can remove the subscription.

Figure 4 shows just a snapshot of the resource event mechanism. The subscriptions require a periodic renewal, where multiple subscriptions can be summarized as one request. Nonetheless, the number of messages to renew a subscription is below the number of polls continuous resource updates would require. Moreover notifications contain hash keys for the current resource representation to enable the comparison of notified resource change and later retrieved representation. The event push concept is designed for implemen-

Figure 4. Integration of HTTP and an out-band event protocol

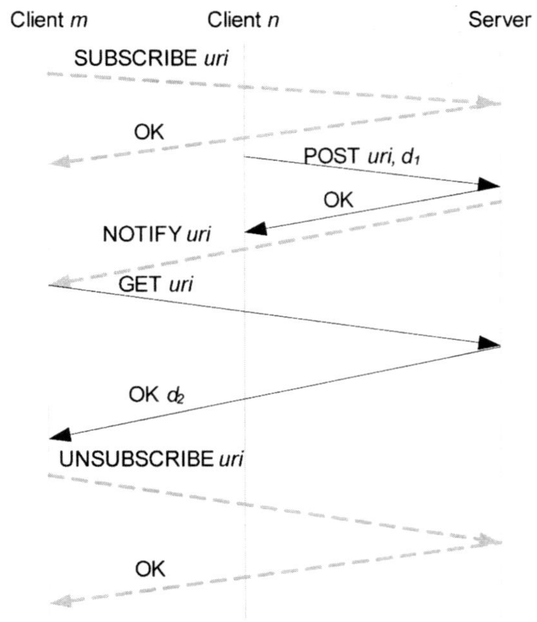

tation with the Session Initiation Protocol (SIP) and its extension for event notifications described in (XMPP.org, 2009). A working group for a new version of HTML, the Web Hypertext Application Working Group (WHATWG) and a community headed by the IETF called HyBi (IETF, 2009) follow the same direction with long-polling and similar techniques. For this purpose they build on efforts like Comet (Russel, 2006). The basic differences of these efforts to the approach we chose for MWR are the focusing on TCP as transport protocol and the integration of data communication and signaling. For instance, in Comet the resource update is communicated directly, and not through a "resource update" message.

Generic Host Interface and Device Services

While platform independence and uniformity are regarded as two of the central reasons for the success of the Web, recent efforts push for stronger platform integration and specialization. BONDI is only one indicator for this trend. Web application developers should be allowed to make use of the distinguishing features of mobile phones such as the camera, acceleration sensor, GPS, power management, contact list, local file system or call control. But such specialization inevitably leads to platform fragmentation, as resources found on one device may not be available on another.

However, the Web architecture is well prepared for the unavailability of resources. For instance, application developers may prepare their client-site code for a server response with status message *not found*. The Generic Host Interface ports the semantics of accessing remote resources to the means of accessing local resources. All resources the host environment of MWR exposes to a Widget are accessed through Ajax and Resource Event API. Thus, the Generic Host Interface does not solve the platform fragmentation problem that arises from the exposure of local services to the script scope, but it unifies the handling of these services.

For example, a built-in GPS receiver for the positioning of the client can be accessed at the local address *http://localhost/GPS*. A subscription to this resource allows the application to react if the terminal position changes. A request to this resource returns a tuple of longitude and latitude. If the GPS device is not available at the current hardware platform, an application request to the respective resource is responded with a common *not found* status message.

Widget Management

The User Interface provides user access to the Application Manager and the Security Framework. The Application Manager allows the user to download a Widget from an URL, and start, suspend or terminate Widgets. Different from traditional Web applications, the capability to control or at least monitor the application life-cycle is mandatory for Widgets. The application model of Widgets is designed to preserve the execution state at the client. When the application is deactivated or temporarily suspended, for instance on an incoming phone call or a battery running empty, the latest state achieved by the user needs to be saved. For example, this user state may be a certain game level or a written text. MWR provides a Life Cycle API that enables application developers to define actions for life-cycle events at runtime. The covered events are activation, deactivation, suspension or reactivation. Thus, an application developer can design the Widget to store all data needed to preserve the state on reception of the deactivation event.

The Security Framework is required to cope with the security lack created through opening services of the host environment for access by arbitrary code downloaded from the Internet. Legacy Web browsers ensure security by executing code only in a secure container, a so called sandbox. The Security Framework uses policies to control the degree of freedom of a Widget in accessing services of the host environment. The policies are supposed to help protecting user data and misuse of resources.

Also, mobile applications consume a significant amount of hardware resources like processing and storage capacity or battery power. The Security Framework supports the user to define for each application separately which resources can be used and to which extent. Policy enforcement is realized by intercepting each communication of the Runtime Environment Core to resources like network interface or device services.

Show Cases

A prototypical realization of the FOKUS Mobile Widget Runtime for J2ME served as testing platform for the implementation of several applications to check concept, integrity of specifications as well as correctness and handling of runtime environment, application framework and available APIs. In the following two selected implementations of mobile location-based community applications are presented. The applications demonstrate how the ubiquity, personalization and rich hardware equipment of mobile devices can be utilized by the mobile Web platform to enable a new generation of applications which merge the virtual and the physical world.

The sample applications utilize the following MWR features:

- Scalable Vector Graphics (SVG 1.1 Tiny) rendering
- DOM Scripting (ECMAScript / JavaScript)
- Asynchronous server requests (AJAX)
- Server event notification (via SIP)
- Satellite-based positioning (via GPS)
- Packaged deployment (compressed tar archives, TGZ)
- Credit-based cost control (users can set a limit on the maximally allowed data transfer caused by a Widget)
- Keypad and Touch-Screen access
- Telecommunication services

The first and simpler application is a virtual Geo-Caching Widget. Originally, Geo-Caching is a GPS-based real world game. Players hide a so-called "cache", which can be a place, box or another real-world object and publish its coordinates on the Web. Other players obtain the cache position as a coordinate pair of longitude and latitude, and try to find the cache equipped with a GPS receiver and a map only. While in classical Geo-Caching the cache is a real-world object, in the realized virtual Geo-Caching the cache can be a digital message, picture or a riddle displayed on the device when the user moves close enough into the range of the cache.

The virtual Geo-Caching Widget running on the MWR renders a radar-like interface that shows direction and distance of caches nearby, as depicted on the left side of Figure 5. Also other players nearby are shown. If the player approaches a cache and the distance is less than 25 meters a message

Figure 5. Sample applications

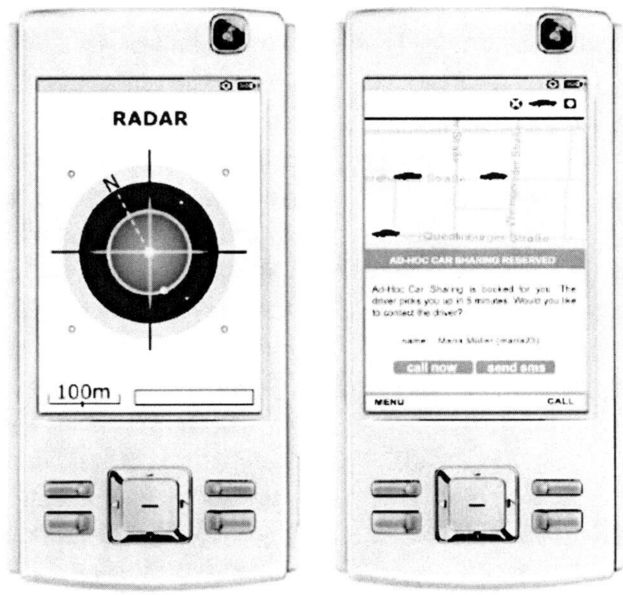

(l.h.) Geo Caching Widget – On a radar like screen distance and direction of caches are presented to the user. (r.h.) Mobile Car Sharing Widget – User receive the notification about the successful arrangement of a shared ride and can call the driver or alternatively send an SMS.

or a picture pops up depending on the type of the cache. The Widget runs e.g. on the Nokia N95 mobile phone, which is equipped with built-in GPS. The user interface is completely designed with SVG instead of HTML; the utilized scripting language is JavaScript. Each virtual cache is requested from the server when the player physically approaches the position of the cache. The Widget has a total size of 26kB only.

The second application for MWR that we realized is a mobile ad-hoc car sharing service. People who search for a car ride can spontaneously use the application to post their request to a list of potential drivers. To receive requests for a ride, drivers utilize the same application and just enter their destination right before they move off. During the ride the drivers are localized and tracked, therefore allowing true ad-hoc matches between potential riders and drivers depending on their both real-time physical positions. Car sharing requests and offers are compared on the server, and both parties are notified through the application if a match is found. If both parties agree, driver and requester for the ride receive additional information, for example about the color of the car, or the place for pickup. Additionally, route derivations and pickup points are calculated by the server and send to the mobile clients.

The interface of the car-sharing Widget is also based on SVG. The maps that serve as orientation for driver and requester are SVG graphics provided by a 3rd party map service. Map information is fetched progressively via the Ajax interface from the Web when the user changes position or scrolls/zooms the map on the Widget. The SVG based presentation simplifies zooming and reduces the amount of data to be communicated. According to storage capacity of the mobile phone and user settings the tiles of the map are kept or discarded when not in use. The Widget can also make use of a Call Control API provided by MWR. The realization of call functionality is realized through VOIP services. The distinguishing characteristic of the Call Control API is the support for handling calls and conference calls in the background and just notify the Widget about events (e.g. hang up). Thus, Call Control is integrated with the Widget in a completely seamless fashion. In the application the driver may anonymously call the passengers from the Widget, e.g., to clarify details of the pick-up.

FUTURE RESEARCH DIRECTIONS

In this chapter we described the potential of the Mobile Web as a sophisticated platform for mobile applications. In general, still a better understanding of the properties of future user interaction with mobile applications is needed, for example new interaction models, and efficient methods for developing mobile applications. Also security issues and the fragmentation problem are future research challenges.

Additionally, current development of next-generation SIM cards (Universal Integrated Circuit Cards; UICCs) paves the way towards new business models and technological approaches in the mobile field: UICCs are Web-enabled through a TCP/IP stack and an on-card Web server, multi-application capable, security enhanced, and are equipped with growing processing and storage capabilities. UICCs are managed by the operator with Over The Air (OTA) technology from remote, which e.g. allows for automatic software updates. When residing on the UICC, operator controlled applications (or third party applications which can "rent" space on the UICC) and the user's personal information are not tied to a particular device, that is, are portable between devices. The integration of UICCs into mobile Web applications is part of our future research.

CONCLUSION

Mobile Web applications reduce time to market, encourage innovation and enable a larger target market. They offer a better value proposition to application developers, which profit from the low learning curve of the applied technologies, by offering the possibility to develop for both the Desktop Web and the mobile Web at the same time. In this chapter we have shown that there are a number of activities on the way to extend the Mobile Web platform towards a "hybrid" platform, which can compete with platforms for locally installed "fat" applications. We also presented our prototype of a hybrid platform, the FOKUS Mobile Widget Runtime and sample applications to demonstrate how these future hybrid applications may look like. In the future we will continue research in this area according to the requirements derived in this chapter and the potential research directions described above.

REFERENCES

W3C (2008). *W3C WebApps Working Group*. Retrieved April 11, 2009, from http://www.w3.org/2008/webapps/

W3C (2009). *About W3C*. Retrieved April 11, 2009, from http://www.w3.org/Consortium/

W3C (2009). *HTML5*. Retrieved April 11, 2009, from http://www.w3.org/TR/html5/

W3C (2009). *Mobile Web Initiative*. Retrieved April 11, 2009, from http://www.w3.org/Mobile/

W3C (2009). *Widgets 1.0: Packaging and Configuration*. Retrieved April 11, 2009, from http://www.w3.org/TR/widgets/

Fielding, R. T., & Taylor, R. N. (2000). Principled design of the modern Web architecture. In *Proceedings of the 22nd international Conference on Software Engineering (Limerick, Ireland, June 04 - 11, 2000)* (407-416), ICSE '00. New York:ACM

IETF. (2009). *hybi: Bidirectional communication for hypertext*. Retrieved April 11, 2009, from http://trac.tools.ietf.org/bof/trac/wiki/HyBi

Kesteren, A. V. (2008). *The XMLHttpRequest Object*. Retrieved April 11, 2009, from http://www.w3.org/TR/2006/WD-XMLHttpRequest-20060405/

Linner, D., Krüssel, S., & Steglich, S. (2008). CAPgets: *Mobile Web Runtime Environment for Community Applications*. 1st International Workshop on Next Generation Networks: Open Platforms & Services (NGNOPS 2008), September 2008, Wales, UK.

OMTP BONDI. (2009). *Home - BONDI*. Retrieved April 11, 2009, from http://bondi.omtp.org/default.aspx

OpenAjax Alliance. (2009). *OpenAjax Alliance*. Retrieved April 11, 2009, from http://www.openajax.org/index.php

Rajapakse, D. C. (2008). *Fragmentation of Mobile Applications*. Retrieved April 11, 2009, from http://www.comp.nus.edu.sg/~damithch/df/device-fragmentation.htm

Russel, A. (2006). *Comet: Low Latency Data for the Browser*. Retrieved April 11, 2009, from http://alex.dojotoolkit.org/2006/03/comet-low-latency-data-for-the-browser/

WHATWG. (2009). *Web Hypertext Application Technology Working Group*. Retrieved April 11, 2009, from http://www.whatwg.org/

ADDITIONAL READING

W3C (2009). *Web Sockets API*. Retrieved March 27, 2009, from http://dev.w3.org/html5/websockets/

Duhl, J. (2003). *White Paper: Rich Internet Applications*. Retrieved March 27, 2009, from http://www.adobe.com/platform/whitepapers/idc_impact_of_rias.pdf.

Lentczner, M. (2009). *Reverse HTTP*. Retrieved March 27, 2009, from http://tools.ietf.org/html/draft-lentczner-rhttp-00

Rabin, J., & Nevile, C. (Eds.). (2008). *Mobile Web Best Practices 1.0 – Basic Guidelines – W3C Recommendation 29 July 2008*. Retrieved March 27, 2009, from http://www.w3.org/TR/mobile-bp/

Russell, A., Wilkins, G., Davis, D., & Nesbitt, M. (2007). *Bayeux Protocol -- Bayeux 1.0draft1*. Retrieved March 27, 2009 from http://svn.cometd.org/trunk/bayeux/bayeux.html

WHATWG. (2009). *HTML5 - Draft Standard*. Retrieved March 27, 2009, from http://www.whatwg.org/specs/web-apps/current-work/.

XMPP.org. (2009). XEP-0124: Bidirectional-*streams Over Synchronous HTTP (BOSH)*. Retrieved March 27, 2009, from http://xmpp.org/extensions/xep-0124.html

Chapter 6
A J2ME Mobile Application for Normal and Abnormal ECG Rhythm Analysis

Qiang Fang
RMIT University, Australia

Xiaoyun Huang
RMIT University, Australia

Shuenn-Yuh Lee
National Chung Cheng University, Taiwan

ABSTRACT

Cardiovascular disease has become the world's number one killer. The prevalence of cardiovascular disease has caused many unnecessary premature deaths and imposed substantial burden to healthcare systems. Many continuous heart monitoring systems have been proposed with the aim to issue early stage warning for a possible forthcoming heart attack by utilising advanced information and communication technologies. Nevertheless, there is still a significant gap between the usability and reliability of those systems and the requirements from medical practitioners. This chapter presents our recent development of a mobile phone based ECG real-time intelligent analysis system. By fully employing the computational power of a mobile phone, the system provides local intelligence for ECG R wave detection, PQRS signature identification and segmentation, and arrhythmia classification. Because those processing can be performed on realtime, an early status warning can be issued promptly to initiate further rescue procedures. As an application of e-commerce in healthcare, a telecaridiology system like this is of great significance to support chronic cardiovascular disease patients.

INTRODUCTION

Recently, the patients suffering from cardiovascular disease (CVD) have been undergoing a rapid increase world widely due to the lifestyle change and the aging of population. For many nations such as USA, Australia, European nations, Canada and China, various CVDs are the number one killer while the cerebrovascular diseases (CBD) such as stroke are the number two killer (Roberts, 2006). The

DOI: 10.4018/978-1-61520-761-9.ch006

prevalence of CVD has risen by 18% over the last decade and is expected to continue to rise over the coming decades due to an expected increase in the elderly population (AIHW, 2004a). In Australia, 37% of the total death in 2001 was caused by cardiovascular diseases and CVD affected total 3.67 million Australian people which is about 18% of national population (AIHW, 2004b). Among Australians having heart attacks, about 25% die within an hour of their first-ever symptoms and over 40% will be dead within a year (Access Economics Report, 2005). CVD together with CBD are also the leading causes of long term disability in adults (Access Economics Report, 2005). They impose a big burden on patients' families as well as the national healthcare system due to the high costs of care, the resulted lower quality of life, and the premature death. Chronic CVD patients are at high risk of having heart attacks and the majority of such heart attacks take place in out-of-hospital environment where the emergence services cannot be available immediately. Therefore, there is an urgent need to develop a personal monitoring and alarming system which can effectively detect early indications of a heart attack and issue timely warning signals for calling for rescue efforts.

Since last decade, we have witnessed the explosive expansion of the use of mobile phone. Now, mobile phone is one of the most pervasively used single electronics devices in the world. For example, Australia has more subscribed mobile phone handsets than its total population (ACMA Report, 2009). The ever increasing computation power of a mobile phone plus its great mobility make it an ideal pervasive computing platform for telehealth monitoring. Although handheld devices such as mobile phone have been widely proposed to use in various telemedicine applications, they are generally utilized as the wireless data transmission tools. The power from the mobile computing devices has not been fully harnessed. On the other hand, many ambulatory and medical monitoring systems need the acquired vital physiological data be processed in real-time so as to generate the much needed precaution and alarming signals. For such applications, the acquired physiological data should be processed locally, rather than sent to a remote server via a GPRS or 3G mobile telephony network systems, to avoid transmission delay and reduce transmission cost. In order to develop the local intelligence, some limitations pertinent to handheld computing devices need to be addressed. Those limitations include the bottleneck of the bandwidth for large amount of stream data continuous transmission, the partial support of the full extended ASCII set, the limited hardware resources such as processor speed and memory amount, and the restricted programming environment such as no directly support of floating point and no multi-dimensional array support for many mobile phone handsets (Sufi et al., 2006).

This chapter presents a realtime stream data mining system for one human vital physiological signal, the electrocardiogram (ECG) on compact mobile phone handsets. This lightweight data mining system is able to extract information which is important to medical practitioners to make clinical diagnosis and treatment decision. The first part of the chapter is a brief introduction of ECG signal and the general steps of clinical analysis of this crucial electrophysiological signal. The current mobile cardiac monitoring development efforts are also briefly reviewed in this section. It is followed by an introduction to J2ME, the development platform used in this research. The key analysis techniques employed in this research which include the time series analysis, the discrete wavelet transform (DWT) and a naive Bayesian classifier are elaborated in this section. The frequency domain analysis of the recorded heart rhythm such as power spectrum density can be performed by implementing the fast Fourier transform (FFT) algorithm on a mobile phone handset. However, it is the discrete wavelet transform, a time-frequency analysis method that shows it superiority over Fourier transform to display the high frequency details of an ECG waveform in different scales and suppress the low frequency baseline wander. Moreover, the fast pyramid algorithm of DWT has been explained in this Section. In the System Design section, the requirement of a realtime analysis, which is one key design requirement in this research, is proposed at first. Then the Record Management System, a unique database-like permanent storage system of J2ME is discussed. The detailed analyses steps are introduced in the following subsections. A further endeavor to complete this lightweight data mining system is to incorporate a Bayesian classifier to realize the arrhythmia classification. As a matter of fact, the proposed system is a hybrid system combining time series analysis, DWT and Bayesian classification to identify the R wave, the QRS complex and then compute the RR interval, the PR interval, the ST interval, and finally to determine the arrhythmia types. In order to ensure all analysis performed on mobile handsets are in realtime, a real-time measurement metric is defined.

The experiment results, which are given in Section 4, suggest that with the choice of fast algorithms and optimized coding, most of the analyses including the time series analysis, the frequency domain analysis and the time-frequency analysis can be implemented, deployed and executed at a realtime speed on a plain mobile phone handset. The implemented system provides satisfactory normal sinus rhythm classification accurate rate. However, further efforts are required to incorporate more advanced classifiers such as support vector machine (SVM) and wavelet de-noising techniques to improve the classification accuracy rates for real clinical applications.

With the rapid expansion of the computational power of the mobile phone handsets, it is possible to carry out relatively complicated data analysis tasks on such lightweight "tiny" computing platform. The presented work successfully implemented several major time domain, frequency domain, and time-frequency data analysis techniques. The successful results suggest that a more sophisticated real-time data mining system can be developed on the mobile computing platform to form a human vital physiological signal acquisition, analysis and alarming system for critically ill patients, such as CVD patients, as a life saver. Telemedicine can be regarded as an example of e-commerce in healthcare. Thus, this presented mobile phone based system demonstrates an important application area of mobile devices in e-commerce.

BACKGROUND

Electrocardiogram (ECG)

The Electrocardiogram (ECG) is the recorded electrical signal generated by the contractile activities of the heart. The heart is composed of myocardial cells and conductive neural-like cells. Those neural like cells which include the sinuatrial (SA) node, the atrioventricular (AV) node, the bundle of His, and the Purkinje fibers, have the capability to initiate electrical impulses spontaneously. While this electrical impulse passes through the conductive pathway of the heart, the ECG is formed. ECG is widely used in cardiovascular disease diagnosis. A normal ECG has a unique signature pattern containing P, Q, R, S and T waves (see Fig. 1). The P wave characterizes the atrial depolarization and the QRS complex signifies more vigorous ventricular depolarization. The T wave represents the ventricular redepolarization, however the redepolarization of atria is too small in amplitude and it also occurs within the QRS Complex area. Thus the atrial redepolarization is not individually identifiable. The typical duration of a normal QRS is 0.06-0.1 seconds. The time interval between two consecutive R peaks is called RR interval. The RR interval and the QRS complex reveal crucial information of a heart condition, thus these two parameters are important for ECG analysis. The normal range of heart rate (HR), the reciprocal of RR interval, is from 60-120 beats per minute (BPM). The cardiologic abnormality is

Figure 1. A typical ECG waveform. The units for the horizontal axis and the vertical axis are second and millivolt respectively

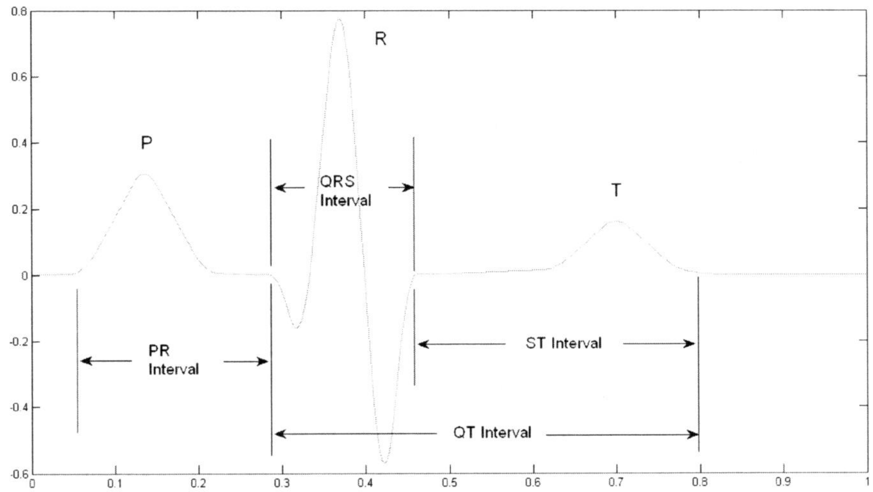

termed as bradycardia or tachycardia respectively if the heart rate is lower or higher than the normal rate. The heart rate variability (HRV), which is generally believed as an indicator of the balance of sympathetic and vagus nerve activities, can be further derived from RR intervals.

ECG Analysis on Mobile Computing Platform

The electrocardiogram one major human electrophysiological signal and its analysis has been extensively conducted. The ECG waveform morphology characterizes different cardiovascular diseases and abnormalities such as myocardial ischemia, myoinfarction, ventricular fibrillation (VF) and ventricular tachycardia (VT).

For a clinical ECG trace analysis, a 5-step approach can be generally adopted (Becker, 2006). These 5 steps are to:

1. Check if the rhythm is regular or irregular
2. Check if all QRS complexes are similar and narrow in width
3. Check if all P waves are similar and PR intervals are normal
4. Check if the rate is normal
5. Check if waves and complexes proceed in normal sequence

Among those steps, it is essential to determine the QRS complex and its width as well as the RR interval. This is because QRS complex is the most significant feature of an ECG recording. If no QRS complex can be identified or its morphology is severely distorted, then there must exist life-threatening arrhythmia, such as asystole or VF. Based on the RR interval, the type of rhythm can be determined. Based on these two observations, all basic arrhythmias can be classified into different groups (Becker, 2006). Further analysis can be done by examining the existence of the P waves and the T waves as well as their relationships with the QRS complexes. For example, the atrial fibrillation can be characterized by a complete missing of P wave and an irregular heart rate. For a normal ECG, the P wave occurs early than QRS complex and the time interval between these two structures is less than 0.2 second.

In recent years, many new computer based automatic ECG identification and classification algorithms have been proposed with satisfactory specificities and sensitivities (Chuah & Fu, 2007; Jiang, 2007; Chen, 2007; Thomas et al., 2007; Castells et al., 2007; Ercelebi, 2004). Those developments greatly alleviate the work burden of cardiologists as well as improve the analysis efficiency and accuracy. Nevertheless, those algorithms adopting either nonlinear methods such as chaos or complex pattern recognition algorithms such as Hidden Markov Model or artificial neural network, are not targeting mobile telecardiology applications. They require heavy computational overheads and are difficult, if not completely impossible, to be implemented on a mobile device. In the mean time, many mobile or ambulatory ECG systems are also proposed together with the development of body area sensor network (BASN) systems. These systems can be classified into two broad categories: one transmits the compressed or uncompressed raw ECG data to a central server and the data are analyzed at the server side (Kail, 2004) and the other processes the raw ECG locally, i.e., within the patient-side hand held devices or ambulatory devices (Jimena, 2005; Helfenbein, 2006). For the practical perspective, the second category approaches are of great importance as it can avoid the transmission delay and issue prompt warning messages. It also can save the transmission cost by avoiding sending large volume of ECG data through GPRS, GSM, or 3G mobile phone network systems. Another advantage for doing so is to reduce the power consumption as the wireless transfer of large amount of data consumes substantial battery power.

Several telecardiology systems have already been proposed emphasizing on local processing. Most of those proposed systems adopt Personal Digital Assistant (PDA) (Goh et al., 2005; Fensli et al., 2005; De Capual et al., 2006) or high end SmartPhone (Leijdekkers & Gay, 2006) rather than plain J2ME based mobile phone handsets. They target on specific ECG condition or arrhythmia patterns only, e.g., the PDA-based ECG beat detector for home cardiac care proposed by Goh et al. (2006) detects the ECG beat only while the personal heart monitoring system proposed by Leijdekkers and Gay (2006) is focusing on Ventricular Fibrillation (VF) and Ventricular Tachycardia (VT). More importantly, those systems don't address whether their implemented ECG analyses match the real-time criteria. The correct classification rates from those systems are often not shown explicitly. In addition, many proposed systems need dedicated hardware such as System-On-a-Chip (SOC) or Field Programmable Gate Array (FPGA) to realize the realtime arrhythmia detection algorithms (Zhou, 2005; Wu, 2005).

Mathematically, the ECG recordings can be treated as a time series, a signal in time domain. Therefore, many time series analysis and frequency analysis methods, such as segmentation, moving averaging, filtering, and spectrum analysis, can be adopted. Because the frequency contents change with the time, the ECG is a non-stationary signal for which the Fourier transform based spectrum analysis is not the optimal analysis candidate which was actually superseded by the time-frequency analysis, such as short-time Fourier transform and wavelet transform. In recent years, many sophisticated data mining techniques have been introduced to analyze ECG traces (Haghighi, 2009; Hu, 2008). However, most of them have complex algorithms which are not suitable for execution in a compact computation environment.

Due to the limited computational resources that a mobile device can have, it is not practical to expect a full scale of ECG trace analysis can be performed on a mobile computing platform. On the other hand, a mobile device provides the much needed mobility and flexibility for ambulatory monitoring. Thus, the research presented here focuses on developing a solution for realtime ECG rhythm analysis. In particular, the most important goal is to determine normal sinus rhythm (NSR) with high sensitivity and specificity. Most of those identified NSR ECG segments will be discarded except last 5 recordings each with eight seconds duration. To achieve high sensitivity and specificity it is important to minimize the false alarming. One important requirement for a wearable or ambulatory ECG monitoring and alarming system is the ECG processing speed. Unless an ECG abnormality can be identified in realtime or quasi realtime, the system cannot be accepted as a life saver for chronic CVD patients. So, the realtime analysis is another requirement of this investigation. Though clinical ECG from a cardiograph or a bedside monitor with 12-lead data provides a better analysis, one single lead ECG (Lead II) is used in this investigation as our goal is to differentiate the normal sinus rhythm and other basic arrhythmias. The classification of mixed and complex arrhythmias caused by complex syndromes such as the coexistence of myocardial ischemia, coronary artery diseases, injury and myocardial infarction are not considered.

J2ME

In order to expand the potential user group, the ECG monitoring system presented here targets on plain medium level mobile phone rather than high end PDA, iPhone or SmartPhone. The core communication tasks of this ECG monitoring system are conducted by the mobile phones carried by both users (patients) and telemonitoring service providers (medical doctors). In this initial investigation, a pair of popular Nokia91 handsets are chosen. Most recent mobile phones support the execution of miniature programs that utilize the mobile processing power. Java 2 Micro Edition (J2ME), .Net Compact Framework, Binary Runtime Environment for Wireless (BREW), Carbide C/C++ are some of the popular programming environments for mobile phone application development. Among these development environments, J2ME is pervasively used since the compact Java runtime environment, Kilobyte Virtual Machine (KVM), has been supported by a wide range of mobile phone handsets already. J2ME is basically a subset of the standard Java platform (J2SE) and is designed to provide Java APIs for applications on tiny, small and resource-constrained devices such as cell phones, PDAs and set-top boxes. One major advantage of choosing Java is that a single program written in J2ME can be executed on a variety of mobile phones that support Java. Apart from the basic computation framework provided by KVM, each of the mobile phone also supports additional Java libraries for supporting additional functionalities such as Bluetooth connectivity, camera functionality and messaging services, etc. These additional libraries expose Application Programming Interfaces (APIs) to the programmer of the handset. J2ME architecture is composed of configuration and profile. Connected Limited Device Configuration (CLDC) defines the minimal functionalities required for a range of wireless mobile devices, e.g., mobile phone, PDA, Pocket PC, home appliances etc. Mobile Information Device Profile (MIDP) further focuses on a specific type of device like mobile phone or pager. MIDP also describes the minimum hardware or software requirement for a mobile phone. To the mobile application developers, both CLDC and MIDP expose Application Programming Interfaces (APIs) and functionalities supported by the KVM.

Since the computational powers of the mobile phone handsets are expanding rapidly, current mobile phones possess considerable computational powers which can perform runtime complex tasks such as 3D games, data compression, MP3 and MPEG encoding and decoding. Even Optical Character Recognition (OCR) software was tested on current mobile phones (Graham-Rowe, 2004). It is feasible to utilize the processors inside the mobile phone to process, compress, and transmit data in realtime for various telehealth applications. The realtime availability is of great importance for the sake of life saving. In principle, by careful design or selection of the proper computational algorithms, many complicated medical data processing and analysis tasks such as compression, decompression, encryption, correlation and transformation, feature extraction, and pattern recognition, can be implemented. However, the mobile phone platform supporting Java™ language is subject to some software and hardware specific limitations. Unlike a Java runtime for PC, the KVM on mobile devices is a miniature version that can only run a subset Java APIs. Compared with a desktop PC, mobile phones based CLDC and MIDP restrict the usage of floating point operations, which means all the floating point must be removed before performing any operations on the mobile devices or a set of custom floating point supporting APIs need to be developed. Multi dimensional arrays are not supported as well; hence, any algorithm performing matrix based calculation needs to find an alternative approach.

Wavelet Transform

The wavelet transform has been proven to be a powerful time-frequency analysis tool for non-stationary biological signal (Akay, 1995). The wavelet transform can map a time domain signal $s(t)$, such as an ECG trace, into a two-dimensional representation of scale and time. The decomposition elements for the wavelet transform are a family of wavelets rather than a set of sinusoids with different frequencies in the Fourier transform. Wavelets are a family of translations and dilations of a single function that is called the mother wavelet. The wavelet name is from the fact that the function always has some localized oscillation (Daubechies, 1988; Mallat, 1989; Chui, 1992).

The wavelet transform can be viewed as an inner product operation that measures the similarity or cross-correlation between the signal and the dilated and translated wavelets. Since the scale has strong relation to frequency, the wavelet transform also leads to a time-frequency analysis (Daubechies, 1992; Strang, 1996). The continuous wavelet transform of $s(t)$ is defined as

$$cwt(a,b) = \int s(t)\frac{1}{\sqrt{a}}\psi(\frac{t-b}{a})dt \quad (1)$$

where $s(t)$ is the analyzed signal, $\psi(t)$ is the basic (or mother) wavelet and $\psi((t-b)/a$ are the wavelet basis functions, sometimes called baby wavelets.

Figure 2 shows the continuous wavelet transform of a non-stationary signal, the chirp signal. The small scales representing high frequencies are arranged at the bottom. The increasing trend of the frequency is also shown clearly.

The continuous wavelet transform is not an orthogonal transform because it contains many redundant transform coefficients which cause a heavy computational overhead and a large storage space. Therefore it is not a good candidate to be used on a mobile device. The discrete wavelet transform (DWT) can be achieved with discretized parameters a and b and a discretized wavelet $\psi(t)$ (Chan, 1995). A particular sampling scheme which also allows a perfect reconstruction is an octave time scaling for a and a dyadic translation for b, i.e., $a_0=2$ and $b_0=1$ (Mallat, 1989). Using this sampling scheme, the wavelets become:

$$\psi_{mn}(k) = 2^{-\frac{m}{2}}\psi(2^{-m}k - n) \quad (2)$$

where m, n are integers. Then the DWT is

$$DWT(m,n) = 2^{-\frac{m}{2}}\sum_{k} s(k)\psi(2^{-m}k - n) \quad (3)$$

Using Equation 3 to compute DWT is slow and inefficient. Mallat (1989) has discovered that the pyramid algorithm can be applied to the discrete wavelet transform under a multi-resolution analysis framework as long as a set of wavelet coefficients is used as the filter coefficients of a Quadrature Mirror Filter (QMF) pair (Mallat, 1989). A QMF pair combines an in-phase symmetric filter (low pass) and an in-quadrature anti-symmetric filter (high pass). He uses a two-channel subband filtering with two filter sequences, hn, the smoothing or scaling filter, and gn, the detail or wavelet filter (Mallat, 1989). Later, Daubechies constructed the compactly supported orthogonal wavelet transform (Daubechies, 1992). The fast

Figure 2. Continuous wavelet transform of a chirp signal, $s(t)=sin(0.6\,t^2)$. The instantaneous frequency increases linearly for a chirp signal. The Morlet wavelet is used

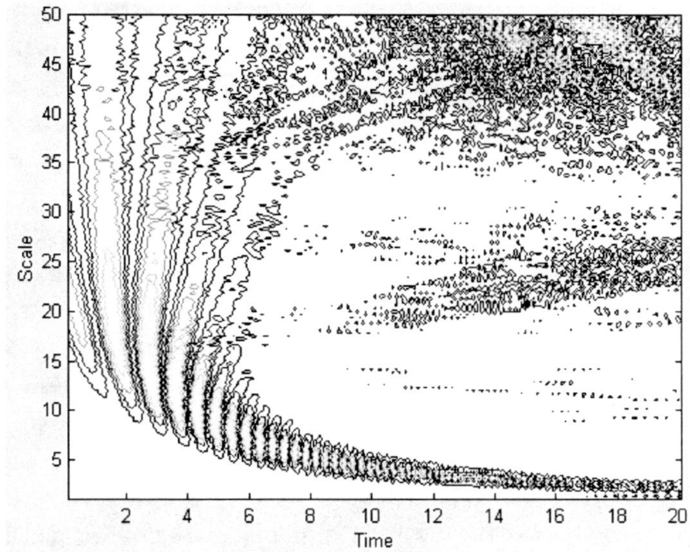

orthonormal wavelet decomposition of a discrete signal is obtained by a pyramid-filtering algorithm, which also allows exact reconstruction of the original data from the new coefficients. The decomposed signals are orthonormal and uncorrelated. It can be seen that the subband component of a signal *s(t)* obtained by the multiresolution analysis is just the orthonormal discrete transform of *s(t)* (Mallat, 1989).

An original signal *s(t)* which is measurable and has a finite energy can be considered to be a sum of a low frequency part and a high frequency part. The low frequency part preserves the overall characteristics of a signal while the high frequency part gives it local characteristics. Therefore, the low-frequency component is called the approximation of *s(t)* and the high-frequency component is called the detail of *s(t)*. The approximation of *s(t)* can be obtained by filtering the *s(t)* using the scaling filter, *hn* and the detail can be obtained by filtering the *s(t)* using the wavelet filter *gn*. The *hn* is a low-pass filter associated with the scaling function and the *gn* is a high-pass filter associated with the wavelet function. The approximation and the detail at the resolution or scale level 2^j in the dyadic sampling grid is denoted as $A_{2^j}^{s(t)}$ and $D_{2^j}^{s(t)}$. The detail signal represents the difference of information between two successive approximations. For the sake of convenience, the resolution level for the original signal is set to 1, i.e., $j=0$.

The approximation of a signal at the 2^j resolution contains all the necessary information to compute the same signal at a smaller resolution 2^{j+1}. When computing an approximation of *s(t)* at resolution 2^{j+1} some information about *s(t)* is lost from the finer resolution 2_j but is still stored in its detail signal at resolution 2^{j+1}. The approximation and the detail operation are similar at all resolutions through the down-sampling or up-sampling (for reconstruction). So from $A_1^{s(t)}$, the approximation at the resolution 1 that is the same as the original signal *s(t)*, all approximations $A_{2^j}^{s(t)}$ and the details $D_{2^j}^{s(t)}$ for $j=2,3,4,\ldots$, could be obtained through the pyramid algorithm. Following the pyramid algorithm for multiresolution analysis, the original discrete signal $A_1{}^{s(t)}$ measured at the resolution 1 can be represented by

$$A_1{}^S = \sum_j D_{2^j}{}^S + A_{2^J}{}^S \quad 1 \le j \le J \qquad (4)$$

where J is the coarsest decomposition level.

This set of discrete signals, $D_{2^1}S$, $D_{2^2}S$, ..., $D_{2^J}S$ and $A_{2^J}S$ is called an orthogonal wavelet representation of the originally measured signal $A_1 s$. This representation has the coarsest approximation $A_{2^J}{}^s$ at the resolution of 2^J and the detail signals at the resolutions from 2 to 2^J. It also can be viewed as a decomposition of the signal into a set of independent frequency channels.

The orthonormal discrete wavelets, e.g., the famous Daubechies wavelets, are not linear phase filters. This drawback can be alleviated by using the nearly symmetric Symlet wavelet proposed by Daubechies (1992). Another frequently used wavelet is the biorthogonal wavelet, which has the linear phase property (Daubechies, 1992). The biorthogonal wavelet analysis is achieved by using two QMF pairs, one for decomposition and the other for reconstruction. One of the problems with the Fourier transform is its nonlocality. All components of a signal *s(t)* in time domain contribute to its spectrum in frequency domain. That is to say Fourier transform (FT) has only good localization in frequency domain but not in time domain. Thus Fourier transform has difficulty with functions having transient components, i.e., components well localized in time such as the QRS complex. Another problem is that the Fourier transform of a signal does not convey any information pertaining to translation of the signal, although this drawback can be corrected slightly by short-time Fourier transform. The third main limitation of FT is the requirement of an infinite length of the studying signals otherwise a periodic assumption has to be adopted. The wavelet transform can overcome these problems. This transform could be regarded as the natural alternative and further development of Fourier transform (Chui, 1992; Strang, 1996).

Bayesian Classifier

A variety of classification techniques such as artificial neural network, fuzzy logic, support vector machine, independent component analysis, have been applied for ECG arrhythmia classification. The particular classifier we incorporated into our system is one of the most popular machine learning methods, the Bayesian classifier (Tan et al., 2006). The Bayesian classifier is a statistical approach to the problem of pattern recognition which aims to recognize a particular class from a measurement vector. Different pattern classes with different measurement vectors can be denoted as different points in the measurement space and patterns with similar properties tend to cluster together. Thus a mapping relationship can be established from the measurement space into the decision space. The Bayesian classifier is based on the Bayes rule. For a set of N measured ECG waveforms w_1, w_2, ..., w_N, with the associated m ECG types t_1, t_2, ..., t_m, each measured ECG waveform is represented by a n-dimensional feature vector $w=f_1, f_2, ..., f_n$ where f_i is the *i-th* measured feature. Let $P(w_i|t_i)$ be the class-conditional probability for a measured ECG waveform whose distribution depends on the type t_i. Then $P(t_i|w_i)$, the *a posteriori* probability that waveform w_i belongs to class t_i can be computed from $P(w_i|t_i)$ by Bayes rule:

$$P(t_i|w_i) = P(w_i|t_i)\, P(t_i) / P(w_i) \qquad (5)$$

The Naïve Bayesian classifier applies "naïve" conditional independence assumptions which state that all n feature $f_1, f_2, ..., f_n$ of the measured ECG waveform w_i are all conditionally independent of one another for a given t_i. This assumption significantly simplifies the representation of $P(w_i|t_i)$, and the problem of estimating it from the training data. In our case, the measured ECG waveform w_i belongs to the known ECG type t_i with the highest probability $P(t_i|w_i)$. Since $P(w_i)$ is fixed for every class t_i, it is sufficient to choose the class that maximizes $P(t_i)\prod_{i=1}^{n} P(f_i \mid t_i)$.

In other words,

$$t_{\max} = \arg\max P(t_i)\prod_{i=1}^{n} P(f_i \mid t_i) \qquad (6)$$

Despite of their simplified assumptions of independence, a Naïve Bayesian classifier often competes well with more sophisticated classifiers (Zhang, 2004). Thus, it is chosen in this investigation for ECG arrhythmia classification on mobile handsets.

SYSTEM DESIGN AND IMPLEMENTATION

Realtime ECG Signal Acquisition and Analysis

As shown in Fig. 3, the developed ECG signal acquisition and analysis system has an ECG sensing unit and a mobile phone processing unit. The ECG sensing unit developed in our group comprises an ECG analogue front end which contains a pre-amplifier, a low-pass filter, a notch filter, an Analogue-Digital convertor, a block of buffer memory, a transmission control unit and a Bluetooth module. An Altera CycloneII FPGA is employed as central processing unit in digital part which contains a HDL designed sampling circuit, an asynchronous FIFO core, and a NiosII soft processor. The FPGA uses 2 clock domains, one 10MHz clock for the sampling circuit and another 50MHz for NiosII processor. The asynchronous FIFO acts as a data exchange pool which uses 10MHz as the write clock signal and 50MHz as the read clock signal. The NiosII processor system was built upon the SOPC function of QuatusII design software with which the hardware resources of the microprocessor can be configured as demand. In this design, NiosII was constructed to have one processor core, one 16KB ram, 3 GPIOs and one UART.

Based on American Heart Association (AHA)'s recommendation (Rijnbeek, 2001), the ECG sampling rate is set to 500 Hz (500 samples per second). The amplified, filtered and digitized ECG raw data is saved into a dedicated buffer memory. Because we consider 8 seconds of 1 lead ECG as one trace of recording, the minimum size of the buffer memory is 8 Kbytes while a 16-bit ADC is used. The buffer full flag is set while 8 seconds of recording is done. Then the content of the buffer

Figure 3. Block diagram of a realtime ECG signal acquisition and analysis

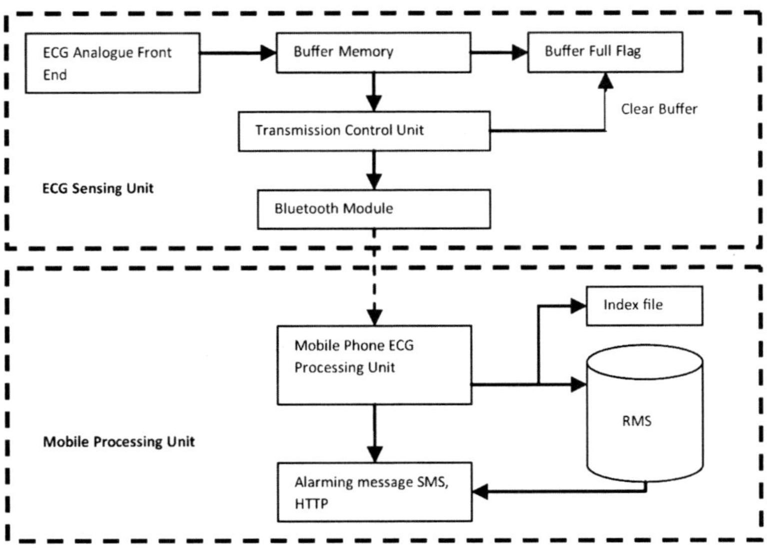

memory will be copied to the transmission unit which will transfer this trace of acquired ECG data to the mobile unit wirelessly and the buffer memory will start to accept data for next trace immediately. The mobile processing unit is basically a Bluetooth enabled mobile phone handset (Nokia 91). Once an ECG rhythmic abnormality is detected, an alarming message will be sent out via a SMS message while the raw pathological ECG recording will be sent out using the file uploading function via HTTP POST method. The raw ECG can be optionally compressed before it is transmitted.

In order to ensure the realtime analysis, the time used for Bluetooth data transmission T_t and analysis T_a should be less than the time used for acquisition T_q, that is $T_t + T_a < T_q$. Because Nokia 91 supports Bluetooth SPP (Serial Port Profile), the highest supported baud rate is 38400 bps. However, 19200 bps is used in this investigation. This is because the highest possible baud rate causes the Java Phone crash frequently due to a possible hardware bug. So for 1 trace of ECG data, the transmission time is 64, 000 bits / 19,200 bps = 3.33 seconds. Consequently, the maximum allowed analysis time is $T_q - T_t = 8.00 - 3.33 = 5.27$ seconds. Some low end mobile handsets only support a maximum 9600 bps baud rate, thus the maximum allowed analysis time is set to 1.34 seconds (8.00 seconds – 6.66 seconds).

Java Mobile Phone and Record Management System

Unlike many proposed telecardiology solutions which only utilize mobile phone as a data wireless transmission tool (Jasemian, 2005), our developed system processes the received ECG recordings locally. After an ECG trace is analyzed, it will be saved into a permanent data store. The permanent data store follows a *First In First Out (FIFO)* rule, i.e., whenever the latest trace is saved, the first trace will be deleted. If the latest trace is identified as a Non-NSR ECG trace, then the alarming routine will be triggered and this ECG trace will be sent immediately to the Telemonitoring Centre for further actions.

For a personal computer, both the file system mounted on the hard disks and the database management system if installed can be used as a permanent data repository. The later provides advanced data management and manipulation functions and supports standard query languages, such as SQL. However, there is no hard disk and file system are readily available to mobile phone handsets. The software platform for mobile Java application development provides a totally different approach which is called Record Management System (RMS) for MIDP applications to persistently store data across multiple invocations. This persistent data store is based on nonvolatile memory such as flash memory and is created in platform-dependent locations for multiple MIDlets access. The RMS classes call into the platform-specific native code that uses the standard OS data manager functions to perform the actual database operations.

Within the RMS, each record store can be viewed as a collection of records with each record has its own unique integer identifier. The running MIDlets can persistently store data and retrieve data from selected collections (see Fig. 4). The RMS can be treated as a database management system though the relationship among records cannot be readily defined. The RMS is responsible for data synchronization, serialization, and integrity. It also manages data concurrency access from multiple MIDlets. The timestamp mechanism is implemented to denote the last modification time and a simple serial data versioning mechanism is also implemented to record content modification history.

The javax.microedition.rms package contains APIs for MIDlet to communicate with the persistent store on a mobile phone. In our implementation, an rmsTable class is defined with a RecordStore class (ecgRecordStore) as its key member variable as well as a Vector class (ecgIndexVector)

Figure 4. MIDlets and RMS interfacing

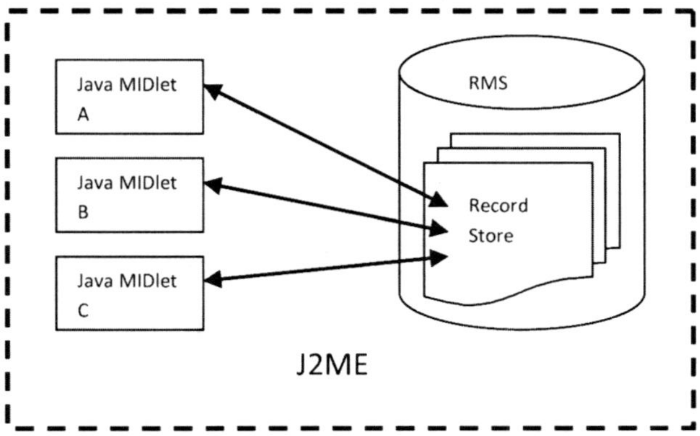

for indexing ECG records. The ecgRecordStore can be created as follows:

```
    RecordStore ecgRecordStore =
null;
    Vector ecgIndexVector =
null;
    Try {
      ecgRecord-
Store = RecordStore.
openRecordStore("myEcgStore",
true);
              ecgIndexVector = new
Vector();
    } catch (RecordStoreExcep-
tion ex) {

ex.printStackTrace();
    }
```

The rmsTable class also contains methods to manipulate the ecgRecordStore and the ecgIndex-Vector such as addIndexElement(), removeRe-cordElement(), getRecordId(), setRecord().

SMS and HTTP Client

The Hypertext Transfer Protocol (HTTP) and the Short Message Service (SMS) are employed for sending out the detected abnormal cardiac rhythm and the warning signal respectively. The HTTP client deployed on the handheld device is responsible for abnormal ECG trace uploading to a remote server.

The J2ME code snippet below shows a typical HTTP client connection between a CLDC device and a server for multipart file uploading using HTTP POST method. A J2ME HttpConnection Object is instantiated. The ecgOut is an Output-Steam object which contains the detected abnormal ECG trace. The ECG data series can be option-ally compressed to reduce the size as well as the transmission cost (Sufi et al., 2009).

```
// www.myecgcarecenter.com/up-
loadecg.php/ is the server side
script // processing the upload-
ed ECG data and performing fur-
ther analysis
String url = "http://www.myecg-
carecenter.com/uploadecg.php/";
HttpConnection myHttpConnection
= null;
Byte[] postData = null;
myHttpConnection = (HttpConnec-
tion) Connector.open(url);
//boundaryString is required for
processing multipart file up-
loading
String boundaryString;
```

```
// getBoundaryString is a method
returning the current boundary
string
boundaryString = getBound-
aryString();
myHttpConnection.setRequestProp-
erty(
"Content-Type", "multipart/
form-data; boundary=" + bound-
aryString;
// use POST method
myHttpConnection.
setRequestMethod(HttpConnection.
POST);
OutputStream ecgOut = myHttpCon-
nection.openOutputStream();
// createPostData is a method
creating the data stream ready
to be
// sent to the server using POST
method
postData =createPostData(ecgOut,
boundaryString);
ecgOut.write(postData);
ecgOut.close();
myHttpConnection.close();
```

SMS is used to send alarming messages. Mobile SMS supports two operations, Message Originating (MO) and Message Terminated (MT). MO is for sending SMS and MT is for receiving SMS. Once an SMS is sent from a mobile phone, the message arrives at Short Message Service Center (SMSC). SMSC generally follows Store & Forward rule, which means message is stored inside SMSC until it reaches the recipient. Hence, an SMSC constantly tries to transmit the SMS until it is received by the recipient. Some SMSCs are guided by Forward & Forget rule which means after sending the SMS, the SMSC deletes the message from the server. Both text and binary data of limited length can be transmitted by SMS.

Table 1 shows the message format used during our implementation of patient to doctor communication via SMS. Unlike HTTP implementations, SMS can only accommodate a short message restricted by 160 characters. Therefore, SMS is not suitable for ECG signal transmission. But, the length of a SMS message is long enough to transmit a single alarming message including the major abnormal ECG features, e.g., the abnormal QRS duration and the HR. In our design, one alarming SMS message contains 4 major fields. The first field is the Message Identifier (MID) which contains 5 bytes. Since each of the byte can have maximum 160 different values, the proposed MID can contain as many as 160^5 different combinations. This enormous range of combination ensures the unique identifying of messages sent from different patients at different times. Health Index (HI) is a 2-byte field which can uniquely differentiate 25600 ECG conditions, such as Right bundle branch block (RBBB), Left bundle branch block (LBBB), Tachycardia, Bradycardia, Arrhythmia etc. The Timestamp field records the time when the ECG abnormality was detected. The remaining bytes are used for storing calculated features, such as heart rate and QRS duration. The following J2ME code highlights the key steps implementing the alarming message sending.

```
// assume the mobile phone num-
ber of the medical professional
is
// 0412345678
String addr =
"sms://0412345678";
String alarmingMessage;
MessageConnection conn= (Mes-
sageConnection) Connector.
open(addr);
```

Table 1. A SMS frame for alarming message

MID	HI	Timestamp	Parameters
(5 bytes)	(1 byte)	(14 bytes)	(up to 140 bytes)

Figure 5. ECG trace analysis steps

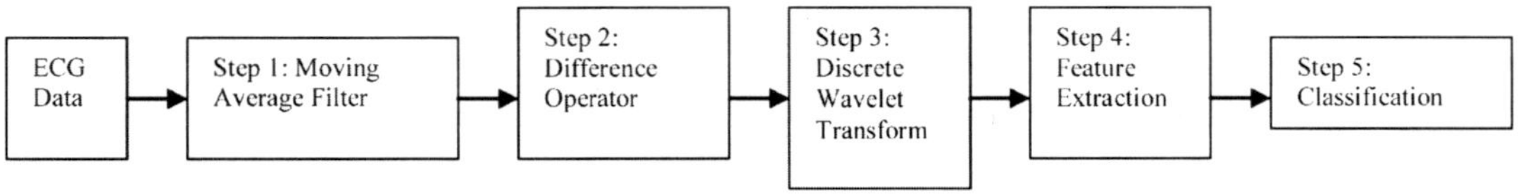

Figure 6. (a) A normal ECG waveform and an ambulatory ECG waveform with baseline wander. (b) The 1st order differences of the two waveforms shown in (a). The two difference traces are almost overlapped each other

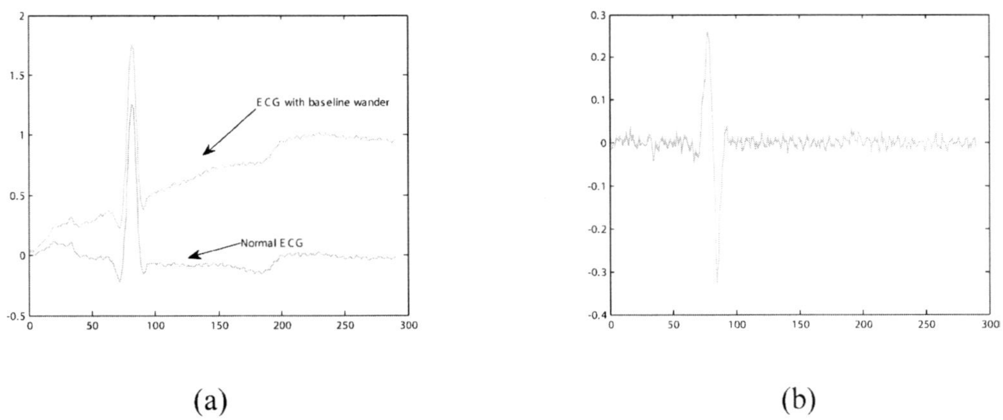


```
TextMessage msg= (Text-
Message) conn.
newMessage(MessageConnection.
TEXT_MESSAGE);
// the content of alarmingMes-
sage is set by method setMes-
sage();
alarmingMessage = setMessage();
msg.setPayloadText(alarmingMess
age);
conn.send(msg);
```

ECG Trace Analysis

Comparing the rest ECG recordings, the realtime ambulatory ECG signal is more susceptible to noise and has much fewer leads. For example, the ambulatory ECG may contain severe high frequency electromagnetic interference from the surrounding environment as well as the increased electromyographic (EMG) noise due to the body motion. Although the baseline wandering caused by the respiration has a typical frequency band between 0.15 and 0.3 Hz, the body movement can cause baseline wandering at much higher frequencies. As discussed in Section 3.1, many existing ECG analysis algorithms are not suitable for ambulatory applications. The suitable processing algorithms are required to be able to handle streamed incoming ECG data and also should be executed fast enough to avoid dropping packets.

As shown in Figure 5, our mobile phone based ECG trace analysis system comprises 5 processing steps, namely the high frequency noise filtering using a 6-point moving average filter, low frequency motion artifact removing using a difference operator, discrete wavelet transform, feature extraction and selection, and a Bayesian based classifier.

Figure 6 illustrates an example of the removing of baseline wander by using a 1st order difference operator. The x-axes for both Figure 6 (a) and 6 (b) are sampled data points with a sampling rate of 500 Hz and the unit for y-axes is millivolt. The obtained two difference traces are almost identical and the R peak is now the zero crossing point from which the QRS complex can be easily identified. The J2ME programs implementing the moving average filter and the difference operator have rather low computational overhead and are readily executed on a mobile device. Thus the remaining part of this chapter will focus on other two core analysis steps, i.e., the discrete wavelet transform and the Bayesian classifier. Their relative more complex algorithms need more lines of code and more memory usage.

We mentioned early that the CLDC-1.0 device, the current most popular mobile device, doesn't support the floating point operation. However, the floating point operation is unavoidable for fast Fourier transform, moving average calculation, and wavelet transform. In order to solve this problem, we tried two approaches. The first approach is to remove the floating points, carry out the numerical operation in integer, carry out the operation of the floating points, then restore the floating point to the numerical operation results. This is virtually equivalent to construct a floating operation library. The second approach is to use a third party floating point library. The particular floating point library we chose is called Microfloat. MicroFloat is a Java software library for doing IEEE-754 floating-point math on small devices which don't have native support for floating-point types. We compared these two approaches and found though the first approach is slightly faster than the second approach, it is error prone. Thus, we decided to choose the third party Microfloat package to solve the floating point operation problem.

For the fast Fourier transform (FFT), the radix-2 Cooley-Tukey fast algorithm is implemented. The J2ME code snippet below shows the FFT implementation. Note the Micrfloat package is used.

```
   public Complex[]
fftEcg(Complex[] ecgRaw, int
powerOf2) {

     Complex[] y = new
Complex[powerOf2];
     // Point to exit the re-
cursive loop
          if (powerOf2 == 1) {
               y[0] =
ecgRaw[0];
               return y;
          }

     // radix 2 Cooley-Tukey
FFT
          Complex[] even = new
Complex[powerOf2 / 2];
          Complex[] odd = new
Complex[powerOf2 / 2];

     for (int k = 0; k < pow-
erOf2 / 2; k++) {
               even[k] =
ecgRaw[2 * k];
          }
          for (int k = 0; k <
powerOf2 / 2; k++) {
               odd[k] =
ecgRaw[2 * k + 1];
          }

     Complex[] q = fftEcg(even,
powerOf2/2);
          Complex[] r =
fftEcg(odd, powerOf2/2);
          for (int k = 0; k <
powerOf2 / 2; k++) {
               long kth = Mi-
croDouble.mul(MicroDouble.intT-
oDouble(-2),
```

```
MicroDouble.intToDouble(k));
                kth = MicroDou-
ble.div(MicroDouble.mul(kth,
                    MicroDou-
ble.PI),
                    MicroDou-
ble.intToDouble(powerOf2));

         Complex wk = new
Complex(MicroDouble.cos(kth),
                MicroDouble.
sin(kth));
                y[k] = q[k].
plus(wk.times(r[k]));
                y[k + pow-
erOf2 / 2] = q[k].minus(wk.
times(r[k]));

     }
         return y;
    }
```

DWT for Baseline Wander Removal and QRS Complex Identification

Figure 7 depicts the DWT of a normal ECG trace (a) and an ECG trace containing baseline wander artifact (b) by using a Daubechies 4 wavelet and the decomposition level is up to 6. The Matlab Wavelet Transform Toolbox is used for this computation. Previous research (Thakor, 1984) indicates that most energies of a typical ECG QRS-complex are concentrated at scales 2^3 and 2^4, especially at scale 2^3. Thus, 2^6 is chosen as the largest scale. From Figure 7 (b), it can be shown that the baseline wander is revealed at a_6 and the QRS complex is clearly displayed in d_2, d_3 and d_4. Thus by combining the wavelet transform modulus maxima (WTMM) and zero-crossing points at these 3 scales, it can efficiently and accurately identify the location of R waves. The fast Mallat pyramid algorithm for DWT is successfully implemented on J2ME platform and deployed on a Nokia N91 mobile phone handset.

RESULT AND DISCUSSION

ECG Signal

A Fluke MPS450 Multiparameter Patient Simulator is used to generate the normal sinus rhythm as well as a variety of arrhythmia ECG traces such as atrial fibrillation, atrial flutter, atrial tachycardia, ventricular tachycardia, ventricular fibrillation, heart block, and asystole. Those normal and abnormal ECG waveforms can be used not only for testing arrhythmia-detection system, but also for training medical personnel, hospital administrators, and research staff. The generated ECG signal is then acquired by the hardware ECG acquisition unit that we built and is further transmitted via Bluetooth to a Nokia 91 handset for local ECG processing. A total of 10 traces of normal sinus rhythm ECG are collected and another 50 traces of different arrhythmia ECG traces are also collected as training data to construct a Naïve Bayesian classifier. Each of those ECG traces is of 8 seconds. Then the realtime collected and digitized ECG data are continuously fed into a Bluetooth enabled Nokia 91 handset and are tested by this Bayesian classifier.

System Implementation and Functions

Screen Menus

The Netbean 6.0 IDE is used to develop the J2ME application. Figure 8 shows the snapshots of our MIDlet implementation of this ECG identification and classification system. Figure 8(a) is the main menu of the GUI. The "Connect ECG Device" command establishes the Bluetooth connectivity between the mobile phone handset and the ECG acquisition unit. The "Show ECG Graph" continuously displays the acquired realtime ECG signal. The "Display ECG List" command displays previously stored ECG traces in RMS and the results are listed in Figure 8 (b). Those

Figure 7. (a) The DWT decomposition of an ECG trace. The Daubechies 4 wavelet is used and the decomposition level is 6. (b) The DWT decomposition of an ECG trace with a motion wander artifact. The motion wander is clearly revealed in a6

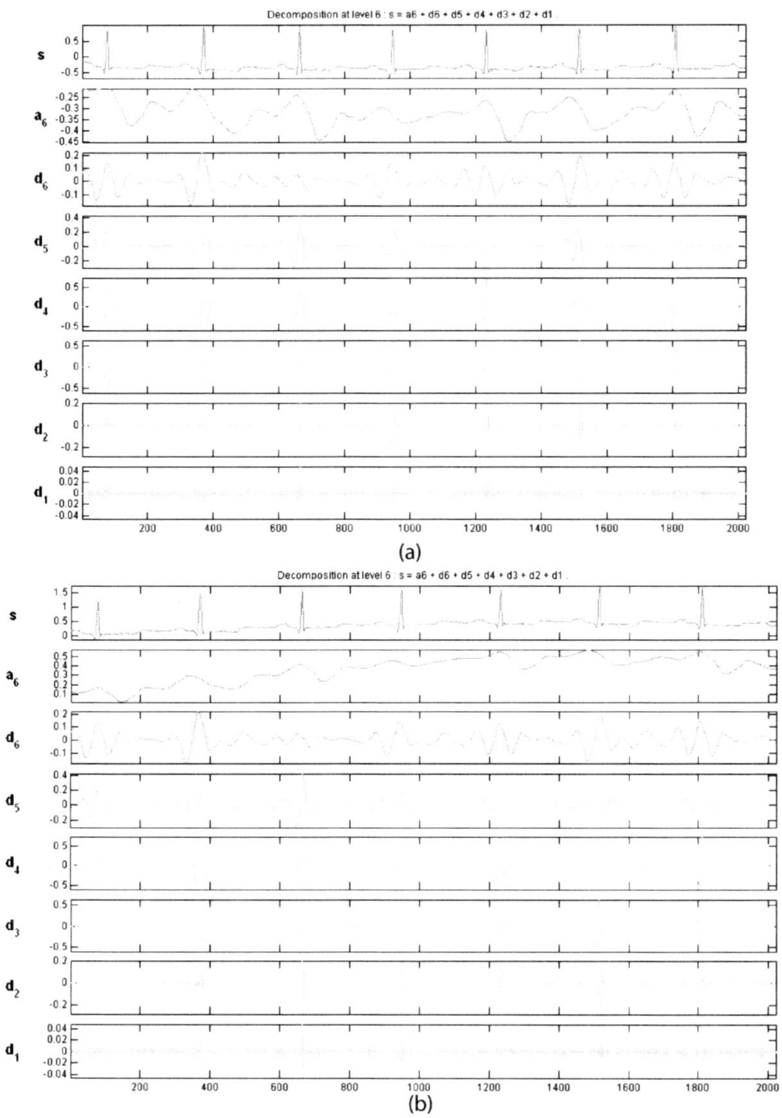

recorded ECG traces are timestamped with a versioning option. The version number 0 as shown in Figure 8 (b) indicates it is an original version without any modification. Figure 8 (c) is the ECG analysis submenu. The user can choose "Display ECG", "ECG Segmentation", "Show ECG Data", "Spectrum Analysis", Daubechies 4 discrete wavelet transform ("Daub4 DWT"), inverse discrete wavelet transform ("Daub4 IDWT"), and "Classification".

DWT Function

Figure 9 shows the snapshots of the DWT for ECG traces using a Daubechies 4 wavelet and the decomposition level is 6. After the DWT is performed for a chosen ECG trace, the user can select to display each component, i.e., 6 levels of details (d_1 to d_6) and 1 level of approximation (a_6). The *d2*, *d3* and *d4* are used by the background classification algorithm to determine the QRS complex.

ECG Segmentation

Figure 10 shows the result of ECG segmentation and interval calculations. The identified ECG intervals are rendered in different colors in Figure 10 (b) and (c), e.g., the red segment is the P wave, the green segment is the R wave and the yellow segment is the ST wave. Those intervals are further supplied into a Bayesian classifier to determine whether the current ECG wave is a normal sinus rhythm or is an arrhythmia.

Performance Evaluation

The Naïve Bayesian classifier is chosen to classify the acquired and segmented waveforms due to its simple implementation and fast computational speed. In this study, total 50 normal sinus rhythm waveforms, and 42 waveform from 5 different

Figure 8. The implementation of a realtime ECG analysis system on a mobile handset. (a) The main menu in Patient's mobile handset. (b) The ECG traces stored in RMS are listed. (c) The sub-menu of ECG analysis. (d) A realtime displayed ECG trace

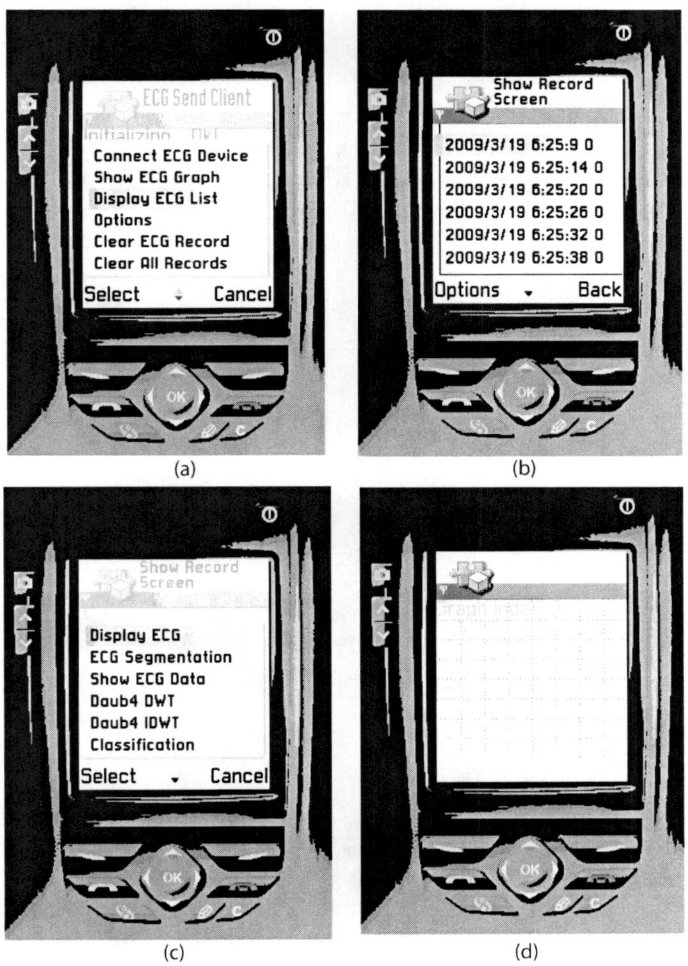

Figure 9. The discrete wavelet transform for ECG analysis using a Daubechies 4 wavelet

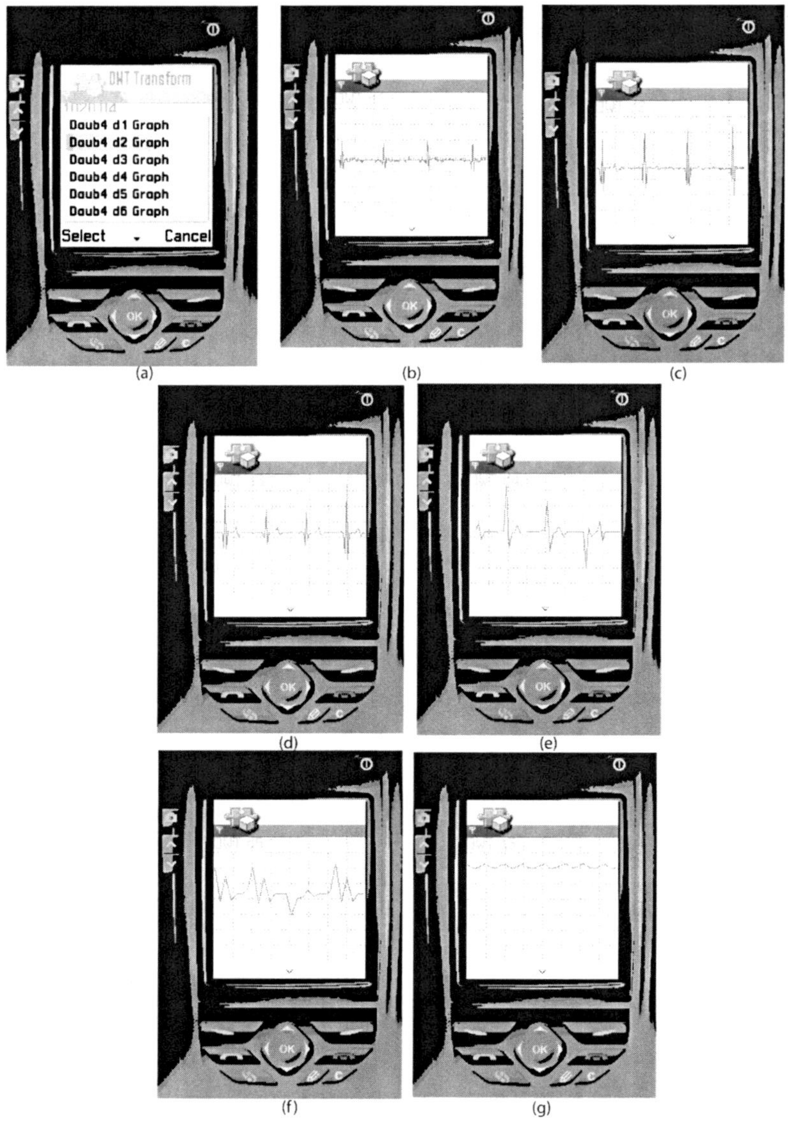

Figure 10. ECG segmentation and intervals calculation

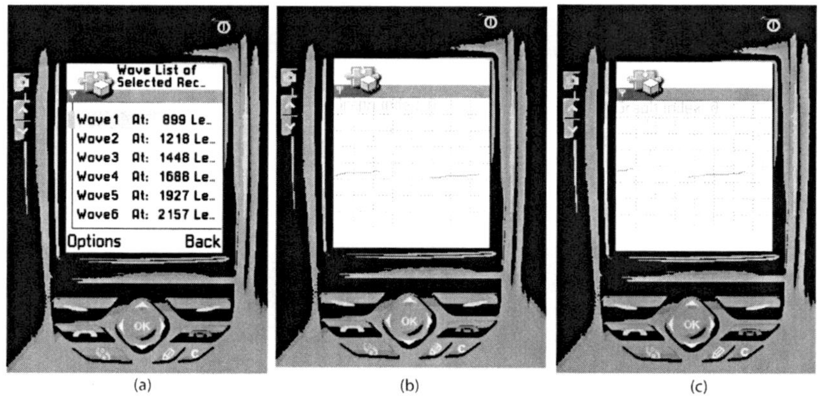

abnormal ECG types (10 atrial flutter waveforms, 10 atrial tachycardia waveforms, 10 ventricular tachycardia waveforms, 10 ventricular fibrillation waveforms and 2 asystole waveforms) are acquired in realtime to test the system performance. Table 2 shows the performance. A false positive result is an allocation of an ECG waveform from one type into its complement set while a false negative result is an allocation of one ECG type from one complement set to its type. It can be seen form the table that this classifier can achieve a rather high overall correct classification rate. The correct rate is defined as:

$$CorrectRate = \frac{FalsePositives + FalseNegatives}{TotalSamples} \quad (7)$$

FUTURE RESEARCH DIRECTIONS

The aim of this investigation is to develop and implement an automated abnormal ECG alert system based on mobile phone technology to activate medical support and possible care after a patient collapses. The presented system integrates time series analysis as well as time-frequency analysis. By adopting a simple Bayesian classifier, total 6 different ECG waveform groups including 5 different types of arrhythmia are tested in this study. The result indicates that different groups have different correct classification rates. Some arrhythmia types such as ventricular fibrillation, atrial flutter, and ventricular tachycardia have correct rates lower than 90%. However the normal sinus rhythm waveform which is the focus of this study can be identified at a high correct rate of 95.6%. This result also shows that the realtime requirement, the total analysis time $T_a < 1.34$ seconds, has been very well attained. The observed average analysis time, from receiving 8 seconds of ECG data in a patient mobile handset to sending alarming message, is around 600 ms which is much less than 1 second.

The correct rate is crucial for providing services with quality because every false positive will trigger a false alarm and every false negative might lead to life loss. It can be noted from Table 2 that the current correct rates are largely affected by the false negative numbers. Thus by reducing the false negative numbers, which were caused by the noises that smear the characteristics of arrhythmia patterns, it is possible to improve the correct rate. The wavelet de-noising algorithm that utilizes the already implemented wavelet decomposition results will be adopted in future work to replace the currently used simple moving average method which is less efficient to process high order motion artifacts.

Table 2. Realtime classification results of 6 different ECG waveform types

ECG types	No. Waveforms	False Positives	False Negatives	Total False	Correct Rate (%)
Normal Sinus Rhythm	50	1	3	4	95.6
Atrial Flutter	10	2	10	12	87.0
Atrial Tachycardia	10	2	6	8	91.3
Ventricular Tachycardia	10	2	10	12	87.0
Ventricular Fibrillation	10	2	13	15	83.7
Asystole	2	0	0	0	100

CONCLUSION

As an e-commerce application in healthcare, telecaridiology is of great significance to support the chronic cardiovascular disease patients. In this chapter, we present a novel, but low cost and relatively equitable ECG signal analysis and alert system for telecardiology. This system fully harnesses the computational power of a plain mobile phone to perform realtime data mining tasks. The evaluation results not only prove it is a feasible approach but also show its potential for future practical applications. The future work will focus on the development of simplified and fast algorithms of other advanced classifiers such as SVM and wavelet packets to improve the arrhythmia classification correct rate.

REFERENCES

Access Economics Report. (2005). *The shifting burden of cardiovascular disease in Australia, A report of Heart foundation*. Retrieved March 20, 2009, from http://www.heartfoundation.com.au/media/nhfa_shifting_burden_cvd_0505.pdf

ACMA. 2009. *Australian Communications and Media Authority Report (2009): Convergence and Communications*. Retrieved March 20, 2009, from http://www.acma.gov.au/webwr/_assets/main/lib100068/convergence_%20comms_rep-1_household_consumers.doc

AIHW. (2004a). *Indigenous Australians carrying heaviest burden of cardiovascular disease*. Retrieved March 20, 2009, from http://www.aihw.gov.au/mediacentre/2004/mr20040505.cfm

AIHW. (2004b). Heart, stroke and vascular diseases—Australian Facts 2004. *AIHW Cat. No. CVD 27. Canberra: AIHW and National Heart Foundation of Australia (Cardiovascular Disease Series No. 22).*

Akay, M. (1995). Wavelets in biomedical engineering. *Annals of Biomedical Engineering*, *23*, 531–542. doi:10.1007/BF02584453

Becker, D. (2006). Fundamentals of electrocardiography interpretation. *Anesthesia Progress*, *53*(2), 53–64. doi:10.2344/0003-3006(2006)53[53:FOEI]2.0.CO;2

Castells, F., Cebrián, A., & Millet, J. (2007). The role of independent component analysis in the signal processing of ECG recordings. *Biomedizinische Technik. Biomedical Engineering*, *52*(1), 18–24. doi:10.1515/BMT.2007.005

ChanY. T. (1995). Wavelets basics. Boston: Klumer Academic Publishers.

Chen, S. W. (2007, Nov-Dec). A nonlinear trimmed moving averaging-based system with its application to real-time QRS beat classification. *Journal of Medical Engineering & Technology*, *31*(6), 443–449. doi:10.1080/03091900701234267

ChuahM.FuF. (2007). ECG *Anomaly detection via time series analysis*. Lecture Notes in Computer Science: Frontiers of High Performance Computing and Networking ISPA 2007 Workshops, 2007 (pp. 123–135). Springer.

ChuiC. K. (1992). An introduction to wavelets. New York: Academic.

DaubechiesI. (1992). Ten lectures on wavelets. Philadelphia: Society for Industrial and Applied Mathematics.

De Capual, C., De Falco, S., & Morellol, R. (2006). *A Soft Computing-Based Measurement System for Medical Applications in Diagnosis of Cardiac Arrhythmias by ECG Signals Analysis*. 2006 IEEE International Conference on Computational Intelligence for Measurement Systems and Applications. pp: 2-7.

Ercelebi, E. (2004). Electrocardiogram signals de-noising using lifting-based discrete wavelet transform. *Computers in Biology and Medicine*, *34*, 479–493. doi:10.1016/S0010-4825(03)00090-8

Fensli, R., Gunnarson, E., & Gundersen, T. (2005). A Wearable ECG-recording System for Continuous Arrhythmia Monitoring in a Wireless Tele-Home-Care Situation. In *Proceedings of the 18th IEEE Symposium on Computer-Based Medical Systems (CBMS'05)*.

Goh, K., Lavanya, J., Kim, Y., Tan, E., & Soh, C. (2005, September 1-4). A PDA-based ECG Beat Detector for Home Cardiac Care. In *Proceedings of the 2005 IEEE Engineering in Medicine and Biology 27th Annual Conference*(pp: 375-378). Shanghai, China.

Graham-Rowe, D. (2004*). Cameraphones will be high-precision scanners, NewScientist.com news service*. Retrieved Oct 10, 2008, from http://www.newscientist.com/article.ns?id=dn7998.

Haghighi, P. D., Zaslavsky, A., Krishnaswamy, S., & Gaber, M. M. (2009). *Mobile Data Mining for Intelligent Healthcare Support*. 42nd Hawaii International Conference on System Sciences, 2009, pp: 1-10.

Helfenbein, E., Zhou, S., Lindauer, J., Field, D., Gregg, R., & Wang, J. (2006). An algorithm for continuous real-time QT interval monitoring. *Journal of Electrocardiology*, *39*, 123–S127. doi:10.1016/j.jelectrocard.2006.05.018

Hu, F., Jiang, M., Celentano, L., & Xiao, Y. (2008). Robust medical ad hoc sensor networks (MASN) with wavelet-based ECG data mining. *Ad Hoc Networks*, *6*, 986–1012. doi:10.1016/j.adhoc.2007.09.002

Jasemian, Y., & Arendt-Nielsen, L. (2005). Evaluation of a realtime, remote monitoring telemedicine system using the Bluetooth protocol and a mobile phone network. *Journal of Telemedicine and Telecare*, *11*(5), 256–160. doi:10.1258/1357633054471911

Jiang, W., & Kong, S. G. (2007, Nov). Block-based neural networks for personalized ECG signal classification. *IEEE Transactions on Neural Networks*, *18*(6), 1750–1761. doi:10.1109/TNN.2007.900239

Kail, E., Khoor, S., & Nieberl, J. (2005). *Ambulatory Wireless Internet Electrocardiography: New concepts & Maths*. 2nd International Conference on Broadband Networks (pp: 1001-1006).

Kranen, P., Kensche, D., Kim, S., Zimmermann, N., Muller, E., Quix, C., et al. (2008). *Mobile Mining and Information Management in HealthNet Scenarios*. 9th International Conference on Mobile Data Management (pp: 215-216).

Mallat, S. (1989). A theory for multiresolution signal decomposition: The wavelet representation. *IEEE Transactions on Pattern Analysis and Machine Intelligence*, *11*, 674–693. doi:10.1109/34.192463

Mannila, H., Tikanmki, J., Himberg, J., Korpiaho, K., & Toivonen, H. (2001) Time series segmentation for context recognition in mobile devices. In *First IEEE international conference on data mining* (pp:203–210).

Rijnbeek, P., Kors, J., & Witsenburg, M. (2001). Minimum Bandwidth Requirements for Recording of Pediatric Electrocardiograms. *Circulation*, *104*, 3087–3090. doi:10.1161/hc5001.101063

Roberts, R. (2006). *Use of Remote Monitoring Devices Increases, Telemedicine Information Exchange, (Original Source: Wall Street Journal, April 18, 2006).* Retrieved Feb 20, 2008, from http://tie.telemed.org/legal/news.asp

Rodríguez, J., Goñi, A., & Illarramendi, A. (2005). Real-Time Classification of ECGs on a PDA. *IEEE Transactions on Information Technology in Biomedicine, 9*(1), 23:34.

StrangG.NguyenT.(1996). Wavelets and filter banks. Wellesley, MA: Wellesley-Cambridge Press.

Sufi, F., Fang, Q., Khalil, I., & Mahmoud, S. (2009). Novel Methods of Faster Cardiovascular Diagnosis in Wireless Telecardiology. *IEEE Journal on Selected Areas in Communications*, *27*(4), 537–553. doi:10.1109/JSAC.2009.090515

Sufi, F., Fang, Q., Mahmoud, S., & Cosic, I. (2006). A mobile phone based intelligent telemonitoring platform. In *The Proceedings of 3rd IEEE EMBS International Summer School on Medical Devices and Biosensors*(pp: 101–104).

TanP. N.SteinbachM.KumarV. (2006). Introduction to Data Mining. Boston: Pearson Education, Inc.

Thakor, N.V., Webster, J.G., & Tompkins, W.J. (1984). Estimation of QRS complex power spectra for design of a QRS filter. *IEEE Transactions on Biomedical Engineering. 31*(11), 702:706.

Thomas, J., Rose, C., & Charpillet, F. (2007). A support system for ECG segmentation based on Hidden Markov Models. In . *Proceedings of Annual Conference of IEEE Eng Med Biol Soc., 2007*, 3228–3231.

Wu, B., Zhuo, Y., Zhu, X., Yan, Q., Zhu, L., & Li, G. (2005). A Novel Mobile ECG Telemonitoring System. In *Proceedings of 27th Annual International Conference of the IEEE Engineering in Medicine and Biology Society* (pp: 3818 – 3821).

Zhang, H. (2004). The optimality of naive bayes. In Barr, V., Markov, Z., Barr, V., and Markov, Z., editors, *FLAIRS Conference*. AAAI Press.
Zhou, H., Hou, K., Ponsonnaille, J., Gineste, L., & De Vaulx, C. (2005). *A Real-Time Continuous Cardiac Arrhythmias Detection System: RECAD* (pp: 875 – 881).

Chapter 7
Factors Facing Mobile Commerce Deployment in United Kingdom

Ziad Hunaiti
Anglia Ruskin University, UK

Daniel Tairo
University of Greenwich, UK

Eliamani Sedoyeka
Anglia Ruskin University, UK

Sammi Elgazzar
Anglia Ruskin University, UK

ABSTRACT

This chapter discuss the challenges facing mobile commerce deployment in the United Kingdom. Although the number of mobile phone users is increasing and the technology is available for successful implementation of m-commerce, only a small number of users utilise m-commerce services. At the same time, mobile phones are becoming smarter, and most of latest phones are capable of connecting to the Internet. This chapter looks at the background of m-commerce as well as the technological development of mobile phones to their current stage. Also, technical and non technical issues which hinder the adoption of m-commerce are discussed and solutions and recommendations are given.

INTRODUCTION

The discovery of "radio waves" - electromagnetic waves and radio communication was behind the birth of new era of transfer and exchange information. That has been emerged in a number of technologies and applications such as; radio and TV broadcasting, satellite communications and mobile communications. Hence, nations became interconnected across the globe and made it as one village. That has been strengthening with birth of the Internet; to complement other wire and wireless networks. As a result of this major advancement, new forms of lifestyle have evolved online applications; which evolves conducting tasks using the power of the Internet and networking infrastructure.

DOI: 10.4018/978-1-61520-761-9.ch007

Online shopping which as focus Electronic commerce (E-Commerce) is one of main platforms for trading and shopping. E-Commerce has become popular for both business and consumers. After then, the adaptation of E-Commerce over mobile networks (m-commerce) has evolved, particularly after the deployment of new mobile generations with high speed Internet access capabilities, which is seen as a key factor for fostering this kind of applications. That was expect to generate a big opportunities; but it was not the case in many countries including the UK, where the growth of m-commerce remind very slow and far below the expected penetration rate. In this chapter reasons behind the hindering of m-commerce will be discussed and recommendation for future m-commerce industry will be presented.

BACKGROUND

Internet is growing very fast and millions of users are connected worldwide. It has changed the way people communicate, socialize, and live in general. Different businesses are also using Internet in the way they are doing their business, from selling of products and services to online banking. Business – to – Business (B2B) which involves trading between business and Business-to-Consumer (B2C) which facilitates trading between commercial organizations and consumers and Consumer-to-Consumer (C2C) are all part of Electronic commerce (E-Commerce). Now people can use mobile devices (Mobile phones and PDAs) to perform electronic commerce, and the term used for this is M-commerce, and its best described as

m-commerce = E-Commerce + mobile

M-commerce is possible when mobile phones can be connected to the Internet, but the use of Mobile phones for M-commerce (shopping) has not been popular in UK as predicted. M-commerce has been in use since the end of twentieth century and has developed a lot from then, but in UK it is not as popular for conducting electronic commerce as it has been with Personal Computers (PCs) and compared with other countries like China and Japan where it is popular (Sadeh, 2002).

M-commerce started unsuccessfully with the introduction of Wireless application Protocol (WAP). This technology enables mobile devices to browse the Internet because it support extensible markup language (XML) and hypertext mark-up language (HTML) which are key languages used for Internet content. WAP enabled devices to run a micro browser. These are applications that suit the small memory size of handheld devices and the bandwidth constraints of a wireless handheld network. Another important M-commerce technology, which is used every day with mobile phone users, is short message service (SMS). This popular service allows short text messages to be sent from and to mobile devices at a low cost. This has a wide application in the use of M-commerce technology (Lewis D, 2004).

Mobile commerce was then coined in the late 1990s during the dot-com boom. The idea that highly profitable mobile commerce applications would be possible through the broadband mobile telephony provided by 2.5G and 3G mobile phone services was one of the main reasons for hundreds of billions of dollars in licensing fees paid by European telecommunications companies for UMTS and other 3G licenses in 2000 and 2001. PDAs and Mobile phones have become so popular that many businesses are beginning to use m-commerce as a more efficient method of reaching the demands of their customers. Although technological trends and advances are concentrated in Asia and in Europe, North America (Canada and the United States) is also beginning to experiment with early-stage of m-commerce. The recent alliance between Sprint Nextel and Clear wire for WiMAX networks being built for completion by 2008 will accelerate the more data-intensive 4G networks that will provide a turning point in m-commerce in North

America. Google, irking Verizon Wireless and AT&T are pushing rules changes that are geared for more consumer options and less control by telecom operators.

With the high pressure to change, telephone companies were forecasting a systemic decline in the use of voice and decided to add new capabilities to their networks. The telephone companies reengineered and upgraded the network to connect machines better. These companies also sponsored the opportunity for a new device. By 1997, a little company called Unwired Planet worked with telephone companies to install, free of charge, a micro browser in mobile phones so voice subscribers could cruise the Internet. Sprint underwrote a lot of the initial development. A number of European telecommunications companies convinced the Unwired Planet to recast the technology as the Wireless Application Protocol. In 1998, the Wireless Application Protocol (WAP), which is standard that enable access to the Internet for Mobile phones, was the basis for the small-screen web phone found in European and some U.S. phones. The computer industry had its PDA, and the telephone companies' industry had its WAP. Although Unwired Planet and WAP Mobile phones failed to take off, the web phone found its time in the year of 1999 where the popularity increased. In 2000 and 2001 hundreds of billions of pounds in the licensing fees were paid by the European telecommunications companies for the universal mobile telecommunications and other 3G license reason being high expectations of the highly profitable M-commerce applications. These M-commerce applications would be delivered through broadband mobile telephony, which is provided by the 2.5G and 3G mobile phone services.

Throughout the 1990s, the telephone companies upgraded 1G analogue mobile phone to 2G digital Mobile phones and what we currently use now 3.75G, making the networks ready for data (Shi 2003). Unlike the wired network, which is primarily filled with data, the mobile network is filled with voice. According to the FCC, only two percent of wireless mobile traffic is currently data. But rather than pioneering the inevitable trends toward digital traffic, the Telephone companies simply just added subscribers. The mobile voice

Figure 1. a) Nokia 6681 b) Dopod 818 Pro c) Apple iPhone 3G

Table 1. Smart phones and PDA

Mobile phone	a) Nokia 6681 (Smartphone)	b) Dopod 818 Pro (PDA)	c) Apple iPhone 3G (Smartphone)
Manufacturer	Nokia	Dopod International (HTC)	Apple
Network	**2G** Network	**2G** Network	**2G** + 3G Network
Display size	2.1 inches	2.8 inches	3.5 inches
OS	Symbian OS	Windows **Mobile** 5.0	Mac OS X v10.4.10
Browser	**WAP** 2.0/xHTML, HTML	**WAP** 2.0/xHTML, HTML (Pocket IE)	HTML (Safari)
Announced (in UK)	2005	2006 as i-mate JAMin, 2006 as O2 XDA Neo	2008
Operator (for these phones)	O2 UK (Pay as you go)	T **Mobile** UK (Pay as you go)	O2 UK (Contract)

market is still the primary 'dollar' attractor, and it is how most people continue to use the wireless networks. Most telephone companies still consider data an *up sell* to their phones (Mark B, 2001). With time, Mobile phones and PDAs have become smarter, performing more complex tasks (Figure 1 and Table 1).

Technology and Characteristics of Mobile Phones

Mobile phones in the UK have taken a number of changes; The Mobile phones in the UK have gone through a number of changes in the standards that they use which are.

First-generation wireless telephone technology: Starting in 1981. This generation is 1G, the first for using Mobile technology that let users place their own calls and continue their conversations seamlessly as they moved from Mobile to Mobile. AMPS use what is called FDM or frequency division multiplexing. Each phone call uses separate radio frequencies or channels.

Second-generation wireless telephone technology: In mobile telephony, second-generation protocols use digital encoding and include GSM, D-AMPS (TDMA) and CDMA. 2G networks are in current use around the world. These protocols support high bit rate voice and limited data communications. They offer auxiliary services such as data, fax and SMS. Most 2G protocols offer different levels of encryption.

2.5 generation wireless telephone technology: In mobile telephony, 2.5G protocols extend 2G systems to provide additional features such as packet-switched connection (GPRS) and enhanced data rates (HSCSD, EDGE), which was the start of real internal mobile networks.

Third-generation wireless telephone technology: In mobile telephony, third-generation protocols support much higher data rates, measured in Mbps, intended for applications other than voice. 3G networks trials started in Japan in 2001. 3G networks are expected to be starting in Europe and part of Asia/Pacific by 2002 and in the US later. 3G will support bandwidth-hungry applications such as full-motion video, video-conferencing and full Internet access (Korhonen J, 2004).

Key Features of M-Commerce

- **Ubiquity and Accessibility:** The use of wireless device enables the user to receive information and conduct transactions anywhere, anytime.
- **Convenience:** The portability of the wireless device and its functions from storing data to access to information or persons unlike the use of E-Commerce with desktop computers.
- **Localisation:** The emergence of location-specific based applications will enable the user to receive relevant information on which to act i.e. sales at a local shop,

special offers where ever you are. Instant Connectivity (2.5G): Instant connectivity or "always on" is becoming more prevalent with the emergence of 2.5 G networks, GPRS or EDGE. Users of 2.5G services will benefit from easier and faster access to the Internet (3G is even much faster)

- **Personalisation:** The combination of localization and personalization will create a new channel/business opportunity for reaching and attracting customers. Personalization will take the form of customized information, meeting the users' preferences, followed by payment mechaniSMS that allow for personal information to be stored, eliminating the need to enter (payment) credit card information for each transaction.
- **Time Sensitivity:** Access to real-time information such as a stock quote that can be acted upon immediately.

Main Issues, Controversies, Problems

With the characteristics of mobile phones, one might expect to see users and businesses embrace m-commerce but according to latest report, out of 73 million mobile phone users in UK, only 17 million users connect their Mobile phones to the Internet. The results which were collected at the end of the year 2007 show that only 23.29% are connected to the Internet which is very low percentage. This low percentage of users connected to the Internet will surely affect m-commerce and study was conducted to see the reason behind this low penetration rate.

There are number of issues which in one way or another contribute the usability of connection of mobile phones to the Internet. Frequently m-commerce is represented as a "subset of all E-Commerce" thus implying that any E-Commerce site could and should be made available for a wireless device. We believe that such conclusions are misleading. M-commerce should be recognized as a unique business opportunity with its own unique characteristics and functions, not just an extension of an organization's Internet-based E-Commerce channel. Of course there are similarities between E-Commerce and m-commerce from being able to purchase a product or service in a "virtual" vs. a build and mortar environment (Herness newsletter, 2001). Although M-commerce is a mobile E-Commerce, it still needs a different approach from the initial design stages, which is how is to be used and what will be the content, to the deployment stages.

Technical and non technical issues can affect the deployment of mobile commerce. These issues can affect the usability of mobile phone and suitability of using m commerce. Some of the technical issues are: the communication over the air interface between mobile device and network introduces additional security threats e.g. eavesdropping which is the capability that the attacker to listen in on signal and data connections associated with other users, mobile devices offer limited capabilities, such as limited (small) display, small processors and limited memory, which is a big setback when using the Internet as this will slow the loading time. Bluetooth, GPRS, 3G and Wi-Fi are Smartphones and PDA features that consume a lot of battery, and using them for a long time will shorten the battery life.

Non technical issues include theft, cost and viruses/worms. Mobile phones are more prone to theft and destruction, Government reports show that more than seven hundred thousand mobile phones are stolen in the UK each year. The cost of buying phone with required technical specification can be very expensive and sometimes more than the price of a new laptop. Since Smartphones and PDAs have their own OS and other applications installed, there is easy possibility of getting warms and viruses when connected to the Internet. Also, there is a threat of a user being tricked and install a snoopware program without knowing.

Figure 2. Mobile phone have internet connection

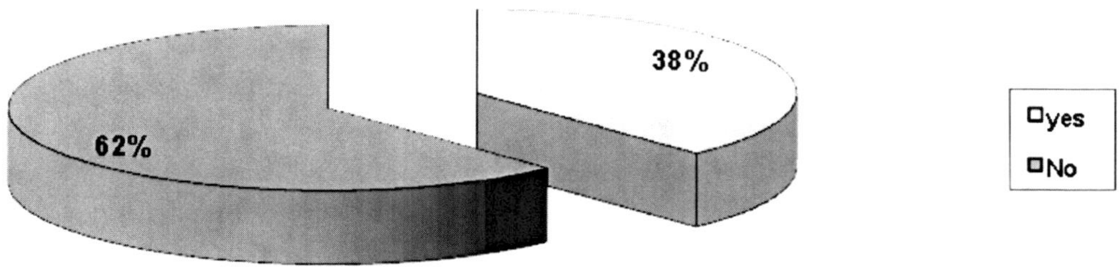

Snoopware programs are capable of turning your mobile phone into a remote monitoring device (activating microphones and/or cameras to record you or your communications)

People believed that the mobile Internet would give them full access to the wired web. But when customers tried to access a site like Amazon.com from a mobile phone, they found it could take as much as 50 minutes to place an order unlike using E-Commerce from a computer which wouldn't be any way near this time (Everson E, 2007)

Data Collection: The Survey

The study to evaluate the current status of m-commerce use in UK was carried out and the questionnaire was used as a primary research method. The questionnaire was chosen because of its easiness in reaching more users and the aim was to find the use of m commerce. The questions were designed in a way that the respondents could provide accurate answers in an easier and quicker way. Participants were asked about their experience in using mobile phones especially for E-Commerce activities. Also, information was collected from previous studies carried out on m-commerce. The questions were asked to 200 people both students and members of the public from 16 years old and above.

The Results

Different questions were asked the following section provides the results and discussion from some of the questions. The first question asked was either respondent has an email address or not reason being to assess the knowledge of respondent in using computers in general. This question showed that 87% of people asked have an e-mail address and only 13% do not currently have an e-mail address. Again, people were asked if they have Internet access and the outcome of the questionnaire shows that 83% of the public have currently got Internet access and 17% has not got access to the Internet. When asked where they get Internet access, the results showed that 60% o usually connect to the Internet at home, 17% connected in the library, 9% connected in the office, 6% connected at their place of work, 4% at a Internet café, only 3% using their mobile phones and finally 1% connected using other ways of connecting to the Internet.

Figure 2 show that 62% of the people asked do not have Internet connection in their mobile phones and 38% of the people asked were able to connect to the Internet using their mobile. Although later they were asked the type of Mobile phones they have, the reason 62% are not able to connect to the Internet is not known if its lack of technical specification, or other reasons.

The outcome of this question shown in figure 3, shows that a huge 84% of the people asked did not use their mobile or PDA to connect to the Internet, only 9% used their mobile phone to connect and only 7% used their PDA. Out of 38% who have Internet connection in their Mobile phones, only 16% are using their Mobile phones or PDAs to do m commerce.

Figure 3. Using mobile phone or PDA to connect to the Internet for online shopping

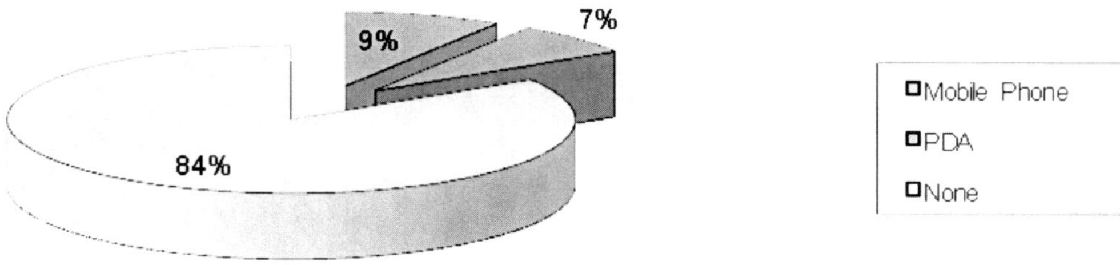

Figure 4. Easy to connect to the internet using my mobile

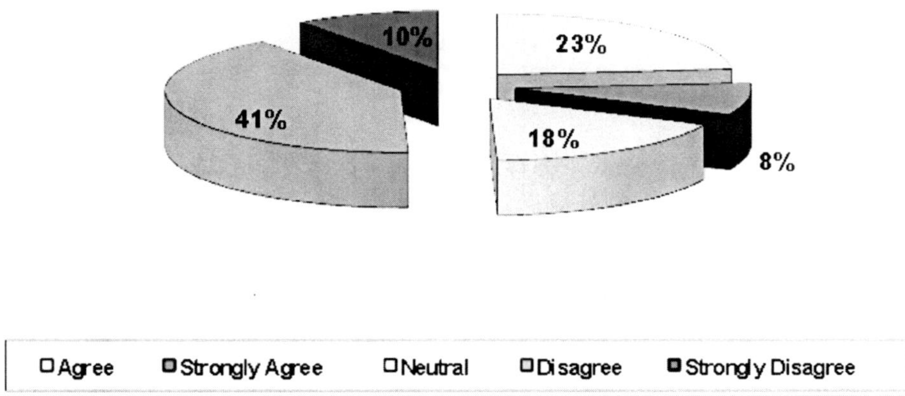

Service providers in UK have almost equal shares of users and the results show that 27% were connected to Vodafone, 21% were connected with O2, 18% with T-mobile, 15% with Orange, 12% with Virgin and finally 7% with 3G. Different service providers offers different mobile phone packages through contract and pay as you go and this can influence the Internet connection from mobile phone. When asked what type of service they have, 68% of the respondents use the pay as you go service and 38% use contract services. Type of service used, will give an understanding on price. Compared to pay as you go some contract deals give a number of minutes or free Internet access. This will also answer in a way if Internet cost is the problem. For contract phone, user who take the 'browsing' package don't care about the cost of Internet since it is part of the package, even with fair usage policy, and those without contract, will use the Internet only when they need to do so. The cost can be the problem only if someone really need to connect mobile phone to the Internet and cannot afford to do so at all.

The study shows that most used mobile type was the Nokia phone with 40% of the people were asked use it, 18% use the Sony Erickson, 13% used the different Samsung models, 10% used the LG, 5% used the Siemens mobile and only 2% use the new apple iPhone which was released at the end of 2007 in the UK. Although types of Mobile phones were given, it is still difficult to determine the technical specifications of each model in order to determine if they are capable of connect to the Internet. Although most new phone models can connect to the Internet, still each model design offer a different usability in terms of model size, keyboard type, and other specifications as shown in figure 1.

As shown in the pie chart in Figure 4, nearly half of the participants do not find it easy to connect to the Internet using their mobile phone as 23% agreed that it was easy, 9% strongly agreed,

Figure 5. The cost of connecting to the Internet by mobile is reasonable

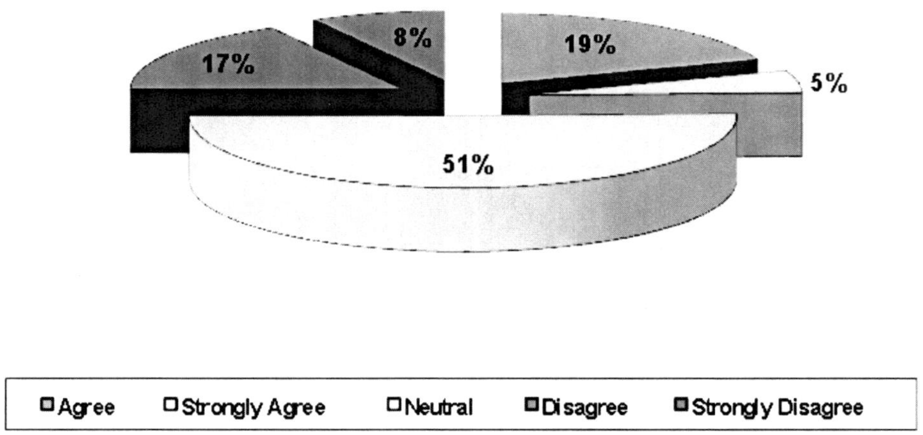

18% neutral which means they do not know if it easy or not, 40% disagreed and 10% strongly disagreed. When asked if they can use their Mobile phones to do Online shopping, the results show that the majority of people asked, do not know. 35% agreed they did not know how to use their phone for Online shopping.

The outcome of this question shows that exactly 51% answered neutral so they did not find price as reasonable but they didn't find it reasonable so this may show that either they do not use their mobile phone at all to connect to the Internet or the price is not an issue to them or even they have not bothered to compare with other service providers. 20% of the answers did agree that it was a reasonable price, 5% strongly agreed, 17% disagreed and 8% strongly disagreed. Also, when asked if the overall cost of Internet is a concern, 17% agreed that the cost of Internet access was expensive, 22% strongly agreed, 32% neutral, 23% disagreed and 6% strongly disagreed

A majority of the participants agreed that using the Internet on your mobile phone does use a lot of battery, as a result to the question above 30%

Figure 6. Safe to use mobile phone in public

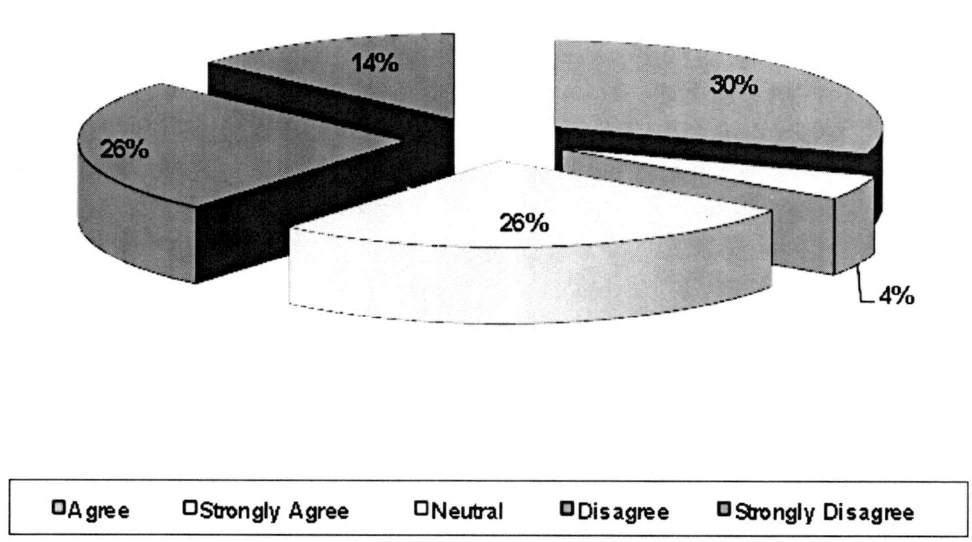

agreed, 10% strongly agreed, 45% neutral, 12% disagreed and 3% strongly disagreed. Although it is a known fact that GPRS and Wi-Fi use a lot of battery, still the question was asked to see if this might be the actual thing concerning users during Internet connectivity.

The results to this questionnaire was a mixed response as 30% agreed, 4% strongly agreed, 26% neutral, 26% disagreed and 14% strongly disagreed. This shows that some people feel safe and some people do not feel safe to use their phone in public. Although when later asked if they feel safe to use their mobile phones to shop Online while in public, 13% agreed, 4% strongly agreed, 34% neutral, 40% Disagree and 9% strongly disagree. This means that, its normal and safe to use mobile phone in public for other things like talking or sending/receiving short message, but when it come to shop Online in public, people don't feel safe. Also, when asked if they feel its secure enough to provide their personal details and bank details through Mobile phones, majority of people doesn't feel secure as, 22% agreed, 6% strongly agreed, 23% neutral, 38% disagreed and 11% strongly disagreed that is wasn't safe enough. But when asked if they will be willing to show in future, there were mixed results but slight high percentage of respondents answered that they are not willing. The question was 'will use m commerce in future' and results were 22% agreed, 5% strongly agreed, 27% Neutral, 37% Disagree and 9% strongly disagree.

The main concern for me to use mobile shopping is time consuming was another question asked and the outcome of this question was that 29% agreed that their concern was it was time consuming to access the Internet through their mobile phone, 19% strongly agreed, 26% neutral, 19% disagreed and 7% strongly disagreed.

A big concern to the users is the usability of the mobile phone as 48% agreed that the usability of the mobile phone was a big concern, 19% strongly agreed, 19% neutral, 9% agreed and 5% strongly disagreed.

When asked if comparing prices is a problem, majority of answers that was received from the participants did agree that their main concern was comparing prices this maybe because of the display size. The result that was received was that 21% agreed that it was difficult to compare prices, 18% strongly agreed, and 43% Neutral, 10% disagreed and 8% strongly disagreed.

Figure 7. Mobile phone usability for online shopping

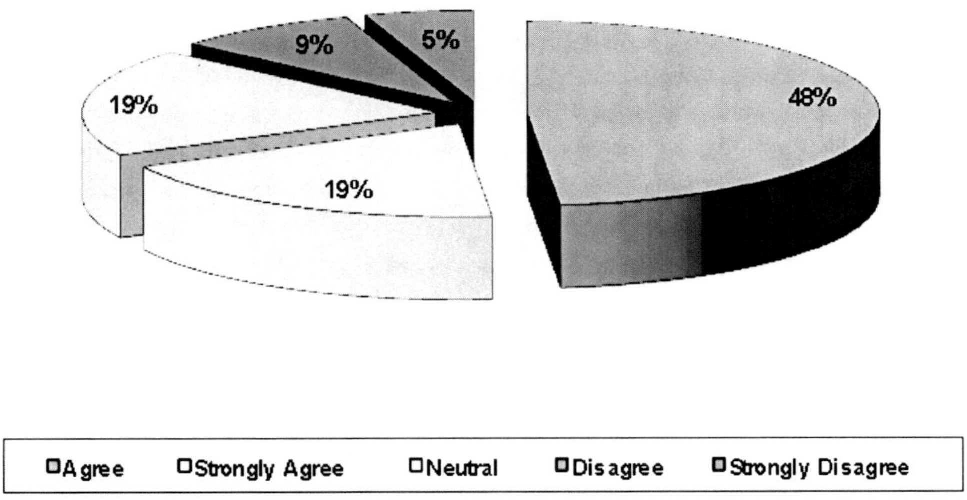

DISCUSSIONS

From the results, it's clear that the percentage of mobile phone users who are using their Mobile phones to connect to and do shopping Online is very low. Different reasons can be inferred from the results and some of them are people do not feel safe to shop Online in public, people are not satisfied with the security provided by wireless connection, it is not easy to connect to the Internet through mobile phone, even if it were easy to connect still people don't know how to connect, usability of mobile phone, connection cost and it takes time to use mobile phone to do m commerce.

There are other issues which affect the deployment of m commerce in UK and which were not captured in the questionnaires questions. There are 'normal' mobile phones, Smartphones and PDA. They are normal in a sense that they provide basic original intention of mobile phone which is voice and short message services. Users who are still using normal Mobile phones might not need to upgrade to Smartphones or PDA if their original reason was voice communication and short message services, and if they are still satisfied with the capability of their Mobile phones.

Smartphones and PDA are far more advanced than the normal mobile phones as they have operating system which allows other applications to be installed. This makes these types of Mobile phones to have a lot of features which will lead to wider customer choice. Example, the Nokia 6681 shown in figure 1 apart from doing basic voice and short message service, it also have a powerful camera, can open PDF, Word, Excel files, can play music of MP3, AAC format and other more features, the Dopod shown is like a small computer with windows media player, Internet explorer, and other features and iPhone with even more features. This means user will be tempted to use simple and cheaper (in terms of cost) features, that is, for a mobile phone with voice, SMS, camera, music and browsing capabilities, users will use those features which are common and easy to use and probably browsing will be the least used. People are connected to their Mobile phones in a personal way. They want to show of that they have latest or beautiful Mobile phones. This means sometimes technical specifications are not main factor in users' choice of mobile phone and maybe fashion or 'feel good factor' are key factor. But it can be argued that, if same users are informed about the full ability of their Mobile phones, they might start to use rip the benefits by trying to use all features as much as possible.

There is difference between surfing the Internet using mobile phone and using the same mobile phone to do m commerce. Surfing the Internet using mobile phone, if one is capable to do so, involves visiting the mobile or normal website for various purposes and it does not involve any transaction. The surfing of Internet only require a particular content to be present, example news, sports, entertainment etc and there are enough mobile news and sport website and portals, and the good example is each service provider provides a portal where one can browse such info.

Figure 1 shows three different mobile phones with display size of 2.1 inches, 2.4 inches and 3.5 inches. All phones have opened www.google.co.uk website. From the figure, the phone with 3.5 inches is clearer and shows more lines at the same time and therefore will be the best choice, but there are two simple factors which will make it not so. First is, majority of users find that even the phone with 2.1 inches of display is too big for comfortable mobility and that means some users prefer small mobile phone. This shows that even the normal browsing will be difficult as this 2.1 inches phones is already too big. Second factor is price, and brand new unlocked iPhone is more than £600, and O2 pay as you go cost more than £300 and this is only purchasing cost and does not include 'running' cost.

Why would anyone with Internet access in a personal computer use a mobile phone to shop? In order to imitate e commerce in mobile phone,

one need clear display and easy to use navigation system and keyboard. From figure 1, Nokia 6681 is using phone pad and 5 way navigation button, Dopod 818 is using stylus and mini keyboard and iPhone is touch screen (finger) and all of them are not easy to use as compared to a full size personal computer. Depending on the type of transaction one intends to perform using mobile phone, it is still a setback for users conducting m commerce.

Another issue which might not be very strong but is important is, there are too many 'players' in the mobile phone industry, and this is end to end connection. The service providers (example O2) provide connection to a machine (example Nokia) with OS from different developer. Although there are standards and although it is successful in computer industry, from the general view, this is the problem. Apple iPhone is trying to solve this by developing its own machine, powering it up with its own operating system and browser and selecting specific operators to provide the connection.

With 73 million mobile phone users in UK, with some of users have more than mobile phone, some of users are under 14, some of users are using 'normal' Mobile phones, some of users buy Mobile phones because of 'looks' and not specifications, its fair to say 17 million users who connects to the Internet is not a very bad number although it can be increased and therefore probably increase the overall users who would do m commerce.

Solutions and Recommendations

The type of Internet content available for mobile phone should be made for mobile phone and this is not only for m commerce but for normal Internet browsing. There should be enough mobile websites for mobile since there are still Mobile phones with smaller display than 2.1 inches. To have a successful deployment of m commerce, there should be a new approach in the manner which business is to be done using Mobile phones. Although websites like Face book, Flickr, CNN, Yahoo, BBC, eBay, YouTube and others have mobile websites, still more mobile websites are required. To encourage the development of mobile websites, all participating parties, starting with World Wide Consortium W3C, Mobile phones manufacturers and developer, businesses to push for mobile website standards, focusing on content presentation and accessibility. Also there should be a rule to force website developers to develop a mobile website for any full size website they are going to develop.

There should be a certain choice of goods or services which can be purchased Online and some of these are digital products like digital media and application software since they don't require one to see the shape of size of the product. But there are also products which required being seen properly and other require comparing prices. These kinds of products need redesign of how Internet content is presented in a mobile phone.

The normal process of purchasing something Online is selection, registering (if it's first time), dispatch, payment, confirmation. The registration is done only once where by a user account will be created and select a user name and a password. Again, during payment process, user is required to provide credit/debit card details. This process is repeated every time a user wants to purchase an item in a new e shop. This to a problem of people getting tired of registering every time and this can be overcome by using the services type of PayPal or Google checkout. With Pay Pal or Google check out, user is registered only once and can purchase items Online easily from Pay Pal of Google checkout merchant. Currently, both Pay Pal and Google checkout have mobile services, that is mobile Pay Pal and mobile Google check out and with these services customers can access sites designed for mobile devices on their mobile and purchase products and services using their accounts without having the problem of getting out their credit card in the public which was part of the security fears. Moreover, the new pay pal system will encourage more users to use the M-commerce as their bank details will be secure and are also covered on insurance for Online fraud. What can be done now is to encourage more e merchants to embrace the payment method offered by the likes of mobile Pay Pal and mobile Google check out so as to capture the new customers who would like to use this type of technology.

Even with mobile website and easy payment method for mobile phone users, if there will not be a way to access these sites and services easily, then it will be useless. Display size and the overall usability of Mobile phones are setback in m commerce deployment. To solve this setback, Apple released a mobile phone with 3.5 inches display called iPhone and BT using Smartphone HTC S620 released BT Total Broadband Anywhere. Both the approach of Apple and BT was focused on developing a mobile phone or a mobile phone package with Internet (display size, usability) and speed in mind.

Apple iPhone shown in figure 1 C offer a lot of benefits and solve some of the issues which hinder successful deployment of m commerce in UK. Some of the advantages it has over other Smartphones and PDAs are

- The large multi-touch touch screen display and innovative software, the iPhone let's you control everything using your fingers.
- Allows you to type using only your fingers which will be much easier than using normal mobile phone pads which are much smaller and several letters just on one key pad.
- Automatically finds and connects to trusted Wi-Fi networks so you can surf the web at faster speeds, so this mobile phone is now acting similar in the way desktop computers and laptops do.
- Possible to determine the location of a switched-on mobile within about 500 meters. This allows users to search for local hotels and restaurants with live menu availability and pricing.

The BT broadband anywhere allows for mobile phone (the model they offer is HTC S620 Smartphone with full qwerty keyboard and 2.4 inches display) to be connected through your Internet. From the BT broadband you can connect, download, and surf the Internet at broadband speeds when you are in the Wi-Fi spot but even if you are not I the spot you can still browse the Internet and download pictures, music and information onto your phone while you are out. The phone is designed to make browsing easier on the Internet and the 'BT to go' screen has been designed to look similar to a PC desktop. Also, the phone allows connecting to the Internet only with one touch of the button. (Phones review, 2008).

Marketing still plays a major role in bringing awareness to people. Mobile phone technology is growing fast, and normal users if are not informed properly, will not know what possibilities are around them. Business should embrace and advertise new technologies and show clearly the benefits to users.

FUTURE RESEARCH DIRECTION

The security when using M-commerce is a big issue to the users. Biometrics can be used as security approach, for example, the integration of signature verification into mobile phones or others like voice and fingerprint verification. This allows the capture of not only static but also dynamic parameters of signatures such as the speed of writing, pressure applied, letter shape and the rhythm of the writing process. This way of security used from biometrics may encourage more users of M-commerce both for security and usability. Biometrics will encourage users to make payments, as they will need to sign for permission for payment. Biometrics will be much widely used not just for Online access to accounts etc; E-Commerce websites will soon be using biometrics to log on to accounts and make more secure payments which worries both m-commerce and E-Commerce users. If M-commerce would use different types of biometric security it may attract more m commerce users in the UK (Moody, 2007)

CONCLUSION

This chapter presented the outcome of study conducted to identify the main factor/challenges behind the low penetration rate of using mobile commerce in UK. Its is clear from the outcome of this study presented that unless a complete framework for Mobile commerce has been established the view of tackling M-commerce has been established with the view of tackling M-commerce identified shortcomings, the growth will remain slow and might not reach targeted bred, which will make it risky for future investment of M-commerce industry.

REFERENCES

Andace, D. (2004). *E-CommerceandM-commercetechnologies.* London: IRM press.

Balan, E. (2007). *8,000iPhones Sold in the UK on First Day.* Retrieved April 15, 2008, from http://news.softpedia.com/news/8-000-iPhones-Sold-in-the-UK-on-First-Day-70696.shtml

Bell, A. (2007). *The latest apple**iPhone**.* Retrieved February 10, 2008, from http://www.ianbell.com/2007/09/26/iPhone-mania-persists-despite-apples-cold-shoulder

Everson, E. (2007). *Holiday shopping scams targetingMobile.* Retrieved March 18, 2008, from http://community.zdnet.co.uk/blog/0,1000000567,10006581o-2000440756b,00.htm

Herness newsletter. (2001). *Mobile commerceand itsfuture.* Retrieved February 21, 2008, from http://www.netmode.ntua.gr/courses/postgraduate/edi/material/11th_Hermes_Newsletter(Mobicom).pdf

KorhonenJ. (2004). *Introduction to 3Gmobilecommunications.* Norwood, MA: Artech house.

Lynch, I. (2000). *Mobile commerce- big or REALLY big?* Retrieved February 2, 2008, from http://www.vnunet.com/vnunet/news/2113522/mobile-commerce-big-really-big

M-commerce. (2006). *SeparatingMobile commercefrom Electronic Commerce.* Retrieved March 20, 2008, from http://www.mobileinfo.com/Mcommerce/differences.htm

MitchellC. (2004). Security for mobility. London: IET.

Moody, S. (2007). *Biometrics in the Here and Now.* Retrieved May 25, 2008, from http://www.technewsworld.com/story/59728.html

NCC. (2008). *M-commerce: more money less hype.* Retrieved May 20, 2008, from http://www.nccmembership.co.uk/pooled/articles/BF_WEBART/view.asp?Q=BF_WEBART_113234

Payment Processing Expert. (2007). *PayPalM-commerce.* Retrieved May 13, 2008, from http://paymentprocessingnews.blogspot.com/2007/10/paypal-m-commerce.html

Phones review. (2008). *BT launches BT total broadband anywhere with free smart phone.* Retrieved May 21, 2008, from http://www.phones-review.co.uk/2008/05/20/bt-launches-bt-total-broadband-anywhere-with-free-Smartphone

Public technology. (2006). *Cheltenham launches coin free parking with newmobilephone payment system*. Retrieved April 22, 2008, from http://www.publictechnology.net/modules.php?op=modload&name=News&file=article&sid=5185

Regan, K. (2008). *Amazon Aims to LightM-commerceFire with TextBuyIt*. Retrieved May 21, 2008, from http://www.ecommercetimes.com/story/62417.html

SadehS. (2002). *M-commerce: Technologies, Services and Business Models*. New York: John Wiley and Sons.

Section 2
Handheld Computing Research and Technologies

Chapter 8
UbiWave:
A Novel Energy-Efficient End-to-End Solution for Mobile 3D Graphics

Fan Wu
Tuskegee University, USA

Emmanuel Agu
Worcester Polytechnic Institute, USA

Clifford Lindsay
Worcester Polytechnic Institute, USA

Chung-han Chen
Tuskegee University, USA

ABSTRACT

Advances in ubiquitous displays and wireless communications have fueled the emergence of exciting mobile graphics applications including 3D virtual product catalogs, 3D maps, security monitoring systems and mobile games. Current trends that use cameras to capture geometry, material reflectance and other graphics elements mean that very high resolution inputs are accessible to render extremely photorealistic scenes. However, captured graphics content can be many gigabytes in size, and must be simplified before they can be used on small mobile devices, which have limited resources, such as memory, screen size and battery energy. Scaling and converting graphics content to a suitable rendering format involves running several software tools, and selecting the best resolution for target mobile device is often done by trial and error, which all takes time. Wireless errors can also affect transmitted content and aggressive compression is needed for low-bandwidth wireless networks. Most rendering algorithms are currently optimized for visual realism and speed, but are not resource or energy efficient on a mobile device. This chapter focuses on the improvement of rendering performance by reducing the impacts of these problems with UbiWave, an end-to-end framework to enable real time mobile access to high resolution graphics using wavelets. The framework tackles the issues including simplification, transmission, and resource efficient rendering of graphics content on mobile device based on wavelets by utilizing 1) a Perceptual Error Metric (PoI) for automatically computing the best resolution of graphics

DOI: 10.4018/978-1-61520-761-9.ch008

content for a given mobile display to eliminate guesswork and save resources, 2) Unequal Error Protection (UEP) to improve the resilience to wireless errors, 3) an Energy-efficient Adaptive Real-time Rendering (EARR) heuristic to balance energy consumption, rendering speed and image quality and 4) an Energy-efficient Streaming Technique. The results facilitate a new class of mobile graphics application which can gracefully adapt the lowest acceptable rendering resolution to the wireless network conditions and the availability of resources and battery energy on mobile device adaptively.

INTRODUCTION

Motivations

Computer graphics is an exciting and rapidly growing field. It has influenced many aspects of our daily life, such as games, movies, advertisements, and education. Traditionally, 3D computer graphics can only be achieved on high performance computers with dedicated graphics hardware. This limits their applications. Recently, two major technology developments have made mobile graphics become possible. One catalyst is the wide adoption of high-bandwidth wireless networks in universities, hospitals, hotels, and other working environments. A second catalyst is the emergence of affordable graphics hardware. Driven by the multi-billion computer game market, graphics hardware has become more and more powerful, cheap, and portable. As a result, many mobile devices are now equipped with dedicated graphics hardware and graphics on mobile devices is becoming popular because untethered computing is convenient and increases the productivity of workers. The following scenario demonstrates how mobile graphics applications can be used.

Motivating real estate mobile graphics use scenario: Ann is an architect who works for Ulo corporation. Ulo corporation is a multi-national architectural film with clients and workers in 50 countries across the world. Ulo maintains a large database of high-resolution 3D architectural drawing of various types of buildings. In order to accommodate workers with PDAs, laptops and cell phones with graphics capability, different teams of architects work on different projects that are maintained in Ulo's database. Initially, an Ulo team visits a client and after preliminary discussions, retrieves possible design solutions and shows them to the client. These serve as starting points of the design process. After the client selects a viable option and requests modifications, the architects annotate the diagrams and return to Ulo's office to make necessary amendments. Periodically, the architects return to the client to show progress and seek more feedback, towards a mutually agreeable design. Some of Ulo's clients are not connected to the Internet. In such cases, Internet hotspots can serve as valuable affordable meeting locations.

In the scenario above, mobility in the home viewing software allowed teams to retrieve new architectural designs for clients on the spot after the client rejected the first one and it was convenient to avoid driving back to their office with clients. Although videos of the homes could have been used in this scenario, graphics allows teams to modify the drawings to answer clients' what if questions. Clients could also interact with the homes and take a closer look at aspects that were important to clients. Indeed, mobile graphics is exploding and new applications are emerging. Computers can reduce boredom on long commutes by playing mobile versions of their favorite games during commutes. Other mobile graphics applications include telesurgery, security monitoring systems, 3D maps, and educational animations. Mobile graphics applications offer a new commercial opportunity especially considering that the total number of mobile devices sold annually far exceeds the number of personal computers sold. The mobile gaming industry already reports revenues in excess of $2.6 billion worldwide annually, and is expected to exceed $11 billion by the year 2010 (Mobile Games, 2005).

More and more 3D computer graphics models are now available on the web for educational and commercial use. Mobile graphics will further expand the applications of computer graphics and make 3D graphics models and applications more widely available. In the following, we list some interesting and promising applications of mobile graphics.

- **Pervasive medical visualization environment:** Mobile graphics will enable doctors to work with 3D patient data anywhere, anytime, and using various devices
- **Navigation of virtual environment:** Mobile graphics is making remote virtual tours become reality. For example, we can remotely walk through photorealistic virtual museums worldwide without expensive travel costs
- **Remote diagnosis:** Remote diagnosis will greatly facilitate medical treatment
- **3D advertisement:** 3D advertisement on TV is a very successful application of computer graphics. Mobile graphics will lead to the wide adoption of 3D graphics models for advertisement on mobile devices, such as PDA and mobile phone.
- **Remote education:** Many 3D graphics models have already been available on the web for education
- **Collaborative learning:** Recent research has demonstrated that using wirelessly interconnected handheld computers is an effective way for collaborative learning
- **Mobile 3D gaming:** Playing games on computers has been overtaken by game playing on mobile devices like cellphones and PDAs. Mobile graphics makes mobile 3D gaming possible in mobile device with limited resources.

Challenges

Mobile graphics, which involves running networked computer graphics applications on mobile devices across wireless network, is a fast growing segment of the networks and graphics industries. The quest for visual realism in graphics is endless. A trend has emerged whereby real world scenes are now digitized to capture scene geometry, lighting, textures and material properties that can be used later to generate visually stunning graphics scenes. However, rendering 3D graphics on mobile devices still faces some fundamental problems including:

High-precision capture of graphics content is creating massive data: The quest to make graphics scenes indistinguishable from the real word is endless. Today, movies and computer games have become extremely realistic because more precise geometry, materials and lighting are used. Classic techniques such as modeling object geometry using software packages and rendering (drawing) using Phong's shading equation are now rarely used when extreme realism is desired. The current trend in graphics is to place cameras around real objects and digitize scene attributes that can be used later for rendering. Today, almost every scene attribute can be captured from the real world including scene geometry (meshes) (Stanford 3D Scanning Repository, 2006), object reflectance properties (Reflectance Data, 2006), object texture (Bidirectional Texture Function, 2006) and scene lighting (Paul Debevec's Light, 2006). Several graphics research groups focus entirely on capturing elements of real graphics scenes. However, the increased precision of cameras today yields captured graphics content that is extremely large. For instance, in 1999, a team of 30 researchers from Stanford and the University of Washington spent a year in Italy, and digitized Michelangelo's statue to create a mesh representation. This geometric model can be obtained from their website and used to create highly realistic images. The largest models they captured had 2 billion faces and would require hundreds of gigabytes of memory to render. Even powerful desktop personal computers do not have enough memory to load a model of that size. Many such models are available on the Stanford group's website (The Digital Michelangelo Project Archive). High-resolution creates many issues including:

- **Different mathematical representations:** Each captured element, such as scene geometry (meshes), object material properties and scene illumination (lighting) is stored in a different mathematical representation. Each representation has many different file formats and each file format is only supported by certain graphics tools. This leads time-consuming conversion between different content's formats.
- **Manual LoD selection:** Since there is no metric for automatically determining the best resolution for each mobile device configuration, the scaling process is currently manual, involves trial-and-error to determine the best resolution for the specific mobile device. This manual approach is limiting with the hundreds of mobile device available in numerous configurations, such as the memory, battery energy and screen size. Essentially, there is no automatic sizing feature that makes it possible for two users to access the same graphics scene with a cell phone and headmounted displays respectively and automatically download content at the best resolution for their devices.
- **Low wireless bandwidth and high error rate:** Wireless channels can have low bandwidth and high Bit Error Rate (BER). Users experience long transmission times on low bandwidth wireless network links and some latency due to retransmission of damaged packets. These sometimes affect the usability of interactive graphics applications such as Internet multiplayer games.
- **Limited mobile resources:** Mobile devices tend to be limited in resources such as memory, CPU power, disk space, screen resolution and battery energy while graphics applications require large amount of these resources. Mobile devices also do not have adequate hardware support of graphics. These limitations make it difficult to process high resolution meshes and textures, or run sophisticated rendering algorithms that are necessary for visual photorealism. While the area of LoD management is rich, previous approaches focussed on controlling frame rates, but did not consider energy conservation on mobile devices.

In summary, the graphics content must be simplified before they can be used on small mobile devices. Scaling and converting graphics content to a suitable rendering format involves running several software tools, converting between mathematical representations and selecting the best resolution for a target mobile device is often done by trial and error, which all takes time. Wireless errors can also affect transmitted content and aggressive compression is needed for low-bandwidth wireless networks. At the mobile device, most rendering algorithms are currently optimized for visual realism and speed, but are not resource or energy efficient on mobile devices.

Therefore, this chapter focuses on the improvement of rendering performance by reducing the impacts of these problems with UbiWave, a novel energy-efficient end-to-end solution for Mobile 3D Graphics to enable real time mobile access to captured graphics using wavelets. The solution tackles the issues including simplification, transmission, and resource efficient rendering of graphics content on mobile device based on wavelets. The results facilitate a new class of mobile graphics application which can gracefully adapt the lowest acceptable rendering resolution to the wireless network conditions and the avail-

Figure 1. The overview of our mobile graphics approach

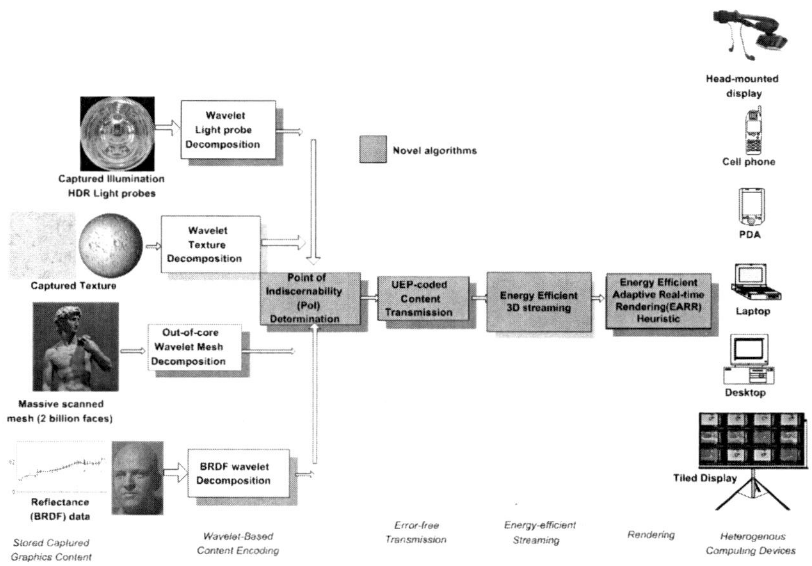

ability of resources and battery energy on mobile device adaptively.

Ubiwave

We have created a framework to scale and transmit high resolution graphics content to mobile devices at various scales. The chapter presents our approach, UbiWave, a novel energy-efficient end-to-end solution that encompasses all stages including retrieving captured content, transmission and rendering on the mobile device. This wavelet-based solution ties in current trends in the capture of graphics content with our directions in mobile graphics.

Figure 1 is an overview of our approach. Captured content is encoded using wavelets (on the left of figure). When retrieved, the content is tailored to the resources of a mobile device and wireless network, transmitted wirelessly to the mobile device where it is rendered (right of the figure). The realism of rendering on the mobile device can be varied to accommodate mobile device constraints on the screen size and battery energy adaptively. Essentially, small devices such as a cell phone on a GPRS cellular data network or a laptop on a broadband WiMax network, can render the same scene, access the same rendering parameters from the same content databases, but automatically achieve the best resolutions for their configuration with less energy consumption. The network can be used in several ways. It can be programmed into a software download tool that downloaders of scanned content can use offline. In a more ambitious scenario, the quality of rendered images in mobile graphics applications would be varied dynamically based on available resources. For instance, the geometry of rendered objects and the quality of shading of a mobile flight simulator could be gracefully degraded as the devices's battery dies.

To achieve our end-to-end vision in UbiWave, we developed several novel algorithms. The shaded boxes in Figure 1 are novel algorithms and techniques in UbiWave (Wu *et al*, 2006). Our UbiWave has following benefits and solved the problems introduced in this section.

1. **Uniform Representation and Increased Productivity:** Captured content will be more

Figure 2. Proposed system framework

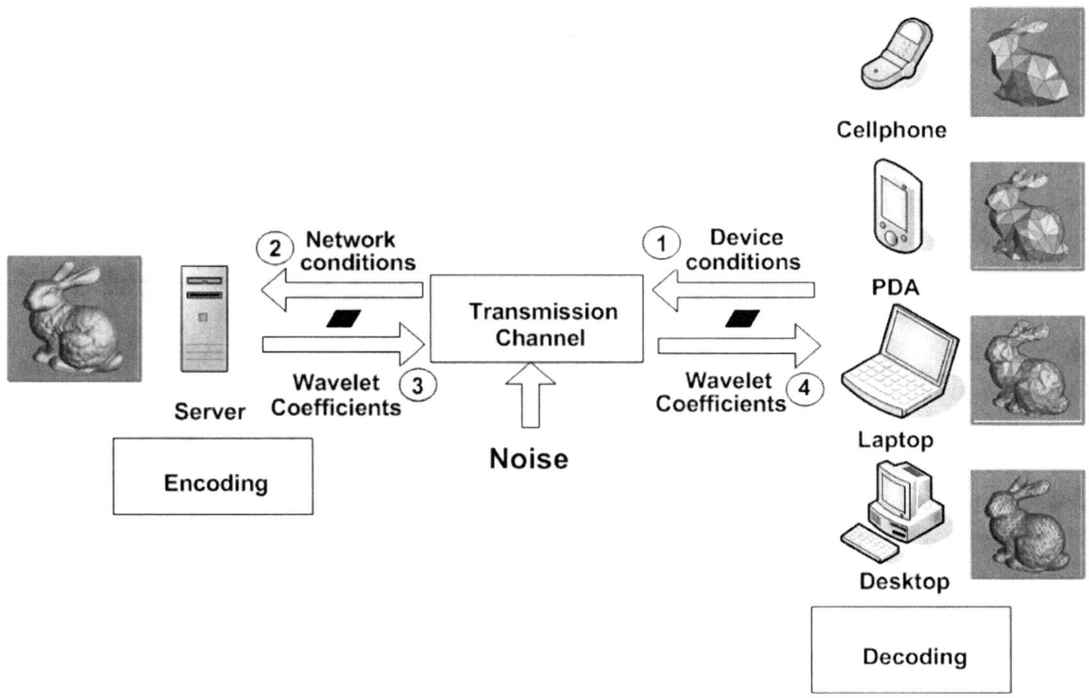

accessible to many heterogeneous devices with minimal effort, speeding up prototyping of mobile graphics applications. Groups that spend months capturing content would just need a few extra hours to run software that converts captured content to a predetermined wavelet representation. Our envisioned framework takes the wavelet-encoded content as input and virtually eliminates manual processes currently required to scale and size graphics content for a target device.

2. **Pareto-based perceptual error metrics for different mobile device's display (Wu et al, 2007):** To save scarce mobile device, we proposed a perceptual error metrics for automatically rendering at the lowest level-of-detail that does not show visual artifacts, called the *Point of Imperceptibility (PoI)*.
3. **Forward error correction scheme (UEP) to make wavelet-encoded graphics content more resilient to wireless errors (Wu & Agu, 2006):** We propose a coding scheme that assigns redundant FEC bits to wavelet data prior to transmission for different parts of the transmitted wavelet content, depending on how important the content is.
4. **An energy-efficient adaptive real-time rendering (EARR) heuristic (Banerjee et al, 2005; Wu et al, 2008):** To balance energy consumption, rendering speed and image quality, we proposed the heuristic adaptively changes LoDs or CPU allocation to compensate for the changing demands of application elements in order to maintain a constant real time rendering frame rate.
5. **An energy-efficient 3D streaming:** We present an energy-efficient 3D streaming technique to enable scalable 3D content streaming in wireless network and avoid data transmission which cannot maintain real-time rendering speed in mobile device.

Figure 2 shows our proposed system framework. The server only needs to send basic mesh connectivity information and corresponding wavelet coefficients to mobile devices, saving bandwidth and memory. The system works using the following four steps:

1. **Mesh preprocessing:** To speed up rendering, we perform wavelet decomposition as a preprocess at a server. In this pre-processing step, The server processes the original high resolution mesh to generate the base connectivity file and coefficient files for different levels of detail and calculates our perceptual metric for different screen sizes using different mesh and image LoDs. This computed data (or plot) is stored along with the corresponding meshes or images.
2. **Receiving mobile parameters:** At run-time, the mobile device transmits certain parameters to the server, so that the server can decide what LoD to transmit to a given mobile device. The transmitted mobile parameters include two parts: mobile device specification and wireless channel conditions. The mobile device specification includes its screen size, CPU, memory and battery energy. The wireless channel parameters include measured bandwidth and error rate measured in the area around the mobile device.
3. **Server decision on what wavelet LoD to send:** After the server receives the mobile parameters, it decides which level of wavelet coefficients will be sent to the mobile device using our perceptual error metrics for simplification to render the lowest level-of-detail that is just adequate for each type of mobile device. And unequal error protection coding scheme can protect the most important package in the high error rate wireless network.
4. **Client decision on what wavelet LoD to render:** After the client mobile device receives mesh data, it decides which level of wavelet coefficients will be rendered to the mobile device using energy-efficient adaptive real-time rendering heuristic. Typically, this decision is based on mobile device screen size, available CPU resources and user requirement. It can be expressed in the general form:

f (CPU,energy,screensize,bandwidth,error rate...) = level of coef. (1)

Roadmap

The remainder of this chapter is organized as follows: *Background and Related Work* section provides wavelets background and related research in the areas of UbiWave; *Pareto-Based Perceptual Error Metric* section describes our perceptual error metric (PoI); *Unequal Error Protection for Wavelet-Based 3D Transmission* section describes our Forward Error Correction scheme(UEP); *Energy-efficient Adaptive Real-time Rendering Heuristic* section describes our Energy-efficient Adaptive Real-time Rendering (EARR) heuristic; *Energy-efficient 3D Streaming* section describes wavelet-based multiresolution 3D streaming in UbiWave; *Future work* section outlines possible future work; and finally *Conclusion* section summarizes this chapter and draws conclusions.

BACKGROUND AND RELATED WORK

Background on Wavelets

This section reviews basic concepts of wavelets and its current applications in computer graphics.

Wavelets, which originated from the work of Fourier, are a mathematical tool that can represent input functions at multiple resolution (Graps, 1995). Figure 3 shows the process of wavelet

Figure 3. Wavelet-based multi-resolutions

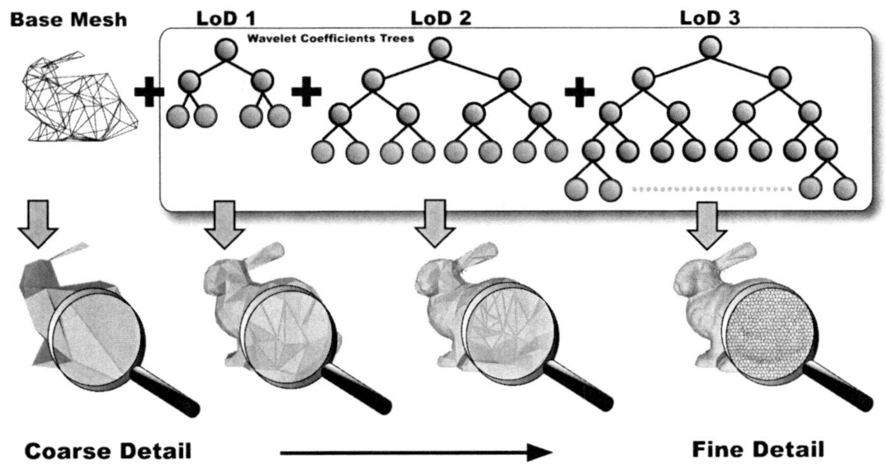

transformation.

Wavelets can *decompose* input functions to yield a coarse (rough) base function, plus a tree of detail coefficients, as shown in Figure 3. *Reconstructing* the original function starts from the coarse base function. Its resolution is then successively improved by adding more levels of the detail coefficient tree. In UbiWave, our system for ubiquitous graphics all rendering inputs such as meshes (Lounsbery, 1994), textures (Christopoulos *et al*, 2000) and material reflectance properties (Schroder & Sweldens, 1995) are converted and distributed as decomposed wavelets (base + coefficient tree) to facilitate scalable rendering on heterogeneous computing devices even when inputs are extremely large captured files.

While wavelets has been applied to many diverse fields, we limit our review here to research that uses wavelets in computer graphics. Wavelets have been used in a wide range of applications including graphics and image processing, ray tracing (Clarberg *et al*, 2005), information retrieval (Park *et al,* 2005), FBI fingerprint storage (Bradley & Brislawn, 1994), and geographic modeling. Today, published work has shown that almost all aspects of a graphics scene can be decomposed using wavelets including meshes, textures, material and reflectance properties. Schroeder (Schroeder, 1992) was one of the first to use wavelets in computer graphics and used wavelets to compress geometric and evaluate global illumination rendering equations.

- **Meshes:** Lounsbery proposed wavelet-based 3D compression (Derose *et al,* 1997; Lounsbery, 1994) by applying wavelet transforms to an arbitrary 3D mesh at several detail levels. During wavelet decomposition of meshes, a mesh is subdivided and deformed to make it fit the surface to be approximated. The original high resolution mesh is processed to generate a base connectivity file along with a sequence of smooth and detail cofficients that express the difference between successive levels of detail. Reconstruction starts with the base mesh. As more wavelet coefficients are included, a higher resolution mesh will be rendered. These steps can be repeated at the required resolution levels. A hierarchy of meshes is obtained from the simplest one M^0, called base mesh, to the original mesh M^{∞}. The wavelet transform of meshes removes a large amount of correlation between neighboring vertices. This hierarchy of meshes at different resolutions is the basis of *multiresolution analysis* (Lounsbery, 1994). To make the mesh approximation M^{j-1}as close as possible to the original mesh M^j, the lifting scheme (Sweldens, 1996) is used. Valette's scheme (Valette & Prost, 2004) tries to convert the connectivity simplification to 1:4 subdivision as much as possible. If 4:1 simplification is not possible, other groups of three or two faces are chosen, or some faces are left unchanged. Several methods for performing wavelet transforms on meshes are based on interpolating subdivision schemes such as the Butterfly (Dyn *et al,* 1990) that defines both interpolating and smoothing parts. The Loop (Loop, 1987) wavelet transform is an approximating scheme that has the advantage that the inverse transform uses Loop subdivision and produces the smoothest surfaces. After wavelet decomposition, adaptive arithmetic coding is often used to compress the size of the transmitted mesh and coefficients. In wavelet decomposition, a mesh is subdivided and deformed to make it fit the surface to be approximated. It consists of basic smooth coefficients and wavelet detailed coefficients. As more and more wavelet coefficients are included, a mesh of better resolution will be rendered. These steps can be repeated at the required resolution levels. We obtain a hierarchy of meshes from the simplest one M^0, called base mesh, to the original mesh M^j. Following (Derose *et al,* 1997), wavelet decomposition can be applied to the geometrical properties of the different meshes that are linked by the following matrix relations:

$$C^{j-1}=A^jC^j \tag{2}$$

$$D^{j-1}=B^jC^j \tag{3}$$

$$C^j = P^jC^{j-1}+Q^jD^{j-1} \tag{4}$$

where C^j is the $v^j \times 3$ matrix representing the coordinates of the vertices of M^j, v^j is the number of vertices for each mesh M^j. D^{j-1} is the $(v^j - v^{j-1}) \times 3$ matrix of the wavelet coefficients at level j. A^j and B^j are the analysis filters, P^j and Q^j are the synthesis filters. To ensure the exact reconstruction of M^j from M^{j-1} and D^{j-1}, the filter-bank must satisfy the following constraint:

$$\left[\frac{A^j}{B^j}\right] = \left[P^j \mid Q^j\right]^{-1} \tag{5}$$

To make the mesh approximation M^{j-1} as close as possible to the original mesh M^j, the lifting scheme (Sweldens, 1996) is used. Valette's scheme (Valette & Prost, 2004), (Valette & Prost, 2003) tries to convert the connectivity simplification to 1:4 subdivision as much as possible. If 4:1 simplification is not possible, other groups of three or two faces are chosen, or some faces are left unchanged.

- **Textures and images:** Techniques to compress images and textures using wavelets have also been proposed. A 2D wavelet transform that can be obtained by a separable decomposition in the horizontal and vertical directions (Lemarie & Meyer, 1986). Image compression based on the Discrete Wavelet Transform (DWT) is used in the JPEG2000 image standard (Christopoulos *et al,* 2000). Wavelet decomposition of textures and images is slightly different from that of meshes. In a preprocessing step, a nonstandard 2-D Haar wavelet decomposition of images is performed, which involves one step of horizontal pair wise averaging and differencing on the pixel values in each row of the image, followed by applying vertical pair-wise averaging and differencing to each column of the result.

- **Material reflectance and BRDFs:** Wavelets have been used to represent material reflectance or Bidirectional Reflectance Distribution Functions (BRDFs). In (Schroder & Sweldens, 1995), reflections were encoded from one incident direction using a spherical wavelet representation, which can represent a slice of the BRDF with several hundreds of coefficients. (Lalonde, 1997) extended this work and represents 4D reflectance functions using 4D basis wavelet functions stored in a compact wavelet coefficient tree that keeps only the highest coefficients to reconstruct the BRDF and thresholding the rest to zero.

If captured content is available as decomposed wavelets, heterogeneous mobile devices can retrieve resolutions suitable for their use. Wavelets achieve aggressive compression, which is also useful for low cellular network bandwidths. Wavelets also support progressive refinement since users can view the increasingly improved intermediate results after receiving coefficients. Finally, using wavelets for graphics content facilitates integration of emerging mobile graphics standards with existing MPEG4 video and JPEG2000 image standards, where wavelets are already used.

Related Work

This section reviews the research work related to the work in this chapter. Five relevant research areas are covered including scalable graphics systems, perceptual error metrics for simplification, error protection coding schemes for wireless transmission of wavelet-encoded meshes, heuristic for energy-efficient rendering and 3D streaming technique.

Systems for Scalable Graphics

Previously proposed techniques to reduce the bandwidth and resource usage of graphics applications but do not use wavelets include image based simplification (Lindstrom & Turk, 2000), geometry compression (Alliez & Desbraun, 2001; Gumbold & Straber, 1998; Touma & Gotsman, 1998), and progressive transmission (Chen & Nishita, 2002; Fogel *et al,* 2001; Hoppe, 1998). Alternate scalable graphics representations such as the use of points (Chen & Nguyen, 2001; Duguet & Drettakis, 2004; Rusinkiewicz & Levoy, 2000) has also been proposed. Points supports scalable rendering and transmission but does not achieve aggressive compression rates. Spherical harmonics (Lindsay & Agu, 2005; Ramamoorhthi & Hanrahan; 2002) can also be used to factorize low frequency lighting and speed up rendering, but not geometry or high frequency lighting. A few related graphics systems are also worth mentioning because they do try to adapt resource usage of graphics applications to the host machines. The ARTE system (Martin, 2000) implements primarily vertex-based techniques such as polygon simplification and LoD techniques, but does not use wavelets or consider error-resilience techniques against wireless channel errors. Repo3D (Macintyre & Feiner, 1998) is a distributed graphics library that proposes an object-oriented framework for distributing input graphics models, but does not use wavelets or compress graphics content. The remote rendering pipeline (Schmalstieg, 1997) uses polygonal LoD techniques, progressive transmission and incremental encoding but not wavelets. (Lamberti & Zunino, 2003; Zunino & Lamberti, 2002) have also proposed other graphics architectures for mobile devices, Yang combines multiple compression techniques to improve performance. We adopt wavelet-based multiresolution analysis for simplification because in addition to facilitating simplification, wavelets also achieve extremely aggressive compression ratios.

We present on a system solution for wavelets-based multiresolution. Our scheme only sends a base mesh and corresponding coefficients tree from the server side.

Perceptual Error Metric for Simplification

This section reviews the research work in error metric. The two most related bodies of work are surface-to-surface geometric simplification metrics and perceptual metrics.

A. Surface-to-Surface Geometric Simplification Metrics

Typically geometric metrics measure the deviation of the surface of a simplified version of a mesh from the original mesh. The Simplification envelopes algorithm (Cohen *et al*, 1996) imposes a bound on the maximum geometric deviation between the original and simplified surface. Gueziecs approach to simplification (Gueziec, 1999) uses a bounding volume approach to measure simplification error. Ronfard and Rossignac (Ronfard & Rossignac, 1996) measures for each potential edge collapse, the maximum distance between the simplified vertex and each of its supporting planes. Bajaj and Schikores plane mapping algorithm (Bajaj & Schikore, 1996) uses a priority queue of vertex removal operations to simplify a mesh while measuring the maximum point wise mapping distance at each step of the simplification. Garland and Heckbert (1997) modify the error metric of Ronfard and Rossignac and propose a quadric error metric. Appearance-preserving simplification by Cohen, Olano and Manocha (Cohen *et al,* 1998), tries to bound the pixel-level shift of a particular point on the simplified objects surface.

In summary, previous mesh simplification error metrics quantify how much a simplified mesh deviates from the original high-resolution mesh, but these metrics did not factor in the mobile screen dimensions. These simplification error metrics are insensitive to changes in screen size and using them unmodified would wrongly select the compute the same optimal mesh resolution for a tiny cell phone screen as it would for a larger laptop screen.

Tools such as METRO (Cignini *et al,* 1998) and MESH (Aspert *et al,* 2002) have been proposed to directly measure simplification errors, but do not factor in the device screen and behavior of human vision.

B. Perceptual Simplification Metrics

While surface-to-surface metrics focus mainly on the distortion of mesh geometry, visual effects such as lighting, shading and texturing also affect how perceivable simplification artifacts are. To account for these effects, elements of human vision have to be incorporated. A number of simplification metrics based on human perception have been developed. Rather than measure simplification errors in object space, perceptual metrics focus on how different mesh and image LoDs affect the contrast and frequency of pixel color changes. This theory is formalized as the Contrast-Sensitivity Function (CSF). Reddy (1997) describes early work to guide LoD selection using a perceptual model. Reddy (1997) analyzed the frequency content of objects and their LoDs in images rendered from multiple viewpoints. Reddy (2001) presented a version of this approach for terrains. Lindstrom and Turk (2000) describe an imaged-driven approach for guiding the simplification process itself. Luebke and Hallen (2001) use the CSF to guide local view-dependent simplification based on the worst-case contrast and spatial frequency of features the simplification would induce in the rendered image. Williams *et al* (2003) extends the work of Luebke and Hallen to 3D texture deviation.

In summary, The look of objects after rendering on a screen is considered by some proposed perceptual metrics that model human vision, but also do not account for differences in mobile display sizes.

We focus on producing a closed form expression that can be computed easily, while accounting for geometric distortion, lighting effects and screen resolution.

Unequal Error Protection for Wavelet-Encoded Meshes

Recent research efforts in the transmission of 3D objects over unreliable links have mostly focussed on still images and video sequences (Mohr *et al,* 2000). The compression and simplification of meshes is another active area of research. Very little research has attacked the issue of transmitting 3D graphics models over wireless networks. This is partly due to the fact that popular applications such as multiplayer games, which require this service have only recently emerged. Existing techniques for mitigating error while transmitting graphics models ranges from robust error coding to retransmission schemes for damaged network packets.

Two popular strategies for handling transmission errors are retransmission (Automatic Repeat-request, ARQ) and Forward Error Correction (FEC). ARQ schemes retransmission is used in many popular network protocols such as TCP/IP (Transmission Control Protocol, 1981) and the IEEE 802.11 Wireless LAN standard (IEEE 802.11, 2001). However, when using ARQ techniques, a receiver has to wait one roundtrip delay every time a packet is retransmitted. In the worst case, the IEEE 802.11 standard will retransmit a packet up to 7 times. This retransmission delay is inappropriate for real time applications such as video streaming and mobile online games where latency can affect the user. For such real time applications or along satellite links where retransmission can take too long (Tobagi *et al,* 1984), FEC is preferred. FEC adds extra bits to transmitted data such that a receiver can detect and correct a small amount of bit errors. A retransmission-based error-resilient technique has been proposed by Bischoff and Kobbelt (2002). In their scheme, the base mesh is re-transmitted along with every Level-of-Detail (LoD) to guarantee that it is correctly received at the mobile client. However, the overhead of transmitting the base mesh can be significant, making this scheme inefficient when packet loss rate is low.

The Hamming code (Hamming, 1950) and Reed-Solomon (Reed & Solomon, 1960) codes are two popular FEC schemes that perform well for most applications. However, wavelet-specified FEC schemes frequently outperform these codes for content that is encoded using wavelet. Wavelets-specific FEC techniques for images (Cosman *et al,* 2000) and video transmission (Sohn *et al,* 2001) have been proposed, but not for meshes. Al-Regib *et al* (1999) previously applied Unequal Error Protection (UEP) to Compressed Progressive Meshes (CPM), but did not use wavelet encoding. We propose applying UEP for wireless transmission of wavelet-encoded meshes.

Bajaj *et al* (1998) proposed several robust source coding methods for meshes. Even though this method adds a level of protection to the transmitted mesh, it does not adapt well to different ranges of channel packet loss rate. Yan *et al* (2001) propose partitioning a 3D model into several segments that are then transmitted independently. However, they use experimental calibration to determine the number of error-protection bits assigned to different segments before transmission, which can be time-consuming. Our proposed technique applies an analytic distortion metric to determine the number of bits assigned per segment and does not require experimental calibration. MPEG-4 also uses error-resilient coding of 3D models that is similar to that proposed by Yan *et al* (2001). UEP is an error coding paradigm that assigns FEC bits based on the amount of information a given segment contains. Al-Regib *et al* (2005) applies UEP to the Compressed Progressive Mesh (CPM) (Pajarola & Rossignac, 2000), a popular mesh representation in order to increase its resilience to transmission errors. As our main contribution, we apply UEP method to meshes that have been encoded using wavelets to make them more resilient to wireless errors. We note that UEP encoding of any content closely depends on a) the underlying structure of the content to be encoded and b) the ability to determine the relative importance of different parts of the mesh.

Heuristic for Energy-Efficient Rendering

The two bodies of work that are most related to our work are the areas of Level of Detail (LoD) management to achieve real-time rendering speeds, and energy management techniques for mobile devices.

A. LoD Selection to Achieve Real-Time Frame Rates

Funkhouser and Sequin (1993), and Gobetti (1999) both implement systems that bound rendering frame rates by selecting the apprioprate Level-of-detail (LoD). While Funkhouser and Sequin used discrete LoDs, Gobetti extended their work by using multiresolution representations of geometry. Wimmer and Wonka (2003) investigated a number of algorithms for estimating an upper limit for rendering times on graphics hardware. The problem of maintaining a specified rendering speed has also been addressed in the Performer system (Rohlf & Helman, 1994), which reacts to changes in frame rate by switching LoDs. A model for predicting the time budget for rendering on mobile devices can be found in Tack et al (Tack *et al,* 2004).

B. Application-Directed Energy Management Techniques

Application-specific energy management schemes use either Dynamic Voltage and Frequency Scaling (DVFS) (Liu *et al,* 2005; Yuan *et al,* 2004) or trade off the application's quality to increase energy efficiency (Flinn *et al,* 2001; Tamai *et al,* 2004). For instance, energy consumption can be reduced by intelligently reducing video quality (Tamai *et al,* 2004) or document quality (Flinn *et al,* 2001). DVFS techniques save energy by dynamically reducing the processor's speed (or voltage) when possible and does not change the application's quality. GRACE-OS (Yuan & Nahrstedt, 2004) proposes a DVFS framework for periodic multimedia applications. The GRACE-OS framework probabilistically predicts the CPU requirements of periodic multimedia applications in order to guide CPU speed settings. Chameleon (Liu *et al,* 2005) proposes CPU scheduling policies for a diverse applications including soft real-time (multimedia), interactive (word processor) and batch (compiler) applications. However, to the best of our knowledge, dynamic CPU scheduling to conserve energy has not previously been applied to graphics applications. Moreover, our approach saves energy savings while maintaining acceptable frame rates and image quality.

3D Streaming

Streaming geometry involves piece-wise incremental transmission of mesh geometry from a server to a client. Streaming of multiresolution geometry is closely related with multiresolution representation and compression. Any type of multiresolution representation can be naturally extended to a view-independent geometry streaming framework. Moreover, using streaming, we can reduce the required network bandwidth between a server and a client with compressed multiresolution representation.

Progressive meshes was the first algorithm for progressive representation on meshes and was introduced by Hoppe (1996). This progressive representation is based on successive mesh simplification by edge contractions, which remove one vertex at a time. The inverse, that is, the reconstruction, is achieved by vertex splits. Khodakovsky *et al.* (2000) presented a compression technique for semi-regular meshes. Valette and Prost (2004) proposed a wavelet-based progressive compression scheme for irregular meshes.

Figure 4. Mobile graphics scenario

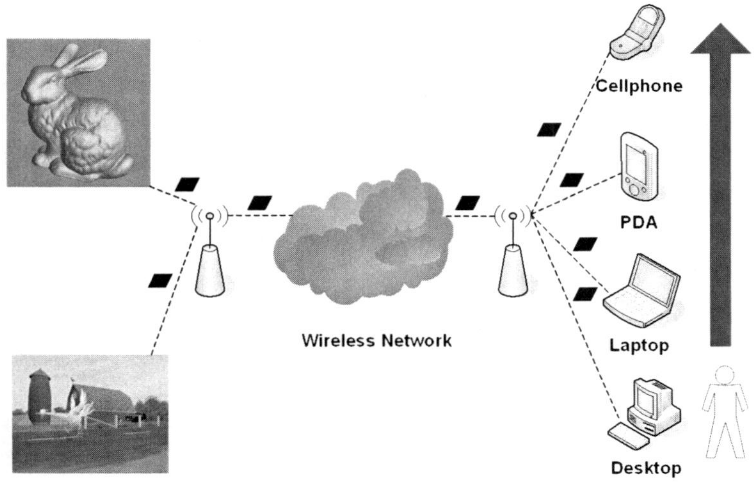

Rusinkiewicz and Levoy proposed a new view-dependent streaming based on QSplat (Rusinkiewicz & Levoy, 2000). They provide a network based visualization for every dense polygon meshes but the splatting approach is not suitable when the client requires the full mesh connectivity. Therefore, a small number of errors during communication does not affect the global shape of the reconstructed mesh on the client side. However, loss of mesh connectivity can occur, since the technique ignores the original mesh connectivity.

Yang *et al.* (2004) introduced a patch-based viewdependent streaming technique. They divide a mesh into several patches and compress each patch offline. In the streaming of a mesh, the entire connectivity information of the mesh is first transmitted to the client and then the compressed patches are selected and streamed with respect to the client viewing information. With this approach, the resolution of the mesh cannot be changed smoothly on the client side.

Kim *et al.* (2004) introduce a framework for view-dependent streaming of multiresolution meshes. The transmission order of the detail data can be adjusted dynamically according to the visual importance. This approach has to send the operation packets, which increases the network overhead. So it is not suitable for wireless network with low bandwidth.

PARETO-BASED PERCEPTUAL ERROR METRIC

This section presents the research work for Pareto-Based Perceptual Metric (PoI) for Simplification on Mobile Displays (Wu *et al,*2006; Wu *et al,*2007).

Overview

Our work focusses on a typical mobile graphics scenario shown in figure 4. Very high resolution graphics meshes and textures are stored on a server, and then simplified when requested by a mobile client. Meshes and textures are simplified on mobile devices for several reasons. First, mobile devices have limited battery energy, memory and disk space and lower resolution meshes and textures consume less of these scarce system resources. Secondly, increasing mesh and texture resolutions generally increases visual realism. However, above a certain Level-of-Detail (LoD), users cannot perceive these increases in mesh and texture resolution. We call this LoD the *Point of Imperceptibility (PoI)*. Essentially, increasing LoD above the PoI wastes mobile resources since users cannot perceive improvements in image quality.

In order to minimize wasting mobile resources, a mobile client should render meshes and textures that are as close as possible to its PoI. Our preliminary experiments showed that the level of detail users can perceive depends on the screen size: smaller screens show less detail and hence have a lower PoI. For instance, we found that for a given mesh, a laptop's display had a PoI of 20K faces, while a cell phone's PoI was 5K faces for the same mesh. This represents a 4x change in the acceptable LoD level based on screen size. Previous work has neglected to directly relate selected LoD levels with target screen size. Other factors such as distance of the rendered object from the screen, object details and whether the user zooms in all affect the perceptibility of simplification artifacts. However, we focus primarily on how PoI changes with screen resolution.

In the scenario in figure 4, we need metrics that enable the server to compute the PoI that corresponds to a mobile device's screen size. Since so many different mobile display resolutions exist, experimentally determining PoI for each mobile display resolution would be impractical. Thus, we would prefer a closed form expression that can be easily computed on-the-fly to determine PoI. Walkthrough applications that are dynamically scaled down for mobile clients would benefit from a PoI metric. Follow-me applications graphics applications (Wang *et al*, 2004) in mobile environments are another class of applications that emphasize the need for a PoI that can be quickly computed based on screen resolution. In follow-me mobile applications, mobile users physically move between mobile devices but can access the same applications using these devices from different locations. The server would need to easily compute the PoI of the user's current device and then transmit graphics files that correspond to the PoI of that mobile display's resolution. We adopt wavelet-based multiresolution analysis for simplification because in addition to facilitating simplification, wavelets also achieve extremely aggressive compression ratios that are suitable for ultra-low bandwidth wireless links such as wide-area cellular phone networks.

Metrics for LoD selection while accounting for different target screen resolutions is a general problem that is addressed by this paper. We develop a metric that can be used to find the PoI of both meshes and textures (images). Our metric is developed in two distinct stages. First, we consider the geometry of test meshes without considering the effects of lighting. In addition to the influence of screen size, visual effects such as lighting and antialiasing make simplification artifacts less perceive-able and hence further reduce the PoI. Luebke and Hallen (Luebke & Hallen, 2001) showed that mesh lighting can reduce the perceptibility of simplification errors by a factor of 2-3. To account for the effects of lighting, we then extend our geometry-only metric using results from work on perceptual simplification metrics. In summary, our metric determines the mesh (and texture) LoD that corresponds to the PoI and takes as input 1) the original mesh (or texture) LoD 2) mobile device screen size and 3) lighting that will be applied to the mesh. We validate our proposed metric through extensive user studies.

Our metric generates a pareto distribution that corresponds to meshes or images at various LoDs. We use this pareto shape to determine thresholds on the perceptibility of mesh distortion on various mobile screen sizes. Since our metric explicitly factors in screen size, a family of slightly shifted pareto plots are generated for mobile displays at different resolutions. To account for reductions in error perception when meshes are lit and shaded, we use Contrast Sensitivity Function (CSF) curves that have become the basis for many perceptual metrics in graphics. As a contribution, we are able to easily determine mesh undulation frequencies during our wavelet decomposition of meshes and

Figure 5. Sample pareto plots of our final PoI metric

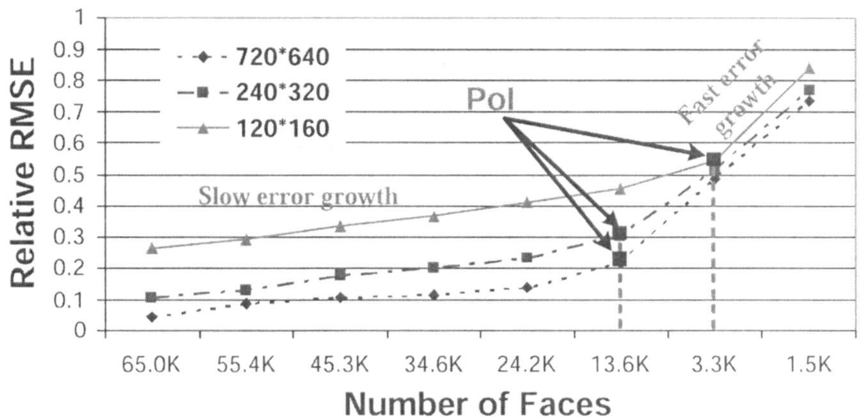

use these frequencies as inputs to the CSF curve. The results studied in this section are used in our Energy-efficient Adaptive Real-time Rendering Heuristic section and 3D streaming technique section.

Our Approach for Perceptual Simplification

In this section, we give an overview of our approach with an emphasis on building intuition and presenting our hypotheses. Our proposed metric for imperceptible simplification extends the work of Tack *et al* (2005). Tack *et al* expressed the surface-to-surface L_p norm error due to mesh simplification but did not explicitly address how perceptible these errors were on different screen resolutions, or consider the effects of lighting on the final rendered mesh. We integrate the original mesh LoD, the target display size, and the effects of scene lighting on error perceptibility into a single expression that can easily be computed. We develop our PoI in two distinct phases. First, in Geometry-only PoI Metric section only distortion in mesh geometry is considered without considering the effects of lighting. Next, in Perceptual Metric section, we extend our PoI metric by integrating perceptual elements (using the CSF) to account for scene lighting.

At this point, we preview some of our final results and give a qualitative description of our general direction. Figure 5 shows sample pareto plots generated by the version of our metric that considers only distortions in mesh geometry (no lighting). Three plots are shown corresponding to three different target screen resolutions (laptop:640x720, PDA:240x320, cellphone:120x160). Starting with an original high-resolution mesh, we generate fourteen levels of detail. We then use our PoI metric to compute the root mean square error generated by an LoD on each of our three target screens and plot them. Essentially, our metric produces a family of plots, one for each target screen resolution. Based on figure 5, we hypothesize that:

- **Hypothesis 1:** Each of the curves in figure 5 follows a pareto distribution. Starting with the original mesh on the left of the plots, relatively low errors are generated as LoD is reduced up until a *knee point*. Beyond the knee point, reducing LoD levels result in sharp increases in error. We conjecture that a) users will be unable to perceive simplification errors to the left of the knee point b) the knee point corresponds to the Point of Imperceptibility (PoI); and c) To the right of the PoI (knee point), errors rise quickly and users easily perceive simplification errors.

- **Hypothesis 2:** Based on the results of Luebke and Hallan, we conjecture that lighting will further reduce the perceptibility of errors, essentially lowering PoI. Referring to figure 5, lighting will essentially shift our pareto plots to the right (knee point occurs at higher LoDs).

The original metric proposed by Tack *et al* and other previously mentioned surface-to-surface metrics are oblivious of the perceptibility of simplification errors when rendered on various screen sizes. As a further note, Tack's original expression would generate the same pareto plot (and not a family of plots) for all three target screen resolutions. Essentially, our goal is extend Tack's expression to account for changes in the pareto distribution plots to account for different mobile screen sizes and then factor in the effects of lighting on error perception.

PoI Error Metrics

Geometry-Only PoI Metric

This section derives the first part of our metric that considers only the distortion of mesh geometry without factoring in the effects of lighting. Our derivation has three steps: 1) Calculate mesh distortion due to simplification; 2) Render the simplified mesh to a large virtual screen M_1; 3) Minify blocks of pixels of M_1 to a pixel of the mobile device's screen M_2. We can magnify if $M_2 > M_1$ as in a large tiled display. For screen-aligned images, only step 3 is performed; Figure 6 summarizes the steps to derive our metric. Equation 6 is our PoI metric for geometry only.

$$l^p\left(S_1,S_2\right)=\underbrace{\left(1-\frac{F_2}{F_1}\right)\frac{\sum_{i=0}^{F}A\left(T_i\right)l^p\left(T_i,S_2\right)}{\sum_{i=0}^{F}A\left(T_i\right)}}_{Object-space}+\underbrace{E_p}_{Screen-space} \quad (6)$$

where F_1 is the number of triangles in the surface S_1, F_2 is the number of triangles in the surface S_2.

Figure 6. Steps in deriving our PoI metric

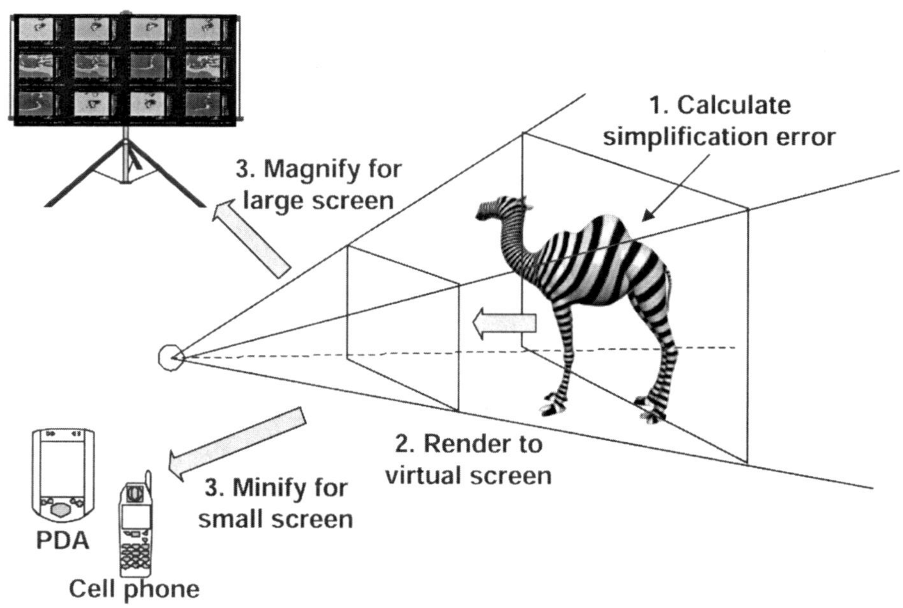

Figure 7. Minifying virtual screen pixels onto mobile screen

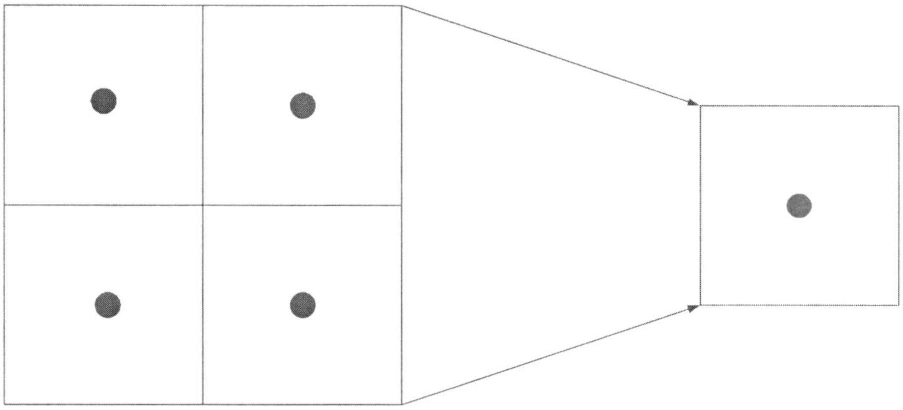

If $F_1 < F_2$, we can rewrite the factor $1 - \frac{F_2}{F_1}$ as $1 - \frac{F_1}{F_2}$.

The first part of Equation 6 deals with surface-to-surface LoD simplification errors in object space and the second term (E_p) deals with pixel-level minification errors as a result of rendering to different screen resolutions (see figure 7). A high-resolution mesh that is rendered to a small screen potentially incurs errors in both terms. A screen-aligned texture incurs errors only due to the second (E_p) term. Likewise, if the same mesh LoD (no surface simplification) is rendered to two different screen sizes, the error due to the first term is zero and the error due to the second term is calculated. For a target mobile display width, W (in pixels) and height H (in pixels), the term E_p is defined as:

$$E_p = \sqrt[p]{\frac{1}{W_2 \times H_2} \sum_{i=1}^{W_2 \times H_2} \left(\frac{W_2 \times H_2}{W_1 \times H_1} \sum_{j=1}^{\frac{W_1 \times H_1}{W_2 \times H_2}} \sqrt[p]{S_p}\right)^p \textit{where}}$$

$$S_p = \frac{1}{3}\left[\left(\frac{R_{i2} - R_{j1}}{256}\right)^p + \left(\frac{G_{i2} - G_{j1}}{256}\right)^p + \left(\frac{B_{i2} - B_{j1}}{256}\right)^p\right] \quad (7)$$

where W_1 and H_1 are the width and height of screen M_1 and W_2 and H_2 are the width and height of screen M_2. We assume that $W_1 > W_2$. Otherwise, W_1 and W_2 should be interchanged. In our system, we use relative Root Mean Square Error (RMSE) ($p = 2$). In Equation 7, we calculate the screen space *RGB* error pixel by pixel and normalize it. As shown in figure 7, S_p calculates the average relative mean square error of *RGB* values between one pixel on the smaller screen and the corresponding group of pixels on the larger screen. If the screen sizes are not same, we compare the $\frac{W_1 \times H_1}{W_2 \times H_2}$ pixels on the screen with one corresponding pixel on small screen and calculate the relative root mean square error between them.

We calculated and averaged our final error metric in equation 8 for all pixels on a target screen while considering four different meshes. Three different screen sizes were considered: 640x720 pixels for laptop, 240x320 pixels for PDA and 120x160 pixels for the cellphone. Figure 8 shows the computed errors which when plotted resemble a pareto distribution with a knee point. One way to calculate the knee point of the pareto plots, the slope of segments of the could be calculated. The point between two consecutive segments with the highest change in slope is the knee point (PoI).

Figure 8. Our metric plotted for meshes at different LoDs

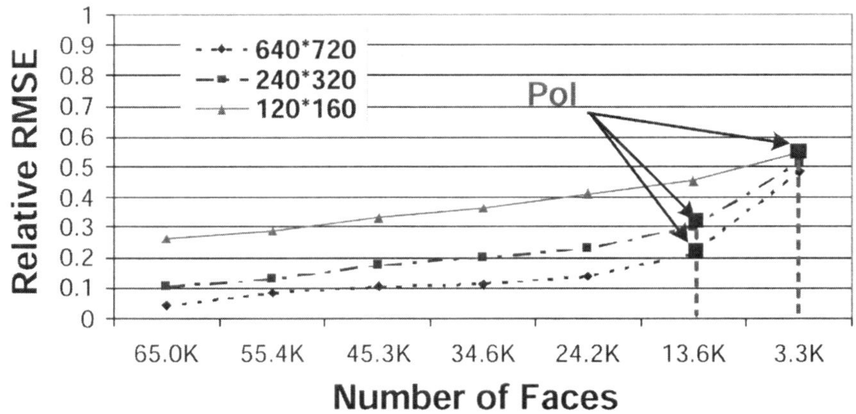

Perceptual Metric

In this section we extend our PoI metric to account for lit meshes using the Contrast Sensitivity Function(CSF). First, we note that effects such as lighting and shading can reduce the perceptibility (sharpness) of mesh edges, hiding differences in detail between LoDs. Essentially, lighting and shading makes geometric distortion less visible. We can model this reduction in the perceptibility of errors as passing the rendered mesh image (sharp) through a filter that removes some of the distortion. To account for the error masking caused by lighting, we multiply our geometry-only expression (equation 6) by a factor $M_p(S_1,S_2)$. As before, this $M_p(S_1,S_2)$ factor considers the perceptibility of errors when rendering our lit mesh onto a large virtual screen of size $S1$ and minifying the image onto a target mobile display of size S_2. Thus our new PoI expression takes the form:

$$I^p(S_1,S_2) = \left[\left(1-\frac{F_2}{F_1}\right)\frac{\sum_{i=0}^{F} A(T_i) I^p(T_i,S_2)}{\sum_{i=0}^{F} A(T_i)} + E_p\right] \times M_p(S_1,S_2) \qquad (8)$$

Next we derive an expression for $M_p(S_1,S_2)$. The human visual system is often modeled as a linear system and its response to visual excitation is expressed as a convolution of the input stimulus with the visual cortex's impulse response. Equivalently, to determine the perceptibility of a lit mesh, we can determine the eye's visual response by multiplying the wavelet transform of the mesh by the CSF. The CSF measures the response of human vision at different spatial frequencies. Mannos and Sakrison, after conducting a series of psychophysical experiments on human subjects, found that the CSF can be modeled by the function in the equation 9. Here f_s is spatial frequency in cycles per degree.

$$C_s(f_s)=[0.0499+0.2964f_s]\times\exp[-(0.114f_s)^{1.1}] \qquad (9)$$

where f_s is spatial frequency in cycles per degree. To integrate the CSF into our metric, during wavelet decomposition we determine the frequency ranges corresponding with each LoD. We then multiply each mesh frequency range with the CSF's response curve in that range. Figure 9 shows the CSF function mapped to frequency ranges obtained during wavelet decomposition of a mesh. This curve essentially defines how sensitive the human eye is to frequency ranges generated during wavelet transformation of the original mesh. Thus, for each frequency band a *sensitivity weight*, C_m can be computed by integrating the CSF curve in figure 9 over that frequency band. The weight measures the *average contrast sensitivity value* of the CSF curve in each band. We then multiply the wavelet coefficients at each LoD (frequency level) by the CSF sensitivity weights *Cm* corresponding to that frequency range. Wavelet transformation involves the iterative application of two mirror filters, *L*, a low-pass filter and *H*, a high-pass filter. Thus, by applying *H* to a discrete input with bandwidth (0,π), a level of coefficients with bandwidth (π/2,π) is acquired. Thus, after *m* iterations, the weight for level *m* is:

$$C_m = \frac{\int_{F_m} CSF(\omega)\, d\omega}{A(F_m)} \quad (10)$$

Where F_m is the frequency subband $\left(\frac{\pi}{2^m}, \frac{\pi}{2^{m-1}}\right)$ and $A(F_m)$ is the width of the band.

We now describe how the *sensitivity weight, Cm* can be incorporated during wavelet transformation of meshes. Wavelet decomposition of a mesh yields a coarse mesh and a tree of wavelet coefficients. To determine the perceptibility of a mesh LoD, all wavelet coefficients at that tree level are multiplied by the CSF sensitivity weight corresponding to that level. When a given mesh LoD is rendered to a screen, each wavelet coefficient *i* in that level of the wavelet tree refines (modifies) a mesh face at that LoD which in turn maps to a block of pixels when rendered. For each mesh LoD, we need to track which group of pixels are modified by each wavelet coefficients at that level. A brute force approach would be to render all LoD levels to a screen and determine what pixels each face maps to. However, the following method to track this relationship requires only one rendering of the original mesh. At the lowest level of the wavelet tree (finest LoD), each wavelet coefficient maps to a triangle in the original mesh which in turn maps to a group of pixels after rendering. By rendering the original mesh, we can track what group of pixels each triangle (wavelet leaf node) maps to. At any higher (coarser) level in the wavelet coefficient tree we can calculate what screen pixels each coefficient in that level maps to as the union

Figure 9. Contrast sensitivity function curve

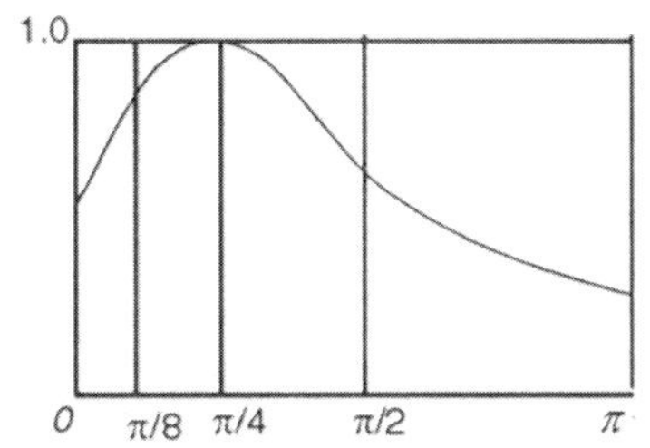

of all pixels corresponding to all leaf nodes that are its children in the tree. Thus, for each pixel *(i,j)* on the target mobile device, we can multiply the wavelet coefficients in a given frequency band with the contrast sensitivity weight corresponding to that frequency band giving:

$$D_l(m,i,j) = C_m W(1,m,i,j) \quad (11)$$

Here C_m is the contrast sensitivity weight and $W(m,i,j)$ is the wavelet coefficient at level *m* and pixel location (i, j). Essentially, we quantify the perceptibility of error to the frequency input at pixel (i, j) in the mth sub-band frequency. Our perceptual comparison metric is then computed as:

$$M_p(S_1, S_2) = \frac{\sum_{m,i,j} \left| D_1(m,i,j) - D_2(m,i,j) \right|^2}{N_h \times N_v} \quad (12)$$

where *D*1 and *D*2 are error values of pixel *i,j*, when considering level *m* of the wavelet coefficients. N_h and N_v are the number of pixels in horizontal and vertical directions on the small screen. If the screen sizes are not the same, we calculate the screen error between one pixel on the smaller screen and the corresponding group of pixels on the larger screen (minification) as shown in figure 7. Figure 10 shows our final results using equation 8. The errors with lighting and shading are clearly smaller than the errors without light-

Figure 10. Curves with shading and without shading

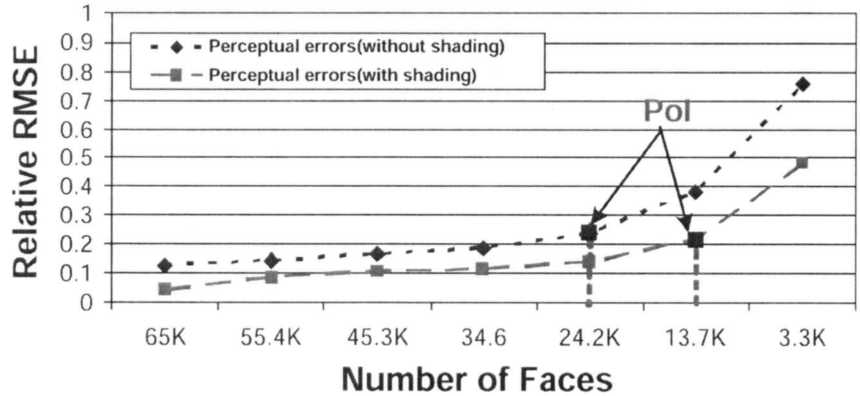

*Figure 11. File size and relative RMSE A: 640*720, B:240*320, C: 120*160 23*

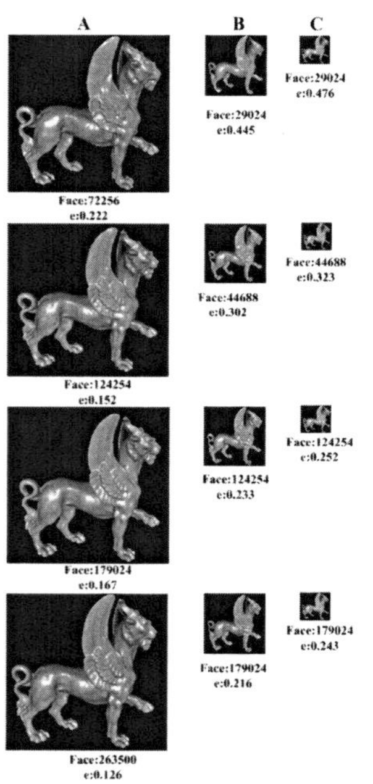

ing and shading.

Figure 11 shows meshes of the different LoD levels of the model. This demonstrates the visual depiction of the results of using our perceptual error metric.

Metric Validation and Analysis

User Studies

Having derived a metric that can be computed to automatically determine the PoI of a given mesh or image, we needed to validate that it works for real users. Specifically, we needed to ascertain that our metric accurately selects the LoD at which users stop perceiving increases in mesh or image resolution. Our approach was to generate a series of mesh and image LoDs and use our metric to determine the PoI LoD. We then asked real users to visually inspect the actual rendered meshes and images that correspond to those LoDs. Our metric worked correctly if it correctly determined the same PoI chosen by real users. Our user studies involved 84 participants.

In our study, several LoDs of a bunny model were rendered at three different screen sizes (laptop:640x720, PDA:240x320 and cellphone:120x160). Figure 12 shows one set of bunny images for screen size 240x320 pixels, ordered from highest(left) to lowest (right) resolution. Each LoD level is placed beside the original and shown to the user in pairs. For instance, images 1 and 2, 1 and 3, 1 and 4, and 1 and 5 in figure 12 are presented to the user in pairs. For each pair of images, users are required to respond in one of three ways: a) A is more detailed than B; b)

Figure 12. An example of rendered meshes of seven different LoDs in user study

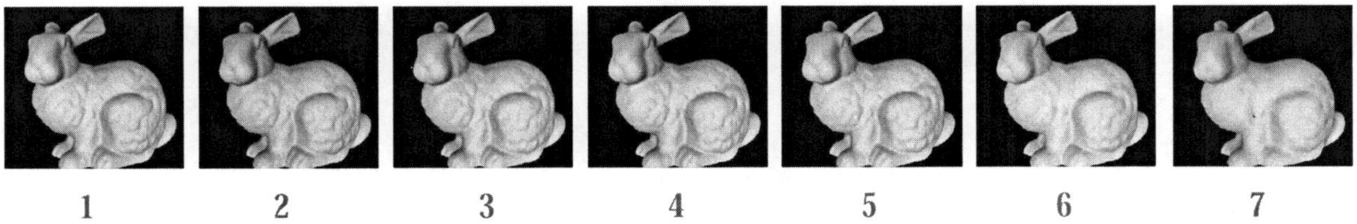

A and B are approximately same; c) B is more detailed than A. The permutations of the two mesh models and three different screen sizes generate eighteen different image pairs that we randomly show the user as questions 1-18. For example, in figure 14, Q05 presents two images to the user with the option of responding with a, b or c as described above.

For each screen size, as we processed from high resolution pairs to low resolutions pairs, as long as the user is able to correctly distinguish between pairs of images, the PoI has not been reached. Once the number of incorrect answers (or user answers 'approximately same') becomes significantly less than the number of correct answers, the lower resolution of the pair is regarded as the PoI that we are looking for. We then compare this experimentally determined PoI with PoI calculated by our metric.

The relative positions of each image pair are also randomized so that the user does not use the image position as a cue to guess which one is more detailed. For instance, if we always placed the high resolution image to the right (image B) and the user happened to notice this, she may always guess that B is more detailed even if she visually cannot see this. To minimize the effect of ambiguities in our phrasing of our questions or problems due to language barriers (English was not the first language of some participants), the users are first shown sample images along with the correct answers.

Figure 13 is the screen shot of the survey pages.

Figure 14 shows sample results of user study. Each question corresponds to a pair of images at a particular screen size. For instance Q05 refers to images 1 and 6 rendered to a PDA screen size. In Figure 8, the resolutions employed in our user studies are shown as black dots. The red line shows where the PoI computed by our metric lies. Comparing the result in figure 14 and figure 8, we observe that users indeed begin to wrongly distinguish the models or answer incorrectly at the PoI. Our calculated error metric is shown with each image to assist the reader in visually mapping calculated error values to an actual image quality.

Figure 13. Screen shot of survey pages

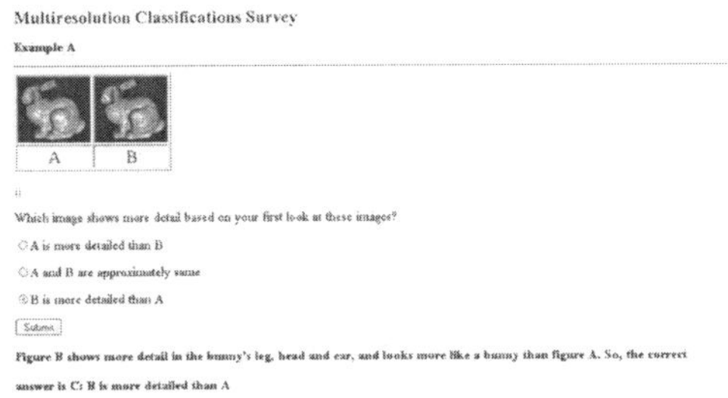

Figure 14. Sample results of the user study

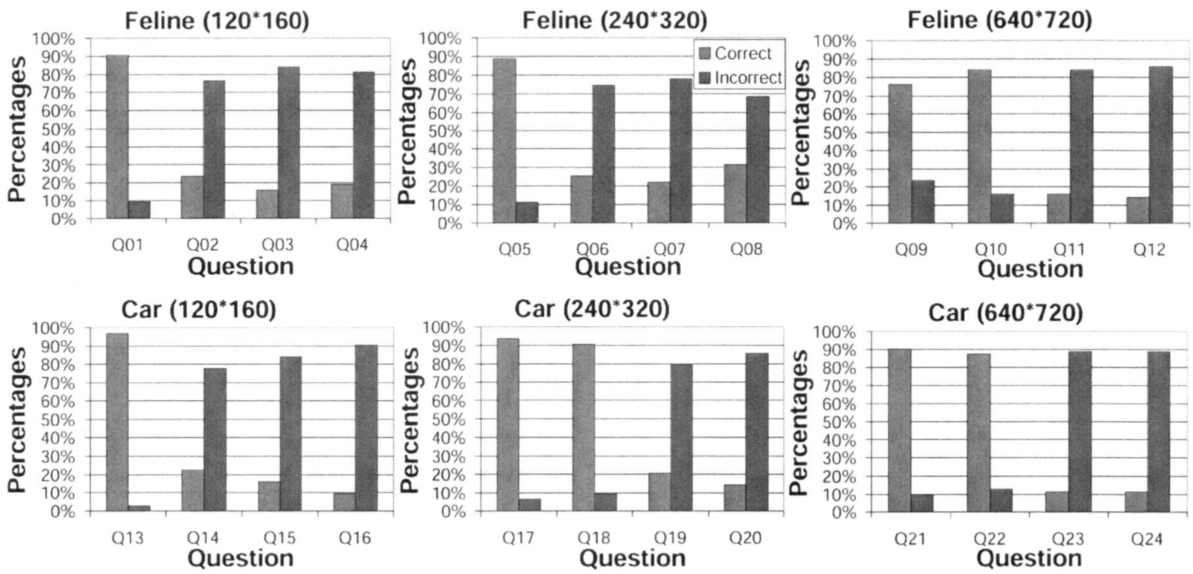

Resource Saving Using PoI

Battery energy, CPU cycles, memory and disk space are all resources that are scarce on mobile devices. Using a mesh or image at the PoI instead of its original resolution can improve usage of these resources. We measure encoding, transmission and decoding times, and quantify potential battery energy savings by using a lower resolution mesh. We measure the energy consumption of receiving, decoding and rendering a given mesh resolution on the mobile client by using our tool (Banerjee *et al*, 2005). PowerSpy is a software tool that tracks the energy consumption of MS Windows applications at the thread and I/O device levels. We calculate the total energy consumption, *E*, by summing up the energy consumption of the CPU, disk and network cards giving $E = E_{CPU} + E_{Disk} + E_{Network\ Card}$.

For a screen size of 640x720, our metric yields a PoI of 13654 faces for a bunny mesh, meaning that there is no significant perceptual difference if the number of faces is greater than 13654. If we use 13654 faces instead of the original mesh, the difference in resource usage is saved. Table 1(a) summarizes the saved transmission time, decoder time and energy consumption in the mobile device.

Thus, using our perceptual metric, in this example we save over 80% of the transmission time, 44% of the decoding time and 61% of the total battery energy. Similar results for an image are tabulated in Table 1(b). For the image in this example, it is possible to save over 60% of transmission time, 35% of the decoding time and 38% of the total battery energy. The above numbers on savings clearly depend on how large the original mesh (or image) is compared with the PoI. Our numbers are included above mainly for illustration purposes.

It is important to note that in calculating our PoI metric, the mesh is rendered from a single viewpoint. However, as an object is moved, different viewpoints will lead to different screen errors and PoIs. To make our metric view independent, in the server pre-processing step, the PoI can be calculated from multiple view points around the mesh. The PoI's generated from multiple viewpoints

Table 1. Resource savings

Faces Number	13654	64951	Saved
ttrans.	1.23ms	7.03ms	82.5%
rt	463ms	832ms	44.4%
Energy Consumption	**12865mwh**	**33298mwh**	**61.4%**
(a) Saved Resources for mesh			
Size of Coef. File	64KB	173KB	Saved
ttrans.	47.6ms	120.5ms	60.5%
rt	340ms	530ms	35.8%
Energy Consumption	**7467mwh**	**12156mwh**	**38.6%**
(b) Saved Resources for image			

can then be averaged or the minimum used in a conservative scheme.

Section Summary

This section presents a wavelet-based multiresolution framework for scalable graphics content transmission and rendering. We present a Point of Imperceptibility (PoI) error metric that accurately picks the lowest acceptable mesh resolution based on the target mobile device's screen size. We develop versions of our PoI that considers only mesh geometry without considering lighting, as well as an extension that considers the effects of lighting on the perceptibility of distortion. We present LoD selection heuristics based on our proposed metric and analyzed the relative Root Mean Square Error (RMSE) our metric. We perform user studies to validate our metric, employed our metric in a heuristic to save mobile device resources and quantized resulting resource savings.

UNEQUAL ERROR PROTECTION FOR WAVELET-BASED 3D TRANSMISSION

This section presents the research work for Unequal Error Protection(UEP) for Wavelet-Based Wireless 3D Mesh Transmission (Wu & Agu, 2006).

Overview

To minimize transmission times on low-bandwidth network links, several compression (Chow, 1997; Rossignac, 1999; Touma & Gotsman, 1998) techniques have been developed to reduce transmitted mesh sizes. Additionally, the wireless channel is well known to have significantly high error rates. Retransmission of damaged packets or Forward Error Correction (FEC) are two strategies that are frequently used to mitigate wireless channel errors. However, the roundtrip delays caused by retransmissions in network protocols such as TCP/IP and the IEEE 802.11 Wireless LAN protocol appear as latency to users, which sometimes affects the interactivity of networked graphics applications. For such applications, FEC is preferred to retransmissions. FEC schemes add redundant bits to the original meshes before transmission such that minor errors can be corrected by the receiver, hence avoiding retransmissions.

As one of our main contributions, we propose a FEC scheme to protect wavelet-encoded meshes from wireless errors. The Hamming code (Hamming, 1950) and Reed-Solomon (Reed & Solomon, 1960) codes are two popular FEC schemes that mitigate error well for most applications. However, FEC schemes that consider the underlying structure of wavelet-encoded content frequently outperform more general schemes that

Figure 15. The importance of different level

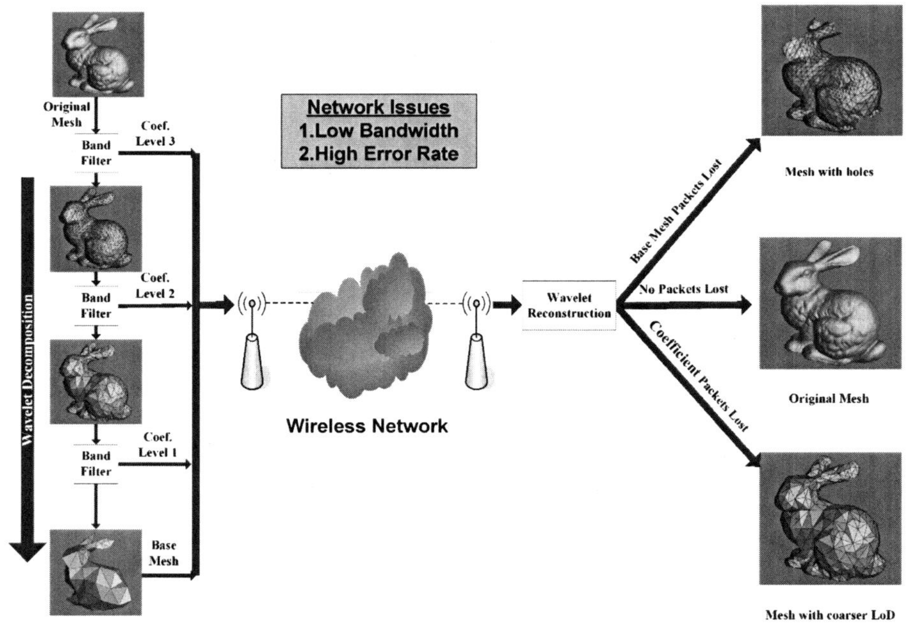

do not. Wavelet-specific FEC techniques for image (Cosman *et al*, 2000) and video transmission (Sohn *et al*, 2001) have been proposed, but not for wavelet-encoded meshes. We propose FEC scheme based on the principle of Unequal Error Protection (UEP). In UEP (Al-Regib *et al*, 2005), the number of FEC bits alloted to each part of the mesh is proportional to the amount of information it contains: more bits are added to parts with more information. Thus, areas of a mesh such as a human face that has many fine details are allocated more FEC bits than areas such as the back with less details.

Unequal Error Protection of Wavelet-Encoded Meshes

Unequal Error Protection

Approaches to mitigate wireless channel errors packets losses can be network-oriented solution such as retransmissions in TCP, post-processing solutions such as error concealment, or preprocessing solutions such as Forward Error Correction (FEC) codes. The roundtrip delay incurred make retransmissions unsuitable for interactive graphics applications. In multicast environments, retransmissions would also flood the sender with acknowledgements and performance could suffer. We consider the use of FEC. FEC strategies include Equal Error Protection (EEP) and Unequal Error Protection (UEP). EEP methods apply the same FEC code to all parts of the mesh's bit stream and is suitable when the channel has a low packet loss rate. However, at higher packet loss rates, considerable degradation on the decoded model quality may occur because of the high possibility that important parts might be lost. In this case, UEP is more suitable since important parts of the decoded mesh get more assigned more FEC bits. Figure 15 shows if the information in base mesh lost, the holes will happen after rendering. But if some coefficients lost, the LoD will decrease after rendering.

In our approach, after applying wavelets decomposition to a mesh, the base mesh as well as wavelet coefficients are assigned an FEC code rate depending on their contribution to the decoded mesh quality. The distribution of these FEC codes is calculated using a statistical distortion measure. Based on this measurement, we determine the number of error-protection codes to be assigned to the base mesh and each level of detail. The FEC codes used in this paper are Reed-Solomon (RS) codes. These error codes are perfect for error protection against bursty packet losses because they are maximum distance separable codes. An (n,k) RS-code encodes k information symbols where each symbol is represented by q bits. These k symbols are encoded into a codeword of n symbols, which is restricted by $n \leq 2^q - 1$. As soon as k symbols are received, all lost symbols can be reconstructed.

UEP in Wavelet-Based Multiresolution

After wavelet decomposition, the base mesh and first few levels of wavelet coefficient tree should be strongly protected to prevent packet loss. We examine several strategies for adding Forward Error Correction (FEC) bits to the base mesh and wavelet coefficients. First, we apply Equal Error Protection (EEP) where an equal number of FEC bits are applied to all parts of the base mesh and to all levels of the wavelet coefficient tree. That is, $S^1 = S^2 = \ldots = S^{M+1}$, where S^k is the number of FEC bits added to on the kth level of wavelet coefficients. Next, we propose applying Unequal Error Protection (UEP) where bits in the encoded mesh are classified based on their contributions to the final look of the reconstructed mesh. Each class is then protected by a number of FEC bits that can provide a certain level of protection against channel losses. In our research, each level of the wavelet coefficient tree and the base mesh, is assigned an FEC code based on amount of distortion that would be introduced into the reconstructed mesh if that portion of the bit stream is lost. Parts of the bitstream that distort the look of the reconstructed mesh most when they are lost are the most important and hence we apply the largest portion of the FEC bit budget. Wavelet coefficients with large absolute values contain the most detail receive more error bit budget, since this level of coefficients contains more information (e.g. fine details such as eyes and nose of a face) compared to other levels. The FEC codes used are the Reed-Solomon (RS) codes. Reed-Solomon codes are block-based error correcting codes with a wide range of applications for error protection against burst packet losses. We also adapt our encoding order of our bitstream to further increase resilience to burst errors. The output bitstream is encoded in blocks of packets, where the data is placed in horizontal packets and then RS is applied across the block of packets vertically. Each block of packets is protected with a FEC code that is proportional to the importance of the corresponding base mesh or coefficients.

Since all types of error protection add extra bits to the original mesh bitstream prior to transmission, both EEP and UEP incur overheads that reduce the number of actual data bits sent compared with NEP. However, since reconstruction starts from the base mesh, loss of the base mesh or parts of it are particularly devastating. Essentially, the base mesh as well as coarser wavelet coefficients are more important than detail coefficients. At high packet loss rates, losing the base mesh or coarser wavelet coefficients degrades the decoded mesh quality significantly even if the detail coefficients are received correctly. EEP distributes error correction bits equally to the base mesh, and all levels of detail coefficients.

Distortion Measure

To determine the level of channel coding associated with each level of the wavelet coefficient tree, we need to evaluate the importance of those coefficients. In this section, we develop a distortion metric that evaluates the relative importance of the various levels of a wavelet coefficient tree. After we determine the importance of each level of the wavelet coefficient tree, we can then assign a fraction of the total FEC bits that is proportional

Figure 16. Wavelet coefficient tree for a mesh with three LODs. C_i^j is the wavelet coefficient at level j

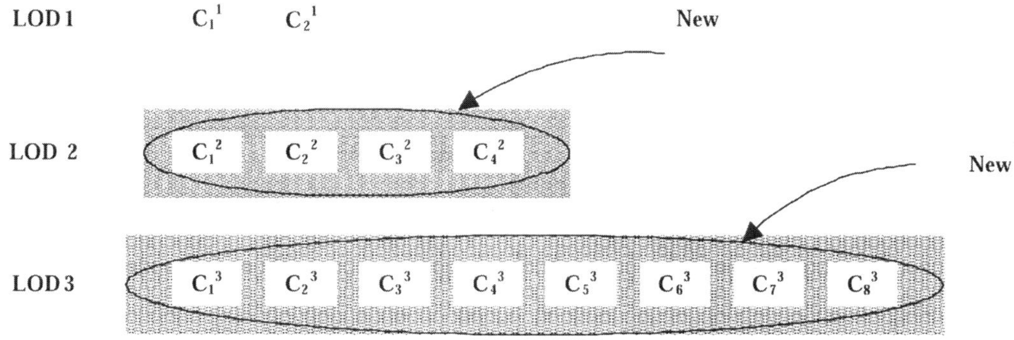

to their importance. The main factors integrated into this distortion measure are: 1) The amount of information contained in the wavelet coefficient, 2) the total number of error-protection bits. As figure 17 shows, in each LoD, some new coefficients are added to the mesh, which provide more detailed information to the final rendered mesh.

To calculate the importance of each level of the wavelet coefficient tree, we evaluate the distortion that would be present in the final decoded mesh if all the coefficients in that level of the tree were lost. We associate a coefficients distortion quantity, $D_{wLOD}^{(j)}$ with the jth LOD, which is defined as the average distortion (per coefficient) added when all coefficients that are added by this LOD are lost. The $D_{wLOD}^{(j)}$ is given by:

$$D_{wLOD}^{(j)} = \frac{1}{N_j} \sum_{1}^{N_j} \left| c_i^j \right| \tag{13}$$

where N_j is the number of coefficients added by $LOD^{(j)}$. This distortion measure estimates the error between the meshes with the jth LOD and the $(j + 1)^{th}$ LOD. We use this distortion measure to calculate the fraction of the total error protection bit budget that is assigned to each level in UEP. In EEP, the available error protection bit-budget can be calculated as follows:

$$S = \sum_{j=1}^{M+1} \left(n - k\right) \times q \times B^j \tag{14}$$

where q is the codeword size. B^j is the number of codewords in each horizontal packet. In the case of UEP, the bit-budget, S, and the total packet size, n, are provided. Therefore, the RS code rates for all M layers need to be computed. Let α^j be the portion of the total bit-budget to protect j^{th} level of decoded mesh. That is, $\alpha^j = \frac{S^j}{S}$. So the jth level bit-budget is given by:

$$\left(n - k_j\right) = \frac{\alpha^j \times S}{q \times B^j} \tag{15}$$

From Equation 15, we know α^j is the main factor to determine the RS code rate. We set α^j to be equal to the coefficients distortion quantity, $D_{wLOD}^{(j)}$ which was given in Equation 13. In this way, we can calculate RS code $(n\text{-}k_j)$ using Equation 15 for each part of decoded mesh.

Block-Based Encoding

To further increase the error-resilience of our transmitted meshes, we apply block-based encoding after UEP encoding, before transmission. A simple example of our approach to block-based error correcting is described. Consider a 3D model that has been decomposed into a base mesh and three levels of wavelet coefficients (L_1, L_2 and L_3). Applying RS codes, the resulting packets are shown in Figure 17. The base mesh consists of five data packets with five error protection packets. The wavelet coefficients corresponding to level one, L_1, consists of six data packets with four error protection packets. Wavelet coefficient level L_2 consists of eight data packets with two error protection packets and level L_3 consists of ten data packets with no error protection packets. The base mesh and its associated RS packets are transmitted first, followed by the coarse wavelet coefficients, until the finest one. As shown in Figure 17, more FEC codes are assigned to the coarser level of coefficients than the finer one. Such an allocation of FEC codes is calculated by a distortion quantity that is described above. At a certain packet loss rate, some of the packets will be lost. Taking an example of three packets for each block being lost. Since the base mesh uses (10,5) error correction codes, when the number of lost packets is not more than five, the client can recover all lost packets. Therefore, in this example, it can recover all three lost packets. For the same reason, all three lost packets in L_1 can be recovered. But the lost packets in L_2 and L_3 can not be recovered by the assigned RS codes. At the client, the base mesh and L_1 level of coefficients have adequate protection but L_2 and L_3 levels of coefficients get lost. Therefore, the more important parts of the mesh are protected, are correctly received by the client and decoded even when the wireless channel loses a significant number of packets.

Result

In this section, we describe tests that we conducted using meshes to evaluate the performance of our method. In particular, the performance of the UEP, EEP and NEP are compared. First we describe a two-state Markov model known as the G-E model (Pimentel & Blake, 1998) for the wireless channel.

Channel Model

We use a Markov model with only two states to model a wireless channel with high bit error rates (Pimentel & Blake, 1998). We shall now briefly describe its main characteristics.

G-E models are defined by the distribution of error-free intervals, which are called gaps. The gap is defined as the interval of length $v - 1$ packets between two consecutive received error packets. This model is illustrated in figure 19 and the probability density function (pdf) $g(v)$ and cumulative distribution function (cdf) of the gaps greater than $v - 1$ packets $G(v)$ are defined as equation 16 and equation 17, respectively.

$$g(v) = \begin{cases} 1 - P_{BG}, & v = 1 \\ P_{BG}\left(1 - P_{BG}\right)^{v-2} P_{BG}, & v > 1 \end{cases} \quad (16)$$

Figure 17. Example of transmitted packets in unequal error protection methods

Data Transmission →

Base Mesh	1	2	3	4	5	RS	RS	RS	RS	RS
L_1	1	2	3	4	5	6	RS	RS	RS	RS
L_2	1	2	3	4	5	6	7	8	RS	RS
L_3	1	2	3	4	5	6	7	8	9	10

Figure 18. G-E two state Markovian Channel Model. P_{GB} is the transition probability from the good state to the bad state while P_{BG} is the transition probability from the bad state to the good state

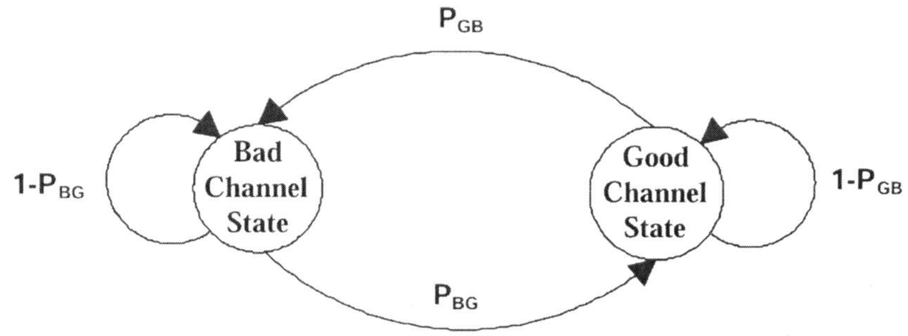

$$G(v) = \begin{cases} 1, & v = 1 \\ P_{BG}\left(1 - P_{BG}\right)^{v-2}, & v > 1 \end{cases} \tag{17}$$

Let $R(m,n)$ denote the probability of having $m-1$ packet losses within the $n-1$ packets following a lost packet. Then $R(m,n)$ is given by:

$$R(m,n) = \begin{cases} G\left(n\right), & m = 1 \\ \sum_{v=1}^{n-m+1} g(v)R(m-1, n-v), & 2 \le m \le n \end{cases} \tag{18}$$

So, the probability of losing m symbols, each of which is of q bits in length, within a block of n symbols can be written as:

$$p(m,n) = \begin{cases} 1 - \sum_{m=1}^{n} p(m,n), & m = 0 \\ \sum_{v=1}^{n-m+1} P_B g(v)R(m, n-v+1), & 1 \le m \le n \end{cases} \tag{19}$$

Simulation Results

We applied the proposed unequal error protection (UEP) method on several models and here we report the results for the small bunny mesh. We consider three cases: encoding the original mesh into a base mesh and 5 levels of detail, 10 levels of detail and 15 levels of detail. In general, the more levels of detail we use, the less information each layer contains. We use the Hausdorff distance to measure the amount of distortion in our received mesh. The Hausdorff distance expresses the geometric distance.

Figure 19 depicts the distortion as a function of the packet loss rate for the small bunny model. Three curves in this figure represent the cases of EEP, UEP, and NEP with level 5. As can be seen from these curves, for an error-free channel no packets are lost and the distortion in the transmitted mesh is zero. As the packet loss rate increases, the performance of EEP and NEP become closer to each other since neither technique can recover when packets of the base mesh or coarse level of coefficients are lost. However, UEP manages to protect the base mesh and coarse wavelet coefficients by assigning more error-protection bits and therefore improving the quality of the decoded mesh quality is better compared to other two methods. When the packet loss rate $P_{LR} \geq 0.2$, the base mesh information is lost and only UEP is able to protect the base mesh.

Figure 20 shows the distortion as a function of the packet loss rate for the small bunny mesh. Three curves in this figure represent the cases of 5, 10, 15 levels of detail. The figure shows a slow increase in the Hausdorff distance up till a knee point at which the Hausdorff distance (or distortion) increases quickly. Before the knee

Figure 19. Maximum Error (Hausdorff distance) between the transmitted and the decoded mesh when the RS code used for EEP is a: (n,k) = (63,45) and b: (n,k) = (63,51). NEP: no error protection is applied, EEP: equal error protection is applied, and UEP: unequal error protection is applied

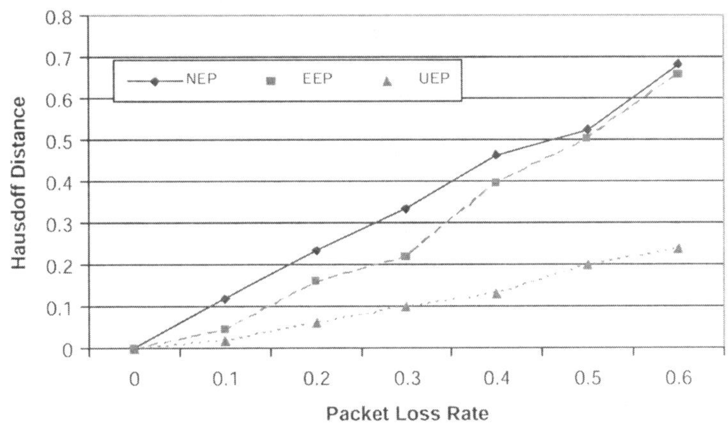

a

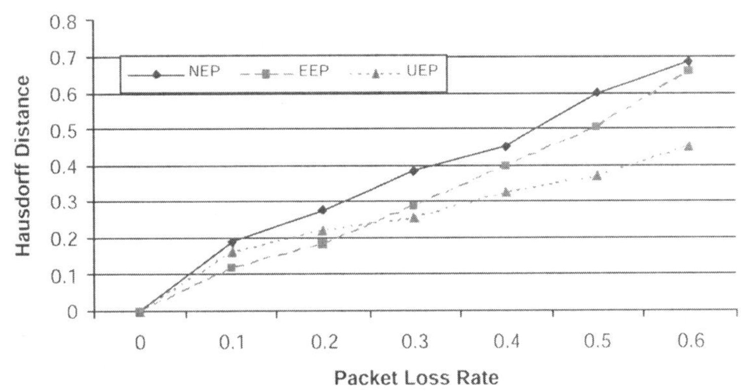

b

point, only wavelet coefficients are lost while the base mesh is correctly received. Beyond the knee points the high error rates cause the base mesh to get lost, causing a large increase in distortion (Hausdorff distance). The knee point of the 5-level LoD is larger (more resilient to errors) than that of the 10-level and 15-level LoDs. This is intuitive since as the mesh is encoded into more LoD levels, each level of the wavelet coefficient tree as well as the base mesh all receive fewer error protection bits. Hence, meshes that are encoded into more LoD levels will lose the base mesh information easier than meshes encoded with fewer LoD levels. Thus for a fixed UEP bit budget, we find an inverse relationship between the number of mesh LoDs used and the error resilience of the wavelet-encoded mesh. Before the knee points, the base mesh is received and only wavelet coefficients are lost. As the mesh is encoded into more LoDs, the importance of each level of the wavelet tree level is reduced and the degradation introduced when wavelet coefficients are lost are also reduced. Therefore, before the knee point, the distortion of the meshes encoded with more LoDs is slightly lower than that of meshes that use fewer LoDs.

Figure 20. Maximum error (Hausdorff distance) between the transmitted and the decoded mesh when different level of detail (5,10,15) are used with RS code (n,k) = (63,45)

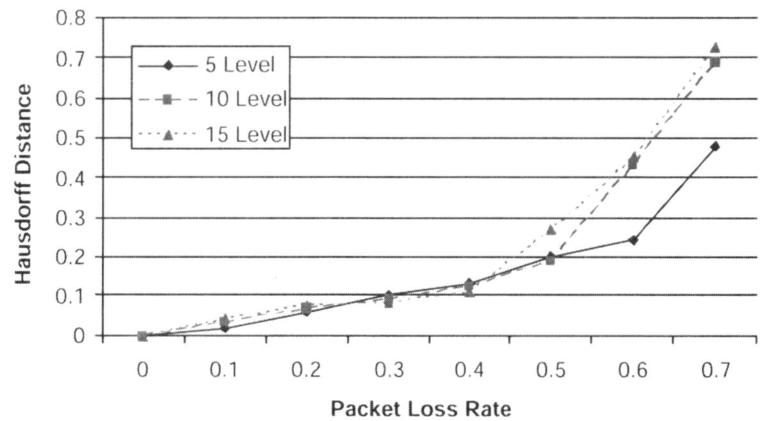

Objective results have been presented above. We also compare the three methods, NEP, EEP, and UEP, subjectively by looking at images of the final rendered mesh after passing them through a simulated wireless channel. Figure 21 shows the experimental results for the small bunny mesh. The first column on the left shows the decoded mesh in the NEP case for different packet loss rates. Similarly, the second and the third columns show the decoded meshes for EEP and UEP respectively. As shown, UEP maintains a reasonable decoded mesh quality as the packet loss rate increases. We have encoded the mesh into 5 Levels of Detail. As the error rate increases, UEP loses some detail coefficients but the base mesh and coarse coefficients are adequately protected and correctly received. Hence, only minor artifacts can be observed on the UEP as error rates increase. We can thus conclude that using our proposed UEP method on wavelet multiresolution, the quality of the decoded meshes is better as the packet loss rate increases.

Section Summary

This section presents Unequal Error Protection (UEP), a Forward Error Correction (FEC) scheme for the error-resilient transmission of meshes that have been encoded using wavelets, to increase decoded mesh quality. Error-protection bits are allocated according to the importance of parts of the wavelet-encoded mesh. The importance of each level is determined by a distortion measure that reflects the information the coefficients contain. Theoretically, the UEP method increases the resilience of wavelet-based mesh transmission to high error rates. By simulating mesh transmission using our proposed scheme on two different channel models, we compare the performance of the proposed UEP, EEP and NEP methods.

ENERGY-EFFICIENT ADAPTIVE REAL-TIME RENDERING HEURISTIC

This section presents the research work for Energy-efficient Adaptive Real-time Rendering (EARR) heuristic (Wu *et al*, 2008).

Overview

The most limiting resource on a mobile device is its short battery life. While mobile CPU speed, memory and disk space have grown exponentially over the years, battery capacity has only increased 3-fold in the past decade. Consequently, the mobile

Figure 21. Subjective results of applying no error protection (NEP), equal error protection (EEP), and unequal error protection (UEP) methods on the SMALL BUNNY model. The caption under every image gives the error protection method and the packet loss rate of the channel. RS code (n,k) = (63,45)

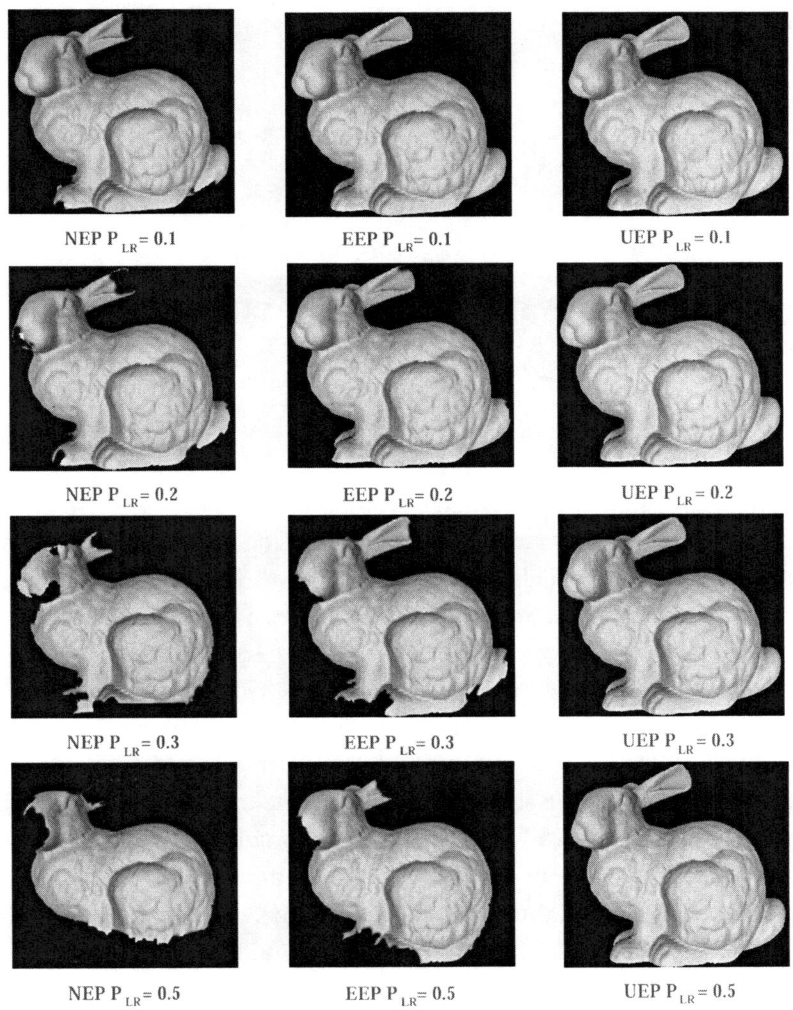

user is frequently forced to interrupt their mobile graphics experience to recharge dead batteries.

Application-directed energy saving techniques have previously been proposed to reduce the energy usage of non-graphics mobile applications. Our main contribution is the introduction of application-directed energy saving techniques to make mobile graphics applications more energy-efficient. The main idea of our work is that energy can be saved by scheduling less CPU timeslices or lower the CPU's clock speed (Dynamic Voltage and Frequency Scaling (DVFS)) for mobile applications during periods when its requirements are reduced.

In order to vary the CPU timeslices alloted to a mobile application, we need to accurately predict its workload from frame to frame. Workload prediction is a difficult problem since the workload of real-time graphics applications depends on several time-varying factors, such as user interactivity level, the current Level-of-Detail (LoD) of scene meshes and mid-mapped textures, visibility and distance of scene models, and the complexity of animation and lighting. Without dynamically

Figure 22. Application running at high real-time frame rate

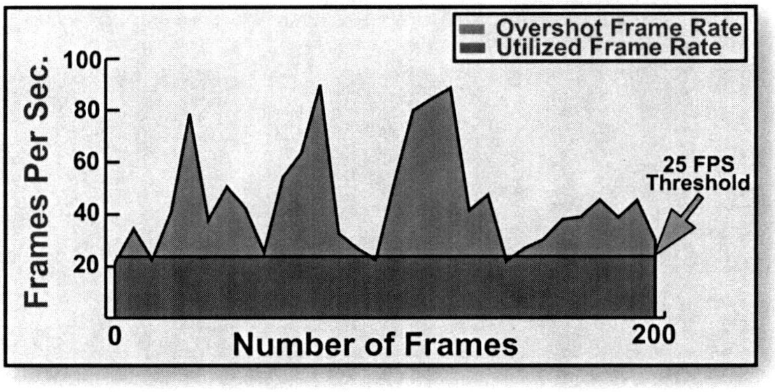

changing the application's CPU allotment to correspond to its needs, the mobile application's frame rate fluctuates whenever there is a significant change in scene LoD, animation complexity, or other factors that affect its workload. Such spikes above 25-30 Frames Per Second (FPS) drain the mobile device's battery and increased energy consumption by up to 70% in our measurements (see figure 23). We propose an accurate method to predict the mobile application's workload and determine what fraction of the CPU's cycles it should be alloted to maintain a frame rate of 25 FPS. As the application's workload changes, we update its CPU allotment at time intervals determined by a windowing scheme that is sensitive to applications with fast-changing workloads and prudent for applications with slow-changing workloads. Our adaptive CPU scheduling scheme dampens frame rate oscillations and saves energy.

Many techniques have been proposed to achieve three desirable qualities of mobile graphics: photorealism, real-time rendering and energy efficiency. For instance, Level-of-Detail (LoD) management allows scenes to be rendered at real-time speeds while maximizing visual realism. Also, intelligent scheduling and application-directed Dynamic Voltage and Frequency Scaling (DVFS) have been proposed to save energy on mobile devices. While these techniques work if applied separately, they can create conflicts when they are integrated into the same graphics framework. Specifically, techniques that improve one attribute can degrade another. For instance, improving image quality requires increasing mesh LoD, which need more CPU cycles and memories accesses which kills (degrades) the mobile devices battery. Essentially, we can think about these three attributes as orthogonal axes. Ideally, we would like to make progress along all three axes. However, in practice, proposed techniques have fundamental limitations that allow them to only make progress along one or two axes but typically not all three axes (See Table 2 for examples).

Since the application's workload changes and should be re-estimated whenever LoDs are

Table 2. Proposed techniques improve one or two desirable mobile graphics attributes while degrading the third one

Technique	Realism	Rendering Speed	Energy Efficiency
LoD Reduction	⇓	⇑	⇑
Voltage Scaling	⇓	⇓	⇑
Frequency Scaling	⇓	⇓	⇑
CPU Scheduling	⇓	⇓	⇑
Ray tracing	⇑	⇓	⇓
Complex (HDR) lighting	⇑	⇓	⇓
Complex material (BRDF)	⇑	⇓	⇓

switched, we have coupled our CPU scheduler with the application's LoD management scheme. When switching scene LoD, we minimized energy consumption by selecting the lowest LoD at which the user does not see visual artifacts, also known as the Point of Imperceptibility (PoI) (Wu *et al*, 2007). Although our primary goal was to minimize the mobile application's energy consumption, we also ensured that the frame rates and visual quality of the rendered LoD were acceptable. In summary, our integrated EARR (Energy-efficient Adaptive Real-time Rendering) heuristic minimizes energy consumption by i) selecting the lowest LoD that yields acceptable visual realism, ii) scheduling just enough CPU timeslices to maintain real-time frame rates (25 FPS). EARR also switches scene LoD to compensate for workload changes caused by animation, lighting, user interactivity and other factors outside our control. To the best of our knowledge, this is the first work to use CPU scheduling to save energy in mobile graphics. Our results on animated test scenes show that CPU scheduling reduced energy consumption by up to 60% while maintaining real time frame rates and acceptable image realism.

Our Approach

Heuristic Architecture

Our framework includes components for monitoring application frame rate and the rendered appearance of a selected mesh LoD, as well as a component for allocating CPU resources to our mobile graphics application. Our adaptation algorithm balances desired attributes using these components, which is shown along with our system architecture in figure 23. The energy monitor measures system-wide energy consumption.

Overview of EARR Heuristic

Our approach is a generalization of the predictive strategy. We predict the LoDs that will be rendered at the speed threshold of 25 frames per second. Within a real-time application such as a game, LoD is just one of many factors that affect the application's frame rates. Other aspects include lighting, texturing, system animation, artificial intelligence and networking of the application. In fact, in complex real-time graphics application such as a game or flight simulator, it is difficult to accurately model and predict all factors that affect observed frame rates including when the user will interact with the scene or to anticipate the animation paths of meshes. We can not hope to consider all of these complex factors that can be computed efficiently. However, using efficient workload predict model, we have developed approximate heuristics that are both efficient to compute and accurate enough to be useful. Our algorithm takes actions such as switching mesh LoD or CPU allocation to compensate for the demands of game components outside its control,

Figure 23. Heuristic architecture

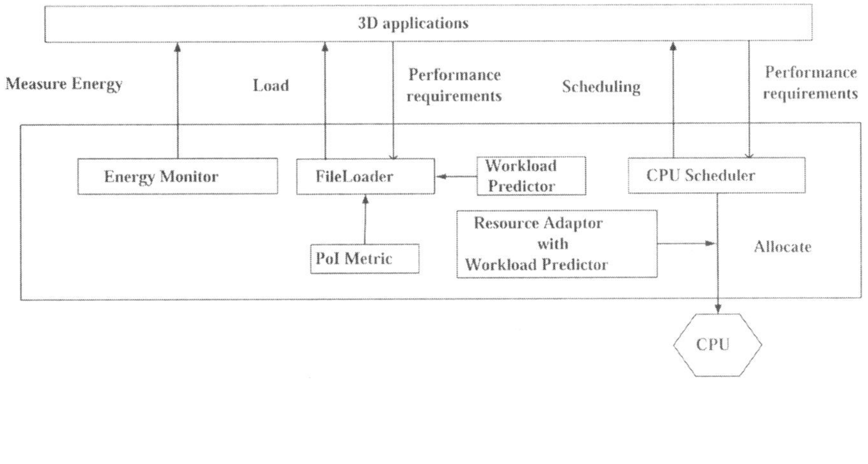

such that the frame rate of the entire application remains within the threshold frame rate.

More formally, we define *Energy(O,S,R)*, to be the energy required to render an instance of a mesh or object *O*, rendered in the scene *S*, with adaptive algorithm *R*. Our approach can be stated as:

Minimize: *Energy*(*O,S,R*)

Subject to: *Rendering frame rate* ≥ *Threshold* (20)

Subject to: *Visual Realism* ≥ *Threshold* (21)

This formulation captures the essence of 3D graphics rendering on mobile devices with real-time constraints. Verbally stated, our goal is to reduce mobile device energy consumption as much as possible, while rendering the lowest LoD that meets the PoI (visual realism) within the target frame rate.

Workload Predicting Model

Overview

The workload predicting model predicts what fraction of available CPU timeslices should be alloted to our mobile application in order to render a given mesh LoD or scene at our target frame rate of 25 FPS. We derive our predicting model in two parts. The first part predicts the workload of a single mesh object. Since most real world scenes consist of multiple objects, as a next step, we extend our workload predicting model to estimate the workload of complex scenes with multiple objects.

In general, as a given mesh is rendered faster, more CPU timeslices are consumed per unit time, and more battery energy is expended with no improvement in visual realism. Thus, to minimize energy consumption, the goal of the CPU scheduler is to allot *just enough* CPU cycles to finish rendering each frame *just before* its deadline expires. We strived to maintain a frame rate of 25 FPS, which means that each frame should finish rendering within a deadline of 40 milliseconds. Based on this deadline, if the rendering time of each frame using a particular LoD is estimated to be 20 milliseconds when 100% of CPU resources are alloted to our mobile graphics application, then the alloted CPU resources (and rendering speed) can be halved without exceeding the frame's deadline. The optimal (fewest) CPU resources C_{opt} to meet our task's deadline can be expressed as:

$$C_{opt} = \frac{\tau}{r_{\max}} \times C_{\max} \qquad (22)$$

where C_{max} is the maximum available allotment of the processor's timeslices, C_{opt} is a reduced allotment of CPU timeslices generated by our algorithm, which just meets the frame's deadline. r_{max} is the rendering time of a mesh with all available processor cycles alloted to our application and τ is the deadline for the frame. Since our target frame rate is 25 Frames Per Second, we set τ, the deadline for each frame, to 40 ms.

We apply our workload predictor as follows. At runtime, given a frame rendering deadline, τ, we use equation 22 to calculate the optimal CPU processor allotment, C_{opt}. We then use our pre-generated statistics to estimate the mesh LoD that corresponds to C_{opt}.

For our workload predictor to be successful, we derive our predicting model in two parts. The first part predicts the workload of a single mesh object. Since most real world scenes consist of multiple objects, as a next step, we extend our workload predicting model to estimate the workload of complex scenes with multiple objects.

Workload Predicting Model for a Single Object

Given a scene, we would like to use certain observable features to predict its rendering time. Funkhouser and Sequin (1993) previously suggested that the number of triangles in a mesh

Figure 24. Sample meshes and their correlation coefficients

a. Bunny model (r_{xy}=0.9343) b. Feline model (r_{xy}=0.8974) c. Venus model (r_{xy}=0.9438)

was a good predictor of its rendering time. To examine how accurately the number of triangles in a mesh predicts of its rendering time, we set up experiments to study how correlated rendering times are with mesh LoD. In a offline calibration pre-process, various meshes (bunny, feline, venus) were rendered at different LoDs and statistics were collected on their rendering times and corresponding processor demand for each LoD. To formally establish the degree of correlation between mesh LoDs and their rendering times, we calculated the first and second order statistics of measured rendering times and triangle counts.

Let x and y be two random variables corresponding to the mesh size(number of triangles) and rendering time, respectively; and let μx and σ_x be the mean and standard deviation of the mesh size; and also let μ_y and σ_y be the mean and standard deviation of the rendering time, respectively. Thus the theoretical correlation coefficient ρ_{xy} between x and y is given by:

$$\rho_{xy} = \frac{E\left[\left(x-\mu_x\right)\left(y-\mu_y\right)\right]}{\sigma_x \sigma_y} \quad (23)$$

Now assume we have N experimentally measured pairs of x and y values. The correlation coefficient ρ_{xy} may be estimated from these N pairs of data as:

$$r_{xy} = \frac{\sum_{i=1}^{N}\left(x-\overline{x_i}\right)\left(y-\overline{y_i}\right)}{\left[\sum_{i=1}^{N}\left(x-\overline{x_i}\right)^2 \sum_{i=1}^{N}\left(y-\overline{y_i}\right)^2\right]^{1/2}} \quad (24)$$

In general, a correlation coefficient of 1.0 is the highest achievable value and implies that given a value of x, the corresponding value of y can be predicted with 100% accuracy. Figure 24 shows three meshes (bunny, feline and venus) with calculated correlation coefficients, respectively. These results show strong correlation between mesh LoD and rendering time. In fact, there is a linear relationship between the number of triangles and rendering time, which corroborates corroborate the results of Funkhouser and Sequin (1993). The slope of this linear relationship depends on the mesh features and how powerful the machine on which it is rendered is. Thus, for a particular mesh and rendering machine, the slope and intercept of the linear function can be determined during pre-processing by rendering the same model at n LoDs, and graphing measured rendering time versus the number of triangles. Depending on its features, different meshes produce functions of different slopes. For instance, increasing the LoD of a complex model by 1000 triangles yields a larger increase in its rendering time than if the LoD of a simpler mesh were increased by 1000 triangles. Hence, complex models have steeper slopes than simple models. For example, the feline model is more complex than the bunny mesh, and thus yields a steeper slope.

Finally, using observed data points of rendering times for different LoDs, we can use linear regression to generate a line of best fit. Let s_i and rt_i denote the number of triangles and rendering time of the *i*th LoD, respectively, with all available CPU cycles alloted to our mobile graphics application. Thus the slope *(b)* and intercept *(a)* of the line of best fit are given as:

$$\bar{s} = \frac{\sum_{i=1}^{n} s_i}{n}$$

$$\overline{rt} = \frac{\sum_{i=1}^{n} rt_i}{n}$$

$$b = \frac{\sum_{i=1}^{n} (s_i - \bar{s}) rt_i}{\sum_{i=1}^{n} (s_i - \bar{s})^2}$$

$$a = \overline{rt} - b\bar{s} \qquad (25)$$

Figure 25 shows a sample best fit line. This line of best fit is used in our workload predictor. To charactize the overall accuracy of our workload predictor, the relative error between actual measured rendering times and predicted rendering times produced by our workload predictor, was calculated for various LoDs. Figure 26 is a plot of calculated relative error corresponding to different LoDs. The figure shows that our workload predictor is reasonably accurate where all relative errors are less than 5%.

Workload Predicting Model for Multiple Objects

Note that in a real game or application, there are typically many objects at various LoDs. Our pro-

Figure 25. Sample best fit line

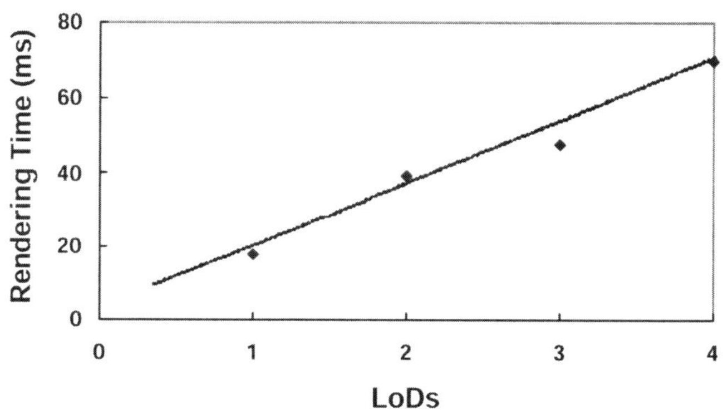

Figure 26. Relative error between actual and estimated rendering times

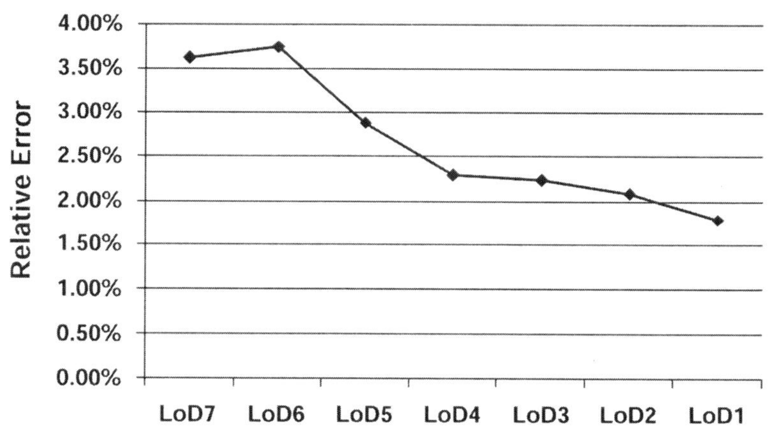

posed method should also predict the workload of multiple objects in a game or applications.

In complex scene with multiple objects, the workload for rendering the scene depends on the visibility of objects in the scene, which can vary over time as objects and the camera move. Tens of thousands of polygons might be simultaneously visible from some observer viewpoints, whereas just a few can be seen from others. Thus, the rendering effort for a dynamic scene is proportional to the triangles that are visible. We used an *eye-to-object* visibility algorithm described in (Teller, 1992) to determine a set of potentially visible objects to be rendered in each frame. Thus, the workloads of all visible objects (as determined in *Workload Predicting Model for a Single Object* section), are then linearly combined to generate the workload of the complex scene.

Next, we considered changes in the application's workload over time. Since application workload changes only slightly from one frame to the next (milliseconds), the workload of successive frames is highly correlated. Thus, we use the current frame's workload to predict the workloads of next *n* frames. We define the *window size* as the number of frames in the future *n* for which current frame's workload is used as an estimate. The choice of *n* affects the performance of our algorithm. If *n* is too small, then we need to updated workload value too often and it will increase the computation overhead; If it *n* is too large, then the variance between the predicted and actual workload will be high, and the variance could be too high to be accepted. Therefore, in our predicting model, this window size (*n*) is updated adaptively at run-time. Figure 27 shows how the window size is updated in our predicting model, which is inspired by the Transmission Control Protocol (TCP) in networking. It starts from 2. At the end of window size time point, we check the workload error. If it is smaller than the threshold, then the window size is doubled or increased by 1. Normally, there will not be much workload change within 8 frames. Therefore, if

Figure 27. Window size updating

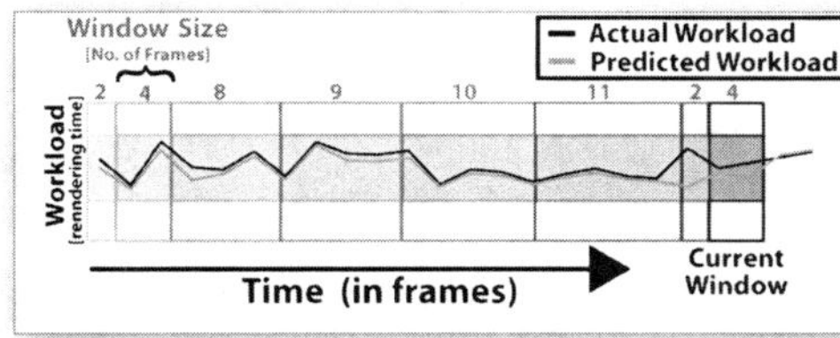

the window size is less than 8, then we double window size, otherwise increase window size by 1. If the workload error is larger than threshold, which means the workload predicting value is not accurate, we reset the window size to 2 and update the workload predicting value with current actual workload. Figure 28 shows the working flow.

The adaptive workload predictor is then used to estimate the workload of each frame at full processor speed, so that we can get the fraction of available CPU timeslices required to render a frame at our target frame rate. We tested it with two scenes provided by the Benchmark for Animated RayTracing(BART)(Lext *et al*, 2001), The results are shown in figure 30. As we can see, the relative errors are both bounded in 0.18.

CPU Scheduler

To conserve battery energy, our CPU scheduler runs a three-phase algorithm. The phases of the scheduler algorithm are workload estimation, estimating processor availability and determining processor resource allocation. More detail is now given on each of these steps. (1) *Estimated workload:* In this step, our workload predictor in *Workload Predicting Model* section is used to estimate how many CPU timeslices running the mobile graphics application will consume. (2) *Estimating processor availability:* Since the system may be running other applications or performing system house-keeping functions, the amount of CPU cycles available to our mobile graphics application varies over time. In this step, the amount of CPU

Figure 28. Flow chart for workload predicting model. In our work load predicting model, N=8

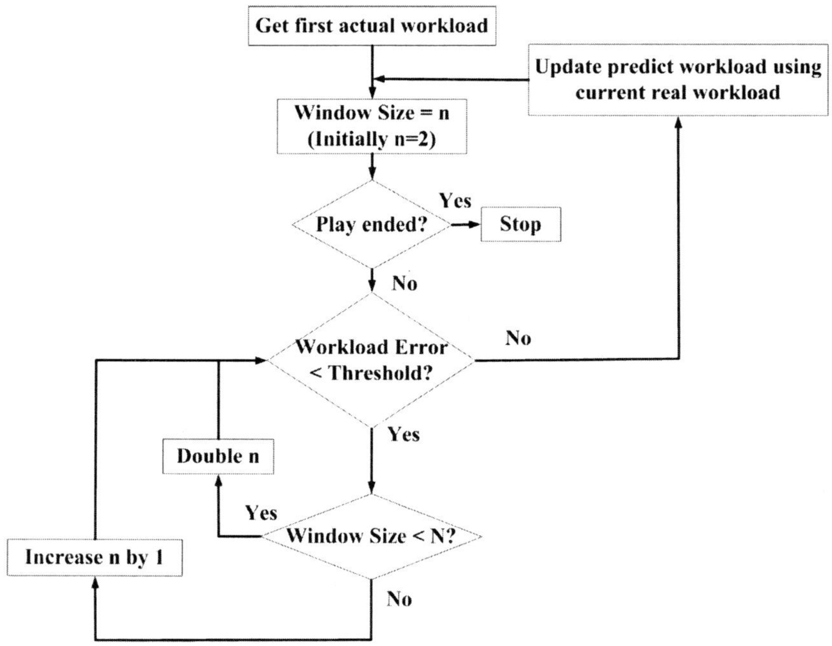

resources currently available for applications is estimated. (3) *Determine processor resource allocation:* The last step chooses what fraction of available CPU resources is alloted based on the predicted workload and processor availability. For instance, if the predicted workload is only one third of the CPU resources available, then the mobile graphics application can save energy by using one third of available CPU resources. Likewise, if the predicted workload exceeds available CPU cycles, all available CPU cycles are allocated to the mobile graphics application

Figure 29. Workload predicting model

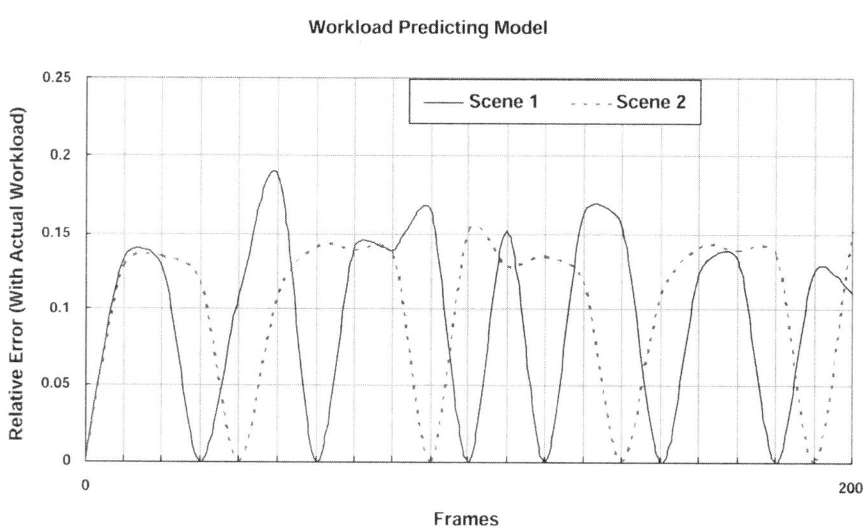

and a lower mesh is selected to maintain a frame rate of 25 FPS.

We shall now formalize our CPU scheduling algorithm. For each real-time task T, let us denote its start time by t_s and its deadline as t_d. Let C_{max} denote the maximum fraction of CPU timeslices that are currently available for running applications. It is important to note that without the intervention of our scheduling algorithm, all tasks will run with 100% allocations of all available CPU timeslices, C_{max}. The number of processor timeslices required by T will be denoted by p. We note that the execution time of the task T is inverse proportional to p. In summary, a feasible schedule of the task guarantees that the task T receives at least a fraction, A, of the maximum available CPU cycles such that it receives $A * C_{max}$ CPU cycles before its deadline, where $A \leq 1$.

Given the application workload p, maximum processor availability C_{max} and interactivity deadline t_d, as shown in figure 31, our policies to allocate processor resources fall into two distinct cases that are now described.

Case 1: If $C_{max} < p$, then the application's demand for CPU timeslices exceeds CPU availability. In this case, the CPU schedule has allocated 100% of all available CPU resources to the task and cannot meet the task's deadline while using the current mesh LoD. Our scheduling algorithm shall allot all available CPU timeslices to the task and additionally reduces mesh LoD to lower the offered workload p.

Case 2: If $t_s + p < t_d$, the task can complete before its deadline. If all available CPU resources are alloted to this task, the rendering speed achieved is larger than 25 frames per second. In this case, the algorithm reduces the fraction of CPU timeslices alloted such that the demanded workload p is just adequate to complete the task before its deadline. The percentage of CPU resources alloted should be:

$$A = \frac{p}{\min\left(C_{\max}, t_d - t_s\right)} \tag{26}$$

In the equation 26, the deadline $t_d - t_s$ is known. In our case, we choose $t_d - t_s$ as 40 ms, p is determined by using our workload predictor. The maximum CPU resources currently available, C_{max} can be monitored by our resource adaptor.

Given an estimated $\hat{p}$ of the demanded workload and the maximum processor availability, $\hat{C}_{\max}$, the optimal CPU resource allocation is computed as:

$$C_{opt} = \begin{cases} C_{\max} : \hat{C}_{\max} < \hat{p} \\ \min\left(C_{\max} \times \frac{\hat{p}}{\min(\hat{C}_{\max}, t_d - t_s)}, C_{\max}\right) : \ otherwise \end{cases} \tag{27}$$

Figure 30. Symbols illustration

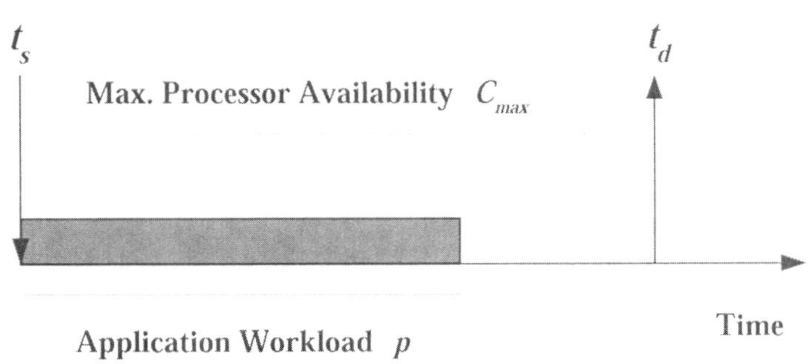

EARR Heuristic

Building on our workload predictor and CPU scheduling policy, we now describe our complete optimization algorithm to balance application frame rate, visual realism and energy consumption constraints. Our algorithm monitors predicted frame rate and the rendered appearance of meshes and takes corrective action such as switching mesh LoD or changing the CPU resource allocation, when frame rate or LoD changes considerably.

Our optimization algorithm works as follows. At the start of the algorithm, the LoD of meshes corresponding to their PoI is selected for rendering. As the mesh moves during an animation, the algorithm reallocates CPU resources using the CPU scheduling policy and the workload predicting model. If the predicted frame rate becomes less than 25 FPS, the algorithm chooses a lower LoD that increases application frame rate to 25 FPS. The optimal CPU allotment that minimizes energy consumption without affecting frame rate is then computed. The algorithm chooses the PoI LoD of the mesh for rendering when the adequate CPU resource can be alloted to render meshes at our speed threshold of 25 FPS.

There are three cases to which our heuristic is required to adjust the application parameters, each require different action. If we let d denote the current LoD of a mesh and d_p denote its PoI LoD. Let f denote the frame rate at which that mesh is currently being rendered. Essentially, there are three cases that our algorithm reacts to:

> **Case 1, predicted frame rate drops such that fi < 25, current LoD i = minimum LoD possible, and 100% of CPU cycles already alloted to this task:** In such a case, since we are at the limits of the factors under our control (minimizing LoD and maximizing CPU cycles), we conclude that it is impossible to meet the rendering speed threshold of 25 FPS. Essentially, the resources of mobile devices are not enough to render the mesh and animation and we cannot rectify the situation. In such a scenario, we simply choose the minimum possible LoD and set the CPU cycles to a maximum and achieve the highest frame rate possible with this setting (best effort).
>
> **Case 2, predicted frame rate drops such that fi < 25, current LoD i = PoI, dp:** In such a case, the algorithm will allocate more CPU resources to increase the rendering frame rate. If the rendering speed is still less than 25 FPS, the algorithm will choose a lower LoD level that can be rendered at 25 FPS and allocate the optimal fraction of CPU cycles, C_{opt} accordingly. We note that in this case, to achieve 25 FPS, we are forced to use a LoD below the mesh PoI, which introduces simplification artifacts.
>
> **Case 3, predicted frame rate increases such that fi >> 25, current LoD i = PoI, dp:** the algorithm continues to use the PoI LoD but tries to save energy by reducing the percentage of CPU timeslices scheduled for our application to the minimum required to maintain a frame rate of 25 FPS. Figure 31 is the flow chart of our algorithm and the complete pseudocode of the algorithm is shown in algorithm 1.

Algorithm 1 Balancing Heuristic in Animation.

Choose the PoI {Rendering the possible lowest LoD without perceivable difference}
if Mesh Move **then**
if ($f_{predicted}$ < 25) and (d_p is the lowest LoD) **then**
Break
else
if d_p is not the lowest LoD **then**
Choose the suitable lower LoD within current CPU resource constraint by predicting model.

Figure 31. Algorithm flow chart

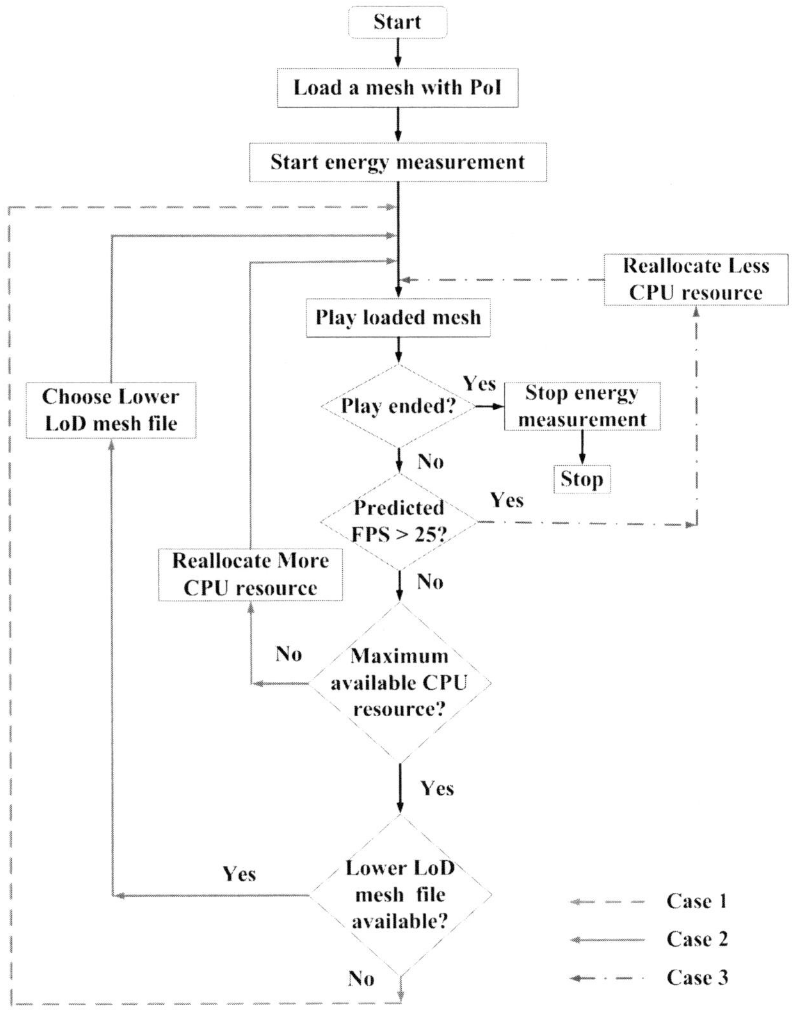

8: **if** Cannot find such kind of d_p **then**
9: **Break**
10: **end if**
11: **end if**
12: **end if**
13: **if** ($f_{predicted}$ > 25) **then**
14: Do CPU scheduling using our CPU scheduling policy to maintain rendering speed almost 25 FPS
15: **end if**
16: **if** ($f_{predicted}$ < 25 in some point) **then**
17: Increase allocated CPU resource to the maximum available CPU resource
18: **if** still $f_{predicted}$ < 25 **then**
19: choose the suitable lower LoD within current CPU resource constraint by predicting model.
20: choose PoI to render until CPU resource is enough to maintain frame rate of 25
21: **end if**
22: **end if**
23: **end if**
24: Choose the LoD nearest to PoI when it reaches the destination .

Experiment and Results

Experiment

We extensively evaluated the performance of our proposed algorithm both a laptop and PDA. The laptop used was a Gateway 3040GZ laptop equipped with an Intel Celeron 1.5GHz processor and 512MB RAM. The laptop's operating system is Linux. The PDA is a HP iPAQ Pocket PC h4300 with a 400 MHz intel XScale processor and 64MB RAM. The operating system of the PDA is windows CE. We repeated all experiments eight times, eliminated the minimum and maximum values before averaging all other values. We animated a mesh bunny along a pre-determined animation path in a scene provided by the Benchmark for Animated RayTracing (BART). The test animation path was chosen because it is representative of typical behavior of real applications. We ran a three-set of experiments using the bunny mesh animated along a sample path in the museum scene, applying three levels of adaptations:

- **Simple (No LoD switching, no CPU scheduling:** The bunny model is rendered at the highest LoD all the time. No LoD changes are made throughout the application's running time and no dynamic CPU scheduling for energy conservation is done. The measured performance of this level of adaptation provides a baseline for establishing how much performance improves with our adaptations.
- **LoD Selection (LoD switching, no CPU scheduling):** The bunny model is rendered, switching mesh LoD as necessary either to react to significant frame rate deviations from 25 FPS, or to react to significant deviations in mesh appearance from acceptable visual realism (PoI). However, no dynamic CPU scheduling for energy conservation is employed in this case.
- **Our Complete Optimization (LoD selection with CPU scheduling):** LoD Selection is done to satisfy achieve a frame rate of 25 FPS and also to satisfy the visual realism constraint. Additionally, the CPU scheduling policy is also applied. Essentially, this is our complete algorithm to balance visual realism, frame rate and energy conservation.

We now present more details about our experiments. First, we generated a series of mesh LoDs and found the LoD corresponding to the PoI of each mesh.

Next, we calibrated our workload predictor for the bunny mesh. The rendering times for three different LoDs were measured. These measured values were then used to generate a line of best fit that predicts rendering time with error rates of less than 5%.

Our goal was to minimize energy consumption of the CPU excluding peripherals and other system components. Thus, to track how well our algorithm worked, we needed to measure the energy consumption of the CPU alone. Measuring the exact energy consumption of the CPU alone is a fairly hard problem. We use a subtractive technique for estimating CPU energy consumption. First, we measured the power consumption of the entire laptop while running our test application. We then measured the base power consumption of the laptop while running just the operating system in idle mode. Finally, we subtracted this base idle power from measured application energy values. In our experiment, the base power consumed by the laptop in idle mode is 7.19W.

Discussion

During our experiments, we set 20 check points along the animation path of the mesh. Figure 32 is a plot of measured frame rates at these check points along the test path with different algorithms tested. Three different plots are used to compare the a) Simple rendering; b) LoD selection and c) optimization algorithms.

In the experiments called *simple*, the mesh is always rendered at the highest LoD. In such a case, the rendering speed is low, as figure 33.a shown. The straight dashed line is the target minimum frame rate of 25 FPS. Without appropriate LoD selection in the simple experiment, the target frame rate of 25 FPS cannot be achieved.

In the experiments called *LoD selection* algorithm, LoD selection is performed but no CPU scheduling is done to conserve energy. In this case, the mesh does not show visual artifacts due to LoD reduction and the application frame rate is always above 25 FPS. However, since no CPU scheduling is done, 100% of all available CPU cycles are always alloted (C_{max}) to the application, and at many points during the application's lifetime, the LoD selected can be rendered much faster than (overshoots) 25 FPS. Figure 32.b shows an example. At frame 45 and frame 210, since the frame rate drops, we choose the lower LoD to render and the frame rate goes up. However, this lower LoD will show some visual artifacts since it is below the PoI LoD. At frame 120 and frame 255, since the available CPU resource is enough maintain a frame rate greater than 25 FPS, we choose render the PoI LoD since visually there is little noticeable difference between PoI and original LoD.

In contrast, in addition to performing LoD selection, *our complete optimization algorithm* reduces alloted CPU resources when the frame rate is far above 25 FPS to save the energy. As a result, the frame rate generated using our optimization algorithm is much more uniform with less fluctuations, as shown in figure 33.c. As in the LoD selection algorithm, at frame 45 and frame 210, the frame rate drops. Our complete optimization algorithm first tries to increase alloted CPU timeslices while using the PoI LoD. Since the frame rate continues to drop, the optimization algorithm selects a lower LoD and runs

Figure 32. Frame rates at check points along animation path

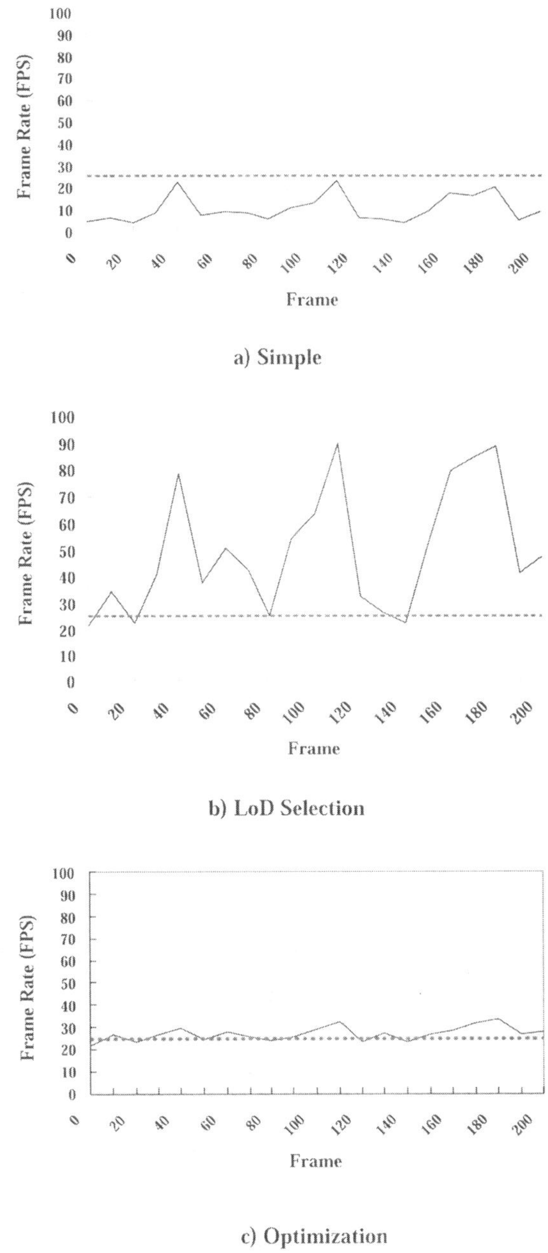

the CPU scheduler algorithm, which reduces the CPU resources alloted to 40% of the maximum available. An application frame rate of 25 FPS is maintained while energy is saved.

Figure 33 shows screenshots of our test applications. In figure 34.a, the simple algorithm is used with the bunny at its original LoD. The achieved frame rate is only 4.43 FPS. In figure 34.b, our complete optimization algorithm is used with the bunny at the PoI LoD. Visually, there is no noticeable difference between the original and PoI LoD. However, the PoI LoD can be rendered at up to 27.01 FPS. In figure 34.c, since the target frame rate cannot maintained even when all available CPU resource are allocated to the application, a lower LoD is chosen. This LoD is lower than the PoI and introduces some visual artifacts. However, the target frame rate of 25 FPS is maintained. Figure 34 shows screenshot of our test applications on a PDA.

Table 3 summarizes the energy saved before and after employing the simple, LoD selection and optimization algorithms. The LoD selection algorithm saves 27.4% of the energy, while our complete Optimization algorithm saves around 62.3% of the energy consumption.

Section Summary

This section presents our heuristic to balance energy consumption, rendering speed and image quality. In summary, our integrated EARR (Energy-efficient Adaptive Real-time Rendering) heuristic minimizes energy consumption by i) selecting the lowest LoD that yields acceptable visual realism, ii) scheduling just enough CPU timeslices to maintain real-time frame rates (25 FPS). EARR also switches scene LoD to compensate for workload changes caused by animation, lighting, user inter-activity and other factors outside our control. To the best of our knowledge, this is the first work to use CPU scheduling to save energy in mobile graphics. Our results on animated test scenes show that CPU scheduling reduced energy consumption by up to 60% while maintaining real time frame rates and acceptable image realism.

Figure 33. Screenshots on laptop using a) simple algorithm with bunny at original LoD; b) Our Optimization algorithm with bunny at the PoI LOD, and c) optimization algorithm with bunny at LoD lower than PoI

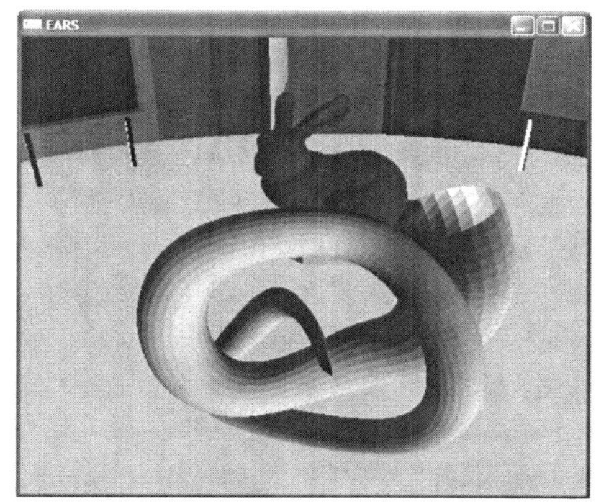

a) Simple Algorithm (4.43 FPS)

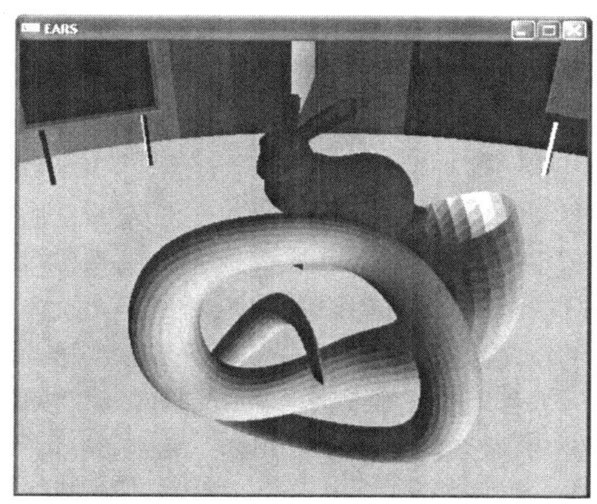

b) Optimization algorithm (27.01 FPS)

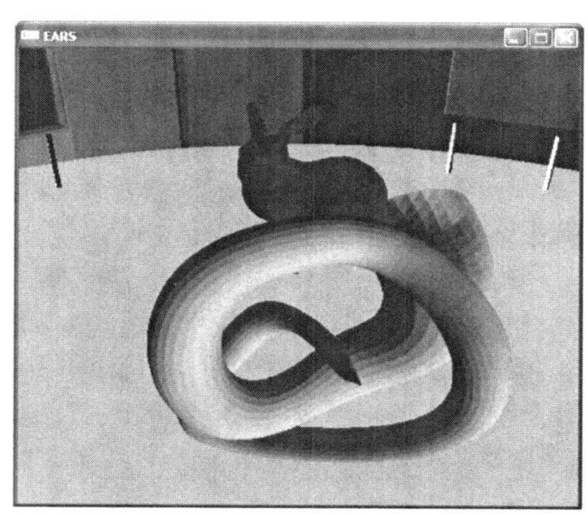

c) Optimization algorithm (40.77 FPS)

Figure 34. Screenshot on a HP iPaq Pocket PC

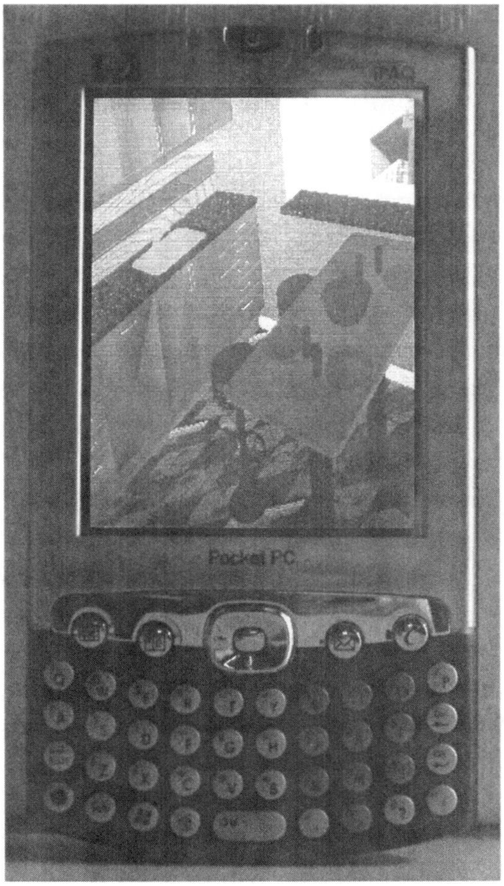

ENERGY-EFFICIENT 3D STREAMING

This section describes a wavelet-based multiresolution mesh streaming technique in UbiWave that utilizes PoI perceptual error metric, unequal error protection coding scheme and energy-efficient adaptive real-time rendering heuristic.

Overview

In this section, we will outline our proposed technique for 3D streaming in UbiWave.

Normally 3D objects are stored and maintained by either a central or distributed servers. A client sends requests to the server for model retrieval, and the requested models are transmitted accordingly by available communication channels from the server to the client. This scenario is typical when using wireless PDA or cell phone as a tool for Internet access. Since the storage of these mobile device tends to be very limited so that it is difficult to store a lot of 3D data locally. And the size of high resolution 3D data causes long download times in low bandwidth wireless channels, making it is difficult to maintain real-time rendering speeds. UbiWave uses wavelet as the uniform representation for 3D content, which forms the basis to 3D streaming from servers to clients, making rendering 3D data without a complete download.

Figure 35 depicts the proposed mesh streaming technique in UbiWave.

To maintain real-time rendering of 3D graphics model, UbiWave decompose 3D meshes into a base mesh and a coefficient tree that is stored in the server so that only base mesh and a certain level of coefficient in coefficient tree need to be encoded and transmitted. Once a mobile device successfully establishes a connection to a server, the parameters of mobile device and network, such as the resources of mobile device and network conditions will be sent to the server. The server determines the PoI of 3D meshes and sends back the data accordingly. The received data is stored at the mobile device side for rendering. Mobile devices use energy-efficient adaptive real-time rendering heuristic to guide rendering so that real-time rendering speed is maintained with minimum energy consumption, while not degrading image quality on mobile devices. Although mobile device has enough resources to maintain real-time rendering speeds, if the network condition is poor, the mobile device still need to wait to receive the whole data so that the real-time rendering speed cannot be maintained. A Level of Detail selection algorithm in the server is needed to avoid wasted transmission energy consumption in mobile device.

Table 3. Energy savings

Algorithm	Before(mwh)	After(mwh)	Saved
Simple	9690	9690	0.00%
LoD Selection	9690	7035	27.4%
Our Optimizations	**9690**	**3653**	**62.3%**

3D Streaming in UbiWave

From the mobile devices' perspective, the most important qualities of a mesh streaming technique are battery energy, rendering speed and visual quality. Our EARR heuristic in the mobile device will balance these three factors.

From the server's perspective, it is preferable if the LoD that is just adequate for each type of mobile device is sent through wireless network with as little data lost as possible. Mesh streaming has two stages: the selection of LoD of the meshes (as determined by the specifications of mobile device) and the efficient transmission of selected data. The first stage involves our PoI perceptual error metrics, while the second stage involves an optimal transmission strategy, unequal error protection coding scheme.

The proposed mesh streaming in Ubiwave consists of the following three steps which described in the following subsection.

Streaming Generation

Streaming data of the model were generated offline in a preprocess stage in the server. The streaming data has two features: (1) the availability of finer granularity, which can provide a more flexible data organizing structure during transmission; and (2) the remarkable reduction of the size of the base mesh and refinement data, which can dramatically decrease the transmission time.

Figure 36 shows the transmission time for meshes and coefficient files of bunny model with wireless network speed 11Mbps. It can be observed that the time required to transfer time coefficient files is significantly less than the transfer time for the actual mesh. This demonstrates that the use of wavelets to encode meshes can save transmission time and network bandwidth.

Figure 37 shows the transmission time for images and coefficient files with wireless network speed 11Mbps. Again, it can be observed that the time required to transfer coefficient files

Figure 35. The proposed mesh streaming technique in UbiWave

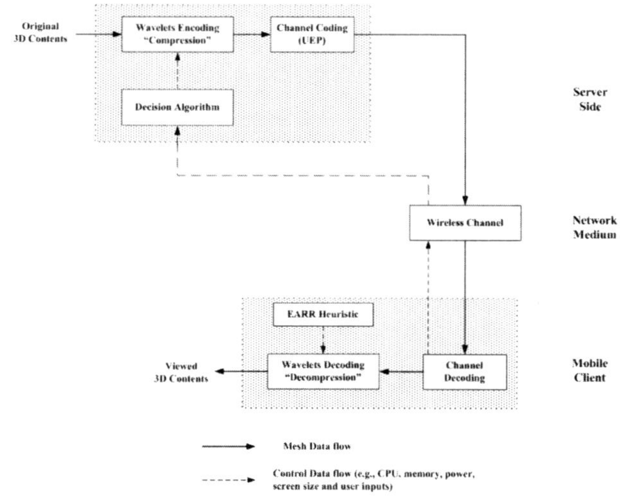

is significant less than the transfer time for the actual images. This demonstrates that the use of wavelets to encode images can save transmission time and network bandwidth.

In UbiWave, wavelet transform decomposes the 3D mesh into base mesh (structural data) and coefficients (geometric data). The coefficients (geometric data) are then decomposed into different levels. Each level of coefficients is related to one level of detail mesh. After preprocessing, the 3D data is stored as structural and geometric levels. Note that the pre-processing needs to be performed only once offline for a given 3D data stream. All 3D content are initially stored at a server, and mobile devices obtain them through a streaming process from the server.

Server Decision Algorithm

As mentioned, mesh streaming has two stages: the selection of LoD of the meshes and the efficient transmission of selected data. In this section, we describe these two stages in server decision algorithm in detail.

1. Level-of-Detail Selection

The Level of Detail of each mesh is determined on the basis of the three factors: (1) Human perceptual error in different mobile devices; (2) Configurations of mobile device, such as display size, CPU resource and battery energy; (3) Network conditions, such as bandwidth and package loss rates.

The flow of data in our system is illustrated in figure 36. The Level-of-Detail Selection algorithm can be illustrated as three basic steps, as shown in figure 39:

1. Once a mobile device establishes a connection to a server, the server will immediately transmit the base mesh to the mobile device and the configurations of the mobile device are periodically sent to the server. The transmitted configurations information from mobile device is used to determine the level of coefficients to be streamed, which includes the resolution of displayer of mobile device and the current available resources.
2. After the server receives the configurations information of a mobile device, it can calculate the PoI for the mobile device and record the LoD, d_{sent}, which has been sent to the mobile device, starting with base mesh. This information is organized in the following format in the server: [Device ID][Model ID][PoI][Level of Detail]
3. The server monitor the channel information when the configuration information of mobile device is received by the server. In mobile devices, it predicts the real-time rendering time, rt_i for possible acceptable LoD i, d_i within the current available resources and sends them to the server. The server calculates the transmission time, t_{trans} for the

Figure 36. Transmission time of bunny model

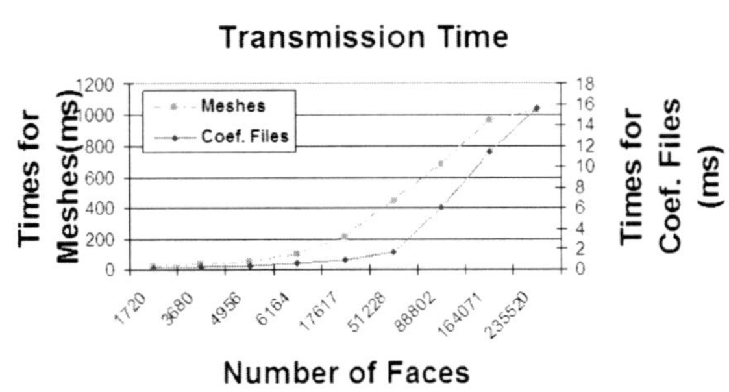

Figure 37. Transmission time for images

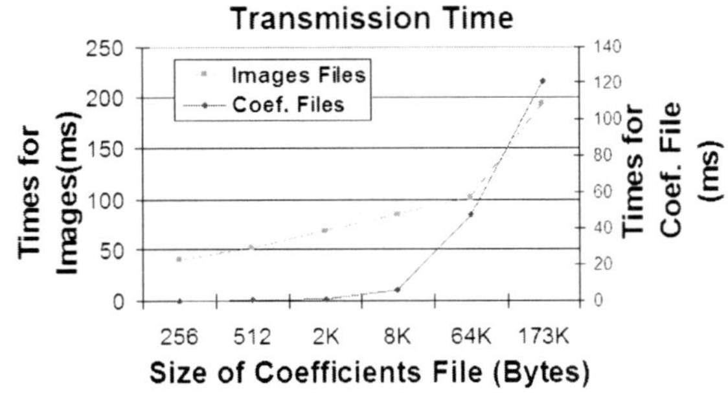

mesh data of LoD i, d_i. Then we have three cases:

1. if d_i is lower than d_{sent}, there is no need to stream it.
2. if d_i is higher than d_{sent}, but the transmission time, $t_{trans.}$ is larger than the real-time rendering time, rt_i, there is pointless to stream it to the mobile device, since mesh data can not be transmitted to the mobile device on time.
3. if d_i is higher than d_{sent}, and the transmission time, $t_{trans.}$ is smaller than the real-time rendering time, rt_i, the difference will be streamed to the mobile device.

Figure 38. The communication process of level of detail selection algorithm

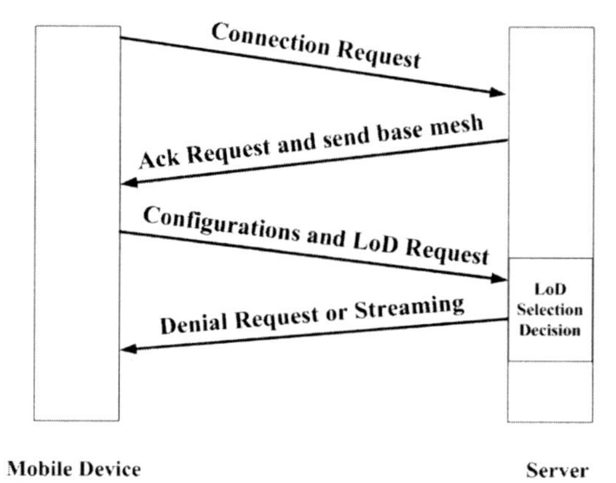

Figure 39 is the flow chart of the Level of Detail selection algorithm.

Note that there is no significant visual error for the Level of Detail above the PoI for the specific model and mobile device. So the highest Level of Detail for a model sent to a specific mobile device is its PoI.

2. Efficient Data Transmission

We then use the UEP coding scheme to protect data from being corrupted. First, to guarantee the same connectivity of the decoded mesh as the original mesh, we assign more FEC bits on the base mesh. Next, we consider the importance of the coefficients and assign more FEC bits. Since the loss of the coefficient data only affects the quality of the decoded mesh and will not make it crash, we can assign less FEC bits to them.

Figure 39. Flow chart of level of detail selection algorithm

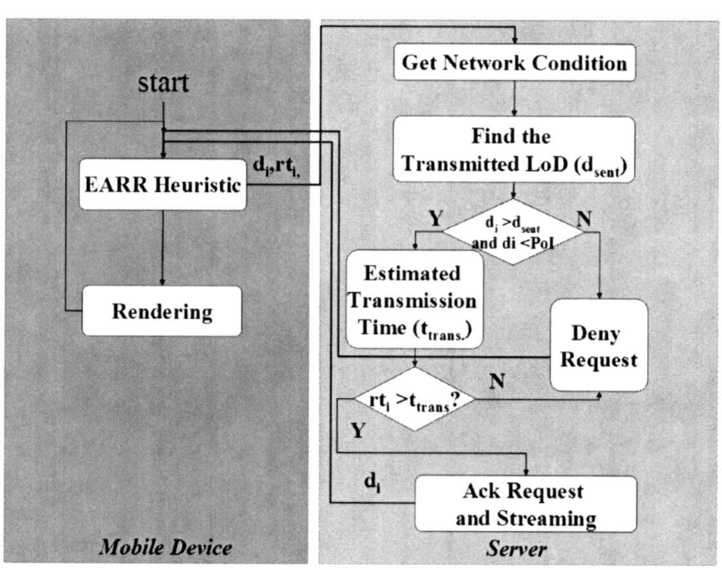

Rendering

Once the connection between mobile device and server is established, the server sends the base mesh to the mobile device. Normally the size of base mesh is small enough for most of the mobile device to render. Then the mobile device predicts the possible acceptable LoD using EARR heuristic and sends the request to the server. If the requested LoD satisfies the Level of detail selector requirement, the server streams the additional requested coefficients to the mobile device. Since the available resources may change, when the coefficients arrive the mobile device, the mobile device decides whether to render the new LoD or not based on EARR heuristic.

Results

The bunny model in the kitchen scene is transmitted over low bandwidth, high error rate wireless channel. We compare the performance without mesh streaming decision technique in terms of rendering speed, image quality and energy consumption. Table 4 summarizes rendering speed, image quality and energy consumption in both wireless network channel.

From this table, we know that the rendering speed and image quality are almost the same since our EARR heuristic will maintain the real-time rendering speed around 25fps. There are two advantages of our approach:

1. **Streaming Latency:** With our streaming technique, the server will not send the requested data to mobile device and deny the request when the network conditions are not good, although the mobile device has enough resources to render the 3D models. The mobile device does not need to wait for the data sending from the server. Our streaming technique achieves a better streaming latency.

Table 4. Performance comparison

	Without our streaming technique	**With our streaming technique**
Rendering Speed	31fps	27fps
Image Quality (faces)	7328	7328
Energy Consumption	**16432 mwh**	**10387 mwh**

2. **Energy Efficiency:** Without our proposed mesh streaming, the server keep sending mesh data with higher LoD. Because of the additional transmission time the mobile device cannot maintain real-time rendering speed, our EARR heuristic will lower the LoD rendered in mobile device automatically. The received data from server with higher LoD is not useful in mobile device. Therefore the transmission energy is wasted. With our proposed mesh streaming technique, if the transmission time is longer than the real-time rendering time, the server will deny the request from the mobile device, and the mobile device will not waste energy receiving data with higher LoD from the server. From the above table, the energy could be saved by 36.8%.

Section Summary

This section presents our wavelet-based energy-efficient streaming technique in UbiWave. Our streaming technique includes three steps: 1) Streaming Generation; 2) Server Decision Algorithm and 3) Rendering in mobile devices. Our streaming technique is useful in wireless network with low bandwidth. It reduces the wasted energy for data transmission. Our experiment results show that Level-of-Detail selection in our steaming technique achieves better streaming latency and saves energy consumption up to 36.8% in low bandwidth wireless networks.

FUTURE WORK

Some possible future work can be extended from UbiWave.

- Even though our Point of Imperceptibility (PoI) error metric works well for meshes, we could make our metric view independent. We propose calculating our PoI metric for each object from multiple view points around the object, and then combines these values. This approach is similar to the image-driven simplification approach of Lindstrom and Turk (2000). We intend to investigate the behavior of the average, minimum and maximum of the PoI calculated from these different views. Texture is another factor which affects on human perception. The future work should consider texture mapping and how it affects on human perception and make the Point of Imperceptibility (PoI) more accurate.
- Texture is another factor which affects on human perception. The future work should consider texture mapping and how it affects on human perception and make the Point of Imperceptibility (PoI) more accurate.
- We analysis the performance of Unequal Error Protection (UEP) scheme and compare the performance with Equal Error Protection (EEP) and None Error Protection (NEP). Comparing the performance of the proposed UEP scheme when applied to wavelet-encoded meshes to UEP on Compressed Progressive Meshes could also be considered.
- We also could investigate the benefits of zero-tree coding. In zero-tree coding, coefficients with values greater than some appropriate threshold value are kept and low-valued coefficients (little information) are replaced by zero.
- Currently, we only did simulations on simple G-E two states Markovian Channel Model. A more complicated channel model, like noise channel model could be applied in the simulations.
- Improve energy saving by integrating Dynamic Voltage Scaling (DVS) and Dynamic Frequency Scaling (DFS). DVS and DFS are popular to be used in graphics hardware. We expect our heuristic will yield further savings after integrating DVS or DFS.
- Improve PoI by integrating eye's gaze pattern. Eye's gaze pattern is another important factor affecting human visual perception. With cues about the eye's gaze pattern, we can increase the LoD of objects that user focuses on while reducing the LoD of objects outside of the focus area. In this way, even more rendering costs can be saved.
- Accurately measuring CPU energy usage. We currently estimate CPU energy usage using a subtractive technique, which can be improved in accuracy. We plan to develop more accurate methods to more accurately measure CPU energy consumption on mobile devices.

CONCLUSION

This chapter presented UbiWave, an end-to-end framework using wavelets to transmit and render graphics content at various resolutions on mobile devices. Ubiwave improves the performance of mobile graphics applications by balancing energy consumption, rendering speed and image quality. Ubiwave includes four parts: 1) a perceptual error metric to guide the scaling of mobile graphics scenes to the lowest LoD at which users do not perceive distortion due to simplification (called the PoI); 2) a novel Forward Error Correction (FEC) scheme based on the principles of Unequal Error Protection (UEP); 3) an Energy-efficient Adaptive Real-time Rendering (EARR) heuristic to balance energy consumption, rendering speed and image quality and 4) an energy-efficient 3D streaming technique. By combining PoI, UEP, EARR and our streaming technique, the rendering speed and image quality of mobile graphics applications in wireless networks can be maximized, while minimizing energy consumption.

In this chapter, our main results were:

- Our Point of Imperceptibility (PoI) error metric accurately picks the lowest acceptable mesh (or image) resolution based on the target mobile device's screen size, which is validated by our user studies. By using our perceptual metric, up to 61% of the total battery energy, can be saved.
- Our Unequal Error Protect Scheme allocates more Forward Error Correction (FEC) bits to important parts of the decoded mesh (parts that show more details). Our UEP scheme performs better than Equal Error Protection and No Error Protection.
- In order to efficiently allot CPU cycles to a running graphics application, its workload needs to be estimated. Our window-based workload predictor can predict the workload of a running graphics application dynamically over time with an error rate of at most 20%.
- Our integrated Energy-efficient Adaptive Real-time Rendering (EARR) heuristic reduces energy consumption by up to 60% while maintaining acceptable image quality at a real-time frame rate of 25 FPS.
- Our energy-efficient 3D streaming technique enables scalable rendering on mobile devices with low streaming latency and maintains real-time frame rates while reducing energy consumption by up to 36%.

REFERENCES

Al-Regib, G., Altunbasak, Y., & Rossignac, J. (2005). An unequal error protection method for progressively transmitted 3D models . *IEEE Transactions on Multimedia, 7*(4), 766–776. doi:10.1109/TMM.2005.850981

Alliez, P., Desbraun, M., (2001). Compression for Lossless Transmission of Triangle Meshes. In Proceeding of SIGGRAPH 2001 (pp. 195–202). Progressive.

Aspert, N., Santa-cruz, D., & Ebrahimi, T. (2002). Mesh:measuring errors between surfaces using the hausdoff distance. In *Proceeding of IEEE Int'l Conf. on Multimedia and Expo* (pp. 705–708)

Bajaj, C., Cutchin, S., Pascucci, V., Zhuang, G., (1998). *Error Resilient Transmission of Compressed VRML* (Tech. Rep.). Austin, TX: TICAM, The University of Texas at Austin.

Bajaj, C., & Schikore, D. (1996). Error-bounded reduction of trianges meshes with multivariate date. *SPIE, 2656*, 34–45.

Banerjee, K., Wu, F., & Agu, E. (2005). Estimating Mobile Memory Requirements and Rendering Time for Remote Execution of the Graphics Pipeline. In*Proceeding of Eurographics 2005.*

Bischoff, S., & Kobbelt, L. (2002). Toward robust broadcasting of geometry data. *Computer Graphics, 26*(5), 665–675. doi:10.1016/S0097-8493(02)00122-X

Bonn, U. (2006). *Bidirectional Texture Function Database Bonn. Retrieved* 2006, from http://btf.cs.uni-bonn.de/index.html

Bradley, J. N., & Brislawn, C. M. (1994). The wavelet/scalar quantization compression standard for digital fingerprint images. In*Proceeding of IEEE International Symposium on Circuits and Systems(ISCAS).*

Chen, B., & Nguyen, M. (2001). Pop: A Hybrid Point and Polygon Rendering System for Large Data, In *Proceeding of IEEE Visualization 2001.*

Chen, B. Y., & Nishita, T. (2002). Multiresolution Streaming Mesh with Shape Preserving and QoS-like Controlling. In *Proceeding of Web3D 2002* (pp. 35-42)

Chow, M. (1997). Optimized geometry compression for real-time rendering. In *Proceeding of IEEE Visualization 97* (pp. 347–354)

Christopoulos, C., Skodras, A., & Ebrahimi, T. (2000). The JPEG2000 still image coding system: an overview. In *Proceeding of IEEE Trans. on Consumer Electronics* (pp. 1103–1127), Vol. 46, Issue 4, Cignini, P., Rocchini, C., & Scopigno, R. (1998). *Metro: measuring error on simplified surfaces.* Computer Graphics Forum, pp. (167–174).

Clarberg, P., Jarosz, W., Akenine-Moller, T., Jensen, H. W., (2005). Efficiently Evaluating Products of Complex Functions. In Proceeding of ACM SIGGRAPH 2005 (pp. 1166–1175). Wavelet Importance Sampling.

Cohen, J., Olano, M., & Manocha, D. (1998). Appearance Preserving Simplification. In *Proceeding of ACM SIGGRAPH 1998* (pp. 115-122).

Cohen, J., Varshney, A., Manocha, D., Turk, G., Weber, H., Agarwal, P., et al. (1996) Simplification Envelopes. In *Proc. of ACM SIGGRAPH 1996.*

Cosman, P. C., Rogers, J. K., Sherwood, P. G., & Zeger, K. (2000). Combined Forward Error Control and Packetized Zero TreeWavelet Encoding for Transmission of Images over Time-Varying Channels. *IEEE Transactions on Image Processing, 9*(6), 982–993. doi:10.1109/83.846241

Data, R. (2006) *Cornell University, Program of Computer Graphics*, Re trieved 2006, from http://www.graphics.cornell.edu/online/measurements/reflectance/index.html.

Debevec, P. (2006). *Paul Debevec's Light Probe Image Gallery.* Retrieved (n.d.)., from http://www.debevec.org/Probes/

Derose, T., Lounsbery, M., & Warren, J. (1997). Multiresolution analysis for surfaces of arbitrary topological type. *ACM Transactions on Graphics, 16*, 34–73. doi:10.1145/237748.237750

Duguet, F., & Drettakis, G. (2004). Flexible Point-Based Rendering on Mobile Devices. *IEEE Computer Graphics and Applications*, (July/August): 57–63. doi:10.1109/MCG.2004.5

Dyn, N., Levin, D., & Gregory, J. A. (1990). *A* butterfly subdivision scheme for surface interpolation with tension control. *ACM Transactions on Graphics*, pp160–pp169.

Flinn, J., deLara, E., Satyanarayanan, M., Wallach, D., & Zwaenepoel, W. (2001). Reducing the energy usage of office applications. In *Proceeding. of Middleware '01.*

Fogel, E., Cohen, D., Revital, I., & Zvi, T. (2001). A Web Architecture for Progressive Delivery of 3D Content. *In Proceeding of ACM Web3D 2001* (pp. 35-41)

Funkhouser, T., & Sequin, C. (1993). Adaptive display algorithm for interactive frame rates during visualization of complex virtual environments. In *Proceeding of ACM SIGGRAPH'93* (pp. 247–254)

Games, M. (2005). *Mobile Games Industry Worth US $11.2 Billion by 2010.* Retrieved 2005 from http://www.3g.co.uk/PR/May2005/1459.htm

Garland, M., & Heckbert, P. (1997). Surface Simplification using Quadric Error Metrics. In*Proceeding of ACM SIGGRAPH 1997* (pp. 209-216)

Gobbeti, E., & Bouvier, E. (1999). Time-Critical Multiresolution Scene Rendering. In *Proceeding of IEEE Visualizatoin* (pp. 123–130).

Graps, A. (1995). A friendly guide to wavelets. *IEEE Computational Science & Engineering, 2*(2).

Gueziec, A. (1999). Locally toleranced surface simplification. *IEEE Transactions on Visualization and Computer Graphics, 5*(2), 168–189. doi:10.1109/2945.773810

Gumbold, S., & Straber, W. (1998). Real Time Compression of Triangle Mesh Connectivity. In *Proceeding of ACM SIGGRAPH 1998* (pp. 133-140).

Hamming, R. W. (1950). Error Detecting and Error Correcting Codes. *The Bell System Technical Journal, 29*, 147–160.

Hoppe, H. (1996) Progressive meshes. In *Proceeding of ACM SIGGRAPH* (pp. 99–108).

Hoppe, H. (1998). Efficient Implementation of Progressive Meshes. *Computers & Graphics, 22*(1), 27–36. doi:10.1016/S0097-8493(97)00081-2

IEEE 802.11 Wireless LAN Standard, 2001.

Khoakovsky, A., Schroder, P., & Sweldens, W. (2000). Progressive geometry compression. In *Proceeding of SIGGRAPH 2000* (pp. 271–278)

Kim, J., Lee, S., & Kobbelt, L. (2004). *View-dependent streaming of progressive meshes.* In *IEEE Trans Circuits and Systems for Video Technology* (2004).

Lalonde, P. (1997). *Representations and uses of light distribution functions*, PhD Dissertation, Vancouver, BC: The University of British Columbia.

Lamberti, F., Zunino, C., Sanna, A., Fiume, A., & Maniezzo, M. (2003). An Accelerated Remote Graphics Architecture for PDAs. In*Proceeding of ACM Web3D 2003* (pp. 55-61).

Lemarie, P., & Meyer, Y. (1986). Ondelettes et bases hilbertiennes. *Rev. Mat. Iberoamericana, 2*, 1–18.

Levoy, M. (2006). *The Digital Michelangelo Project Archive.* Retrieved (n.d.)., from http://graphics.stanford.edu/data/mich/

Lext, J., Assarsson, U., & Moller, T. (2001). *A* Benchmark for Animated Ray Tracing. *IEEE Computer Graphics and Applications, 21*(2), 22–31. doi:10.1109/38.909012

Lindsay, C., & Agu, E. (2005). Wavelength dependent Rendering Using Spherical Harmonics. In *Proceeding of Eurographics* (2005).

Lindstrom, P., & Turk, G. (2000). Image-driven simplification. *ACM Transactions on Graphics, 19*(3), 204–241. doi:10.1145/353981.353995

Liu, X., Shenoy, P., Corner, M., (2005). Application level power management with performance isolation. In Proceeding. of ACM MM'05. Chameleon.

Loop, C. (1987). *Smooth subdivision surfaces based on triangles. Master's thesis.* Salt Lake City, UT: Department of Mathematical, University of Utah.

Lounsbery, M. (1994). *Multiresolution analysis for surfaces of arbitrary topological type.* PhD Dissertation, Seattle, WA: University of Washington.

Luebke, D., & Hallen, B. (2001). Perceptually driven simplification. for interactive rendering. In *Proceeding of Eurographics Rendering Workshop* (pp. 7–18)

Macintyre, B., & Feiner, S. (1998). A Distributed 3D Library. In . *Proceedings of SIGGRAPH, 2005*, 361–370.

Martin, I. M. (2000). ARTE: An Adaptive Rendering and Transmission Environment for 3D Graphics. In *Proceeding of 8th ACM International Conference on Multimedia* (pp. 413-415).

Mohr, A., Riskin, E., & Ladner, R. (2000). Unequal loss protection: Graceful degradation of image quality over packet erasure channels through forward error correction. *IEEE Journal on Selected Areas in Communications, 18*(6), 819–828. doi:10.1109/49.848236

Pajarola, R., & Rossignac, J. (2000). Compressed pregressive meshes . *IEEE Transactions on Visualization and Computer Graphics, 6*(1), 79–93. doi:10.1109/2945.841122

Park, L., Ramamohanarao, K., & Palaniswami, M. (2005). A novel document retrieval method using the discrete wavelet transform. *Trans. on Graphics (TOG), 23*(3).

Pimentel, C., & Blake, I. (1998). Modeling burst channels using partitioned fritchman's markov models. *IEEE Trans Veh Tech, 47*(3), 885–899. doi:10.1109/25.704842

Transmission Control Protocol (TCP), Request For Comment (RFC) 793, Internet Engineering Task Force (IETF), September 1981.

Ramamoorhthi, R., & Hanrahan, P. (2002). Frequency Space Environment Map Rendering. In *Proceeding of ACM SIGGRAPH 2002* (pp. 517-526).

Reddy, M. (1997). *Perceptually modulated level of detail for virtual environments*. PhD Dissertation, Edinburgh, UK: University of Edinburgh, UK.

Reddy, M. (2001). Perceptually optimized 3D graphics. *IEEE Computer Graphics and Applications, 21*(5), 68–75. doi:10.1109/38.946633

Reed, I.S., & Solomon, G. (1960). Polynomial Codes over Certain Finite Fields, *Journal of the Society for Industrial and Applied Mathematics.*

Rohlf, J., Helman, J., (1994). A High Performance Multiprocessing Toolkit for Real-Time 3D Graphics. In Proceeding of ACM SIGGRAPH'94 (pp. 381–395). IRIS Perfromer.

Ronfard, R., & Rossignac, J. (1996). Full-range approximation of triangulated polyhedral. *Computer Graphics Forum, 15*(3), 67–76. doi:10.1111/1467-8659.1530067

Rossignac, J. (1999). Edge breaker: connectivity compression for triangle meshes. *IEEE Transactions on Visualization and Computer Graphics, 5*(1), 47–61. doi:10.1109/2945.764870

Rusinkiewicz, S., Levoy, M., (2000). A Multiresolution Point Rendering System for Large Meshes. In Proceeding of ACM SIGGRAPH 2000 (pp. 343–352). Qsplat.

Schmalstieg, D. (1997). *The Remote Rendering Pipeline*, PhD Dissertation, Vienna University of Technology, Austria.

Schroder, P., & Sweldens, W. (1995). Spherical wavelets: Efficiently representing functions on the sphere. In *Proceeding of ACM SIGGRAPH . Computer Graphics, 29*, 161–172.

Schroeder, W. (1992). Decimation of triangle meshes. In *Proceeding of ACM SIGGRAPH* (pp. 65–70).

Sohn, K., Lee, C., Ryou, J., & Jang, W. (2001). *Error-resilient Zerotree Wavelet Video Coding. SPIE Journal of Optical Engineering*, (pp. 2480-2488).

Stanford 3D Scanning (2006). Stanford 3D Scanning Repository. Retrieved 2006, from http://graphics.stanford.edu/data/3Dscanrep/

Sweldens, W. (1996). The lifting scheme: A custom-design construction of biothogonal wavelets. *Applied and Computational Harmonic Analysis, 3*, 186–200. doi:10.1006/acha.1996.0015

Tack, N., Lafruit, G., Catthoor, F., & Lauwereins, R. (2005). Pareto based optimization of multiresolution geometry for real time rendering. In *Proceeding of ACM Web 3D* (pp. 19–27).

Tack, N., Moran, F., Lafruit, G., Lauwereins, R., (2004). 3D Rendering Time Modeling and Control for Mobile Terminals. In Proceeding of ACM Web3D (pp. 109–117). Synposium.

Tamai, M., Sun, T., Yasumoto, K., Shibata, N., & Ito, M. (2004). Energy-aware video streaming with QoS control for portable computing devices. In *Proceeding of ACM NOSSDAV'04* (pp. 68–73).

Teller, S. (1992). *Visibility Computations in Densely Occluded Polyhedral Environments.* PhD Dissertation.

Tobagi, F. A., Binder, R., Leiner, B., (1984). Packet Radio and Satellite Networks (pp. 24–40). IEEE Communications.

Touma, C., & Gotsman, C. (1998). Triangle Mesh Compression. In *Proceeding of Graphics Interface* (pp. 26-34).

Valette, S., & Prost, R. (2003). Wavelet-based progressive compression scheme for triangle meshes: Wavemesh. *IEEE Transactions on Visualization and Computer Graphics, 10*(2).

Valette, S., & Prost, R. (2004). Multiresolution analysis of irregular surface meshes. *IEEE Transactions on Visualization and Computer Graphics, 10*, 113–122. doi:10.1109/TVCG.2004.1260763

Wang, X., Silva, F., & Heidemann, (2004). J. Demo abstract: Follow-me application—active visitor guidance system. In *Proceedings of the 2nd ACM SenSys Conference*.

Williams, N., Luebke, D., Cohen, J., Kelley, M., Schubert, B., (2003). Perceptually guided simplification of lit, textured meshes. In Proceeding of Interactive 3D (pp. 113–121). Graphics.

Winmmer, M., & Wonka, P. (2003). Rendering time estimation for Real-Time Rendering. In *Proceeding of the Eurographics Symposium on Rendering* (pp. 118–129).

Wu, F., Agu, E., (2006). Unequal Error Protection for Wavelet-Based Wireless Mesh Transmission. Boston, MA: ACM SIGGRAPH.

Wu, F., Agu, E., & Lindsay, C. (2007). Pareto-Based Perceptual Metric for Imperceptible simplification on mobile displays. In *Proceeding of Eurographics 2007*, Prague, Czech Republic.

Wu, F., Agu, E., & Lindsay, C. (2008), Adaptive CPU Scheduling to Conserve Energy in Real-Time Mobile Graphics Applications. In *Proceeding of ISVC 2008*, Las Vegas, NV.

Wu, F., Agu, E., & Ward, M. (2006). Multiresolution Graphics on Ubiquitous Displays using Wavelets. *International Journal of Virtual Reality, 5*(3).

Yan, Z., Kumar, S., & Kuo, C. (2001). Error resilient coding of 3-D graphic models via adaptive mesh segmentation. *IEEE Transactions on Circuits and Systems for Video Technology, 11*(7), 860–873. doi:10.1109/76.931112

Yang, C. K., & Chiueh, T. (2005). An Integrated Pipeline of Decompression, Simplification and Rendering for Irregular Volume Data. In *Proceeding of 4th International Workshop on Volume Graphics* (pp 147-237)

Yang, S., Kim, C., Kuo, C., (2004). A progressive view-dependent technique for interactive 3D mesh transmission. In IEEE Trans. Circuits and Systems for Video Technology.

Yuan, W., & Nahrstedt, K. (2004). Practical voltage scaling for mobile multimedia device. *In Proceeding Of ACM MM'04* (pp.924–931).

Zunino, C., Lamberti, F., Sanna, A., Montrucchio, B., (2002). A Wireless Architecture for Performance Monitoring and Visualization on PDA Devices. In Proceeding of SCI 02 (Vol. XV, pp. 143–148). Proceedings.

Chapter 9
Peer-to-Peer Service Sharing on Mobile Platforms

Maria Chiara Laghi
University of Parma, Italy

Michele Amoretti
University of Parma, Italy

Gianni Conte
University of Parma, Italy

ABSTRACT

True ubiquitous computing requires peer-to-peer service sharing on mobile platforms, with application entities communicating and providing services to each other and to users. In order to enforce this paradigm to devices with limited processing and storage resources, lightweight middleware components are required. In this chapter, we define a theoretical model for autonomic and altruistic computational entities, and we use it to build a framework for peer-to-peer service-oriented infrastructures, focusing on three key aspects: overlay scheme, dynamic service composition and self-configuration of peers. Based on this framework, JXTA-SOAP Mobile Edition is a software component that completes the Sun MicroSystem's JXTA platform, supporting peer-to-peer sharing of Web Services.

INTRODUCTION

The emergence of compact albeit powerful devices is giving users the ability to access, anytime and anywhere, globally available applications. For challenging contexts such as ambient intelligence and emergency management, requiring highly efficient, pervasive and dependable solutions, we envision a synergetic approach based on ubiquitous computing models and service-oriented technologies. Moreover, to improve scalability, we support the shift from traditional client/server architectures to systems based on the peer-to-peer (P2P) paradigm, completed by the self-organization and the self-adaptation principles.

The peer-to-peer paradigm enables two or more entities to collaborate spontaneously in a network of equals (peers) by using appropriate information and communication systems without the necessity for central coordination. Furthermore, a peer-to-peer system is a complex system, because it is composed of several interconnected parts that as a whole ex-

DOI: 10.4018/978-1-61520-761-9.ch009

hibit one or more properties (i.e. behavior) which cannot be easily inferred from the properties of the individual parts.

At the beginning of the P2P era, Barkai (2002) proposed the following requirements for a general-purpose P2P middleware:

- portability
- interoperability
- security
- local autonomy
- persistence
- scalability
- extensibility

With these objectives in mind, in recent years some researchers have focused on designing robust overlay schemes (with respect to bootstrapping, connectivity, message routing) and distributed security / trust mechanisms, while others have targeted application-specific problems. Next step is to create decentralized and self-organizing infrastructures, being able to provide services to users according to their availability and the network status, and also supporting the spontaneous creation of services provided by heterogeneous nodes, such as mobile devices interacting through ad hoc connections without any prior planning (Gaber, 2007).

This chapter introduces the *Networked Service-oriented Autonomic Machine (NSAM)*, which is a theoretical model of an hardware/software entity that is programmed to be altruistic in sharing its resources. In particular, we focus on special kinds of resources, i.e. services, offered to and by mobile devices. Based on NSAM, we present a framework for peer-to-peer service sharing, based on three key aspects: overlay scheme, dynamic service composition and self-configuration of peers.

In section 2 we provide a survey on mobile devices and platforms. In section 3 we focus on peer-to-peer service-oriented infrastructures, discussing design issues, defining the NSAM theoretical model, and illustrating the formal framework. In section 4 we illustrate its implementation in the mobile edition of JXTA-SOAP, a software component that completes JXTA middleware in order to support peer-to-peer sharing of Web Services in mobile networks. In section 5 we illustrate the objectives for future work. Finally, in section 6 we conclude the chapter with a summary and a discussion of the achieved results.

BACKGROUND

Due to digital convergence, mobile industry is facing a significant disruption in these years. Multifunctional products are emerging for consumers, and diversification is introducing a new set of requirements for architectures and platforms, such as flexibility, scalability and modularity. Mobility is considered a strategic component of enterprise business, and deploying mobile applications provides great productivity improvements. Mobility is complex, because it involves multiple back-end systems, some legacy, some newly deployed, and a collection of mobile devices with an increasing number of mobile operating systems (BlackBerry OS, Windows Mobile, Symbian OS, Mac OS X, Palm OS, Android and mobile Linux). A great variety of wireless technologies is also available in a global workplace, from current cellular networks with CDMA and GSM standards, to WiFi, WiMax, and future next-generation 4G networks.

Mobile Platforms

Most portable devices (PDAs, smart phones, digital media and music players, handheld gaming units, and calculators) are built on ARM[2], a 32-bit RISC processor architecture developed by ARM Limited that is widely used in embedded designs. Because of their power saving features, ARM CPUs are dominant in the mobile electronics market, where low power consumption is a critical design goal. Prominent branches in this family include Marvell's XScale, the ST-Ericsson's

NOMADIK series and the Texas Instruments OMAP (Open Multimedia Application Platform), a proprietary microprocessor for multimedia applications, used by many mobile phones (i.e. Nokia's N-series).

An alternative to ARM architecture is AMD Geode, a series of x86-compatible System-on-a-chip microprocessors and I/O companions targeted at the embedded computing market. Geode processors are optimized for low power consumption and low cost while still remaining compatible with software written for the x86 platform. The processor family is best suited for thin client, set top box, tablet PCs and embedded computing applications, and is typically found in industrial control systems.

Finally, INTEL Atom is a line of x86 and x86-64 CPUs intended for use in MIDs, smart phones and ultra-mobile PCs meant for portable and low-power applications. Actually, it is the most diffused processor for netbooks.

Mobile enterprise platforms are developing to build and deploy mobile applications, and to keep a consistent synchronization between back-end sources and applications on the mobile device. An extensible platform can speak to the variety of applications, devices, and wireless technologies and at the same time deliver key management and security components and meet the demand and expectations of both IT and the mobile user. Different platforms and operating systems are available for mobile devices, most of them offering integrated development environment and emulators. Examples are BlackBerry[3], Windows Mobile[4], Symbian[5], LiMo[6], Android[7].

In addition to the development platforms offered by these OSs, mobile applications can be implemented using toolkits like J2ME, .NET or BREW. Java ME (commonly referred to by its previous name: Java 2 Platform, Micro Edition or J2ME), designed by Sun Microsystems, is a specification of a subset of the Java platform aimed at providing a certified collection of Java APIs for the development of software for resource-constrained devices; the Microsoft .NET Compact Framework (.NET CF) is a version of the .NET Framework that is designed to run on Windows CE based mobile/embedded devices such as PDAs, mobile phones, factory controllers, set-top boxes, etc. BREW (Binary Runtime Environment for Wireless) is an application development platform created by QUALCOMM for mobile phones. It was originally developed for CDMA handsets, but has since been ported to other air interfaces including GSM/GPRS. BREW is a software platform that can download and run small programs for playing games, sending messages, sharing photos. The main advantage of BREW platforms is that the application developers can easily port their applications between all supported devices.

Service-Oriented Architectures on Resource-Constrained Devices

Besides hardware constraints, mobile devices introduce many other specific challenges which make difficult the deployment of Web Services on top of them (Berger, 2003). Unlike dedicated servers, mobile devices will typically have intermittent connectivity to the network. As a result, the services offered on a mobile device may not be accessible all the time. An application that uses or composes such Web Services needs to operate in an opportunistic manner, leveraging such services when they become available. On the server side, Web Services on mobile devices should also attempt to keep messages as short as possible. Another issue to be addressed is the change of IP address which may arise when a mobile device moves between different locations, and from one administrative domain to another. However, with the P2P in place, the need for the Public IP can be eliminated and the mobiles can be addressed with unique peer ID. Each device in the P2P network is associated with the same peer ID, even though the peers can communicate with each other using the best of the many network interfaces supported by the devices like Ethernet, WiFi, etc. (Srirama, 2006).

Since the WS message protocol, namely SOAP, introduces some significant overhead, few toolkits support the deployment of Web Services on limited devices, such as PDAs, smart phones, etc. One is gSoap (van Engelen, 2002), which provides a WS engine with run-time call de-serialization. Unfortunately, gSoap is written in C/C++, thus requiring a priori stub/skeleton generation by means of a specific compiler, which also means lack of portability.

.NET Compact Framework[15] is a subset of the .NET platform, targeting mobile devices. Its class library enables the development of Web Service clients, but does not allow to host Web Services.

Looking at the Java Micro Edition (J2ME) platform, most libraries are only for client side functionality. The Java Wireless Toolkit (WTK) provides J2ME Web Services API (WSA)[16], based on JSR 172[17], which specifies runtime ServiceProvider interface to allow the generation of portable stubs from WSDL files. The specification contains some notable limitations, most of them due to the requirement for WS-I Basic Profile compliance. Conforming to the profile ensures interoperability, but also prevents using alternative methods. Another widely used solution is the kSoap2[18] open source component, which is a parser for SOAP messages (with RPC/literal or document/literal style encoding), not supporting the generation of client side stubs. kSoap2 is compliant with devices lacking JSR 172 support, and allows to access non WS-I conformant services.

To the best of our knowledge, the unique solution enabling J2ME applications (CLDC, CDC) as service endpoints is the Micro Application Server (mAS)[19]. It can be considered a lightweight version of Axis, by which it is inspired. For this reason we have chosen it to implement the J2ME version of JXTA-SOAP.

P2P SERVICE-ORIENTED INFRASTRUCTURES

In a ubiquitous computing environment, a service-oriented infrastructure must be enabled with service discovery protocols (SDPs) to find the most appropriate services, either upon direct request from the users or proactively. Moreover, mobility and resource scarcity introduce two dimensions that service-oriented infrastructures for wired networks don't take into account: location awareness and physical proximity between the service provider and the user. In a broader vision, to find the most appropriate services, the service-oriented infrastructure should exploit context.

To improve decentralization, scalability, robustness, and to avoid single points of failure, the peer-to-peer paradigm is a viable solution for such advanced service-oriented infrastructures. In contrast with the client/server approach, in which resource providers and resource consumers are clearly distinct, peers usually play both roles. The key concept of the peer-to-peer paradigm is leveraging idle resources to do something useful, like cycle sharing or content sharing.

Design Challenges

The operation of any peer-to-peer system relies on a network of peer software/hardware nodes, and connections (links) between them. This network is formed on top of - and independently from - the underlying physical computer (typically IP) network, and is thus referred to as an *overlay network*. The topology, structure, and degree of centralization of the overlay network, and the message routing and location mechanisms it employs for messages and resources are crucial to the operation of the system, as they affect its scalability, security, fault tolerance, and self-maintainability.

The scalability of an overlay scheme measures its effectiveness and efficiency, with respect to the target application(s), when applied to large situations (e.g. large workloads or large number of participating nodes). A distributed system should be inherently more scalable if using the P2P paradigm, rather than the client/server approach. But some P2P overlay schemes scale better than others, with respect to resource discovery effectiveness and performance, bandwidth occupancy, etc. For example, a message routing protocol is considered scalable with respect to network size, if the number of message propagations that are necessary to find a resource grows as *O(log N)*, where *N* is the number of nodes in the network.

The effectiveness and efficiency of P2P protocols for resource sharing usually depends on how peer are connected. One key operation is *bootstrapping*, the initial discovery of other nodes participating in the network. Nascent peers need to perform such an operation in order to join the network. Bootstrapping usually includes operations needed to repair overlays that have split into disconnected subgraphs (GauthierDickey & Grothoff, 2008). Another important operation is *connectivity management*, i.e. the maintenance of connections or exchange of topology information for peers that are already connected to the network at large.

Search performance and *consistency* are two important measures for the sharing of dynamic contents (e.g. in P2P storage systems). Search performance concerns how fast the users locate and obtain copies of requested resources (time complexity) and how many nodes must be involved in that process (space complexity). Consistency concerns how old the acquired data (resource descriptions or shared content) are with respect to the actual available resources. There are several studies on *replication strategies* to improve the search efficiency of unstructured P2P networks (Lv, 2002). To optimize network-wide search performance given limited storage capacity, more replicas are preferred for more frequently accessed objects. The search time and traffic under random walk search is minimized when the number of replicas for each object is proportional to the square root of its query rate (Cohen & Shenker, 2002). With controlled flooding search, the search traffic is minimized under the same square root replica distribution, whereas the search time is minimized when the number of replicas for each object is linearly proportional to its query rate (Tewari & Kleinrock, 2006). However, none of the above work has considered keeping the replicas consistent with the authoritative contents. In general, there are two classes of methods to maintain *consistency*: push-based and pull-based. In push-based methods, the content owners keep track of the replica locations and send invalidation messages or updated contents to the replicas whenever the contents are modified. In contrast, pull-based methods are replica-driven. The replicas, when considered outdated, are validated before serving new requests. A recent work (Tang, 2008) proposes to assign each replica an expiration time (time-to-live, TTL) beyond which the replica stops serving new requests unless it is validated.

Peer-to-peer architectures present a particular challenge for providing high levels of *availability*, *privacy*, *confidentiality*, *integrity*, and *authenticity*, due to their open and autonomous nature. Network nodes cannot be considered trusted parties, and no assumptions can be made regarding their behavior. Preserving integrity and authenticity of resources means safeguarding the accuracy and completeness of data and processing methods. Unauthorized entities cannot change data; adversaries cannot substitute a forged document for a requested one. Privacy and confidentiality mean ensuring that data is accessible only to those authorized to have access, and that there is control over what data is collected, how it is used, and how it is maintained. A malicious node might give erroneous responses to requests, both at the application level, returning false data, or at the network level, returning false routes and partitioning the network. Moreover, the P2P system must be robust against a conspiracy of a *malicious collective*, i.e. a group of nodes acting in concert to attack reliable ones. Attackers may have a number of goals, including traffic analysis against systems that try to provide anonymous communication, and censorship against systems that try to provide high availability.

Security attacks in P2P systems can be classified into two broad categories: passive and active (Govoni, 2002). Passive attacks are those in which the attacker just monitors activity and maintains an inert state. The most significant passive attacks are *eavesdropping*, which involves capturing and storing all traffic between some set of peers searching for some sensitive information (such as personal data or passwords), and *traffic analysis*, where the attacker not only captures data but tries to obtain more information by analyzing its behavior and looking for patterns, even when its content remains unknown. In active attacks, communications are disrupted by the deletion, modification or insertion of data. The most common attacks of this kind are: *spoofing*, in which one peer impersonates another; *man-in-the-middle*, where the attacker intercepts communications between two parties, relaying messages in such a manner that both of them still believe they are directly communicating; *playback or replay*, in which some data exchange between two legitimate peers is intercepted by the attacker in order to reuse the exact data at a later time and make it look like a real exchange; *local data alteration*, which goes beyond the assumption that attacks may only come from the network and supposes that the attacker has local access to the peer, where he can try to modify the local data in order to subvert it in some malicious way.

There are several other issues that potentially can hinder the deployment of large-scale P2P applications. For example, asymmetric bandwidth in the access network, in particular the uploading capability of each peer, can become a bottleneck in the system. This can significantly impact ISP (Internet service providers) and how ISP perform traffic dimensioning. Moreover, a large portion of the Internet bandwidth is occupied by P2P applications, where many ISP have enforced traffic engineering mechanisms, in particular for inter-domain traffic. For file sharing, this implies considerable slowdown in performance; but for streaming applications, this can be fatal. Finally, NAT and firewalls can impose fundamental limitations on the pair-wise host connectivity in the overlay network. It is well-known that a significant portion of broadband users experience NAT or firewall problems, and this requires particular attention.

NSAM Model

A *Networked Service-oriented Autonomic Machine (NSAM)* is a theoretical model of an hardware/software entity that is programmed to be completely altruistic, providing atomic services to other NSAMs, and cooperating with other NSAMs to build composite services. A system of NSAMs is a peer-to-peer system, in which each node can act both as service consumer and service provider, and contributes to the effective and efficient functioning of the whole system.

NSAMs can be of different types and complexities, depending on the device and on the characteristics of the offered services. Several kinds of devices are considered: PCs and workstations, notebooks, PDAs, smart-phones, as well as sensors and actuators. Devices can be classified on the basis of their system characteristics (OS, processor type, memory, I/O type, battery, connectivity) or their functionalities (camera, communicating, processing, sensors...). The software layer of each NSAM includes a lightweight control module (implementing a peer-to-peer overlay scheme) and services.

Formally, a NSAM node is a tuple

$$NSAM = \langle URI, CTRL, R \rangle \quad (1)$$

where *URI* is a unique identifier, *CTRL* is the control layer (modeled, for example, as a finite state machine), and *R* is a set of resources.

Resource attributes describe the device characteristics; some of them may change with time, others are fixed:

Figure 1. NSAM basic ontology

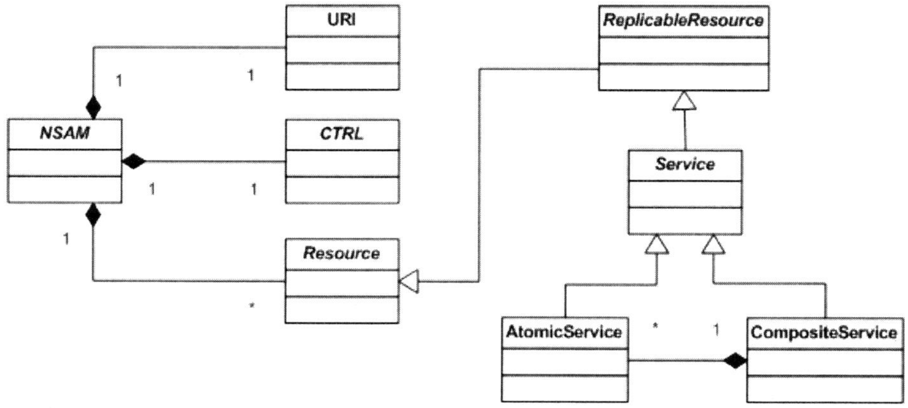

$R = \{r_1, r_2, ..., r_n\}$ (2)

Example of resource:

r_1 = device HW = <CPU, memory(t), battery(t), connectivity>

representing the hardware of the device. Of course, resources can be defined with finer granularity. Each resource property has a name and a range. In the example, battery is represented by a percentage, connectivity is a string in {wired, wireless} or in a more rich enumeration, CPU is an integer value, etc. Some properties are time-dependant.

A service $s \in S$ is a resource consisting in a unit of work executed by a service provider to achieve the results desired by a service consumer. Formally, a service is a tuple

$s = <I, O, P, E>$ (3)

where I is a set of input parameters, each one being characterized by type and semantics, i.e. for each $i \in I$, *type(i)* and *sem(i)* are defined. The O set includes the output parameters of the service. They also have associated type and semantics *type(o)*, *sem(o)*. It is important that service consumers and service providers share the same domain ontologies in order to have a common understanding of shared services. Semantic descriptions of services are used to organize service advertisements in centralized or distributed repositories, allowing to efficiently retrieve and use services in the NSAM network. P and E are the precondition and effect sets, respectively. Such optional parameters are expressed in the form of logical conditions which can assume the true or false value. Preconditions must be verified in order to invoke the service, while an execution effect may become a precondition for the successive invocation in a composition scenario. For example, in an ambient intelligence scenario, if we need a service that assigns the value "*ON*" to the "*status*" property of a "*living room light*", we specify an invocation effect very precisely.

An atomic service is defined as the minimal executable function unit, that cannot be decomposed and whose execution can transform a given state to another state. It is represented as a tuple:

$a = <I, O, P, E, Q>$ (4)

where Q is the set of quality of service attributes, depending on the device characteristics and on the amount of resource required to process inputs and generate outputs. Each node can provide different atomic services. The number of concurrent service instances and the quality of service (QoS) of each instance at a certain time depends on the current availability of hardware resources on the node.

Atomic services provided by different peers can be statically or dynamically aggregated (proactively or on-demand) to realize new complex tasks. A composite service is a tuple:

$$c = <I, O, P, E, Q, G_w> \quad (5)$$

where G_w is the rule that allows to combine atomic services; this rule is represented as a directed workflow graph

$$G_w = <S, L_w> \quad (6)$$

where S is a set of services (both atomic and composite) and L_w is a set of links that represent transitions (i.e. *I-O* connections) among services.

Framework for P2P Service-Oriented Infrastructures

The NSAM model is particularly suitable to characterize service-oriented peers, interacting with complex environments (figure 2). In this framework, among the resources of the peer (R set, according to the NSAM model), functional modules are implemented as services. For example, the overlay scheme mechanisms are implemented as atomic services that each NSAM runs. Considering a group of NSAMs, the peer-to-peer interaction of their overlay services leads to the emergence of a composite overlay service. Self-configuration works similarly, with atomic services that adapt the configuration of the NSAM, based on information that in general is both local and external, for which a composite service spanning the whole network drives a global adaptation process. Thus, service composition mechanisms are embedded in the implementation of atomic services. A detailed discussion is postponed to section 3.3.2.

The *CTRL* component of the NSAM is basically a lightweight resource manager, that configures the NSAM at startup, deciding which resources must be run, and manages their runtime allocation. In particular, it defines and implements the instantiation policy of atomic services that are requested by multiple consumers at the same time.

Overlay Scheme

In section 3.1 we introduced the concept of overlay network, as one of the distinguishing features of P2P systems. The *overlay scheme* defines how

Figure 2. The structure of a service-oriented peer, supporting ubiquitous computing for mobile users in highly dynamic and heterogeneous environments

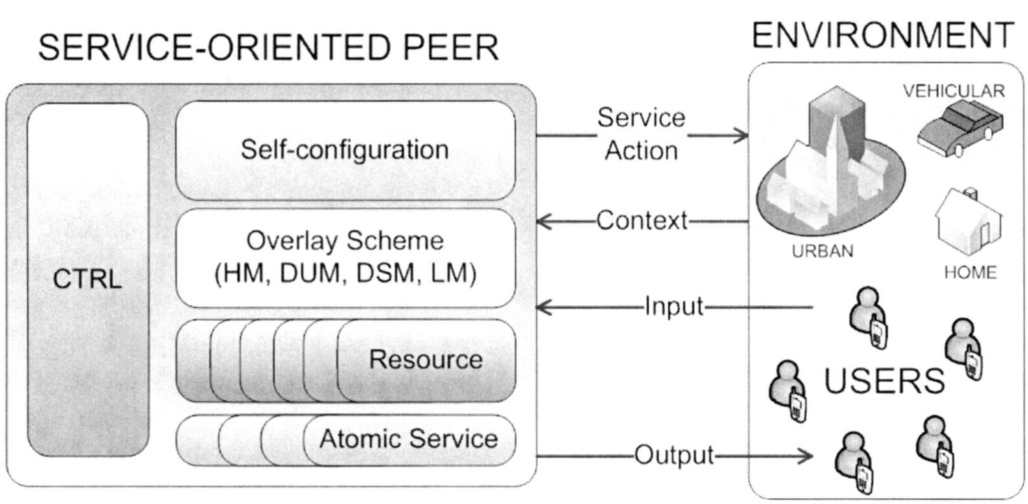

peers are connected, how messages are propagated among nodes to share resources and information about them, and which security mechanisms are adopted.

In our opinion, *the placement of information about shared resources* plays an important role in the characterization of an overlay scheme. Information about shared resources can be:

- published to a central server
- or published to other peers
- or locally stored by resource owners and not published

The first approach leads to *hybrid overlay schemes* (based on the **Hybrid Model - HM**), so called because they are based on the client/server paradigm in resource publication and discovery, while the peer-to-peer approach is used for resource consumption. Centralized servers can also be used to support trust among peers, for example by playing the role of Certification Authorities (CAs) (Amoretti, 2005).

The other approaches lead to *decentralized overlay schemes*, only relying on local information available at each node (such networks are often referred as "pure" P2P systems). Decentralized P2P systems can be divided in two groups, depending on the topology awareness of peers. A decentralized P2P overlay is *unstructured* (based on the **Decentralized Unstructured Model - DUM**) if links among peers (being them actual or potential connections) can be represented by a random graph, whose characteristics are unknown to the peers, and not relevant to their message routing strategies. On the contrary, a decentralized P2P overlay is *structured* (based on the **Decentralized Structured Model - DSM**) if its topology is controlled and shaped in a way that resources (or resource advertisements) are placed at appropriate locations.

To improve the performance (with respect to scalability, lookup performance and stability) of P2P networks, *layered overlay schemes* (based on the **Layered Model - LM**) have been studied and implemented (Garces-Erice, 2003; Peng, 2007). Such overlays are characterized by interacting layers, each one being organized according to one of the "flat" models (HM, DUM, or DSM).

Dynamic Service Composition

In a service-oriented infrastructure, user and application requests typically need to combine the functionality of several services and resources spread over the networked environment. The mechanism of combining two or more services together to form a complex service is known as service composition. Typically, a service composition system accepts a complex user task as an input and attempts to meet the needs of the task at hand by appropriately matching the task requirements with the available services. Such composite services enable users (applications) to reach their goal without having to discover and coordinate among a number of services on their own.

Service composition is highly desirable in peer-to-peer (P2P) systems where application services are naturally dispersed on distributed peers. However, it is challenging to provide high quality and failure resilient service composition in P2P systems due to the decentralization requirement and dynamic peer arrivals/departures. Moreover, in pervasive computing environments peers are hosted on a number of devices with heterogeneous functionality sets. In the presence of such variety, it is desirable to dynamically combine available basic services (as building blocks) to create composite services.

Dynamic composition mechanisms built using *graph techniques* (Kalaspur, 2007) provide support to user tasks in the face of dynamic challenges such as heterogeneity, resource restrictions, user and resource mobility, locality of service provisioning, and so forth. It is also necessary to dynamically capture information regarding the state of a device while the device is operational. Such a dynamic mechanism will ensure uniform

Figure 3. The three network layers: P2P service overlay network, peers' overlay network, and physical network

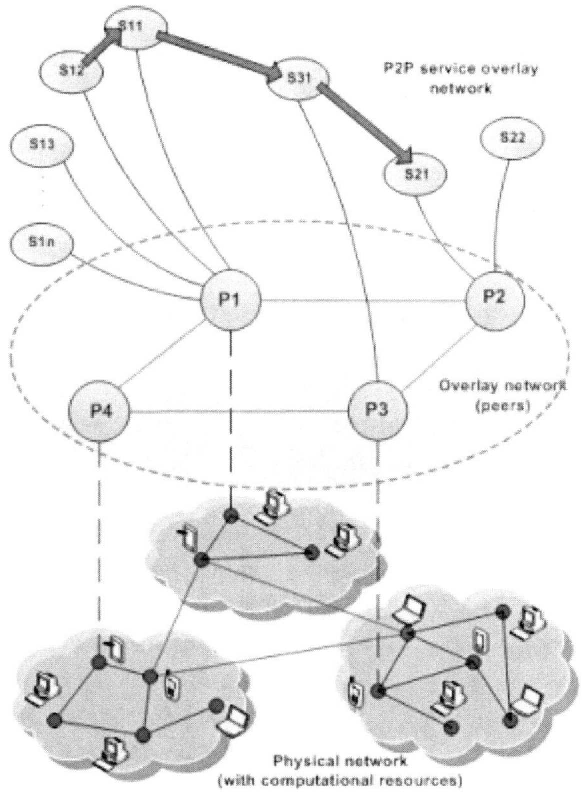

resource consumption, timely support, and fairness in resource utilization. A **P2P service overlay network** (figure 3) may be defined, over which service consumers send requests to service providers and new services can be flexibly composed from available service components based on the user's function and quality-of-service (QoS) requirements. However, in general, mobile devices still have difficulties in fully satisfying users' requirements, due to shortcomings in system resources, especially limited battery life. Restrictions in battery capacity prohibit the use of fully functional applications for satisfactory durations. In addition, the mobile computing environment requires applications to adapt dynamically to their context, including the user's role, capability, and current environment, while maintaining the constant functionality of applications. When an application invokes a complex task that can be performed by a combination of services, that application is resolved to a service composition or a service flow that is represented as a service composition graph. QoS control for applications running in a mobile peer must be invoked in a way that does not exhaust the resources of the device, including residual battery energy.

The metadata that represents the service includes descriptions about service capabilities. By employing semantics, formal declarative descriptions are attached to services. Semantic descriptions of services are used to organize services in a repository, retrieve the appropriate services and use them correctly. A domain ontology may used to conceptualize domain knowledge with commonly accepted vocabulary and to provide semantics to service descriptions. The syntactic parameters of a service define input, output, QoS parameters, pre-conditions and post-conditions, if present.

According to the NSAM model, a composite service is defined as aggregation of atomic and composite services (recursion). This allows the definition of increasingly complex applications by progressively aggregating components at higher levels of abstraction.

Creating a complex process requires not only a clear definition of collaboration patterns of all its components, but also a way of depicting service interactions. Task resolution is performed firstly deriving several different compositions at the semantic level, then identifying the underlying services that can take part in the composite results. Service composition mechanisms are classically treated as extensions to service discovery strategies (that are usually implemented as atomic services hosted by all peers). Service discovery is achieved by matching service requests with the ontology-based service descriptions of shared services. According to the proposed NSAM model, I and O attributes are used as parameters for discovery mechanisms. When a peer receives a service request that cannot process by itself, either partially or completely, it searches for other peers able to process the request. For this reason, it should be able to locate peers that provide any type of service and to send messages to a fraction of its neighbours in order to propagate the requests.

The requirements for service composition are that the output produced by a service S_1 can be consumed by S_2, i.e. for each $o \in O_1$, $\exists\, i \in I_2$ so that *type(o)= type(i)* and *sem(o)=sem(i)*.

Figure 4 illustrates an example in which a service with input set I and output set O is composed using two alternative, semantically matching, flows:

S1 → S2 since I≡I1, O1=I2, O2=O

S1 → S3 → S4 since I≡I1, O1≡I3, O3≡I4, O4≡O

The quality parameter Q in the definition of the requested service is used to select a composition among all the possibilities, and to stop the discovery process when at least a composition with the required quality of service is discovered.

A service providing system is considered self-adaptive if it can dynamically adjust its service structure so as to reflect the changing demand and improve user's satisfaction. To make a system self-adaptive an effective coordination mechanism can be created in which peers are considered to be cooperative in nature.

Some NSAMs may also act as orchestrators for service composition, offering a service that

Figure 4. Example of service compositions

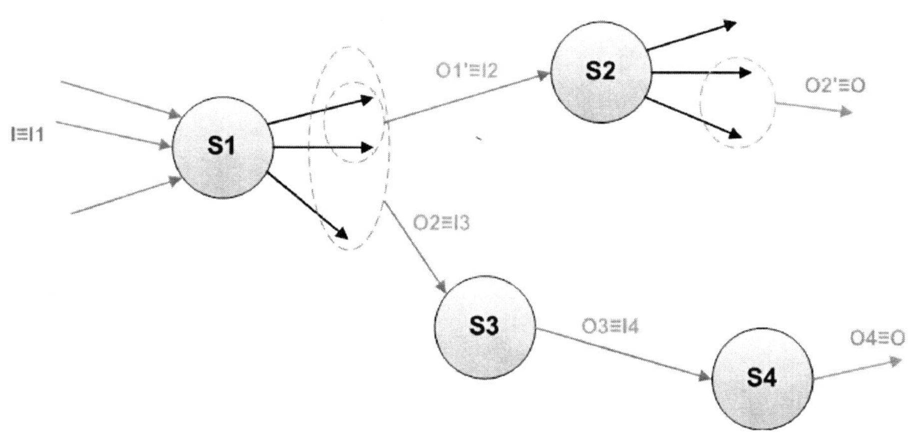

collects service information (by triggering discovery processes), and creates a combination of available services that meet user requirements. Such coordinators manage both the discovery process and the service invocation once a satisfying composition is found.

Finally, one of the major challenges in pervasive computing applications is the issue of mobility. In any pervasive computing environment, once the initial composition is identified and a service session is established, the mobility of the peer can change the composed solution. In such situations, the challenge is to reconfigure the session under progress as quickly as possible by considering the current resource availability around the user. Within the P2P service overlay, it is possible to ensure that the request can be recomputed with minimal interruption of the session under progress. The effect of user mobility while a service session is in progress can lead to a complete dynamic recomposition of the service.

Self-Configuration of Peers

A peer-to-peer system is a complex system, because it is composed of several interconnected parts (the peers) that as a whole exhibit one or more properties (i.e. behavior) which are not easily inferred from the properties of the individual parts. The reaction of a peer to direct or indirect inputs from the environment is defined by its internal structure, which can be either based on static rules shared by every peer (protocols), or on an adaptive plan which determines successive structural modifications in response to the environment, and turns the P2P network in a complex adaptive system (CAS).

Many considerable peer-to-peer protocols have been recently proposed. They can be grouped in few architectural models, taking into account basically two dimensions: the dispersion degree of information about shared resources (centralized, decentralized, hybrid), and the logical organization (unstructured, structured). The behavior of a peer-to-peer system based on protocols follows a pre-established pattern.

On the other side, there is a lack of common understanding about adaptiveness. In our view, peers' internal structure may change in order to adapt to the environment. For example, consider a search algorithm whose parameters' values change over time in a different way for each peer depending on local performance indicators. The evolution of a structure can be based on memoryless transformations that are applied to the structure to modify it, or based on learning and knowledge transmission. In general, adaptive peer-to-peer networks emulate the ability of biological systems to cope with unforeseen scenarios, variations in the environment or presence of deviant peers.

In a recent work (Amoretti, 2009B), we proposed the Adaptive Evolutionary Framework (AEF) for peer-to-peer architectures. According to the AEF, the internal structure of the peer is based on an adaptive plan τ which determines successive structural modifications in response to the environment. The adaptive plan, in the AEF framework, is based on an evolutionary algorithm, which utilizes a population of individuals (structures), where each individual represents a candidate solution to the considered problem. To show the potential of AEF, we used it to define a resource sharing scheme in which the evolutionary aspect is driven by a genetic algorithm.

Evaluation of Service Discovery and Aggregation Strategies

Due to the complexity of NSAM interactions, analytical studies may give some insights into the behavior of a NSAM system, but are inadequate in practice. Unstructured overlays are usually more complex to study than structured ones. Moreover, a network of peers is highly dynamic, since joins and departures occur continuously. Also for these reasons, we usually complete our studies with simulations carried out with the Discrete Event Universal Simulator (DEUS) (Amoretti, 2009A).

This tool provides a simple Java API for the implementation of nodes, events and processes, and a straightforward but powerful XML schema for configuring simulations.

Service composition strategies rely on resource discovery mechanisms (as we explained in section 3.3.2). Here we propose a *search cost analysis* that considers different overlay schemes. The search cost SC is the number of steps until approximately the whole network is revealed.

Decentralized unstructured overlay schemes are usually explored with the following strategies (Adamic, 2001; Zhang, 2007):

- random walk
- flooding
- probabilistic flooding

We remind that the probability generating function (PGF) of a network with degree distribution $P(k)$ is

$$G(z) = \sum_k P(k) z^k$$

With the random walk strategy, each message is forwarded to a randomly chosen neighbor, at each step, until the time-to-live (TTL) expiration. The average degree of a randomly chosen node is

$$<k> = G' \quad (1)$$

It has been demonstrated that

$$SC = N / <k>_{2B} \quad (7)$$

where N is the total number of nodes and $<k>_{2B}$ is the average number of second neighbors (Adamic, 2001). The latter can be computed as

$$<k>_{2B} = [G'_1(1)]^2$$

where $G_1(z)$ is the PGF that gives number of new neighbors encountered on each step of a random walk. Many unstructured peer-to-peer networks are *scale-free*, i.e. their separation degree grows sublinearly with respect to N. Such networks are characterized by a power-law distribution in the node degree

$$P(k) = ck^{-\tau}$$

Assuming $k_{max} \sim N^{1/\tau}$, then

$$SC = N^{3(1-2/\tau)} \quad (8)$$

Thus, if $\tau < 3$, random walk strategies in scale-free networks have search costs that scale *sublinearly* with the size of the network. The scale-free feature is a consequence of growth and preferential attachment (i.e. the probability with which a new node connects to the existing nodes is not uniform). Without preferential attachment, the resulting node degree distribution would be exponential

$$P(k) \sim e^{-\beta k}$$

In such networks, random walk strategies have search cost that scale *linearly* with N.

Flooding strategies are usually very effective (much more than random walks), but too expensive in terms of network bandwidth usage. To tackle this problem, probabilistic forwarding of query messages is a viable solution. The forwarding probability is varied according to the popularity of the resource being searched and the node degree. Peers estimate the popularity of the resource in the network based on feedback from previous searches. Such search mechanisms balance the volume of control traffic and the search performance.

Currently we are working on a different strategy, whose novelty is given by the integration of service semantics with the rigid constraints of a structured overlay scheme. The detailed description of the strategy is out of the scope of this chapter, and will be the subject of a future paper. Here we would like to emphasize that the advantage of using a DSM-based architecture is that lookups take *O(logN)* time with high probability, also in presence of high churn.

JXTA-SOAP MOBILE EDITION

Sun MicroSystem's JXTA is mainly the specification of a set of open protocols for building overlay networks, independent from platforms and languages (Traversat, 2003). Such protocols are implemented as services (e.g. Discovery Service, Peer Information Service, etc.) that are locally executed by each peer, leading to emergent global behaviors. The mapping with the framework we introduced in section 3 is quite immediate, considering local service instances as atomic services, and global instances as composite services.

Currently, there are three official JXTA implementations: J2SE-based, J2ME-based and C/C++/C-based. In particular, an almost complete version of the JXTA Java Micro Edition (JXTA-J2ME, a.k.a. JXME) has been recently released. It provides a JXTA compatible platform on resource constrained devices using the Connected Limited Device Configuration (CLDC) with Mobile Information Device Profile 2.0 (MIDP), or Connected Device Configuration (CDC). Supported devices range from smart-phones to PDAs.

Within JXTA developers community, we are responsible for the development and maintenance of the component called JXTA-SOAP[8], which is currently the sole open source project supporting peer-to-peer sharing of Web Services both on fixed and mobile platforms. Each JXTA peer provided with JXTA-SOAP is able to deploy its own Web Services, advertise them in the network, discover and invoke those provided by other peers. Advertising and discovery is based on JXTA core protocols (Traversat, 2003), and SOAP messages for request/response interaction with Web Services are carried by JXTA pipes. The JXTA-SOAP component has been implemented in Java, in two editions (J2SE-based and J2ME-based) that are completely interoperable. In the following of this section we focus on the mobile edition, since the standard edition has been already presented in Amoretti (2008).

JXTA-SOAP may be compared to the Mobile Web Services Mediation Framework (Srirama, 2006; Srirama, 2008), whose code unfortunately is not publicly available. MWSMF provides a hybrid solution, since it must be configured as JXTA-J2SE peer and established as an intermediary module between Web Service clients and mobile hosts, being these configured as JXME peers. Web Service clients may invoke the services deployed on mobile hosts via the MWSMF, which encodes SOAP messages to BinXML format, and sends them through JXTA pipes. The MWSMF also manages message persistence, guaranteed delivery, failure handling and transaction support.

Architecture

JXTA-SOAP Mobile Edition (ME) supports J2ME's Connected Device Configuration (CDC) Profile. This JVM configuration does not allow to use Apache Axis as SOAP engine. In general, most WS-oriented APIs for J2ME only support client-side functionalities, *i.e.* service inspection and invocation. For example, the Java Wireless Toolkit (WTK) provides J2ME Web Services API (WSA)[9], based on JSR 172[10], which specifies runtime service provider interfaces to allow the generation of portable stubs from WSDL files. The specification contains some notable limitations, most of them due to the requirement for WS-I Basic Profile compliance. Conforming to the profile ensures interoperability, but also prevents using alternative methods. Another widely used solution is the kSoap2[11] open source component, which is a parser supporting the generation of client side stubs. kSoap2 is compliant with devices lacking JSR 172 support, and allows to access non WS-I conformant services. To the best of our knowledge, the unique solution enabling J2ME applications (CLDC, CDC) as service endpoints is the Micro

Figure 5. Internal architecture of a peer based on JXTA-SOAP Mobile Edition

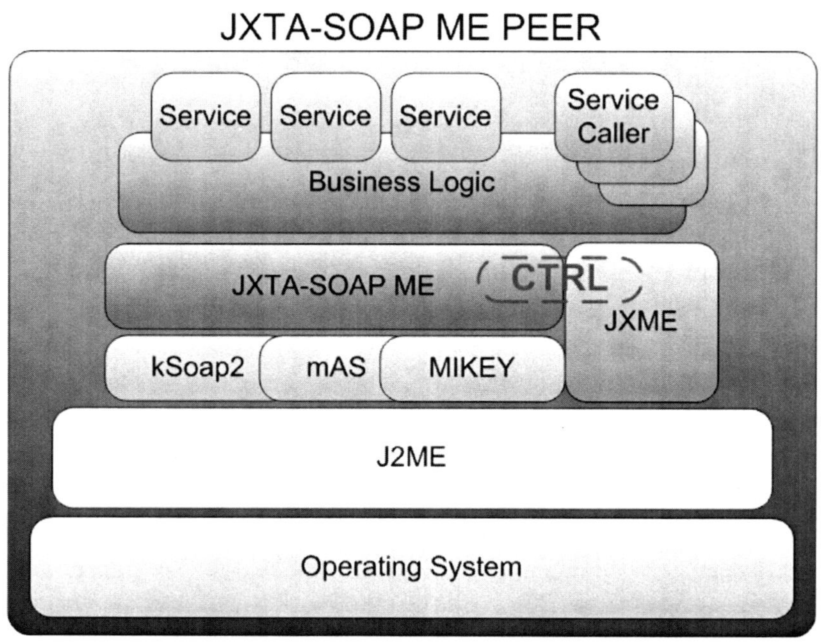

Application Server (mAS)[12], that can be considered a lightweight version of Axis.

The layered architecture of a JXTA-SOAP ME peer is illustrated in figure 5. In this framework, to create a service-oriented application, the developer must perform the following steps: 1) define the WSDL interface of the services, 2) implement the service code, 3) implement remote service callers (if needed), 4) implement the application logic (i.e. the main loop of the peer). Local service activation, as well as remote service discovery and invocation, are managed by JXTA protocols.

In details, service invocation is allowed by a kSoap2-based implementation of the Call Factory class. The latter instantiates a kSoap2's Soap Object, and sets all the properties for message exchanging through JXTA pipes. Soap Object is a highly generic class which allows to build SOAP calls, by setting up a SOAP envelope. JXTA-SOAP defines a Call Factory class that is used to create a Call object, passing the reference to a Service Descriptor, a public pipe advertisement of the service and the peergroup as parameters for the creation. The Call Factory class also allows to create an instance of kSoap Pipe Transport, the class we implemented to manage the transmission of SOAP messages using service pipes. The kSoap2 API provides a Transport class that encapsulates the serialization and deserialization of SOAP messages, but does not manage communication with the service. The HTTP Transport subclass allows service invocation over HTTP, setting up the required properties, but it uses URLs as absolute references of remote services, and it is not suitable for usage in JXTA-SOAP, where services (as every resource) are identified by JXTA-IDs and must be invoked through JXTA pipes. Thus, we extended the Transport class with the implementation of a call functionality that configures a JXTA pipe and creates the messages to be sent over it. After instantiating the transport using the Call Factory class, the consumer peer creates the request object, indicating the name of the remote method to be assigned to a Soap Serialization Envelope, as the outbound message for the soap call. Soap Serialization Envelope is a kSoap2 class that extends the basic Soap Envelope, providing support for the SOAP Serialization format specification and simple object serialization. The same class provides a getResponse method that extracts the parsed response from the wrapper object and returns it.

For service provision, we integrated the Server class of the Micro Application Server (mAS) into the basic service class of the JXTA-SOAP API. mAS implements the Chain of Responsibility pattern (Gamma, 1995), the same used in Axis. It avoids coupling the sender of a request to its receiver by giving more than one object a chance to handle the request; receiving objects are chained and the request passed along the chain until an object handles it. Moreover, mAS allows service invocation by users and service deployment by the owner, and suppurts browser management of requests, distinguishing whether the HTTP message contains a Web page request or a SOAP envelope.

Security

To cope with malicious attacks, security policies adopted at the overlay P2P network level usually consist of key management, authentication, admission control, and authorization. These are the strategies we took into account for securing consumer-to-service communication in JXTA-SOAP. Currently, JXTA-SOAP supports secure service invocation by means of two orthogonal mechanisms. The first one, transport-level security, allows to create a secure channel which guarantees the integrity and confidentiality of exchanged information, by means of mutual authentication between parties (using certificates) and data encoding. The other approach is WSS-based message-level security, for which SOAP messages sent by service consumers contain security parameters (tokens) which are extracted by service providers to check for consumers' compliance with the security policy of the invoked service.

In JXTA, the default Membership Service is PSE, which stands for Personal Security Environment. This service is the only one that is considered secure and the one that will be analyzed. PSE provides credentials based on X.509 certificates. Any number of such certificates may be included as Certificate elements in the PSE credential, together with the Peer Group ID and the subject's Peer ID. The credential itself is also signed.

Implementing secure invocation mechanisms in the mobile version of JXTA-SOAP required the porting of PSE membership classes from JXTA-J2SE to JXTA-J2ME, for peergroup authentication, the implementation of a new type of JXTA pipe, by which it is possible to cipher message contents, and the definition of a new security policy, suitable for J2ME's Connected Device Configuration (CDC) and Personal Profile. We introduced Multimedia Internet KEYing (MIKEY)[13] protocol to create the key pair and all the required parameters for encryption and decryption operations. Although memory and processing power have dramatically improved for handheld devices, encryption remains a resource-intensive task that requires consideration when designing protocols. MIKEY is a schema for management of cryptographic keys which can be used in real-time and peer-to-peer applications; it has been developed with the intention to minimize latency when exchanging cryptographic keys between small interactive groups that reside in heterogeneous networks. The protocol is defined in RFC 3830 and in JXTA-SOAP project we introduced an implementation with RSA-R algorithm[14].

Applications and Performance Evaluation

Ubiquitous computing, for its nature, has an extremely wide range of applications. Here we consider two important and challenging fields, i.e. ambient intelligence and emergency management, for which we are developing solutions based on JXTA-SOAP.

Ambient Intelligence (AmI) refers to digital environments that proactively support people in their daily lives, based on the convergence of three key technologies: Pervasive Computing, Artificial Intelligence, and Intelligent User Friendly Interfaces (Ramos, 2008). AmI represents a step beyond the current concept of a "User Friendly Information Society", bacause the technologies should be fully adapted to human needs and cognition. Indeed, AmI should be orientated towards community and cultural enhancement, helping citizens to build knowledge and skills, and to achieve better quality of life. At the same time, AmI should inspire trust and confidence, working in a seamless, unobtrusive and often invisible way.

One of the most challenging AmI services is User Activity Monitoring, which may be transversal to every AmI scenario. The framework illustrated in section 3, is able to provide the flexibility required to deal with highly dynamic environments where devices continuously change their availability and (or) physical location (e.g. those which are carried or worn by the user). This complex problem of composing and decomposing connections among nodes is abstracted in an overlay network where the Activity Monitor (AM) component subscribes for raw context events coming from other distributed components (sensors, specialized data filters, etc.), searches for remote services which may provide useful information for its reasoning function, and publishes context events which describe indoor and outdoor activity of the user, taking into account different contour information such as medical prescriptions, planned agenda, etc.

Emergency Management (or disaster management) is the discipline of dealing with and avoiding risks (Haddow, 2004). It involves preparing for disaster before it happens, disaster response (e.g. emergency evacuation, quarantine, mass decontamination, etc.), as well as supporting and rebuilding society after natural or human-made disasters have occurred. ICT support is very important during the disaster response (DR) phase of an emergency, which may commence with search and rescue, but in all cases the focus will quickly turn to fulfilling the basic humanitarian needs of the affected population. This assistance may be provided by national or international agencies and organizations. Effective coordination of disaster assistance is often crucial, particularly when many organizations respond and local emergency management agency capacity has been exceeded by the demand or diminished by the disaster itself. Tracing missing people, coordinating donor groups, recording the locations of temporary camps and shelters are examples of problems in the immediate post-disaster period that can be effectively addressed by using ICT.

Using JXTA-SOAP mobile, we developed a GUI-based application that allows to join a JXTA-based P2P network to share services for supporting disaster response activities. The application has several overlapping panels (or tabs), each one being related to a specific function. As illustrated in figure 6, the Remote panel shows discovered remote services. It is possible to search for services in the P2P network (offered by other rescue operators), and to select one of them from the resulting list, in order to see all the operations it offers, which are shown in the Operation tab. The user puts a description of the desired service in the search field, and all the matching services are listed in the table. Some services from the back-end are assumed to be always available, such as the one that provides photos taken by a satellite. The Operation management panel (figure 7) shows all the functionalities provided by the selected service; the operator can choose a particular operation and fill the input parameters table in the invocation panel.

We tested the performance of the application with respect to several point-to-point configurations, combining different settings for each participant. JXTA-J2SE peers have been deployed on laptops and desktop computers running either Windows XP, Linux or Mac OS X, equipped with 1GB RAM and 1.6GHz processors. A JXTA-J2ME peer has been hosted on an I-Mate JASJAR

Figure 6. Disaster Response GUI: remote service selection panel

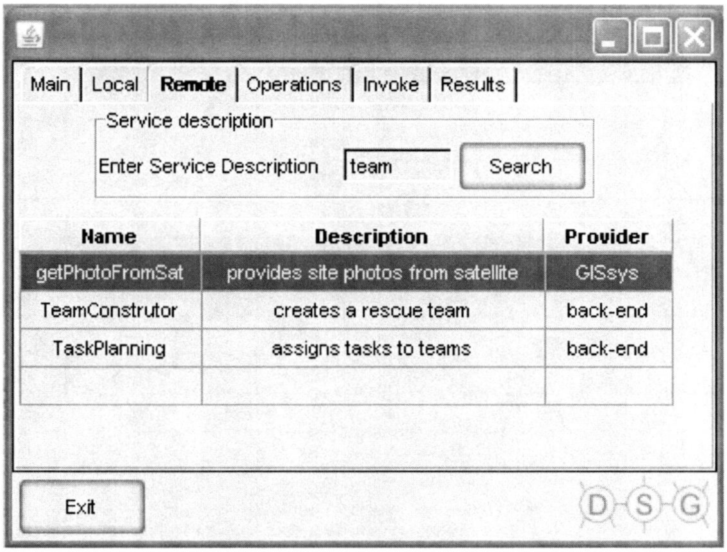

Pocket PC PDA, equipped with 64MB RAM and 520MHz processor. The memory footprint of a Java program is predominantly due to objects, classes, and threads that the users create directly, and native data structures (like the constant-pool, the string-table, etc.), native code, and the virtual machine (JVM) itself that are loaded indirectly by the user. For JXTA-SOAP peers running on laptops and desktop computers with J2SE v1.5, we measured a 22.5MB footprint (at least 10MB are needed by the sole JVM). On the other side, the peer installed on the Pocket PC with J2ME

Figure 7. Disaster Response GUI: operation management panel. A photo of the disaster location is taken, and a short description written, both ready to be sent to the back-end upon request, or proactively by the rescue operator

Table 1. Evaluation of JXTA-SOAP performance: rendezvous peer discovery, service discovery, service invocation

Overlay peers	Data link	multicast	tr (s)	ts (s)	ti (s)
edge J2SE c, rdv J2SE p	Ethernet	Off	2.5	0.5	0.1
edge J2SE c, rdv J2SE p	Ethernet	On	n.n.	2.0	0.1
edge J2SE c, edge J2SE p	Ethernet	On	n.n.	0.5	0.1
edge J2SE c, edge J2SE p, rdv J2SE b	Ethernet	Off	2.5	0.5	0.1
edge J2SE c, rdv J2SE p	WiFi	Off	2.0	0.5	0.7
edge J2SE c, rdv J2SE p	WiFi	On	n.n.	2.0	0.7
edge J2SE c, edge J2SE p	WiFi	On	n.n.	1.0	0.4
edge J2ME c, rdv J2SE p	WiFi	On	n.n.	3.1	2.0
edge J2ME c, edge J2ME p	WiFi	On	n.n.	2.5	2.0
adhoc J2SE c, adhoc J2SE p	adhoc	On	n.n.	1.0	0.4
edge J2ME c, rdv J2SE p	adhoc	On	n.n.	1.0	4.4
adhoc J2ME c, adhoc J2SE p	adhoc	On	n.n.	1.0	0.4
adhoc J2ME c, adhoc J2ME p	adhoc	On	n.n.	1.0	0.5

(personal profile v1.1) had a 7MB RAM footprint (with 3MB for the JVM). All tested configurations are listed in table 1.

We configured peers as service providers (p), consumers (c), or bridge nodes (b) which store advertisments and route messages but do not provide or consume Web Services. At the data link layer we considered Ethernet, WiFi and ad-hoc mode. Experimental results refer to the following sequential actions performed by the consumer peer:

- elapsed time for rendezvous peer discovery (t_r)
- elapsed time for service discovery (t_s)
- elapsed time for service invocation (t_i)

Rendezvous peer discovery is not necessary (n.n.) when multicast is active (on), but service discovery requires much time with respect to the multicast off case. Without multicast, a list of rendezvous hosts must be used to allow peers join the network at bootstrap. Once an edge peer is connected to its rendezvous, if the service has been advertised (and replicated among rendezvous peers) the discovery process is very fast. Test results are encouraging, being performance significant in almost all examined cases. It appears that, when hosts are connected in ad-hoc mode, best performance is achieved if also at the application level peers are configured in ad-hoc mode.

FUTURE RESEARCH DIRECTIONS

In the near future, our research activity on peer-to-peer service sharing will go ahead, focusing both on the refinement of the NSAM model and on middleware development. We are studying novel distributed strategies for service composition. Moreover, we are studying alternative solutions to genetic algorithms, in order to implement the Adaptive Evolutionary Framework (AEF) illustrated in section 3. In this context, we are considering complex environments and applications (in particular mobile ones), for which adaptiveness is not a plus but a fundamental requirement.

CONCLUSION

In this chapter we introduced the *Networked Service-oriented Autonomic Machine (NSAM)*, which is a theoretical model of an hardware/software entity that is programmed to be altruistic in sharing its resources. We focused on NSAMs whose hardware resources can be classified as mobile devices, offering and consuming services. In this context, we presented a framework for peer-to-peer service sharing, based on three key aspects: overlay scheme, dynamic service composition and self-configuration of peers. This framework is suitable to characterize many existing platforms and to define new ones. In particular, JXTA and JXTA-SOAP fit well with the NSAM concept of emerging composite services, aggregating atomic service instances deployed by peers. We described the Mobile Edition of JXTA-SOAP, showing its good performance and proposing some interesting and challenging applications.

REFERENCES

Adamic, L. A., Lukose, R. M., Puniyani, A. R., & Huberman, B. A. (2001). Search in power-law networks. *Physical Review E: Statistical, Nonlinear, and Soft Matter Physics*, *64*(4), 1842–1845. doi:10.1103/PhysRevE.64.046135

Amoretti, M. (2009B). A Framework for Evolutionary Peer-to-Peer Overlay Schemes. In *European Workshops on the Applications of Evolutionary Computation*, Tubingen, Germany.

Amoretti, M., Agosti, M., & Zanichelli, F. (2009A). DEUS: a Discrete Event Universal Simulator. *In 2nd ICST/ACM International Conference on Simulation Tools and Techniques (SIMUTools 2009)*, Roma, Italy.

Amoretti, M., Bisi, M., Zanichelli, F., & Conte, G. (2005). Introducing Secure Peergroups in SP2A. In *2nd IEEE International Workshop on Hot Topics in Peer-to-Peer Systems, co-located with Mobiquitous,* 2005, San Diego, California.

Amoretti, M., Bisi, M., Zanichelli, F., & Conte, G. (2008). *Enabling Peer-to-Peer Web Service Architectures with JXTA-SOAP*. In IADIS International Conference e-Society 2008, Algarve, Portugal.

BarkaiD. (2002). Peer-to-Peer Computing: Technologies for Sharing and Collaborating on the Net. Santa Clara, CA: Intel Press.

Baset, S. A., & Schulzrinne, H. G. (2006). An Analysis of the Skype Peer-to-Peer Internet Telephony Protocol. In *25th IEEE International Conference on Computer Communications (INFOCOM 2006)*, Barcelona, Spain.

Berger, S., McFaddin, S., Narayaswami, C., & Raghunath, M. (2003). Web Services on Mobile Devices - Implementation and Experience. In *5th IEEE Workshop on Mobile Computing Systems & Applications*, Monterey, CA.

Cohen, E., & Shenker, S. (2002). Replication Strategies in Unstructured Peer-to-Peer Networks. In *ACM SIGCOMM '02*, Pittsburgh, PA.

GaberJ. (2007). GLOBECOM Workshop 07. Washington, DC: Spontaneous Emergence Model for Pervasive Environments. In IEEE.

GammaE.HelmR.JohnsonR.VlissidesJ. (1995). Design Patterns. Reading: Addison-Wesley.

Garces-Erice, L., Biersack, E. M., Felber, P. A., Ross, K. W., & Urvoy-Keller, G. (2003). Hierarchical Peer-to-Peer Systems. In *International Conference on Parallel and Distributed Computing (Euro-Par 2003)*, Klagenfurt, Austria.

GauthierDickey. C., & Grothoff C (2008). Bootstrapping of Peer-to-Peer Networks. In *International Workshop on Dependable and Sustainable Peer-to-Peer Systems*, Turku, Finland.

Govoni, D., & Soto, J. C. (2002). JXTA and security. In *JXTA: Java P2P Programming*. Indianapolis, IN: Sams Publishing.

HaddowG. D.BullockJ. A. (2004). Introduction to Emergency Management. Amsterdam: Butterworth-Heinemann.

Kalaspur, S., Kumar, M., & Shirazi, B. A. (2007). Dynamic Service Composition in Pervasive Computing. *IEEE Transactions on Parallel and DistributedSystems*, *18*(7), 907–917. doi:10.1109/TPDS.2007.1039

Kleis, M., Lua, E. K., & Zhou, X. (2005). Hierarchical Peer-to-Peer Networks using Lightweight SuperPeer Topologies. In *10th IEEE Symposium on Computers and Communication (ICSS05)*, La Manga del Mar Menor, Cartagena, Spain.

Lv, Q., Cao, P., Cohen, E., Li, K., & Shenker, S. (2002). Search and Replication in Unstructured Peer-to-Peer Networks. In *ACM International Conference on Supercomputing (ICS '02)*, New York.

Peng, Z., Duan, Z., Qi, J. J., Cao, Y., & Lv, E. (2007). HP2P: A Hybrid Hierarchical P2P Network. In *1st International Conference on the Digital Society*, Gaudeloupe.

Ramos, C., Augusto, J. C., & Shapiro, D. (2008). Ambient Intelligence - the Next Step for Artificial Intelligence . *IEEE Intelligent Systems*, *23*(2), 15–18. doi:10.1109/MIS.2008.19

Srirama, S. N., Jarke, M., & Prinz, W. (2006). *A* Mediation Framework for Mobile Web Service Provisioning. In *10th IEEE International Enterprise Distributed Object Computing Conference Workshops (EDOCW 2006)*, Hong Kong, China.

Srirama, S. N., Jarke, M., & Prinz, W. (2008). MWSMF: a Mediation Framework Realizing Scalable Mobile Web Service. In *Mobilware 2008*, Innsbruck, Austria.

Tang, X., Xu, J., & Lee, W. C. (2008). Analysis of TTL-Based Consistency in Unstructured Peer-to-Peer Networks. *IEEE Transactions on Parallel and Distributed Systems*, *19*(12), 1683–1694. doi:10.1109/TPDS.2008.44

Tewari, S., & Kleinrock, L. (2006), Proportional Replication in Peer-to-Peer Networks. In *25th Annual Joint Conference of the IEEE Computer and Communications Societies (INFOCOM)*, Barcelona, Spain.

TraversatB.AroraA.AbdelazizM.DuigouM. HaywoodC.HuglyJ.-C. (2003). *Project JXTA 2.0 Super-Peer Virtual Network* (Tech. Rep.). Sun MicroSystems.

van Engelen, R. A., & Gallivan, K. (2002). The gSOAP Toolkit for Web Services and Peer-To-Peer Computing Networks. In *2nd IEEE International Symposium on Cluster Computing and the Grid (CCGrid 2002)*, Berlin, Germany.

Zhang, H., Zhang, L., Shan, X., & Li, V. O. K. (2007). Probabilistic Search in P2P Networks with High Node Degree Variation. In *IEEE International Conference on Communications (ICC 2007)*, Glasgow, Scotland.

ENDNOTES

1. BitTorrent official site http://www.bittorrent.org
2. ARM http://www.arm.com
3. RIM (Research in Motion) http://www.rim.com
4. Windows Mobile http://www.microsoft.com/windowsmobile/en-us/default.mspx
5. SYMBIAN http://www.symbian.org/
6. LiMo http://www.limofoundation.org/

7. ANDROID http://www.android.com
8. JXTA-SOAP https://soap.dev.java.net/
9. WSA http://java.sun.com/products/wsa/
10. JSR 172 http://jcp.org/en/jsr/detail?id=172
11. kSoap2 http://ksoap2.sourceforge.net
12. mAS https://sourceforge.net/projects/masproject
13. MIKEY http://www.ietf.org/rfc/rfc3830.txt
14. MIKEY-RSA-R http://www.ietf.org/rfc/rfc4738.txt
15. Microsoft's .NET Compact Framework http://msdn.microsoft.com/en-us/netframework/aa497273.aspx
16. Sun MicroSystems's J2ME Web Services APIs (WSA) http://java.sun.com/products/wsa/
17. Sun MicroSystems's JSR 172: J2ME Web Services Specification http://jcp.org/en/jsr/detail?id=172
18. kSoap2 project http://ksoap2.sourceforge.net
19. mAS project https://sourceforge.net/projects/masproject

Chapter 10
Scripting Mobile Devices with AmbientTalk[1]

Elisa Gonzalez Boix
Vrije Universiteit Brussel, Belgium

Christophe Scholliers
Vrije Universiteit Brussel, Belgium

Andoni Lombide Carreton
Vrije Universiteit Brussel, Belgium

Tom Van Cutsem
Vrije Universiteit Brussel, Belgium

Stijn Mostinckx
Vrije Universiteit Brussel, Belgium

Wolfgang De Meuter
Vrije Universiteit Brussel, Belgium

ABSTRACT

This chapter is about programming mobile handheld devices with a scripting language called AmbientTalk. This language has been designed with the goal of easily prototyping applications that run on mobile devices interacting via a wireless network. Programming such applications traditionally involves interacting with low-level APIs in order to perform basic tasks like service discovery and communicating with remote services. We introduce the AmbientTalk scripting language, its implementation on top of the Java Micro edition platform (J2ME) and finally introduce Urbiflock, a pervasive social application for handheld devices developed entirely in AmbientTalk.

INTRODUCTION

For the past five years, we have been researching coordination abstractions to structure mobile computing applications. These applications are typically deployed on mobile devices (e.g. cellular phones, PDAs, …) equipped with wireless communication technology (e.g. WiFi, Bluetooth,…) (Mascolo, Capra, & Emmerich, 2002). Such devices form

DOI: 10.4018/978-1-61520-761-9.ch010

so-called *mobile ad hoc networks* which have two discriminating characteristics: the connectivity between devices is often intermittent (connections drop and are restored as devices move about) and there is little or no fixed support infrastructure, such that devices can often communicate *only* with physically proximate devices, favouring a peer-to-peer architecture rather than a client-server approach.

Traditionally, developing, testing and deploying mobile computing applications is laborious. One of the major reasons for this difficulty is that the programming languages that are commonly used for this task (e.g. C, C++, Java) have not been designed to deal with the hardware characteristics of mobile ad hoc networks. Especially on runtime platforms for handheld devices such as J2ME or the .NET compact framework, programmers have little more than a low-level socket API to work directly on top of supported networking protocols. Consequently, more high level abstractions such as service discovery, remote messaging, failure handling, asynchronous event handling, etc. must all be dealt with manually by the programmer.

In this chapter, we will describe AmbientTalk: an experimental scripting language for mobile devices (Dedecker, Van Cutsem, Mostinckx, D'Hondt, & De Meuter, 2006; Van Cutsem, Mostinckx, Gonzalez Boix, Dedecker, & De Meuter, 2007). To the best of our knowledge, AmbientTalk is the first high-level distributed object-oriented programming language that specifically targets mobile devices connected via an ad hoc wireless network. While the language features the standard toolbox of any object-oriented scripting language (similar to popular languages such as Ruby, Python or Groovy), it also integrates built-in support for service discovery (built on top of UDP), remote messaging (built on top of TCP/IP), failure handling, asynchronous event processing and publish/subscribe coordination between remote services. AmbientTalk is implemented entirely in Java and thus benefits from the platform-independence of the Java Virtual Machine. In addition, AmbientTalk can interoperate with Java applications. This allows concerns related to distribution (service discovery, asynchronous communication, failure handling) to be handled in the scripting language, while still enabling the reuse of existing Java libraries (e.g. for XML parsing, GUI construction, encryption etc.)

After having presented AmbientTalk, we introduce Urbiflock, an application that we have built using the language. Urbiflock is a framework for the development of so-called "pervasive social applications": applications that allow people to interact by means of handheld devices (such as their cell phones). Such applications aim to extend the so successful web-based social network services (e.g. Facebook, MySpace, etc.) to mobile services, opening new possibilities for mobile commerce. They enable spontaneous interaction between groups of people: people may broadcast announcements to each other, they can browse one another's profile, launch interactive polls, etc.

BACKGROUND

The hardware characteristics of mobile devices introduce certain phenomena that must be dealt with when writing mobile computing applications. In this section, we summarize these hardware phenomena. Subsequently, we discuss related work in the field of programming languages and middleware that has influenced the design of AmbientTalk.

Hardware Phenomena

There are two discriminating properties of mobile networks, which clearly set them apart from traditional, fixed computer networks: applications are deployed on *mobile* devices connected by *wireless* communication links with a limited communication range. Such networks exhibit two phenomena which are rare in their fixed counterparts:

Volatile Connections. Mobile devices equipped with wireless media possess only a limited communication range, such that two communicating devices may move out of earshot unannounced. The resulting disconnections are not always permanent: the devices may meet again, requiring their connection to be re-established. Often, such *transient* network partitions should not affect an application, allowing it to continue its collaboration transparently upon reconnection. Partial failure handling is not a new ingredient of distributed systems, but these more frequent transient disconnections do expose applications to a much higher rate of partial failure than that which most distributed languages or middleware have been designed for. In mobile networks, disconnections become so omnipresent that they should be considered the rule, rather than an exceptional case.

Zero Infrastructure. In a mobile network, devices that offer services spontaneously join with and disjoin from the network. Moreover, a mobile ad hoc network is often not manually administered. As a result, in contrast to stationary networks where applications usually know where to find collaborating services via URLs or similar designators, applications in mobile networks have to find their required services dynamically in the environment. Services must be discovered on proximate devices, possibly without the help of shared infrastructure. This lack of infrastructure requires a *peer-to-peer* communication model, where services can be directly advertised to and discovered on proximate devices.

Any application designed for mobile networks has to deal with the above phenomena. Because the phenomena are universal, an appropriate computational model can and should be developed which eases distributed programming in a mobile ad hoc network by taking these phenomena into account from the ground up. Moreover, because the effects engendered by partial failures or the absence of remote services often pervade the entire application, the above phenomena are not easily hidden behind traditional library abstractions. Therefore, distribution is often dealt with in dedicated middleware or programming languages.

Distributed Languages and Middleware

In what follows we describe how a number of programming languages and middleware deal with the above mentioned hardware phenomena. We will not only focus on approaches specifically designed for mobile ad hoc networks because some approaches outside this domain provide interesting features for this context and because their discussion further illustrates the differences between systems developed for traditional, fixed computer networks on the one hand and mobile networks on the other hand.

Distributed Languages

Most of the distributed languages that have been designed for local area networks (LAN), like Emerald (Jul, Levy, Hutchinson, & Black, 1988) and Obliq (Cardelli, 1995), are based on a synchronous communication model (Remote Procedure Call or RPC). To abstract over temporary disconnections, objects either remain blocked waiting for an outstanding RPC to a disconnected object (making the application unresponsive), or the RPC fails which requires cumbersome failure handling code for each remote call. Other distributed languages, such as ABCL/f (Yonezawa, Briot, & Shibayama, 1986), are based on the actor model (Agha, 1986). In this model, actors refer to one another via *mail addresses*. When an actor sends a message to another actor, the message is placed in a mail queue and is guaranteed to be eventually delivered by the actor system. Asynchronous communication via mail addresses decouples actors in time and synchronisation, making the actor model in itself almost suitable for mobile networks. However, it lacks means to perform service discovery, i.e. to acquire the mail address of a remote actor via anonymous communication. The ActorSpace model (Callsen & Agha, 1994) extends the actor model to solve this issue: messages can be sent to a *pattern* rather than to a mail address, and they will be delivered by the actor system to an actor with a matching pattern. The ActorSpace model, however, was conceived for traditional networks, as it relies on infrastructure to manage the matching of the patterns.

Some distributed languages designed for open networks (such as the Internet), have adapted the RPC model, like Argus (Liskov, 1988), while others are based on the asynchronous message passing model of actors, like Salsa (Varela & Agha, 2001) and E (Miller, Tribble, & Shapiro, 2005). Many of those languages introduce pure asynchronous communication in order to cope with higher latency of communication and failures. Argus and E make use of *futures* (also known as promises) to avoid forcing programmers to rely on explicit, separate callback methods to obtain the result of an asynchronous computation. An asynchronous send immediately returns a *future* object: a placeholder object (i.e. a proxy) which is eventually *resolved* with the return value. Most future abstractions support synchronisation by suspending a thread that accesses an unresolved future. The E language pioneered the use of callbacks on futures to express synchronisation on the resolution of a future in a non-blocking, event-driven manner (Miller, Tribble, & Shapiro, 2005).

In terms of failure handling, Argus features built-in support for atomic transactions to cleanly deal with unfinished computations resulting from partial failures. E allows to monitor the connection with a remote object by registering observers on remote references which are triggered upon failure.

Middleware

An alternative to distributed languages is middleware. In the past few years, many middleware platforms to support mobile computing have been proposed (Mascolo, Capra, & Emmerich, 2002). Approaches like the Rover toolkit adapted the RPC model to support volatile connections by queueing RPCs (Joseph, deLespinasse, Tauber, Gifford, & Kaashoek, 1995). This works well for temporary disconnections, but does not address long-lasting disconnections. In order to deal with this issue, the Jini architecture for network-centric computing (Waldo, 2001) was built from the ground up with the notion of *leasing* (Gray & Cheriton, 1989). A *lease* denotes the right to access a resource (e.g. an object) for a finite amount of time. Leases were introduced in Jini to allow clients and services to leave the network gracefully without affecting the rest of the system.

Several approaches have been proposed (Davies, Friday, Wade, & Blair, 1998; Mamei & Zambonelli, 2004; Murphy, Picco, & Roman, 2001) for mobile computing based on tuple spaces (Gelernter, 1985). In the tuple space model, processes communicate by inserting and removing tuples from a shared tuple space, which acts like a globally shared memory. Because tuples are anonymous, they are extracted by means of pattern matching on their content. Communication is decoupled in both time and space: processes can insert and remove tuples independently and the publisher of a tuple does not necessarily specify, or even know, which process will consume the tuple. This decoupling makes the tuple space model suitable for mobile ad hoc networks. Most research on tuple spaces for mobile computing extended this model by distributing the tuple space over a set of devices. In LIME (Murphy, Picco, & Roman, 2001), the tuples in the local tuple space of all devices in range are conceptually merged into a *federated tuple space*. Nodes can post and read tuples to and from this federated tuple space by means of the typical tuple space operations. However, when devices move out of range their tuples are no longer shared and removed from the federated tuple space. TOTA (Mamei & Zambonelli, 2004) improves on this model by allowing tuples to be replicated from location to location loosening the restriction that the sender and the receiver of a tuple have to be connected at the same time, i.e. decoupling devices in time.

The publish/subscribe communication paradigm (Eugster, Felber, Guerraoui, & Kermarrec, 2003) has also proven to be a fruitful basis for mobile computing middleware because it supports decoupling in time, space and synchronisation. For example, Jini uses such an approach to allow clients and services to spontaneously join an unadministered network and pass along a remote reference to the service. However, this paradigm has the disadvantage of requiring callbacks to handle results. The main difference between traditional, centralised publish/subscribe architectures and those for mobile networks is the incorporation of geographical constraints on event dissemination and subscriptions. For example, in the location-based Publish/Subscribe (LPS) (Eugster, Garbinato, & Holzer, 2005) architecture, a publisher defines a *publication range* and a subscriber defines a *subscription range*. Only when the publication range of the publisher and the subscription range of the subscriber overlap is an event disseminated to the subscriber. The Scalable Timed Events and Mobility (STEAM) middleware (Meier, Cahill, Nedos, & Clarke, 2005) even introduces geographical locations as first-class entities named *proximities*.

Summary

Although there has been a lot of active research with respect to mobile computing middleware (Mascolo, Capra, & Emmerich, 2002), there has been little innovation in the field of programming language research to tackle the issues raised by mobile networks. Although distributed programming languages are rare, they form a suitable development tool for encapsulating many of the complex issues engendered by distribution (Bal, Steiner, & Tanenbaum; Briot, Guerraoui, & Lohr, 1998). However, none of the distributed programming languages developed to date have been explicitly designed for mobile networks. They lack the language support necessary to deal with the radically different network topology. In the following section, we describe AmbientTalk: a scripting language that has been designed for mobile ad hoc networks from the ground up.

AMBIENTTALK: SCRIPTING FOR AD HOC NETWORKS

Mobile ad hoc networks are complex environments because of the lack of (server) infrastructure and because of the volatile connections between devices. We have designed AmbientTalk as a small object-oriented scripting language to ease the development of programs for these types of networks. We have chosen to implement AmbientTalk as a scripting language rather than as a library or toolkit for an existing language (e.g. Java or C#) because it fosters rapid application development, and because AmbientTalk introduces a reactive event loop application model that is not easily integrated with the typical multithreaded application model offered by mainstream languages. AmbientTalk's primary goal is to serve as an experimental language for our research on mobile ad hoc networks and is also used for teaching distributed computing at our university. However, we also wanted AmbientTalk to be a practical language. Therefore, AmbientTalk scripts can access any Java library or communicate with Java applications, as we will explain in more detail later.

In order to deal with the latency of wireless connections and the intermittent connectivity of devices due to transient network partitions, remote communication in AmbientTalk is entirely built around the concept of asynchronous message passing (as in the actor model) and smart buffering of messages (as in the Rover toolkit). As a result, AmbientTalk's asynchronous communication model allows objects to abstract over temporary network failures without blocking the control flow. In traditional languages, dealing with asynchronous communication is complicated because it does not integrate well with multithreading, which is the standard model to support concurrency in most programming languages. AmbientTalk solves this issue by replacing multithreading with reactive event loop concurrency. This model maps well onto the inherently reactive nature of distributed applications that must react to all kinds of network events. Devices may join or leave the network and messages can be received from remote devices at any point in time. It is similar to the model used by GUI frameworks (reacting to user events) and Web servers (reacting to incoming HTTP requests). The particular event loop model of AmbientTalk is based on that of the E programming language (Miller, Tribble, & Shapiro, 2005) and of Twisted Python (Fettig, 2005), an asynchronous network programming library for Python.

In order to deal with the fact that services have to be discovered in the environment without relying on an intermediary lookup service that may not be available when they meet, the language has a built-in publish/subscribe engine that allows objects to discover one another in a peer-to-peer manner, without depending on any centralised infrastructure.

In this section, we explain the language and illustrate some of its key features by means of a toy advertising application. In this application, advertisers can broadcast advertisements, which are printed on the screen of the cell phones of nearby potential customers that have announced their interest in these advertisements.

AmbientTalk Objects

Even though AmbientTalk is a scripting language for distributed programming in mobile ad hoc networks, it remains a full-fledged object-oriented language in its own right. AmbientTalk is dynamically typed and prototype-based. Computation is expressed in terms of objects sending messages to one another. Objects are not instantiated from classes. Rather, they are either created *ex-nihilo* or by cloning and adapting existing objects. The code snippet below defines a prototypical advertisement object. The code defines a new anonymous object and binds it to a variable named Advertisement. This object serves as a prototypical advertisement object, defining a number of fields to store the advertisement's state and a number of methods to define useful behaviour, e.g. to get a description of the content of the advertisement.

```
def Advertisement:= object: {
  def category; // a type clas-
sifying the advertisement
  def title;  // a string de-
scribing the subject of the ad-
vertisement
  def content; // a string de-
scribing the content advertise-
ment
  def advertiser; // a string
describing contact details
  // this method serves as the
"constructor"
  def init(aCategory, aTitle,
aText, anAdvertiser) {
    category:= aCategory;
    title:= aTitle;
    content:= aText;
    advertiser:= anAdvertiser;
  };
  def getDescription() {
    "Advertiser: " + advertiser.
getContactDetails() + "\n" +
    "Title: " + category + "\n"
+ text;
  };
};
// instantiate a new advertise-
ment
def anAdvertisement:= Advertise-
ment.new(Leisure,
  "Cheap drinks",
  "Cheapest bar in the neigh-
bourhood!"
  sender);
```

The last four lines of code show how to create a leisure advertisement in the advertising application. Sending the message new to the prototypical Advertisement object creates a clone (a shallow copy) of itself which is initialised using its init method with the arguments passed to new.

When an object receives a message it does not understand, it *delegates* the message to its parent object. Delegation is an object-based alternative to class-based inheritance (Lieberman, 1986). A declarative syntax is provided for specifying that a new object delegates to an existing prototype by means of extend:with. In the code excerpt below, a new prototype MovieAdvertisement is created which delegates to the Advertisement prototype. When a MovieAdvertisement is cloned, the clone has its own Advertisement parent object with its own copies of the category, title, content and advertiser slots.

```
def MovieAdvertisement:= extend:
Advertisement with: {
  def director;
  def imdbURL;
  def isSameMovieAndDirector(aMo
vieTitle, aDirector) {
    (self.title == aMovieTitle).
and: { director == aDirector };
  };
}
```

AmbientTalk uses block closures to represent *delayed* computations, such as implementing the branches of an if:then:else: control structure or nested event handlers, as will be described later. Block closures are constructed by means of the syntax {|args|body}, where the arguments can be omitted if the block takes no arguments. The following code excerpt shows a typical use of blocks to iterate over an array of advertisements, to show all advertisements on the screen.

```
myAdvertisements.each: { |ad|
ad.show() }
```

AmbientTalk supports both traditional canonical syntax (e.g. ad.show()) as well as keyworded syntax (e.g. myAdvertisements.each: block) for method definitions and message sends. As a general rule, keyworded syntax is used for control structures (e.g. while:do:) or object declarations (e.g. object:) while the canonical syntax is used for expressing application-level behavior.

Distributed Programming in AmbientTalk

In AmbientTalk, concurrency is not spawned by means of threads but rather by means of actors (Agha, 1986). AmbientTalk actors are not represented as active objects, but rather as *communicating event loops*, as is done in the E programming language (Miller, Tribble, & Shapiro, 2005). An actor is an event loop encapsulating regular objects which can communicate with one another using either synchronous method invocations (expressed as o.m()) or asynchronous message passing (expressed as o<-m()). Asynchronous messages are enqueued in an actor's queue of incoming messages, called its *mailbox*. An actor perpetually removes the next message from its mailbox and executes the corresponding method on the receiver of the message. Actors process messages from their message queue serially, i.e. one by one, to avoid race conditions on the state of regular objects.

In AmbientTalk, each object is said to be *owned* by exactly one actor. Only an object's owning actor may directly execute one of its methods (ensuring thus exclusive access to its mutable state). It is possible for objects owned by an actor to refer to objects owned by other actors. Such references that span different actors are named *far references* (the terminology stems from the E language) and only allow asynchronous access to the referenced object. Performing a method invocation via a far reference provokes a runtime exception. Asynchronous messages sent via far references are enqueued in the message queue of the actor that encapsulates the receiver

Figure 1. AmbientTalk actors as event loops

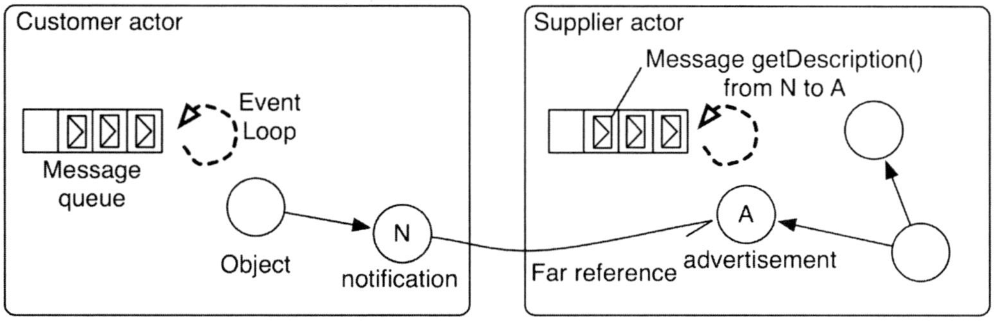

object. Figure 1 illustrates AmbientTalk actors as communicating event loops. The dotted lines represent the event loop processes of the actors that perpetually take messages from their message queue (represented as a sequence of boxes containing messages) and synchronously execute the corresponding methods on the actor's owned objects. An event loop process never "escapes" its actor boundary. When communication with an object in another actor is required, a message is sent asynchronously via a far reference to the object. For example, when a notification object N sends a message getDescription() to advertisement object A to request the title of the advertisement, the message is enqueued in the message queue of A's actor which eventually processes it.

Asynchronous Message Passing

In AmbientTalk, asynchronous messages can be sent between objects owned by the same actor (via a local reference) or by different actors (via a far reference). When sending an asynchronous message to an object that is encapsulated within the same actor, the message's parameters are passed *by reference*, exactly as is the case with regular synchronous message sending. When sending a message across a far reference, objects are instead parameter-passed *by far reference*: the parameters of the invoked method are replaced by far references to the original objects. Objects that have declared themselves to be *isolates* form an exception. Isolate objects are serializable objects and instead passed by (deep) copy. This allows the recipient actor to operate on the copy synchronously, without additional inter-actor communication and without violating the exclusive state access property.

To illustrate asynchronous message passing more concretely, consider the advertisement application described previously. Customers can use their mobile phone to receive advertisements of nearby services and can get extra information regarding the advertisement. Each cellular phone runs an advertisement application written in AmbientTalk. This application consists of a single actor. Given that advertisement denotes a far reference to the advertisement broadcasted by another actor, the description of the advertisement can be requested as follows:

```
def descriptionFut:=
advertisement<-getDescription();
```

The variable descriptionFut contains a *future*, which is a placeholder for the return value that will be computed asynchronously. Once the return value is computed, it "replaces" the future object; the future is then said to be *resolved* with the value. In AmbientTalk, futures are objects which can in turn be sent asynchronous messages. Those messages are accumulated within the future as long as it is unresolved. When the future is resolved, accumulated messages are forwarded to the resolved value. It is also possible to register a block of code with a future, which is executed asynchronously when the future becomes resolved. Such "in-line event handlers" are very useful when access to the actual return value of a message send is required. For example, the description that supplied the advertisement can only be printed to the screen when the descriptionFut future is resolved to a string value:

```
when: descriptionFut becomes: {
|description|
  // execution is postponed un-
til future is resolved
  system.println("New advertise-
ment received: " + description);
} catch: { |exception| ... };
// code following when: is pro-
cessed immediately
```

The when:becomes:catch: function takes a future and two block closures as arguments, and registers the functions as *observers* on the future. If the future is resolved to a proper value, the becomes: function is applied with the resolved value as parameter. If the asynchronously invoked method raises an exception, rather than returning a value, the corresponding future is resolved with the exception and the catch: function is applied to the exception. This enables applications to catch asynchronously raised exceptions in a way similar to the well-known try-catch abstraction. The execution of either of the above block closures is always scheduled in the owning actor's message queue, such that their execution is serialised w.r.t. other messages processed by the actor.

Far References and Partial Failures

In AmbientTalk, two objects are said to be *local* when they are owned by the same actor. Objects are considered *remote* when they are owned by different actors, even if those actors are hosted by the same device. By design, AmbientTalk abstracts from the physical location of actors and considers actors as the unit of distribution. Because objects residing on different devices are necessarily owned by different actors, the only kinds of object references that can span across different devices are far references. This ensures that all distributed communication is asynchronous.

By allowing far references to cross virtual machine boundaries, we must specify their semantics in the face of partial failures. AmbientTalk's far references are by default resilient to network disconnections. When a network failure occurs, a far reference to a disconnected object starts buffering all messages sent to it. When the network partition is restored at a later point in time, the far reference flushes all accumulated messages to the remote object in the same order as they were originally sent. Hence, messages sent to far references are never lost, regardless of the internal connection state of the reference. Making far references resilient to network failures by default is one of the key design decisions that make AmbientTalk's distribution model suitable for mobile ad hoc networks, because temporary network failures have no immediate impact on the application's control flow. This behavior is desirable in mobile ad hoc networks since they exhibit more frequent transient network partitions than traditional computer networks. However, not all network partitions are transient. Some of these failures will be permanent (e.g. a device moving out of wireless communication range that never returns) and require application-level failure handling.

To preserve the resilience of far references to transient failures while still being able to deal with permanent failures, AmbientTalk employs *leasing* (Gray & Cheriton, 1989). A far reference only provides access to a remote object for a limited period of time (the lease period). At the discretion of the owner of the resource a lease can be renewed, prolonging access to the resource. Figure 2 summarizes the different states a far reference can be in. When the far reference is connected and

Figure 2. States of a far reference

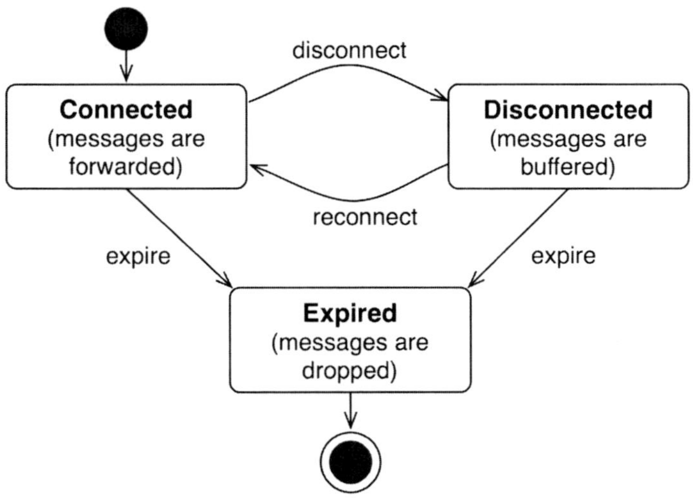

active, i.e. there is network connection and the lease has not yet expired, it forwards the buffered messages to the remote object. While disconnected, messages are accumulated as previously explained. When the time period has elapsed, the access to the remote object is terminated and the far reference is said to *expire*. Any attempt in using it will not result in a message transmission since an expired far reference behaves as a *permanently* disconnected remote reference.

A far reference can expire either because the lease cannot be renewed if a disconnection outlasts the lease period, or simply because the reference is not actively being used (and thus not renewed). When the reference expires, both client and service objects can schedule clean-up actions. This allows client and service objects to treat a failure as permanent (i.e. to detect when the reference is permanently broken) and to perform appropriate compensating actions. At server side, this has important benefits for memory management. Once all leased references to a service object have expired, the object can be taken offline, becoming subject to garbage collection once it is no longer locally referenced. Without such a mechanism, a single disconnected far reference could keep an object online forever.

Exporting Objects as Services

Objects can acquire far references to objects by means of parameter-passing or return values from inter-actor message sends. However, it remains to be explained how objects can acquire an *initial* far reference to an object owned by a remote actor. In this section we explain how objects can be made available to remote actors, an actor can explicitly *export* objects that represent certain services. In most distributed systems, exported objects are identified by means of a simple name or UUID in a name server or by a URL. However, in a mobile ad hoc network, name servers are impractical due to the limited infrastructure and the URL of a service may not be known to other actors.

In AmbientTalk, service objects are exported by means of a *type tag*. Type tags are a lightweight classification mechanism, used to categorise objects explicitly by means of a nominal type. One use of type tags in AmbientTalk is to provide a description of what kinds of services an object provides to remote objects. In AmbientTalk, a type tag can be a subtype of one or more other type tags, and one object may be tagged with multiple type tags. Although type tags are not used for static type checking, they are best compared with empty Java interface types, like the typical "marker" interfaces used to merely tag objects (e.g. java.io.Serializable and java.lang.Cloneable). One assumption we make is that all devices in the network attribute the same *meaning* to each type tag, i.e. we assume they define a common ontology to classify services.

Recall again the example of the advertising application where users receive advertisements from nearby services. Advertisements need to be *exported* to be made available on the network. The code snippet below shows how an advertisement object can export itself by means of the type tag stored in the advertisement's category field.

```
def pub:= export: self as: self.
category;
```

From the moment an object is exported, it is discoverable by objects owned by other actors by means of its associated type tag. The export:as: function returns an object that can be used to take the exported object offline again, by invoking pub.cancel(). How remote objects can acquire a reference to the exported object is explained in detail in the following section.

Service Discovery

AmbientTalk employs a publish/subscribe service discovery protocol. A publication corresponds to exporting an object by means of a type tag. The type tag serves as a *topic* known to both publishers and subscribers (Eugster, Felber, Guerraoui, & Kermarrec, 2003). A subscription takes the form of the registration of an event handler on a type tag, which is triggered whenever an object exported under that tag has become available in the ad hoc network.

In the advertising application, a user can be notified whenever a leisure advertisement is received as follows:

```
whenever: Leisure discovered: {
|advertisement|
  when: advertisment<-getDe-
scription() becomes: { |descrip-
tion|
    system.println("New leisure
advertisement received: " + de-
scription);
  }
};
```

The whenever:discovered: function takes as arguments a type tag and a block closure that serves as an event handler. Whenever an actor is encountered in the ad hoc network that exports a matching object, the handler function is scheduled for execution in the message queue of the owning actor. An object matches if its exported type tag is a subtype of the type tag argument of whenever:discovered:. The advertisement parameter of the handler function is bound to a far reference to the exported advertisement object of another actor. The function can then start sending asynchronous messages via this far reference to communicate with the remote object. Similar to the export:as: function, the whenever:discovered: function returns an object whose cancel() method cancels the registration of the handler function.

Interoperability with the JVM

AmbientTalk has been built in Java and thus runs on top of the Java Virtual Machine (JVM). AmbientTalk has been designed so that it can interoperate with the underlying JVM. The interoperability with the JVM is similar to that of other dynamic languages implemented on top of the JVM such as Groovy, Jython and JRuby. This means that all Java libraries available on the underlying platform are accessible to the AmbientTalk programmer. Hence, AmbientTalk scripts can call upon Java for standard tasks like XML parsing, GUI construction, encryption etc. We describe the interoperability of AmbientTalk with Java by means of the implementation of a GUI for the advertising application. The small AmbientTalk script shown below constructs a graphical user interface using the Java Swing framework. The GUI consists of a simple input field for the title of the advertisement, a text area used for the description of the advertisement and a button to publish the advertisement.

```
def swing:= jlobby.javax.swing;
def JFrame:= swing.JFrame;
def JTextField:= swing.JText-
Field;
def JTextArea:= swing.JTextArea;
def JButton:= swing.JButton;
// instantiate classes by send-
ing them the "new" message
def frame:= JFrame.
new("Advertisement");
def titleField:= JTextField.
new(20);
def textArea:= JTextArea.new();
def advertiseButton:= JButton.
new("Advertise!");

// static Java fields appear as
fields on the class object
frame.
setDefaultCloseOperation(JFrame.
EXIT_ON_CLOSE);

// these are all Java methods
invoked from AmbientTalk
def pane:= frame.getContent-
Pane();
pane.setLayout(jlobby.java.awt.
GridLayout.new(1,3));
pane.add(titleField);
pane.add(textArea);
pane.add(advertiseButton);
// the anonymous object is an
AmbientTalk object that
// masquerades as a Java Action-
Listener object
advertiseButton.
addActionListener(object: {
  def
actionPerformed(actionEvent) {
    def title:= titlefield.get-
Text();
    def content:= textArea.get-
Text();
    def advertisment:= Adver-
tisement.new(theCategory, title,
content, self);
    def pub:= export: advertise-
ment as: advertisement.category;
  };
});
frame.setVisible(true);
```

Accessing Java Objects in AmbientTalk

In order for AmbientTalk objects to interact with Java objects, they first need to gain access to Java classes. From classes, objects can then be referenced via static fields or by instantiating the referenced classes. Java classes are organised hierarchically by means of packages. We have chosen to mimic this structural hierarchy by means of simple objects whose public slot names correspond to nested Java package or class names. The root of this hierarchy is named jlobby (ordinary AmbientTalk programs use an object called the lobby to load external objects, hence the name jlobby for loading Java classes). As shown in the definition of swing in the code example, package objects can be created by selecting the slot with the appropriate name from jlobby.

Java classes are AmbientTalk objects whose fields and methods correspond to public static fields and methods in the Java class. Hence, these fields and methods can be accessed or invoked using regular AmbientTalk syntax. Java classes can be instantiated in AmbientTalk similar to how AmbientTalk objects are instantiated, i.e. by sending new to the object, which returns a new instance of the class. Arguments to new are passed as arguments to the Java constructor. For example, in the advertising application above, a new instance of a JFrame is created with the title of the frame passed as an AmbientTalk string. Java objects are AmbientTalk objects whose fields and methods correspond to public instance-level fields and methods in the Java object.

There are built-in conversions between the primitive data types of Java and AmbientTalk. For example, AmbientTalk strings are converted into Java Strings and vice versa. These predefined conversions make the interoperability between Java and AmbientTalk highly transparent in most cases.

Accessing AmbientTalk Objects in Java

When AmbientTalk code invokes a Java method that expects an argument typed as an *interface*, any AmbientTalk object can be passed to that method. The interoperability layer automatically generates a Java proxy object that implements the appropriate interface. When a Java object invokes a method on the proxy, the proxy forwards it to the AmbientTalk object. In the above example, the call to addActionListener requires a parameter of type ActionListener, which is an interface type. Instead of passing a wrapped Java object implementing this interface, one can pass any AmbientTalk object; the object is not even required to implement all declared interface methods. The anonymous object passed in the above code properly implements the actionPerformed callback, and will be notified by Java code whenever the user presses the advertiseButton. Method invocations like the actionPerformed callback are scheduled in the actor's message queue, to make sure that there can be no race conditions on AmbientTalk objects that are made accessible to Java threads. The details of mapping Java threads onto AmbientTalk events can be found in earlier work (Van Cutsem, Mostinckx, & De Meuter, 2008).

Deployment and Platform Constraints

AmbientTalk has been implemented entirely in Java and requires a regular J2SE Java Virtual Machine supporting version 1.3 or higher. The language is available at http://prog.vub.ac.be/amop/at/download. The implementation also runs on the Java 2 micro edition (J2ME) platform, under the connected device configuration (CDC). This means that AmbientTalk runs on PDAs and high-end cellular phones. Our current experimental setup consists of a number of HTCP3650 Touch Cruise phones that communicate by means of a wireless ad hoc WiFi network. Furthermore, AmbientTalk also runs on the JamVM Java virtual machine (cfr. http://sourceforge.net/projects/jamvm), which can be installed on the Apple iPhone and can make use of JNI libraries. Currently, AmbientTalk does not run on J2ME/CLDC (Connected Limited Device Configuration) devices because the AmbientTalk virtual machine relies on the Java Reflection API, which is not supported by the CLDC configuration. The AmbientTalk VM requires this API to implement the interoperability layer between AmbientTalk and Java. This is not a strict dependency, however: a preprocessor could be used to avoid the use of Java reflection by generating the proxies and conversion methods necessary for the interoperability between AmbientTalk and Java ahead of time. This would allow AmbientTalk to run on CLDC phones, but currently remains an area of future work.

At the implementation level, AmbientTalk interpreters communicate with one another by means of sockets via a TCP/IP network. AmbientTalk's topic-based publish/subscribe service discovery mechanism is peer-to-peer and does not require a centralised repository. AmbientTalk interpreters discover one another by means of the network's support for multicast messaging using UDP. After a successful discovery, the two interpreters exchange discovery information (e.g.

registered subscriptions and exported objects) in order to find a match.

As described previously, the naming and discovery of services happens via type tags. We make the underlying assumption that the name of such tags represents a unique service and is known by all participating services. This discovery mechanism also does not take versioning into account explicitly, e.g. if a certain service is updated, older clients may discover the updated service, and clients that want to use only the updated service may still discover older versions. Clients and services are thus themselves responsible to check versioning constraints.

AmbientTalk has currently not been optimized for computational performance. This is mainly because the bulk of applications written in AmbientTalk are communication-bound rather than computation-bound. Computation-intensive parts of an application can be written in Java through the strong interoperability with the Java VM. The performance of these parts will thus be limited by the performance of the underlying Java VM. Therefore, AmbientTalk's performance should mainly be focused on its network layer. Justin & Rajive (2008) have benchmarked the network efficiency of AmbientTalk against that of LIME and Spatial Views (Yang, Ulrich, Adrian & Liviu, 2005). The latter two are both state of the art frameworks that aim to tackle similar problems associated with mobile ad hoc networks. These benchmarks compare the network overhead for client-server throughput, group communication and connection-reestablishment in the face of frequent disconnections. From these results we can conclude that AmbientTalk performs better than LIME and Spatial Views in terms of network throughput and similar to LIME for group communication and connection-reestablishment. Spatial Views performed worse than AmbientTalk and LIME for communication over wireless links for all benchmarks. For a full account of these benchmarks we refer the interested reader to (Justin & Rajive, 2008).

PERVASIVE SOCIAL APPLICATIONS

Social networking applications such as Facebook, MySpace and Flickr, have gained tremendous popularity in the last few years. Despite being used every day by an enormous amount of users worldwide, social networking applications are still poorly integrated into the real world. Users need to manually upload content to a website (e.g. what they did over the weekend) and during social events these applications are out of the picture. Nowadays web-based social networking applications are still more a reporting tool of social events than a tool used to engage in social interactions. With the increasing miniaturisation of computing devices, we believe these limitations will be overcome in the near future, thus enabling a brand new form of social networking by means of *pervasive social applications*: social networking applications that allow people to interact by means of handheld devices (such as their cell phones). While social networking applications allow users to come in touch with people with similar interests, pervasive social applications emphasize the social aspect of interactions with friends and social links (Ben Mokhtar & Capra, 2009). Pervasive social applications also open the door to a new sort of mobile commerce service based on user' social preferences and social links.

We have developed a prototype framework for the development of such applications called *UrbiFlock* that allows people to meet and interact using their cell phones. Recently, a number of middleware solutions have been designed targeting pervasive social applications (Beale, 2005; Kalofonos, Antoniou, Reynolds, Van-Kleek, Strauss & Wisner, 2008; Ben Mokhtar & Capra, 2009). Kalofonos, Antoniou, Reynolds, Van-Kleek, Strauss & Wisner (2008) propose a secure platform which enables end-users to easily organize their social networks. Ben Mokhtar & Capra (2009) explore a number of social-based matching algorithms to reason about user preferences and their social links. In contrast to those approaches, UrbiFlock aims to be a platform for the rapid prototyping of pervasive social applications. In this regard, UrbiFlock has a stronger resemblance to BT Communities (Beale, 2005), a framework aimed to ease the development of pervasive social applications that communicate over Bluetooth. In this framework, a number of applications have been developed (e.g. a dating application) which all have in common that the Bluetooth connections (or the lack thereof) between different devices denote the different *communities*. Unlike BT Communities, in Urbiflock communities may not be related to hardware phenomena such as Bluetooth or WiFi connectivity. In the remainder of this section we describe the Urbiflock framework and how to prototype a simple rating application called *I rate you* (IR8U).

UrbiFlock

UrbiFlock is a framework sculpted for the development of applications that enable spontaneous interactions of people and exploit new technologies such as wireless networks and mobile devices. As in Facebook, users that join Urbiflock (called *flockrs*) can meet other users and interact with them, for example by sending each other messages. Flockrs have a profile which can be browsed by other flockrs. The Urbiflock framework takes care of managing a flockr's friends lists, called *flocks*. A flock can be compared to a Facebook group (for example, a group of your old classmates), but it additionally allows for the definition of groups of *proximate* flockrs (for example, a group of all of your friends that are currently nearby). Unlike current social network sites, Urbiflock allows the *specification* of flocks both in terms of *physical* proximity (defined by for example the bluetooth communication range of the flockr's cellular phones) and *semantic* proximity (e.g. in terms of being friends of someone).

Similar to Facebook, users can build applications and plug them into the Urbiflock framework. Several core applications are currently available in the Urbiflock framework, such as flock creators and profile viewers. In the remainder of this section, we describe the main concepts of the Urbiflock framework. Subsequently, we describe how a programmer can define his own plug-in application.

Figure 3 shows an UML diagram of the relevant parts of the UrbiFlock framework. Flockr plays a central role in this design. A flockr has exactly one profile and can be registered to multiple Flocks. In addition, a Flockr can have multiple installed applications. Applications need to be explicitly added to a Flockr before they can be used. From then on the flockr sees the application in a launch screen (similar to the Home screen on the Apple iPhone). Running applications have controlled access to flockr information via the framework such as the flockr profile. This is a common functionality found in a range of social networking applications. In addition, Urbiflock applications have access to the user's flocks (so they can talk to nearby flockrs) and the flockrs who have installed the same application on their devices.

Applications can register listeners that are notified when other flockrs enter or leave communication range, when they change their profile (e.g. when they update their status) or when flockrs running the same application appear in the proximate environment (e.g. to enable application-specific interaction). The latter event can be detected by calling the registerApplication-Listener method on an Application.

Profiles in UrbiFlock are highly extensible and besides a number of mandatory fields, flockrs can add as many custom fields as they like (for example, they could add their year of graduation). The fields of a profile can be used to match other users in the proximity by grouping them in flocks. For example, a flockr could create a flock of nearby flockrs which graduated the same year. When adding custom fields to the profile, the user can specify the type of the field (e.g. a number, a

Figure 3. UrbiFlock design diagram

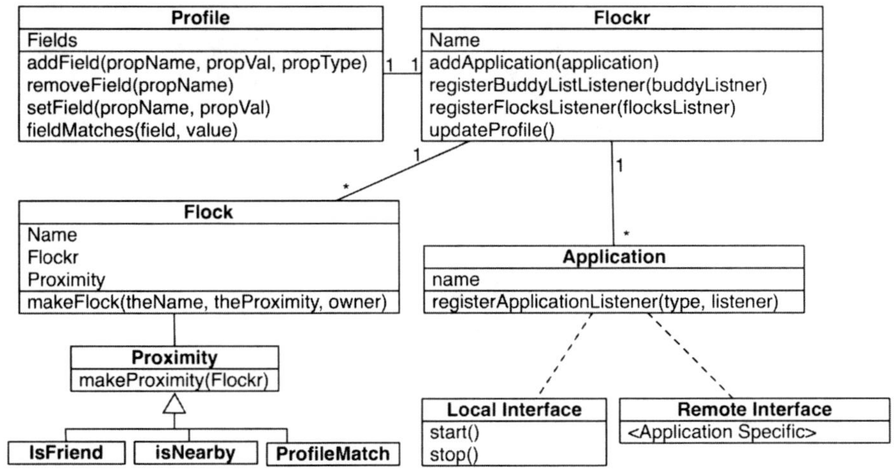

piece of text, a date, a choice etc.). Furthermore, the framework provides some infrastructure that makes it easy to add new custom types without having to write too much boiler plate code.

A Flock consists of a list of flockrs and a proximity function that determines whether a certain Flockr belongs to that list. There are several predefined proximities in the Urbiflock framework: isFriend encodes a friendship relationship (i.e. if a flockr is a friend of another flockr), isNearby encodes physical proximity relationship (currently defined by the communication range of their cellular phones) and doesProfileMatch tests an attribute of a flockr's profile. The operator used for such a test can be specified by the user and depends on the type of the field that is being compared (e.g. comparison operators for numbers and dates work differently and are different altogether for plain text). Users can define their own proximity functions as combinations of existing proximities by combining them using "and" and "or" operators. For instance, a user could specify a flock consisting of all male people in the neighbourhood who like to drink Belgian beers. This can be encoded in Urbiflock with a proximity function that combines the physical proximity to discover nearby flockrs and matches their profile to select the nearby flockrs that are male and like drinking Belgian beers.

A proximity function is recomputed whenever there is an event that alters the type of encoded relationship. These events become visible to the user as an addition or removal of a flockr in a flock. For example, if a flockr moves out of communication range, the proximity function in the nearby flock will be recomputed removing the disconnected flockr. The same happens when any of the connected flockrs adapts his or her profile, since this change may cause a flockr to enter or leave the proximity as defined by the corresponding proximity function.

The Urbiflock framework provides programmers with the necessary infrastructure to deal with the highly dynamic environment to which pervasive social networking applications running in mobile ad hoc networks are exposed. Programmers do not need to manually track the appearance and disappearance of flockrs in their environment (by means of the AmbientTalk service discovery constructs described in the previous Section), or changes in their profiles. In addition, plug-in applications themselves can be notified of the appearance or disappearance of nearby flockrs running the same application, making it easy to have small applications interact with each other.

As shown in Figure 3 every application has two interfaces: a local and a remote one. This distinction between local and remote interfaces has

Figure 4. Screenshot of IR8U application in Urbiflock

been introduced for reasons of security: methods defined in the local interface can only be called by local objects. Remote objects can only invoke methods defined in the application's remote interface. This allows the application to enable local objects, which can be trusted, to call different operations on the application than remote objects (e.g. changing the application's settings).

Writing a Simple Application in Urbiflock

UrbiFlock is a toolkit for the rapid development of pervasive social network applications. When making use of our toolkit programmers do not have to be concerned about discovery of services or failures in the network layer, but can instead work with different notions of proximity that make sense for pervasive social networking applications. To plug additional applications into the framework that make use of its offered infrastructure, they only have to implement a small set of methods.

In this section we explain the implementation of a simple application called *I rate you* (IR8U). This application allows users to ask proximate users to rate them on a certain subject. Figure 4 shows a screenshot of the IR8U application in Urbiflock. It depicts the Urbiflock screen launcher for a flockr called Andoni, which consists of buttons to access its profile, defined flocks and the installed applications (IR8U and Guanotes, a pre-installed application described in the next section). The figure also shows the flock viewer (launched when the user clicks the flocks button) with the two predefined flocks (corresponding to the isNearby and isFriend proximities). The bottom part of the figure 4 shows the GUI for IR8U which consists of a list of ratings in progress. In this example, the flockr has an ongoing rate about his level of English (with one reply from a flockr called Tom) and he is going to launch another rate about his latest Urbiflock application. Other users can rate the subjects by giving a rating between 0 and 5 stars.

The first step in the creation of the IR8U application is to extend the prototypical application with custom infrastructure as shown in the code snippet below. We define the needed data structures to keep track of who is connected in the proximity and who rated certain subjects. A vector connectedRaters stores of the people who are connected in the proximity while a hashmap ratingSubjects stores of the subjects (as keys) and their ratings (as values). Each rating itself consists of a pair of a far reference to the flockr who rated the subject and an integer between 0 and 5 representing the flockr's rating. In order to identify applications in Urbiflock, every application is associated to a type tag. Therefore, we create one for IR8U with the same name as the application itself. Last we define a variable to contain a reference to the GUI, this variable is not initialized here yet.

```
def localInterface:= extend:
makeApplication("IR8U", aFlockr)
with: {
  def connectedRaters:= Vector.
new();
  def ratingSubjects:= HashMap.
new();
  deftype IR8U;
  def ui;
  ...
}
```

The next step is to implement two mandatory methods start and stop which are called by the framework when the user starts and stops the application. The main purpose of these functions is to initialize and clean up listeners and required data structures. The code snippet below shows the start method for the IR8U application.

```
def start() {
  ui:= jlobby.at.urbiflock.
ui.ir8u.IR8U.new(self);
  self.export(IR8U);
  subscription:= self.registerAp
plicationListener(IR8U,
    object:{
      def
notifyApplicationJoined(flockr,
profile, ir8uApp){
          connectedRaters.
add(ir8uApp);
      };
      def
notifyApplicationLeft(flockr,
profile, ir8uApp){
          connectedRaters.
remove(ir8uApp);
      };
  });
};
```

First the GUI is initialized after which the application is exported to the network by calling the export method with its type tag. The framework takes care of exporting the application in the network and notifying listeners for this application. Finally, a listener is registered that updates the vector when connected IR8U users enter or leave the neighbourhood. This is done by calling the registerApplicationListener method with the type tag IR8U and a listener object. This listener object implements two methods notifyApplicationJoined and notifyApplication-Left which are called when another application in the proximity is discovered or leaves, respectively. Both of these methods are called with a reference to the flockr in the proximity, a copy of his profile, and a reference to the remote interface of the IR8U application of the remote flockr. The code snippet below shows the implementation of the stop method.

```
def stop(){
  super^stop();
  if: (subscription != nil)
then: {
    subscription.cancel();
    subscription:= nil;
    connectedRaters:= nil;
  };
};
```

The stop method is responsible for cleaning up the IR8U application. It first issues a super-send to invoke the default cleanup code defined in the prototypical application (which takes the application offline by unexporting it). Application listeners are then removed (by invoking subscription.cancel()) and its data structures are set to nil such that they can be garbage collected.

Now that we have explained how to start and stop the application and what data structures are used, we can explain the implementation of the basic functionality of IR8U. A user can ask all proximate users to rate a certain subject as implemented in the askRatingFor method. When this method is called it first creates a new subject and adds it to the ratingSubjects hashmap. Next, it sends an asynchronous message rateMe to all connected raters asking them to give a rating on the subject. A flockr can give this rating by calling the rateFlockr method. This method sends the asynchronous message rate to the remote application. Note that rateMe and rate must be defined in the remote interface as they are called from remote devices.

```
def askRatingFor(subject) {
  ratingSubjects.put(subject,
[]);
  connectedRaters.each: { |ir8u-
app|
  ir8uapp<-
rateMe(remoteInterface, aFlockr.
getProfile(), subject);
};
def rateFlockr(ir8uApp, subject,
rating) {
  ir8uApp<-rate(aFlockr,
aFlockr.getProfile().username,
subject, rating);
};
```

All of the previously shown method definitions are contained in the local interface as these methods and data structures should not be accessible from remote devices. Now we describe the remote interface by which remote devices can interact with the applications in the proximity. As mentioned earlier, this remote interface defines the methods rate and rateMe. The rate method is invoked when a remote Flockr issues a rating on this flockr's subject. It simply records the rating given by the Flockr and updates the GUI. The rateMe method is called when a remote Flockr asks this Flockr to rate him on a subject. The method sends a message to the GUI to ask the user for a rating.

```
def remoteInterface:=
  extend: localInterface.remote-
ApplicationInterface with: {

    def rate(ratingFlockr, rat-
ingFlockrName, subject, rating)
{
      ratingSubjects.
put(subject, prevRatings +
[[ratingFlockr, rating]]);
      ui.updateRating(ratingFloc
krName, subject, rating);
    }
    def rateMe(ir8uApp, profile-
ToRate, subject) {
      ui.askToRate(ir8uApp, pro-
fileToRate, subject);
    };
};
```

Guanotes: An Advanced UrbiFlock Application

In addition to the IR8U application, we have implemented an application called *Guanotes*, inspired by the "Wall" plug-in of Facebook where people can post notes on somebody else's "wall". In Urbiflock, Guanotes allows flockrs to post notes to any flockr present in their surroundings or belonging to a particular flock. This allows users to define a target group of note receivers in addition to target individuals (as done in the Wall plugin). For example, one could imagine sending price reductions only to people who have a birthday while the reduction is active. This information is retrieved by the corresponding proximity from the flockrs' profiles and is used to update the corresponding flocks, which can be used by different applications, such as in this case the Guanotes application.

A guanote thus consists of a message and a receiver list (a flock or individual flockrs). Similar to IR8U, Guanotes keeps track of the connected

Figure 5. Guanotes usage scenario in Urbiflock

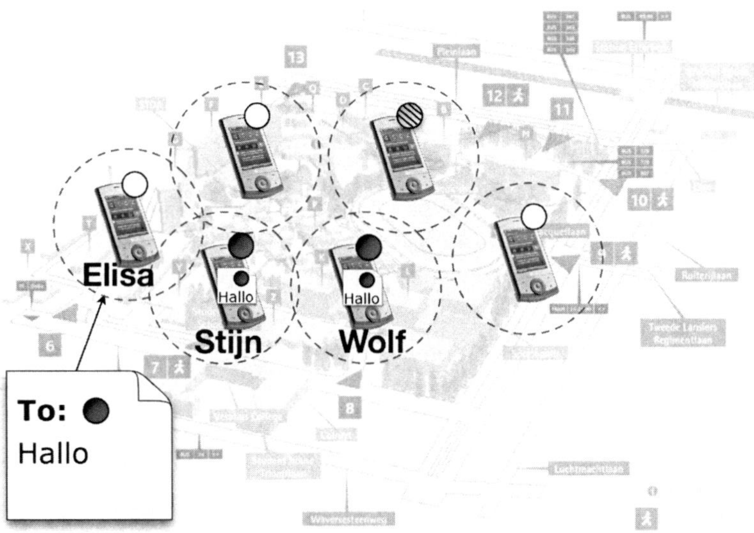

flockrs which are running the guanotes application. Guanotes applications communicate with each other (by means of the Urbiflock framework) to interchange guanotes they carry. A guanote is propagated to another device only if a flockr belongs to the receiver list. A guanote is thus propagated through the network hopping from device to device in order to ensure that it gets received by as much targeted flockrs as possible.

Figure 5 illustrates the propagation of a guanote in Urbiflock. It shows six different flockrs connected in Urbiflock and running the Guanotes application. The communication range of their devices is depicted with a dotted line while the colored dot on their devices denotes the gender of the flockr (white for girls, black for males, and grey with stripes if the gender is not set in the flockr's profile). In particular, the figure 5 shows how a hallo guanote is sent by Elisa to the blueFlock which is defined as "all nearby male flockrs". This guanote is transitively propagated until all connected flockrs belonging to the blueFlock are reached. The guanote is not propagated to flockrs that do not adhere to the definition of blueFlock. The guanotes is first propagated to Stijn and then to Wolf. Elisa's device is in communication range of two devices (corresponding to a white flockr and Stijn), but since the guanote is only meant for male nearby flockrs, it is only propagated to Stijn. Similarly, Stijn's device is nearby to Elisa, Wolf and a third white flockr, but it only propagates the guanote to Wolf. It is important to note that Wolf receives the hallo guanote without being in direct communication range of Elisa's device.

Guanotes, like IR8U, builds upon the services offered by the Urbiflock framework, such as the discovery of remote flockrs, the communication between applications, accessing and comparing profiles, etc. Thanks to Urbiflock, the Guanotes application only needs to concern itself with the storage and propagation of notes to nearby flockrs.

FUTURE RESEARCH DIRECTIONS

There are concrete plans for the deployment of Guanotes as an advertisement application on the Brussels public transportation system. With this application commuters will be able to exchange advertisements for items they wish to sell or buy as they encounter each other on trams or buses.

This first city-wide experiment will allow us to assess the scalability of our language when it is used by a massive amount of users.

We are currently also implementing an ambient game in Urbiflock that will be deployed on the campus at the Free University of Brussels. Players are divided into two teams competing for virtual items around the campus (e.g. "capture the flag"). By making use of virtual weapons players will be able to hinder players of the opposite team. This game is more complex than the current applications developed in Urbiflock and it will require a number of external input devices like a GPS receiver and possibly an RFID reader to pick up virtual items. The implementation of such an advanced dynamic application will help us to further identify general repetitive patterns that can then be integrated into the UrbiFlock framework.

AmbientTalk is also used for teaching distributed systems in the computer science master program at the Free University of Brussels. The language has proven to be the appropriate framework to get familiar with the fundamental issues of programming mobile devices deployed on wireless ad hoc networks. It also allows us to incorporate the latest developments of our research into the teaching material. For example, we intend to use Urbiflock next year to make students write their own pervasive social networking applications. We also aim to enhance our development support for AmbientTalk which is currently limited to a simple plugin (for TextMate) supporting syntax coloring, autocompletion for statements and running AmbientTalk scripts. In particular, we are working on an integrated development environment for AmbientTalk in Eclipse focusing on debugging support.

CONCLUSION

We have described AmbientTalk, a distributed object-oriented scripting language specifically designed to deal with the hardware characteristics inherent to mobile ad hoc networks. What makes AmbienTalk a suitable scripting language for the implementation of mobile computing applications are its event-driven application model, its automatic buffering of messages to deal with intermittent connectivity and its built-in peer-to-peer service discovery abstractions to discover nearby applications.

We have introduced Urbiflock, a framework written in AmbientTalk, designed to ease the development of so-called pervasive social applications. Urbiflock enables the spontaneous interaction of people by means of handheld devices, bringing social networking applications one step closer to become tools used during social events rather than merely tools to report on past social activities. The IR8U and Guanotes applications illustrate that Urbiflock is an ideal platform to prototype new pervasive social applications without having to deal explicitly with many low-level issues such as the appearance and disappearance of users in the ad hoc network, tracking changes in the users' profiles or performing group management.

REFERENCES

Agha, G. (1986). *Actors: a Model of Concurrent Computation in Distributed Systems*. Cambridge, MA: MIT Press.

Bal, H. E., Steiner, J. G., & Tanenbaum, A. S. (1989). Programming Languages for Distributed Computing Systems. *ACM Computing Surveys*, *21*(3), 261–322. doi:10.1145/72551.72552

Beale, R. (2005). Supporting Social Interaction with Smart Phones. *IEEE Pervasive Computing / IEEE Computer Society [and] IEEE Communications Society*, *4*(2), 35–41. doi:10.1109/MPRV.2005.38

Ben Mokhtar, S., & Capra, L. (2009). From Pervasive To Social Computing: Algorithms and Deployments. *To Appear in the ACM International Conference on Pervasive Services (ICPS '09).*

Briot, J.-P., Guerraoui, R., & Lohr, K.-P. (1998). Concurrency and Distribution in Object-Oriented Programming. *ACM Computing Surveys*, *30*(3), 291–329. doi:10.1145/292469.292470

Callsen, C. J., & Agha, G. (1994). Open Heterogeneous Computing in ActorSpace. *Journal of Parallel and Distributed Computing*, *21*(3), 289–300. doi:10.1006/jpdc.1994.1060

Cardelli, L. (1995). A Language with Distributed Scope. In *Proceedings of the 22nd ACM SIGPLAN-SIGACT Symposium on Principles of Programming Languages*, 286-297.

Davies, N., Friday, A., Wade, S. P., & Blair, G. S. (1998). L2imbo: a distributed systems platform for mobile computing. *Mobile Networks and Applications*, *3*(2), 143–156. doi:10.1023/A:1019116530113

Dedecker, J., Van Cutsem, T., Mostinckx, S., D'Hondt, T., & De Meuter, W. (2006). Ambient-oriented Programming in AmbientTalk. In *Proceedings of the 20th European Conference on Object-oriented Programming (ECOOP), 4067*, 230-254.

Eugster, P., Felber, P., Guerraoui, R., & Kermarrec, A. (2003). The many faces of publish/subscribe. *ACM Computing Surveys*, *35*(2), 114–131. doi:10.1145/857076.857078

Eugster, P., Garbinato, B., & Holzer, A. (2005). Location-based Publish/Subscribe. *Fourth IEEE International Symposium on Network Computing and Applications*, 279-282.

Fettig, A. (2005). Twisted Network Programming Essentials. Cambridge, MA: O'Reilly Media, Inc.

Gelernter, D. (1985). Generative communication in Linda. *ACM Transactions on Programming Languages and Systems*, *7*(1), 80–112. doi:10.1145/2363.2433

Gray, C., & Cheriton, D. (1989). Leases: an efficient fault-tolerant mechanism for distributed file cache consistency. *SOSP '89: Proceedings of the twelfth ACM symposium on Operating systems principles*, 202-210.

Joseph, A. D., deLespinasse, A. F., Tauber, J. A., Gifford, D. K., & Kaashoek, M. F. (1995). Rover: a toolkit for mobile information access. In *Proceedings of the 15th ACM Symposium on Operating Systems Principles (SOSP '95)*, 156-171.

Jul, E., Levy, H., Hutchinson, N., & Black, A. (1988). Fine-Grained Mobility in the Emerald System. *ACM Transactions on Computer Systems*, *6*(1), 109–133. doi:10.1145/35037.42182

Justin, C., & Rajive, B. (2008). Programming in Mobile Ad Hoc Networks. *The Fourth International Wireless Internet Conference (WICON).* 10.4108/ICST.WICON2008.4932

Kalofonos, D. N., Antoniou, Z., Reynolds, F. D., Van-Kleek, M., Strauss, J., & Wisner, P. (2008). MyNet: a Platform for Secure P2P Personal and Social Networking Services. *Sixth Annual IEEE International Conference on Pervasive Computing and Communications (PerCom)*, 135-146.

Lieberman, H. (1986). Using prototypical objects to implement shared behavior in object-oriented systems. *Conference proceedings on Object-oriented Programming Systems, Languages and Applications*, 214-223.

Liskov, B. (1988). Distributed programming in Argus. *Communications of the ACM, 31*(3), 300–312. doi:10.1145/42392.42399

Mamei, M., & Zambonelli, F. (2004). Programming Pervasive and Mobile Computing Applications with the TOTA Middleware. *PERCOM '04: Proceedings of the Second IEEE International Conference on Pervasive Computing and Communications*, 263-276.

Mascolo, C., Capra, L., & Emmerich, W. (2002). Mobile Computing Middleware . In *Advanced lectures on networking* (pp. 20–58). New York: Springer-Verlag New York, Inc.doi:10.1007/3-540-36162-6_2

Meier, R., Cahill, V., Nedos, A., & Clarke, S. (2005). *Proximity-Based Service Discovery in Mobile Ad Hoc Networks* (pp. 115–129). Distributed Applications and Interoperable Systems.

Miller, M., Tribble, E. D., & Shapiro, J. (2005). Concurrency among strangers: Programming in E as plan coordination. *Symposium on Trustworthy Global Computing, 3705*, 195-229.

Murphy, A., Picco, G., & Roman, G. C. (2001). LIME: A Middleware for Physical and Logical Mobility. In *Proceedings of the The 21st International Conference on Distributed Computing Systems*, 524-536.

Van Cutsem, T., Mostinckx, S., & De Meuter, W. (2008). Linguistic Symbiosis between Actors and Threads. *Computer Languages, Systems & Structures*, *1*(35).

Van Cutsem, T., Mostinckx, S., Gonzalez Boix, E., Dedecker, J., & De Meuter, W. (2007). AmbientTalk: object-oriented event-driven programming in Mobile Ad hoc Networks. In *Proceedings of the XXVI International Conference of the Chilean Computer Science Society (SCCC 2007)*, 3-12.

Varela, C., & Agha, G. (2001). Programming dynamically reconfigurable open systems with SALSA. *SIGPLAN Not.*, *36*(12), 20–34. doi:10.1145/583960.583964

Waldo, J. (2001). Constructing Ad Hoc Networks. *IEEE International Symposium on Network Computing and Applications (NCA'01), 9.*

Yang, N., Ulrich, K., Adrian, S., & Liviu, I. (2005). Programming ad-hoc networks of mobile and resource-constrained devices. *SIGPLAN Not.*, *40*(6), 249–260. doi:10.1145/1064978.1065040

Yonezawa, A., Briot, J. P., & Shibayama, E. (1986). Object-oriented concurrent programming in ABCL/1. *Conference proceedings on Object-oriented programming systems, languages and applications*, 258-268.

ENDNOTES

i This work is based on an earlier work: AmbientTalk: Object-oriented Event-driven Programming in Mobile Ad hoc Networks, in Proceedings of the XXVI International Conference of the Chilean Computer Science Society (SCCC 2007) © IEEE Computer Society Proceedings, 2007.

i i Elisa Gonzalez Boix is funded by the prospective Research for Brussels program of the Institute for the encouragement of Scientific Research and Innovation of Brussels (IWOIB-IRSIB). Christophe Scholliers and Andoni Lombide Carreton are funded by a doctoral scholarship of the Institute for the Promotion of Innovation through Science and Technology in Flanders (IWT-Vlaanderen), Belgium. Tom Van Cutsem is a Postdoctoral Fellow of the Research Foundation - Flanders (FWO).

Chapter 11
Interrupt Handling in Symbian and Linux Mobile Operating Systems

Ashraf M.A. Ahmad
Princess Sumaya University for Technology, Jordan

Mariam M. Biltawi
Princess Sumaya University for Technology, Jordan

ABSTRACT

Handling interrupts is at the heart of a real time operating systems, such operating systems are the Mobile OS. The most commonly used Mobile OS are the Symbian and RT-Linux operating systems. This paper will introduce the differences of interrupt handling in many different aspects to measure these differences effect on mobile applications performance and throughput. The major contributions to this chapter are first to introduce the interrupt handling mechanism in mobile system with through elaboration on the types of interrupt handling that a Mobile OS may use. Then a deep analysis for both interrupt handling mechanisms used by the Symbian and RT-Linux OS is presented. A comprehensive conclusion will be explained about the major differences in all aspects among both Symbian and RT Linux mobile OS.

INTRODUCTION

The production and usage of handheld computers in the form of smart phones have been growing rapidly the last few years, at the same time as the market share of PDAs (Personal Digital Assistants) in their pure form have declined. Due to the development of mobile industry from the hardware perspective, the production and programming of softwares that control the interaction with the hardware must be taken into consideration; this software is called the "mobile operating system." All the mobile operating systems are considered real time operating systems. A real time operating system is a system that requires the computing result to be correct and produced in a specified deadline period. It should also be single purposed having a small size with inexpensive mass-production and specified timing requirements.

Mobile operating system puts constraints on a suitable operating system similar to those of advanced PDAs. The operating system has to have a low memory footprint and a low dynamic memory usage, an efficient power management framework,

DOI: 10.4018/978-1-61520-761-9.ch011

Figure 1. Layered architecture of mobile OS

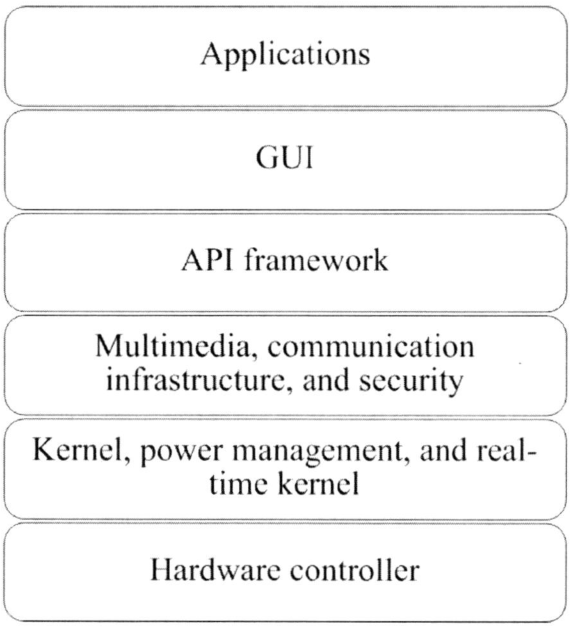

and real-time support for communication and telephony protocols. Furthermore, users often have a more cavalier attitude to mobile phones than to PCs. For instance, when removing the battery while the phone is still switched on a user still expects device and data integrity. Thus the mobile operating system needs a completely new architecture and different features to provide adequate services for handheld devices which can be illustrated in six layers as in figure 1:

The dominant mobile CPU market in respect to cores and architectures is the one designed by Cambridge-based ARM Holdings Ltd. The features of the mobile processors must include: high performance, low power consumption, multimedia capability, and real-time capability.

The goal of this paper is to compare the interrupt handling for two mobile operating systems; the Symbian OS and the Linux OS, the palm OS would have been a part of the comparison if it did not switch to Linux. The Palm OS is a single threaded operating system unlike the Symbian OS and the Linux OS which are multi-threaded. Due to the fact that developers are leaning towards the production of bigger applications for the mobile operating systems, it will be a problem for the Palm OS to stay single threaded because bigger applications need multi-threading. With the Linux-based operating system, Palm hopes to enhance the everyday mobile user's experience with an OS, by making it more reliable and performing better than the previous generation of Palm OS.

Handling interrupts is at the heart of a real time operating system. Managing the interaction with external systems through effective use of interrupts can dramatically improve system efficiency and the use of processing resources. Numerous actions are occurring simultaneously at a single point and thus have to be handled efficiently and in a fast manner. Interrupts are a pinnacle point in the architecture of modern CPU's, to illustrate this point further: The basic mechanism for interrupts is as follows: the CPU hardware has a wire called the interrupt-request line, this line is pulled after each instruction the CPU finishes. If any device has "pulled" the wire, the CPU performs a state save and jumps to the interrupt handler routine. The interrupt handler determines what raised the interrupt, performs what has to be done, does a state restore, and executes a return from interrupt instruction to return the CPU to the state prior to the interrupt.

As a real time operating system, Symbian OS is pretty new (Mäkeläinen & Di Flora & Mikkonen, 2008). A real time kernel was first introduced with version 8.0. Symbian Ltd. was started in 1998 by Psion, Ericsson, Nokia and Motorola. One of the more recent advancements has been the rapid movement of tailoring Linux for suitability in the embedded systems market. This started with kernel and compiler support for all the popular 32-bit microprocessors being designed into embedded systems today, including Intel x86, ARM, Motorola/IBM PowerPC, NEC MIPS and Hitachi SH. Several fast-growing commercial embedded Linux software distributions have popped up, with support for features required in embedded systems designs. In addition, the Linux is an open source project, using the GNU General Public License (GPL). This license allows all code developed for the Linux kernel to be used freely by others, for personal or commercial use, and specifically disallows distribution of the system without also having accompanying source code, including all kernel modifications which have contributed to a big part of its success.

BACKGROUND AND RELATED WORK

Symbian OS and RT-Linux OS use a strategy known as system-on-chip (SOC). Here, the CPU, memory (including cache), memory-management-unit (MMU), and any attached peripheral ports, such as USB ports, are contained in a single integrated circuit. This feature reduces the cost of any real time operating system. One of the important features of real-time operating system is that it should respond to a real time process as soon as that process requires the CPU. Thus the scheduler for real time operating system must support a priority based algorithm with preemption, this algorithm is supported by both Symbian OS and RT-Linux OS.

Symbian OS is designed for the mobile phone environment. It addresses constraints of mobile phones by providing a framework to handle low memory situations, a power management model, and a rich software layer implementing industry standard for communications, telephony and data rendering. Symbian OS has a lightweight 32-bit pre-emptive kernel. This multi tasking operating system runs the kernel in a privileged mode while other tasks run in a non-privileged mode; therefore, access to memory and memory mapped hardware is protected. A Non-privileged mode is the user mode, and a non-privileged program can only access memory allocated to it directly. On the other hand the privileged mode is kernel mode and can access all the memory belonging to any program. The process is the unit of memory protection in the Symbian OS and the thread is the unit of execution as well as the unit that gets scheduled. A process may contain more than one thread. For example the kernel in Symbian OS has two threads: the 'kernel server' and the 'NULL' threads. The kernel server is the highest priority thread while the NULL thread is the lowest priority thread which is run when there is nothing suitable to run on the processor putting the system into the various power saving and sleeping modes. It is also responsible for loading the file server and for booting the kernel on start up.

On the other hand the RT-Linux is also a light-weight 32-bit pre-emptive kernel (Brown, 2007). It includes support for memory management, thread processing and thread creation, inter-process communications mechanisms, interrupt handling, execute-in-place ROM file systems, RAM file systems, flash management, and TCP/IP networking. Real time processes are light-weight threads executing each in their own address space and have the highest priorities. But the Linux kernel has low priority and can be preempted by a real time thread or task. Eventually both operating systems use Round Robin scheduling for threads with equal priorities (Golatowski & Hildebrandt & Blumenthal & Timmermann, 2002).

Interrupt Mechanism

Interrupts are caused either by software and called software interrupts or by hardware and called hardware interrupts. Software interrupts are synchronous interrupts and are caused by events triggered in user mode. One example of software interrupts is a real time clock or timer, periodically setting the interrupt pin high. On the other hand, hardware interrupts are asynchronous interrupts and are caused by a hardware occurring events. An example of hardware interrupts is a button press setting the interrupt pen on the processor high. A signal is created when an interrupt occurs, for instance, a typical interrupt signal could be a button pressed signal or a real time clock signal.

Figure 2. Interrupt controller

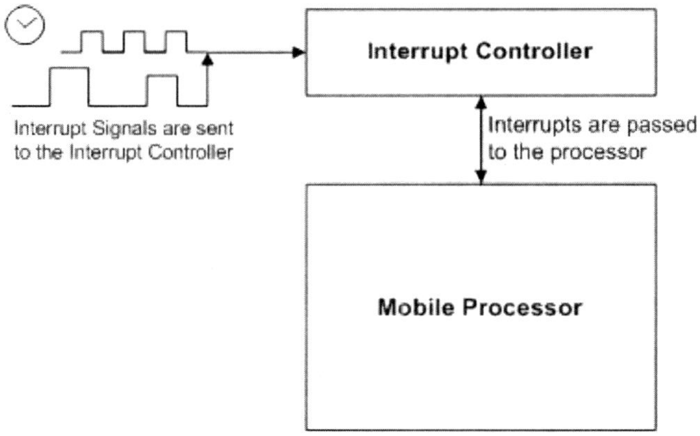

Then all these signals are sent to the interrupt controller as stated in figure 2.

If the interrupt controller disables this interrupt then it will not be passed to the processor. If the interrupt is passed to the processor, the interrupt handler is executed, and the state of the current process is saved, then the interrupt is served using a specific interrupt service routine (ISR), after servicing the interrupt, the context of the interrupted process is restored (See figure 3). The period of time from the arrival of the interrupt at the CPU to the start of the interrupt service routine is referred to as the interrupt latency. Interrupt handler is the routine that is executed when an interrupt occurs. An Interrupt service routine is a routine that acts on a particular interrupt.

Current processor status register (CPSR) is a store to set a bit for enabling and disabling interrupts and also controls the processor mode (SVC, System, User etc.). In a privileged mode the program has full read and write access to CPSR register but in a non-privileged mode the program can only read the CPSR register (Etsion & Tsafrir & Feitelson, 2003). When an interrupt occurs the processor will go into the corresponding interrupt mode and by doing so a subset of the main registers will be swapped out and replaced with a set of mode registers. In privileged modes there is another register for each mode called Saved Processor Status Register (SPSR). The SPSR is used to save the current processor state register CPSR before changing modes.

Several mobile processors have two interrupt inputs. The first is called Interrupt Request (IRQ) and the second is called a Fast Interrupt Request (FIQ). The system needs to determine which interrupt source caused the interrupt and dispatches the relevant handling routine accordingly. In software interrupts; the User Mode is the only mode that is a non-privileged mode. If the CPSR register is set to User mode then the only way for the processor to enter a privileged mode is to execute a software interrupt. The software interrupt call is normally provided as a function of the RTOS. The software interrupt will have to set the CPSR mode to SVC or SYS and then return to the halted program. When any type of interrupt (IRQ or FIQ) occurs and the interrupts are enabled in the CPSR register, the processor will continue executing the current instruction before servicing the interrupt. In general, FIQ's are reserved for high priority interrupts that require short interrupt latency and IRQ's are reserved for more general purpose interrupts. It is recommended that RTOS's do not use the FIQ so that it can be used directly by an application or specialized high-speed driver. The following example will demonstrate the IRQ and FIQ in brief.

Example on IRQ: Assuming that, at the beginning both FIQ and IRQ are set to Zero allowing

Figure 3. Interrupt handling

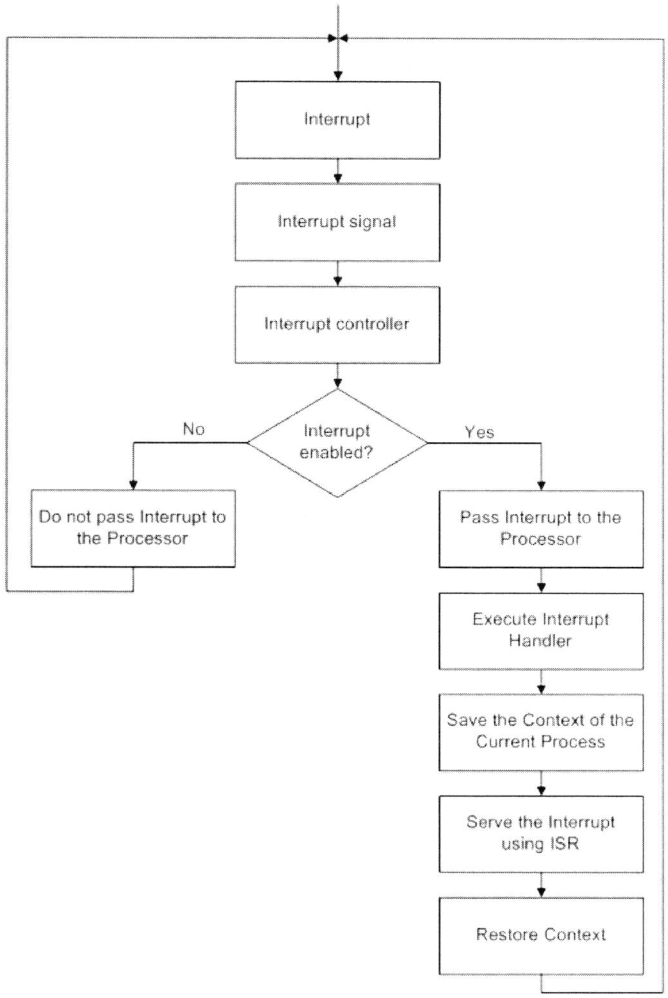

both an IRQ and FIQ to interrupt the processor. When an IRQ occurs the processor will automatically set the I-bit to 1, disabling any further IRQ. The F-bit remains set to 0, allowing an FIQ to interrupt the processor. FIQ are at a higher priority to IRQ, therefore, they should not be disabled. When the mode changes to IRQ mode the CPSR, of the previous mode, in this example User mode is automatically copied into the SPSR register. The interrupt handler then takes over. An example on when FIQ occurs: The processor goes through the same procedure as an IRQ interrupt but instead of just disabling further IRQ (I-bit) from occurring, the processor also disables FIQ's (F-bit). This means that both I and F bits will be set to 1 when entering the interrupt handler. This illustration is shown in figure 4.

In the previous example, after the processor disables any further interrupts, it should vector to the appropriate interrupt handler, this is done via the vector table. A vector table consists of a set of instructions that manipulate the Program Counter (PC). These instructions cause the PC to jump to a specific location that can handle a specific interrupt. Chaining interrupt handlers means saving the existing vector entry and inserting a new entry. If the new inserted handler cannot handle a particular interrupt source this handler can return control to

Figure 4. **(a)** *IRQ interrupt* **(b)** *FIQ interrupt*

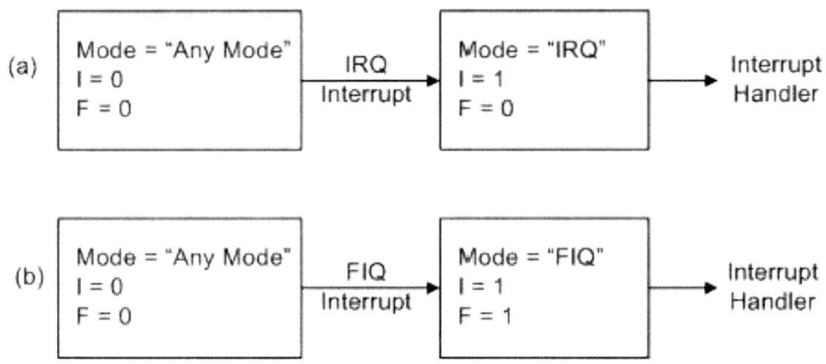

Figure 5. Vector table and interrupt handler

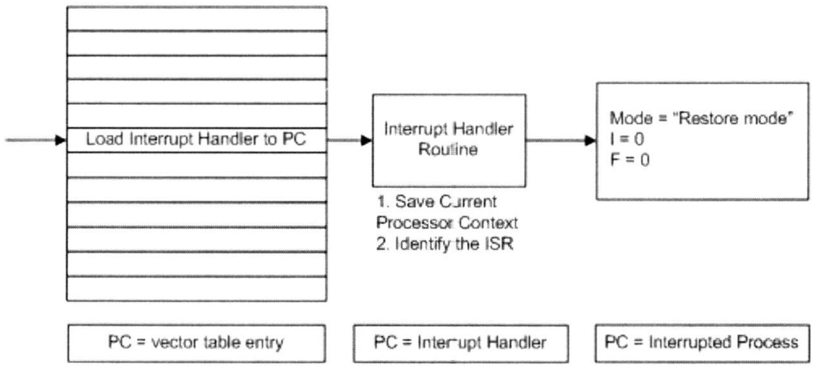

the original handler by the called saved vector entry. Once the new handler has been chained and an interrupt occurs, this new handler will identify the source. If the source is known to it then the interrupt will be serviced. If not, the previous handler will be called. Chaining can be used to share an interrupt handler. Therefore, the interrupt handler saves the current processor context and identifies the interrupt service routine (ISR) to serve the interrupted process, finally after the interrupt is served the context is restored and interrupts are enabled, both FIQ and IRQ are set to Zero.

Interrupt Handling Mechanisms

There are several methods for interrupt handling, any mobile OS may use:

1. **Non-nested interrupt handler:** in this handler interrupts are disabled until control is returned back to the interrupted process. It also services a single interrupt at a time. When the IRQ interrupt is raised the processor will disable further IRQ interrupts occurring. Then the processor will set the PC to point to the correct entry in the vector table and executes that instruction. This instruction will alter the PC to point to the interrupt handler. Once in the interrupt code, the interrupt handler has to first save the context, so that the context can be restored upon return. The handler can now identify the interrupt source and call the appropriate Interrupt Service Routine (ISR). After servicing the interrupt the context can be restored and the PC manipulated to point back to next the instruction prior to the interruption. This is elaborated in figure 6.
2. **Nested interrupt handler:** allows for another interrupt to occur within the

currently called handler. This is achieved by re-enabling the interrupts before the handler has fully serviced the current interrupt as elaborated in figure 7. For a real time system this feature increases the complexity of the system and to be designed carefully.

3. **Re-entrant interrupt handler:** is a method of handling multiple interrupts where they are filtered by priority. The basic difference between a re-entrant interrupt handler and a nested interrupt handler is that the interrupts are re-enabled early on in the interrupt handler to achieve low interrupt latency.
4. **Prioritized interrupt handler:** associates a priority level with a particular interrupt source. A priority level is used to dictate the order in which the interrupts will be serviced. This means that a higher priority interrupt will take precedence over a lower priority interrupt, which is a desirable characteristic in an embedded system. There are several techniques for the prioritized interrupt handling, these technique are: *Simple, Standard, direct,* and *grouped.* The simple and nested interrupt handler services interrupts on a first-come-first serve basis. A *simple priority interrupt* handler tests all the interrupts to establish the highest priority. An alternative solution is to branch early when the highest priority interrupt has been identified this is how the *standard priority interrupt* functions, it follows the same entry code as for the simple prioritized interrupt handler and both has the same start, but the *standard priority handler* intercepts the interrupts with a higher priority earlier. A *direct prioritized interrupt* handler branches directly to the interrupt service routine (ISR), each ISR is responsible for disabling the lower priority interrupts before modifying the CPSR register so that interrupts are re-enabled. This type of handler is relatively simple since the disabling is done by the service routine; it also causes minimal duplication of code since each service routine is effectively carrying out the same task. The *grouped priority interrupt handler* is assigned a group priority level to a set of interrupt sources. This is important when there is a large number of interrupt sources. It tends to reduce the complexity of the handler since it is not necessary to scan through every interrupt to determine the priority level. This may improve the response times.

Figure 6. Non-Nested interrupt handler

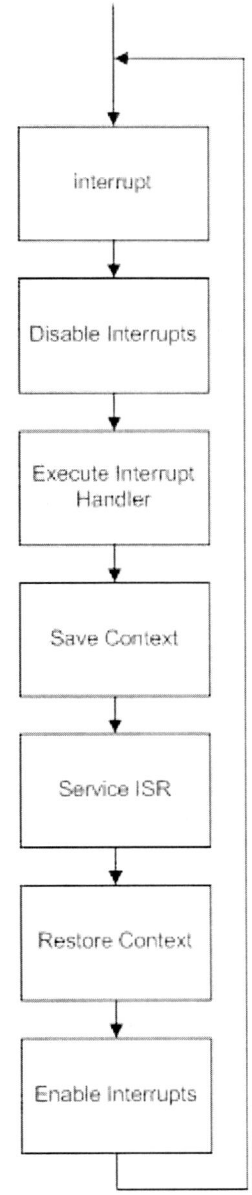

Main Differences Among Interrupt Handling Mechanisms

1. **Simple non-nested interrupt handler:** Handles and services individual interrupt sequentially. Its Interrupt latency is high, and it is easy to implement and debug, but cannot be used to handle complex embedded systems with multiple priority interrupts.
2. **Nested interrupt handler:** Handles multiple interrupts without a priority assignment. Its Interrupt latency is Medium to high. This type of handler can enable interrupts before servicing an individual interrupt is complete, reducing interrupt latency, but it does not handle prioritization of interrupts, so lower priority interrupts can block higher priority interrupts.
3. **Re-entrant interrupt handler:** Handle multiple interrupts that can be prioritized. Its interrupt latency is Low and it can handle interrupts with different priorities. But the Interrupt handler tends to be more complex.
4. **Prioritized interrupt handler:**
   a. **Simple:** Handles prioritized interrupts. Its Interrupt latency is Low and deterministic because the priority level is identified first and then the service is called after the lower priority interrupts are disabled. But the time taken to get to a low priority service routine is the same as for a high priority routine (Rengnier & Lima & Barreto, 2008).
   b. **Standard:** Handles higher priority interrupts in a shorter time to lower priority interrupts. Its Interrupt latency is Low. It treats higher priority interrupts with greater urgency with no duplication of code. But this handler suffers from time penalty because it requires two jumps resulting in the pipeline being flushed each time a jump occurs.

Figure 7. Nested interrupt handler

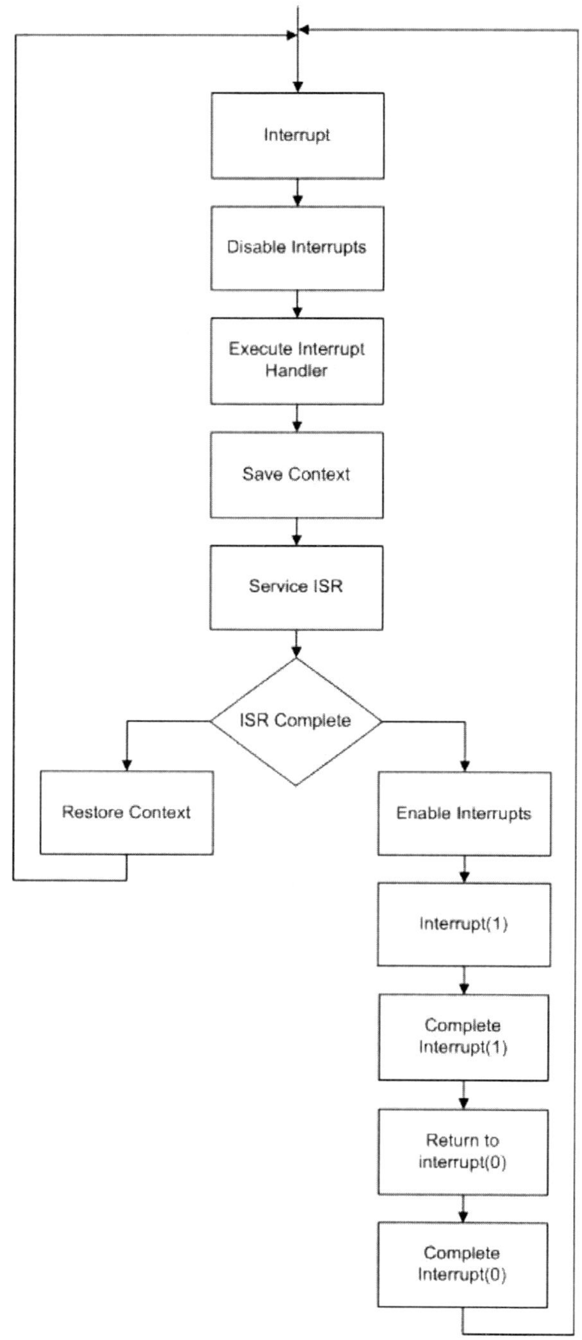

   c. **Direct:** Handles higher priority interrupts in a shorter time goes directly to the specific service routine. Its interrupt latency is Low. It uses a single jump and saves valuable cycles to go to the service, but each service routine has to have a mechanism to set the external interrupt mask to stop lower priority interrupts from halting the service routine.

d. **Grouped:** handles interrupts that are grouped into different priority levels. Its interrupt latency is Low and this handler is Useful when the embedded system has to handle a large number of interrupts. It also reduces the response time since the determining of the priority level is shorter. But determining how the interrupts are grouped together is the main disadvantage this type suffers from.

The following sections will explain the interrupt handling mechanisms for each RT-Linux OS and Symbian OS:

RT-Linux Interrupt Mechanism

The RT-Linux uses the prioritized interrupt handler.

The interrupts in RT-Linux are divided into two groups: soft and hard interrupts. Soft interrupts (Etsion & Tsafrir & Feitelson, 2003) which on average offer a good latency are those under the control of Linux. However, hard interrupts are those controlled by RT-Linux. When a hard interrupt occurs the processor enters IRQ mode, disabling any further IRQ interrupts and dispatching to the appropriate real time interrupt handler, and then the IRQ interrupts are enabled before exiting the real time interrupt handler (Terrasa & García-Fornes, 1999).

All interrupts are initially handled by the Real-Time kernel, then, passed to the Linux task but only when there are no real-time tasks to run. A layer of emulation software between the Linux kernel and the Interrupt Controller Hardware is provided. Thus, when Linux has "disabled" interrupts, the emulation software will queue interrupts that have been passed on by the Real-Time kernel. Linux uses three functions to handle interrupts: The cli macro executes the x86 machine instructions, which clears the enable interrupt bit in the processor control word. The sti macro executes the x86 instructions that set the interrupt flag bit, enabling interrupts. The iret function saves and restores the CPU state before and after the interrupt handler is called. All occurrences of these functions are replaced with emulating macros: S_CLI, S_STI, and S_IRET. This routes all hardware interrupts through the RT interrupt handler (Wang & Lin, 1998).

For disabling interrupts, interrupt state variable in the emulator is reset. When an interrupt occurs, the emulator checks that variable, if it is set; Linux has interrupts enabled, and the Linux interrupt handler is invoked immediately. On the hand, if the Linux interrupts are disabled; the handler is not invoked, instead a bit is set in that variable that holds the information about all pending interrupts. When Linux re-enables interrupts, the emulation software causes control to the Linux handler for the highest priority pending interrupt. This is how the soft interrupts are handled. And because Linux has no direct control over the Interrupt Controller, it does not affect the processing of real time interrupts that do not pass through the emulator.

The S_CLI routine clears the interrupt state variable in the emulator. When the Linux kernel executes the S_STI macro, data is pushed onto a stack (emulating a trap) and then calls the S_IRET routine. The S_IRET routine saves the contents of the registers and initializes the data segment register to point to the kernel. This ensures the kernel data address spaces is accessible, thus making global variables accessible. The bit string variable that contains pending interrupts is scanned. If a set bit is not found, the interrupt state variable is set and control is returned from the interrupt via the iret instruction. If a set bit is found, control is shifted to the Linux handler. The handler ends with an S_IRET call so other pending interrupts will be serviced. During the execution of the iret routine, the Linux kernel also examines the contents of the stack to determine if the interrupt occurred in kernel mode or user mode. If it determines the interrupt originated from the kernel it will not use its own scheduler. Because of this, the routines that prepare the stacks make it appear as if control has been passed directly from the hardware interrupt controller. Linux handlers examine the stack to find out whether it was the user or the kernel code that was interrupted and make decision based on it (Hong & Zhang & Jin-Long Hu, 2006; Momtchev & Marquet, 2002; Rengnier & Lima & Barreto, 2008).

Symbian OS Interrupt Mechanism

Symbian OS uses the Simple Non-nested interrupt handler. The following sections will introduce the software and hardware interrupts in Symbian OS.

Software interrupts in the Symbian OS.

In the Symbian OS programs never link to kernel directly but interface to it through a shared library called euser.dll. This library is located in a known address and contains the necessary instructions to interface to the OS and request its services, thus to access the processor in a privileged mode an executive call (or system call) must be executed. Executive calls switch control to the kernel executives. This means that when programs call user library functions, the user library is pre-programmed to cause a software interrupt, therefore causing the processor to branch to the interrupt handler routine at the processor's interrupt (exception) vector. The interrupt handler checks the type of the executive call and branch to the correct kernel function accordingly (Gao & Hope, 2008; Morris, 2006).

There are two kinds of executive calls designed in Symbian OS: slow and fast executive calls. *Fast executive calls*, operate with the interrupt requests (IRQs) disabled (I-bit = 1) but the fast interrupt requests (FIQs) are enabled (F-bit = 0), thus they are designed to be so short as not to impact interrupt latency, while they usually carry zero to one parameters. Such executive calls are mostly used to gain access to kernel-side objects or to hardware resources. Fast executive calls run in the context of the calling thread (although the processor is switched to supervisor mode); thus they use the heap of the calling thread itself. Nevertheless in order to avoid faulting the system (remember user threads have entered privileged mode now) because of a lack of space on their stack, they make use of a predefined re-entrant stack. Following a fast executive call, the kernel does not try to reschedule any threads, so execution continues from the calling thread.

Slow exec calls, operate with all interrupts enabled (both I-bit and F-bit are set to Zero) and thus can be interrupted by both FIQs and IRQs. Such executive calls are usually for operations that make use of more parameters (up to four), need to save more state and in general need more time for processing (for example when looking up a dll's entry point or ordinal). Slow exec calls run in the context of the calling thread and make use of either the kernel server or null thread stack. Some slow executive calls may also call fast executive calls from the user library. After a slow executive call the scheduler will get the opportunity to switch if necessary to the highest, in priority, ready-to-run thread. Indeed before such a re-schedule takes place the kernel scheduler will attempt to sequentially execute any queued DFCs (deferred function calls, i.e. top half of interrupt handling routines). Thus slow executive calls are called slow; because they need to do more work, they can be interrupted and they may lead to context switching. See table 1.

Executive calls, may access and even modify certain kernel-side objects, as well as offering privileged access to hardware. One thing they are not allowed to do though is to create and/or destroy such kernel-side objects (and in general perform allocations or de-allocations on the kernel heap).

Hardware interrupts in Symbian OS

Only the kernel can access hardware directly and device drivers are used to provide user-side code with a mechanism to access hardware services. A device driver is effectively an add-on to the kernel. It resides on the kernel side and therefore has the same access rights, uses the kernel heap and links to the kernel so that it can call kernel functions. Due to Symbian OS pre-emptive scheduling, there is no known context when an interrupt occurs. In the device driver architecture, the Interrupt Service Routine (ISR) called at interrupt time can schedule a Delayed Function Call (DFC) that runs when the kernel is in a known state (Harrison & Shackman, 2007; Morris, 2006). The ISR/DFC mechanism allows the device driver developer to choose where to perform specific tasks in order to minimize the thread latency response to hardware. Device driver architecture is elaborated in figure 8.

As stated earlier most of Mobile Processors have two input lines, the FIQ and IRQ. Symbian OS has been designed to reuse an interrupt line between multiple source devices. To achieve this, Symbian OS makes use of *interrupt chaining* where the ISRs that correspond to the same interrupt signal but different sources are chained together to form a single linked list of ISRs. When an interrupt occurs, the interrupt service routine provided by each interrupt service object in the list is called in the order in which it was chained. At the same time, the Symbian OS interrupt handling framework prevents from assigning an ISR to multiple signals (source devices). An ISR runs on the kernel side, while the context of the system is unknown to the ISR. At the point of the interrupt, the state of the kernel is undefined, which imposes restrictions on what can be done in the service routine.

When the interrupt request signal is asserted,

1. The processor enters IRQ mode and,
2. branches to the interrupt handler from the vector table then,
3. the interrupt handler immediately looks to discover the source of the interrupt by checking the specific hardware's interrupt controller register for pending interrupts that

Table 1. Differences between fast and slow executive calls

Fast Executive Call	Slow Executive Call
Operate with IRQs disabled and FIQs enabled (I-bit = 1) (F-bit = 0)	Operate with all interrupts enabled I-bit = 0 F-bit = 0)
Designed to be so short	Need more time for processing
Usually carry zero to one parameters	Usually carry more parameters up to four
Mostly used to gain access to kernel-side objects or to hardware resources	---
Run in the context of the calling thread and uses the heap of the calling thread.	Run in the context of the calling thread and make use of either the kernel server or null thread stack
Make use of a predefined re-entrant stack.	---
Following a fast executive call, the kernel does not try to reschedule any threads, so execution continues from the calling thread.	After a slow executive call the scheduler will get the opportunity to switch if necessary to the highest, in priority, ready-to-run thread. but before such a re-schedule the kernel scheduler will attempt to sequentially execute any queued DFCs
Creating a message in SMS application	Playing music in the background while doing other things.

Figure 8. Device driver in symbian OS

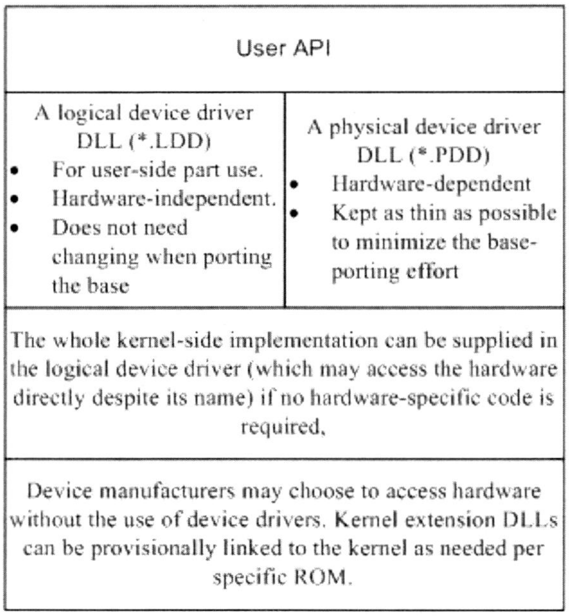

have not been disabled by linearly checking every bit in that register that corresponds to an interrupt signal source.

4. For every pending IRQ source found, the interrupt handler dispatches to the ISR(s), by looking into an interrupt vector table.
5. In the interrupt vector table usually lies a service chain of ISRs; which are offered the id of the source of the IRQ in a FIFO manner.

Previously mentioned process is elaborated in Figure 9.

ISRs have no priorities and nested interrupts are not permitted, therefore, they have to be very short in order to avoid blocking other interrupts for too long; hence, Interrupt handling in Symbian OS is separated into two levels: Interrupt Service Routines (ISRs) and Deferred Function Calls (DFCs), ISR is responsible for:

- Checking whether the interrupt source has a pending interrupt. This is important when several interrupt sources share an interrupt signal.
- Clearing the interrupt bit in the CPSR register.
- Acknowledging the device that its interrupt request has been received
- Doing any necessary I/O.
- Queuing a DFC to continue processing any data if necessary.

When an interrupt happens, the state of the kernel is undefined, which imposes restrictions on

Figure 9. Hardware interrupt in symbian OS

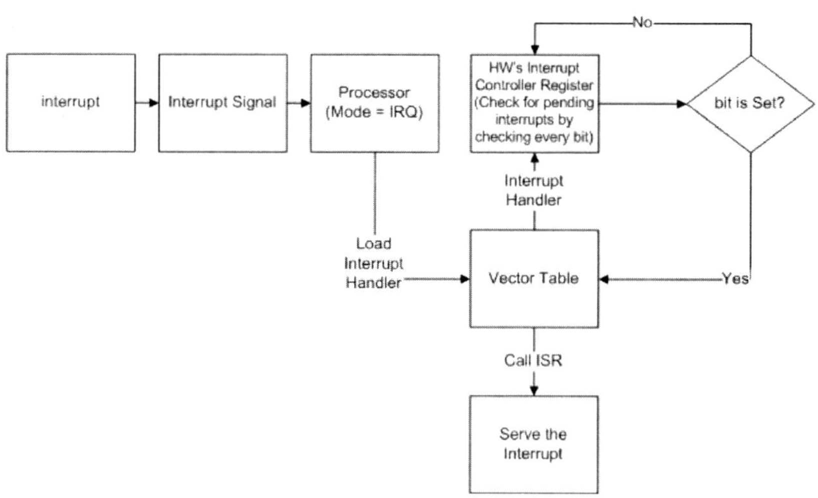

what can be done in the ISR. While a service routine can access the kernel heap and can, therefore, access certain kernel member variables and any memory areas previously allocated, but it cannot allocate or free memory, read from or write to user memory space and signal a thread.

Because nesting of interrupts is not permitted, as stated earlier, any interrupt signal must be fully handled before other interrupt signals can be serviced and in order to perform processing that would otherwise be impossible or inappropriate inside the service routine, ISRs need to queue DFCs. Because the kernel is guaranteed to be in a known state prior to scheduling any DFCs, it means that a DFC can call general kernel functions, signal a thread and access any previously allocated memory and existing data structures. But DFCs cannot allocate or free memory (on the kernel heap) just like ISRs.

During the execution of a DFC, interrupts (IRQs and FIQs) are enabled so that execution time is not as critical as the execution time of an ISR; and interrupt latency is kept to a minimal. Nevertheless, it is still important to keep processing time short because control cannot return to user threads until all DFCs have run. This is because DFCs are scheduled after all ISRs have been called, but just before the kernel reschedules any user threads. Although DFCs are quick, nevertheless they operate with both IRQs and FIQs enabled; which means that they can be interrupted by some other ISR.

Symbian OS interrupt architecture allows for multiple interrupt service routines to be bound to an interrupt signal. This makes shared interrupt lines easier to handle. Interrupt service routines can be added and removed dynamically at runtime; this allows device drivers to add and remove ISRs when they are loaded or unloaded. Furthermore, interrupt service routines can be dynamically enabled and disabled. Simply, if a service routine is disabled, it is not called when the interrupt signal to which it is bound, occurs. DFCs are normally allocated early when a device driver is loaded. Adding them to the kernel's queue of DFCs or removing them from the queue, which simply involves manipulating pointers, requires no further memory allocation or de-allocation. As with ISRs, Symbian OS imposes no limit to the number of DFCs which can be queued.

CONCLUSION

This paper introduced a survey on differences among interrupts in the Linux and Symbian Mobile operating systems; we concluded that both interrupt mechanisms are similar in some ways and different in another, especially in organizational. In Symbian OS the pending interrupts are handled in a FIFO order but in the RT-Linux they are handled in a prioritized order.

REFERENCES

Brown, G. N. (2007). Linux: a platform for innovation in converged mobile handsets. *BT Technology Journal*, *25*(2), 126–132. doi:10.1007/s10550-007-0036-2

Etsion, Y., Tsafrir, D., & Feitelson, D. G. (2003). Effects of clock resolution on the scheduling of interactive and soft real-time processes. In Joint International Conference on Measurement and Modeling of Computer Systems: *Proceedings of the 2003 ACM SIGMETRICS international conference on Measurement and modeling of computer systems: Operating systems* (pp. 172 - 183). New York: Association for Computing Machinery.

Gao, F., & Hope, M. (2008). Collaborative middleware on Symbian OS via Bluetooth MANET. *WSEAS TRANSACTIONS on COMMUNICATIONS*, *7*(4), 300–310.

Golatowski, F., Hildebrandt, J., Blumenthal, J., & Timmermann, D. (2002). Framework for Validation, Test and Analysis of Real-Time Scheduling Algorithms and Scheduler Implementations. In RSP, *Proceedings of the 13th IEEE Intl. Workshop on Rapid System Prototyping* (RSP'02), (pp. 146).

Harrison, R., & Shackman, M. (2007). *Symbian OS C++ for Mobile Phones*. Hoboken, NJ: Wiley Publishing.

Hong, X., Zhang, L., & Jin-Long, Hu. (2006). New Scheme of Implementing Real-Time Linux. In icsea, *(pp.67), International Conference on Software Engineering Advances* (ICSEA'06).

Mäkeläinen, R., Di Flora, C., & Mikkonen, T. (2008). Enhanced integration of Java to symbian OS using smart pointers. In *ACM International Conference Proceeding Series; Vol. 343. Proceedings of the 6th international workshop on Java technologies for real-time and embedded systems.* Real-Time JVM implementation issues (pp. 38-47).

Momtchev, M., & Marquet, P. (2002). An Asymmetric Real-Time Scheduling for Linux. In *ipdps, vol. 2, (pp.0096), Intl. Parallel and Distributed Processing Symposium: IPDPS 2002 Workshops.*

Morris, B. (2006). *Symbian OS Architecture Sourcebook*. Hoboken, NJ: John Wiley & Sons.

Rengnier, P., Lima, G., & Barreto, L. (2008). Evaluation of interrupt handling timeliness in real-time Linux operating systems. *ACM SIGOPS Operating Systems Review*, *42*(6), 52–63. doi:10.1145/1453775.1453787

Terrasa, A., & García-Fornes, A. (1999). Real-Time Synchronization Between Hard and Soft Tasks in RT-Linux. In rtcsa, pp.434, *Sixth International Conference on Real-Time Computing Systems and Applications* (RTCSA'99).

Wang, Y. C., & Lin, K. J. (1998). Enhancing the Real-Time Capability of the Linux Kernel, In rtcsa, (pp.11), *Fifth Intl Conference on Real-Time Computing Systems and Applications* (RTCSA'98).

ADDITIONAL READING

Campbell, A., Aurrecoechea, C., & Hauw, H. (1996). *A survey of qos architectures*. New York: Multimedia Systems.

Divakaran, D. (2002). *RTLinux HOWTO. Internet FAQ Archives Online Education.* Retrieved August 8th, 2002, from http://www.faqs.org/docs/Linux-HOWTO/RTLinux-HOWTO.html

Forin, A., Forin, R., Raffman, A., Raffman, A., & Aken, J. V. (1998). *Asymmetric Real Time Scheduling on a Multimedia Processor*. (Technical Report MSR-TR-98-09). Redmond, WA: Microsoft Research.

Franke, M. (2007). *Seminar Paper: A Quantitative Comparison of Realtime Linux Solutions.* Chemnitz, Germany: Chemnitz University of Technology, Department of Computer Science.

Graf, A., Dabrunz,O,, Assmann, S. (2009). *Interrupt Handling on x86 (RT) and Boot Interrupt Quirks*. Nürnberg, Germany: Maxfeldstr

Higel, S. (2003). *Towards an Intuitive Interface for Tailored Service Compositions*, Compositions, - DAIS 2003 . *Lecture Notes in Computer Science*, *2893*, 17–21.

Iannello, G., Pescapè, A., Ventre, G., & Vollero, L. (2004). *Experimental analysis of heterogeneous wireless networks. WWIC 2004, Wired/Wireless Internet Communications 2004*. LNCS.

Iftode, L., Borcea, C., Ravi, N., Kang, P., & Zhou, P. (2004). Smart phone: An embedded system for universal interactions. In *Proceedings of the tenth International Workshop on Future Trends in Distributed Computing Systems* (pp. 88-94).

Kagami, S. (2001). *Humanoid robot h7 for autonomous and intelligent software research.* In Real Time Linux Workshop, Milan, Italy, 2001. ftp://ftp.realtimelinuxfoundation.org/pub/events/rtlws-2001/proc/k02-kagami.pdf

Kirste, T. (1995). An infrastructure for mobile information systems based on a fragmented object model. *Distributed Systems Engineering Journal*, *2*, 161–170. doi:10.1088/0967-1846/2/3/004

Ledvinam, B., Mota, F., & Kintner, P. M. (2000). A coming of age for gps: A rtlinux based gps receiver. In *Proceedings of the Workshop on Real Time Operating Systems and Applications and Second Real Time Linux Workshop (in conjunction with IEEE RTSS 2000)*, Orlando, Florida, 2000.

Mantegazza, P., Bianchi, E., Dozio, L., & Papacharalambous, S. (2000). Rtai: Real time application interface. *Linux Journal, 72*, 1-1. Retrieved April 2000, from http://noframes.linuxjournal.com/lj-issues/issue72/3838.html

Micheal, J. (2007). *Smart Phone Operating System Concepts with Symbian OS. West Sussex PO19 8SQ*. England: John Wiley & Sons Ltd.

Morris, B. (2007). *The Symbian OS Architecture Sourcebook. West Sussex PO19 8SQ*. England: John Wiley & Sons Ltd.

Pagonis, J., & Sinclair, M. C. (1999). Initial Considerations . In *IEE Colloquium on Lost in the Web: Navigation on the Internet, Digest No. 1999/169*. Evolving Personal Agent Environments to Reduce Internet Information Overload.

Pomiers, P., & Noel, T. (2000). *SynDEx Communications Under Linux*. INRIA Rocquencourt.

Proctor, F. M., Damazo, B., Yang, C., & Frechette, S. (1993). *Open architectures for machine control*. Technical report, National Institute of Standards and Technology, Gaithersburg . *MD Medical Newsmagazine*, (December): 1993.

Proctor, F. M., & Shackleford, W. P. (2001). Timing studies of real-time Linux for control. In *Proceedings of DETC 01 ASME 2001 Design Engineering Technical Conferences & Information in Engineering Conference*, Pittsburgh, PA, September 9-12 2001. ASME.

Proctor, F. M., & Shackleford, W. P. (2002). Embedded real-time Linux for cable robot control. In *Proceedings of DETC'02 ASME 2002 Design Engineering Technical Conf. & Computers & Information in Engineering Conference*, Montreal, Canada, September 29-October 2 2002.

Retrieved Julyt 30th, 2000, from http://www-rocq.inria.fr/syndex/doc/U/SynDExCommsLinux.html

Roe, P., & Chan, S. Y. (1999). *I/O in the gardens non-dedicated cluster computing environment*. IEEE Press.

Rohs, M. (2005). Camera Phones with Pen Input as Annotation Devices. In *Proceedings of the Workshop PERMID* (pp. 23-26).

Schreier, P. G. (2001). *Interfacing DA Hardware To Linux (Technical report)*. United Electronic Industries.

Terziyan, V. (2001). *Architecture for Mobile P-Commerce: Multilevel Profiling Framework*. In Workshop Notes for the IJCAI01, Workshop on E-business & the Intelligent.

Yodaiken, V., Cloutier, P., Schleef, D., Daly, P. N., Rajkumar, R., & Kuhnm, B. (2000, November 27-30). Development of RTOSes and the position of Linux in the RTOS and embedded market. In *Proceedings of the 21st Symposium on Real-Time Systems (RSS-00)*, (pp. 8-8), Los Alamitos, CA: IEEE Computer Society.

Zhang, H., & Arora, A. (2004). All-IP wireless networks. *IEEE Journal on Selected Areas in Communications*, *2*, 613–616.

Zhang, J., Chen, X., Yang, J., & Waibel, A. 2002. A PDA-based sign translator. In Proc. the 4th IEEE Int. Conf. on Multimodal Interfaces.

Chapter 12
Web Page Adaptation and Presentation for Mobile Phones

Yuki Arase
Osaka University, Japan

Takahiro Hara
Osaka University, Japan

Shojiro Nishio
Osaka University, Japan

ABSTRACT

According to the explosive growth of mobile phones, mobile Web has been a part of our life. People can access the Web with their mobile phones and obtain information anywhere and anytime. This trend will stimulate the coming of mobile commerce, where people look for and purchase products on the Web whenever they want. Mobile Web is one of the key technologies for mobile commerce. However, since mobile phones have to be handheld, their interface is strictly limited. Users have to browse large-sized Web pages designed for large displays with a small screen and poor input capability of mobile phones. Additionally, considering mobile users browse Web pages in various situations, users' needs towards presentation functionalities may different depending on their browsing situations. To provide comfortable Web browsing experience under these constraints, we have proposed two systems for mobile phone users. One system provides various presentation functions for Web browsing so that users can select appropriate ones based on their browsing situations. The other system provides functions to navigate users within a Web page so that they can find the information of their interest without getting lost in the page. In this chapter, we briefly introduce designs of these systems and introduce results of user experiments, through which we show that our systems can reduce users' burden on mobile Web by enabling to select appropriate presentation functions adapted to their situations and by navigating them on a large Web page with the entertaining interface.

DOI: 10.4018/978-1-61520-761-9.ch012

INTRODUCTION

We are witnessing the explosive growth of mobile devices. The number of mobile subscribers in the world is projected to be over 4 billion by 2010 from 2.7 billion at the end of 2006.According to this trend, Web access using mobile phones has been also getting popular. In some countries, such as Japan and India, the number of users who access the Web using their mobile phones has exceeded that of PC users. The mobile Web is already a part of our life. At the same time, electronic commerce has got popular as well. Considering these facts, we can expect that the next decade will be the decade of mobile commerce. As a key technology of mobile commerce, mobile Web browsing is important, since people find something to purchase on the mobile Web anywhere and anytime.

However, the current usability of mobile Web is still far from comfortable standard. The problems are twofold: the one comes from a low-bandwidth and the other does from the poor interface of mobile phones, i.e., a small screen and poor input capability. As for the bandwidth, the situation is getting better according to the improvement of the communication facilities, which is apparent from the launch of the advanced connection services, such as 3G and CDMA. On the other hand, the limited interfaces are difficult to improve, since mobile phones have to be handheld. In this chapter, we focus on conventional mobile phones, which only have an ordinary (non-touch) screen and a telephone keypad. As represented by the iPhone from Apple, some advanced smart phones with a touch-screen of comparatively larger size are released; however, the majority of mobile phones in the world still follow the conventional style. Such conventional mobile phones are especially suffered by their limited interfaces on Web browsing.

To solve the problems on Web browsing using mobile phones, we have proposed two browsing systems to provide following functions:

1. Selectable presentation functions based on multimodal mobile user situations
2. Navigation within a Web page

The rest of this chapter is organized as follows. We firstly review prior works related to Web page presentation on mobile devices. Next, we introduce the first and the second system and report the user evaluation. Finally we describe future direction and conclude our chapter.

RELATED WORK

To solve problems of Web browsing using mobile phones, many studies have been conducted. Power Browser (Buyukkokten, Garcia-Molina & Paepcke, 2000; 2001) summarizes text contents within a Web page and then creates an index of the page, deleting all images within the page. When users select a content from the index of the page, it is fully displayed. By doing so, it can reduce the size of the Web page and display more contents on the small screens of mobile phones. RSVP Browser (Bruijin, Spence & Chong, 2002) extracts and sequentially displays important images from a Web page. Doing this allows users to grasp the outline of the page without being bothered by operations. However, it is effective only for pages that contain many meaningful and large images associated with the content.

Some commercial Web browsers for mobile phones, such as the NetFront (NetFront) and the Opera for Mobile (Opera for Mobile), are initially installed in recently released mobile phones. Among them, restructuring Web pages is standard so that users can read pages using only vertical scrolling. However, it is difficult to properly restructure a complicated Web page, e.g., one containing nesting tables.

These prior works have a significant drawback in which they have to change the layouts of Web pages by simplifying or deleting contents of the pages. If the layout of a Web page is changed, users cannot refer to their past Web browsing experience on desktop PCs. For example, users may be used to the presence of a menu list on the left side of a Web page. However, if the layout is changed and this usual feature is removed or is different, users might not be able to comfortably browse the page. In addition, while most prior approaches use HTML tag analysis to change the layout, HTML tags determine the layout of a page but cannot semantically describe the content. Therefore, changing the layout of a Web page might go against the intention of a Web page's author. For example, if an author writes "See the left figure" in a Web page's text and the layout is different, readers may not understand which figure the author means. Additionally, in the user studies of our previous work (Arase, Maekawa, Hara, Uemukai & Nishio, 2007), we confirmed that the number of operations of the browser linearly restructuring pages was not significantly different from that of a conventional browser presenting pages as they are, in the same manner as desktop PCs. The subjects said that it was really bothersome to scroll through Web pages even if the browser linearizes the contents. Furthermore, all subjects deemed more effective our approach preserving the original layouts and providing functions to present them comfortably. Based on this drawback and previous experiment results, we believe that it is best to preserve the original layouts of Web pages in Web browsing using mobile phones, as well as to reduce users' scroll operations. In the following, we present several prior systems that keep the layouts of Web pages.

WebThumb (Wobbrock, Forlizzi, Hudson & Myers, 2002) first displays the overview of a Web page, which is a scaled down image of the page. When a user selects a content area from the overview, this content is displayed in a new application window at its original size. Collapse-to-Zoom technique (Baudisch, Xie, Wang & Ma, 2004) allows users to collapse areas deemed irrelevant from the overview of a Web page. Collapsing a content causes all of the remaining contents to be redrawn with more detail, which increases the users' chance of identifying a relevant content. On the other hand, Baluja (2006) proposed a system that divides a Web page into nine regions so that users can select and zoom in a region on the overview by pressing a corresponding key. The Minimap proposed in (Roto, Popescu, Koivisto & Vartiainen, 2006) changes the widths of text paragraphs and scales down images and tables, while preserving the layout as close as possible to the original layout of a Web page. Our previous system (Maekawa, Hara, & Nishio, 2006b) presents a entire Web page using auto-scrolling to show users the entire structure of the page, which can effectively reduce the number of operations.

Our two systems, which we introduce in this chapter, also follow the policy to preserve the original layouts of Web pages. Furthermore, they provide functions to adapt multimodal mobile user situations and to navigate users in a Web page.

WEB BROWSING SYSTEM FOR MULTIMODAL USER SITUATION

In this system, we aim to adapt to users' Web browsing situations. Web browsing styles using mobile phones are much different from those using PCs. PC users generally browse Web pages sitting in front of their computers, thus, their browsing style is basically static. On the other hand, mobile phone users browse Web pages in various situations, e.g., while shopping in a department store, walking down a street, sitting in a train to commute, or eating meals. Accordingly, appropriate Web browsing styles are different based on users' situations. Although many browsing systems have been proposed as we already described in the previous section, such variety of mobile users' situations has not been considered.

It is usually difficult to precisely detect or predict users' situations by using sensing devices, and improper operations on mobile phones are quite stressful for users because the poor interface

Table 1. Functions assigned to keys on a telephone keypad

key	Function	Outline of the function	
		Menu	**Outline**
1	Overview	Scaled-down view	A scaled-down page that fits the screen size is displayed.
		Tile view	The screen is divided into four sub-screens and a part of components is displayed on each sub-screen.
2	Scrolling per display size	--	User can scroll a page at the unit as the mobile phone's display size.
3	Jump to the previous component	--	User can jump to the previous component (block of related information).
4	Jump to the next component	--	User can jump to the next component.
5	Jump to an image	--	User can jump to an image within a component sequentially.
6	Fisheye view	--	User can browse content with the fisheye view on the overview of a page.
7	Word search	Same word search	User can search words that match with a link that he/she now focuses on.
		Synonym search	User can search synonym words of a link that he/she now focuses on.
		Antonym search	User can search antonym words of a link that he/she now focuses on.
		Input word search	User can search words that match with a word that he/she inputs.
8	Jump to a relevant component	--	User can jump to a relevant component.
9	Auto-scrolling	--	User can browse a component using auto-scrolling.

costs many operations to recover to a proper condition. Therefore, we think that it is reasonable to provide functions so that users can easily select an appropriate presentation style by themselves according to their situations. We propose a novel Web browsing system, called *OPA Browser*, in which the keys of the telephone keypad of a mobile phone have different functions for presenting Web pages. This system enables users to select an appropriate presentation style adapted to their situations.

System Design of OPA Browser

Table 1 summarizes OPA Browser's functions. Users can refer the allocation of these functions on their telephone keypad by pressing a softkey.

In our previous works, we confirmed the effectiveness of the overview(Arase, Maekawa, Hara, Uemukai & Nishio, 2007), jumping to the previous/next components, and auto-scrolling (the display automatically scrolls following a path determined by the system) functions (Maekawa, Hara, & Nishio, 2006b). Most of the other functions listed in Table 1 are those users thought effective for Web browsing, which became apparent in the informal user interviews in the user studies. Thus, we chose these 12 functions and integrated into a system so that users can select one by pressing a single key.

Structure of Web Pages

Before we explain each function of OPA Browser, let us explain the structure of Web pages, which is the basis of the functions. Generally, a Web page is composed of a large number of different components, each of which can be viewed as an information block, such as a site directory and news located on the top page of a portal site. Figure 1 shows an example of components, where each block enclosed with dashed rectangle is a component. Many prior studies have addressed component extraction from a Web page (Chen, Ma & Zhang, 2003; Embey, Jiang & Ng, 1999; Yang, Tan, Mukherjee, Ramakrishnan & Davulcu, 2003). They analyze the structure of HTML tags and perform image processing and text analysis to precisely extract components. However, these studies do not take sizes of components into account.

Figure 1. Example of components

OPA Browser extracts components from a Web page so as to adjust to the size of mobile phone's display because it presents components on the display, and an excessively large component would require a long time for users to read. Specifically, OPA Browser extracts components based on the method proposed in our previous work (Arase, Maekawa, Hara, Uemukai & Nishio, 2007), which uses the DOM (Document Object Model) tree so that the sizes of all components are within the objective size (width and height), from 1 to 5 times the size of the mobile phone's display. We also consider HTML tags to enhance the accuracy of the extraction.

In the following subsections, we briefly introduce OPA Browser's functions. The detailed explanation of each function can be found in (Arase, Hara, Uemukai & Nishio, 2007).

OPA Browser Functions

Function 1: Overview

For mobile phone users, it is difficult to grasp the entire structure of a Web page since a mobile phone displays only a small part of the page. Users usually recognize the role of each content area (e.g., a main content, a menu of the page, and an advertisement) based on the structure of the page. Users also decide in which direction to scroll from the structure of the page, and thus, they often lose their way on the page if they cannot grasp the page structure. To solve this problem, presenting an overview of the page is effective. OPA Browser provides the following two styles of overview.

Scaled-Down View

A scaled-down page that fits the screen size of a user's mobile phone is displayed, so that he/she can grasp the structure of the entire page. Figure 2 (a) shows an example of a scaled-down page. OPA Browser provides a function to zoom up a component. On the bottom of Figure 2 (a), there is a menu represented by "+" mark. When the user presses a softkey, the component that he/she now focuses on is consecutively zoomed up.

Tile View

A mobile phone's display is divided into four parts (sub-screens) and part of each component of a Web page is displayed on each sub-screen. Compared to the scaled-down view, a user cannot grasp the structure of the entire page from this tile

Figure 2. Two styles of overview

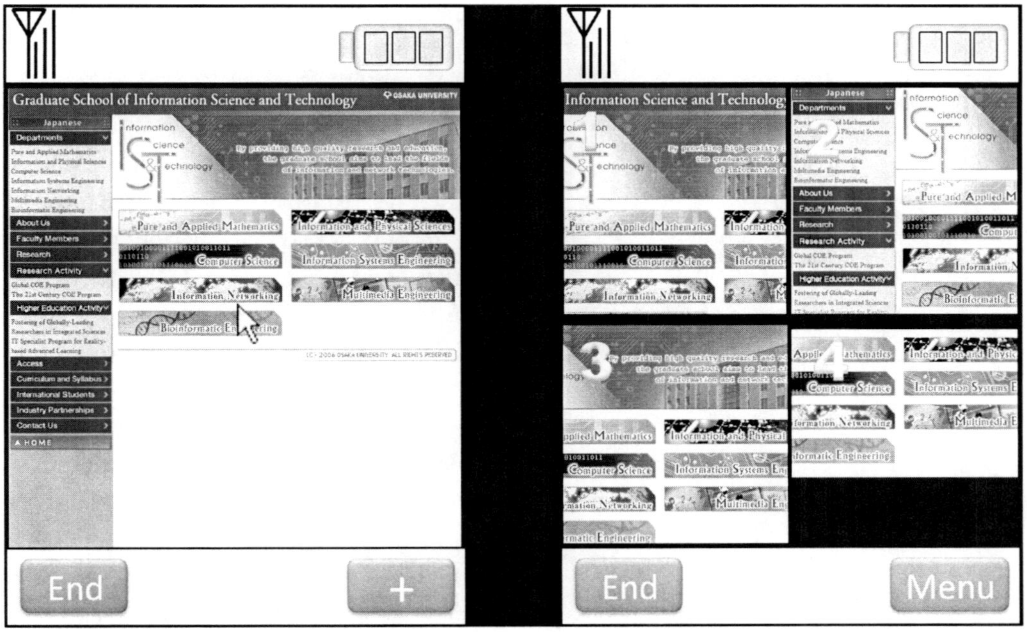

view but can browse each component with the original size, and compare some components all at the same time. Figure 2 (b) shows an example of the tile view, where each sub-screen is allocated a number. If the user selects a certain sub-screen's number by pressing the corresponding key, the entire component is displayed on the full screen with its original size.

Function 2 / Functions 3 and 4

Function 2 (scrolling per display size) aims to decrease the number of users' operations by enlarging the unit of scrolling to the display's width/height.

Functions 3 and 4 (jump to the previous/next components) enable users to jump to the previous and next components from the currently read one by only pressing a single key. In this manner, we can reduce users' scrolling operations. The order of jumping components is determined based on their appearances in the HTML file. Users can move within the page being aware of information blocks.

Function 5: Jump to an Image

Users can sequentially jump to an image within a component. A Web page is generally composed of three contents, i.e., texts, links and images, and images are usually used to attract people's attention. Therefore, the function to jump to an image within a component is useful for finding important information.

Function 6: Fisheye View

While users can grasp the structure of a Web page using the overview, the zoom ratio of the overview is often too small, and thus, users have difficulty to view details of the page. To solve this problem, this function provides a fisheye view. Figure 3 shows an example of the fisheye view, where users can browse contents with the original size on the overview and read them.

Figure 3. Fisheye view

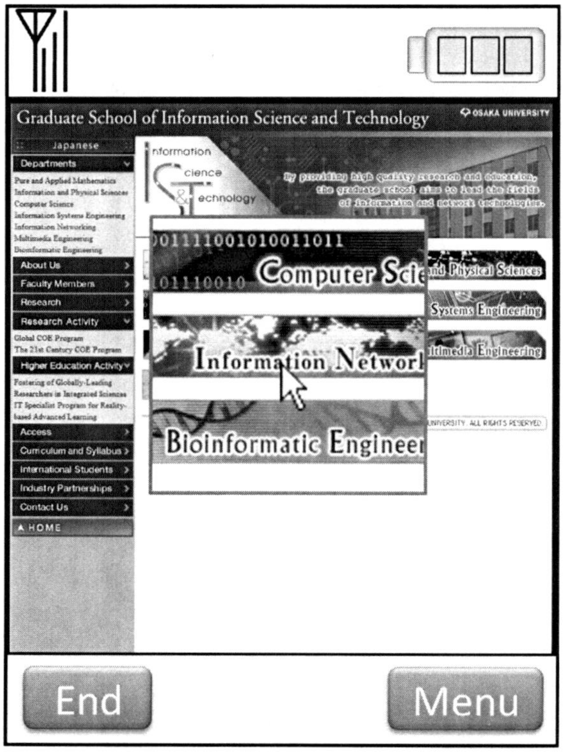

Function 7: Word Search

This function provides four kinds of word search functions that enable users to directly find the information they need. In addition to ordinary word search functionality, it also enables users to find synonyms and antonyms of a queried word. Here, due to a matter of implementation, OPA Browser can use only link strings to word search but cannot use those of plain text. When users input a word or specify a link with the pointer, the display automatically scrolls to links containing the corresponding results.

Function 8: Jump to a Relevant Component

As mentioned before, components are information blocks, and thus some of them seem to have relevant components that share common information within a page. This function enables users to jump to such relevant components from one to another. We assume that this function is effective when users need to collect some relevant information of a specific topic. For example, if a user has concrete aims such as to read articles about a newly released mobile phone in a Web page, the user can easily find the needed article using the word search function (function 7). On the other hand, if a user wants to collect information of other mobile phones as well, the user can find more articles using this function after reading one of the articles of interest.

To determine relevant components, OPA Browser uses all text (both plain text and link strings) and number of images within a component to model it as a *feature vector*. It calculates the relevance score between two components using the cosine similarity measure based on the following equation.

$$relevance = \frac{\vec{v}_1 \cdot \vec{v}_2}{\left|\vec{v}_1\right| \cdot \left|\vec{v}_2\right|}$$

Here, $\vec{v}_1$ and $\vec{v}_2$ are feature vectors of two components. OPA Browser determines two components as relevant if they have high enough relevance score.

Function 9: Auto-Scrolling

Users can view a component by auto-scrolling, without conventional hand scrolling. We confirmed in our previous works (Arase, Maekawa, Hara, Uemukai & Nishio, 2007; Maekawa, Hara, & Nishio, 2006b) that auto-scrolling can reduce users' scrolling operations and enables them to browse Web pages comfortably.

OPA Browser determines the path and speed of auto-scrolling based on our previous work (Maekawa, Hara, & Nishio, 2006b). At first, OPA Browser determines the path according to the shape of a component. Specifically, when the component's height is higher than that of a mobile phone's display and the component's width is narrower than that of the display, the scroll path is set to the vertical direction. On the contrary, when the component's height is lower than that of the display and the component's width is broader than that of the display, the scroll path is set to the horizontal direction. When both the height and width of the component are larger than those of the display, the scroll path is set to zigzag.

After determining the path, OPA Browser calculates the speed [pix/msec] of auto-scrolling based on the following equation, which is based on our previous work (Maekawa, Hara, & Nishio, 2006b).

$$v = \frac{c \cdot Attribute\ Area}{A01 \cdot Breadth}$$

Attribute is a role of the component, such as "HEADER," "FOOTER," "LEFT(RIGHT)SIDE," and "BODY," which is determined based on its shape and location within the Web page. We defined these attributes based on Web page design theory and our previous observation of Web pages. OPA Browser sets the speed faster for minor components, i.e., HEADER and FOOTER. *Area* [pix^2] is the dimension of the component, and *AoI* [msec] means the amount of information within the component, i.e., the time period for users to read the component, which is estimated based on a general insight; average humans need 1 [min] to read 280 words and 100 [msec] to view an image. *Breadth* [pix] is set as component's width for vertical scrolling, as component's height for horizontal scrolling, then as screen's height for zigzag scrolling.

User Experiment of OPA Browser

We conducted a user experiment to verify the effectiveness of OPA Browser. We designed the experiment to be as close as possible to the actual situations of using mobile phones. We asked 30 participants in their twenties to browse Web pages for three days using both OPA Browser and NetFront (for comparison purpose) in the same situations as their own phones. NetFront is a pre-installed commercial Web browser for mobile phones, which restructures Web pages so that users can browse them only by vertical scrolling. This presentation style is one of the standards among commercial Web browsers for conventional mobile phones. We chose NetFront to compare as a representation of the commercial Web browsers for conventional mobile phones.

The 30 participants were volunteers from our laboratory; 8 women and 22 men. Among them, 12 participants have used another commercial browser several times, and they knew the basic operations of commercial Web browsers for mobile phones. The remaining 18 participants had no experience of using such browsers.

The participants used an SH902iS phone for browsing over W-CDMA connection. The display size of the SH902iS is [pix], however OPA Browser and NetFront can use only a [pix] area. The main input control is a direction pad and a center action button for selecting, and it also has two softkeys.

We sent each participant an experimental task by e-mail, instructing which browser to use for the task. We selected 12 goal-oriented tasks (6 tasks for each browser) that access many different types of Web pages, textual and graphical, simple and crowded, small-sized and large-sized, and different page structures. We tried to select tasks that would be somehow interesting to participants, thus most pages were major ones and the contents were recent, such as "Please find a CD you want to listen to on the Sony Music page." In addition, we selected each set of two tasks from twelve tasks at two Web sites that have the same kind of contents and a similar structure and page size, and used them for OPA Browser and NetFront in turn so as to avoid fixing browsers and experimental Web sites for the fairness of the experiment. Furthermore, we requested participants to browse three Web sites freely using OPA Browser. Before starting the experiment, we explained to all participants how to use both browsers and gave them time to get used to the browsers.

When participants finished each browsing task, they sent the feedback via e-mail to report their browsing situations, subjective amount of operations, difficulty of the task, and comments. We also recorded participants' operation logs on OPA Browser. These logs contain information on each operation (the selected functions, the keys pressed by the participants and the time) and the position in a Web page that was displayed on the mobile phone at the rate of 0.1 [sec] intervals to examine participants' orbits of browsing. We could not record logs on NetFront because it is impossible to modify commercial products. Additionally, participants used our experimental mobile phones installed both browsers in a train, at home, etc. as well as their own phones. Therefore, we could not collect logs even manually. In compensation, we could learn users' impressions using them in real situations. Also, orbits of use in NetFront become only straight lines because of the alteration of Web pages, and thus we might not obtain useful insights.

Selected Functions According to Users' Situations

Figure 4 shows the ratio of the selected functions in each situation. The bars are composed of ten cells, each of which corresponds to overview (function 1) to auto-scrolling (function 9) from the bottom and indicates the percentage of the selected times against the total number of selected functions.

In the situation of lying, it is apparent from Figure 4 that jumping to the previous/next component functions (functions 3 and 4) were mainly used (the ratio of functions 3 and 4 was 54%). Participants said that they could easily find the information they need by using jumping to the previous/next component functions because the presented upper-left areas of components were enough informative to grasp what information the components contained at a glance. The posture of using a mobile phone while lying causes numb on arms, therefore, participants felt reluctant against operations. Accordingly, they selected jumping to the previous/next component functions to find the information required by the experimental tasks with simple operations.

On the other hand, in the situation of using public transportation, participants tended to use various functions more frequently than in other situations to view a Web page for fun, such as using jumping to an image and relevant component functions. To understand their motivation of selecting functions, we verified their browsing orbits on the experimental Web pages. As a result, we confirmed that participants in public transportation first looked for the information required by a task, and then, freely browsed the Web page for fun using above functions. This is because participants could concentrate on their display, and in most cases, their motivation of browsing was to kill time while commuting. Therefore, in such situations, it is effective to provide functions to entertain users as well as to decrease operations.

Effects of Page Type

Generally, Web pages can be classified into three categories based on their contents; graphical pages mainly containing images such as top pages of online shopping sites, text-based pages mainly containing text and one or two illustrations or pictures related to the text, such as detailed reports of news, and intermediate pages mixture of text and images, such as top pages of portal sites. Specifically, we classified the experimental pages into graphical pages if images occupy more than half of a page, text pages if it contains text and less than two images except for a logo of the page, and intermediate pages otherwise.

Figure 5 shows the ratio of selected functions in each situation when participants browsed the graphical pages. It is apparent that the ratio of jumping to an image function increased in all situations. As we verified the participants' browsing orbits, we confirmed that they used the function to find images with fewer operations. On the other

Figure 4. Ratio of the selected functions in each situation. The bars are composed of ten cells each of which corresponds to overview (function 1) to auto-scrolling (function 9) from the bottom and indicates the percentage of the selected times against the total

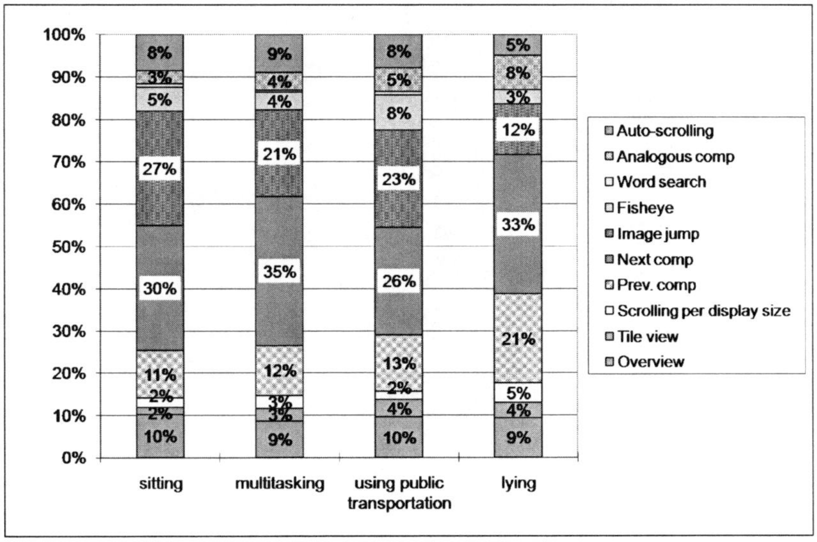

hand, the selected ratio of jumping to an image function decreased on the intermediate pages, and the ratio was lowest on the text-based pages.

In addition, tables are the most difficult part to view on small displays. Figure 6 shows the ratio of selected functions when browsing tables in the text-based pages in each situation. It is apparent that the ratio of the fisheye view function increased. Basing on participants' browsing orbits, we confirmed that participants used the fisheye view function to view tables, which is an unexpected usage for us. They fixed the fisheye view on a row and scrolled it in the horizontal direction to the aiming cell looking the entire table on the background overview. By doing so, they could easily follow rows of a table, even though they protruded away from their screens.

These results show that users can select functions from our OPA Browser adapted not only to their situations but to characteristics of Web pages.

Users' Subjective Impression

In the experiment, participants gave their feedback via e-mail after executing each task. They selected one from "very much," "much," "average," "little," and "very little" according to their subjective amount of operations on Web pages. Figure 7 shows the ratio of participants' subjective amount of operations in each situation. It shows that the ratio of "little" and "very little" was larger in all situations on OPA Browser compared to NetFront, that is, participants felt operations decreased when using OPA Browser.

As for OPA Browser, participants' subjective amount of operations was decreased in the situation of using public transportation (participants felt operations was "little" or "very little" in 60% browsing, the second highest rate among the four situations). This is because participants browsed Web pages to kill time using various functions to find new and interesting information. Consequently, the actual number of operations could not help increasing, however, participants got liberalized against operations, and thus, the

Figure 5. Ratio of the selected functions on the graphical pages

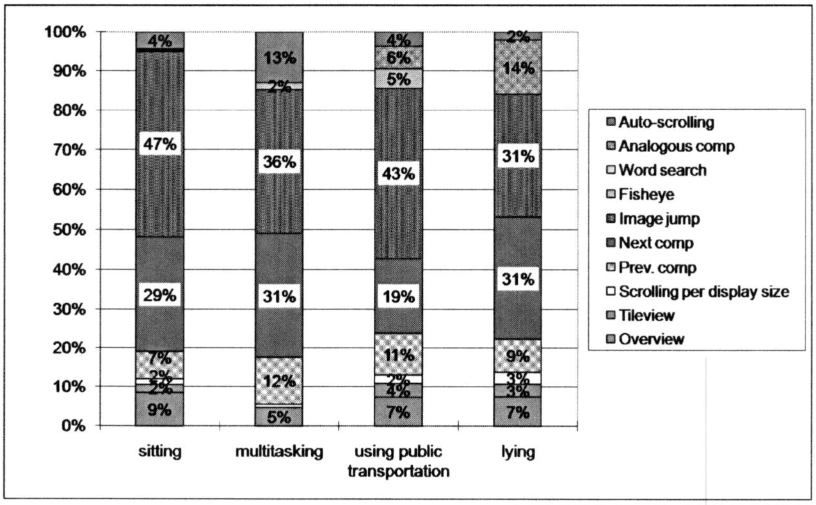

subjective amount of operations decreased. On the contrary, in the situation of sitting, their subjective amount of operations was increased (the ratio of "little" or "very little" was only 38%, the lowest rate among the four situations.) We infer that participants conducted the experimental tasks more seriously because they took time only for the experiment (contrary to multitasking and using public transportation situations) and their posture was more courteous than lying, and thus, they were sensitive to the amount of operations.

As for NetFront, participants' subjective amount of operations especially increased in the situation of multitasking (the ratio of "very little" or "little" was only 14%, the lowest rate among the four situations.) In this situation, participants could not fully concentrate on browsing, and thus, they were more likely to feel burden to operations. In addition, NetFront requires more concentration on display since users cannot expect when the information they need appears due to the alteration of Web pages. Therefore, they have to fix

Figure 6. Ratio of the selected functions on the text-based pages containing tables

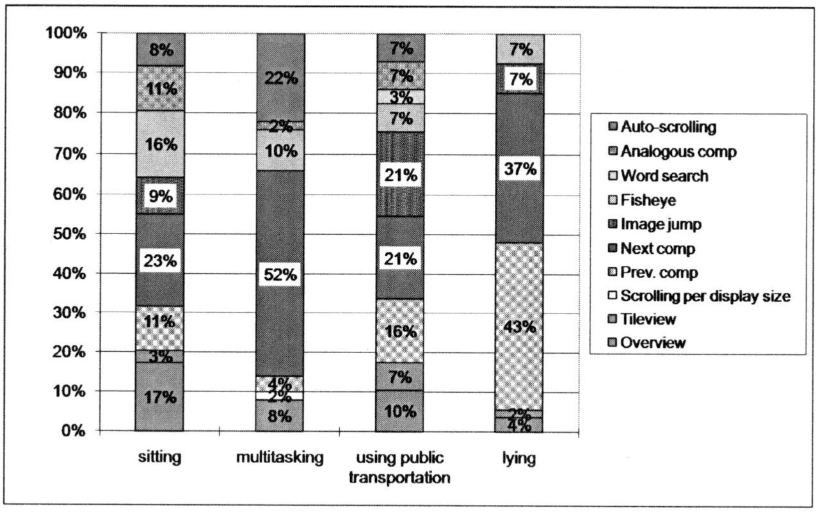

Figure 7. Subjective amount of operations

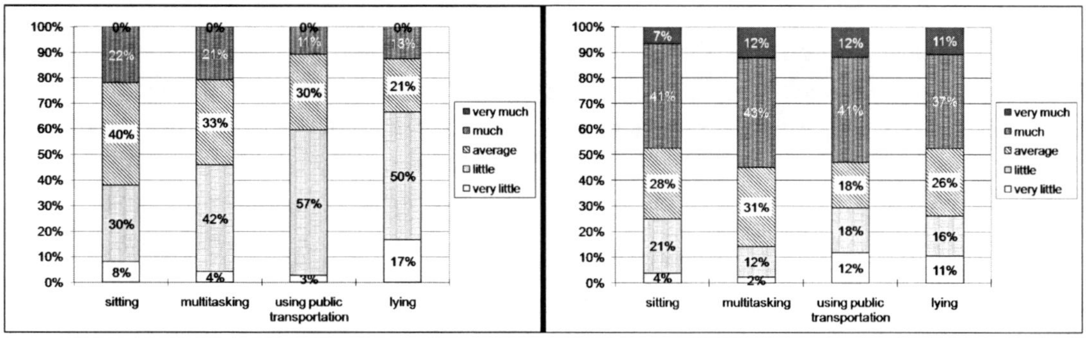

their eyes on the displays while scrolling. Thus, participants' subjective amount of operations increased in the multitasking situation. On the other hand, OPA Browser could decrease participants' burden compared to NetFront by jumping to the previous/next component and auto-scrolling functions, which do not require stubborn concentration on the display.

As a whole, functions that enable users to browse Web pages with simple operation, such as jumping to the previous/next component and overview functions were effective on any situations as basic functions. Additionally, functions that adapt to Web pages' characteristics and users' situations, such as jumping to an image, relevant component, and fisheye view functions, enhanced efficiency and enjoyment on Web browsing. Users can select these functions only by pressing a single key on OPA Browser. We think that users can browse Web pages comfortably by using such basic and other functions in combination by pressing a single key in other situations which we could not observed in this experiment.

WEB BROWSING SYSTEM FOR USER NAVIGATION ON A WEB PAGE

In this section, we describe another system that aims to navigate users within a Web page. Web pages are well structured in nature, where information of the same sort is aggregated into components. Users can unconsciously understand routes to follow when browsing pages with large screens. However, mobile phone users cannot recognize the routes, since a displayed portion of Web pages is too small. As a result, users have troubles to decide a direction to scroll. To enable users to browse pages smartly without getting lost, we propose a system that navigates users within Web pages, which we named as *MotoBrowser*. We adopt an analogy of motoring to a destination on the MotoBrowser.

System Design of MotoBrowser

Browsing large-sized Web pages with small screens has common characteristics with motoring unknown cities, e.g., although people try to get their destinations as efficiently as possible, their available information is strictly limited. However, in the motoring case, drivers usually achieve their goals more easily than mobile users browsing Web pages. The keys are traffic signs, maps, and most importantly, limitation of moving direction due to roads in a city. The roads limit drivers' choices of moving directions, and thus, prevent them from getting lost very often. Too much choices mess people up and disturb making a correct judgment. Additionally, traffic signs navigate drivers, and maps make it possible for them to understand where they are and which direction they are going.

Therefore, MotoBrowser models a Web page as a city by paving roads and presenting traffic signs on it. It provides *Drive mode* that enables users to "drive" the city by auto-scrolling and *Overview mode* that shows them the map of the city.

In the following subsections, we first show a scenario of using MotoBrowser, and next, describe details of methods to pave roads and generate traffic signs. Then, we describe the interface of Drive and Overview modes.

USAGE SCENARIO OF MOTOBROWSER

We show a scenario example of using MotoBrowser to explain its interface. At first, a user launches the MotoBrowser stored in his/her mobile phone, and inputs the URL of a Web page which he/she wants to browse or select it from the bookmark list. Then, the requested page on which roads are paved is presented on the screen of the user's phone, and the user can select Drive mode by pressing the key "1". By doing so, the page is automatically scrolled along with the roads on the page (see Figure 8). While auto-scrolling, an information sign and a speed limit sign are presented, which show what information the neighboring content area contains and the amount of information within it, respectively (see Figure 8 (b) and (c)). When it reaches an intersection, MotoBrowser stops auto-scrolling and waits for the user's selection of which direction (route) he/she wants to go. Additionally, at the intersection, an information sign with a white arrow is displayed to annotate what kinds of contents are on the route ahead (see Figure 8 (a)). The user can choose one route by using his/her direction pad considering the information provided by the information sign. Then, MotoBrowser restarts auto-scrolling. If the user finds a content of interest while auto-scrolling, he/she can stop auto-scrolling by pressing any keys, and can read the content in detail by manual scrolling. The user presses key "2" when he/she wants to go back to the route and restart auto-scrolling.

If the user wants to view the entire page structure, e.g., a map of the page, he/she can select Overview mode by pressing key "3". Then, MotoBrowser presents a scaled-down page that fits the mobile phone's screen. Moreover, it presents an information sign on the content area specified by the pointer to show what kind of information the area contains (see Figure 9). The user presses key "0" to end each mode.

Road Paving Phase

To pave roads on a Web page, MotoBrowser first divides a Web page into components based on the same method with the OPA Browser. After component extraction, MotoBrowser paves roads between components so that users can drop by any components while scrolling.

MotoBrowser paves roads on the Web page based on attributes of the components. First, the attribute of each component is determined based on its location in the page and shape, as "HEADER," "FOOTER," "LEFT(RIGHT)SIDE," and "BODY". Next, roads are paved on the page based on components' attributes so that users can easily access every component from roads. Specifically, roads are paved along the bottom edge of HEADER, the upper edge of FOOTER, the right edge of LEFTSIDE, the left edge of RIGHTSIDE, and the bottom edge of BODY. If a BODY component locates on the left (right) half of the page, a road is also paved along its right (left) side edge. After that, MotoBrowser merges overlapping roads and joins roads with neighboring ones so as to avoid isolated roads.

Annotation Generation Phase

We explain how MotoBrowser generates information signs. MotoBrowser presents two kinds of information signs; one shows detailed information about neighboring components' topics while driving along with routes and the other shows categories of components' topics at each intersection, which are located along the route ahead.

Figure 8. Drive mode (auto-scrolling is proceeding in order of (a), (b), then (c))

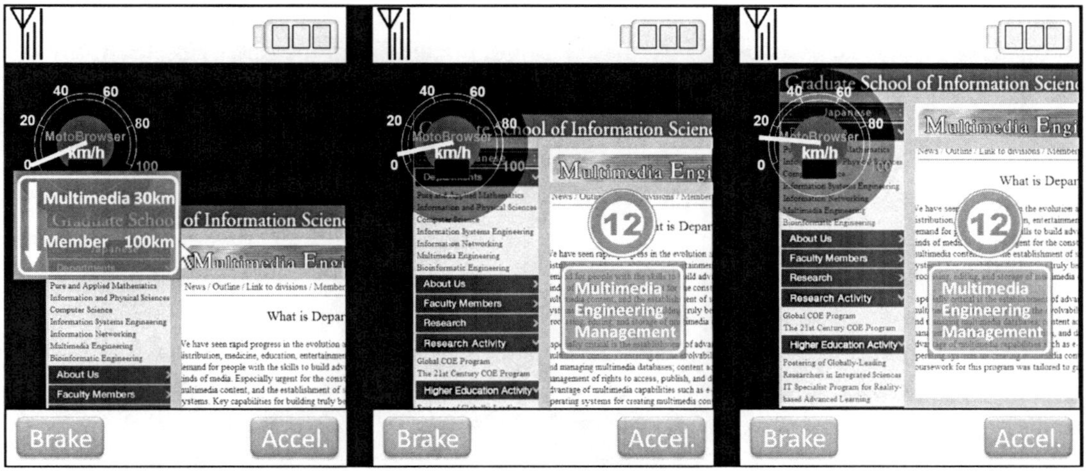

MotoBrowser presents annotations on the information signs. However, it is not easy to automatically create annotations from the source of a Web page, since a Web page usually does not contain enough words to extract annotations using conventional text processing approaches, such as TF/IDF and Lexical Compounds (Anick & Tipirneni, 1999). Some prior studies took into account characteristics of Web pages such as a link structure and layout to summarize them. The method proposed by Shen et al. (2004) extracts a main topic of a Web page by page-layout analysis, and then, uses the sentences within the main topic as the summary of the page. The InCommonSense system (Amitay & Paris, 2000) looks for pages that have a link to the page which is the target of summarization. Then the system extracts sentences around the link to the target page because these sentences are likely the descriptions of the page. However, these approaches cannot be used in MotoBrowser to extract annotations for each component because a component contains much fewer words compared with the entire page, and thus they are not enough to extract appropriate annotations. Additionally, the approach in (Amitay & Paris, 2000) that uses links to the target page is difficult to apply in MotoBrowser since links are usually directed to pages but not to components.

Therefore, in MotoBrowser, we use link structures of components to extract annotations, and also use HTML tags to extract kinds (categories) of topics within the components.

Figure 9. Overview mode

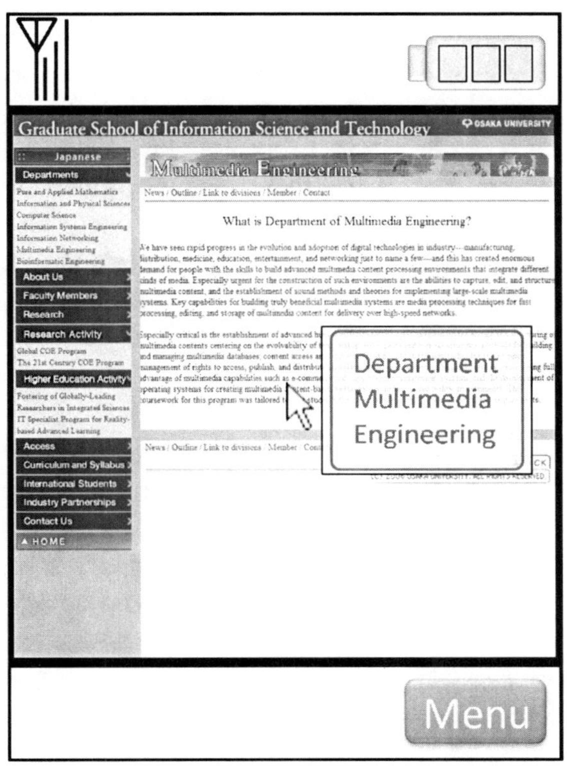

Annotation Extraction

MotoBrowser uses Web pages linked from a target component to extract annotations about topics within the component. Since linked pages usually contain the information related to the topics of the component in most cases, it is reasonable to use them for annotation extraction.

Here, BODY components can be further classified into two types based on the amount of link and text; "Link" components if the total number of characters in link texts is larger than that in plain texts, and "Text" components, otherwise. A Link component can be regarded as a directory, which is a set of links to the same topic, while a Text component can be regarded as a topic itself. MotoBrowser adopts different ways to extract annotations for each component type because they have different characteristics. Specifically, as for Link components, linked pages are more important to extract annotations, while as for Text components, containing texts are important. In the following, we explain the details.

As for Link components, MotoBrowser fetches pages that are linked from the target component (see Figure 10), and then, conducts morphological analysis to pick up only nouns from the entire text within all these pages because verbs and adjectives are not suitable for annotations. After that, MotoBrowser computes each noun's importance using TF/IDF and selects the top three nouns as annotations of the component. To calculate IDF, MotoBrowser uses all nouns derived from the linked pages.

As for Text components, an inside text usually represents their topics, however, its amount is not enough to precisely extract annotations. Therefore, MotoBrowser also makes use of their linked pages if available. If a Text component contains links, MotoBrowser fetches the linked pages, extracts nouns from them, and computes the nouns' importance using TF/IDF, in the same way with that of the Link component. Then, MotoBrowser also conducts a morphological analysis over the plain text within the target component and extracts nouns. If a noun extracted from the linked pages duplicates with the one extracted from the plain text of the component, the importance of the noun is increased. After that, MotoBrowser selects the top three nouns as annotations of the component. If the target component does not contain any links, MotoBrowser uses only nouns within the component and computes their importance using TF/IDF, then, selects the top three nouns as annotations.

Category Extraction

MotoBrowser presents categories of topics contained in components on the route ahead at an intersection. However, it is difficult to detect categories using the method described in the previous section because annotations extracted by the method are too specific as categories. Therefore, we use another feature of Web pages; HTML tags. HTML tags determine the layout of pages, and additionally, they are used to emphasize words and sentences. We checked 50 Web sites of various kinds (news, corporate, and web shopping sites), and confirmed that most components in each page have titles emphatically written using particular HTML tags, such as <H1>, <B>, and <BIG>.

Therefore, MotoBrowser extracts sentences and words that are emphasized by particular tags, i.e., from <H1> to <H6>, <EM>, <STRONG>, <B>, <I>, <U>, <SMALL>, <BIG>, and <FONT> tags, and then, conducts morphological analysis to extract nouns as the categories. If the component contains more than three emphasized nouns, MotoBrowser gives higher priorities to nouns enclosed by the tags listed in higher positions in the above tag list, which emphasize words more strongly than latter ones.

Drive Mode

Drive mode automatically scrolls Web pages along with the paved roads. While auto-scrolling,

Figure 10. Linked pages from a component

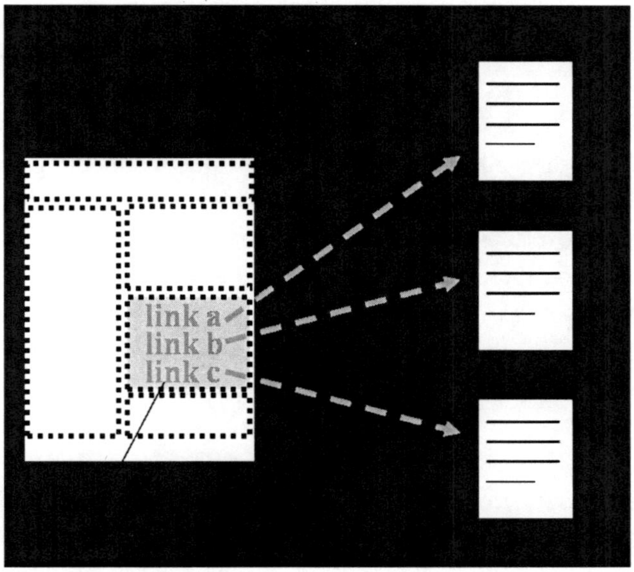

MotoBrowser presents the following traffic signs to help users to browse pages. All these signs are translucently presented not to disturb browsing.

- An information sign with an arrow (Figure 8 (a)) is presented at each intersection to show users what kinds of topics are expected within components on the route ahead. The categories are extracted based on emphasizing HTML tags. For this information sign at an intersection, MotoBrowser selects and shows only one category each from three components on the road ahead. It also shows the distance to the component, where the distance is defined as the Euclid distance from the current center position of the display to the coordinate of the component's upper-left corner. Although the actual unit of distance is pixel, we use "kilometer (km)" in mimicry of actual information signs.
- An information sign (Figure 8 (b) and (c)) is presented while auto-scrolling to show users what information (topics) the neighboring component contains. The annotations are extracted based on the link structure of the component.
- A speed limit sign (Figure 8 (b) and (c)) is presented on a component to show users how much information the component has. The speed limit is calculated based on the time needed to read a component (Maekawa, Hara, & Nishio, 2006b), as the same way with the OPA Browser. The speed limit represents the maximum speed of auto-scrolling that enables users to read the component. Therefore, its value is inversely proportional to the amount of information within the component.
- During Drive mode, a speed meter (Figure 8) is presented to show users the speed of auto-scrolling. Users can adjust the speed using softkeys of mobile phones (the left softkey corresponds to a brake pedal and the right one corresponds to a gas pedal).

Overview Mode

MotoBrowser shows users a "map" (the scaled-down page that fits the mobile phone's screen) of a Web page in Overview mode so that users can grasp the entire page structure and understand where they are viewing and which direction they are going. In addition, MotoBrowser presents an information sign on the component specified by the pointer, to show him/her what kind of information the component contains as shown in Figure 9. The categories are extracted based on emphasizing HTML tags. Here, the information sign shows all three extracted categories of topics within the component.

MotoBrowser Evaluation

We conducted two experiments to verify the effectiveness of our MotoBrowser. At first, we evaluated how accurately it can extract annotations and categories. Then, we conducted a user experiment comparing with a commercial Web browser for mobile phones.

Accuracy of Annotation and Category Extraction

We selected thirty Web sites of various kinds (news, corporate, and web shopping sites; each of them is a popular site such as Amazon) and tested how accurately the two methods in MotoBrowser (link structure based and HTML tag based) can extract annotations and categories. To calculate accuracy, one of the authors of this chapter judged whether the extracted annotations and categories were correct or not. Then, "accuracy" is defined as the ratio of the number of correct annotations (categories) extracted by MotoBrowser to that of all the extracted annotations (categories).

Table 2 shows the accuracy (%) of the two methods for both Link and Text components. As for the method based on the link structure, we think that these results are acceptable considering the limitation of available information to extract annotations. Here, we found an interesting fact between the accuracies of Link and Text components. For Text components, most of the linked pages contain information directly related to the components, such as detailed information of headline news, which resulted in high accuracy of annotation extraction. On the other hand, although Link components have more links than Text components, there are sometimes noisy links, e.g., those connecting to advertising pages, which resulted in lower accuracy than Text components. We have to eliminate these noisy links to achieve higher accuracy. Moreover, some noisy (general) words, such as month and date, were also extracted as annotations. We plan to exclude these noisy words considering their meanings determined by a morphological analysis.

Although the HTML tag based method is very simple, it achieved high accuracy for extracting components' categories. To further improve accuracy, we can apply conventional techniques of detailed HTML source analysis, such as HTML tag pattern recognition. Such an extension is open to our future work.

MotoBrowser User Experiment

We asked 16 participants in their twenties to use both our MotoBrowser and NetFront again for comparison. We should note that the MotoBrowser aims to navigate users on a Web page, while the OPA Browser we introduced in the previous sections aims to provide various presentation functions on a telephone keypad of mobile phones so that users can select appropriate ones by only pressing a single key. Thus, the goals of these two browsers are different. Therefore, we here compare our MotoBrowser with NetFront, as a representation of commercial Web browsers for conventional mobile phones.

The participants used an SH902iS phone for browsing over a W-CDMA connection. The display size of the SH902iS is [pix], however MotoBrowser and NetFront can use only a [pix] area. The main input control is a direction pad and a center action button, as well as two soft keys.

The participants were volunteers from our laboratory; 3 women and 13 men. As a result of the precede questionnaire survey, we found that the seven participants browse Web pages using mobile phones more than once a week (we refer them as "mobile Web experts" in the next subsection). They have been browsing news and portal sites by using mobile phones. On the other hand, the nine other participants rarely browse Web pages using mobile phones (we refer them as "beginners" in the next subsection). Before starting the experiment, we explained all participants how to use both browsers, and gave time to get used to operations of both browsers.

We set six tasks and allocated three tasks for each browser because successive experiments on the same site would affect the results. Tasks are to access top pages of news sites and select ordered links within the pages, such as to access CNN Web site and select "Science" link. After that, we asked participants freely browse linked pages and select links that interest them. The participants repeated this free browsing twice, that is, they browse 3 pages for each task and 18 pages in total.

Here, users' browsing styles can be classified into "Net-surfing" and "search". The former is the style that browses Web pages looking for something new and interesting without specific goals. The latter is the style that browses pages to find target information. We designed this experiment to verify the effectiveness of MotoBrowser in these two styles. Therefore, the tasks given to participants correspond to "search" and the free browsing after the tasks corresponds to "Net-surfing".

The participants were given different combinations of browsers and tasks, and different orders to use browsers for generality and fairness of results. Through the experiment, we recorded operation logs on MotoBrowser. For NetFront, since it is impossible to modify commercial products, we manually record participants' number of operations. Additionally, we asked participants to estimate their subjective amount of operations on each page by selecting one from "very much," "much," "average," "little," and "very little" to investigate users' actual impressions on browsers. As the precede questionnaire survey, we checked whether participants have an experience to browse the experimental pages before.

When finishing the experiment, we conducted a questionnaire survey to compare the MotoBrowser with the NetFront. Specifically, the participants were asked to respond to each of following questions by scoring from -2 (strongly disagree) to 2 (strongly agree).

i. Providing enjoyable browsing situation
ii. Easy to operate
iii. Suitable for Net-surfing
iv. Suitable for search

We analyzed the experimental results from three different aspects; comparing MotoBrowser and NetFront, mobile Web experts and beginners, and already-read pages and unread pages. We describe the details in the following.

MotoBrowser vs. NetFront

We first analyze the results by comparing MotoBrowser with NetFront to understand their effects on users.

For this aim, we collected 275 browsing logs. We confirmed that collected logs have great differences among individuals, and thus, it is not suitable to conduct parametric tests over them. As

Table 2. Accuracy of annotations and categories [%]

	Link component	Text component
Link structure based method	73	82
HTML tag based method	86	83

a result of fit tests, number of operations turned out not to follow a normal distribution. Therefore, we conducted a Mann-Whitney U test with a significance level of 5% for number of operations of both browsers. The result showed that there were significant differences between MotoBrowser and NetFront (p=0.0001). The median value of number of operations was 66.0 on MotoBrowser and 46.0 on NetFront. This result is not surprising because we observed that when using NetFront, the number of operations could be reduced by keeping pressing the direction pad.

On the other hand, Figure 11 shows participants' subjective amount of operations on the experimental pages using both browsers, i.e., how much operations did participants "feel" to need for each browser. As opposed to the actual number of operations, as for MotoBrowser, participants felt the amount of operations was "very much" or "much" in only 13% browsing, while as for NetFront, they felt so in 38% browsing. Furthermore, they felt that their operations were "little" or "very little" in 59% browsing when using MotoBrowser, while only 18% when using NetFront. This result shows that MotoBrowser could decrease participants' actual burden to view pages. As we confirmed in our previous work (Arase, Hara, Uemukai & Nishio, 2007), users' subjective amount of operations does not always correspond to the actual number of operations because it is affected by other various elements. For example, users feel the amount of operations as less when they could find the information of interest soon, even if the actual number of operations was large.

Consequently, although scroll operations can be reduced by keeping pressing a direction pad in NetFront, even such operations were burdensome for participants. Additionally, participants had to focus on the display until information of interest would appear. These are the reasons why users felt much more burden than expected from the actual number of operations. On the contrary, as for MotoBrowser, participants could be relaxed while browsing because they were able to predict what information would appear due to the MotoBrowser's information signs. Furthermore, MotoBrowser's interface could entertain the participants, and thus the burden of operations could be reduced. Indeed, five participants said that they could enjoy the experiment due to the attractive interface of MotoBrowser.

Mobile Web Experts vs. Beginners

Next, we verify the effects of past experience of Web browsing using mobile phones. As we described above, we refer seven participants who browse Web pages using mobile phones more than once a week as "mobile Web experts," while refer nine other participants who rarely browse Web pages using mobile phones as "beginners." We classified logs into ones of mobile Web experts and beginners, and then conducted a Mann-Whitney U test with a significance level of 5% for the number of operations on both browsers. As a result, there were no significant differences stemming from the past experiences both on MotoBrowser and NetFront.

As for NetFront, it is interesting that there was no significant difference of number of operations between mobile Web experts and beginners, since experts are used to using such Web browsers for mobile phones. This is because operations on NetFront are quite simple, and thus, beginners could use it as experts do. As for MotoBrowser, this result is expectable because operations on MotoBrowser are totally different from that of NetFront, and thus, past experiences on commercial Web browsers for mobile phones did not affect the result.

We also investigated the subjective amount of operations of mobile Web experts and beginners on both browsers. As for NetFront, although the actual number of operations was not significantly different between them, Mobile Web experts felt more burden (they felt their operations were "much" or "very much" on 46% browsing) than beginners (they felt their operations were "much" on 32% browsing and none was "very much"). On the other hand, there was no outstanding dif-

Figure 11. Subjective amount of operations: MotoBrowser vs. NetFront

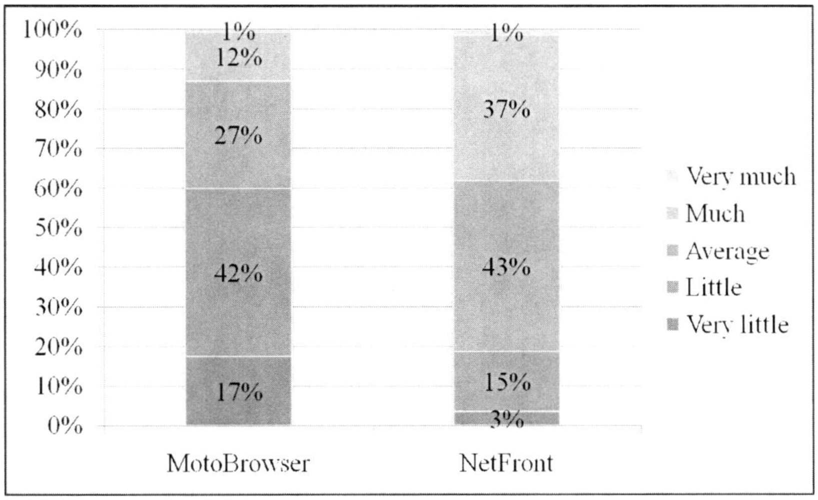

ference on MotoBrowser. This result shows that as users get used to NetFront, they feel more burden even though the actual number of operations does not change because operations on NetFront are monotone and its interface is too simple to attract users.

Already-Read Pages vs. Unread Pages

Finally, we classified participants' logs based on whether or not they have browsed the experimental pages before. We conducted a Mann-Whitney U test with a significance level of 5% for the number of operations of the already-read pages and the unread pages on both browsers. Although we expected that the already-read pages would result in a smaller number of operations and browsing time than the unread pages, there were no significant differences.

However, as Figure 12 shows, participants' subjective amount of operations was different. As for MotoBrowser, participants felt their operations were "little" or "very little" in 72% browsing on already-read pages and felt "much" in only 4% browsing. On the other hand, participants felt their operations were "little" or "very little" in 53% browsing on unread pages and felt "much" or "very much" in 17% browsing. It is apparent that participants could view already-read pages more comfortably than unread pages because MotoBrowser keeps original layout of Web pages, so the appearance of the pages are the same with when using desktop PCs. On the other hand, as for NetFront, there was no big difference between the result on the already-read pages and the unread pages. This result shows the effectiveness of preserving the original layout of Web pages in MotoBrowser.

Questionnaire Survey

Figure 13 shows the scores (each number next to a bar indicates the score) for questions obtained by the questionnaire survey; we present total scores and scores classified into mobile web experts and beginners.

It is apparent from the score of question (i) that the MotoBrowser could entertain participants while browsing. As we confirmed the reduction of the participants' subjective amount of operations, the MotoBrowser can make Web browsing more enjoyable. As for question (ii), NetFront took an advantage for simplicity of operations against the MotoBrowser because NetFront's operations are only scrolling in vertical direction. However, as we confirmed in the previous section, users' burden

Figure 12. Subjective amount of operations: already read pages V.S. unread pages

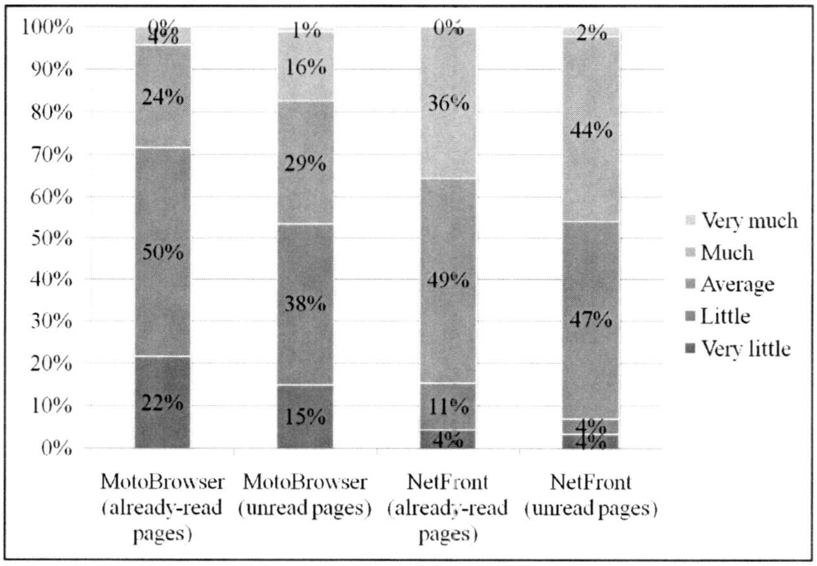

cannot be decreased. This is because NetFront's interface is simple but not really entertaining. On the other hand, even though MotoBrowser's operations are complicated than NetFront, none of participants complained about its difficulty and all of them could use it comfortably as we observed. The positive answer of question (i) also shows the operationality of the MotoBrowser.

Moreover, the MotoBrowser got very high scores for questions (iii) and (iv) which asked participants the effectiveness of MotoBrowser for Net-surfing and searching the target information, respectively. Two participants pointed out that the MotoBrowser's information signs really helped them find the target information, and they could browse the pages effectively. Therefore, improving the accuracy of annotations and category extraction would increase the usability of MotoBrowser. An interesting remark is that mobile Web experts rated NetFront worse for question (iv) although they were used to using NetFront. This shows that as users get familiar with Web browsers for mobile phones, just simple operations are not enough to browse pages comfortably.

CONCLUSION AND FUTURE WORK

The world is witnessing the prologue of mobile commerce, where mobile Web browsing is one of the key technologies, as people look for something to purchase at first on the Web. To provide comfortable mobile Web experience, we have proposed two Web browsing systems for conventional mobile phones.

As for the first system, we proposed a system that offers various presentation functions on each key of the telephone keypad of a mobile phone. In the user experiment, we asked participants to use our system for three days as the same way with their own phones in their real lives. The results showed that they could choose appropriate presentation functions according to their situations and type of browsing Web pages. As a future work, we plan to investigate the effectiveness of functions on each situation. Additionally, we also plan to conduct larger and longer period experiments to collect more logs on various mobile usage situations, where we can expect to obtain useful knowledge for designing mobile applications.

As for the second system, we adopt an analogy of driving a car in a city aiming at navigating users

Figure 13. Total scores on questionnaire survey

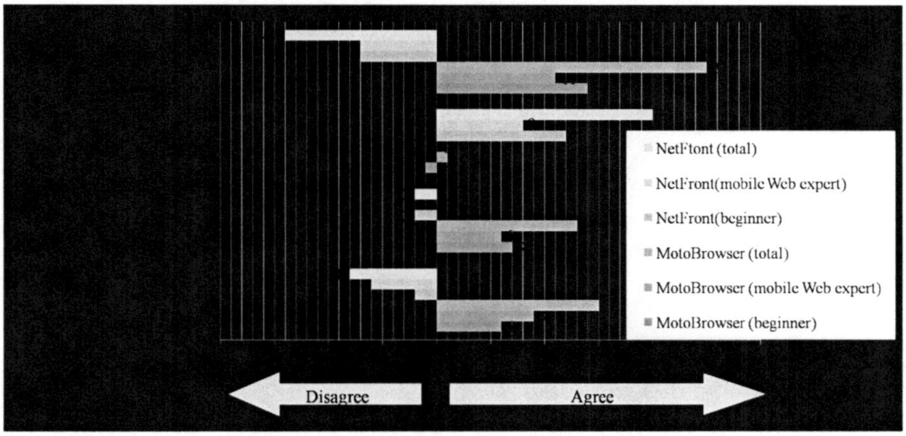

in a Web page. We conducted experiments and confirmed that the MotoBrowser could decrease users' burden while browsing. Additionally, users could enjoy their browsing due to the attractive interface of MotoBrowser. This is the first attempt to adopt "entertainment" aspect to a Web browser for mobile phones. The experimental results showed that it made a good first impression. As a future work, we plan to conduct a follow-up study about the entertainment aspect of MotoBrowser, where users use the MotoBrowser for a longer period.

Considering that the number of mobile subscribers is rapidly increasing and more and more variety of mobile devices would appear, it is interesting to use multiple mobile devices collaboratively to browse Web pages. As we have proposed the concept of "collaborative browsing" (Maekawa, Hara, & Nishio, 2006a), such trend is getting real. By doing so, although each mobile device's capability is limited, users can benefit their advanced features.

ACKNOWLEDGMENT

The authors wish to thank Dr. Shigeyuki Akiba, President & CEO of KDDI R&D Laboratories Inc. for his continuous support for this study. This research was partially supported by "Global COE (Centers of Excellence) Program" and Grant-in-Aid for Scientific Research on Priority Areas (18049050) of the Ministry of Education, Culture, Sports, Science and Technology, Japan.

REFERENCES

Amitay, E., & Paris, C. (2000). Automatically summarizing Web sites - Is there a way around it? In *Conference on Information and Knowledge Management* (pp. 173-179).

Anick, P. G., & Tipirneni, S. (1999). The paraphrase search assistant: Terminological feedback for iterative information seeking. In *ACM SIGIR Conference* (pp. 153-159).

Arase, Y., Hara, T., Uemukai, T., & Nishio, S. (2007). OPA Browser: A Web browser for cellular phone users. In *ACM Symposium on User Interface Software and Technology* (pp. 71-80).

Arase, Y., Maekawa, T., Hara, T., Uemukai, T., & Nishio, S. (2007). A Web browsing system for cellular phone users based on adaptive presentation. *Universal Access in the Information Society*, *6*(3), 259–271. doi:10.1007/s10209-007-0088-6

Baluja, S. (2006). Browsing on small screens: Recasting web-page segmentation into an efficient machine learning framework. In *International World Wide Web Conference* (pp. 33-42).

Baudisch, P., Xie, X., Wang, C., & Ma, W. Y. (2004). Collapse-to-zoom: Viewing web pages on small screen devices by interactively removing irrelevant content. In *ACM Symposium on User Interface Software and Technology* (pp. 91-94).

Bruijin, O., Spence, R., & Chong, M. Y. (2002). RSVP browser: Web browsing on small screen devices. *Personal and Ubiquitous Computing*, *6*(4), 245–252. doi:10.1007/s007790200024

BuyukkoktenO.Garcia-MolinaH.PaepckeA. (2000). Efficient web browsing for PDAs. In CHI conference (pp. 430–437). Power Browser.

Buyukkokten, O., Garcia-Molina, H., & Paepcke, A. (2001). Seeing the whole in parts: Text summarization for web browsing on handheld devices. In *International World Wide Web Conference* (pp. 652-662).

Chen, Y., Ma, W.-Y., & Zhang, H. J. (2003). Detecting web page structure for adaptive viewing on small form factor devices. In *International World Wide Web Conference* (pp. 225-233).

Embey, D. W., Jiang, Y., & Ng, Y. K. (1999). Record-boundary discovery in web documents. In *ACM SIGMOD Conference* (pp. 467-478).

Maekawa, T., Hara, T., & Nishio, S. (2006a). A Collaborative Web Browsing System for Multiple Mobile Users. In *IEEE International Conference on Pervasive Computing and Communications* (pp. 22-33).

Maekawa, T., Hara, T., & Nishio, S. (2006b). Two approaches to browse large web pages using mobile devices. In *International conference on Mobile Data Management*.

NetFront. (n.d.). *NetFront*Retrieved (n.d.), from http://www.access-netfront.com/

Opera for Mobile. (n.d.) *Opera for Mobile*. Retrieved (n.d.), from http://www.opera.com/products/mobile/

RotoV.PopescuA.KoivistoA.VartiainenE. (2006). A Web page visualization method for mobile phones. In CHI conference (pp. 35–45). Minimap.

Shen, D., Chen, Z., Yang, Q., Zeng, H. J., Zhang, B., Lu, Y., & Ma, W. Y. (2004). Web-page classification through summarization. In *ACM SIGIR Conference* (pp. 242-249).

Wobbrock, J., Forlizzi, J., Hudson, S., & Myers, B. (2002). WebThumb: Interaction techniques for small-screen browsers. In *ACM Symposium on User Interface Software and Technology* (pp. 205-208).

Yang, G., Tan, W., Mukherjee, S., Ramakrishnan, I. V., & Davulcu, H. (2003) On the power of semantic partitioning of web documents. In *Information Integration on the Web* (pp. 39-46).

Chapter 13
Technologies and Systems for Web Content Adaptation

Wen-Chen Hu
University of North Dakota, USA

Naima Kaabouch
University of North Dakota, USA

Hung-Jen Yang
National Kaohsiung Normal University, Taiwan

Weihong Hu
Shandong Sport University, China

ABSTRACT

The world has witnessed the blossom of mobile commerce in the past few years. Traditional Web pages are mainly designed for desktop or notebook computers. They usually do not suit the devices well because the pages, especially the large files, cannot be properly, speedily displayed on the microbrowsers due to the limitations of mobile handheld devices: (i) small screen size, (ii) narrow network bandwidth, (iii) low memory capacity, and (iv) limited computing power and resources. Therefore, loading and visualizing large documents on handheld devices become an arduous task. Various methods are created for browsing the mobile Web efficiently and effectively. This chapter investigates some of the methods: (i) page segmentation, (ii) component ranking, and (iii) other ad hoc methods. Though each method employs a different strategy, their goals are the same: conveying the meaning of Web pages by using minimum space. The major problem of the current methods is that it is not easy to find the clear-cut components in a Web page. Other related issues such as mobile handheld devices and microbrowsers will also be discussed in this chapter.

INTRODUCTION

Mobile commerce has drawn great attention these days and people start using mobile handheld devices such as smart cellular phones to perform all kinds of activities such as mobile Web browsing and instant messaging. According to Gartner, Inc., a market research company, the number of units of PCs, smartphones, and cellular phones shipped in 2008 are

DOI: 10.4018/978-1-61520-761-9.ch013

- 302.2 million PCs including desk-based PCs, mobile PCs, and X86 servers (Gartner, Inc., 2009a)
- 139.3 million smartphones, which are mobile phones with advanced functions such as PC-like functions (Gartner, Inc., 2009b)
- 1.22 billion mobile phones (Gartner, Inc., 2009c)

The number of smartphones shipped is increased fast in recent years and it is a little less than half of the number of PCs shipped. It is expected the number of smartphones shipped will surpass the number of PC shipped in the near future. When people started using handheld devices to browse the mobile Internet about ten years ago, Webmasters usually created two versions of their Web pages. One version using HTML is for desktop browsers and the other one using WML, cHTML, or other languages is for microbrowsers. However, this approach has been proved futile and time-consuming and most Web sites have only one version in HTML for both desktop browsers and microbrowsers today. Most Web pages are mainly designed for desktop or notebook computers. They usually do not suit the devices well because the pages, especially the large files, can not be properly, speedily displayed on the microbrowsers due to the limitations of mobile handheld devices: (i) small screen size, (ii) narrow network bandwidth, (iii) low memory capacity, and (iv) limited computing power and resources. Therefore, loading and visualizing large documents on handheld devices become an arduous task.

A wide variety of methods have been used for Web content adaptation for mobile handheld devices. This chapter gives the challenges faced by these methods. It includes three themes:

- **Internet-enabled mobile handheld devices:** Mobile users browse the mobile Internet by using mobile handheld devices, which include six major components: (i) mobile operating systems, (ii) mobile central processing units, (iii) microbrowsers, (iv) input and output components and methods, (v) memory and storage, and (vi) batteries.
- **Microbrowsers:** Microbrowsers are a small version of desktop browsers such as Microsoft Internet Explorer and Firefox. They usually apply one of the four approaches to access the mobile Internet: (i) wireless language direct access, (ii) HTML direct access, (iii) HTML to wireless language conversion, and (iv) error.
- **Web content adaptation:** Various methods are used to browse the mobile Web and none of them is dominant. Most of them use the segmentation-and-ranking approach, that is, they display the page components in the order of their importance. This chapter investigates some of the methods:
  - **Page segmentation:** which is used to segment Web pages
  - **Component ranking:** which is used to rank page components after segmentation
  - **Other ad hoc methods:** such as text summarization, transcoding, and Web usage mining

Though each method employs a different strategy, their goals are the same: *conveying the meaning of Web pages by using minimum space.* The major problem of the current methods is that it is not easy to find the clear-cut components in a Web page.

A related survey of Web content adaptation is also given by Alam & Rahman (2003).

INTERNET-ENABLED MOBILE HANDHELD DEVICES

Mobile users interact with mobile commerce applications by using small wireless Internet-enabled devices, which come with several aliases

Figure 1. A system structure of mobile handheld devices

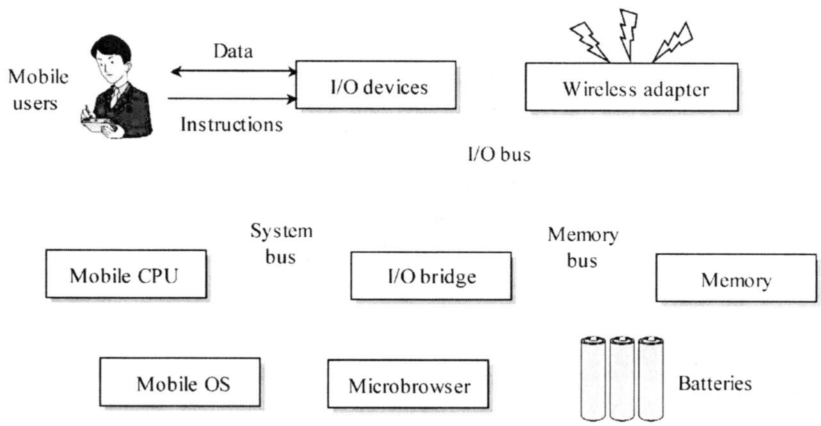

such as handhelds, palms, PDAs, pocket PCs, and smartphones. To avoid any ambiguity, a general term, mobile handheld devices, is used in this book. A mobile handheld device is small enough to be held in one hand and is a general-purpose, programmable, battery-powered computer, but it is different from a desktop PC or notebook due to the following three special features:

- **Limited network bandwidth:** This limitation prevents the display of most multimedia on a microbrowser. Though the Wi-Fi and 3G networks go some way toward addressing this problem, the wireless bandwidth is always far below the bandwidth of wired networks.
- **Small screen/body size:** This feature restricts most handheld devices to using a stylus for input.
- **Mobility:** The high mobility of handheld devices is an obvious feature that separates handheld devices from PCs. This feature also makes possible many new applications such as mobile recommendations that normally cannot be done by PCs.

Short battery life and limited memory, processing power, and functionality are additional features, but these problems are gradually being solved as the technologies improve and new methods are constantly being introduced. The limited network bandwidth prevents the display of most multimedia on a microbrowser. Though the Wi-Fi and 3G networks go some way toward addressing this problem, the wireless bandwidth is always far below the bandwidth of wired networks. The small screen/body size restricts most handheld devices to using a stylus for input. Figure 1 shows a typical system structure for handheld devices, which includes the following six major components: (i) mobile operating systems, (ii) mobile central processing units, (iii) microbrowsers, (iv) input and output components and methods, (v) memory and storage, and (vi) batteries, wich will be detailed next (Hu, et al, 2005). Synchronization connects handheld devices to desktop computers, notebooks, or peripherals to transfer or synchronize data. Without needing serial cables, many handheld devices now use either an infrared (IR) port or Bluetooth technology to send information to other devices.

Mobile Operating Systems

Simply adapting desktop operating systems for handheld devices has proved to be futile. A mobile operating system needs a completely new architecture and different features to provide adequate

Figure 2. A generalized mobile operating system structure

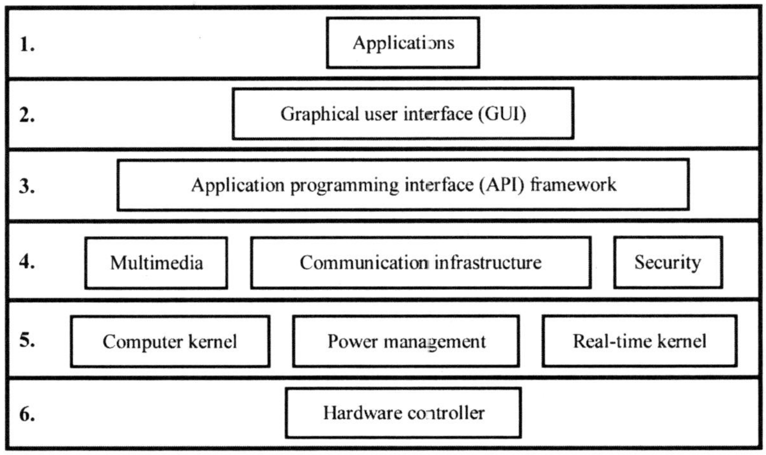

services for handheld devices. A generalized mobile operating system structure as shown in Figure 2 can be visualized as a six-layer stack: (i) applications, (ii) GUI, (iii) API framework, (iv) multimedia, communication infrastructure, and security, (v) computer kernel, power management, and real-time kernel, and (vi) hardware controller.

Mobile Central Processing Units

The core hardware in mobile handheld devices is the mobile processor, and the performance and functionality of the devices are largely dependent on the capabilities of their processors. There used to be several brands available, but recently mobile processors designed by ARM Ltd. have begun to dominate the market. ARM is the industry's leading provider of 32-bit embedded RISC microprocessors, with almost 75% of the market. Handheld devices are becoming more sophisticated and efficient every day and mobile users are demanding more functionality from the devices. To achieve this advanced functionality, in addition to the obvious feature, low cost, today's mobile processors must have the following features:

- **High performance:** The clock rate must be higher than the typical 30 MHz for Palm OS PDAs, 80 MHz for cellular phones, and 200 MHz for devices that run Microsoft's Pocket PC.
- **Low power consumption:** This prolongs battery life and prevents heat buildup in handheld devices that lack the space for fans or other cooling mechanisms.
- **Multimedia capability:** Audio/image/video applications are recurring themes in mobile commerce.
- **Real-time capability:** This feature is particularly important for time-critical applications such as voice communication.

Microbrowsers

Microbrowsers are miniaturized versions of desktop browsers such as Netscape Navigator and Microsoft Internet Explorer. They provide graphical user interfaces that allow mobile users to interact with mobile commerce applications. Microbrowsers usually use one of the following four approaches to return results to the mobile user: (i) wireless language direct access, (ii) HTML direct access, (iii) HTML to wireless language conversion, and (iv) error. Details of microbrowsers will be given in the next section.

Table 1. A comparison of the four kinds of storage available for handheld devices

	Capacity	*Erasable*	*Price Per Unit*	*Speed*	*Volatile*	*Writable*
Flash Memory	~ 5 GB	Yes	3rd	3rd	No	Yes
Hard Disks	~ 100 GB	Yes	4th	4th	No	Yes
RAM	~ 2 GB	Yes	1st (highest)	2nd	Yes	Yes
ROM	~ 1 GB	No	2nd	1st (fastest)	No	No

Input and Output Components and Methods

Because of their size, handheld devices necessarily use different input and output components, methods, and strategies from those used by PCs:

- **Input components and methods:** Entering data into handheld devices is never an easy task because the devices are so small. Various input methods for handheld devices have been developed, the most important of which are: (i) keyboards, (ii) navigator, (iii) touch screens, (iv) writing areas on screens, and (v) speech recognition. Another input option that is often used is to receive data and files directly from PCs.
- **Output components and methods:** Although several alternative input devices and methods are available for handheld devices, the options for output devices and methods are more limited, with the main output component for a handheld being its screen. Handheld devices normally use synchronization technology to print data and files via PCs; handheld printers are available, but they are not common.

Memory and Storage

Desktop PCs or notebooks usually have between a few hundred Mbytes and a few Gbytes of memory available for users, whereas handheld devices typically have only few tens or hundreds of Mbytes. PDAs normally have more storage space than smart cellular phones, with the former commonly having 64 Mbytes, and the latter a memory size that may be as low as a few Mbytes. Four types of storage are usually employed by handheld devices: (i) flash memory, (ii) hard disks, (iii) random access memory (RAM), and (iv) read-only memory (ROM). Hard disks, which provide much more storage capacity, are likely to be adopted by handheld devices in the near future. Table 1 compares these four types of storage; a comprehensive survey of storage options can be found in Scheible (2002). Today's wireless devices demand higher memory throughput for more advanced features, such as Internet browsing, e-mail, data streaming, and text messaging.

Batteries

Replaceable, rechargeable lithium-ion batteries are most commonly used in handheld devices. In smartphones using this kind of battery, the talking time, standby time, and full recharging time currently take a couple of hours, a few hundred hours, and a couple of hours, respectively, and the browsing time will be slightly shorter than the talking time. In the future, it should be possible to use handheld devices without the need to recharge them frequently by replacing the lithium-ion batteries with fuel cells, which although they are not yet practicable are likely to represent the best choice in the long-term. Table 2 provides a comparison between lithium-ion batteries and fuel cells, and detailed descriptions are given below.

Table 2. A comparison between Lithium-Ion batteries and fuel cells

	Contents	*Output*	*Type & Method*
Lithium-Ion Battery	Lithium ions	Electricity	Rechargeable using a power outlet
Fuel Cell	Natural gas	Electricity and water	Refuelable using fuel such as natural gas

MICROBROWSERS

Microbrowsers are miniaturized versions of desktop browsers such as Netscape Navigator and Microsoft Internet Explorer. They provide the graphical user interfaces that enable mobile users to interact with mobile commerce applications.

Features

Due to the limited resources of handheld devices, microbrowsers differ from traditional desktop browsers in the following ways:

- smaller windows
- smaller footprints
- fewer functions and multimedia features

Several microbrowsers, such as Microsoft Mobile Explorer and Wapaka Java Micro-Browser, are already available. America Online, Inc., the parent company of the Netscape Network, and Nokia are developing and marketing a Netscape-branded version of Nokia's WAP microbrowser with AOL enhanced features for use across a wide variety of mobile handheld devices. Figure 3 shows a microbrowser, NetFront Browser v3.5 from ACCESS, which supports Visual Bookmarks—a pan & zoom navigation tool for the desktop-like presentation of web pages on mobile devices with limited screen size (ACCESS Co., Ltd., 2006).

Technologies

Several markup languages are used to present mobile content on microbrowsers. These may not be able to handle all the languages currently used, therefore some content will not be displayed by some microbrowsers. Microbrowsers usually take one of the following four approaches, as shown in Figure 4, to display mobile content (Lawton, 2001):

1. **Wireless language direct access:** Here, a microbrowser supports some wireless languages, such as WML, CHTML, and XML, and directly displays any content written in a wireless language supported by that microbrowser
2. **HTML direct access:** This approach displays the HTML contents directly, with no intervention, but may distort the content. For example, large images cannot be displayed on the small screens of microbrowsers

Figure 3. NetFront Browser v3.5 © 2008 ACCESS Co., Ltd.

Figure 4. Four approaches used by microbrowsers to display mobile content

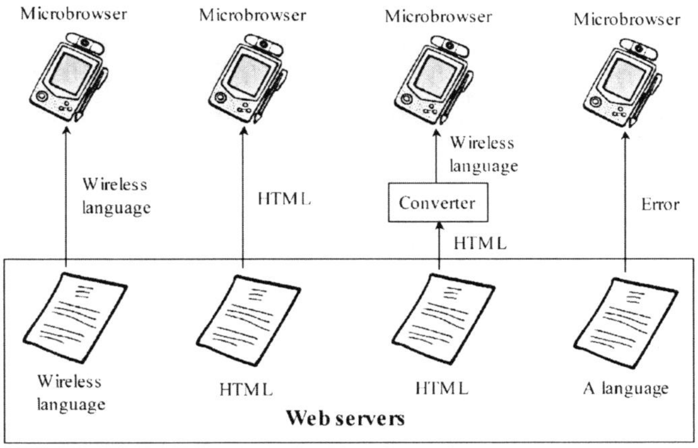

3. **HTML to wireless language conversion:** Some mobile middleware provides conversion software that converts an HTML script to the script of a wireless language supported by that microbrowser. For example, i-mode includes a Corporate Conversion Server that converts existing HTML files into i-mode-compatible HTML, the CHTML.
4. **Error:** If a microbrowser is not able to handle the content, it displays an error code such as "Invalid WML code."

Some microbrowsers, like most desktop browsers, can automatically send and receive information with the aid of a cache, which is known as Web caching (Davison, 2001). Web caching offers significant advantages, such as reduced bandwidth consumption, server load, and latency. Taken together, these advantages make accessing the Web less expensive and improve performance. These three components unique to mobile handheld devices, namely mobile OSs, mobile CPUs, and microbrowsers, result in a significant difference between the performance of handheld devices and desktop PCs; the remaining components do not play such a crucial role.

Major Microbrowsers

A number of microbrowsers are currently available commercially. Four popular microbrowsers are: (i) Opera 8.65, (ii) Openwave Mobile Browser, Mercury Edition, (iii) Access NetFront Browser 3.5, and (iv) Microsoft Pocket Internet Explorer. Table 3 compares these four microbrowsers and detailed descriptions of the microbrowsers are given below. Some companies also provide microbrowser emulators/simulators such as Opera Mini Simulator that enable developers to test their products on desktop computers because small devices are not convenient for mobile application development.

Two Examples of Built-in Web Content Adaptation

This sub-section gives two examples of how microbrowsers display Web content from the industry:

- **ACCESS:** ACCESS' NetFront Browser includes Smart-Fit Rendering technology (n.d.), which intelligently adapts standard web pages to fit the screen width of any mobile device enabling an intuitive and rapid vertical scrolling process, without

Table 3. A comparison of the four leading Microbrowsers

	Mobile Browser 9.5	Mobile Browser, Mercury Edition	NetFront Browser 3.5	Internet Explorer Mobile
Vendor	Opera	Openwave	Access	Microsoft
Support HTML?	Yes	Yes	Yes	Yes
Support WML?		Yes	Yes	Yes if extra software installed
Major Technologies	Small-Screen Rendering	Progressive rendering of content	Smart-Fit Rendering™	Fit-to-Screen menu
Special Features	Flash	Ajax	Ajax	JScript

degrading the quality or usability of the pages being browsed. Concretely, the following process is performed:

- Images larger than the screen width are scaled down to fit the screen width.
- Tables larger than the screen width are split and laid out vertically as shown in Figure 5.

Figure 5. A Web page table split by ACCESS' NetFront Browser

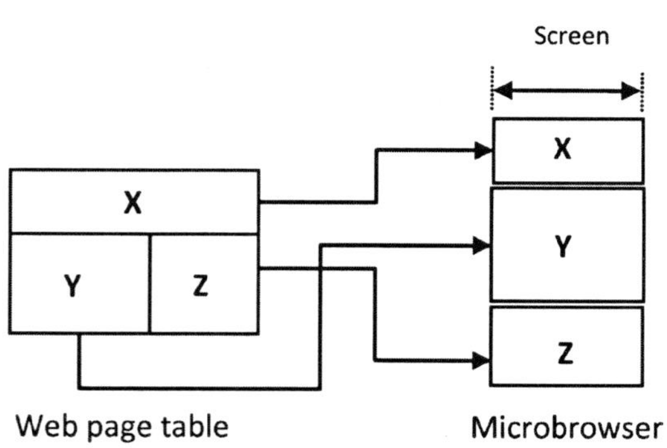

- **Opera:** Opera's Small-Screen Rendering technology (n.d.) reformats Web page to fit it inside the screen width and eliminate the need for horizontal scrolling. All the content and functionality is still available, it is only the layout of the page that is changed. Figure 6 shows an example of the Opera's method.

WEB CONTENT ADAPTATION

Most Web pages are designed for the use of desktop or notebook browsers like Microsoft Internet Explorer and Firefox in mind. When the pages are

Figure 6. Screen shots of Opera' Small-Screen Rendering™: (a) before rendering and (b) after rendering

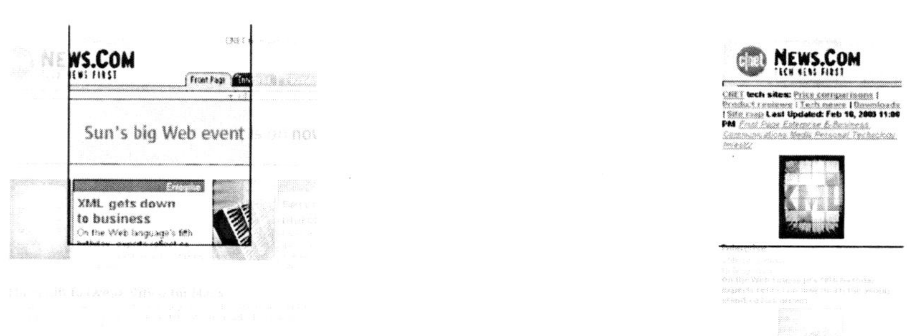

(a) (b)

accessed from microbrowsers, they are distorted or not functioning fully or properly because many of their features such as images and Ajax are removed or disabled. Various methods are created to try to solve or relieve the problems. This section divides the methods into three categories: (i) page segmentation, (ii) component ranking, and (iii) other ad hoc methods, and each category will be detailed next. Though each method employs a different strategy, their goals are the same: *conveying the meaning of Web pages by using minimum space.*

Page Segmentation

Mobile users usually are not interested in every detail of a Web page. For example, a typical commercial Web page usually includes three columns: the left navigation, the main content, and the right navigation/advertisements. Most mobile users would like the main content being displayed first if they have choices. The main idea behind the method of page segmentation is to display parts of Web pages instead of the whole pages when using microbrowsers. In order to realize this idea, Web pages need to be segmented. Several methods are designed to be used to segment Web pages (Gupta, Kumar, Mayank, Tripathi, & Tapaswi, 2007; Hua, Xie, Liu, Lu, & Ma, 2006); Xie, Miao, Song, We, & Ma, 2005; Chen, Ma, & Zhang, 2003). The most popular methods are to analyze the HTML source code and segment the pages according to the HTML tags. For example, Figure 7 shows a typical Web page and its corresponding HTML source code. The method studies the HTML code and re-organizes the columns, which use the HTML tags <table>, <tr>, <td>, and <th>. The central column is usually displayed first. Figure 8 shows the sample page after being re-organized and its corresponding HTML source code.

However, this method suffers two major disadvantages:

- **The segmentation inaccuracy:** Page segmentation is to find Web page components. The problem is most page components are not clean-cut. Therefore, the segmented components may not be ideal.
- **The ranking problem:** This method tries to display the important parts of a Web page first. The question is how to define the importance of a Web component, which is ambiguous. Therefore, page segmentation is usually followed by component ranking, which will be explained next.

Component Ranking

Page segmentation is usually followed by component ranking, which is used to rank the page components. So they can be displayed in the order of their importance. The following Web page features can be used to rank components:

- **Audio/figure/flash/table/video caption:** A caption is usually a description of the subject.
- **Content:** Web page content provides the most accurate and full-text information. However, it is also the least-used information for a search engine since content extraction is still far less practical.
- **Description:** Web page descriptions can either be constructed from the meta tags or submitted by webmasters or reviewers. A metatag is an HTML tag that provides information such as author, expiration date, a list of keywords, about a web page.
- **Distance:** Components closer to the central point of a page are usually more important than components far away from the central point.
- **Hyperlink text:** Hyperlink text is normally a title or brief summary of the target page.
- **Hyperlink:** Hyperlinks contain high-quality semantic clues to a page's topic. A

Figure 7. (a) A sample Web page and (b) the corresponding HTML source code

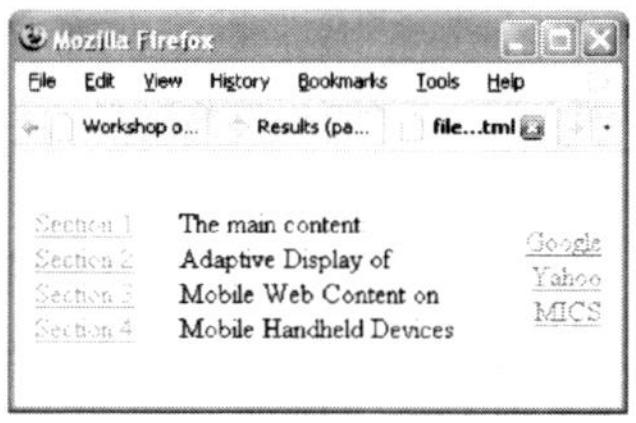

(a)

```
<html><body>
 <table width="100%" valign="middle"><tr>
  <td width="15%">
   <a href="Sec_1/">Section 1</a><br />
   <a href="Sec_2/">Section 2</a><br />
   <a href="Sec_3/">Section 3</a><br />
   <a href="Sec_4/">Section 4</a><br />
  </td>
  <td width="30%">
   The main content: Adaptive Display of Mobile Web Content
      on Mobile Handheld Devices
  </td>
  <td width="15%" align="right">
   <a href="http://www.google.com/">Google</a><br />
   <a href="http://www.yahoo.com/">Yahoo</a><br />
   <a href="http://www.micsymposium.org/">MICS</a><br />
  </td>
 </tr></table>
</body></html>
```

(b)

hyperlink to a web page represents an implicit endorsement of the page being pointed to.

- **Keyword:** Keywords can be extracted from full-text documents or metatags. Filtering operations are applied to a document before retrieving keywords from the full-text document. Typical operations include the removal of common words using a list of stopwords, the transformation of upper-case letters to lower-case letters, etc.
- **Page structure:** HTML source code has a tree structure. Important information may be revealed from the structure. For example, the central column of a three-column table usually contains more important information than other two columns do.
- **Page titles:** The title tag defines the title of an HTML document.
- **Size:** Large-size components are usually more important than small-size ones.
- **Text with a different font, style, color, or size:** Emphasized text is usually given a different font to highlight its importance.
- **The first sentence:** The first sentence of a Web page is usually an introduction or an abstract.

Figure 8. (a) The sample page of Figure 7.a after being re-organized and (b) the corresponding HTML source code

(a)

```
<html><body>
 <table width="100%" valign="middle">
  <tr><td>
    The main content: Adaptive Display of Mobile Web Content
       on Mobile Handheld Devices
  </td></tr>
  <tr><td>
   <a href="Sec_1/">Section 1</a><br />
   <a href="Sec_2/">Section 2</a><br />
   <a href="Sec_3/">Section 3</a><br />
   <a href="Sec_4/">Section 4</a><br />
  </td></tr>
  <tr><td>
   <a href="http://www.google.com/">Google</a><br />
   <a href="http://www.yahoo.com/">Yahoo</a><br />
   <a href="http://www.micsymposium.org/">MICS</a><br />
  </td></tr>
 </table>
</body></html>
```

(b)

Many methods are created to rank Web page components and each method is quite different from the others (Borodin, Mahmud, & Ramakrishnan, 2007; Hattori, Hoashi, Matsumoto, & Sugaya, 2007). The following example shows one of the methods and readers can check the references to find other methods. The example uses the PageRank algorithm, which is used by Google search, to rank page components (Yin & Lee, 2004). It performs the following tasks in sequence:

1. Segment a Web page and collect the page components.
2. Convert the Web page into a graph, whose nodes are page components and edges are relationships among components. Each edge is associated with a weight. For example, each paragraph could be a component and two consecutive paragraphs have an edge between them. Figure 9 shows a segmented page and its corresponding graph/tree. The root of this tree is the <html> element, which has three children: the left column, the central column, and the right column.
3. The algorithm "PageRank" is then applied to the graph to find the ranks of page components or graph nodes. It analyzes the edges to uncover two types of pages:
    - *authorities*, which provide the best source of information on a given topic and
    - *hubs*, which provide collections of links to authorities.

Two major steps are used to find the authorities and hubs and their weights:

a. **A sampling component:** which constructs a focused collection of several components likely to be rich in relevant authorities; and
b. **A weight-propagation component:** which determines numerical estimates of hub and authority weights by an iterative procedure:
    - **Authority weight update:** If a component is pointed to by many good hubs, we would like to increase its authority weight x_p, for a component p, by the sum of y_q over all components q that link to p.
    - **Hub weight update:** In a strictly dual fashion, if a component points to many good authorities, we increase its hub weight.

The authority and hub weights are then used to decide the component ranks.

Other Ad Hoc Methods

A wide variety of methods are created for Web content adaptation for microbrowsers. This chapter is not possible to cover all of the methods. Other than the above two methods: page segmentation and component ranking, this sub-section lists some of the methods: (i) page summarization, (ii) transcoding, and (iii) Web usage mining. Some other methods can be found in the related articles, for example, multimedia adaptation (Maekawa, Hara, & Nishio, 2006; Laakko & Hiltunen, 2005), context-aware adaptation (Pashtan, Kollipara, & Pearce, 2003), RSS feeds method (Blekas, Garofalakis, & Stefanis, 2006), and grammar induction method (Kong, Ates, Zhang, & Gu, 2008).

Page Summarization

This method tries to display a summary of a large document on a microbrowser. Text summarization gives a short version of a document without losing its meaning. It has been a research topic for a long time and no major breakthrough was made in many years because of its high difficulty. Yang and Wang (2004) propose a fractal summarization for large documents. It is based on the theory of

Figure 9. (a) A sample page with eight components labeled with letters from A to H and (b) the graph corresponding to the page

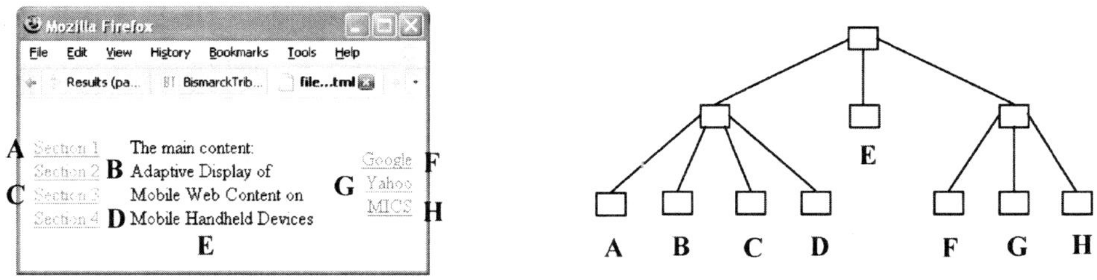

(a) (b)

fractal, which is a geometric shape that is repeated itself under several levels of magnification. It generates a brief skeleton of summary at the first stage, and the details of the summary on different levels of the document are generated on demands of users. Otterbacher, Radev, and Kareem (2006) use the method of hierarchical summarization, which displays the most important sentences in an article first. If the reader finds the initial summary interesting or relevant, he/she may "drill down" the details of the story by expanding the message. The hierarchical summarization includes two stages. First, it identifies the salience of each sentence in a document and ranks the sentences accordingly. Second, it builds a tree of all sentences such that its root is the sentence with the highest salience.

Transcoding

Transcoding is to convert one document to another. For mobile Web browsing, transcoding tries to translate a Web document to another and expects the latter document will be better displayed on handheld devices compared to the former document. Hwang, Kim, and Seo (2003) develop a syntax-based Web transcoding system that allows universal access to Web pages without manual reauthoring. It is based on *structure-aware* transcoding heuristics, which preserve the original Web page's underlying layout as much as possible. The proposed heuristics extract the relative importance of Web components from an intelligent syntactic analysis and display them in the order of their importance. Hsiao, Hung, and Chen (2008) propose an architecture of versatile transcoding proxy (denoted by VTP) for Web content adaptation. In their framework, the proxy can accept and execute the *transcoding preference script* provided by the client or the server to transform the corresponding data or protocol according to the user's specification.

Web Usage Mining

World Wide Web Data Mining includes content mining, hyperlink structure mining, and usage mining. All three approaches attempt to extract knowledge from the Web, produce some useful results from the knowledge extracted, and apply the results to certain real-world problems. The first two apply the data mining techniques to Web page contents and hyperlink structures, respectively. The third approach, Web usage mining, is the application of data mining techniques to the usage logs of large Web data repositories in order to produce results that can be applied to many practical subjects, such as improving Web sites/pages, making additional topic or product recommendations, user/customer behavior studies, etc. Zhou, Hui, and Chang (2006) try to enhance mobile-browsing experience by using Web recommendations. Each user is observed as a unit of unknown identity, although some properties may be accessible from demographic data. A runtime component dynamically inserts recommended or related links into the top of each requested page. Therefore, their system can generate recommendations even for a new mobile user with no historical access records. A related research can be found in the article from Hu, et al. (2008).

SUMMARY

Mobile commerce is a promising trend of commerce and mobile handheld devices are the mandatory tools for performing mobile commerce transactions. It uses microbrowsers to access the mobile content. However, many problems are associated with mobile Web browsing. This chapter discusses various issues related to mobile Web browsing. The first issue is the study of mobile handheld devices, which include six components: (i) mobile operating systems, (ii) mobile central processing units, (iii) microbrowsers, (iv) input and output components and methods, (v) memory and storage, (vi) batteries. The component closely tied to mobile Web browsing is microbrowsers, which usually apply one of the four approaches to access the mobile Internet: (i) wireless language direct access, (ii) HTML direct access, (iii) HTML to wireless language conversion, and (iv) error. The last issue is the difficulty of mobile Web browsing. Various methods are created for browsing the mobile Web efficiently and effectively. Each method has its own advantages and disadvantages and none of them is dominant. Though each method employs a different strategy, their goals are the same: *conveying the meaning of Web pages by using minimum space.* This chapter investigates some of the methods:

- **Page segmentation:** which is used to segment Web pages
- **Component ranking:** which is used to rank page components after segmentation
- **Other ad hoc methods:** such as text summarization, transcoding, and Web usage mining

Most methods segment the Web page first and then display the components in the order of their ranks. The major problem of the current methods is that it is not easy to find the clear-cut components in a Web page.

REFERENCES

ACCESS. (n.d.). *Small-Fit Rendering*. Retrieved June 14, 2008, from http://www.access-company.com/products/netfrontmobile/contentviewer/mcv_tips.html#Anchor-Smar-45765

Adipat, B., & Zhang, D. (2005). Adaptive and personalized interfaces for mobile Web. In *Proceedings of the 15th Annual Workshop on Information Technolgies & Systems (WITS)*, Las Vegas, NV.

Alam, H., & Rahman, F. (2003). Web document manipulation for small screen devices: a review. In *Proceedings of the 2nd International Workshop on Web Document Analysis (WDA2003)*, Edinburgh, UK.

Arase, Y., Maekawa, T., Hara, T., Uemukai, T., & Nishio, S. (2006). A Web browsing system based on adaptive presentation of Web contents for cellular phones. In *Proceedings of the 2006 International Cross-Disciplinary Workshop on Web Accessibility (W4A)*, (pp. 86-89). Edinburgh, U.K.

Blekas, A., Garofalakis, J., & Stefanis, V. (2006). Use of RSS feeds for content adaptation in mobile Web browsing. In *Proceedings of the 2006 International Cross-Disciplinary Workshop on Web Accessibility (W4A)* (pp. 79-85). Edinburgh, U.K.

Borodin, Y., Mahmud, J., & Ramakrishnan, I. V. (2007). Context browsing with mobiles—when less is more. In *Proceedings of the 5th International Conference on Mobile Systems, Applications, and Services (MobiSys '07)* (pp. 3-15). San Juan, PR

Chen, Y., Ma, W. Y., & Zhang, H. J. (2003). Detecting Web page structure for adaptive viewing on small form factor devices. In *Proceedings of the 12th International Conference on World Wide Web* (pp. 225-233).Budapest, Hungary.

Davison, B. D. (2001). A Web caching primer. *IEEE Internet Computing*, *5*(4), 38–45. doi:10.1109/4236.939449

Gartner, Inc. (2009a). *Gartner Says in the Fourth Quarter of 2008 the PC Industry Suffered Its Worst Shipment Growth Rate Since 2002*. Retrieved March 15, 2009, from http://www.gartner.com/it/page.jsp?id=856712

Gartner, Inc. (2009b). *Gartner Says Worldwide Smartphone Sales Reached Its Lowest Growth Rate With 3.7 Per Cent Increase in Fourth Quarter of 2008*. Retrieved March 18, 2009, from http://www.gartner.com/it/page.jsp?id=910112

Gartner, Inc. (2009c). *Gartner Says Worldwide Mobile Phone Sales Grew 6 Per Cent in 2008, But Sales Declined 5 Per Cent in the Fourth Quarter*. Retrieved March 19, 2009, from http://www.gartner.com/it/page.jsp?id=904729

Gupta, A., & Kumar, A. Mayank, Tripathi, V. N., & Tapaswi, S. (2007). Mobile Web: Web manipulation for small displays using multi-level hierarchy page segmentation. In *Proceedings of the 4th International Conference on Mobile Technology, Applications, and Systems* (pp. 599-606). Singapore.

Hattori, G., Hoashi, K., Matsumoto, K., & Sugaya, F. (2007). Robust Web page segmentation for mobile terminal using content-distances and page layout information. In *Proceedings of the 16th International Conference on World Wide Web* (pp. 361-370). Banff, Canada

Hsiao, J. L., Hung, H. P., & Chen, M. S. (2008). Versatile transcoding proxy for Internet content adaptation. *IEEE Transactions on Multimedia*, *10*(4), 646–658. doi:10.1109/TMM.2008.921852

Hu, W. C., Yeh, J. H., Chu, H. J., & Lee, C. W. (2005). Internet-enabled mobile handheld devices for mobile commerce. *Contemporary Management Research*, *1*(1), 13–34.

Hu, W. C., Zuo, Y., Chen, L., & Yang, C. H. (2008). Adaptive mobile Web browsing using Web mining technologies. Memmola M. & Al-Hakim, L., editors, *Business Web Strategy: Aligning the Internet with Corporate Design*, Hershey, PA: Information Science Reference.

Hua, Z., Xie, X., Liu, H., Lu, H., & Ma, W. Y. (2006). Design and performance studies of an adaptive scheme for serving dynamic Web content in a mobile computing environment. *IEEE Transactions on Mobile Computing*, *5*(12), 1650–1662. doi:10.1109/TMC.2006.182

Hwang, Y., Kim, J., & Seo, E. (2003). Structure-aware Web transcoding for mobile devices. *IEEE Internet Computing*, *7*(5), 14–21. doi:10.1109/MIC.2003.1232513

Jindal, A., Crutchfield, C., Goel, S., Kolluri, R., & Jain, R. (2008). The mobile Web is structurally different. In *Proceedings of the 27th Conference on Computer Communications* (pp. 1-6). Phoenix, AZ.

Kong, J., Ates, K. L., Zhang, K., & Gu, Y. (2008). Adaptive mobile interfaces through grammar induction. *Proceedings of the 20th IEEE International Conference on Tools with Artificial Intelligence* (pp.133-140). Dayton, OH.

Lawton, G. (2001). Browsing the mobile Internet. *IEEE Computer*, *35*(12), 18–21.

Maekawa, T., Hara, T., & Nishio, S. (2006). Image classification for mobile Web browsing. In *Proceedings of the 15th International Conference on World Wide Web* (pp. 43-52). Edinburgh, Scotland.

Opera Software ASA. (n.d.). *Opera's Small-Screen Rendering*. Retrieved June 23, 2008, from http://www.opera.com/products/mobile/smallscreen/

Otterbacher, J., Radev, D., & Kareem, O. (2006). News to go: hierarchical text summarization for mobile devices. In *Proceedings of the 29th Annual International ACM SIGIR Conference on Research and Development in Information Retrieval* (pp. 589-596). Seattle, WA.

Pashtan, A., Kollipara, S., & Pearce, M. (2003). Adapting content for wireless Web services. *IEEE Internet Computing*, *7*(5), 79–85. doi:10.1109/MIC.2003.1232522

Xie, X., Miao, G., Song, R., Wen, J. R., & Ma, W. Y. (2005). Efficient browsing of Web search results on mobile devices based on block importance model. In *Proceedings of the 3rd IEEE International Conference on Pervasive Computing and Communications (PerCom 2005)* (pp. 17-26). Kauai, HI.

Yang, C. C., & Wang, F. L. (2003). Fractal summarization for mobile devices to access large documents on the Web. In *Proceedings of the 12th International Conference on World Wide Web* (pp. 215-224). Budapest, Hungary.

Yin, X., & Lee, W. S. (2004). Using link analysis to improve layout on mobile devices. In *Proceedings of the 13th International Conference on World Wide Web* (pp. 338-344). New York.

Zhang, D. (2007). Web content adaptation for mobile handheld devices. *Communications of the ACM*, *50*(2), 75–79. doi:10.1145/1216016.1216024

Section 3

Wireless Networks and Handheld/Mobile Security

Chapter 14
Positioning and Privacy in Location-Based Services

Haibo Hu
Hong Kong Baptist University, China

Junyang Zhou
Hong Kong Baptist University, China

Jianliang Xu
Hong Kong Baptist University, China

Joseph Kee-Yin Ng
Hong Kong Baptist University, China

Location positioning by GPS has become a standard function in modern handheld device specifications. Even in indoor environments, positioning by utilizing signals from the mobile cellular network and the wireless LAN has been intensively studied. This chapter starts with some review of the state-of-the-art technologies. Positioning technologies propel the market of location-based services (LBS). They are mobile content services that provide location-related information to users. However, to enjoy these LBS services, the mobile user must explicitly expose his/her accurate location to the service provider, who might abuse such location information or even trade it to unauthorized parties. To protect privacy, traditional approaches require a trusted middleware on which user locations are anonanonymous ymized. This chapter presents two new privacy-preserving approaches without such a middleware. The first is a non-exposure location cloaking protocol where only relative distances are exchanged. The second is a protocol for nearest neighbor search with controlled location exposure.

INTRODUCTION

With the advent of new-generation smart mobile devices such as Apple iPhone and Google gPhone, location positioning by GPS and aGPS (assisted-GPS) become a standard function in handheld device specifications. Texas Instruments forecasts that by 2012, 34% of mobile handsets will be shipped with GPS modules. Even in indoor environments, positioning by utilizing signals from the mobile cellular network and the wireless LAN has been intensively studied and its accuracy has been sig-

DOI: 10.4018/978-1-61520-761-9.ch014

nificantly improved recently. In the first half of this chapter, we will review some of the state-of-the-art technologies.

Positioning technologies propel the market of mobile value-added services, in particular the location-based services. Location-based services (LBS) are mobile content services that provide location-related information to users. A typical LBS is the nearest neighbor query, in which the user quests for the nearest point of interest (e.g., gas station, restaurant) from where he/she is.

However, in order to enjoy these LBS services, it appears that the mobile user must explicitly expose his/her accurate location to the service provider, who might abuse such location information or even trade it to unauthorized parties. Even worse, the nature of location-based services seems to leave the users with no choice but to see their location privacy compromised in exchange for services. This rising concern is hindering the prosperity of LBS market and the mobile industry as a whole. The research community have identified this issue lately and attempted to solve it using "attribute generalization", a common approach used for privacy protection in RDBMS. The main idea is to blur the user location in a service request, that is, to replace the accurate user location with a cloaked region (usually a circle or a rectangle). This region encloses the accurate user location and satisfies some privacy metric such as k-anonymity (at least k users share the same region so that they are indistinguishable).

However, a centralized and trusted middleware (usually called "anonymizer") is required to form such cloaked regions for users before their LBS requests reach the service provider, and ironically the users have to expose their accurate locations – exactly what they want to hide – to this middleware. In case such middleware is not available or cannot be trusted, the LBS request cannot proceed without privacy compromise. In the second half of this chapter, we will study two approaches that address this issue.

First, we design a non-exposure location cloaking protocol. This protocol consists of a clustering stage and a secure bounding stage, during which only relative distances are exchanged among users to obtain the cloaked region. Furthermore, an anonymizer is not required in this protocol. Second, we take one step further – to allow users to skip the location cloaking phase and request location-based service directly from the (untrusted) server. In particular, we study the nearest neighbor (NN) query, an important location-based service. We learn from computational geometry that the query space forms a Voronoi Diagram, which is composed of Voronoi cells. Each data object corresponds to one Voronoi cell and this object is always the NN in this cell. As such, the essence of an NN query is matching the user location to some Voronoi cell. However, this matching implicitly exposes the user location to the server. We therefore study a client-server protocol that allows the users to learn the cell information in their neighborhood, so that they can resolve NN queries with controlled location exposure to the server.

BACKGROUND: LOCATION POSITIONING TECHNOLOGIES

Satellite-Based Position System

There are several satellite-based navigation systems, including the US's **G**lobal **P**osition **S**ystem (GPS) (Dana, 1998), the Russian Federation's **GLO**bal **NA**vigation **S**atellite **S**ystem (GLONASS) (Russian Space Agency) and the European Union's Galileo navigation system (European Space Agency). They enable receivers or terminals on the earth to gain location information.

Among these systems, the Global Position System is the most widely used one. In the following, we introduce this system in detail in order to have an overview of satellite-based navigation positioning systems. GPS was originally proposed for military applications, and later, in the 1980s, the system started to be available for civilian use (Dana, 1998). It consists of three segments:

- Space Segment. It consists of 24 constellation satellites that orbit the earth in 12 hours. The satellite orbits repeat almost the same ground track every 24 hours. This constellation provides the user with 5~8 satellites visible from any point on the earth. Figure 1 shows the constellation of the satellites.
- Control Segment. It is a system of tracking stations located around the world. These tracking stations are responsible for measuring signals from the satellite, computing precise orbital data (ephemeris) and making clock correction for each satellite.
- User Segment. It consists of the GPS receivers. The GPS receivers receive satellite signals and use the satellite as reference points to compute their positions, which can be accurate to within meters on average (Dana, 1998).

Figure 1. Constellation of GPS satellites (source: http://www.state.gov/cms images/globe2.jpg)

Figure 2. Samples of pseudo random code

Technology Behind GPS

The positioning technology used by the GPS is *triangulation*. To estimate a position, a GPS receiver measures the distances from itself to three or more satellites using the travel time of radio signals. In order to measure the distance between a satellite and the GPS receiver, precise time synchronization is needed. For a decent GPS receiver, it receives at least four timing signals from different satellites. Then, the GPS receiver adjusts its time clock until its time is perfectly synchronized with the timing signals. Each satellite has a unique Pseudo Random Code. Figure 2 shows samples of the pseudo random code. After the synchronization, the GPS receiver compares the received pseudo random codes with its own versions to measure the time delay. By using a simple formula (Distance = Velocity (speed of light) * Time Delay), the GPS receiver can compute the distance between the satellite and itself.

However, the GPS receiver in a mobile station (MS) isnot always on, which would cause the Time to First Fix (TTFF) problem. The TTFF is the time taken for the GPS receiver to lock the satellite signals. The TTFF can be as long as ten minutes or more from switch on which is not acceptable in real-time mobile location services especially if the service is an emergency request. To overcome this problem, a system known as Assisted GPS (A-GPS) has been proposed (Djuknic & Richton, 2001). In A-GPS, the mobile phone network provides the satellite information to the GPS receiver to improve the TTFF. This can be

Figure 3. Positioning method using the cell based algorithm

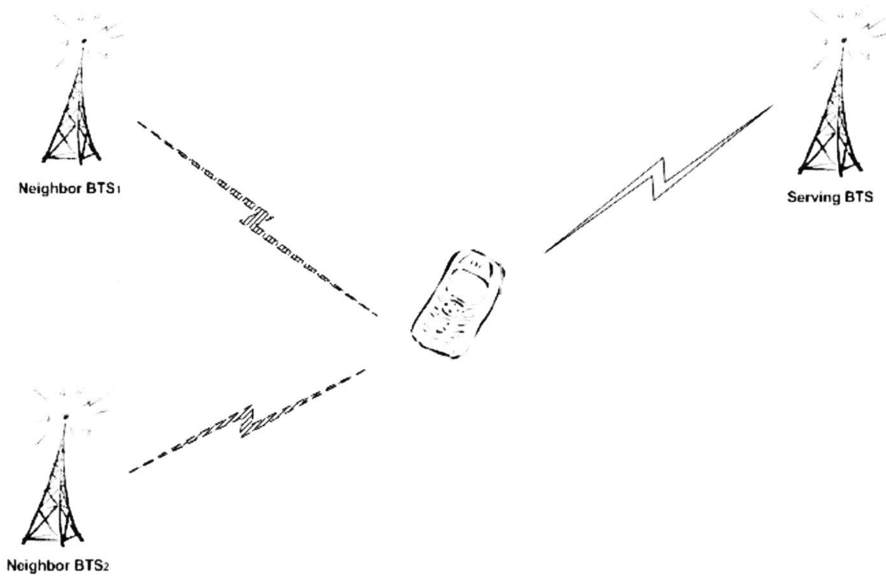

achieved by incorporating a GPS receiver into the base station (BS). In general, the BS is located close to the MS; therefore, the satellite information received by the GPS receiver of the BS is sufficiently accurate to be used by the MS. Although using A-GPS is a feasible positioning solution for the mobile phone network, it cannot provide estimation for indoor environment or places that cannot receive satellite signals.

Mobile Positioning within a Cellular Radio Network

Several mobile positioning methods have been proposed to overcome the deficiency of GPS within a cellular radio network in the literature for positioning a MS (Laitinen et al., 2001; Pahlavan & Krishnamurthy, 2002; Ng, Chan, & Kan, 2002; Kan, Chan, & Ng, 2003; Chu, Leung, Ng, & Li, 2004; Zhou, Chu, & Ng, 2005; Zhou, Chu, & Ng, 2005; Prasithsangaree, Krishnamurthy, & Chrysanthis, 2002; McGuire, Plataniotis, & Venesanopoulos, 2005). These approaches can be generally classified into five categories: cell based approach, time based approach, angular based approach, signal strength based approach and hybrid approach.

Cell Based Approach

A basic positioning approach is to make use of the cellular concept in the cellular radio network. It has been adopted by service providers due to its simplicity and ease of deployment (Laitinen et. al., 2001). This approach works as follows. When an MS roams in the cellular radio network, there exist a serving BS and several neighbor BSs. The serving BS is the one that the MS is currently connected and the neighbor BSs are those the MS can monitor their signal strength levels for possible handoff, as illustrated in Figure 3. The cell based approach uses the location of the serving BS as the estimated location of the MS. Therefore, the accuracy of the cell based approach is dependent on the cell size, which can be up to tens of square kilometers in rural areas. Some enhancements can be employed to improve the cell based approach, for example, further dividing a cell into sectors and estimating positions based on the cell sectors. But even with these enhancements, the

Figure 4. Positioning method using TDOA (hyperbolic multi-lateration)

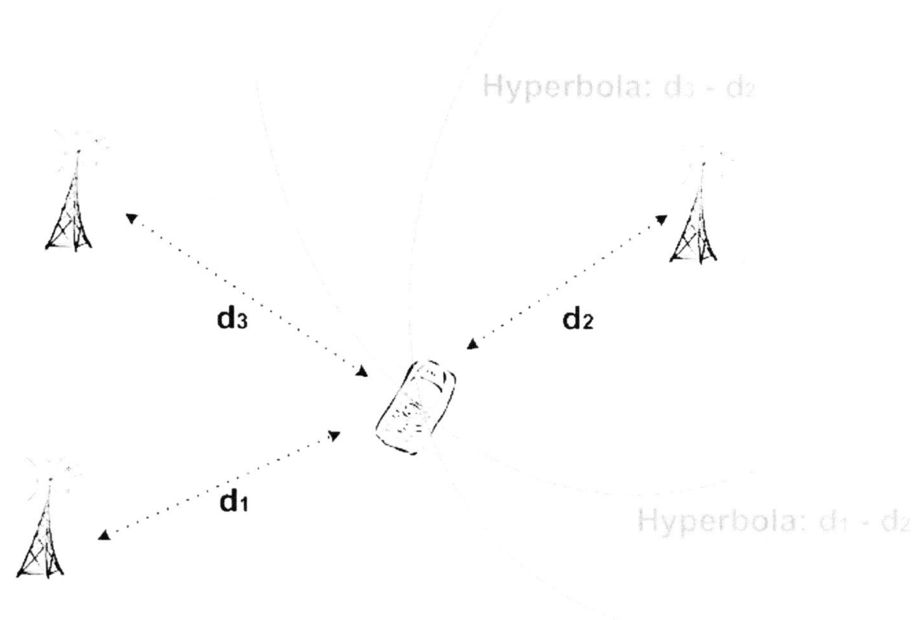

accuracy is still around 150-300m depending on the cell layout of the mobile network (Laitinen el al., 2001).

Time Based Approach

In general, the time-based approaches such as Time-Of-Arrival (TOA), Time-Difference-Of-Arrival (TDOA) and Enhanced-Observed Time Difference (E-OTD) measure the signal propagation time from the BS to the MS and estimate the distance between them. Then, by simple trilateration, it is possible to estimate the location of the MS (Laitinen el. al., 2001). Figure 4 shows the algorithm called Hyperbolic Multi-lateration based on the Time-Difference-Of-Arrival (TDOA) between each BS and the MS, which is measured by the network controller.

Although the time based approaches are simple, they require precise time synchronization between each BS and the MS. Even though a GPS receiver can be used for the synchronization purpose, the extra installation cost would put a heavy burden on the operator.

Angular Based Approach

The angular based approach, Angle-Of-Arrival (AOA), uses the antenna array in the BS to estimate the angle from which the signal arrives, and then estimates of the location of the MS (Laitinen el al., 2001). As shown in Figure 5, each BS is equipped with additional gear to detect the compass direction from which the signal is arriving. In general, for a 3-dimensional geometry space, only three BSs are required to determine a unique location.

The AOA approach has good accuracy in Line-Of-Sight (LOS) situation but it is not the method of choice in densely populated urban areas. This is because LOS to three BSs is seldom present in such areas. In Non Line-Of-Sight (NLOS) situation, the performance of the AOA method can be dropped to 300m (Laitinen el. al., 2001). Figure 6 illustrates the NLOS problem in AOA approach. Also, another major implementation problem of AOA is the need for an antenna array to be deployed at each BS, which is expensive.

Figure 5. Positioning method using AOA

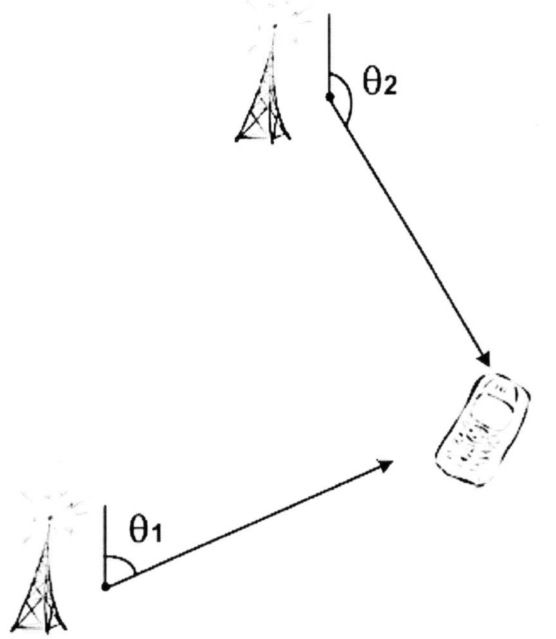

Signal Strength Based Approach

The signal strength is a commonly available parameter in cellular radio networks; therefore approaches based on signal attenuation can be applied to all kinds of cellular radio networks. Several algorithms have been proposed in the literature to estimate the distance based on measured signal strength (Pahlavan & Krishnamurthy, 2002; Ng, Chan, & Kan, 2002; Kan, Chan, & Ng, 2003; Chu, Leung, Ng, & Li, 2004; Zhou, Chu, & Ng, 2005; Zhou, Chu, & Ng, 2005; Prasithsangaree, Krishnamurthy, & Chrysanthis, 2002)

Free Space Propagation Model

The fundamental signal propagation model is the free space propagation model which is given by Pahlavan and Krishnamurthy (Pahlavan & Krishnamurthy, 2002):

$$P_r = \frac{P_t G_t G_r \lambda^2}{(4\pi)^2 d^2} \quad (1)$$

where P_r is the receiver power, P_t is the transmitter power, G_t is the transmitter antenna gain, G_r is the receiver antenna gain, d is the distance in meter between the transmitter and the receiver, and λ is the wavelength in meter.

The free space propagation model is derived from the first principles: it shows that the receiver power decays with a distance at a rate of 20dB/decade when there are no obstacles between the transmitter and the receiver. In real environments especially in urban areas, a signal Line of Sight (LOS) path between the transmitter and the receiver seldom occurs. Thus, the free space propagation model cannot be applied in such environments. Furthermore, the free space propagation model is a non-directional propagation model such that the contour line at the same signal level resembles a circle.

Signal Propagation Model

The signal propagation model is an extension model of the free space propagation model. It requires the estimation of the propagation model parameters. Ng et al. have proposed to choose a Maximum Likelihood method to estimate them (Ng, Chan, & Kan, 2002). The signal propagation model for each BS is assumed as follows:

$$P_r(dB) = K + \alpha \ln(d) \quad (2)$$

where $Pr(dB)$ is the received signal strength (RSS), d is the distance between the BS and the MS, and K and α are two propagation parameters being estimated by the Maximum Likelihood method.

Two algorithms for location estimation based on the signal propagation model have been proposed, namely, the Center of Gravity (CG) algorithm and the Circular Trilateration (CT) algorithm (Ng, Chan, & Kan, 2002; Kan, Chan, & Ng, 2003).

Ellipse Propagation Model

Zhou et al. have proposed an Ellipse Propagation Model (EPM) with the Geometric Algorithm and the Iterative Algorithm (Zhou, Chu, & Ng, 2005; Zhou, Chu, & Ng, 2005). EPM considers the

Figure 6. NLOS problem in AOA

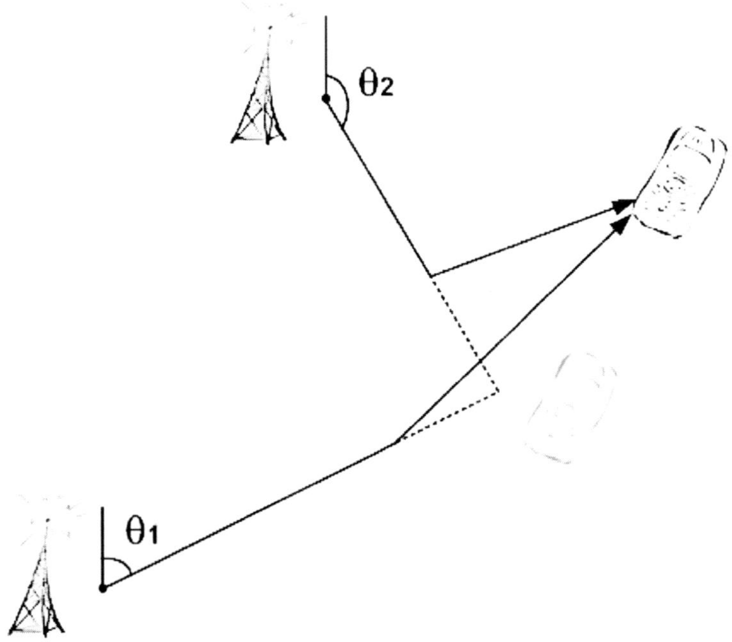

directional transmission property of the antenna and assumes that the contour line of the signal strength is an ellipse where the BS is at one of the focuses. EPM is defined as:

$$d = k(s_0 / s)^{1/\alpha} \frac{1-e}{1-e\cos(\theta)} \quad (3)$$

where d is the distance between the MS and the BS, k is the proportion constant, s_0 is the transmitting power of the BS, s is the signal power received, e is the eccentricity of the ellipse, θ is the deviation between the ellipse principal axis and the line of the MS-BS, and α is called the path loss exponent (Zhou, Chu, & Ng, 2005).

Directional Propagation Model

Chu et al. have designed a Directional Propagation Model (DPM), which takes both the directional effect and the environmental effect into account (Chu, Leung, Ng, & Li, 2004). DPM is defined as,

$$pl=\beta_0+\beta_1 g+\beta_2 log(h)+(\beta_3+\beta_4 log(h)+\beta_5 e)ln(d) \quad (4)$$

where pl is the mean propagation loss (defined as the difference between the received signal strength and the transmit power in decibel), d is the distance in meter between the MS and the BS, g is the directive gain of an antenna type t, e is the environment index and h is the height in meter of the BS.

Fingerprinting Approach

The fingerprinting approach utilizes the received signal strengths (RSSs) on the MS for location estimation as shown in Figure 7. It is divided into two phases, the off-line phase and the on-line phase as shown in Figure 8. In the off-line phase, the RSS vector is collected for a set of sample locations. The RSS vector records the RSS, si, received from each BS identified by the Cell Identify Code, CID*i*. The RSS vector and its corresponding locations are then stored in the database.

During the on-line phase, the RSS vector of the MS is measured and the weighted distance *Lp* between the measured RSS vector and the database entry is computed with the following

Figure 7. The fingerprinting approach

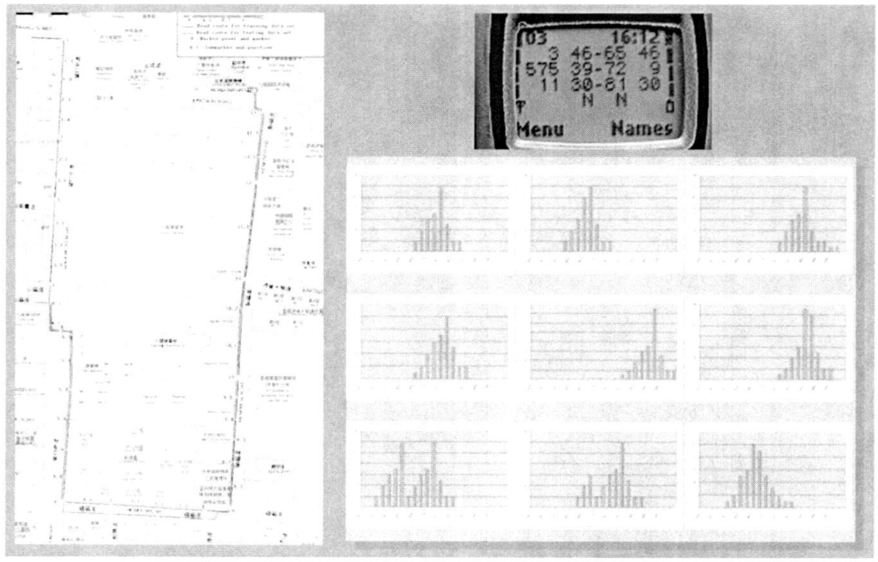

formula (Prasithsangaree, Krishnamurthy, & Chrysanthis, 2002),

$$L_p = \frac{1}{N}\left(\sum_{i=1}^{N}\frac{1}{\varpi_i}\,|\,s_i - s_i^m\,|^p\right)^{1/p} \qquad (5)$$

where N is the number of RSS measurement received from the location, p and ω_i are the scalar factor and the weighted factor of the i-th signal difference respectively. The weighted factor ω_i is used to bias the distance by a factor, which indicates the reliability of the database entry or the RSS measurements. The location of the MS is then estimated in the on-line phase by either of the following two methods:

- Choose the location in the database corresponding to the fingerprint with the minimum distance to the measured fingerprint of the MS. For example, the Manhattan distance L1 and the Euclidean distance L2 with $\omega_i = 1$ for all entries are often been used in this case (Bahl & Padmanabhan, 2000).
- Choose M closest database entries (those with the smallest signal distance) and estimate the location based on average of the coordinates of these M points.

The fingerprinting approach is widely used in indoor location estimation and has been proven for its great accuracy (Prasithsangaree, Krishnamurthy, & Chrysanthis, 2002; Bahl & Padmanabhan, 2000). However, applying the fingerprinting approach in outdoor environments is usually impractical. This is because it is very difficult to collect the RSS measurement for all possible locations in a large area. In addition, the fingerprinting approach is very sensitive to the surrounding environments; thus, re-calibration or re-collection of the data is often required.

Hybrid Approach

A hybrid approach combines two or more approaches mentioned above to achieve a better location estimation (Laitinen el al., 2001; McGuire, Plataniotis, & Venetsanopoulos, 2005). As a case study, we here introduce the hybrid method that combines the time based approach and the signal strength based approach.

Figure 8. Architecture of the fingerprinting approach

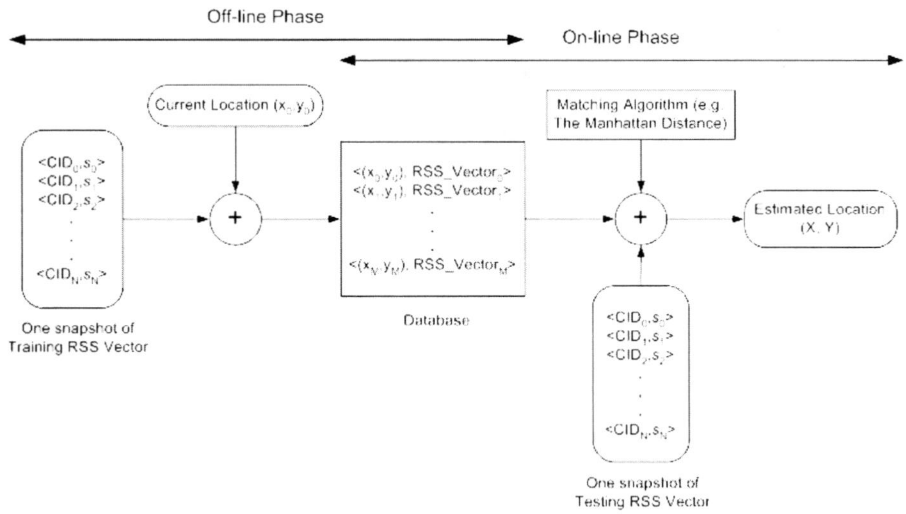

McGuire e. al. proposed a data fusion method based on both the RSS and TDoA measurements, which obtains with a higher positioning accuracy than the approaches using either measurement alone (McGuire, Plataniotis, & Venetsanopoulos, 2005). In this method, nonparametric estimation techniques, which choose the survey data taken from the propagation environment to construct approximate joint probability density functions are employed to compute the location of the MS. They have also demonstrated that the location estimation algorithms using the RSS and ToA or TDoA measurements are robust to additive measurement noise and NLOS propagation.

NON-EXPOSURE LOCATION CLOAKING

Location-based services (LBS) provide dynamic content according to where the user is located. Typical LBS applications include road navigation, nearest point of interest (POI) query, and location-aware advertisement. In order to enjoy such services, the mobile user must explicitly expose his/her accurate location to the server. For example, if the user asks for the nearest restaurant, he/she must provide the LBS server his/her accurate position in terms of GPS coordinates. In this sense, the user's location privacy is compromised in exchange for services. To address this issue, an intuitive solution is to cache the whole dataset of POI on the mobile device, which can then resolve location-based queries locally. However, due to limited resources provided on the mobile device, this solution cannot scale to large POI datasets, neither can it deal with data updates. Therefore, a more sophisticated strategy called *location anonymity* has been proposed and studied (Gruteser & Grunwald, 2003; Gedik & Liu, 2005; Mokbel, Chow, & Aref, 2006). The objective is to allow the mobile user to request services without revealing the accurate location.

Among various approaches proposed along this line, *location cloaking* is predominant. It blurs the accurate user location and replaces it with a well-shaped *cloaked region* (usually a circle or a rectangle), according to some anonymity metric such as *k-anonymity* (the cloaked region must contain at least *k* users) or *granularity* (the size of the cloaked region must exceed a threshold). Most existing location cloaking research focuses on minimizing the size of the cloaked region while still satisfying the anonymity metric. However, to

Figure 9. Two-Phase non-exposure cloaking

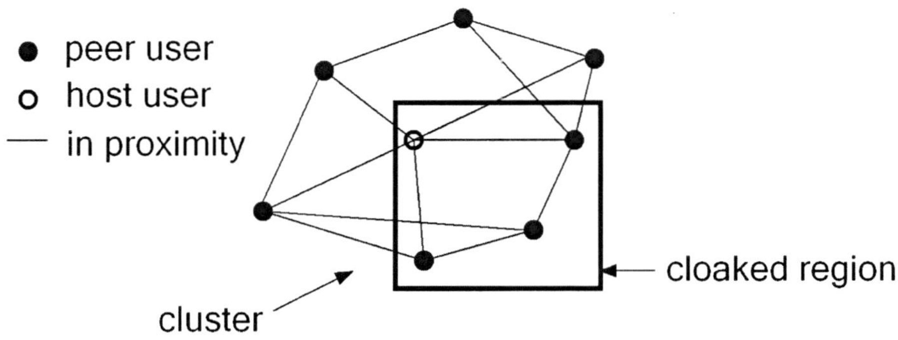

obtain the cloaked region and optimize its size, all existing algorithms require the accurate locations (i.e., the coordinates) of all users. As the accurate locations are exactly what the users want to hide, all existing work essentially imposes an assumption that all parties involved in the cloaking algorithm must be trusted. Typical parties include the *anonymizers* that sit in between the user and the LBS server (Mokbel, Chow, & Aref, 2006; Ghinita, Kalnis, & Skiadopoulos, 2007), and the user peers when the cloaking is performed in a peer-to-peer environment (Chow, Mokbel, and Liu, 2006).

However, in practice any party in the network might be malicious and the exposure of the accurate location information to any party might reveal user's identity or other sensitive information. In this sense, existing algorithms have limited applications and location cloaking without exposing the accurate user location to any party is urgently needed.

In this section, we present such a *non-exposure* location cloaking algorithms (Hu & Xu, 2009). It is designed for k-anonymity and cloaking is performed based on the *proximity* information among mobile users. Proximity information is widely available in practice, using location positioning technologies. For example, a mobile device is able to measure the closeness from its peers, by measuring either the received signal strength (RSS) from its peers (the stronger the closer), or the time difference of arrival (TDOA) of beacon signals from its peers using omni-directional antennas (the shorter the closer).

Figure 9 shows a proximity graph that is based on such RSS information. A vertex in this graph stands for a user, and an edge means that the two users are WiFi neighbors. The proposed non-exposure cloaking process is invoked by the *host user* who wants to request a location-based service. The process involves the surrounding users and is conducted in two phases. In the first phase, k users (including the host user) are identified through the proximity information. They and only they contribute to the resulted cloaked region. Moreover, if they become host users later, they will employ the same cloaked region. The shaded cloud in this figure encloses *4* users for a *4*-anonymity cloaking request from a host user. We will show later that this phase is equivalent to finding a cluster of size at least k in the graph. Furthermore, to minimize the size of the cloaked region, we should minimize the *diameter* of this cluster. This problem is thus called *proximity minimum k-clustering*. It is difficult particularly in the distributed environment because an earlier cluster result for a host user might significantly affect the subsequent cluster result. We tackle this problem by defining an equivalent relation called t-connected. This leads to a nice property called

Figure 10. Weighted proximity graph

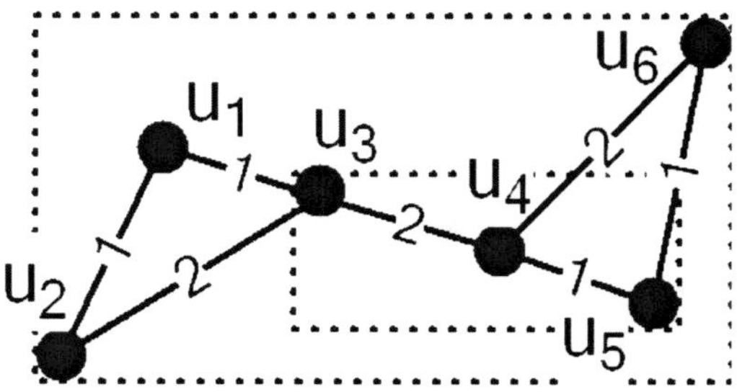

cluster-isolation, where subsequent cluster result is immune to change. Based on this property, we present an efficient distributed algorithm for minimum k-clustering.

In the second phase, the cloaked region --- a bounding box of all users in the cluster --- is obtained without exposing their accurate locations. The solid box in this figure shows a possible bounding box for the cluster of *4* users. Finding this box is equivalent to obtaining lower and upper bounds of users' coordinates without revealing these coordinates. We call this problem *secure bounding*. It is related to secure multi-party computation (SMC) (Goldwasser, 1997). To reduce the size of the cloaked region, the objective of secure bounding is to obtain the bound as tight as possible with the lowest cost. We propose a progressive bounding algorithm in which a bound increases progressively until all users agree with this bound. By developing a sophisticated cost model for the communication cost, we derive the optimal increment value for this algorithm.

Proximity k-Clustering

We are given a dataset *Đ* of all users, and each user in *Đ* has some peer users in proximity. The proximity input can be modeled as an undirected weighted graph where each vertex denotes a user and each edge (u,v) denotes that users u and v are in proximity; and the weight of (u,v) denotes the relative distance between u and v. We call the resulted graph a weighted proximity graph (WPG). Figure 10 shows a WPG where the relative distance is the signal strength. For example, the weights of edges (u_2,u_1) and (u_2,u_3) are 1 and 2, which means that the signal between u_2 and u_1 is stronger than that between u_2 and u_3.

By definition, location k-anonymity is to map a host user u to a set of peer users $S(u) \in$ *Đ* such that the size of $S(u)$ exceeds k, i.e., $|S(u)| \geq k$. As such, location k-anonymity on WPG is equivalent to k-clustering. More specifically, a k-clustering is a partition of *Đ* into a number of groups, each of which has a size of at least k. To minimize the size of the resulted cloaked region, we try to minimize the maximum edge weight (MEW) for each cluster.

It is noteworthy that, in our cloaking problem only the host users need k-clustering. As such, a distributed and local k-clustering algorithm that only finds the cluster for a host user is more desirable. However, intuitive local k-clustering leads to a poor minimum k-clustering result for subsequent host users. Figure 11 shows an example. The dotted line encloses a cluster of 5 vertices; however, removing this cluster isolates vertex g from the rest of WPG. To address this issue, while obtaining a minimum-diameter cluster for the host vertex, the distributed minimum k-clustering algorithm must also guarantee the k-clustering result of the rest WPG is not affected. This property is called *cluster-isolation*, and a distributed k-clustering algorithm that satisfies this property is *cluster-isolated.*

Now present the distributed k-clustering algorithm on WPG that minimizes the MEW in the cluster. It is derived from a centralized k-clustering algorithm and modified to be cluster-isolated. First, we introduce an equivalence relation called *t-connected*, based on which these two k-clustering algorithms are designed.

Definition 1: t-connected relation: Two vertices a and bare t-connected, if there is a path a, v_1, v_2, ..., b in WPG such that no edge weight in this path exceeds t.

Figure 11. Disconnected problem

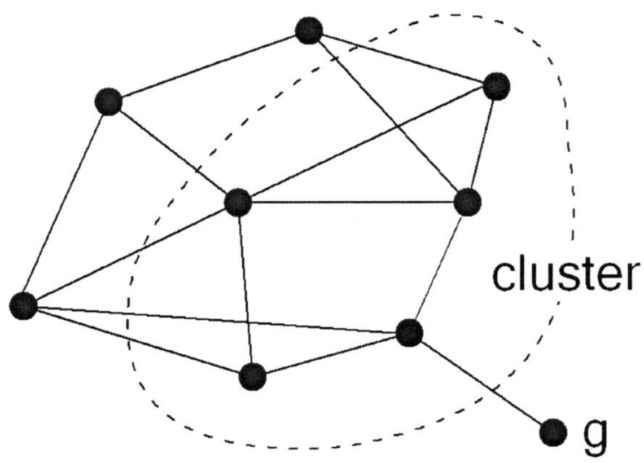

t-connected is an equivalence relation. In general, an equivalence relation partitions all elements in a set into equivalence classes. We therefore obtain a clustering of the vertices in WPG through t-connectivity. In terms of graph notions, an equivalence class corresponds to a connected component in WPG whose edge weights do not exceed *t*. For different *t*, we obtain different clustering results. If *t* is set to the MEW of the whole WPG, the clustering result has only a single connected component — the whole WPG. Then by decreasing *t*, this component is partitioned into smaller connected components, which form another clustering result.

To minimize the cluster size, the centralized clustering algorithm should use the lowest *t* while keeping all connected components valid, i.e., their sizes are no smaller than *k*. It partitions a connected component, i.e., a *t-connectivity* cluster, by removing edges in this cluster in the descending order of their weights, until this cluster is no longer connected and is thus partitioned into some smaller connected components, i.e., clusters. Each of these clusters is partitioned in the same way into even smaller clusters. The recursive partition continues until a further partition will lead to an invalid cluster, i.e., the size is smaller than *k*. As such, the resulted clusters are those that cannot be further partitioned, and we call them the *smallest valid t-connectivity* clusters.

The above algorithm is centralized, as it requires the knowledge of the whole WPG. Now we extend it to a distributed version of *t-connectivity* k-clustering. It finds the cluster Ć for a particular vertex *u*. Intuitively, to minimize the cluster size (i.e., the MEW), the algorithm should find the smallest valid *t-connectivity* cluster of *u* by increasing *t* until the cluster size just exceeds *k*. However, a key observation is that this cluster is not necessarily isolated. To remedy this, we give a sufficient condition of the smallest valid *t-connectivity* cluster being isolated.

Theorem 2: A sufficient condition of the smallest valid t-connectivity cluster being isolated is that, all external border vertices of this cluster (vertices adjacent but not belonging to it) can form a valid t-connectivity cluster in the remaining WPG.

Corollary 3: A distributed algorithm that finds for a host vertex u the smallest valid t-connectivity cluster that satisfies the above condition is cluster-isolated.

Figure 12 details the distributed algorithm in three steps. In the first step (lines 1–6), it obtains the smallest valid cluster of *u* by spanning from *u* through edges with increasing weights until the size reaches *k*. It always chooses the minimum-weight edge from the priority queue of to-be-spanned vertices, the result cluster Ć is guaranteed as the smallest valid cluster. In the second step (lines 7–15), the algorithm checks each external boundary vertex *v* in Ć. If v cannot form a cluster of size *k* with *t-connectivity* in the remaining WPG, *v* is added to Ć and *t* is thus updated. New vertices will then be spanned from Ć using the new *t*. In this process, some vertices become new external boundary vertices. The loop terminates when all external boundary vertices are checked. Finally in the third step (lines 16–17), since the size of C might be well above k and since all the edge weights in Ć are known, the algorithm calls the centralized k-clustering algorithm to obtain the smallest valid cluster for u.

Figure 13(a) and (b) show the distributed t-connectivity 2-clustering process on a WPG, where u is the host vertex. In the first step, the distributed privacy-aware nearest neighbor search algorithm uses 5-connectivity to form a 2-cluster $\{u, v\}$ in solid dots (Figure 13(a)). In the second step, the three external border vertices (the hollow dots) are checked. Among them, only w cannot form a 2-cluster with 5-connectivity. As such, w is added to the cluster and x is added as a new external border vertex (13(b)). Since x can have a 2-cluster with 5-connectivity (shown in dashed line), the algorithm proceeds to the third step and returns $\{u, v, w\}$ as the 2-cluster result.

SECURE BOUNDING

Now that the k-cluster is formed, the next and final step is to hide the true identity of the host user among the users in this cluster. This is achieved by generalizing the identifier or quasi-identifier that is embedded in the service request. Typical examples of such identifier include the geographic coordinates (longitude and latitude) and IP address. Without loss of generality, we assume that the identifier is a scalar and private attribute ξ of the user, and therefore the objective in this section is to obtain tight lower and upper bounds for the ξ values of all users in a cluster without revealing them.

This problem is closely related to secure multi-party computation (SMC) (Goldwasser, 1997). In general, a secure multi-party computation problem is to compute a function from multiple participants in a distributed network, each holding one input. The computation must be secure, that is, no more information is revealed to any participant except for those implied by this participant's input and the function output. Our problem can be reduced to an SMC problem if we want to obtain the tightest bound — the maximum and minimum ξ values. However, this is not favorable due to the following three reasons. First, the theoretical SMC

Figure 12. Distributed t-connectivity k-clustering

```
Input: G: the WPG
       u: the host vertex
       k: the anonymity requirement
Output: C: the cluster of u
Procedure:
C = {u};
push all u's neighbors in G to H;
while |C| < k do
    v = H.pop();
    C = C ∪ {v};
    push all v's neighbors not in C to H;
E is the set of external boundary vertices;
t is the connectivity of C;
while E is not empty do
    v = E.next();
    if v does not have a t-connectivity cluster of size k then
        C = C ∪ {v};
        t=minimum weight of edge between v and C;
        span C with new t;
        insert new external boundary vertices to E;
call centralize_k_clustering(C, k);
C is the cluster that contains u;
```

Figure 13. Distributed t-connectivity 2-clustering

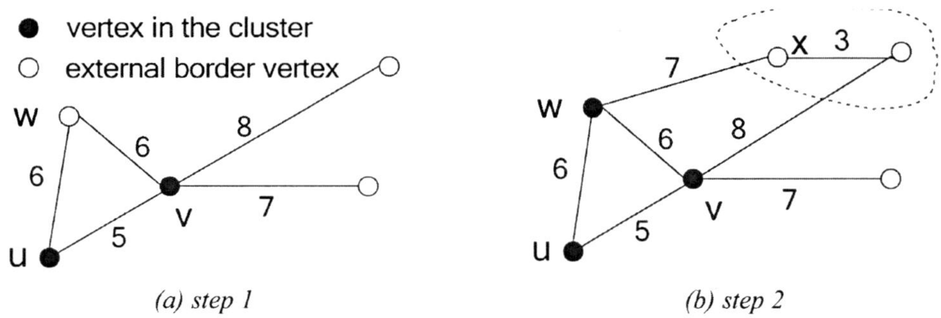

solution to the function of maximum or minimum is impractical. In fact, even for the most primitive SMC problem — the Millionaire problem — the communication complexity of the most efficient protocol (Cachin's algorithm) is polynomial to the number of bits of each input (i.e., the precision of ξ) and the number of participants (i.e., *k* in our problem). Therefore, solving an SMC problem is impractical, in particular in mobile environments where communication is extremely expensive. Second, returning the maximum and minimum ξ values exposes the actual ξ values of some users, which is not fair for them. Third, SMC assumes that participants have no apriori knowledge on the inputs and hence strict security can be guaranteed. However, in our problem, the fact that all participating users are in the same cluster already implies their ξ values are close and even within a certain range. As such, SMC is an overwhelming and inappropriate solution to our problem.

Therefore we propose an alternative secure bounding protocol. For simplicity, we present the protocol for upper bounding, and we assume that users follow a semi-honest model (Du, 2001). In this model, the users follow the protocol properly except that they can record all intermediate results to deduce the ξ values of others. Our protocol is progressive and follows the "hypothesis-verification" paradigm: in each iteration, a hypothetic bound is proposed and verified by all users in the cluster; if not all agree with this bound, a new iteration begins and a larger bound is proposed for those disagreeing users to verify. The protocol terminates when all users agree. Note that the bound can be computed either at a centralized server or at the host user in a distributed environment.

The key factor in this protocol is the increment of the new bound from a disagreed bound. A smaller increment leads to a tighter bound; however, it is at the cost of more iterations and thus higher communication overhead for verification. On the other hand, a larger increment leads to a looser bound, and thus costs higher communication for the subsequent service request. In what follows, we derive the optimal bound to minimize the total communication cost. Let *x* denote the increment from the previous bound. Then for a single user, the optimal *x* can be computed by the following differential equation:

$$P(x)R'(x) = (C_b+R(x))p(x) \qquad (6)$$

where *p(x)* and *P(x)* denote the probability and cumulative density function of the ξ variable, respectively, *R(x)* denotes the communication cost (in terms of *x*) of the service after cloaking, C_b denotes one round-trip communication cost for the bound verification.

We extend this result to *N* users, where the optimal *x* can be computed by the following differential equation:

$$R'(x) = (C^*-R^*)\,Np(x) \qquad (7)$$

where C^* and R^* are the optimal total cost and service cost, respectively, for a single user by Eqn. 1. Let us consider a concrete example. The x variable of a user follows a uniform distribution in the range of $(0, U)$, i.e., $p(x) = 1/U$, and $P(x) = x/U$. The communication cost for the service request is proportional to the area of the bound, i.e., $R(x) = C_r * x^2$. Then Eqn. 2 is reduced to $2C_r x = (C^*-R^*)N/U$, so $x = \frac{N(C^*-R^*)}{2C_r U}$.

PRIVACY-AWARE NEAREST NEIGHBOR SEARCH

Location-based services (LBS) are mobile content services that provide location-related information to users. However, in order to enjoy such services, the mobile user must explicitly expose his/her accurate location to the server. We focus on a typical query type --- *k*-nearest-neighbor (kNN) query shown in Figure 14. The user *o* sends out his/her accurate location and asks for the nearest restaurant. Upon receiving the kNN query, the server returns the name and address of the nearest restaurant (which is *g*), and other up-to-minute information, such as menu, table reservation status and customer reviews.

Mobile users see their location privacy compromised in exchange for services (e.g., finding the nearest restaurant). To address the location privacy issue, researchers have recently been interested in developing online privacy-aware data access techniques. Along this line, location cloaking has been proposed to blur the user locations when requesting services (Gruteser & Grunwald, 2003; Gedik & Liu, 2005; Mokbel, Chow, & Aref, 2006; Ghinita, Kalnis, & Skiadopoulos, 2007). The idea is to replace the accurate user location in the request with a well-shaped cloaked region (usually a circle or a rectangle), according to some privacy metric such as granularity (the area of this region must exceed a threshold) or k-anonymity (this region must contain at least k users). A kNN query with such a cloaked region is called a k-range-nearest-neighbor (kRNN) query (Hu & Lee, 2006), and the server returns to the user the kNNs of all points inside this region. Finally, the user refines the genuine kNNs from the kRNN query results, based on the accurate user location. In other words, to protect location privacy, the user requests a superset of kNN results from the server, thereby trading network bandwidth for location privacy. Figure 15 shows a 1-range-nearest-neighbor (RNN) query (with the dashed-line box as the cloaked region) when location cloaking is applied to the NN query in Figure 14. In this example, the server returns not only the genuine result g for the NN query, but also g_1 and g_2, because they are the NNs for some points in the cloaked region.

Figure 14. Nearest neighbor query

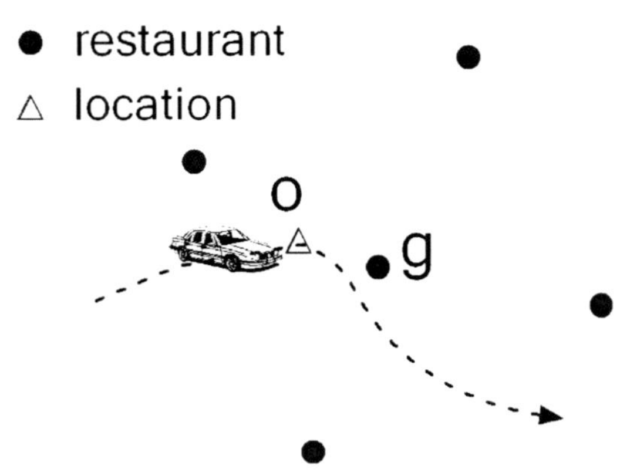

Figure 15. Range nearest neighbor query

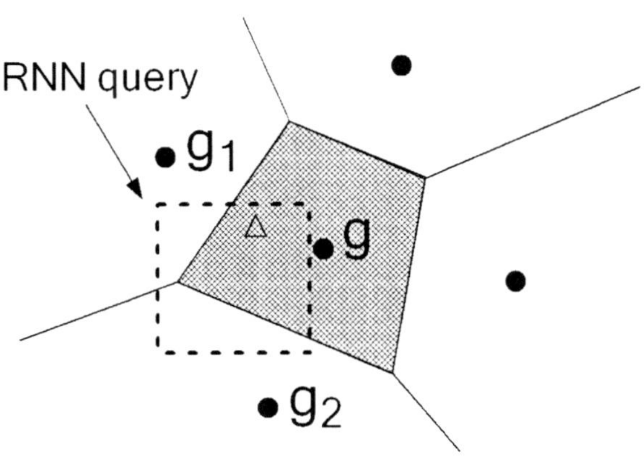

In effect, location cloaking achieves privacy at the cost of requesting non-result objects (e.g., g_1 and g_2) together with their contents (e.g., menu, use review). Since the contents are normally web pages that include texts, images and even videos, their sizes could be significant. Moreover, the larger is the cloaked region, the more privacy is preserved, but the contents of more non-result objects are requested. Requesting these contents waste precious network bandwidth, consume device battery, and charge the user more than necessary. Therefore, an important issue is how to control location cloaking in order to minimize the number of non-result objects.

In this section, we present an innovative result-aware location cloaking approach, called *2PASS* (2-Phase Asynchronous Secure Search), based on Voronoi cells.

Voronoi Cell and Diagram

Given a set of n objects, a Voronoi diagram divides the space into n partitions (Berg, Kreveld, & Overmars, 1997). Each partition is called a Voronoi cell and corresponds to one object. The cell is in such a shape that the nearest neighbor of any point in this cell is the corresponding object. Figure 16(a) shows an example of Voronoi diagram with 6 objects a through f. The solid lines show the borders of Voronoi cells, and the dotted lines connect the objects whose cells are adjacent. These latter lines divide the space into partitions of a special shape — triangles. As such, the set of these lines is called *Delaunay triangulation* of the space.

We design a weighted undirected graph to store the Voronoi diagram and Delaunay triangulation. As shown in Figure 16(b), each vertex in this graph denotes an object, and each edge denotes a line in the Delaunay triangulation. Each vertex i is also assigned a non-negative weight w_i (to be detailed later). We call this graph a *weighted adjacency graph* (WAG) in the sequel. It is noteworthy that a WAG is a special weighted graph, because its vertices, instead of its edges are weighted.

OVERVIEW OF 2PASS

A mobile user wants to request a location-based service (e.g., kNN search) from the LBS server. To protect location privacy, before requesting the service, he/she should invoke location cloaking, which obtains for this user a cloaked region that satisfies certain privacy metric. The user then attaches this region, instead of the accurate location, in the service request. Upon receiving this request, the server processes it and returns the resulted objects. Two predominant privacy metrics are granularity, i.e., the area of the cloaked

Figure 16. Voronoi Diagram and WAG. (a) Voronoi Diagram (solid lines are cell borders, dotted lines are Delaunay triangulation) (b) Weighted Adjacency Graph

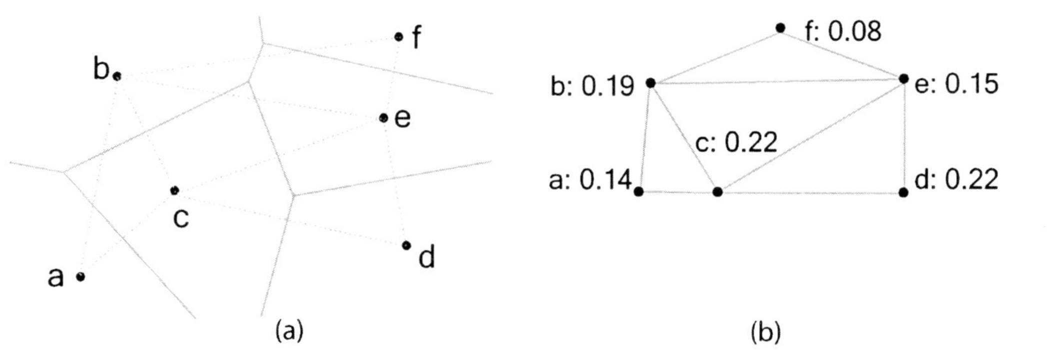

region is no less than a user-specified threshold, and k-anonymity, i.e., the number of users in the cloaked region is no less than k.

Based on the Voronoi cell information, 2PASS requests the objects (including the genuine NN together with other non-result objects) to satisfy the privacy requirement on the cloaked region, which is implied by these requested objects. 2PASS is unique in that the client controls what objects to request from the server so that their total number and thus the overall bandwidth are minimized. To achieve this, 2PASS works in two phases. In the first phase the client requests from the server a WAG of its neighborhood area, where the weight of a vertex is the area of the corresponding Voronoi cell. In the second phase, the client selects objects from this WAG and requests them for their complete contents. To minimize the object number while still meeting the privacy threshold τ, the criteria of object selection are a combination of the following: (1) the sum of the areas of Voronoi cells from the selected objects must exceed τ; (2) the genuine nearest neighbor o^* must be selected; and (3) these Voronoi cells must be connected, i.e., no cell is isolated from the rest of the cells. This last criterion guarantees the cloaked region is a single region, which is a common assumption in all existing approaches. With the introduction of WAG, the above object selection is equivalent to finding a subgraph in the WAG that satisfies the following criteria: (1) the sum of the weights of vertices in the subgraph must exceed τ; (2) o^* must be in the subgraph; and (3) this subgraph must be a connected component. In the sequel, we call such a subgraph a "valid-weight connected component" (VWCC) of a query, and the objective of 2PASS is to find a VWCC with the minimum number of vertices. We formalize this problem as follows.

Minimum Valid-Weight Connected Component (MVWCC) Problem: Given a WAG G, the privacy threshold τ, and the genuine NN object o^*, to find a VWCC that has the minimum number of vertices.

Approximate MVWCC Algorithm

We present an efficient approximation algorithm with a constant bound of approximation ratio. A key observation is that MVWCC problem is very similar to the rooted k-minimum spanning tree (*k-MST*) problem. A k-spanning tree (*k-ST*) is a spanning tree with at least k vertices.

k-MST Problem: Given a weighted undirected graph and a vertex r, to find a k-spanning tree rooted at r that spans at least k vertices with the minimum sum of edge weights.

We show that the MVWCC problem can be reduced to a *k-MST* problem in polynomial time. The key idea is to construct a new edge-weighted graph G' from the WAG G. Initially, G' has the same sets of vertices and edges as G, with each edge assigned a unit weight. Then for each vertex (called "primary vertex") in G', we add a number of auxiliary vertices (called "subsidiary"). Each subsidiary only connects to its primary vertex via an edge with a weight of 0. The number of such subsidiary vertices for each primary vertex i, denoted by p_i, is almost proportional to its weight w_i in G: $p_i = w_i * \Delta - 1$, where Δ is a constant called scaling factor. Figure 17 illustrates the constructed G' from the WAG in Figure 16(b).

Then by any k-MST approximation algorithm, such as Garg's 3-approximation algorithm, we get an approximate k-MST result Γ' for G'. Then we can construct an approximate MVWCC Γ for G from Γ', where $k = \tau * \Delta$: for each primary vertex in Γ', we add it to Γ.

Figure 17. Reduce MVWCC problem to k-MST

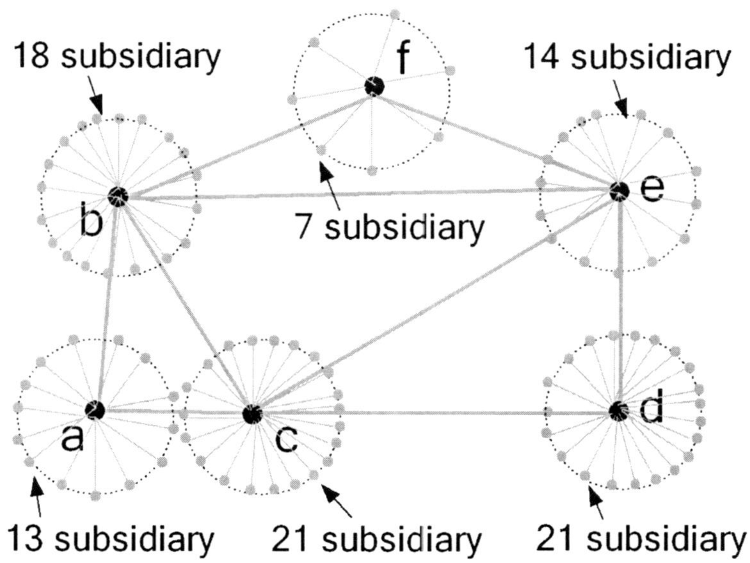

WAG-TREE AND 2PASS

In the first phase of 2PASS, the client requests for the WAG of its neighborhood. However, the definition of neighborhood is not clarified. If the object dataset is not huge, the client can request the WAG of the entire space and cache it to avoid re-request for subsequent queries. However, for a practical dataset with thousands or even millions of objects, it is impractical to request and cache the entire WAG. As such, we partition the entire WAG into *WAG snippets* of reasonable sizes so that the client receives only the snippet(s) surrounding the query location. In essence, a WAG snippet is the WAG of a subspace. In Figure 18, the four snippets are obtained by partitioning the space in Figure 16(a) into four equal-sized subspaces *(A,B,C,D)* and computing their WAG's respectively. It is noteworthy that an object that is outside of a sub-space can still appear in the WAG snippet of this sub-space, as long as the Voronoi cell of this object in the WAG of the entire space overlaps this sub-space, e.g., objects *a* and *c* in snippet *A*.

In order for the client to know which snippet(s) to request in the first phase of the query, we build a hierarchical index called WAG-tree. Like a quad-tree, this index recursively partitions the space into quadrants. Each entry in its leaf node points to a WAG snippet. Note that since the WAG-tree contains no WAG snippets, the size of this index is extremely small. Figure 19 illustrates a WAG-tree and WAG snippets pointed by it

The whole 2PASS procedure is summarized in Figure 20. During the system initialization, the whole WAG-tree is sent to and cached on the client. Upon an NN query, the client looks up the WAG-tree and locates the snippet that contains the query point. If the area of this sub-space is still smaller than the user-specified privacy threshold τ, the client will locate the lowest-level ancestor tree node of this snippet whose sub-space area just exceeds τ and request all snippets rooted at this node. In the sequel, we call all these snippets host snippets. The client then joins the received host snippets into a single WAG and applies the approximate MVWCC algorithm on it. The complete records of the objects that appear in the result VWCC are requested in the second phase.

Figure 18. WAG snippet

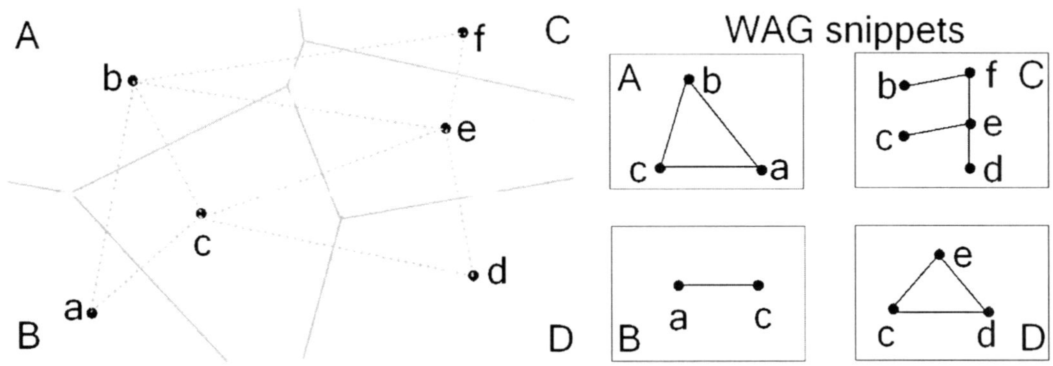

Figure 19. Illustration of WAG-tree

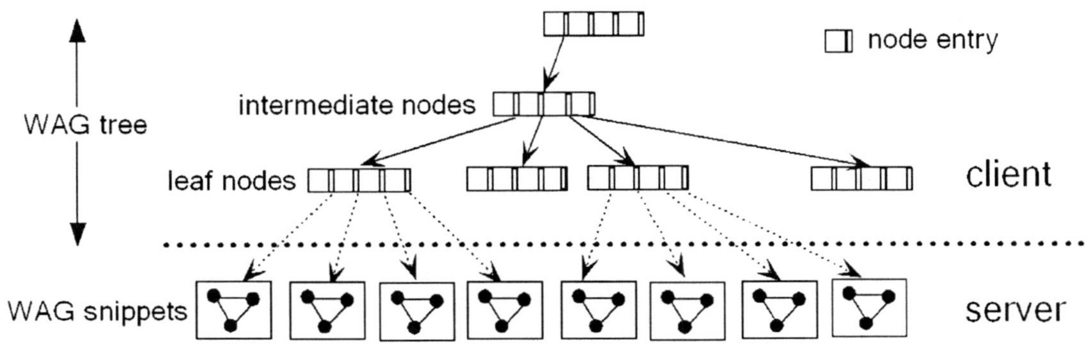

FUTURE RESEARCH DIRECTIONS

There are still a lot of open problems in privacy-aware location-based services. For example, how secure are the proposed techniques? In other words, is it possible for the client or server to "trick" the other side so that the privacy is breached? As another example, how secure are the well-adopted privacy metrics, such as k-anonymity? In location-based services, if the user population is dense, k-anonymity might still lead to privacy compromise.

One problem is receiving particular attention --- how we can design a query processing model that preserves privacy of mutual parties: the server should protect its data privacy and reveal only information that is implied by the query result, and the query user should protect its query privacy so that the server knows nothing about the query and is therefore unable to infer any information about the user. There are some works based on the theoretical results of secure multiparty computation (SMC). However, SMC solutions are in general expensive in terms of computation and communication costs. More practical and efficient solutions are needed for mobile environment where both the computational and the communication resources are limited.

CONCLUSION

Privacy has been an important and active research topic in mobile computing and location-based services. In this chapter, we study how to achieve location privacy during LBS without a centralized

Figure 20. 2PASS procedure for NN query

```
Input: q: the query point
       τ: the privacy threshold
Output: o*: the genuine NN
Procedure:
if WAG-tree is not available then
    obtain WAG-tree from server;
S = {the snippet contains q};
while area(S) < τ do
    move one level up the tree and add snippets to S;
request S from the server;
join the received S into G;
find o* from G;
Γ = Approximate_MVWCC(G, o*, τ);
request complete records of the objects appear in Γ from the server;
return o* and its attributes to the user;
```

and trusted middleware. First, we review the recent progress on location positioning technologies. Second, we investigate how to perform location cloaking without users exposing their accurate locations to a trusted third party. We decompose the problem into two subproblems: proximity minimum k-clustering and secure bounding. Third, we study how to perform nearest neighbor query with guaranteed privacy. We present a framework called 2PASS that allows the client to control what objects to request in order to minimize their number while not compromising location privacy of the user. The core component of 2PASS is a lightweight WAG-tree index from which the client can compute out the objects to request from the server.

REFERENCES

Bahl, P., & Padmanabhan, V. N. (2000). An . In *Building RF-based User Location and Tracking System, in IEEE INFOCOM 2000* (*Vol. 2*, pp. 775–784). Tel-Aviv, Israel: RADAR.

Chow, C. Y., Mokbel, M. F., & Liu, X. (2006). A peer-to-peer spatial cloaking algorithm for anonymous location-based services. In *Proc. of ACM GIS* (pp. 171-178).

Chu, K. M., Leung, K. R. P. H., Ng, J. K., & Li, C. H. (2004), Locating Mobile Stations with Statistical Directional Propagation Model, in *Proc. of the 18th Intl. Conf. on Adv. Info. Networking and Applications (AINA 2004)*, Fukuoka, Japan, pp. 230-235.

Dana, P. H. (1998), *Global Positioning System Overview*, The University of Texas, Retreived (n.d.)., from http://www.colorado.Edu/geography/gcraft/notes/gps/gps.html

de Berg, M., Cheong, O., van Kreveld, M., & Overmars, M. (2008). *Computational Geometry: Algorithms and Applications*. Springer, 3rd edition.

Djuknic, G. M., & Richton, R. E. (2001). Geolocation and Assisted GPS. *Computer*, *34*(2), 123–125. doi:10.1109/2.901174

Du, W. (2001). *A study of several specific secure two-party computation problems*. Ph.D. dissertation, Purdue University.

European Space Agency (ESA). (n.d.). *Galileo*. Retrieved (n.d.)., from http://www.esa.int/esaNA/galileo.html

Gedik, B., & Liu, L. (2005). Location privacy in mobile systems: A personalized anonymization model. In *Proc. of ICDCS*, 2005 (pp. 620-629).

Ghinita, G., Kalnis, P., & Skiadopoulos, S. (2007). Prive: Anonymous location-based queries in distributed mobile systems. In *Proc. of WWW 2007* (pp. 371-380).

Goldwasser, S. (1997). Multi party computations: past and present. In *Annual ACM Symposium on Principles of Distributed Computing*, pp.1-6.

Gruteser, M., & Grunwald, D. (2003). *Anonymous usage of location-based services through spatial and temporal cloaking* (pp. 31–42). In Proc. of MobiSys.

Hu, H., & Lee, D. (2006). Range nearest neighbor query. *IEEE Transactions on Knowledge and Data Engineering*, *18*(1), 78–91. doi:10.1109/TKDE.2006.15

Hu, H., & Xu, J. (2009). Non-exposure location anonymity. *IEEE International Conference on Data Engineering*, 2009, to appear.

Kan, K. K. H., Chan, S. K. C., & Ng, J. K. (2003). A Dual-Channel Location Estimation System for providing Location Services based on the GPS and GSM Networks, in *Proceedings of The 17th International Conference on Advanced Information Networking and Applications(AINA 2003)*, Xi'an, China, pp. 7-12.

Laitinen, H., Ahonen, S., Kyriazakos, S., Lahteenmaki, J., Menolascino, R., & Parkkila, S. (2001). Cellular location technology (Tech. Rep. 007), *VTT Information Technology*.

McGuire, M., Plataniotis, K., & Venetsanopoulos, A. (2005). Data Fusion of Power and Time Measurements for Mobile Terminal Location . *IEEE Transactions on Mobile Computing*, *4*(2), 58–66. doi:10.1109/TMC.2005.24

Mokbel, M. F., Chow, C. Y., & Aref, W. G. (2006). The new casper: Query processing for location services without compromising privacy. In *Proc. of VLDB*, 2006 (pp. 763-774).

Ng, J. K., Chan, S. K., & Kan, K. K. (2002). Location Estimation Algorithms for Providing Location Services within a Metropolitan area based on a Mobile Phone Network, in *Proceedings of The 5th International Workshop on Mobility Databases and Distributed Systems (MDDS 2002)*, Aix-en-Provence, France, pp. 710-715.

Pahlavan, K., & Krishnamurthy, P. (2002). *Principles of Wireless Networks a Unified Approach*. Upper Saddle River, NJ: Pearson Education, Inc.

Prasithsangaree, P., Krishnamurthy, P., & Chrysanthis, P. K. (2002). On Indoor Position Location with Wireless LANs, in *The 13th IEEE International Symposium on Personal, Indoor, and Mobile Radio Communications (PIMRC 2002)*, Lisboa, Portugal, pp.720-724.

Russian Space Agency. (n.d.). *Global navigation satellite system (glonass)*. Retrieved (n.d.)., from http://www.glonass-ianc.rsa.ru/

Zhou, J., Chu, K. M.-K., & Ng, J. K.-Y. (2005). Providing Location Services within a Radio Cellular Network using Ellipse Propagation Model, in *Proceedings of the 19th International Conference on Advanced Information Networking and Applications (AINA2005)*, Taipei, Taiwan, pp. 559-564.

Zhou, J., Chu, K. M.-K., & Ng, J. K.-Y. (2005), An Improved Ellipse Propagation Model for Location Estimation in facilitating Ubiquitous Computing, in *Proceedings of the 11th IEEE International Conference on Embedded and Real-Time Computing Systems and Applications (RTCSA 2005)*, Hong Kong, pp. 463-466.

Chapter 15
Survivability in RFID Systems

Yanjun Zuo
University of North Dakota, USA

ABSTRACT

There has been an increasing popularity of Radio Frequency Identification (RFID) techniques in various applications. A tiny RFID tag is attached to a mobile object, which can be scanned and recognized by a hand-held reader (e.g., a PDA, a mobile scanner). RFID offers opportunities for real-time item tracking, human identification, and inventory management. For applications using low-cost RFID tags and hand-held devices, however, various risks could threaten their abilities to provide essential services to users. In this chapter, survivability issues related to RFID systems are studied. For mission-critical systems empowered by the RFID technology, any interruption of essential services, even for a short period of time, is not acceptable. Hence, survivability must be provided to ensure that the critical services can be continuously delivered, despite malicious attacks and system failures. Our main contribution is a study and survey of survivability enhancing techniques in face of the special challenges that limited computational capacities, high mobility, and sensitive nature of RFID devices pose.

INTRODUCTION

Radio Frequency Identification (RFID) is a wireless technology for automatic item identification and data capture. It uses radio signals to identify a product, animal or person. Many retailers and wholesalers use RFID systems to manage product shipments and inventory tracking. Major retail chains such as Wal-Mart and Target have mandated that all suppliers introduce RFID. RFID have also been used in critical information systems in military, healthcare, and crisis management. The US Department of Defense has ordered that all shipments to its armed forces be equipped with RFID tags (Li & Ding, 2007). The UK armed forces adopted RFID in 2003 (Roberts, 2006).

There are three types of major components in a RFID system: tags, readers, and a backend server.

DOI: 10.4018/978-1-61520-761-9.ch015

A tag is physically attached to an item with a unique identification. A reader is a device that can recognize the presence of RFID tags and read the information supplied by them (Glover & Bhatt, 2006). It can be a PDA, a mobile phone or any kind of devices capable of communicating with an RFID tag. To obtain data from a tag, a reader first queries the tag and then forwards the received identity information to the backend server, which maintains a database of tag entries. After being authorized, the reader can obtain more detailed information about the tag. An RFID reader and the backend server communicate through a secure channel. Since the backend server and the readers can be secured using standard security mechanisms (e.g., public key cryptography), we assume that they are secure and trustworthy. In this chapter, we focus on the survivability enhancing techniques for low-cost RFID tags, which have limited computing and memory resources for security.

Although RFID systems provide numerous benefits for automatically identifying object items in a wide range of applications, they imply security concerns. Military RFID tags could be attacked by enemy forces. Supply chain RFID tags could be scanned by competitors for sensitive logistic information. RFID enabled passports may release personal data if not appropriately protected.

Various mechanisms have been developed to enhance the security of RFID systems. But, no guarantee can be offered for RFID security and privacy. In this chapter we address the issue of survivability for a RFID system and survey the potential survivability enhancing techniques in the literature. The objective of RFID survivability is to ensure that a RFID system provides essential services to users even in presence of malicious attacks and/or system failures. The major challenge for the survivability of an RFID system is the limited resources (e.g., memory, computing power, and area space) of RFID tags for security.

In the following discussions, we first present the background of system survivability and RFID security and privacy. Then, according to the unique features of RFID systems and the threat model, the survivability requirements for RFID systems are specified. Next, the survivability enhancing techniques in the literature are classified and discussed. Those techniques are presented from several perspectives such as preventive, protective and reactive, and recovery-oriented. Finally, future research directions on RFID survivability are discussed.

BACKGROUND

Survivability

System survivability has been studied from different application areas and based on different abilities that a critical system should have. Various definitions ([Deutsch & Millis, 1988; Ellison, *et. al.*, 1997; Knight, *et. al.*, 2003; Hiltunen, *et. al.*, 2000], to cite a few) of system survivability have been proposed and most of them share some common understandings. For instance, survivability is widely understood as a system property, relating the level of services provided to the level of damage present in the system and operating environment; a survivable system must support the system's mission; operating in a hostile environment, a survivable system may offer degraded (but acceptable to users) services to users and have the ability to recover when the environment improves. Tarvainen (2004) identify a set of key properties that a survivable system should have: (1) a survivable system delivers essential services and maintains essential properties of those essential services, e.g., specified levels of integrity, confidentiality, performance and availability; (2) requirements of survivability are often expressed in terms of maintaining a balance among multiple attributes such as security, reliability, and modifiability; and (3) it is crucial to identify the essential services, and the essential properties that support them, within an operational environment.

Survivability requirements can vary substantially depending on the scope of the system, the functionalities of the system, operating environments of the system, and the criticality and consequence of attacks and interruptions of the system. There are two aspects of survivability (Pal, *et. al.*, 2000): survival by protection and survival by adaptation. The former refers to the situation when security mechanisms can effectively protect system from harmful, accidental or malicious changes. The latter refers to the ability of a system to dynamically reconfigure and respond to environment change, thus avoiding the situation when its critical services cannot be provided. We next review research on RFID security and privacy.

RFID Security and Privacy

Most research on RFID security and privacy focuses on tag-reader mutual authentication, tag-reader secure communications, and tag data access control. Techniques were developed to prevent attacks such as message eavesdropping, replay, man-in-the-middle, denial of service, and tag (or reader) impersonations. In the literature, RFID security and privacy protocols can be classified into two broad categories (see [Juels, 2006; Avoine, 2006] for more information): a class of protocols trying to enhance privacy and security without using standard cryptographic primitives (Tsudik, 2006; Weis, *et. al.*, 2003; Lim & Kwon, 2006) and a class of protocols relying on symmetric-key primitives such as block ciphers and hash functions (Vajda & Buttyan, 2003; Jues, 2004; Peris-Lopez, *et. al.*, 2006). In the first category, Tsudik (2006) proposed an RFID authentication protocol where a reader, *R*, shares a key, x_i, with a tag, *T*. *T* also stores an internal timestamp t_i which was the last time interrogated by *R*. To query a tag, *R* sends the current time, t_R. The tag compares t_R with t_i. If $t_R \leq t_i$, then the tag outputs a random response. Otherwise, the tag outputs $Hx_i(t_R)$, where Hx_i represents the HMAC computed with secret key x_i. Finally, the timestamp maintained by *T* is also updated. To validate a response, *R* checks whether the received $Hx_i(t_R) = Hx_j(t_R)$ for any secret key, x_j, in its database. If so, then *R* can be certain that the response comes from a valid tag, T_j. Weis (2003) proposed "hash lock" for private mutual authentication and tag access control. Each tag has two possible states: locked and unlocked states. In a locked state, the tag responds to all queries with only its meta-ID and offers no more functions; in an unlocked state, it performs privileged operations related to security and configuration. Their scheme ensures that a tag enters the unlocked state only if it receives an appropriate command from a legitimate reader. Lim (2006) proposed a strong and robust RFID security protocol with both forward and backward intractability. Their protocol uses a forward hash chain for tag identification and a backward hash chain for reader identification. The protocol allows up to *m* authentication failures between two valid sessions.

There are also solutions for tag-reader mutual authentication using non-cryptographic primitives. Vajda (2003) proposed a set of extremely lightweight challenge response authentication algorithms only using XOR and bit-mapping functions. Juels (2004) proposed a solution based on the use of pseudonyms, without using any hash function. Each tag stores a list of pseudonyms; it rotates them, releasing a different one on each query. After a certain time period, the tag needs to be updated through an out-of-band channel. LMAP (Peris-Lopez, *et. al.*, 2006) and M^2AP (Peris-Lopez, *et. al.*, 2006) are two other ultra-lightweight RFID authentication protocols which only use simple bitwise operations.

RFID THREAT MODEL AND SURVIVABILITY REQUIREMENTS

The Threat Model

RFID tags are a challenging platform from security perspective. The limited resources (both memory and processing power) of a passive tag and the nature of wireless communications between a tag and a reader determine that an RFID system is vulnerable to various passive and active attacks. In the threat model, we consider the following attacks to tag security.

- **Impersonation attacks:** An attacker captures keys or other tag data that allow for impersonation. For instance, an attacker intercepts messages in transmission between a tag and a reader. Or, the attacker can physically compromise a tag and retrieve the information.
- **Replay attacks:** An attacker reuses a tag's response to the reader in a previous session to impersonate the tag in the current session.
- **Denial-of-service attacks:** An attacker causes a tag to assume a state from its normal operation. The tag becomes either temporarily or permanently incapacitated.
- **Relay attacks:** An attacker makes the verifier believe that the prover is in its close vicinity by surreptitiously forwarding the signal between the verifier and an out-of-field prover (Kim, *et. al.*, 2009).
- **Physical attacks:** An attacker physically removes RFID tags, destroy tags, and/or use special devices to read tag memory and obtain/change identification-related information.

RFID System Survivability Requirements

Survivability requirements for RFID system can be summarized as below.

1. **Component identity requirement:** due to its open nature, a RFID system must be able to reliably identify a legitimate tag. If the state of a tag (e.g., its secret keys) is released to an attack, the tag is under the control of the attack which then changes the identity of the tag, put it into sleep, or permanently disable it. All those malicious actions could completely compromise the tag. A RFID system survives only if it can reliably identify self and non-self components even in a hostile environment.
2. **Information access requirements:** a legitimate user must have access to tag information (e.g., tagged item ID or other product information) whenever needed. As a trustworthy label for an item with important data about the item, a RFID tag needs to provide real time data about the state of the tagged item. In order to be reliably used, tag information must be authentic. Tags should have the ability to resist memory tampering and secure communications with readers to ensure data confidentiality, integrity, and availability.
3. **Tag Availability Requirements:** RFID tags must be available when a RFID reader updates their state, e.g., refresh secret keys, reset tag states, or temporarily disable the tags in presence of malicious attacks. Tags should be readily available and searchable when needed.

SURVIVABILITY ENHANCING TECHNIQUES

We are mainly concerned with survivability enhancing techniques at the application layer and also address some issues at the physical layer. Survivability enhancing techniques refer to those technical means and plans of actions that can improve the ability of a RFID system to withstand malicious attacks and continuously support the system mission. Those techniques are classified in two general categories: (1) software based (better protocols suitable for RFID tags and readers); and (2) hardware based (more powerful and efficient hardware to support security and survivability software). Due to the cost pressure of massively deployed RFID tags, however, it is not possible

Figure 1. RFID survivability enhancing techniques

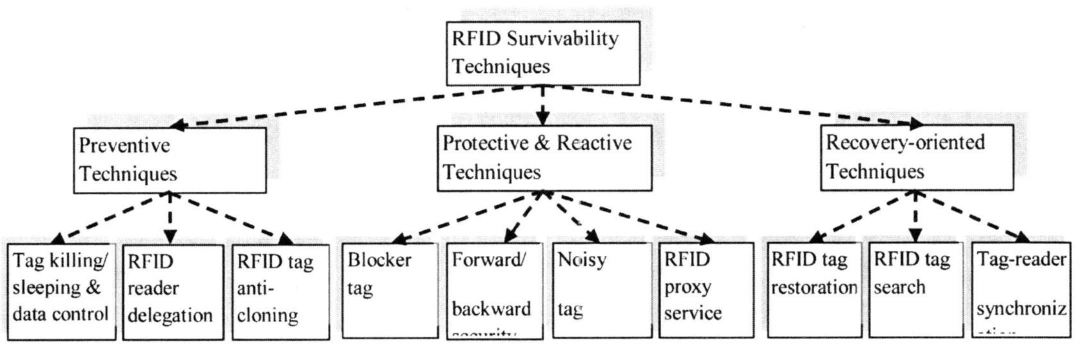

in the near future to produce low-cost RFID tags with sufficient functionalities to support standard security methods (e.g., public-key cryptography and other multi-party security protocols). In the following sections, we focus on the software based approach for RFID survivability. The techniques in this category can be classified as preventive, protective and reactive, and recovery-oriented as shown in Figure 1.

Preventive Techniques

Preventive techniques protect a RFID system from possible attacks by using novel security mechanisms, environment measures, and operational protection means. In the literature, RFID techniques in this category include tag memory protection, tag disabling (temporarily or permanently), reader delegation, and tag anti-cloning.

RFID tags can be rewriteable. Like other security mechanisms, password protection has been incorporated into most tags to eliminate unauthorized data access. The backend server (or the reader) shares a specific PIN with each individual tag. Writing of tag data must be PIN-protected. Furthermore, to protect the integrity of the tag data, digital watermark has been applied to RFID tags.

RFID tag killing has been used by many RFID systems. When a tag is no longer needed (e.g., change of owner or no need to be tracked) or the operating environment is getting too challenging, the tag can be disabled permanently by authorized readers. A tag specific kill PIN is implanted into a tag and the tag can simply disable itself upon receiving an authorized command.

Instead of disabling a tag permanently, an unauthorized reader may put the tag to "sleep", i.e., temporarily render it inactive. To "wake" up the tag when necessary, the reader must transmits another tag specific PIN to authenticate itself. Hence, an RFID tag can be in one of two states: sleep and awake. A physical trigger, like direct touch of a reader probe, might serve as an alternative means of waking a tag (Stajano & Anderson, 1999).

Reader delegation is a way to reduce the exposure of an adversary breaking into an RFID reader. Instead of losing all the secrets for all the tags, the users may lose only what was delegated to that reader (Tan, *et. al.*, 2006). Reader delegation has additional advantage of tolerating poor quality network connections between the reader and the backend server since a reader does not need to communicate with the server for every tag authorization.

Physical security is crucial for tag anti-cloning and anti-counterfeiting. Generally, there are two research directions towards preventions of physical attacks. One is hardware approach, applying tamper resistance property to the system; the other is software approach, focusing on cryptographic ways to solve the problem. These two approaches can be used jointly. For instance, physically unclonable function (PUF) (Tuyls & Batina, 2006) is a technique applied to resist physical attacks on low-cost RFID tags. In order to thwart tag cloning attack, Tuyls (2006) propose a PUF structure to store secret key materials in a tag. In order to make an item unclonable, two components are needed: (1) Physical protection, which uses unclonable physical structure embedded in the package (removal of the structure leads to its destruction). One or more unique fingerprints derived from the physical structure will be printed on the product for the verification of the authenticity of the product; and (2) Cryptography protection, which provides digital signature to detect and prevent tempering with the fingerprints derived from a physical object. It also provides secure identification protocols to identify a product.

A physical unclonable function maps challenges to responses and it is embodied in a physical object (Tuyls & Batina, 2006). It satisfies the following properties: (1) easy to evaluate: the physical object can be evaluated in a short amount of time; and (2) hard to characterize: from a number of measurements performed in polynomial time, an attacker who no longer has the device and who only has a limited (polynomial) amount of resources can only obtain a negligible amount of knowledge about the response to a challenge that is chosen uniformly at random.

Protective and Reactive Techniques

While preventive techniques can be implemented during the system design phase and enforced before any attacks could happen, protective and reactive techniques focus on real-time incident response in face of malicious attacks. They are applied to enhance a RFID system's ability to adapt to the changing environment and respond to attacks correspondingly. The techniques in this category include tag monitoring, tag blocking, fault tolerance, and proxy-based protection.

Shield and blocker of RFID enabled products such as e-passports and credit cards are active techniques to protect tags from unauthorized interrogations. Tag blocking is a measure of operating an RFID system in a degraded model. Juels (2003) propose a special form of RFID tag called a blocker. The blocking depends on the incorporation into tags of a modifiable bit called a privacy bit. A "0" privacy bit marks a tag as subject to unrestricted public scanning; a "1" bit marks a tag as "private." The scheme refers to the space of identifiers with leading '1" bits as a privacy zone. A blocker tag prevents unwanted scanning of tags mapped into the privacy zone.

Noisy tags are designed to secure communication channels between a reader and its tags (by facilitating to establish a secret key) (Castelluccia & Avonine, 2006). A noisy tag is owned by the reader's manager and set out within the reader's field. It is a regular tag that generates noise on the public channel between the reader and the queried tag, such that an eavesdropper cannot differentiate the messages sent by the queried tag from the ones sent by the noisy tag.

Fault tolerance is another way to enhance RFID system survivability. It is achieved by forward and backward security in RFID tag-reader authentication. In many cases, RFID tags often need to change owners (Song, 2008). Forward security refers to the situation that the old owner, say S_{old}, will not be able to identify and control the tag, say, T, after it passes the ownership of T to a new owner, say S_{new}. Forward security can be achieved by simply requiring that S_{new} applies the mutual authentication protocol and updates the currently shared key k' to a new key k (by incorporating some secret materials only known to S_{new}). Since S_{old} cannot figure out k, it has no way to identify any transactions related to T conducted after the key update. Backward security refers to the situation that the new owner will not be able to use any information it has on the tag to track back the past interactions conducted related to tag T. Once again, one simple key update is sufficient to achieve this goal. Before S_{old} transfers the ownership of T to S_{new}, it updates the key k'' to k'. Then S_{old} passes the secret k' to S_{new} via a secure channel.

Figure 2. Updating tag ID

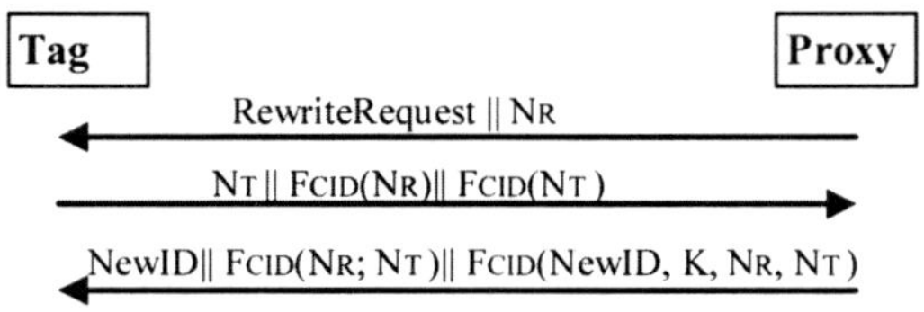

Figure 3. Putting tag to sleep

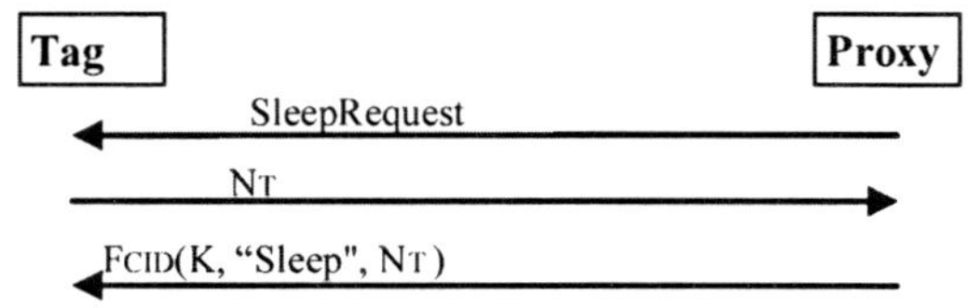

Since S_{new} has no way to calculate *k"* based on *k'* (a one-way function such as a hash function can ensure this), it cannot identify any transactions related to *T* conducted before the ownership transfer. The major challenge of those methods is, however, to securely deliver the secure key from the current owner to the new owner. In the literature, key delivery is often conducted using an off-band secure channel.

Dimitriou (2008) presented a proxy framework for protecting the privacy of users carrying RFID products. This approach uses a mobile phone (or any other similar device) as a proxy for interacting with readers on behalf of the user carrying tagged item. The user can specify when and where information will be released. Essentially, the proxy acts as mediator for tag access to ensure the tag can withstand malicious attacks. The major operations are summarized in Fig. 2-5, where the symbol "||" stands for message concatenation; *CID* stands for the current ID of the tag; *NewID* stands for the new ID of the tag; N_R represents a random nonce generated by the proxy; N_T represents another random nonce generated by the tag; $F_{cid}(.)$ stands for an encryption function using key *cid*; and *K* represents the authentication key shared between the tag and the proxy.

Guardian (Rieback, *et. al.*, 2005) is a device that acts as an intermediary between tags and readers and must always be alert in protecting tag responses from unauthorized read attempts. It allows reader queries, appropriately re-issues queries in encrypted form, or actively blocks tag answers. The RFID Guardian offers granular access control - coordination of security primitives, context-awareness, and tag-reader mediation. For instance, Guardian can control which RFID readers can query which RFID tags under which conditions. Also Guardian can act as a "man-in-the-middle", mediating interactions between RFID readers and RFID tags. Mediation can take either a constructive or destructive form. For a constructive mediation, an untrusted RFID reader first passes a request for a desired query to the RFID Guardian. Upon the successful completion of a possibly complex security negotiation, the Guardian re-issues the query in encrypted form and forwards the cryptographically-protected queries to RFID tags. The Guardian then receives the encrypted tag response, decrypts it, and forwards the response to the RFID reader that requested it. Selective RFID jamming is an example of

Figure 4. Enhancing RFID privacy

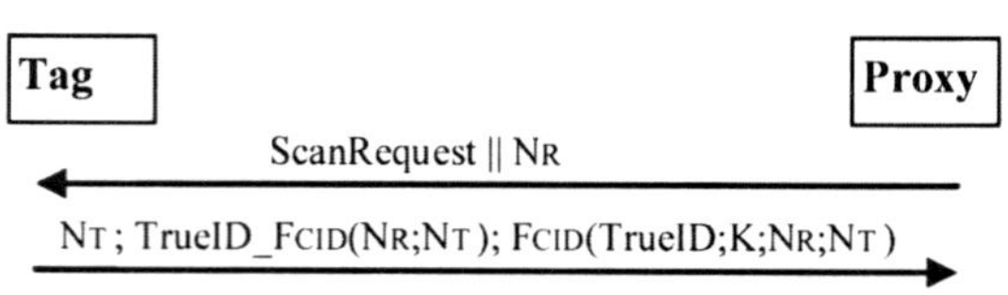

Figure 5. Reverting a tag to its original state

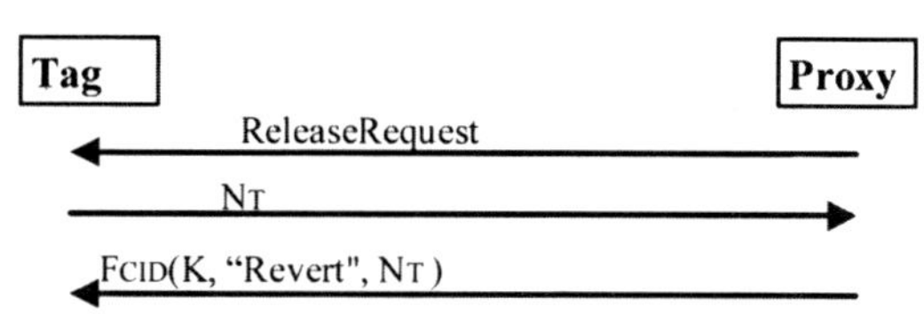

Figure 6. Zuo's tag search protocol

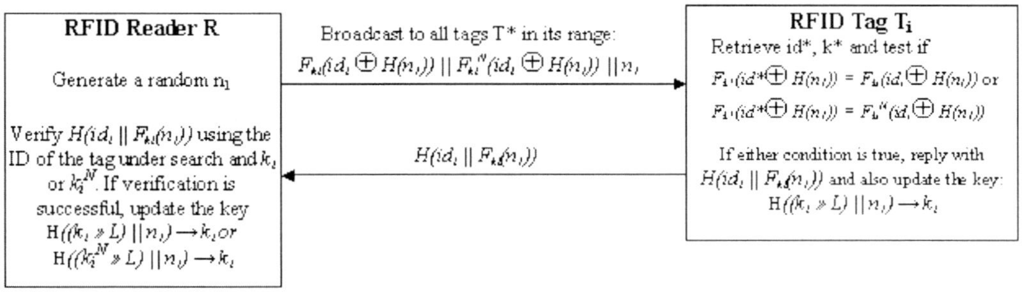

destructive mediation where the RFID Guardian blocks unauthorized RFID queries on the behalf of RFID tags. By filtering RFID queries, Selective jamming provides off-tag access control.

Recovery-Oriented Techniques

A survivable system must be able to recover from damage quickly. We summarize the techniques in this category for RFID systems in three major thrusts: tag restoration, tag search, and tag-read state synchronization.

Tag restoration is to reset the secret of a tag as shared with the reader/backend server whenever it is necessary. It can be achieved by an explicit key reset channel via PIN matching, manual intervention (e.g., physical contact of a RFID tag to trigger tag key reset), or simply normal scan of the tag. Since RFID tag and reader maintain a synchronized state and update their shared secret keys in a coordinated fashion, reading the tag one or two times in a secure environment effectively serves as a tag reset method. Even when an attack has already possessed the tag secret key, reading the tag can trigger a key update using some new materials supplied by the reader. After such secure reading, the attack effect has been wiped out.

Another recovery related technique is tag search. As a highly mobile device, a RFID tag (or even a reader) could be stolen, misplaced, or physically tampered. Secure search for missing tags or detecting tag failures is an important enhancing technique for RFID survivability. If a tag is controlled by an attacker, the corresponding secret information stored in the backend server must be deleted or updated.

Zuo (2009) proposed a set of secure and private protocols to search for tags based on their IDs or certain features that the tags have. One protocol is shown below (also see Figure 6). In this protocol, a reader R broadcasts a search query to all the tags in its read range anonymously. Only the tag under search will reply. Other tags just keep silent. To be robust, the protocol is designed so that R recognizes both the current key shared with a tag, t_i, i.e., k_i, and the next "should-be" secret key, denoted as k_i^N, to encrypt a search query message. k_i^N is used just for the situation when a tag, T_i, has updated the key in the last successful search, but R did not. k_i^N is calculated as $k_i^N = H((k_i \gg L) \,||\, n_i)$, where $H(.)$ presents a one-way hash function; ">>" represents a right cyclic rotation function; and L represent the number of bits to rotate.

1. **Reader:** $R \text{ -> } T^*\!: F_{ki}(id_i \oplus H(n_i)) \,||\, F_{ki}^N(id_i \oplus H(n_i)) \,||\, n_i$, where $\oplus$ represents a bit-wise XOR operation and n_i represents a random nonce generated by R

R broadcasts a search query message to all the tags in its field.

2. **Tag:** Each tag, T^*, which receives the message uses its ID number, id^*, its secrete key,

k^*, and the n_l value to test if $F_{k^*}(id^* \oplus H(n_l)) = F_{ki}(id_i \oplus H(n_l))$ or $F_{k^*}(id^* \oplus H(n_l)) = F_{ki}^N(id_i \oplus H(n_l))$. If either condition is true, then $T^* = T_i$ and T_i responds by sending the following message:
2.1) T_i -> R: $H(id_i \| F_{ki}(n_l))$
2.2) $H((k_i \gg L) \| n_l) \rightarrow k_i$

T_i also updates its secret key k_i.

3. **Reader:** R verifies the received message $H(id_i \| F_{ki}(n_l))$ by applying the ID number of the tag being searched, either the current shared secret key with the tag or the next expected key, and the n_l value transmitted in step 1. If one of the calculated results matches the received message, then R believes that tag T_i has been successfully searched. Otherwise, R believes that either the response is not valid or T_i is not in its range. In addition, if the current shared key, k_i, was used to verify the message, R updates the shared key by doing this: $H((k_i \gg L) \| n_l) \rightarrow k_i$.

Otherwise, if the expected next key k_i^N was used to verify the message, R updates the shared key in this way: $H((k_i^N \gg L) \| n_l) \rightarrow k_i$.

The third technique in lieu of RFID recovery is the system's ability to resynchronize the state of tag and reader if its state is desynchronized possibly by malicious attackers. Desynchronization is a dangerous denial of service attack. If a tag is permanently desynchronized with its owner or the backend server, then the tag is considered completely compromised. Since an attacker has access to the communication channel between a tag and a reader, the attack could block some messages and make the states of the tag and reader out of pace. For instance, an attacker could block the key update message from reaching the tag. Therefore, the backend server updates the key but the tag does not. So, the tag and the backend server will not be able to authenticate each other since their keys are out of synchronization.

Henrici (2004) proposed to solve the synchronization problem by having a tag emit the difference between its current transaction number *TID* and the last successful transaction number *LST*, i.e., $\Delta TID = TID - LST$. So, the reader will be able to determine the current state of the tag. The reader also maintains two entries with each tag, one is for the "should-be" values and the other is for the last successful authentication. A row in the database is never overwritten until the other entry has been addressed by the tag proving that one being currently valid and the one to be overwritten being obsolete. The protocol is summarized below (also see Figure 7):

- Reader R sends a HELLO message to a tag *T* (here we do not consider tag singleton)
- T increments its current transaction number TID by one and then responds with message $A = h(ID), h(ID \| TID), \Delta TID$, where $h(.)$ is a hash function and ID represents the ID of tag *T*
- *R* receives A and forwards it to the backend server *S* for tag data
- S selects the entry with $HID=h(ID)$. The stored last successful transaction number

Figure 7. Henrici and Muller's Protocol

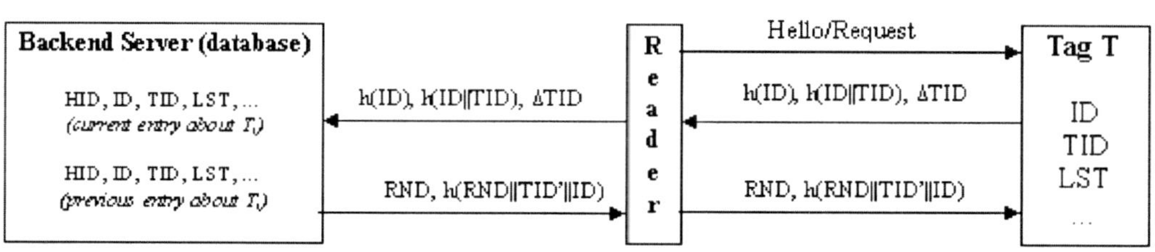

LST and the received *ΔTID* are added to obtain the current transaction ID of tag *T*, i.e., *TID'*. Then *h(ID||TID')* is calculated to verify the received *h(ID||TID)*. If they don't match, the transaction is terminated

- If the *TID'* is not higher than the *TID*, then a replay attack is in process and the message is discarded
- If the above is verified, *S* replaces the stored *TID* with *TID'*. In this way, *S* and *T* synchronize their states
- Next, *S* creates a new *ID' = ID||RND*, where *RND* is a random nonce. *S* creates a reply *h(RND||TID'||ID)* and send it (together with *RND*) to *R*, which forwards the message to *T*
- *T* checks *h(RND||TID'||ID)* and update its ID. In the mean time, *T* sets its last successful transaction number to the current *TID* value

FUTURE RESEARCH DIRECTIONS

RFID survivability is a challenging issue but no existing work in the literature explicitly addresses RFID survivability. We list some research areas which we believe have great potential in enhancing RFID survivability.

There is a need for more lightweight, efficient, and secure cryptographic protocols suitable for low-cost RFID tags. Currently, it is still not realistic to apply sophisticated cryptographic algorithms (e.g., RSA, digital signature and keyed hash function) to low-cost RFID tags. Those limitations prevent a RFID system from benefiting from many of the standard, multi-party, and reliable existing security mechanisms in practice. Although there have been some encouraging experiments in this field, there is still no publicly accepted solution to apply standard cryptographic algorithms to such a resource restricted device as an RFID tag.

Research is highly desired for next generation, intelligent RFID tags. According to Moor's Law, RFID tag prices will be continuously down for the same type of tags today or the tags will have much more functions available if the prices are kept the same. With more functionality available for security, next generation RFID will be more intelligent. They may be self-adaptive to the environment changes, self-robust to misuse or malicious attacks, and self-configurable to resist miss-setting or improve performance. Those new features will greatly increase a RFID system's ability to withstand attacks and provide essential services to users even when operating in a hostile environment.

One important approach to enhance survivability is by adaptation. There is a need to build agile, scalable, and resilient RFID infrastructures which includes RFID communication networks, authentication systems, and auditing mechanisms. This infrastructure is highly distributed but has central functions for intrusion detection, tag monitoring, tag search, and intrusion tolerance. All those features must consider the unique characteristics of resource-restricted RFID tags.

CONCLUSION

In this chapter, we surveyed survivability enhancing techniques for RFID systems. Survivability is a relatively new research area. While traditional security focuses on prevention and protection of computer resources, survivability explicitly deals with the issue that essential services of a system can be provided to users in face of malicious attacks or system failures even after parts of the systems have been damaged. There is little research on survivability of RFID systems. RFID survivability requires innovative techniques to address the limitations of low-cost RFID tags, highly mobile devices, and challenging environment in which an RFID system operates. This chapter summaries the potential survivability enhancing techniques in the literature and provides references for researchers and system developers to develop technologies towards resilient, secure, and survivable RFID systems.

ACKNOWLEDGMENT

This work was supported in part by US AFOSR under grants FA 9550-09-1-0215. The author is thankful to Dr. Robert Herklotz for his support, which made this work possible.

REFERENCES

Avoine, G. (2006). *Security and Privacy in RFID Systems*. Retrieved February 1, 2009, from http://lasecwww.epfl.ch/~gavoine/rfid

Castelluccia, C., & Avoine, G. (2006, April 19-21). Noisy Tags: A Pretty Good Key Exchange Protocol for RFID Tags. *7th IFIP WG 8.8/11.2 International Conference, Smart Card Research and Advanced Applications*, Tarragona, Spain.

DeutschM.WillisR. (1988). Software Quality Engineering: A Total Technical and Management Approach. Englewood Cliffs, NJ: Prentice-Hall.

Dimitriou, T. (2008, January 10-12). Proxy Framework for Enhanced RFID Security and Privacy. *Consumer Communications and Networking Conference*, Las Vegas, NV.

Ellison, R., Fisher, D., Linger, R., & Lipson, H. (1997). *Survivable Network Systems: An Emerging Discipline.* (Technical Report CMU/SEI-97-TR-013), Software Engineering Institute, Carnegie Mellon University.

Feldhofer, M., Dominikus, S., & Wokerstorfer, J. (2004, August 11-13). Strong Authentication for RFID Systems Using the AES Algorithm. *Cryptographic Hardware and Embedded Systems – CHES 2004, 6th International Workshop*, p. 357-370, Cambridge, MA.

Glover, B., & Bhatt, H. (2006). *RFID Essentials*. Norfolk, UK: O'Reilly Publisher.

Henrici, D., & Muller, P. (2004, March 14-17). Hash-based Enhancement of Location Privacy for Radio-frequency Identification Devices using Varying Identifiers. *Second IEEE Annual Conference on Pervasive Computing and Communications Workshops*, Orlando, FL.

Hiltunen, M., Schlichting, R., Ugarte, C., & Wong, G. (2000). Survivability through Customization and Adaptability: The Cactus Approach. *DARPA Information Survivability Conference and Exposition* (pp. 294-307).

Hu, W., Zuo, Y., Wiggen, T., & Krishna, V. (2008, May 18-20). Handheld Data Protection Using Handheld Usage Pattern Identification. 2008 IEEE International Conference on Electro/Information Technology, p. 237-240, Ames, IA

Juels, A. (2004, September 8-10). Minimalist Cryptography for Low-cost RFID Tags. *The Fourth Conference on Security in Communication Networks* (pp. 149-153), Amalfi, Italy.

Juels, A. (2006, February). RFID Security and Privacy: A Research Survey (2006). *IEEE Journal on Selected Areas in Communication, 24*(2).

Juels, A., Rivest, R., & Szydlo, M. (2003). The Blocker Tag: Selective Blocking of RFID Tags for Consumer Privacy. *ACM Conference on Computer and Communication Security*, (pp. 103-111).

Kim, C., Avoine, G., Koeune, F., Standaert, F., & Pereira, O. (2009, December 2-4). The Swiss-Knife RFID Distance Bounding Protocol. *The 12th International Conference on Information Security and Cryptology*, Seoul, Korea.Knight, J., Strunk, E., and Sullivan, K. (2003). Towards a Rigorous Definition of Information System Survivability, *DARPA Information Survivability Conference and Exposition*, Washington, DC.

Li, Y., & Ding, X. (2007, March 20-22). Protecting RFID Communications in Supply Chains. *ACM Symposium on InformAtion, Computer and Communications Security*, Singapore.

Lim, C., & Kwon, T. (2006, December 4-7). Strong and Robust RFID Authentication Enabling Perfect Ownership Transfer. *The 8th Conference on Information and Communications Security*, Raleigh, NC.

Pal, P., Loyall, J., Schantz, R., Zinky, J., & Webber, F. (2000). Open Implementation Toolkit for Building Survivable Applications. *DARPA Information Survivability Conference and Exposition, 2*, 197-200.

Peris-Lopez, P., Hernandez-Castro, C., Estevez-Tapiador, J., & Ribagorda, A. (2006, July 12-14). LMAP: A Real Lightweight Mutual Authentication Protocol for Low-cost RFID Tags. *2nd Workshop on RFID Security*, Graz, Austria.

Peris-Lopez, P., Hernandez-Castro, C., Estevez-Tapiador, J., & Ribagorda, A. (2006, September 3-6). M^2AP: A Minimalist Mutual-authentication Protocol for Low-cost RFID Tags. *International Conference on Ubiquitous Intelligence and Computing* (pp. 912-923), Wuhan, China.

Rieback, M., Crispo, B., & Tanenbaum, A. (2005, July). RFID Guardian: A Battery-powered Mobile Device for RFID Privacy Management. *Australian Conference on Information Security and Privacy, 3574*, 184-194

Roberts, C. M. (2006). Radio Frequency Identification (RFID). *Computers & Security, 25*(1), 1, 18–26. doi:10.1016/j.cose.2005.12.003

Song, B. (2008, July 9-11). *RFID Tag Ownership Transfer*. The 4th Workshop on RFID Security, Budapest, Hungary.

Stajano, F., & Anderson, R. (1999). The Resurrecting Duckling: Security Issues for Ad-hoc Wireless Networks. *7th International Workshop on Security Protocols, 1796*, Lecture Notes in Computer Science, p. 172-194, New York: Springer-Verlag.

Tan, C., Sheng, B., & Li, Q. (2008, March). Secure and Serverless RFID Authentication and Search Protocol. *IEEE Transactions on Wireless Communications*, 7(3).

Tarvainen, P. (2004, November 4-5). Survey of the Survivability of IT Systems. *The 9th Nordic Workshop on Secure IT-systems*, Helsinki, Finland.

Tsudik, G. (2006, March 13-17). YA-TRAP: Yet Another Trivial RFID Authentication Protocol. *The 4th Annual IEEE International Conference on Pervasive Computing and Communications*, Pisa, Italy.

Tuyls, P., & Batina, L. (2006, February 13-17). RFID-Tags for Anti-Counterfeiting. *The Cryptographer's Track at the RSA Conference*, San Jose, CA.

Vajda, I., & Buttyan, L. (2003, October 12). Lightweight Authentication Protocols for Low-cost RFID Tags. *Second Workshop on Security in Ubiquitous Computing*, Seattle, WA.

Weis, S., Sarma, S., Rivest, R., & Engels, D. (2003, March 12-14). Security and Privacy Aspects of Low-cost Radio Frequency Identification Systems. *The 1st International Conference on Security in Pervasive Computing*, Boppard, Germany.

Zuo, Y., Search and Private Tag Search Protocols (2009). *Special Issue on Advanced RFID Technologies, Information Systems Frontier - A Journal of Research and Innovation*, to appear.

ADDITIONAL READING

Avoine, G., & Oechslin, P. (2005, March 8). A Scalable and Provably Secure Hash Based RFID Protocol. *Second IEEE International Workshop on Pervasive Computing and Communication Security*, March 8, Kauai Island, HI.

Dimitriou, T. (2005). A Lightweight RFID Protocol to Protect Against Traceability and Cloning Attacks. *Second IEEE International Workshop on Pervasive Computing and Communication Security*, Kauai Island, Hawaii, USA.

Ghosh, D., Sharman, R., Rao, H., & Upadhyaya, S. (2006). Self-healing Systems – Survey and Synthesis. *Decision Support Systems*, *42*, 2164–2185. doi:10.1016/j.dss.2006.06.011

Hopper, N., & Blum, M. (2001). Secure Human Identification Protocols. In C. Boyd (ed.) *Advances in Cryptology – ASIA CRYPT 2001*, Vol. 2248, *Lecture Notes in Computer Science*, pp. 52-66, New York: Springer-Verlag.

JuelsA.WeisS. (2005). Authenticating Pervasive Devices with Human Protocols. In ShoupV. (Ed.), Advances in Cryptology – Crypto 05, Lecture Notes in Computer Science. New York: Springer-Verlag.

Juels, A., & Weis, S. (2007, March 19-23). Defining Strong Privacy for RFID. *5th Annual IEEE International Conference on Pervasive Computing and Communications*, White Plains, USA.

Katz, J., & Shin, J. (2006). Parallel and Concurrent Security of the HB and HB++ Protocols. In *Advances in Cryptology – EURO CRYPT 2006*, Vol. 4004, *Lecture Notes in Computer Science*, pp. 73-87, New York: Springer.

Medhi, D., & Tipper, D. (2000). Multi-Layered Network Survivability – Models, Analysis, Architecture, Framework and Implementation: An Overview. *DARPA Information Survivability Conference DISCEX 2000*, pp. 173-186, Hilton Head, SC.

Mikic-Rakic, M., Mehta, N., & Medridovic, N. (2002) Architectural Style Requirements for Self-healing Systems. *The First Workshop on Self-healing Systems*, pp. 49-54, Charleston, South Carolina.

Molnar, D., & Wagner, D. (2004, October 25-29). Privacy and Security in Library RFID Issues, Practices, and Architectures. *ACM Conference on Computer and Communications Security*, Washington, DC.

Park, J., & Chandramohan, P. (2004). Static vs Dynamic Recovery Models for Survivability Distributed Systems. *The 37th Hawaii International Conference on System Sciences*, Hawaii, USA.

Peris-Lopex, P., Hernandex-Castro, J., Estevez-Tapiador, J., & Ribagorda, A. (2006, July). A Real Lightweight Mutual Authentication Protocol for Low-cost RFID Tags. *Second Workshop on RFID Security*.

Samyde, D., Skorobogatov, S., Anderson, R., & Quisquater, J. (2002, December 11). On a New Way to Read Data from Memory. In *Proceedings of the First International IEEE Security in Storage Workshop*, Greenbelt, MD.

Chapter 16
Mobile and Handheld Security

Lei Chen
Sam Houston State University, USA

Shaoen Wu
University of Southern Mississippi, USA

Yiming Ji
University of South Carolina Beaufort, USA

Ming Yang
Jacksonville State University, USA

ABSTRACT

Mobile and handheld devices are becoming an integral part of people's work, life and entertainment. These lightweight pocket-sized devices offer great mobility, acceptable computation power and friendly user interfaces. As people are making business transactions and managing their online bank accounts via handheld devices, they are concerned with the security level that mobile devices and systems provide. In this chapter we will discuss whether these devices, equipped with very limited computation power compared to full-sized computers, can make equivalent security services available to users. We focus on the security designs and technologies of hardware, operating systems and applications for mobile and handheld devices.

INTRODUCTION

As mobile and handheld devices are becoming indispensable in this modern world, people are concerned with whether these lightweight and downsized computer systems and mobile networks can achieve the same level of security as in conventional computer and network systems. The purpose of this chapter is to provide readers a perspective of the current mobile and handheld systems through the review of the security designs and technologies of mobile hardware, operating systems (O.S.), and applications. The chapter starts with the background information of handheld devices in the above three areas in Section 2. In Section 3 we discuss the security risks when using mobile devices. Section 4 talks about mobile hardware security. Mobile operating system security and application security are reviewed in Section 5 and 6 respectively. We examine the standards, technologies and tools for layered mobile security in Section 7. The future

DOI: 10.4018/978-1-61520-761-9.ch016

of mobile security is discussed in Section 8 and conclusion is drawn in the last section.

BACKGROUND

A mobile or handheld device is a pocket-sized computing device installed with a mobile operating system supporting various mobile applications. Such devices consist of three main parts: hardware, operating system, and applications. Smartphones and Personal Digital Assistants (PDAs) are the most popular mobile devices which also include Enterprise Digital Assistants (EDAs), ultra-mobile PCs, handheld game consoles, multimedia players and recorders in a broader definition. In this chapter, we will focus our discussion on smartphones, since they correspond to the major market of mobile and handheld devices.

Evolved from a mating of the mobile phone and PDA (Charlesworth, 2009), the smartphone provides not only essential phone features such as calling and receiving calls, but also additional PC-like information accessing services (CEVA, 2009). There is no industrial standard for the definition of a smartphone. "We have between 56 and 85 percent global market share depending on what you say is a smartphone," said Jerry Panagrossi, vice president of U.S. operations for Symbian, the leading provider of mobile operating systems for smartphones. Rick Roesler, vice president of handhelds for Hewlett Packard (HP), considers "Smartphones are computers you talk to," while Jason Langridge, UK mobility business manager at Microsoft says: "For us, smartphones combine traditional communication devices and provide rich data applications." (Needle 2005) Nowadays, smartphones are installed with operating systems that allow users to add applications, such as Word, Excel and games, and hardware, such as Wi-Fi card, GPS card and Secure Digital (SD) card, to enhance connectivity, storage and data processing. Most smartphones support features such as email, Internet browsing, build-in camera, document viewing and editing, media playback and editing, etc. However, compared to conventional desktop applications, mobile applications are often designed and implemented with limited functionality due to the relatively less computation power and low storage space.

The hardware manufacturers of smartphones include Nokia, Research In Motion Limited (RIM), Samsung, Palm, etc. The newly released Nokia N97 (Nokia-N97 2009), as an example, has a 3.5-inch 24-bit colorful screen with resolution of 640 by 360 pixels. N97 runs over the S60 (a software platform runs over Symbian OS) 5th edition platform and supports a wide range of connectivity such as Bluetooth 2.0 Enhanced Data Rate (EDR), USB 2.0, Wi-Fi, GPRS and WCDMA, and applications such as Microsoft Outlook and Lotus Notes. The recent BlackBerry 9000 series runs over Intel XScale 624MHz CPU and supports sending and receiving e-mails wherever it connects a wireless network of certain cellular phone carriers.

Popular mobile operating systems include Symbian OS, iPhone OS, BlackBerry, Windows Mobile, Linux, Palm WebOS and Android (market share shown in Figure 1). Symbian OS, the most popular mobile OS from Symbian Ltd counting for almost half of the world market, is a proprietary operating system that runs exclusively on the Advanced RISC Machine (ARM) architecture which is a 32-bit Reduced Instruction Set Computer, or RISC, processor architecture developed by ARM Limited. These processors are used and equipped in about 98 percent of the mobile phones sold each year. Although Symbian OS has the largest share in the worldwide markets, it falls behind other companies in the North American market. The latest version of Symbian OS 9.5 supports mobile digital television broadcasts, Wi-Fi, Mobile Web Server and lots of open source software.

Users with a smartphone running Windows Mobile operating System will not only be able to use proprietary software such as Microsoft Office Mobile but also a large variety of third-party software. iPhone OS is a close source (with

Figure 1. World market share of Smartphone operating systems (Nov. 2008)

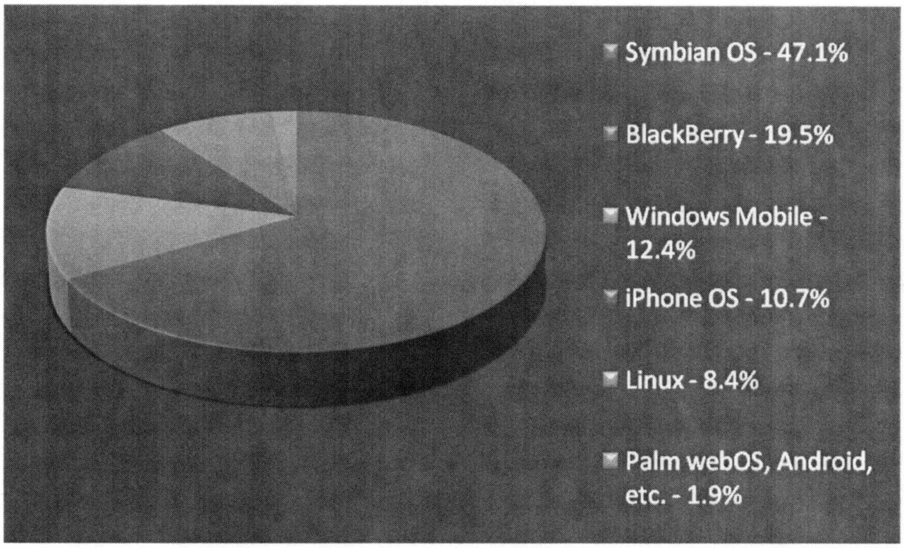

open source components), Mac OS X / Unix-like, operating system developed by Apple Inc. for the iPhone and iPod Touch. The latest version 2.2.1 runs over ARMv6 platform and supports Multi-touch Graphical User Interface (GUI), a set of interaction techniques allowing users to control graphical applications with several fingers. Android is an operating system, based on Linux kernel developed by Google and Open Handset Alliance, that runs over the HTC Dream (marketed as T-Mobile G1) smartphone. Palm webOS is a closed source (with open source components) embedded operating system developed by Palm, Inc.

It is not difficult to discover that the hardware, operating system and applications of smartphones are interrelated and mobile security is based on protecting all these three layers. Before discussing the security of each of the three layers, let us analyze the security risks and find out potential security threats in current mobile systems.

MOBILE SECURITY RISKS AND THREATS

Although smartphones enhance mobility and productivity, they also introduce new security risks and potential security threats. Mobile communication services in the future are expected to expand globally with increasingly diverse functionality, while the mobile handset architecture will become more unified in terms of the operating system, standardization, and open specifications (McAfee-NTT-DoCoMo 2007). As a result of these trends, it is very likely that mobile environments may become increasingly vulnerable to malicious attacks compared to conventional mobile phones systems. The countermeasure to these threats lies not only in securing the mobile network but also handsets, with the two approaches complementing each other.

Year-over-year worldwide mobile phone shipments grew 24 percent to 194.3 million units in Q4 2004 (IDC 2004; Deloitte 2005). Moreover, according to canalys.com Ltd (Reading, England), there were 684 million mobile phones shipped in 2004. More than 2 billion subscribers had used mobile phones by the end of 2005, up from 1.5 billion users in June 2005. As the demand for mobile devices and services dramatically increases, the need for mobile security solutions and enhanced device management is undeniable. In fact, 79 percent of IT managers believe mobile device support disrupts the regular and intended services of the IT department, and 76 percent of IT managers say they have no formal IT management policy in place for mobile devices, according to Gartner (2004a; 2004b) Inc., a Stamford, Connecticut based technology consulting firm.

Mobile users are utilizing wireless technologies every day, but many users are not taking the proper precautions to ensure that they are working in a secure environment. In fact, according to Gartner, about 90 percent of mobile devices lack protection to stay away from hackers. John Girard, research vice president at the firm, notes: "Wireless mobility is the greatest change to occur in corporate data collection and distribution in the past decade ... The solution for enterprises is to institute sound management policies to protect mobile information assets and contain costs."

Mobile Threats: Reality Check

Hackers are increasingly taking aim at mobile devices, even though the majority of security attacks still target conventional PCs and servers. Consider the following seven examples, ranging from mid-2001 to mid-2004 (McAfee-NTT-DoCoMo 2007):

1. **DoCoMo 110 Dialer:** In the middle of 2001, hackers figured out a way to exploit the e-mail applications in several phones. The hacking involved an exploitable bug within messaging software in the NTT DoCoMo system. In consequence, users who exploited the code could direct the phone to dial the "110" emergency number, producing a denial of service attack.
2. **SMS-Bomb:** In the fourth month of 2002, a Windows-based mobile application flooded Short Message Service (SMS) addresses with messages. This cross-platform attack, involving PC malware, resulted in denial of service attacks against SMS addresses and related services.
3. **Nokia 6210:** Hackers discovered the Nokia 6210 exploit in 2003, which involved a bug in the parser for vCard (address book) attachments to SMS messages in this mobile phone model. The result was denial of service attacks that crashed individual phones.
4. **Siemens 35/45:** This exploit in March 2003, similar to the Nokia 6210 exploit, made use of a bug in the SMS handler to cause phones to crash and hang.
5. **Bluejacking:** This short-range spam and prank technique in November 2003 targeted Bluetooth-enabled mobile phones. Hackers proactively beamed "contact" and sometimes wicked information to other phones within Bluetooth range.
6. **Symbian/Cabir Worm:** This worm, running on the Nokia Series 60 (S60) platform, emerged in the mid 2004. It made use of Bluetooth communications to initiate denial of service attacks causing rapid battery drains.
7. **WinCE/Duts Virus:** This virus in July 2004 for Windows CE attached itself to applications in root/current folder. It gave the hacker community a chance to take a closer look at smartphones and handheld devices running Microsoft's mobile OS.

The above seven examples reinforce the fact that mobile devices are already under attack or at least open to attacks. Moreover, experts such as Richard Clarke, former cyber security advisor to President Bush, predicted that mobile devices will become the next major platform for hacker wars. After all, mobile devices are easy, unprotected targets that hackers can exploit. There are many factors making mobile technologies and devices less secure. Here we list the top five of them (Collins & Vile 2007):

1. **50% mobile devices are bought by people themselves:** A recent Freeform Dynamics study showed that about half of the mobile

and handheld devices are bought by the individuals, not by the organization.

2. **New mobile technologies introduce new risks:** Compared to the mobile hardware and software a few years ago, the devices available in the market now are much more improved in processing power, battery life, screen resolution and data process speed. As new applications and new communication methods are introduced, new security risks are also brought in.
3. **Organizations are not prepared well:** Many organizations have not developed up to date policies to control and manage the use and configuration of mobile devices.
4. **Lack of a joined up architecture:** In most cases, mobile technologies and devices are added onto existing systems. The mobile configurations often lack the important security features such as centralized management and encrypted communications as available in desktop scenarios.
5. **Always-on and high speed access:** Many mobile and handheld devices are connected to an always-on, such as Wi-Fi and 3G, network which increases the chance of attacks.

The above listed five insecurity factors are just the tip of iceberg as far as mobile security is concerned. The potential security threats rooted from these factors are analyzed as below (Collins & Vile 2007):

- **Theft or loss of devices:** Some handheld and mobile devices can be expensive and become the targets of theft. In order to provide better mobility, handheld devices are made with small sizes resulting being more likely to be left on the tables in cafés, on buses or in cars. The impact of device loss can be in terms of productivity loss and support overheads plus capital cost.
- **Data confidentiality:** Data stored on mobile devices may be easily access by a third party. The best and simplest way to protect the devices is to use PIN which unfortunately is not very welcome by many users who feel it inconvenient to type in PIN every time when they boot up their devices. Today many smartphones and PDAs support expansion card slots which support various formats of large storage, such as SD cards and Compact Flash (CF) cards. However, few of these devices have built-in capability to encrypt data from the expanded media.
- **Inadvertent publishing:** Mobile devices that are not properly configured may leave data flows open to eavesdropping. A device with Bluetooth capability could be left in discoverable mode which could be open to hijacking. Corporate and business secrets could be "published" in careless mobile phone conversations or video conferences in public area.
- **Fraud:** A message sent from the mobile device looks just like it comes from the owner of the device. If the devices is left unguarded, even for just a little while, a third person can use it to send fraud emails or log onto websites using the owner's ID.
- **Intrusion escalation:** A compromised device could be used as a bridge to gain access to corporate systems. This can be done by misusing the credentials on the mobile device resulting breaches of confidential data such as customer or employee information. This escalating instruction approach is widely used in hacking.

The risks in mobile security are inevitable and the solution is to have sound layered security design and implementation. Each of the next three sections discusses the problems of and possible solutions to mobile security at a different layer.

MOBILE HARDWARE SECURITY

Mobile hardware security deals with the physical access to the mobile handheld devices and the data stored in those devices. To prevent or detect such attacks, multiple approaches are proposed and implemented. The common way is to have stronger encryption, often supported by hardware cryptographic accelerators, hardware or software locking the devices, and blocking unauthorized data reading and writing. Two or multiple-factor authentication is also often used.

Designated Hardware for Stronger and Faster Encryption

Texas Instruments' M-Shield Mobile Security Technology (M-Shield 2009) is a system-level security solution that intimately interleaves hardware and software technologies to provide the highest level of security. According to Texas Instruments (TI), M-Shield hardware security technology is operating system-independent and not sensitive to software attacks. M-Shield's hardware security technology includes:

- Hardware cryptographic accelerators and random number generator
- Public key infrastructure with secure on-chip keys (e-fuse)
- Secure access/restriction to all chip peripherals and memories
- Secure Direct Memory Access (DMA) transfers
- Hardware-based countermeasures against software attacks and cloning
- Secure protection of debug, trace, and test capabilities
- Hardware-reinforced secure execution and storage environment (Secure Environment) embedding:
  - A Secure State Machine
  - Secure RAM for sensitive authorized application execution and secure data storage
  - Secure ROM with 100+ accessible by authorized applications (Protected Applications)
  - Secure storage mechanism

These hardware designs and technologies help protect the data stored and processed on mobile devices, even when the hardware has been compromised by unauthorized parties. In order to allow legitimate users to have the access to resources, reliable and efficient authentication mechanism needs to be designed and implemented.

Two-Factor and Multi-Factor Authentication

Two-factor authentication (T-FA) is an authentication mechanism in which two different factors, or two pieces of information and processes, are used in conjunction for authentication. T-FA normally provides higher level of authentication assurance. Multi-factor authentication uses more than two factors for authentication.

SafeNet smart cards and SafeNet iKey USB tokens (Safenet 2009) are secure authentication devices that can hold users' credentials, such as passwords, keys, certificates, or biometrics in a highly secure fashion. The devices have an open, flexible operating system that can enable other capabilities such as storing personal information or providing physical access credentials securely to the device. The cards/tokens can be used in both PKI and non-PKI environments.

SafeNet's USB authentication tokens are designed to deliver security, interoperability, convenience and performance. SafeNet iKeys support PKCS #11 and MS-CAPI interfaces, allowing for seamless integration and interoperability with identity and access management applications. Used in conjunction with SafeNet's robust client software, SafeNet tokens are widely trusted. The following are main features of SafeNet tokens.

- Three-factor authentication capability
- Strong PKI technology and security

- Secure key generation and storage in hardware
- Enhanced crypto co-processor for improved performance and speed
- Robust hardware and software protection against differential power and timing attacks

MOBILE OPERATING SYSTEM SECURITY

Mobile operating systems serve as the intermediary between mobile subjects (users and programs) and mobile objects (resources). Mobile OS' main job is to manage access control over the resources which can be in the forms of hardware and software. Therefore, mobile operating system security deals with managing accessibility to the system resources.

Symbian Security

Symbian v9.x is the most widely used operating systems on smartphones worldwide. In the security architecture of Symbian v9.x, there are two key design drivers lying behind the security model (Wilce 2001; Dixon & Jakl 2005):

- **Firewall protection of key system servers through the use of capability-based access control:** The capability model is deliberately limited to a small number of capabilities. A capability is an access token that corresponds to a permission to undertake a sensitive action or group of actions. The most important resource that requires access control is the kernel executive and a system capability is required by a client to access certain functionality through the kernel Application Programming Interface (API). System capabilities (permissions) are never exposed to users. In the Symbian OS kernel, the file server including the loader and the software installer are granted this privilege.
- **Data caging creates a protected part of the file system which rogue apps are not able to access.** Data caging involves separating code from data in the file system such that a simple trusted path is created on the platform. A few new directory hierarchies are created, e.g. the new /sys directory contains all executable code residing in the /sys/bin subdirectory. The central idea is that by caging non-Trusted Computing Base (non-TCB) processes into their own part of the file system, the /sys directory becomes hidden to them. The kernel and file server would check that a client process had TCB or AllFiles capability before allowing any access to the /sys sub-tree. In addition, the loader would disallow any attempt to execute code residing elsewhere than /sys/bin. Table 1 below shows the capabilities for data caging in Symbian v9.x mobile OS.

Symbian Signed technology is also used for secure access control (Morris 2008). According to Morris, Signing is the process of encoding a tamper-proof digital certificate into an application. The certificate identifies the application's origin, and grants access to those capability-protected APIs in Symbian OS that the application declared at build-time. Protected APIs are those that allow sensitive operations which include the following:

- access end users' private data, thus potentially breaching privacy
- potentially create billable events, thus costing the end user money
- access the mobile phone network, potentially affecting its operation
- access handset functions that can affect the normal behavior of the phone
- potentially impact the performance of other applications running on the phone

Signing an application is not required if the application uses no capabilities, or if the applica-

Table 1. Capabilities for data caging in Symbian v9.x

Capabilities	/resource		/sys		/private/ownSID		/private/anyOther		/anyOther	
	Read	Write	Read	Write	Read	Write	Read	Write	Read	Write
None	Yes	No	No	No	Yes	Yes	No	No	Yes	Yes
AllFiles	Yes	No	Yes	No	Yes	Yes	Yes	Yes	Yes	Yes
TCB	Yes	Yes	No	Yes	Yes	Yes	No	No	Yes	Yes
AllFiles+TCB	Yes	Yes	Yes	Yes	Yes	Yes	Yes	Yes	Yes	Yes

tion uses "User Grantable" capabilities allowing the user to grant permission for these capabilities during installation or run time, or otherwise the application is targeted for an earlier Symbian version. The software installer SWInstall conforms to an appropriate PKI (Public Key Infrastructure). SWInstall has been designed to support a number of different security policies.

For secure backup and restore, Symbian v9.x considers three different types of files.

- **Public files:** those stored in public directories. They cannot pose any threats to the integrity of the platform, and no special security measures need to be taken from the point of view of Platform Security when they are backed up and restored. They can be backed up using the traditional method of backup, known as passive backup.
- **Private files:** those belonging to particular applications which store them in the /private/ file hierarchy. The information system itself cannot know what these files are or what their sensitivity level might be, so the backup and restore policy regarding them is left under the control of each individual owning application.
- **Executable binaries:** those stored in the /sys/bin/ directory. If these executables were restored from a backup in the same way as a simple restore of public files, this would represent a significant threat to the security of the devices as Software Install would be fully bypassed and there would be no way of guaranteeing that executables had not been interfered with while they were stored off the secure device. The solution to this problem is to use Software Install's own mechanisms to back up executables and ensure their integrity on restore. Each installation package file includes a SISSignedController (SIS: Symbian Installation Source, installation file for Symbian OS) with all the information needed to install the files in the SIS archive and the integrity is guaranteed by its signature. Since the only way to install something in \sys\bin is via Software Install, the stored SISSignedController elements can therefore be used during a backup operation to recreate versions of the SIS file originally used to install every executable present in \sys\bin. During the corresponding restore operation the SISSignedController can once again be used to verify the integrity of the executable being restored and the validity of its digital signature against one of the root certificates always present on the device.

Despite the security technologies and mechanism in Symbian OS, it is reported (S60-Hacked 2008) that the Nokia S60 version 3 (Symbian OS 9.1, 9.2 & 9.3) devices can be hacked to remove the platform security introduced in OS 9.1 onwards thus allowing users to install "unsigned" files (files without certificates validated by Symbian) and allowing access to previously locked system files. This allows changing of how the operating system works, allowing hidden applications to be viewable and possibly increases the threat posed by mobile viruses. A Spanish modder (Jarno 2008) has developed an easy to use privilege escalation hack for Symbian S60 3rd Edition phones. The hack provides unlimited access to the phone's file system. With this access any number of modifications can be made. This hacking was to run a single SISX installation file which contains a simple graphical application to remove the access restrictions of any application that is currently running on the device. The privilege escalation has side effects: the OS is not able to start any new applications until the phone is rebooted. However, whatever is running at the time has total control over the device.

BlackBerry Security

The BlackBerry Enterprise Solution enforces security in the following two categories: Wireless Data Security and Stored Data Security (BlackBerry 2009). For Wireless Data Security, the solution offers two transport encryption options, Advanced Encryption Standard (AES) and Triple Data Encryption Standard (Triple DES or 3DES), for all data transmitted between BlackBerry Enterprise server and BlackBerry smartphones. Private encryption keys are generated in a secure, two-way authenticated environment and are assigned to each BlackBerry smartphone user. Each secret key is stored only in the user's secure enterprise account (i.e., Microsoft Exchange, IBM Lotus Domino or Novell GroupWise) and on their BlackBerry smartphone and can be regenerated wirelessly by the user. Data sent to the BlackBerry smartphone is encrypted by BlackBerry Enterprise Server using the private key retrieved from the user's mailbox. The encrypted information travels securely across the network to the smartphone where it is decrypted with the key stored there. Data remains encrypted in transit and is never decrypted outside of the corporate firewall. BlackBerry Mobile Data System (MDS) Services on BlackBerry Enterprise Server support RSA SecurID authentication, providing organizations with additional authorization when users access application data or corporate intranets on their BlackBerry smartphones.

BlackBerry smartphones support Hyper Text Transport Protocol with Security (HTTPS) communication in one of two modes, depending on corporate security requirements:

- **Proxy Mode:** A Secure Socket Layer / Transport Layer Security (SSL/TLS) connection is created between BlackBerry Enterprise Server and the application server on behalf of BlackBerry smartphones. Data from the application server is then AES or Triple DES encrypted and sent over the wireless network to BlackBerry smartphones.
- **End-to-End Mode:** Data is encrypted over SSL/TLS for the entire connection between BlackBerry smartphones and the application server, making End-to-End Mode connections most appropriate for applications where only the transaction end-points are trusted.

BlackBerry smartphone applications are created using the BlackBerry Java Development Environment (JDE), in which certain functionality, such as the ability to execute on startup or to access potentially sensitive BlackBerry smartphone application data, requires developers to sign and register their applications with RIM. This adds protection by providing a greater degree of control and predictability to the loading and behavior of applications on BlackBerry smartphones.

Additionally, the BlackBerry Signing Authority Tool can help protect access to the functionality and data of third party applications by enabling corporate developers or administrators to manage access to specific sensitive APIs and data stores through the use of server-side software and public and private signature keys.

To help protect BlackBerry Mobile Data Software (MDS) Studio applications from tampering, corporate developers can sign an application bundle with a digital certificate described by an alias. They can use either a trusted certificate authority (CA) or a self-signed certificate. BlackBerry MDS Studio generates and signs applications with certificates that are compliant with the Public Key Infrastructure (X.509) standard.

Windows Mobile Security

Windows Mobile powered devices employ a combination of security policies, roles, and certificates to address configuration, remote access, and application execution (WM-Security 2008). Security policies provide the flexibility to control access to devices. If a user or application is allowed access, security policies then control the boundaries for actions, access and functionality. The following list shows some of the ways security policies can be used on Windows Mobile powered devices:

- control which applications are allowed to run on the device and what they can do
- control who can access specific device settings, and their level of access
- control what desktop applications can do on the device (Remote API control through ActiveSync)
- policies configure security settings that are enforced with the security roles and certificates
- security roles determine access to device resources, based on message origin and how the message is signed
- certificates are used to sign executables, DLLs, and CAB files that run on Windows Mobile-powered devices

Windows Mobile 6 provides the following features to help protect devices against a variety of threats and risks (WM-Security 2008):

- Strong device password protection
- Device lock requires a password or PIN to access the device when it is turned on
- Local device wipe occurs after a specified number of incorrect login attempts
- Remote device wipe erases data and helps to prevent unauthorized use
- Exponential back-off if incorrect passwords are entered
- Local and remote storage card wipe erases data and helps to prevent unauthorized use
- Storage card encryption helps to prevent unauthorized use
- Custom Local Authentication Subsystem (LAS) and Local Authentication Plug-in (LAP) provide the infrastructure for authentication by sophisticated third-party hardware and software methods.
- Password policy enforcement, such as required password for synchronization
- SSL encryption of all data transmitted between the device and the corporate mail server
- AES for SSL channel encryption in 128 and 256 bit cipher strengths
- Encrypted data passes through a single SSL port on the firewall
- Cryptographic implementations are certified by US Federal Information Processing Standard (FIPS) 140-2, and are designed to be protected against a variety of potential threats. Supported algorithms include AES, DES and 3DES, SHA-1, and RSA.
- Flexible client authentication: SSL/TLS, Exchange ActiveSync, Certificate-based, RSA SecurID-protected
- Users can add root certificates without being a manager of the device; user root certificates will not compromise the level of

security established by the device management security settings
- Security policies help to control over-the-air access to device
- Bluetooth discovery mode can be prohibited to help guard device integrity (Supported in Windows Mobile 6 Standard only)
- Security policies help control acceptance of unsigned attachments, applications, or files
- One-tier access for code execution control — executable runs if it is signed.
- Two-tier access for code execution control — executable runs if it is signed; permissions indicate access. (Supported by Windows Mobile powered Smartphone and Windows Mobile 6 Standard only)
- Attachments for download can be denied or size-restricted
- Office Mobile does not support macros, so viruses cannot leverage them to do damage
- Code execution control allows the device to be locked so that only applications signed with a trusted certificate can run

Similar to other mobile operating systems, Windows Mobile manages application execution based on permissions. Windows Mobile powered devices have a two tiered permission model, or applications can be blocked: Privileged, Normal, and Blocked. Applications running at the privileged level have the highest permissions: they can call any API, write to protected areas of the registry, and have full access to system files. Few applications need to run as privileged. In fact, allowing them to run privileged allows them to change the operating system environment, and can threaten the integrity of the device. Most applications run normal. They cannot call trusted APIs, write to protected areas of the registry, write to system files or install certificates to protected stores. They could still install a certificate to MY store, however. Applications do not run if blocked because they are not allowed to execute. An application could be blocked because it is not signed by an appropriate certificate, or because the user blocks it after being prompted.

MOBILE APPLICATION SECURITY

Application Security deals with software applications, including anti-virus, firewall, and Virtual Private Network (VPN) applications, installed over the mobile operating system aiming to protect mobile system from being attacked at the application and network level.

Application Security on Symbian

Symantec Mobile Security 4.3 for Symbian protects Symbian OS Version 9.x Series 60 and User Interface Quartz (UIQ) smartphones from malicious threats and network intrusions. It can automatically allow, deny access, or quarantine upon detecting an infected file. Centralized management enables administrators to configure, lock, and enforce security policies remotely or locally. Its key features include (Symantec_MSS 2009):

- Real-time and prescheduled virus scanning and quarantining of infected files regardless of attack vector (MMS, IR, Bluetooth, email)
- Integrated stateful firewall that is configurable to protect by port and protocol for GPRS, Wi-Fi, and IP connections
- Auto-protect functionality that runs continuously and unobtrusively in the background, providing real-time intrusion prevention and threat alerting
- Advanced on-device event logging that provides extractable details on device activity
- Automatic wireless security and application updates via configurable LiveUpdate™ that keep on-device protection current

There are software solutions for privacy and security on Symbian. For example, Axmor's Symbian (Axmor, 2009) developers implemented an application for Symbian smartphones that allows the user to create, manage, and insure the privacy of personal information, such as private contacts, call logs, SMS and MMS, images and passwords. The private contact list, the most impressive functionality of the developed application, allows the user to:

- create and manage secret contacts in a separate address book on the mobile phone, invisible to those who do not know the password
- make calls to these contacts
- turn off the sound signal when a call from a private contact is received
- forgo saving related call logs in a standard call history
- save SMS and MMS in a separate folder which is not available for unauthorized persons

For less than thirty-five dollars, a user can download and purchase software such as Handy Safe Pro for S60 (Safe-Pro 2009) which is a perfect assistant for secure and convenient data managing. This tool stores passwords, credit card details, user names, codes, accounts, web pages addresses, travel info and so on using strong 448-bit Blowfish data encryption which has no known effective codebreaking so far.

Application Security on Windows Mobile

Symantec Mobile Antivirus 5.1 (Symantec_MAWM 2009) for Windows Mobile enables secure mobile computing by providing comprehensive protection against malicious threats that target Windows Mobile operating systems. On-device, automatic, real-time scanning helps protect against threats downloaded from the Web, sent via email or a Wi-Fi connection, or received via Bluetooth or infrared ports. A Symantec antivirus micro-engine provides both real-time and on-demand scanning functions, as well as wireless Symantec LiveUpdate capability when wireless Internet connectivity is available. The Auto-Protect function constantly scans and detects malicious code without user intervention, at the point of entry, before the handheld device is infected. With Auto-Protect, intercepted threats can't be opened without interaction on the part of the user who has the option of deleting or repairing the infected file.

A user who also needs other security features can choose to purchase Symantec Mobile Security Suite for Windows Mobile. The following shows the key features of this product.

- Antivirus—Real-time and prescheduled virus scanning and quarantining of infected files
- Antispam for SMS
- Stateful firewall functionality to control inbound and outbound network traffic by network address, port, and protocol
- Data protection and loss mitigation using encryption and a file activity log
- Password management and enforcement
- Tamper protection to guard the system against attack and to ensure the integrity of software components
- Administrator-managed phone feature control
- Integration with Symantec LiveUpdate
- Enterprise management

Some organizations may want to apply a "PC-centric approach" to mobile device security. Toward that end, mobile device users may be asked to download, install and configure anti-virus software, software patches and other protective code onto their mobile devices. This approach, however, can be tedious, time consuming and frustrating to users who simply don't have the skills or patience to embrace such processes. In stark contrast, McAfee has employed a smarter approach (McAfee-NTT-DoCoMo 2007) to mobile device security. Working with mobile carriers such as NTT DoCoMo and mobile device manufacturers, McAfee is "climbing inside" mobile devices to offer built-in security services. These services require no extra user action or training. This five-step mobile security strategy calls for:

1. An end-to-end solution for all mobile devices within a carrier's network
2. Mass coverage over all devices and geographic regions
3. Transparent security that requires no end-user involvement
4. Security services maintained by the network operator
5. Cooperation between network carriers, device manufacturers and technology providers, such as software companies that write applications for smartphones.

Layered Mobile Security

Layered security (Wiki_LS 2009), or layered defense, means that in order to maximally secure information and data, various security mechanisms, technologies and solutions are utilized at the same time in the same system. The concept of layered security is similar to Defense in Depth, a term from information assurance and information security meaning multiple layered of defense are placed throughout a system. A simple example would be, a secure mobile system needs both of the following: a reliable operating system with secure access control over the resources, and application level real-time monitoring software which tries to prevent or detect malicious attacks. Layered security also means not only security technologies and tools need to be used, but also security policies, operating planning, and user training should be reinforced.

Consumer level and enterprise level can have different strategy for layered security. For example, consumer layered security strategy includes using extended validation SSL certificates, multi-factor authentication, single sign-on, and enforcing signing and encryption for transactions, web and emails. For enterprises, layered security requires strong network authentication, strong encryption for files, folders, disks and removable media, secure end-to-end communication and enterprise level security policy.

FUTURE OF MOBILE SECURITY

The following security issues (Lamparter & Westhoff 2002) in mobile and handheld systems need to be addressed and discussed in order to provide better mobile security in the future:

1. **Protecting data for privacy and waving trust into data:** New security issues not only appear at all network layers, but also for access rights for data on the individual devices. For example, a person carrying a medical chip storing all his medical data hopes others to access this data when this person is under critical condition. However he does not want this information transparent or to arbitrary persons in non-emergency situations. It is still not clear how this can be managed. One possible solution is to have distributed certification authorities to establish some sort of trust among them.
2. **Integration of mobile wireless networks with wired networks:** When two or more different types of networks are interconnected, problems come. There is no central instance that all devices trust in different types of mobile networks. A number of security protocols are currently standardized, e.g. IPSec, SSL/TLS and Bluetooth security protocol. These protocols use either end-

to-end or hop-by-hop encryption. Without some pre-established security associations over wireless and dynamic connections, building low-cost secure channels is still a challenge. It is also not clear how a public key infrastructure can be managed in a very large scale and dynamic network.

3. **Cryptographic algorithms in constraint systems:** Most modern security protocols require both public key and symmetric key cryptographic algorithms. Public key algorithms require intensive computation which might not perform well on mobile devices considering the equipped very limited computation and battery power.
4. **Cost and reliability of security and privacy applications and services:** Mobile operating systems are not shipped with all security features. It becomes the mobile users' responsibility to purchase and download antivirus and firewall software. However, not every user is willing to spend a few hundred dollars on security software trying to secure a mobile device at about the same cost (of course the data might have much higher value, but not every user understands this thoroughly). Another issue is, if it is too complicated for a common user to download, install, and use these security applications, the user will simply stay away from it. Even if a piece of mobile firewall software can be easily installed and used, can we expect it to provide similar or close security level as the desktops?

CONCLUSION

In this chapter, we have discussed the security issues and possible solutions of mobile security in three layers: mobile hardware, mobile operating system and mobile applications. In order to provide high level security and privacy good for business and daily life, it is essential to strengthen security in all three layers. Robust and reliable security is built on hardware that is initially designed and then implemented with security in mind. Mobile operating systems are expected to have better capability designed and management, while mobile applications need to be standardized and built with reliable quality. Mobile users need to gradually realize the importance of security and privacy on mobile systems and start to learn to utilize secure applications and secure features in the mobile OS to protect their mobile devices.

REFERENCES

S60-Hacked. (2008). *Symbian S60 Hacked.* Retrieved June 10, 2009, from http://www.symbianfreak.com/news/008/03/s60_3rd_ed_feature_pack_1_has_been_hacked.htm

Axmor (2009). *Axmor's Symbian Security Applications*. Retrieved June 10, 2009, from http://www.axmor.com/symbian-development/phone-confidentiality.aspx

BlackBerry. (2009). *BlackBerry Enterprise Solution for Mobile Security.* Retrieved June 10, 2009, from http://na.blackberry.com/eng/ataglance/security/features.jsp

CEVA. (2009). *Glossary of Terms*. Retrieved June 10, 2009, from http://ceva-dsp.mediaroom.com/index.php?s=glossary

Charlesworth, A. (2009). The ascent of smartphone. *Engineering & Technology*, *4*(3), 32–33. doi:10.1049/et.2009.0306

Collins, J., & Vile, D. (2007, May). Mobile Security A primer on the security of mobile devices, and the implications for enterprise IT, *Freeform Dynamics Ltd.*

Deloitte (2005, January), *Worldwide Mobile Phone Subscriber Research*, New York: Deloitte & Touche.

Dixon, J., & Jakl, M. (2005). *Symbian OS v9 Security Architecture*. Retrieved June 10, 2009, from http://developer.symbian.com/main/documentation/sdl/symbian94/sdk/doc_source/guide/platsecsdk/SGL.SM0007.013_Rev2.0_Symbian_OS_Security_Architecture.doc.html

Gartner (2004a). *2004 mobile security research reports*.Stamford, MA: Gartner Inc

Gartner (2004b). Q1 *2004 research report on wireless mobile security and hackers*. Stamford, MA: Gartner Inc.

IDC. (2004). *Worldwide Mobile Phone Shipment Research*. International Data Corp, Press Release 2003 and 2004

Jarno (2008). *Symbian Jailbreak by Spanish modder*. Retrieved June 10, 2009, from http://www.f-secure.com/weblog/archives/00001451.html

Lamparter, B., & Westhoff, D. (2002). *Security Challenges in the future mobile Internet*. PAMPAS'02 Workshop on Requirements for Mobile Privacy & Security. Heidelberg, Germany: NEC Network Laboratories

M-Shield. (2009). *Texas Instruments' M-Shield Mobile Security Technology Solution*. Retrieved June 10, 2009, from http://focus.ti.com/general/docs/wtbu/wtbugencontent.tsp?templateId=6123&navigationId=12316&contentId=4629

McAfee-NTT-. *DoCoMo (2007). The Future of Mobile Security – Here Today. McAfee & NTT DoCoMo*. Retrieved June 10, 2009, from http://www.mcafee.com/us/local_content/case_studies/cs_future_mobile_security.pdf

Morris, B. (2008). *A guide to Symbian Signed* (3rd ed.). London: Symbian Software Ltd.

Needle, D. (2005). *Smartphones Take Center Stage*. Retrieved June 10, 2009, from http://www.wi-fiplanet.com/news/article.php/3551686

Nokia-N97. (2009). *Nokia N97 Tech Specs*. Retrieved June 10, 2009, from http://www.nokiausa.com/find-products/phones/nokia-n97/specifications

Safe-Pro. (2009). *Handy Safe Pro for Symbian*. Retrieved June 10, 2009, from http://www.software.com/downloads/business-applications/review-Handy-Safe-Pro-for-Nokia-9500-9300-521060.html

Safenet (2009). *Two Factor Authentication*. Retrieved June 10, 2009, from http://www.safenet-inc.com/products/tokens/index.asp

Symantec_MAWM (2009). *Symantec Mobile AntiVirus for Windows Mobile*. Retrieved June 10, 2009, from http://www.symantec.com/business/mobile-antivirus-for-windows-mobile

Symantec_MSS (2009). *Symantec Mobile Security for Symbian*, Threat protection for Symbian OS Series 60 and UIQ through integrated antivirus and firewall technologies. California: Symantec.

Wiki_LS (2009). *Layered Security*. Retrieved June 10, 2009, from http://en.wikipedia.org/wiki/Layered_security

Wilce, M. (2001). *High Level Requirements for Release 7.0 of the Symbian Platform v0.04*. London: Symbian.

WM-Security. (2008). *Security Mobdel for Windows Mobile 5.0 and Windows Mobile 6*. Seattle, WA: Microsoft Corporation.

Chapter 17
Design and Performance Evaluation of a Proactive Micro Mobility Protocol for Mobile Networks

Dhananjay Singh
Dongseo University, South Korea

Hoon-Jae Lee
Dongseo University, South Korea

ABSTRACT

This chapter introduces the Proactive Micro Mobility *(PMM) Protocol for the optimization of network load. We present a novel approach to design and analyze IP micro-mobility protocols. The cellular Micro Mobility Protocol provides passive connectivity in an intra domain. The PMM Protocol optimizes miss-routed packet loss in Cellular IP under* handoff*_conditions and during time delay. A comparison is made between the PMM Protocol and the Cellular IP showing that they offer equivalent performance in terms of higher bit rates and optimum value. A mathematical analysis shows that the PMM Protocol performs better than the Cellular IP at 1 MHz clock speed and 128 kbps down link bit rate. The simulation shows that a short route updating time is required in order to guarantee accuracy in mobile unit tracking. The optimal rate of packet loss in the PMM Protocol in a Cellular IP are analyzes route update time. The results show that no miss-routed packets are found during handoff.*

INTRODUCTION

Micro mobility protocols aim to improve the handoff delay and packet loss performance of Mobile IP (Yair A., Claudiu D., Hilsdale M., (2006)). Most micro mobility protocols expose to the home agent, and a single IP address for a mobile node (MN) as long as it remains within a particular foreign domain (Campbell, A. T. & Gomez-Castellanos, J. (2000)). The main losses in mobile communications are of two types: "wireless losses", due to white Gaussian noise in the wireless channel; and "handoff losses", due to the time delay in making a connection to new base station (BS). Handoff losses occur during the allocation of resources

DOI: 10.4018/978-1-61520-761-9.ch017

and packet re-transmission. These losses can be reduced by using an efficient routing protocol on the network layer, in order that a good handover technique minimizes the handoff delay When comparing existing routing protocols, their Mobile IP should be considered first as it provides roaming capability for mobile users in macro-level networks. Problems with conventional tunneling phenomena are faced in small cellular networks, where fast handoff environments can exist due to the high speed of mobile users; in addition mobile tracking consumes lots of signal (Campbell A.T., Gomez J., Kim, Chieh-yih Wan, Zoltán R. T., András G. Valkó. (2002)). To overcome the problem of Mobile IP for fast handoff networks, a new mobile communication technology known as "Micro Mobility" has evolved. Micro Mobility is a field where the domain is divided into pages. The domains can be large WLAN networks such as campuses, etc., and for the best results the domain should be made as large as possible. Micro Mobility Protocols (MMPs) put the responsibility of communication at the page level and Mobile IP operates on the pages so as to extend the scope of macro-level networks.

Background

Lots of work has been done in wireless network in the field of routing protocols. Some initial work went into supporting the roaming of mobile users among cellular areas. Various handoff techniques were proposed and their performance analyzed (Yan Z., Hee S.B., (2004)). Much of the current work is concentrating on intra domain networks, using many *ad hoc* routing protocols for WLAN 802.11b and 802.11a. These *ad hoc* protocols are designed for different physical and mobile environments, such as: DSR for intra domain networks, for slow speed mobile devices and for dynamic networks; AODV for on demand basis networks; and TORA for time dependent networks. Once Macro and Micro Mobility Protocols had been successfully integrated into mobile user systems, various micro mobility protocols were designed on the back bone of the mobile IP. These protocols have been compared, based on their performance and other issues.

Researchers have followed different approaches to give connectivity to a mobile user when the user is roaming. The conventional approach is based on the "prediction of mobile unit in mobile environment". The losses in wireless and mobile environments are very high compared to wired networks; this is one of the main reasons behind the low bit rates used in the wireless and mobile domain. The various different losses existing in wireless and mobile communication networks are:

1. Wireless losses
2. Handoff losses
3. Control message losses

The second types of loss are handoff losses. When a mobile unit leaves the coverage area of the home base station and enters the area of a new base station, the new base station needs to get connected to the network.

In previous work, the Global System for Mobile communications (GSM) provided a roaming facility to mobile units based on a centralized data base, HLR, a switching center, MSC and BSC, and a signaling system, BTS. The process of letting go of the connection of one base station and getting the connection to another base station while roaming is called handoff or handover, depending on the way that connectivity is established (Joachim Tisal, (May 2001)).

- **Network controlled handoff** (NCHO): Delay varies from 200 to 500 ms.
- **Mobile assisted handoff** (MCHO): Provides a handover delay of approximately one second.
- **Softer handoff**: Delay is variable.
- **Mobile controlled handoff**: Provides a delay of the order of 100 ms.

The user can communicate with another user as soon as the user takes handoff from one base station to another. Then the user may encounter a disturbance in the continuous flow of communication. While the user is taking handoff, the user doesn't get any messages and at the same time all packets delivered by the old base station have to be re-transmitted through the new base station. This condition affects real time traffic. Therefore hand off delay should ideally be zero (Campbell, A. T. & Gomez-Castellanos, J. (2000)).

The third kind of loss is protocol oriented. This loss involves a number of control messages that must be queried in order to maintain connection of the mobile user during mobile networking. A number of control messages can be lost as they contain no useful data to transmit. This applies especially to the extra control signaling involved in control messages and required to track the mobile unit. These losses can be optimized with a better routing protocol. If a long period is used for control messages, the probability of accurate tracking of the mobile unit decreases. If a short period is used for control message, the control message loss will be high (Gunnar Heine, Matt Horrer, (1999)).

In this chapter, we show how handoff losses and control losses can be optimized with an efficient routing protocol. Micro Mobility is a field invented to optimize the losses in fast handoff environments. As the mobile IP doesn't suit micro-networks as well as it suits macro-networks, Micro Mobility Protocols (MMPs) are designed so they can be operated on the back bone of a mobile IP, i.e. both MMPs and Mobile IP can be integrated. Among the various MMPs, Cellular IP shows better optimization of losses than other protocols which introduce semi-software handover techniques (Aisha H. A. Hashim, Anwar F., Mohd. S.,Liyakthalikh, H. (2005)).

The aims in designing a new Proactive Micro Mobility Protocol are:

1. To optimize network load by optimizing miss-routed packet losses
2. To optimize control packet losses (even the optimization of control losses serves to optimize the network load)

To solve the above problem, a protocol needs to be designed with:

a. Prediction of handoff
b. Optimization of control messages (i.e. optimization of route update and page update packets in Cellular IP), besides inheriting all the inherent advantages of Cellular IP.

Prediction of handoff is done by modifying the existing Cellular IP (Gunnar Heine, Matt Horrer, (1999)). The prediction of an event (handoff in our case) involving a collision in a mobile environment. Which describes how to predict collisions in directional antennae and how to avoid them. Any collision is a future event in the mobile environment. Another paper entitled showed a directional antenna and how to predict an event. This chapter discusses the handoff of mobile units in a micro mobility environment involving a directional antenna and proposes a Proactive Protocol for Micro Mobility which can optimize losses better than Cellular IP. The performance of the designed protocol is analyzed CMIS software coded on NS-2 simulator.

MICRO MOBILITY

Micro mobility is a field that has evolved to provide high bit rate transmission to mobile users on intra domain networks (figure 1). It works on the back bone of mobile IP and extends the scope of mobile users to the macro-level. Micro mobility protocols need to work in a wide variety of scenarios, such as varied underlying infrastructure support, mobility patterns, MAC and physical layer (Bhaskara,

Figure 1. Micro mobility model

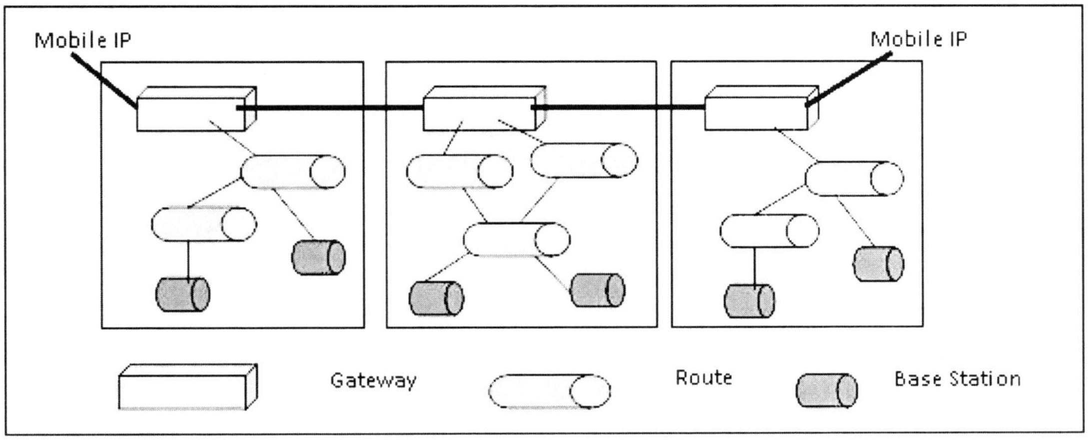

G. Helmy, A. Gupta, S. (2003) & (Vicente C.G., Pablo G.E., Vincent P. (2004)).

Functions Supported by Micro Mobility Protocols

1. **Mobility Management:** Micro Mobility Protocols work within a domain. The mobile user switches from one base station to another base station. It transmits data packets to the mobile user along the shortest path, unlike conventional cellular networks.
2. **Paging:** In micro mobility each domain is organized on a page. Each page contains gateway (GW), router and base station parameters. The router can be on other base stations. Mobile IP works within the gateway to extend the communication to metropolitan networks. The structure of the micro mobility page is shown in Fig. (2).
3. **Handoff:** It supports fast handoffs with the introduction of new handover techniques at the time of cell switching, unlike other cellular structures.
4. **Passive connectivity:** Connection of a user to a new base station is established when the user enters the page without any initiation from the user. When a mobile user roams within a page the connectivity is provided by all the base stations within a cell, without any extra messages for connectivity after every handoff (Kaiduan X., Vincent W.S. Wong, Victor C. M. Leung (September, 2004)).

Functions Not Supported by Micro Mobility Protocols

- **Home Location Register (HLR)**: MMPs do not support any centralized data base system. All the routing operations are performed by updating routes and pages at every BS within a page.
- **Switching center**: It does not support a switching center to control all BSs, although its operation is controlled by the gateway (GW) node.
- **Signaling systems**: It does not require any extra signaling systems like the Base Transmission Station (BTS) and Base Sub Center (BSC) that exist in conventional cellular networks.
- **Notion of 'connection'**: It does not support the sort of conventional connectivity that exists in conventional cellular networks. Instead, it supports its own connectivity depending on its utility.

In the integration of Mobile IP and MMP, problems are handled at the micro- and macro-levels. Two of the major MMPs are Cellular IP and Hierarchical Mobile IP (Xiea B., Kumara A., Agrawal D.P., Srinivasan S. (2006)). Cellular IP supports paging, mobile management, handoff, passive connectivity and avoids a central location data base (HLR), switching centre (MSC), signaling system and, more generally, the type of connection conceived in the standard GSM network. The "semi-software handover technique" introduced by Cellular IP has well optimized handoff losses but fails to optimize network overload due to packets being mis-routed during handoff. Advanced techniques already exist at the micro-level in the Cellular IP for optimizing networks (Gunnar Heine, Matt Horrer, (1999)). Besides inheriting all these positive features of the Cellular IP. The new protocol has been analyzed theoretically and simulated with CIMS software provided by Colombia University Telecommunication Research centre.

Working Model of Micro Mobility Protocols

The structure and a working model of the MMP are shown in Fig. 2. The scope of any MMP is defined by a page. Mobile IP works among pages. Whenever a mobile user wants to communicate with another mobile user in the micro mobility page structure, the IP datagram goes to the mobile user Host Agent (HA). Then, the HA passes the datagram to the page's gateway via conventional mobile IP tunneling. Next, the gateway takes care of the flow of packets within pages using micro mobility packets. The packet flows in a page according to micro mobility protocol rules and reaches the mobile user along the shortest path. The response from the MH reaches the GW in the same path as it was forwarded from GW to MH. Then the GW sends the packet directly to the corresponding host in the network (Aisha H. A. Hashim, Anwar F., Mohd. S.,Liyakthalikh, H. (2005)).

The page update packets are used by the GW to update its page cache. Whenever the BS receives a route update or page update packet, it forwards the packet to the GW via the GW's upward neighbors, by the shortest hop to hop method. The forward and upward neighbors are depicted in Fig. 2. The PMM Protocol nodes maintain the route cache. Packets transmitted by the MH are updated entries in each node's cache. An entry maps the MH's IP address to the neighbor from which the packet arrived to the node. The chain of cached mappings refers to a single MH that consists of a reversed path for downlink packets addressed to the same MH. As the MH migrates, the chain of mappings always point to its current location because its uplink packets are newly created and the old mappings changed. Control packets are ICMP (Internet Control Message Protocol) packets with specific authentic payloads from the MH routed through the PMM Protocol. The MH sends periodic control packets to update its route mappings in the PMM nodes (Magret V., Choyi V.K., (January, 2001)).

Every time that the BS has to send a data packet to the MH, it checks the route validation

Figure 2. Working model of the micro mobility protocol

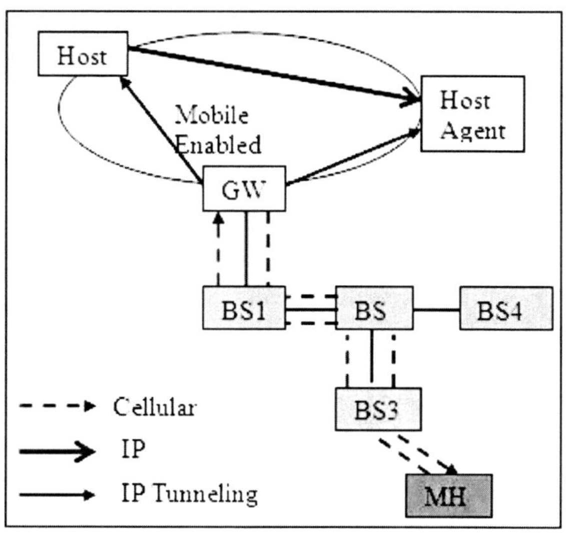

Figure 3. Architecture of PMM network

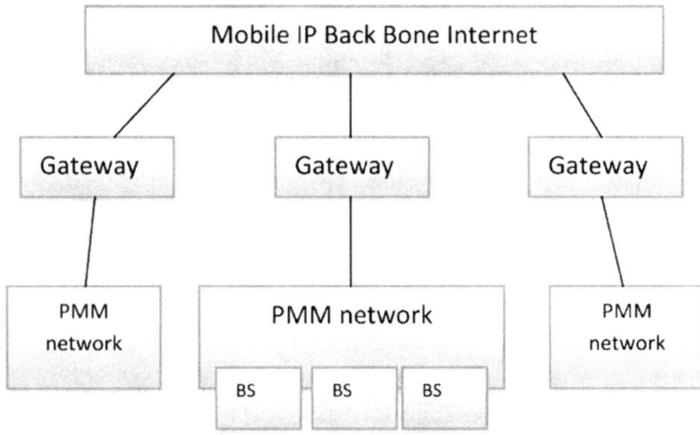

time. If the route is still valid it sends the data packet to the MH. Otherwise it stops sending packets to the MH and sends a message to reset the route in the root cache of the corresponding uplink neighbor.

PROACTIVE MICRO MOBILITY NETWORK

The PMM Network consists of interconnected PMM nodes (figure 3). The role of nodes is two-fold. They route IP packets inside the PMM and IP Network and communicate with mobile hosts via a wireless interface. Referring to the latter role, a PMM node that has a wireless interface is also called a Base Station. PMM node which has wireless interface is called PMM base station. PMM Gateway is a PMM node that is connected to a regular IP network by at least one of its interfaces PMM Mobile Host. A Mobile Host that implements the PMM protocol (Sridhar Jakkula, (June, 2002)).

The Terminology of Active Mobile Host (MH)

A mobile host is in active state if it is transmitting or receiving IP packets. A PMM Network Identifier a unique identifier assigned to PMM Networks, Paging-update, paging-teardown broad cast messages from base station and response messages from MH when it receives broadcast message from base station. The data packets are in IP packet that is not a control packet. Downlink neighbors of a PMM node except its uplink neighbor are referred to as downlink neighbors and a mobile host is in idle state if it has not recently transmitted or received IP packets. A PMM Network provides access to a regular IP network. This IP network in this memo is referred to as "Internet", but it can also be a corporate intranet, for example One PMM node is said to be the neighbor of another if they are connected directly. Neighbors are identified in a PMM node by interface and Medium Access Protocol (MAC) address.

Paging Area is a set of base stations that idle mobile hosts crossing cell boundaries within a Paging area. That does not need to transmit control packets to update their position. A cache maintained by some PMM nodes, used to route packets to mobile hosts. Paging-timeout Validity time of mappings in Paging Caches Paging-update packet (Xie K., Wong W.S. V., Leung C.M.V. (2005)).

The page update packets are used by the GW to update its page cache. Whenever BS receives route update or page update packet it forwards packet to the GW though upward neighbors to the GW, in shortest hop by hop method. The forward and upward neighbors are depicted in Fig.4. PMM nodes maintain Route Cache. Packets transmitted by the mobile host create and update entries in each node's Cache. An entry maps the mobile host's

Figure 4. Forward and upward packet neighbors of BS

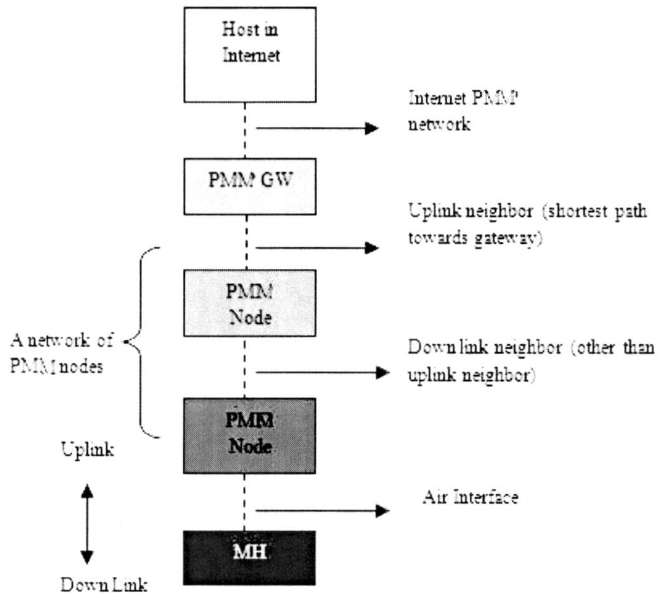

IP address to the neighbor from which the packet arrived to the node. The chain of cached mappings referring to a single mobile host constitutes a reverse path for downlink packets addressed to the same mobile host. As the mobile host migrates, the chain of mappings always points to its current location because its uplink packets create new and change old mappings. Control packets are ICMP packets with specific authentic payloads. Though packets to mobile host routed through the PMM nodes, MH sends periodic control packets to update its route mappings in PMM nodes.

Every time when the base station has to send a data packet to MH it checks the route validation time. If the route is I still valid it sends the data packet to the MH else stop sending packets to MH and sends a message to resets route in corresponding uplink neighbors root cache. (Carli M., Cappabianca F., Tenca A. and Neri A., (2004)).

Location Management

BS broadcasts route query messages periodically to know the information of MH in its coverage area. The response of route queries all MHs send route update packets. Using the time elapsed in control message exchange and antenna lobe angle, in which it receives the max strength of signal for that particular host, BS keeps the track of MH in its circular parameter co ordinates. The route validation time will be set a multiple of control message exchange periodic time, in order to keep the connection though any MH route update packets gets lost due to wireless losses. For the best results all the base stations should align in a page such that there should be less probability in calculating the direction of position of a mobile user.

Routing

All the packets delivered by MH will reach the GW through BS by shortest path hop be hop method. With these packets also the route mapping will updated in PMM nodes. Whenever the base station receives a route update packet for route query packet, it sets the route validity time. It calculates weather the MH going to leave the coverage area by the next hop it sets the route validation time as a fraction of control message exchange time, unlike in general case where the route validation time is a multiple of control message period.

Handoff

PMM network supports two types of hand off.

a) *PMM enabled handoff:* The handoff losses in PMM network, optimized by proactively by predicting handoff. Whenever the BS receives route update message from MH it calculates its positions and computes the velocity of the MH as shown in table 2. Every time before refreshing or updating a route it checks for the route update time. With equation:

$$P(r1,\alpha1)+V(r1,\alpha1)*T>P(R,\alpha)$$

Where P(r1, α1) is current position of the MH, V(rl, α1) is velocity of the MH at that instinct, P(R, α) is point on the boundary of BS coverage area, T is periodic time of broadcasting control message from BS, If any time the position of the MH satisfies the above equation the current base station sets the validity root for that MH to a fraction of T and sends a control message to the BS regarding handover of the particular MH with its sequence number, position, velocity and approximate time of handover calculated as:

$$Tapr = \frac{[P(R,\alpha) - P(r1,\alpha1)]}{V(r1,\alpha1)}$$

Then gate way decides what could be the cross over base station to the MH and sends a packet to that particular BS regarding to give connectivity to a particular MH which is going to be take handoff. Then cross over base station checks its route cache for the route to that particular MH if it is already starts a route to the MH (i.e. already handoff is taken), then discards the control message sent by the GW. If it does not find any route to that particular MH it maps a new route for that particular MH (Gunnar Heine, Matt Horrer, (1999)).

b) *Semi Soft handoff:* If last route update message from MH misses, when it is at the boundary of a BS coverage area, it starts responding to root query message from cross over BS. Then a new route will be established through cross over BS. As the old BS do not know about the handoff, like in cellular IP it starts keep sending packets until the route validation expires. Advantage of PMM enabled Handoff As the current BS station know about the upcoming handoff route validation time for that particular mobile host to a smaller value. So that there will be less number of miss routed packets as compared to cellular IP. The optimization in miss routed packets optimizes the network load. As we can use beacon signals in cellular IP as route query signals there will not be any extra signaling.

Paging in PMM Networks

MH sends page update packets periodically to GW. Every time GW receives page update packet MH it checks the condition of page crossing as like BS boundary crossing. If it finds the crossing of page boundary by the next hop it resets validation of page route to a small value. Whenever the MH enters the new page it sends a page update packet then a new page mapping will be done in the new page. Mobile IP will take care of new connection through the new gate way. Therefore page losses also optimized as GW predicts the time of page crossing (Joachim Tisal, (May 2001)).

Proactive Micro Mobility Protocol Design (PMM)

The structure of the PMM Protocol works as the back bone of the mobile IP. The idea behind the design is to modify the Cellular IP in such a

Table 1. The gateway keeps the location information of all BSs

Base station seq. no.	Location in terms of radius in cm	Location in terms of azimuthal angle in Radian (α)	Shortest path	Minimum no. of hops
BS1	R1	α1	PMMP1, PMMP2	3
BS2	R2	α2	PMMP3	2

Table 2. The base station maintains a route table for the MH

M H no.	Root Validation Time	Current Position (r1, α1)	Previous Position (r1, α2)	Velocity [(r1, α1) -(r1,α2)]/T

Figure 5. Description of tabulated values

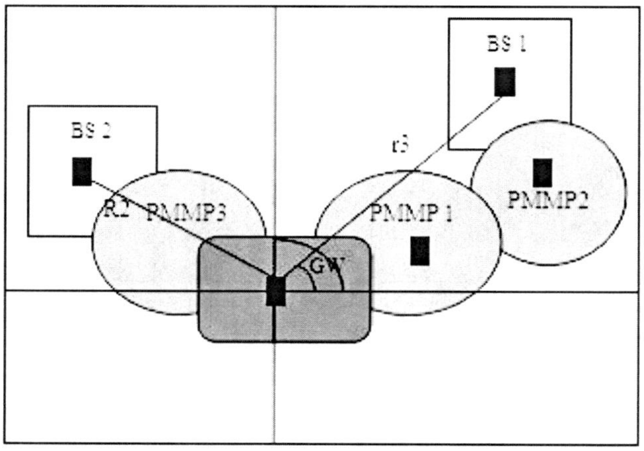

way as to get location information at a particular instant in time and to find the estimated velocity during handoff. To find the location of the MH at a particular instant in time, directional antennae located on the BS are used towards the highest roaming probability areas such as towards roads in a city environment (Shim Y.C., Kim H.A., Lee J. I., (2005)).

The Structure of Micro Mobility Networks

The gateway (GW) keeps the location information of all BSs shown in Table 1. Each BS knows its radius and maintains a routing table for the MH. The intermediate BS also maintains a route cache as in a Cellular IP structure. The Base Station (BS) broadcasts periodic route query messages to detect available MHs in its wireless coverage area. Responding to query messages, all MHs in the coverage area send route update messages. After the time elapsed during the exchange of both control packets, the BS calculates the distance of the MH from the BS and calls this the current BS radial component r1.

The angle of the antenna lobe, in which it receives maximum strength from a particular MH, is taken as approximately equal to the azimuthal angle α1, between the two. The values of the angles are tabulated as the current positions shown in Table 2. After the completion of consecutive control message exchanges, the BS again records the r and α for the MH.

The BS maintains a route table for the MH as shown in Table 2 with its position information. All position entries are taken in circular coordinates.

The Table 2 is updated with r_2, α_2. Using these two position values as well as the time delay between the two entities, the approximate velocity of the MH is calculated and further updated in Table 2. Whenever the BS receives a route update packet from the MH, the BS updates its route cache. If it receives a route update packet for the first time when a new MH enters its area of coverage, a new entry is made for the MH and the route validation time is set. If the BS receives a route update message from an old MH, it refreshes the old route. Besides the route update packet, the MH sends a periodic page update packet to the nearest BS.

Disadvantages in PMM Networks

Complexity of structure will be high due to more number of directive antennas instead of one Omnidirectional antenna. Extra computation will be needed at the base station.

PERFORMANCE EVALUATION

Mathematical Analysis

The performance, Handoff delay is less than for a Cellular IP network; in the ideal case, it is even zero. The small handoff delay depends only on the time required to transmit. For the cross-over signal from BS to MH, the handoff delay depends on the ability to transmit a signal to cross over to the BS from the MH. The BS takes time to create a new route to the GW, and there is time involved in transmitting a signal to cross over from BS to MH along the down link.

Packet loss is proportional to handoff delay or handoff loop time. As handoff loop time is much less than for Cellular IP, the packet loss is also theoretically much less. In the ideal case, the packet loss is zero.

Miss-routed packet loss is equal to a multiple of handoff loop times in terms of downlink bit rate. In the current case, the route validation time is reset at the time of handoff from the current BS and only for a fraction of the cycle time for root query messages to cross over from the BS as it starts sending packets to the MH. Thus, the old BS only keeps sending packets during a fraction $\alpha 1$ of T after crossing the boundary and the crossover BS starts sending packets to the MH during a fraction $\alpha 2$ of T.

Total miss routed packets = $(\alpha 1+ \alpha 2)\, T * r$

Or $(\alpha 1+ \alpha 2) <= 1$

Where: T is the cycle time of route query messages

r is the downlink bit rate in the current case

Whereas, in Cellular IP:

Total miss routed packets $=\alpha T*r$

Or $\alpha>=3$

So, the optimization of miss-routed packets is obviously less in a PMM network than in a Cellular IP.

Route Maintenance Cost

If T_c is the cycle time of control message exchanges, and p is the fraction of Tc required to exchange control messages, then the cost involved in exchanging control messages in one period is $2p\,R_c/T_c$ where R_c is length of control packet.

The total cost of control packets in time T is:

$$\frac{2pTR_c}{T_c} \tag{1}$$

If $(\alpha 1+ \alpha 2).T_c$ is the total handoff loop time, then the cost of miss-routed packets in one handoff is:

$$r\,(\alpha 1+ \alpha 2)\,.\,\frac{T_c}{T_H},$$

where T_H is the handoff time or dwell time.

The cost of miss routed packets in time interval T is:

$$\frac{rT(\alpha 1+\alpha 2)T_c}{T_H} \quad (2)$$

The optimum value of T_c to minimize overall cost is:

$$\text{Optimum } T_c = \frac{\alpha(2R_c pTH)}{r(\alpha 1 + 2)} \quad (3)$$

Equation 3 is derived from by differentiating the losses from Equations 1 and 2 and setting the sum to zero. The total cost of route maintenance is:

$$Ca = \frac{2pRC}{optTC} + \frac{r(\_1 + \_2)optTc}{TH}$$

$$= \frac{\alpha 8prRc(\alpha 1 + \alpha 2)}{TH} \quad (4)$$

SIMULATION

The mathematical analysis is made assuming a down link bit rate of 128 kbps, which is relevant to real-time voice traffic. At higher bit rates, we observe better performance curves similar to Cellular IP. Comparisons of Cellular IP and PMM are made for bit rates of 2 Mbps and 10 Mbps in Figure 6 (A & B), respectively. Figure 6 A, shows that 20 kbps losses correspond approximately to a route update period of 15 s. The losses are more optimized with a 10 Mbps down link bit rate than in a Cellular IP.

The following values are taken for Cellular IP to compare total losses with respect to route update time.

In this simulation, compilation files provided in the CMIS software were used and code was written in Tcl to simulate the PMM Protocol, making some assumptions.

The following assumptions were made in the simulation work:

1. Route update messages from the MH are transmitted as a result of route query messages broadcast by the BS.

Figure 6. Mathematical results for costs

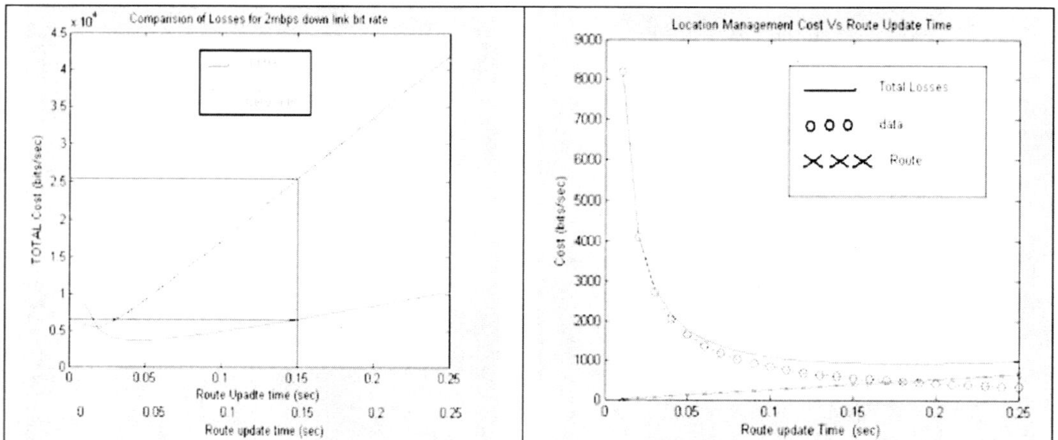

A) Location Management Cost vs. Route Updating Time.

B) Comparisons of Total Loss in Cellular IP and PMM Protocol.

Table 3. Simulation parameters

Parameter	Value
Downloading Bit Rate	128 kbps
Route Update Packet Size	102 bytes
Periodic Route Query Time	0.5 s
Fraction of Period of Route Query Message required for Exchange Control Message	$0.1(\alpha 1+\alpha 2) = 0.6$
Handoff Dwell Time	30 s

Table 4. Simulation parameters

Parameter	Value
Route Update Time	0.5 s
Page Update Time	3 s
Route Validation Time	10 s
Route validation time of old base station when PMM enabled handoff existed	0.5 s
No. of GW nodes	1
No. of Routing nodes	3
No. of Base Stations	4
MH speed	20 m/s
Control Packet Size	50 byte
Handover Delay	0.05
Simulation Time	40 s
MH source and destination points	10.0 and 420.0
X & Y Dimension of the Topography	500

2. As the CMIS software doesn't support multi-directional roaming, instead we predicted the last hop of the MH in the cell coverage area. However, irrespective of time, the MH receives a first beacon from the cross-over base station at the same time as the route update message, when the BS calculates the handoff condition.

The simulation environment used PMM networks like Cellular IPs and tracked the conditions.

An overview of the simulation is as follows:

- The basic contents are: GW with Corresponding Node, Router, BS, MH and modified watchdog agent; Routing table and Route updating; Events.
- Two route updating procedures are followed. The first covers handoff time and the other are for general roaming with a third new global flag for condition checking.
- The following parameters are used for the existing structure of Cellular IP to compare results.

During the simulation, the MH was encountered with 4 handoffs. The performance of the simulation set up for the PMM network in terms of packet optimization is tabulated below in Table 5.

The performance curve for PMM compared with a Cellular IP has been drawn in Fig. 8 and 9, in terms of packet losses. The simulation results show better performance of miss-routed packet optimization in PMM compared to a Cellular IP.

Table 5. Packet optimization is tabulated ←No of misrouted packets, delivered-->

Route update time	1st handoff	2nd handoff	3rd handoff	4th handoff	5th handoff
0.1 s	6	7	9	1	5.75
0.2 s	6	13	10	9	9.5
0.3 s	25	26	17	9	19.25
0.4 s	8	37	29	6	20
0.5 s	23	29	18	11	20.25

Table 6. Simulation conditions are tabulated

Route update time	1st handoff	2nd handoff	3rd handoff	4th handoff	5th handoff
0.1 s	35	26	29	19	27.25
0.2 s	45	47	50	47	47.25
0.3 s	85	86	79	69	79.25
0.4 s	85	108	108	89	97.5
0.5 s	87	107	126	88	102

Comparison of Mathematical and Simulation Results

Mathematical results are plotted in Fig. 6 and simulation results are plotted in Fig. 7. The total number of packets received by the MH in 40 seconds is 3898, assuming no loss on the wireless channel. Therefore, the down link packet rate is calculated to be 97.45 per second. On substituting the down load packet rate into equation (2):

Packet losses = $r(\alpha 1+\alpha 2)\text{opt}Tc$ (6)

While:

$$\text{Packet losses for Cellular IP} = r\left(\alpha - \frac{1}{2}\right)\text{TRu} \quad (7)$$

Figure 7. Simulation results for costs

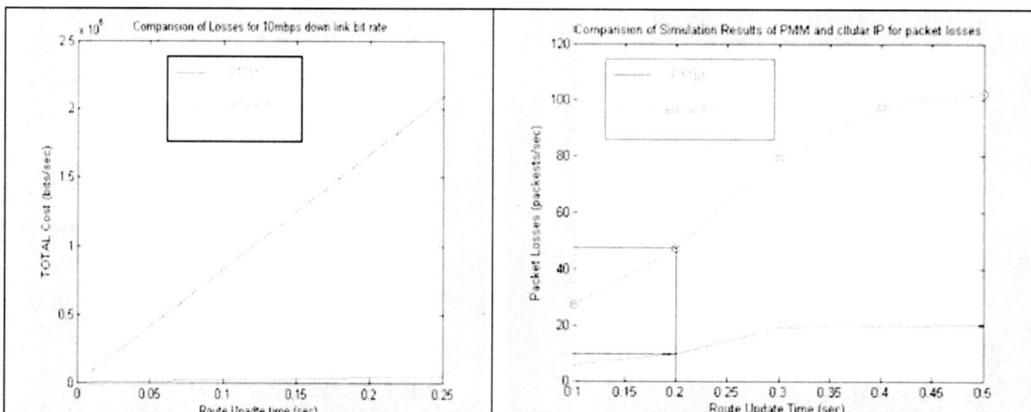

A) Comparison of Losses for 2 Mbps down link bit rate.

B) Comparison of losses for 10 Mbps down link bit rate

Figure 8. PMM and cellular IP, for packet losses simulation

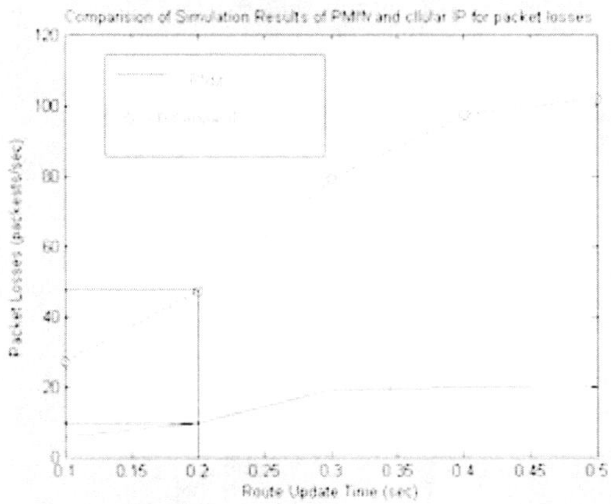

Figure 9. PMM and cellular IP, for packet losses theoretical

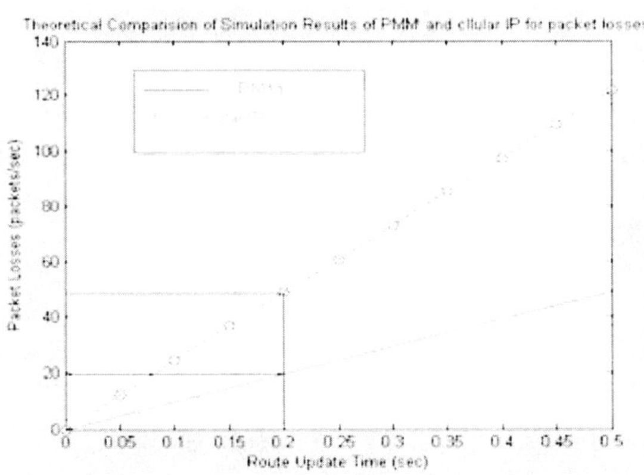

Using equations 6 and 7, the performance of PMM and Cellular IP, in terms of packet losses, are drawn in Fig. 8 as predicted by the mathematical analysis and in Fig. 9 as predicted by the simulation.

CONCLUSION

This chapter deals better design and analyze the performance of the PMM protocol using CBR traffic and UDP agents. Simulation supported only unidirectional moment of MH. A new way should be added to support multi-directional motion of MH. Though, the performance of PMM for optimization of packet loss is better than cellular IP, in both theoretical and simulation analysis, until the simulation supports multi-directional moment of MH the accuracy can be concluded, only within limits. Here, we separate various micro mobility protocols in to distinct, routing, packet forwarding handoff optimization, signaling and paging environment. By using appropriate selection of handoff and widely proactive micro mobility protocol can be designed. It is a new approach for the prediction of handoff conditions in mobile environments. It is based on a Cellular IP and optimizes miss-routed packets in the Cellular IP at the handoff time. The theoretical and simulation results show that the new Proactive Micro Mobility Protocol optimizes miss-routed packet loss in the Cellular IP. Overall, the packet loss for the PMM Protocol in the Cellular IP is optimized mathematically and experimentally for a route update time of 0.2 seconds. By collecting the position of the MH periodically, the velocity of the MH is calculated and handoff predicted. Although we do not claim complete coverage of all scenarios, we showed how our approach helps identify a rich set of design and scenario parameters. The evaluation for such parameters provides better understanding of existing micro-mobility protocols, and a systematic framework for iterative evaluation of further enhancements and modifications of these protocols.

FUTURE WORK

PMM protocol optimization of packet loss is better than cellular IP, in both theoretical and simulation analysis, until the simulation supports multi directional moment of MH the accuracy can be concluded, only within limits. The result of simulation and further needs for better simulation giving encouragement to design complete module for PMM protocol.

REFERENCES

Bhaskara, G., Helmy, A., & Gupta, S. (2003). Micro-mobility protocol design and evaluation: a parameterized building block approach. *IEEE 58th Vehicular Technology Conference* (pp. 2019- 2024.)

Campbell, A. T., Gomez, J., Wan, K. C. y., Zoltán, R. T., & Valko, A. G. (2002). Comparison of IP micro mobility protocols. *IEEE Wireless Communications*, *9*, 72–82. doi:10.1109/MWC.2002.986462

Campbell, A. T., & Gomez-Castellanos, J. (2000). IP micro-mobility protocols. *ACM SIGMOBILE Mobile Computing and Communications Review*, *4*(4), 45–53. doi:10.1145/380516.380537

Carli, M., Cappabianca, F., Tenca, A., & Neri, A. (2004). Mobility Management for the Next Generation of Mobile Cellular Systems . In *Lecture notes of Telecommunications and Networking* (pp. 991–996). Berlin, Heidelberg: Springer.

Hashim, A. H. A., & Anwar, F. Mohd. S., & Liyakthalikh, H. (2005). Mobility Issues in Hierarchical Mobile IP. *3rd IEEE International Conference: Science of Electronic, Technologies of Information and Telecommunications*, pages.

Heine, G., & Horrer, M. (1999). *GSM Networks: Protocols, Terminology and Implementation*. Norwood, MA: Artech House.

Jakkula, S. (June, 2002). *Proactive Micro Mobility Protocol Design*. Unpublished Master's dissertation, IIIT-Allahabad, India.

Kaiduan, X., Wong, V. W. S., & Leung, V. C. M. (2004, September). Support of Micro-Mobility in MPLS-Based Wireless Access Networks. *Oxford Journals IEICE-Transactions on Communications*, *88*(7), 2735–2742.

Magret, V., & Choyi, V. K. (January, 2001). Multicast Micro-mobility Management. *Lecture Notes in Computer Science*, (pages 260-268).Berlin / Heidelberg: Springer.

Shim, Y. C., Kim, H. A., & Lee, J. I. (2005). Design and Evaluation of a New Micro-mobility Protocol in Large Mobile and Wireless Networks . In *Lecture Computational Science and Its Applications* (pp. 9–12). Berlin, Heidelberg: Springer.

Tisal, J. (May 2001). *The GSM Network: The GPRS Evolution: One Step Towards UMTS* Wiley, Forth Worth, TX: John & Sons.

Vicente, C. G., Pablo, G. E., & Vincent, P. (2004). Evaluation of cellular IP mobility Tracking procedures. *The International Journal of Computer and Telecommunication Networking*, *45*(3), 261–279.

Xie, K., Wong, W. S. V., & Leung, C. M. V. (2005). Support of Micro-Mobility in MPLS-Based Wireless Access Networks. Oxford Journal in IEICE Transactions on Communications. E88(B), 2735-2742.

Xiea, B., Kumara, A., Agrawal, D. P., & Srinivasan, S. (2006). Secured macro/micro-mobility protocol for multi-hop cellular IP. *Journal Security in Wireless Mobile Computing Systems*, *2*(2), 111–136.

Yair, A., Claudiu, D., & Hilsdale, M. (2006). Fast handoff for seamless wireless mesh network, *ACM International Conference On Mobile Systems, Applications And Service*, (pp. 83-95).

Yan, Z., & Hee, S. B. (2004). Counting in Hierarchical Cellular System with overflow scheme "Handoff Counting in Hierarchical Cellular System with overflow scheme. *The International Journal of Computer and Telecommunications Networking*, *46*(4), 541–554.

Chapter 18
A Comparative Review of Handheld Devices Internet Connectivity Revenue Models to Support Mobile Learning

Phillip Olla
Madonna University, USA

ABSTRACT

This chapter provides a survey of mobile broadband revenue models deployed by mobile network operators in the UK, USA and Canada. The survey of exiting revenue models highlights the technology adoption trends for handheld devices by consumers and identifies the future impact of these trends on the network operators and content providers with respect to educational content. This article focuses on innovations in consumer propositions that can support the Mobile Learning phenomenon. The study reveals that the various operators aim to differentiate their consumer propositions by branding, technology devices, and flexible pricing structures. From the results of the study it is clear that the current continuous convergence of multimedia applications, information services, digital networks, and devices will likely lead to an increase in adoption of Mobile learning systems in the UK, Canada and the USA especially as the price per bandwidth drops and new innovative connectivity options are deployed such as built in mobile broadband processor in laptops and consumer devices.

INTRODUCTION

There has been a phenomenal evolution of mobile technology over the last decade and the voice capabilities have evolved from a niche technology to an indispensable service. Consumers have adopted mobile voice technology into all facets of their daily lives (Mohr, 2006). In addition to the widespread diffusion of mobile phones, a broadband evolution has occurred in the developed world, due to the proliferation of technologies such as fiber, cable modem, and broadband wireless services. The diffusion of broadband along with the innovation in handheld devices has led to the growth in Mobile learning. There are an abundance of scenarios of learning with mobile technologies. Personal digital Assistants (PDA), Smartphones, and mobile phones are frequently used technologies for mobile learn-

DOI: 10.4018/978-1-61520-761-9.ch018

ing. Mobile Learning can be broadly categorized on the two dimensions of personal vs shared and portable vs static (Naisnith, 2004). This article will focus on personal handheld devices that access learning content while mobile. The advancement in learning technologies to incorporate mobile technology can be viewed as an evolution of distance learning to e-learning (electronic) and know to mobile supported learning. Mobile, wireless, and handheld technologies are being used to re-enact approaches and solutions to teaching and learning used in traditional and web-based formats (Keegan, 2002). Mobile learning facilitates interaction with computer-supported learning environments from mobile devices using a wireless connection.

The next major trend in communication will see the convergence of both mobile and broadband technologies to create a phenomenon called Personal Broadband. Personal Broadband can be viewed as a fusion of the two perpetual markets of mobile technology and broadband, aiming to serve four types of customers: those migrating from mobile voice services and seeking higher speeds for multimedia applications including voice over IP services (Engel, 2007), fixed users who want mobility, Wi- Fi users seeking additional range, and new users who will adopt the new generation of services and applications generated by the high data rates promised by personal broadband technologies. Personal broadband can have a profound effect on deploying learning content over the mobile network.

Due to the proliferation of mobile devices such as smartphones, Blackberry's, and Iphones mobile consumers are dictating that they stay connected ubiquitously irrespective of their location. Users are now accustomed to broadband at home and expect the same connection to be available in their offices, airports, hotels, and other public spaces, similar to the constant convenience of a mobile phone. One factor that is driving this trend is the increase in multimedia content such as learning material available on the internet and mobile networks (Mohr, 2008). Another important factor is the trends towards Web2.0 applications that require users to connect to the Internet to gain access to services and applications.

Handheld devices have evolved to become ubiquitous, networked, and converged devices with enhanced capabilities for rich social interactions, context awareness and internet connectivity, which can have a great impact on learning. Learning will shift outside the classroom into the community and into the learner's environments, both real and virtual, thus becoming more situated, personal, collaborative and lifelong. The challenge that educators face will be to determine how to use mobile technologies to transform learning into a seamless part of daily life to the point that it is not considered learning. One of the challenges that educators will face is ensuring that students will have access to the content once they leave the classroom and educational environment at an affordable cost. There are a variety of networks technologies that can provide connectivity but there is typically a cost, this paper will aim to summarize the potential costs and describe how the revenue models are packaged for consumers.

Most of the reviews of mobile technologies and learning have addressed how mobile technologies have been used to enhance the curriculum. In this review, we take a pragmatic perspective to investigate how student using mobile learning can access the content outside the confines of the classroom of their mobile devices by reviewing revenue models implemented by operators. This article will provide a review of the current approaches being deployed by network operators to provide mobile broadband services. There is a trend of diversification of telecommunications services, which is leading to a wide range of services becoming available in the market (Cha, Jun, Wilson, & Park, 2008). A comparison of revenue models will be presented by surveying the consumer tariff proposition offered by network operators in the United Kingdom, Canada and the USA. This will allow the identifications of contrasting approaches (Alam & Prasad, 2008).

The paper is structured as follows. The first section will provide a review Mobile Learning, this will be followed by a discussion on the technologies that are capable of providing mobile broadband connectivity to handheld devices, followed by a discussion on the evolution of 4G technology. Prior to the conclusion the article provides a review of the mobile broadband revenue models used by operators in the UK, Canada and USA.

LITERATURE REVIEW: MOBILE LEARNING TECHNOLOGIES ON HANDHELD DEVICES

Mobile learning can be conceptualized to describe the technologies such as the devices and networks to support the learning infrastructure. Traxler proposes a definition of m-learning as "learning delivered or supported solely or mainly by handheld and mobile technologies such as personal digital assistants (PDAs) smartphones or wireless laptop PCs" with unique characteristics such as personal, spontaneous opportunistic, informal, pervasive, situated private, context-aware, bite-sized, and portable (Traxler 2007).

Shih and Mills (2007) propose key functionality that enables mobile learning to supports the capability to deliver the phenomena of learning anytime and anywhere through the use of multimedia (text, voice, image, or video) and communication (phone call, voice/text messaging, e-mail Web access). Shih and Mills (2007) also propose that this method of teaching and learning provides "real-time online interaction in a series of short burst learning activities, with features such as voice/ video recording for storytelling or even a mobbloging journal." Lehner and Nosekabel (2007) describe learning as a service that electronically delivers digital content to learners, irrespective of location and time, and provides learners guidance and feedback using new interfaces for diverse learning approaches.

Due to the evolution of the mobile domain, there has been a considerable increase in research that applies handheld devices and mobile technologies to learning (Lai et al, 2007) such as the G1:1 project (Research Center for Science and Technology for Learning 2005; Chan *et al.* 2006) and the M-learning project (ULTRALAB and CTAD 2003). Researchers propose that mobile technologies have the potential to create stimulating opportunities for learning (Curtis *et al.* 2002; Kynäslahti 2003; Ogata & Yano 2004). In the context of educational application, mobile technologies facilitate the delivery of digital multimedia educational content to learners regardless of location and time (Lehner &Nosekabel 2002). Research also shows that Mobile technology will support learners in authentic and seamless learning. Mobile technologies have been shown to impart instant learning guidance, feedback and use innovative interfaces for diverse learning approaches (Liang *et al.*) 2005). Some studies have also revealed that mobile technologies are beneficial on field-trip-based and outdoor learning experiences (Rieger&Gay 1997; Chen *et al.* 2003; Roschelle 2003; Seppälä &Alamäki 2003). There are huge benefits to incorporating computing power and wireless into the learning environment, because these capabilities an make learning expedient, immediate, authentic, accessible, efficient and convenient (Curtis *et al.* 2002; Kynäslahti 2003; Ogata&Yano 2004).

There are a multitude of technologies that can be compartmentalized as mobile learning tools. Naismith (2005) interprets the concept to mean 'portable' and 'movable'; they also imply a 'personal' as opposed to 'shared' context of use. Mobile technologies have been classified in Figure 1 using a two dimensions of personal vs shared and portable vs static. The first arrow describes mobile enabled technologies that offer shareable interactions, examples include interactive classroom whiteboards and video-conferencing facilities. Another scenario of technologies that fits to this group include technologies that can provide learning experiences to users on the move, but the devices themselves are not physically movable are Street kiosks, interactive museum displays

Figure 1. Mobile learning devices

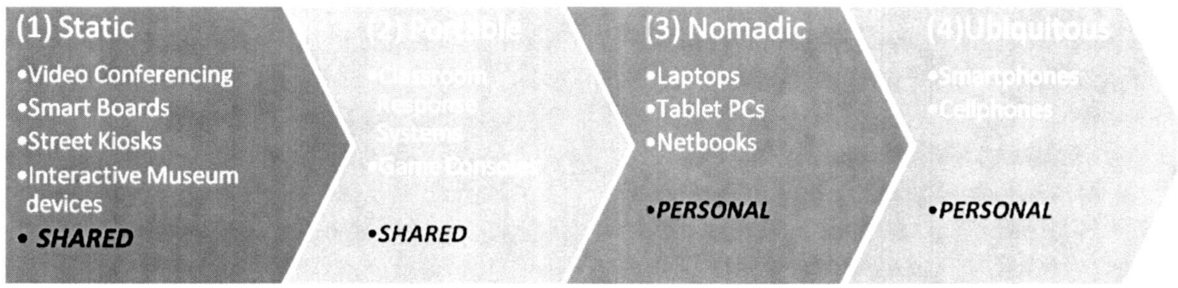

and other kinds of installations offer pervasive access to information and learning experiences, but it is the learner who is portable, not the delivery technology (Naismith 2005). The second arrow describes technologies, less portable than mobile phones and handheld devices that interact with the learning environment to enhance the learning experience such as classroom response systems. These type of describes can be utilized to react to content such as quizzes on an instructors system. This technology is static from the point of view that it must reside at a single location. The third arrow describes devices that can be classified as both portable and personal. These are the devices that people most commonly associate with mobile technologies: mobile phones, PDAs, tablet PCs and laptops. Adding a Nomadic component (arrow 4) distinguished the devices that can be moved around the classroom and campus from the devices that can be taken off campus and still maintain connectivity and access all the content and interact with the learning environment in the same way. The review conducted in the next section is primarily concerned with identifying internet connectivity models for handheld devices mentioned in arrow 4, however it was discovered that some revenue models are emerging that support devices highlighted in arrow 3.

CATEGORIZATION OF EXISTING REVENUE MODELS FOR MOBILE BROADBAND ACCESS

This research reviewed revenue models of mobile data consumer proposition from 12 network operators in three countries, United Kingdom, Canada and the USA to identify the various revenue models being used. Over 100 different consumer propositions were identified and 37 unique revenue models were documented and classified into 7 main groups illustrated in figure 2. The revenue model was deemed to be unique even if they were offered by another operator due to the fact that the network coverage, speeds and terms of usage differ between operators. It was decided to only focus on consumer models because the business proposition were too complicated to decipher and were heavily dependent on factors such as current relationship with operator, number of devices, size of organizations.

The revenue model describes the process by which a mobile broadband service provider intends to generate income by specifying the charging mechanism for a service or the subsidy that will be applied. This is different from a business model (J. Francis, 2007). The main purpose of a business model is to develop a strategy defining what should be done or how to create value (Yamakami, 2006). The revenue model describes the execution that leads to the conversion of the value creation strategy into cash-flow

Figure 2. Mobile broadband revenue models

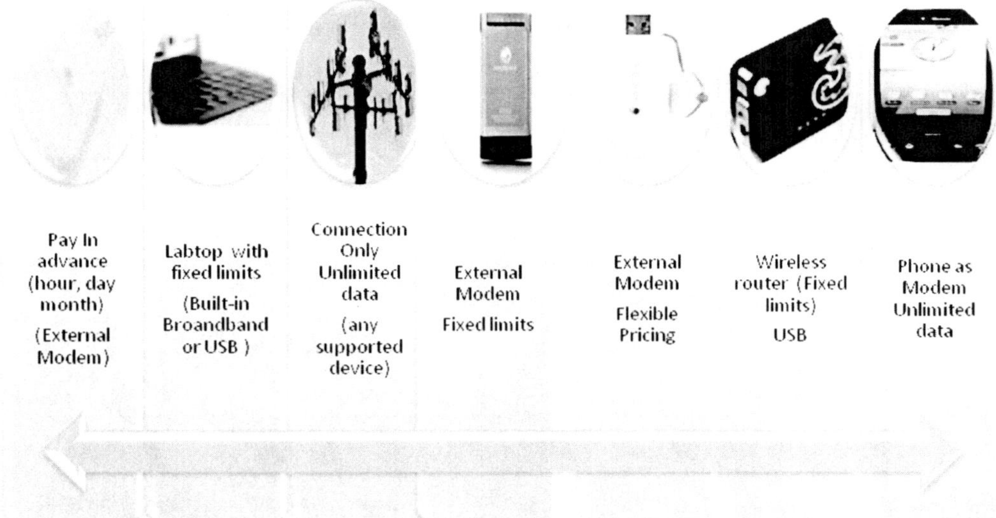

The revenue models from three countries have been studied to understand the various elements and revenue streams along with the documented problems with the approaches. The 7 revenue model classification is illustrated in figure 2. From this study is obvious that network operators are turning to consumer mobile broadband data services as a new source of revenue and means for increasing ARPU (annual revenue per user). In the UK it appears that the operators are marketing the service as a viable alternative to home broadband services, while the USA and Canada the products are more business centric although the propositions are offered to consumers. The revenue models reviewed are all significantly different from the typical cellular networks models which support only per-minute and flat rate charging models, although the equipment subsidization has been maintained along with an allocation for text messaging (Donegan, 2005).

Model 1: Pay in Advance Models

These models have become very sophisticated and users can prepay for services over a multitudes of timeframes as illustrated in table 1. A user can pay in advance on an hourly basis, weekly, month without having to sign up to a contract. With T-Mobile (UK, USA) a user can also access T-Mobile hotspots as well as their 3G network. The main advantage of this is the speeds are likely to be better in a hotspot in some locations. In the UK there has been a steady increase in pre paid mobile broadband due to the lowering of the data charges. There are a variety of options available, a new product launched by 3 UK allows the user to purchase the external modem outright and then purchase top-up data amounts based on what allocation the consumer needs. T-Mobile and Vodafone also offers a rolling contract proposition. With this proposition the consumers can roll a 30 day, or 1 week contract. This type of flexibility that is appealing to consumers who require short term data access.

Model 2: Laptops with Fixed Limits

Network operators are renowned for mobile handset subsidies (Albon & York, 2008), this strategy continued with the smartphones such as blackber-

Table 1. Pay in advance revenue models

Company	Country	Revenue Model	Device Type	Technology	Price	Frequency	Monthly Download (Gig)	Price per additional Gig
XOHM	USA	Pay In Advance	Any	WiMax	$ 30.00	Month	unlimited	n/a
T-Mobile (uk)	UK	Pay In Advance		UMTS 2101	$ 4 (£2 00)	hour	3Gb	20c per mg
T-Mobile (uk)	UK	Pay In Advance		UMTS 2101	**$20 (£10)**	1 day	3 GB	20c per mg
T-Mobile (uk)	UK	Pay In Advance		UMTS 2101	**$40 (£20)**	7 days	3 GB	20c per mg
3	UK	Pay In Advance	USB	HSDPA	$12.50 (£25)	30 days	7GB	20c per mg
3	UK	Pay In Advance	USB	HSDPA	$5 (£10)	31 days	1GB	20c per mg

ries and Iphones, a new phenomenon in the UK involves network operators offering laptops in the place of cell phones. The laptops are free or subsidized and the consumer can use the devices to access the internet on the operator's mobile broadband network. Typically the consumer has a download limit of 1G to 20G and pays a monthly fee for 18 months or 24 months. The longer the contract the less the subscriber has to pay for the laptop. Since 2006 laptop sales out number desktop sales (quote), but not everyone can afford a new laptop. This has led to the creation of a new proposition by Vodafone, Carphone Warehouse, 3, Phones4U, the Link, that allows consumers to receive a free laptops when they purchase a fixed limit mobile broadband package with an 18months 24 month contract. The Mobile Broadband Laptop offer is a UK phenomenon and 3 and Vodafone provide a wide selection of laptops for different demographics such as students or business sectors. At the time of this study, there was no laptop offer available in USA or Canada. Whereas the Carphone Warehouse packages include a free Acer laptop, there was more choice available from 3, with a selection that at the time of our data collection included HP 530, HP 2133 and HP DV6000 laptops. From a student perspective this revenue model can provide a $300 saving a year to a student who requires both broadband and a new laptop. Another benefit is the fact that the cost of the new laptop is spread over 12 or 18 monthly payments. There are two types' connectivity options available. The first option involves using an external modem such as the USB card or PC card, while the preferred option is the built on 3G processors as this requires no software installation. At the time of this study, a survey by Mobile Computing suggests that 30%, of Mobile broadband sales are laptop revenue models, whilst Broadband News states that 27% of all sales.

Mobile broadband operators Orange, 3 and Vodafone are currently running free laptop offers on their mobile broadband, and a similar offer was available from T-Mobile until all stock sold out.

Model 3: External Modems with Fixed Limits

All Internet activities such as browsing the internet, viewing pictures online, music, streaming video or online gaming count towards a users monthly broadband usage. The usage is the amount of this data that a user accesses per month and not the amount that is downloaded onto the computer. This data allowance is typically measured in gigabytes (GB) but in the US some packages exist that use

Table 2. Laptop with fixed limits

Company	Country	Device Type	Technology	Laptop Price	Price	Frequency	Contract	Monthly Download (Gig)	Price per additional Gig
Vodafone	UK	Built-In	UMTS 2100	Dell Inspiron		Monthly	2	3Gb	$ 30 (£15) per GB
3	UK	USB	HSDPA	HP550	$50(£25)	Monthly	2	5GB	$.20 (10p) per mg
Orange	UK	Buit-in	HSDPA	Free (ASUS Eee 901 PC)	$50 ($25)	Monthly	2	3	$.20 (10p) per mg

Table 3.Modems with fixed limits

Company		Device Type	Technology	Price	Frequency	Contract	Monthly Download (Gig)	Price per additional Gig
Verizon	USA	EM	EVDO	$ 59.00	Monthly	1,2,3	5	0.25 per mbyt
Telsus	Canada	EM	EVDO	$ 65.00	Monthly	2	1	$.10 per gig
Telsus	Canada	EM	EVDO	$ 25.00	Monthly	2	4mb	$.12 per mb
AT&T	USA	EM	HSPA	$ 60.00	Monthly	2	5	$0.48 per mb
AT&T	USA	EM	Edge	$ 40.00	Monthly	2	50 mb	$0.97 per mb
3	UK	USB	HSDPA	$ 30 (£15)	Monthly	2	3GB	$.20 (10p) per mb

megabytes MB as the standard for download limits. The GB allocation is the key billing factor that decides the amount a mobile network operator will charge for specific broadband package. Over 90% of the broadband packages reviewed have a download limit (e.g. 1G, 2GB, 5G 15GB, 30GB). This means that a user is only allowed a certain allocation and any additional downloads are charged at a different rate. Heavy internet users require higher data limits and subsequently the more expensive the mobile broadband package. Most of the existing Mobile broadband users access the networks via an external modem which consist of either PC Cards and USB cards. If students are downloading educational content that takes them over their limit then the penalties can vary very high. The external modems are also subsidized based on length of service. The fixed data download vary from country to country. For example

- UK ranges from 1G – 8 GB and prices between $20 to 120$
- Canada ranges from 1G – 5 GB and prices range from $30 to $75
- USA ranges from .5G to 5 GB and prices range from $30 - $ 75

Most consumers who have access to broadband are used to the unlimited access paradigm and this creating some confusion in the mobile arena. Although network operators would like to offer unlimited broadband, the 3G technology implemented does not support cost effective data transfer protocol. The cost of data on a 3G network is more expensive than home broadband technology. Network operators are worried that if all subscribers had unlimited download limits, the system would be overloaded with peer to peer content, bit torrents and other large files, which would overwhelm the network and create havoc

Table 4. Flexible pricing

Company		Device Type	Technology	Price	Frequency	Contract	Monthly Download (Gig)	Price per addl. Gig
Fido	Canada	EM	HSPA	$ 30.00	Monthly	2	0 > 500mb	
Fido	Canada	EM	HSPA	$ 35.00	Monthly	2	500mb > 1gb	
Fido	Canada	EM	HSPA	$ 50.00	Monthly	2	1gb > 2 gb	
Fido	Canada	EM	HSPA	$ 65.00	Monthly	2	2gb> 3gb	
Fido	Canada	EM	HSPA	$ 80.00	Monthly	2	3gb > 5gb	10c/mb
Bell	Canada	EM	CDMA2000 1xEV-DO	$ 65.00	Monthly	2	> 1gb	
Bell	Canada	EM	CDMA2000 1xEV-DO	$75	Monthly	2	1gb > 2 gb	
Bell	Canada	EM	CDMA2000 1xEV-DO	$ 85.00	Monthly	2	2gb >3gb	
Bell	Canada	EM	CDMA2000 1xEV-DO	$ 100.00	Monthly	2	3gb > 5gb	10c/mb

causing user to experience slow networks and poor connections. This will change with the adoption of 4G all IP infrastructures. The release of newer, smaller, faster mobile broadband modems in the form of USB sticks, USB modems (dongles) and data cards are being released onto the UK market on a monthly basis. The largest producer of these devices is Huawei who currently supply USB sticks to 3, Vodafone and T-Mobile.

Model 4: Flexible Pricing

The flexible pricing option is a revenue model that allows a consumer to pay a flat fee for their data usage based on a rolling scale. This model has the benefit of allowing the customer to be automatically upgraded to the higher tariff if their usage pattern exceeds there data limits. Part of the problem with this approach is that a customer is not sure how much they will be paying from month to month, however this is merely a perception because the flat fee pricing will also result in charging that varies from month to month. This model is only available in Canada (at the time of compiling the data). Another problem for consumers is that they are not sure when they will be pushed into the next data scale. The obvious benefit from this type of model is that the consumer avoid potentially hefty penalties. If a consumer exceeds the designated download allowance set by the operator, the fees are charged per Mb or per Gb.

Model 5: Connection Only-Unlimited Data

The connection only approach will allow a subscriber to buy a broadband connection that is not tethered to a particular device. The approach is similar to using a WiFi network with a WiFi en-

Table 5. Connection only with unlimited data

Company		Device Type	Technology	Price	Frequency	Contract	Monthly Download (Gig)	Price per additional Gig
XOHM	USA	Any	WiMax	$ 50.00	Monthly	none	unlimited	n/a
T-Mobile	USA	Any	WiFi	$ 39.99	Monthly by Month	n/a	unlimited	na
T-Mobile	USA	Any	WiFi	$ 6.00	Day	n/a	unlimited	na

Table 6. Wireless modem fixed limit

Company		Device Type	Technology	Price	Frequency	Contract	Monthly Download (Gig)	Price per additional Gig
3	UK	USB + Wireless router	HSDPA	$ 30 (£15)	Monthly	2	3GB	$.20 (10p) per mb

Table 7. Phone as modem revenue model

Company		Device Type	Technology	Price	Frequency	Contract	Monthly Download (Gig)	Price per additional Gig
T-Mobile (uk)	UK	Phone	UMTS 2102	$25 (12GBP)	Day	n/a	Unlimited (exc VOIP)	**n/a**
T-Mobile (uk)	UK	Phone	UMTS 2103	$50 (25GBP)	Day	n/a	Unlimited (Inc VOIP)	**n/a**
AT&T	US	Phone	EDGE / 3G	$50	Monthly	2	5G	**$0.5 MB**
Sprint	US	Phone	EVDO	$15 + Call Plan	Monthly	2	5G	**$0.5 MB**

abled device. Xohlm has launched a service in the USA that will allow a subscriber to use a WiMAX network to access the mobile broadband network via any WiMAX enabled device which could be a set-top box, laptop, smartphone. They are currently offering 2 connections for $50. This service has launched in Baltimore USA and is expected be launched in other cities in 2009. One of the main benefits of this approach is the flexibility of having unlimited data allowance. There are few Mobile Broadband providers that are offering unlimited download. An Unlimited Broadband package is perfect for users who are heavy Internet users, or have erratic usage patterns. Unlimited access will allow users to spend as much time online as they like to download large files, such as videos, music (MP3s), applications and software updates.

Another benefit of an unlimited broadband package is that a monthly bill is guaranteed to be the same each month as consumers pay a fixed fee for unlimited usage as opposed to the fixed limit, which incurs additional costs if your limit is exceeded. Currently T-Mobile in the UK was the only network operator that did not enforce excess fees to users who exceed their limit. In extreme cases, customers that consistently exceed t heir limit are contacted and advised to bring their usage to a level stipulated in the fair usage guidelines. T-Mobile (UK) provides "unlimited broadband" packages with a fair usage allowance of 3Gb on their plus package and 10Gb on their max package. Vodafone has recently launched an unlimited packages which require an external USB dongle, but these are only available to business customers.

Model 6: Wireless Routers-Fixed Limit

Another innovative revenue model involves the user purchasing a wireless router with an external USB card. The user connects the modem to the USB card and creates a WiFi network that will allow the subscriber to share the broadband connection and data limit, this approach is similar to the home broadband.

Model 7: Phone as a Modem-Unlimited Data

Using a 3G phone as a modem is another means for revenue generation. The network operator will typically sign up the customer to a voice plan and add on an additional data plan that will allow the consumer to use their phones to provide data access to their laptops. The laptop is tethered to the phone to access the Internet via a USB cable or Bluetooth. This feature is supported by most of the high end 3G phones such as the Nokia phones, G-phone and the Iphone is expected to support this feature soon. The data plans vary for Phone as Modem range from $49.99/month in the USA to $100 in the UK.

CONCLUSION

This study highlights that for the most part, the bulk of the network operator's revenue is generated from traditional service like voice and related products such as ring tones and SMS and MMS messaging. At the time of this research the seven mobile broadband revenue models identified were focused on providing access to the network and were not service driven. With the growth of Mobile learning network operators must develop capabilities to support the multi-faceted nature of mobile learning services at a cost that is affordable to students while leveraging data services such as telemetry, TV, mobile video, games, and location and navigation applications to generate additional revenue. In the future 4G is expected to support the proliferation of IP enabled device and applications across multiple domains such as personal communications, consumer entertainment, and monitoring; more attention should be paid to identifying educational content to create a unique billing plan that will encourage the proliferation of content onto mobile networks.

The revenue models identified are very diverse and include discounting equipment such as laptops, external USB models, and wireless routers. The concept of discounting laptops and netbooks and bundled with a data connection device is appealing to students and will increase the ability of students to access content, however the data plans are very expensive for downloading multimedia data. The revenue model can be distinguished into those that provide a fixed download limit and models that provide unlimited downloads, the most common model is the fixed download limit with a typical limit ranging from 1G – 6 G of data. With the range of consumer services, price, and availability, there is a likelihood that in the future mobile broadband will someday surpass fixed broadband as mobile phones have surpassed fixed lines in some countries, however the service reliability and coverage must the improved. There is also a trend of embedding 3G/ 4G processing within laptops and consumer devices.

Innovation is driving this fast-moving market and design and technology advancements are working together to fuel uptake of personal broadband services and change people's perceptions of mobile broadband from luxury product to everyday essential required for educational, personal and business needs.

REFERENCES

Ala-Laurila, J., Mikkonen, J., & Rinnemaa, J. (2001). Wireless LAN access network architecture for mobile operators. *IEEE Communications Magazine*, *39*(11), 82–89. doi:10.1109/35.965363

Alam, M., & Prasad, N. (2008). Convergence transforms digital home: Techno-economic impact. *Wireless Personal Communications*, *44*(1), 75–93. doi:10.1007/s11277-007-9380-2

Albon, R., & York, R. (2008). Should mobile subscription be subsidised in mature markets? *Telecommunications Policy*, *32*(5), 294–306. doi:10.1016/j.telpol.2008.02.003

Cha, K., Jun, D., Wilson, A., & Park, Y. (2008). Managing and modeling the price reduction effect in mobile telecommunications traffic. *Telecommunications Policy*, *32*(7), 468–479. doi:10.1016/j.telpol.2008.04.005

Chan, T.W., Roschelle, J., Hsi, S., Kinshuk, Sharples, M., Brown, T., Patton, C., Cherniavsky, J., Pea, R., Norris, C., Soloway, E., Balacheff, N., Scardamalia, M., Dillenbourg, P., Looi, C. K., Milrad, M., & Hoope, U. (2006). One-to-one technology enhanced learning: an opportunity for global research collaboration. *Research and Practice in Technology Enhanced Learning* 1, *3*(29).

Chandrasekhar, V., Andrews, J., & Gatherer, A. (2008). Femtocell networks: A survey. *IEEE Communications Magazine*, *46*(9), 59–67. doi:10.1109/MCOM.2008.4623708

Curtis, M., Luchini, K., Bobrowsky, W., Quintana, C., & Soloway, E. (2002) Handheld use in K-12: a descriptive account. In *Proceedings of IEEE International Workshop on Wireless and Mobile Technologies in Education (WMTE)*, pp. 23–30. Los Alamitos, CA: IEEE Computer Society Press

Dahlman, E., Gudmundson, B., Nilsson, M., & Skold, J. (1998). UMTS/IMT-2000 based on wideband CDMA. *IEEE Communications Magazine*, *36*(9), 70–80. doi:10.1109/35.714620

Donegan, M. (2005). The business case: Can convergence really pay off? Total Telecom, (APR.), 35. Engel, C. (2007). Competition in a pure world of Internet telephony. *Telecommunications Policy*, *31*(8-9), 530–540.

Fabrizi, S., & Wertlen, B. (2008). Roaming in the Mobile Internet. *Telecommunications Policy*, *32*(1), 50–61. doi:10.1016/j.telpol.2007.11.003

Fitchard, K. (2004). *Qualcomm re-imagines mobile media. Telephony, 245(22), 6-7.* Retrieved (n.d.)., from http://www.scopus.com/scopus/inward/record.url?eid=2-s2.0-9744254533&partnerID=40

Forum, W. (2006). 'Mobile WiMAX Part I: A Technical Overview and Performance Evaluation'. *Mobile WiMAX - Part I: A Technical Overview and Performance Evaluation.*

Francis, J. (2007). Techno-economic analysis of the open broadband access network wholesale business case. In 2007 16th IST Mobile and Wireless Communications Summit, 2007 16th IST Mobile and Wireless Communications Summit. Budapest.

Gunasekaran, V., & Harmantzis, F. (2008). Towards a Wi-Fi ecosystem: Technology integration and emerging service models. *Telecommunications Policy*, *32*(3-4), 163–181. doi:10.1016/j.telpol.2008.01.002

Jiang, T., Xiang, W., Chen, H., & Ni, Q. (2007). Multicast broadcast services support in OFDMA-based WiMAX systems. *IEEE Communications Magazine, 45*(8), 78-86. Retrieved (n.d.), from http://www.scopus.com/scopus/inward/record.url?eid=2-s2.0-34548642639&partnerID=40.

Keggan, D. & Fern Univ., H. (2002, September 30). *The future of learning: From e-learning to m-learning. ERIC Document Reproduction* (Service No. ED472435)

Kim, S., Song, S., & Jung, H. (2007). WiBro-based mobile RFID service development. In *IEEE Wireless Communications and Networking Conference, WCNC*, 2007 IEEE Wireless Communications and Networking Conference, WCNC 2007. (pp. 2880-2884). Kowloon.

Kynäslahti, H. (2003). In search of elements of mobility in the context of education . In Kynäslahti, H., & Seppälä, P. (Eds.), *Proceedings of Mobile Learning* (pp. 41–48). Helsinki, Finland: IT Press.

Lai, C., Yang, J., Chen, F., & Chan, T. (2007). (n.d.). Affordances of mobile technologies for experiential learning: the interplay of technology and pedagogical practices. *Journal of Computer Assisted Learning*, *23*, 326–337. doi:10.1111/j.1365-2729.2007.00237.x

Lazaro, O., Gonzalez, A., Aginako, L., Hof, T., Filali, F., & Atkinson, R. (2007). *Enabler for next generation pervasive wireless services. In 2007 16th IST Mobile and Wireless Communications Summit, 2007 16th IST Mobile and Wireless Communications Summit*. Budapest: MULTINET.

Lee, K. (2007). Technology leaders forum - Create the future with mobile WIMAX. *IEEE Communications Magazine, 45*(5). Retrieved (n.d.), from http://www.scopus.com/scopus/inward/record.url?eid=2-s2.0-34249097660&partnerID=40

Lehner, F., & Nosekabel, H. (2002). The role of mobile devices in e-learning – first experience with a e-learning environment. In *IEEE International Workshop on Wireless and Mobile Technologies in Education* (eds M. Milrad, H.U. Hoppe & Kinshuk), pp. 103–106.Los Alamitos, CA: IEEE Computer Society Press.

Mohanty, S. (2006). A new architecture for 3G and WLAN integration and inter-system handover management. *Wireless Networks, 12*(6), 733–745. doi:10.1007/s11276-006-6055-y

Mohr, W. (2006). Strategic steps to be taken for future mobile and wireless communications. *Wireless Personal Communications, 38*(1), 143–160. doi:10.1007/s11277-006-9022-0

Mohr, W. (2008). Vision for 2020? *Wireless Personal Communications, 44*(1), 27-49. Retrieved (n.d.), from http://www.scopus.com/scopus/inward/record.url?eid=2-s2.0-36949011635&partnerID=40.

Naismith, L., Lonsdale, P., Vavoula, G., & Sharples, M. (2004). *Literature Review in Mobile Technologies and Learning REPORT 11: FUTURELAB SERIES.* Retrieved (n.d.), from http://www.google.com/search?q=Literature+Review+in+Mobile&rls=com.microsoft:*&ie=UTF-8&oe=UTF-8&startIndex=&startPage=1 Accessed June 2009

Ogata, H., & Yano, Y. (2004) Context-aware support for computer-supported ubiquitous learning. In *Proceedings of IEEE International Workshop on Wireless and Mobile Technologies in Education (WMTE)*, pp. 27–34. Los Alamitos, CA: IEEE Computer Society Press.

Panken, F., Hoekstra, G., Barankanira, D., Francis, C., & Schwendener, R., Gr°ndalen, O., et al. (2007). Extending 3G/WiMAX networks and services through residential access capacity [Wireless broadband access]. *IEEE Communications Magazine, 45*(12), 62–69. doi:10.1109/MCOM.2007.4395367

Rieger, R., & Gay, G. (1997). Using mobile computing to enhance field study. In . *Proceedings of the Computer-Supported Collaborative Learning Conference: CSCL, 97*, 215–223.

Roschelle, J. (2003). Unlocking the learning value of wireless mobile devices. *Journal of Computer Assisted Learning, 9*, 260–272. doi:10.1046/j.0266-4909.2003.00028.x

Salkintzis, A., Fors, C., & Pazhyannur, R. (2002). WlAN-GPRS integration for next-generation mobile data networks. *IEEE Wireless Communications, 9*(5), 112–124. doi:10.1109/MWC.2002.1043861

Schatz, R., & Egger, S. (2008). Social interaction features for mobile TV services. In *IEEE International Symposium on Broadband Multimedia Systems and Broadcasting 2008, Broadband Multimedia Symposium 2008, BMSB*, IEEE International Symposium on Broadband Multimedia Systems and Broadcasting 2008, Broadband Multimedia Symposium 2008, BMSB. Las Vegas, NV.

Seppälä, P., & Alamäki, H. (2003). Mobile learning in teacher training. *Journal of Computer Assisted Learning, 19*, 330–335. doi:10.1046/j.0266-4909.2003.00034.x

Traxler, J. (2007). Defining, Discussing, and Evaluating Mobile Learning: The moving finger writes and having writ. *International Review of Research in Open and Distance Learning*, *8*(2), 12.

Trkman, P., Jerman Blazic, B., & Turk, T. (2008). Factors of broadband development and the design of a strategic policy framework. *Telecommunications Policy*, *32*(2), 101–115. doi:10.1016/j.telpol.2007.11.001

Yamakami, T. (2006). Lessons in business model development from early mobile Internet services in Japan. In *International Conference on Mobile Business, ICMB 2006*, International Conference on Mobile Business, ICMB 2006. Copenhagen.

Yavuz, M., Diaz, S., Kapoor, R., Grob, M., Black, P., & Tokgoz, Y. (2006). VoIP over cdma2000 1xEV-DO Revision A. *IEEE Communications Magazine*, *44*(2), 88–95. doi:10.1109/MCOM.2006.1593550

Section 4

Handheld Images and Videos

Chapter 19
Mobile Vision on Movement

Lambert Spaanenburg
Lund University, Sweden

Suleyman Malki
Lund University, Sweden

ABSTRACT

In the early days of photography, camera movement was a nuisance that could blur a picture. Once movement becomes measurable by micro-mechanical means, the effects can be compensated by optical, mechanical or digital technology to enhance picture quality. Alternatively movement can be quantified by processing image streams. This opens up for new functionality upon convergence of the camera and the mobile phone, for instance by 'actively extending the hand' for remote control and interactive signage.

INTRODUCTION

The history of technology is one of surprising crossovers, where a technique developed in one area causes subsequently a break-through in another area. For 'movement' the milestone is the invention of the accelerometer for the airbag: the cushion that inflates for large decelerations to protect vehicle passengers from physical damage. Inflating too early may cause an accident, while reacting too late makes it useless. Therefore the sensor needs to perform robust and dependable in real-time to provide the desired safety, while the mass-market requires it to be cheap and mass-producible. The large-scale introduction of this sensor has given credibility for use in other markets (Knivett, 2009).

The most remarkable crossover has happened when the accelerometer is applied to the digital camera. Typically handling causes a jitter of 10 to 20 vibrations per second. The effect is more apparent with the larger pixel sizes and is further magnified by auto-focus and zoom features (Or & Pundik, 2007). Once having the jitter extracted, it can be compensated by optical and mechanical techniques providing image stabilization. This can also be performed by digital techniques and from the movie industry came special hardware boxes to support such functionality. However, initial ex-

DOI: 10.4018/978-1-61520-761-9.ch019

periments to bring the stabilisation function as software into the camera have not been successful. The digital camera, equipped with digital communication channels, has created a new industry with associated printers, photo-finishing kiosks and a variety of on-line services.

The mobile phone quickly caught on and started to integrate the camera onto the mobile platform, called the camera phone. The product philosophy is based on 'convergence' of data, audio and video into a single device. Sharp and J-Phone have introduced the world-first camera-phone (J-SH04) in November 2000 in Japan. Four years later (October 2004) about 75% percent of mobile phones in Japan are camera phones and in 2005 the market penetration saturates around 75 to 85%, i.e. almost all mobile phones in Japan are camera phones. In the meantime the number of cell phones in general is skyrocketing and will break the 1 billion units barrier by 2010 according to the Gartner Group. Though the camera phone overshadows the digital camera in sales volume, its performance still trails because of additional requirements on cost, size and power consumption (Henning, 2008).

A similar development has been in Remote Control, for instance for the Home Amusement Centre. The early devices simply send keyboard pressings over an infra-red channel to the TV. Recently, it has become possible to point or even to copy movements into a cursor position on a screen. By measuring the pulses over the infra-red channel, the Doppler effect of the moving device can be quantified and used for calibration. Such techniques can also be used to free the computer mouse from its wire, and over this application the Gaming industry is entered. Games have been developed first as a PC application and gradually moved to use specialized peripherals. Also here the wiring between computer peripherals and the processor box has always been a cause for irritation. Cables get mixed up and the user is limited in its freedom, even more so with the advent of multi-user games.

(Casual) Gaming has been based on laser-based pointing devices with sensors located in/on the target object. This concept has brought a remarkable spin-off in museums, where one points to art to get a spoken explanation. Recently the accelerometer from the air-bag and the digital camera also made a cross-over to remote control. This way to move a device and let the movement known to a server to be included in a game or service is best known from the Wii technology. The path the accelerometer went from its invention in automotive safety to mass amusement is accompanied by a continuing improvement. According to a recent Gartner study (Savvas, 2008), the market demands further improvement in the mechanical motion sensor technology, gradually adapting the accelerometer better to the requirements of the application areas.

Initially the camera phone is just the addition of a vision sensor to a cell-phone, but with less resolution and at less performance. For the mobile telephone the emphasis has been on image compaction and stabilization by measuring the global motion through electronic means. Alternative to the use of the accelerometer, image stabilization can also been done by digital image processing. This removes the dependence on mechanical stress and temperature effects, making the accelerometer hard to test when the device becomes smaller. In the movie industry, separate boxes where the camera images are post-processed for stable 3D effects, have long been in use. It is not trivial to incorporate this functionality into the camera phone. But when this is done, all kind of pointing and interaction functions can be accommodated on a camera phone. This can be extended to support to capture user directives (Cravotta, 2007).

One may wonder, whether this development path is plausible as it assumes that a technology with limited effect in digital cameras will by nature become better when the market is extended to mobile telephones. For the mobile telephone the emphasis has been on image compaction and stabilization by measuring the global motion through electronic means. This can be extended to support gaming, as to capture such gaming directives a camera model must be trained from global image shifts but differentiate persistent movement from trembling and shaking (Tico & Vehvilainen, 2007). The step from personal gaming to community services seems self-evident, and vision sensors networks with intelligence by assembly lay on the horizon.

In short. Crossovers are a major innovation motor, leading alone or in combination to business disruptions (Christensen, 1997). We have seen here that movement has made such a cross-over and pose that this will create new mobile functions. Therefore we will first review motion capture and illustrate its computational complexity in image stream compaction. Then in camera stabilization the tight restrictions of the mobile platform get introduced. Subsequently we demonstrate that such demands can be alleviated in further applications.

Catching Motion

An image is captured as a matrix of point-wise measured light intensities. Using the technical specifications of the sensor and its objective, values are attached to the physical distance between the matrix elements, turning the matrix into a topographic map. There are many of such maps, of which the image is just one. The difference is the existence of 'blobs', combinations of pixels into uniquely identifiable objects. For natural images these are physically plausible objects, such as a face, but for artificial images any connected component with attached meaning will do.

By lack of generally applicable structure in the overall area of image-based communication, the basic grouping of pixels is the macro block. First introduced for compaction purposes, it is a matrix of 16 by 16 pixels, though also smaller units like 8 by 16 and 8 by 8 have been proposed. The macro block relates to the practice of down-sampling an image for first inspection. This makes for smaller image storage on chip, less image memory access and faster operation. For instance, a large image is first represented by the average intensity value for every 16 by 16 pixel block while only interesting blocks are analysed in more detail by block matching.

For the still image, block-matching is usually performed to compact the representation by communicating only unique macro blocks by all its pixels, and otherwise only transferring a reference to the similar but previously communicated block. In this 'intra' mode, all blocks are compared to one another and a difference measure is computed as for instance the Sum of Absolute Differences $SAD = \sum_{i=0}^{N-1}\sum_{j=0}^{N-1}\left|R_{i,j} - I_{i,j}\right|$, where $R_{i,j}$ and $I_{i,j}$ are luminance values for each pixel of two macro-blocks, and N is the macro-block size. It is the simplest one and only includes addition operations. A viable alternative, but not explored here, is the Sum of Squared Differences $SSD = \sum_{i=0}^{N-1}\sum_{j=0}^{N-1}\left(R_{i,j} - I_{i,j}\right)^2$.

Motion can be found by comparing macro blocks between two subsequent frames in an image sequence (Figure 1). Ideally objects can move at infinite speed, but physical causes such as gravity and friction enforce an upper limit. This is fortunate, as no movements should take place during still image extraction. In that case movement only shows from object displacement between images. One usually distinguishes between (a) foreground, where all the movement is in terms of displacement versus either the previous or a defined reference image, and (b) background, where all commonalities between the images can be found. This distinction between foreground and background is fundamental to movement detection and appears prominently in image stream compaction as the 'inter' mode.

But not all movements need to come to the fore. A well-known example is foliage. Where the wind rustles through the leaves, each leaf will

Figure 1. Illustration for block matching process

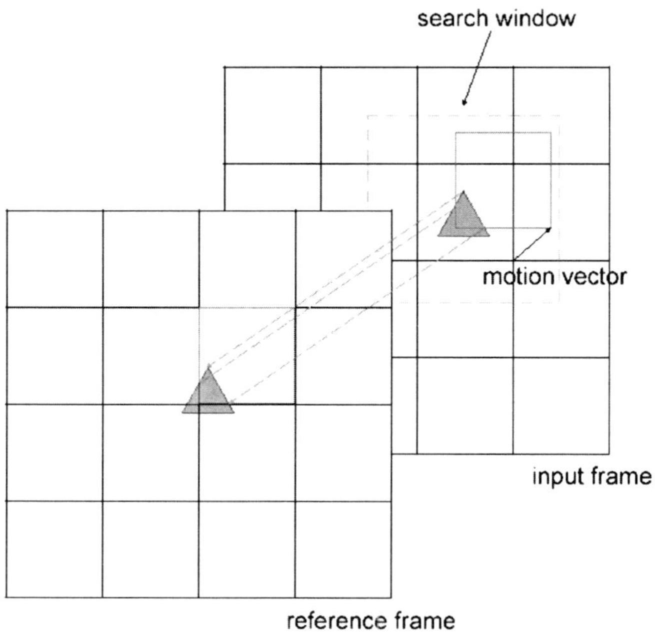

move in its own way. To keep track of the movements of all the individual leaves will increase the foreground. But this has little meaning as motion can be captured through a simple stochastic model, similar to texture models for the non-moving background. All this assumes that the camera does not move. Therefore all objects in the foreground have really been displaced.

As the differences between frames in an image stream can only be found in the foreground, the comparison is not always made with the previous frame. Instead, the reference frame can be used: a frame that is handled as a still image and therefore usable as reference when nothing much has moved. Then, as moving objects are located, these objects can be labelled and a Motion Vector (MV) is determined. In subsequent frames this information can be determined to track the objects by only a local search. This search window around a selected macro block normally determines the complexity of the algorithm in two ways. First is the size of the window SWS in terms of pixels. In addition, the search parameter *p* sets the number of pixels by which the search extends on the current macro block on all sides (Figure 1).

There are various search methods for (macro) block matching algorithms (Vella & Castorina, 2002). The full search is costly and usually not implemented in real-time equipment. Fast searching techniques aim to considerably reduce computational complexity while maintaining good accuracy. These algorithms reduce the full search process to a few sequential steps in which each subsequent search direction is based upon the results of the current step. The algorithms modelled by MATLAB in (Zhang & Chen, 2008) are Full Search (FS), Three Step Search (TSS), Four Step Search (FSS) and Diamond Search (DS). TSS, FSS and DS are all fast block-matching algorithms:

- [Full Search _ FS] This algorithm calculates the SAD cost function for each possible pixel position in the search window. So, it always finds the best match. The obvious disadvantage of FS is that the larger the search window is, the more computations it requires. All further improvements try to achieve the same performance as Full Search but doing as little computation as possible so that a larger area can be covered.

- [Three Step Search _ TSS] TSS first searches the centre location in the search window and sets the 'step size' S (typically 4 for a usual search parameter p of 7). It then searches at 9 locations (8 on the perimeter of a square and 1 in the centre) in a 9*9 window. From these 9 locations searched so far it picks the one giving least SAD cost and makes it the new search origin. It then sets the new step size S=S/2, and repeats the search for two more iterations until S=1. At that point it finds the location with the least cost function and the macro block at that location is the best match.
- [Four Step Search _ FSS] FSS sets a fixed square pattern size of S=2 to start with, no matter what the size of the search window is. Thus it looks at 9 locations in a 5*5 window. If the least weight is found at the centre of the search window, the pattern size is immediately dropped to S=1. If the least weight is at one of the other eight locations, then we continue two times while the search window is still maintained as 5*5 pixels wide. Finally the pattern size is dropped to S=1. The location with the least weight is the best matching macro block.
- [Diamond Search _ DS] For Diamond Search, the search point pattern is changed from a square to a diamond, and there is no limit on the number of steps that the algorithm can take. DS uses two different types of fixed patterns, one is Large Diamond Search Pattern (LDSP) and the other is Small Diamond Search Pattern (SDSP). In this algorithm, we always use LDSP until the least weight is at the centre location; then the last step uses SDSP around the new search origin and the location with the least weight is the best match. As the search pattern is neither too small nor too big and because of the fact that there is no limit to the number of steps, this algorithm can find global minimum very accurately.

Motion capture is a computationally very intensive function. Searching through an image for all the possible objects (areas) that may change place for every potential location requires a lot of calculations. Yet there is only 1/15th to 1/30th of a second to do this before the next frame arrives for processing. This will often force a selection between the alternative methods as listed before. Here we use the 'foreman.yuv' QCIF file as test sequence, with a distance of 2 between reference frame and input frame to implement our algorithms. That means if frame 2 is an input frame then frame 0 is its reference frame, and frame 1 is the reference frame for frame 3 and so on.

The size of the search window has only significant influence on the Full Search algorithm due to the fact that the cost function must be calculated for each pixel in the search window. For the other 3 fast block-matching algorithms, the image quality and the average number of searches required per macro block saturate when the size of the search window is larger than a characteristic value, which is 7 pixels (say half a macro block size) for the 'foreman.yuv' file. In (Zhang & Chen, 2008) it is found that FS takes on average around 225 searches per macro block, while DS and FSS reduce that number by more than an order of magnitude. Nevertheless the signal-to-noise ratio of FSS and DS is pretty close to FS. Also TSS lowers the number of computations required per macro block by almost an order of magnitude, but does not come very close in signal-to-noise ratio to FS. Overall, (Zhang & Chen, 2008) concludes that DS has the best characteristics.

In the different image processing applications that have to do with motion, the basic block matching method appears in slightly different ways.

Figure 2. The original H.264/AVC encoding dataflow

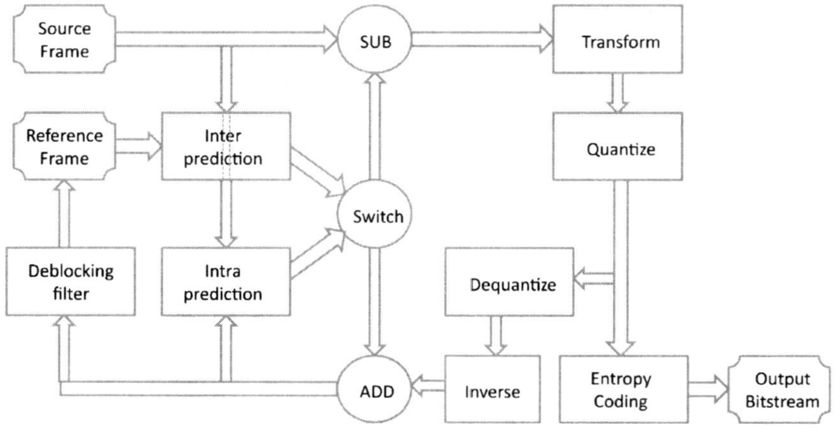

For compaction, the quality of individual motion detection is of prime interest. Instead, image stabilization is more focussed on discriminating between the global (camera) motion and the local movements within the image stream. Mixing both on a mobile platform brings the element of real-time performance and small footprint in. At the end of the chapter we will argue that for camera flocks the requirements can be relaxed by nature of the redundant observation.

In short. The basis of movement detection by image processing is pixel correlation. This complex computational process can be drastically reduced under consideration of the purpose. It is shown in (Postma & vanDartel, 2001) that face recognition in a still picture can be eased by considering that fixation is based on high contrast areas. For movement we have to find the same pixel groups in subsequent images. Again a considerable reduction can be found by applying a heuristic search pattern (Sibiryakov, 2007). Here we plead for the Diamond Search (DS).

Compaction

The fundamental problem in communicating images is stream compaction. Just sending an image as a set of pixels requires a communication channel that can support real-time transport. But where displays get larger, the channel cannot be bettered. Or, even worse, channels have been popularized that are more flexible but slower. This makes compaction a key issue for the channel encoder, requiring in turn a decoder at the receiving side. But even when the channel is fast enough, the move to 3-dimensional pictures and beyond will rapidly become overbearing (Sanchez & Nasiopoulos, 2006). Therefore the central issue in image processing is (and will remain) compaction for the encoder/decoder. To a degree compaction can already be performed in the single image, but taking into account the relation between images in a stream helps to gain further improvements (Figure 2).

The motion estimation compares two or more consecutive frames and determines whether areas of the image have changed or moved between the frames. In many cases an area stays exactly as it was in the previous frame and therefore it is sufficient for the encoder to inform the decoder to display this area as it was in the previous frame. If the area moves in a certain direction, the motion estimation algorithm directs the decoder to use the same piece of image as in the previous frame, but to move it a certain amount in a defined direction. In practice this will be accomplished by sending motion vectors within the MPEG4 bit stream. These vectors will guide the decoder in choosing the appropriate portions of the previously decoded frame to be used in the reconstruction of the current frame. It should be clear that this vastly increases the compression rates. One example is the 'talking head' type of content, such as a newscaster, which results in a very compact MPEG4 stream.

The MPEG standards use different resolutions to find more or less precise motion vectors. MPEG-2 introduces the half pixel motion vector resolution while MPEG-4 adds quarter pixel resolution. This improves the prediction and reduces the error, but it will also make this block more complex. The MPEG-4 standard also introduces intra prediction. The algorithmic complexity has increased in subsequent MPEG standards, but most notably it has risen with the introduction of H.264/AVC. Here motion extraction takes about 95% of the performance, while being responsible for less than 65% of compaction improvement (Chen & Chien, 2006). Therefore it is of interest to see whether and how the algorithmic complexity can be reduced and/or the speed of the algorithm execution can be enhanced. One way for doing this is to implement the image processing in future technology. Unfortunately, it appears that algorithms grow faster in complexity than technology cranks up the speed of processing.

The block diagram of a H.264 encoder is depicted in Figure 2 (Richardson, 2003). The source video is fed into the flow, encoded frame by frame in time order. Each source frame is processed in units of a macro block while ordered in raster scan. Each source macro block is encoded through intra- or inter-prediction. In either case, the source macro block is predicted by former encoded, decoded and reconstructed samples in different ways. For intra prediction, the predicted macro block is built by samples from the current frame. For inter prediction, the macro block is predicted by a macro block from one or several reference frames.

The residual macro block, which is the difference between the source macro block and the predicted macro block, is achieved by subtraction. Then the residual macro block undergoes transformation, quantization and entropy coding to be converted to compressed coefficients. The output bit-stream is assembled from these coefficients and the corresponding encoding decision to inform the recipient that it looks like the other block except for some slight differences.

Beside the forward path mentioned above, there is a reconstruction path, which is shaded in Figure 2. The reconstruction path is a simulation of the decoding process, as will later occur at the decoder side. In order to reconstruct compressed images from the input bit-stream, in a decoder the bit-stream is processed inverse to the way it is generated in the encoder. In detail, the input bit-stream is de-quantized and inverse transformed to form residual macro blocks, and then source macro blocks are reconstructed by adding the residual macro blocks to the macro blocks that are predicted from previously reconstructed macro blocks according to the prediction mode decision extracted from the bit-stream. The reconstructed macro block is passed to intra prediction for the prediction of its neighboring macro blocks, and it is also processed by a de-blocking filter and pieced together with previously reconstructed and de-blocking-filtered macro blocks to form a reference frame. One or several reference frames are stored for the inter prediction of future frames. This certifies that reconstruction is the same at the encoder and the decoder side by having the same reference.

Since data suffers from precision loss during the process of quantization in the encoder, the reconstructed source macro blocks in the decoder are not necessarily identical to the source macro blocks in the encoder. An encoder needs to reconstruct source macro blocks in the same way as a decoder, so that the same basis is used for prediction (in the encoder) and reconstruction (in the decoder). Otherwise, if the encoder uses source macro blocks rather than reconstructed source macro blocks as prediction basis, the difference between source images and decoded source images will be accumulating and rapidly grows out of bounds.

Over the past years a number of studies have been performed to see how image compaction algorithms can be parallelized, see for instance (Jacobs & Chouliaras, 2006). The typical approach is based on a cluster of processors with a shared memory. Each processor is performing the algorithm on part of the image, while reading and writing from a logically integrated storage (Rutten & van Eijndhoven, 2002). The operation is supported by the H.264/AVC notion of a slice: a block of data expected to be manipulated independent of the other slices. Where this cannot be enforced, the option exists to communicate to the decoder that a particular macro-block has not been handled properly. Impressive speed-ups have been noted, but the increase in latency and the need for memory access synchronization takes much potential away (Rodriquez & Gonzalez, 2006). Lately it has been shown that overlapping slices are the solution to this image decomposition problem, allowing for a scaling hierarchy without the shared memory access and synchronization problems.

The storage needs of the image compaction algorithm comprise largely of (1) the input frame, and (2) the reconstructed frame. To identify what is really new, the image predicted from the past and the current one are compared. When the image to be compacted is cut into parts called slices, this mechanism requires that the restored image will fit the partition. Consequently the restored slices will have grown by the amount in which compactable movements are likely to occur. The H.264/AVC standard does not give much detail on the way slicing must be performed. *Simple slicing* destroys the relations found within (intra) and between (inter) the frames for which the decoder must be notified.

Alternative slicing mechanisms are studied in (Wang & Yang, 2007) by modifications of the H264 encoder implementation. Here, the frame partition is limited to horizontal-only slices (i.e. *stripes*) with a relatively equal amount of macro-blocks (e.g. a CIF-frame can be divided into two stripes of 6*18 and two stripes of 5*18 macro-blocks). It is found that the problems of simple slicing can be solved through a separate pre-processing of the left column of macro-blocks in each frame (*pure stripe*). This guarantees that we will have at least a reconstructed left neighbour for every macro-block in a slice. Pre-processing of the stripe is strictly local and therefore does not break the desired parallelization. Still, motion vectors may point outside the stripe. Therefore, as a next measure (Figure 3), each stripe is enlarged with a vertically overlapping border in analogy of the search parameter p in Figure 1 (*bordered stripe*).

Three different versions of the standard algorithm have been benchmarked: (1) the original, non-sliced version, (2) the striped version with borders to take the motion-based growth into account for a search range of (-16, +16), and (3) the pure striped version (Table 1). All lead to output images of about the same quality in each category of quality performance figures: variations are in the order of 10^{-4} dB. Comparing the original and the pure striped version it can be seen that the compaction suffers from slicing, though the effect seems more dependent on the desired quantization step than on the amount of slicing. (Note that a higher Quantisation Step Size QP denotes a lower quality). But due to the additional pre-processing the effect is not as disastrous as generally expected and is actually only a problem for low quality images that do not require a high throughput.

The approach where the reconstructed stripe includes a border of macro-blocks to handle the motion vectors brings a further improvement. For high-quality images (QP=18) no real difference with the non-sliced version can be seen anymore, while a degradation of compression quality can still be observed for low-quality images (QP=34), because the slicing inevitably breaks the exploitation of the spatial correlation at the boundaries

Figure 3. Principle of overlapping slices

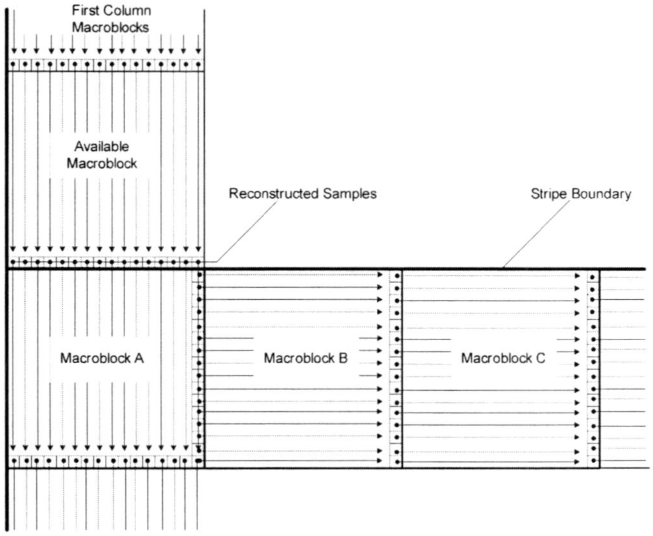

of a stripe. Yet the degradation seems far more acceptable than with simple slicing.

Despite the abundance of scientific literature on image compaction, it is hard to come with a fair comparison, as results tend to be incompletely and inconclusively reported. Some information on the scalability of slice parallelism can be found in (Rodriguez & Gonzalez, 2006). On basis of our observations, we judge that their Foreman experiments on a workstation cluster assume a QP of 34. Their shared memory technology leads to a compression efficiency similar to our use of the bordered stripe, but is burdened with such increases in latency and data synchronization issues that compaction time starts to increase beyond 4 slices. Our alternative using the bordered stripe can be scaled without such a sudden deterioration in performance and is still not limited to 1 dimension. But Foreman is an example of average difficulty. Mobile is much easier, while Highway

Table 1. Benchmark experiments for different slicing and quality performance figures. Shown results are respectively encoded image size in kbytes and Peak Signal Noise Ratio (PSNR) in dB

		CIF Foreman (kbyte / dB))	CIF Highway (kbyte / dB)	CIF Mobile (kbyte / dB)	QCIF Highway (kbyte / dB)	QCIF Carphone (kbyte / dB)
2 Slices QP = 18	No slicing	2,891/43.925	24,795/43.303	6,893/42.523	5,798/43.046	948/44.241
	Brd. stripe	2,893/43.923	24,750/43.295	6,893/42.523	5,801/43.045	947/44.236
	Pure stripe	2,947/43.930	24,790/43.296	6,896/42.522	5,810/43.047	960/44.233
4 Slices QP = 18	No slicing	2,891/43.925	24,795/43.303	6,893/42.523	5,798/43.046	948/44.241
	Brd. stripe	2,896/43.918	24,750/43.291	6,893/42.522	5,806/43.044	949/44.231
	Pure stripe	3,050/43.941	24,943/43.296	6,900/42.524	5,846/43.046	981/44.235
4 Slices QP = 26	No slicing	796/38.438	3,311/39.110	2,754/35.715	1,067/38.469	344/38.549
	Brd. stripe	803/38.480	3,365/39.110	2,757/35.713	1,083/38.470	347/38.553
	Pure stripe	882/38.498	3,454/39.114	2,761/35.710	1,102/38.475	365/38.565

is very difficult and has 14% inefficiency for low quantization step size. Nevertheless, for a QP of 26 and below, even this difficult image has an in-efficiency of maximal 4%. This suggests that not the number of stripes but the demanded Quantization Step Size is dominating the compaction results.

What remains is the question on the significance of these results for compaction hardware architectures. We have shown extensions to the data structures used in intra- and inter-prediction that lead to scalable compaction without significant degradation of the image quality. Its main feature is removal of data hazards, which provides for locality of operation and is therefore leading to a sliced hardware architecture that (a) can be programmed for trading-off compaction quality with encoding throughput and (b) can already produce adequate results without an external network to provide access to shared memory. This makes the approach feasible for multi-core architectures as currently under investigation for mobile applications.

In short. The pixel representation of an image is large. On communication, compaction is required to reduce the size of the pixel file before transport in such a way that de-compaction retrieves the original picture. Simply running compaction on a multi-core architecture gives few computational advantages (Kangas & Hämöläinen, 2006). It seems that building hierarchy into the algorithm is much more rewarding.

Image Stabilization

The captured image is an observation of reality, but only in part. Consequently the picture is placed in the overall landscape, a simple form of hierarchy. Being maps, placements are expressed by coordinates. Coordinate systems can be shared throughout the image hierarchy, but usually every image has its own coordinate system and every placement its own location and origin. Coordinates can be mathematically transformed from the one coordinate system to the other.

As the camera moves, the extracted image moves through the global world. Ideally it can go at any pace, but this is not true in reality. The vision sensor may be hard to turn or slow in capturing. This is the counterpart of moving objects. Of old, the way to obtain sharp pictures of a fast moving object is to move the camera along such that the object always remains in the middle. Then, the moving parts will not be part of the foreground, but of the background. For any case in between, the motion is represented both in foreground and background, and detection will be hard on the compacted image stream. But not all camera movements are deliberate. The camera can be placed on a ship heaving on the waves, can be held by a walking person and can simply tremble in some shaking hands. Usually such problems are met by mechanical image stabilization techniques. Electronic means have had limited application.

The major difference seems to be the point of reference. Intra-image motion can easily be separated from the static background, but reversion can take place when the camera is validated within the global world. Essentially we have here a case of source separation. Two signals are mixed and the only way to separate them is by identifying the characteristics of the respective source. The source separation model covers a 3-dimensional reality and therefore the separation algorithm is more complex than compaction. The camera movement can be evaluated through motion estimation by examining the movement of the background in an image sequence to try to get motion vectors representing the estimated motion. The global motion vectors (GMVs) are extracted in this process. However, the video sequences not only contain still contents but also include some moving objects, which may disturb the accuracy of the extracted GMVs.

In practice, camera motion can be a complex combination of translation and rotation in the 3D space. It has already been investigated how camera stabilization can be accomplished using block matching (Adda & Cottineau, 2003). As we

Table 2. MCPS measurements with different SWS and NSMBs

	Average MCPS	Utilization	Peak MCPS
SWS =14 NSMBs =15	114	55%	204
SWS =14 NSMBs =5	42.2	20%	89.5
SWS =7 NSMBs =8	70.4	34%	108
SWS =7 NSMBs =5	39.8	19%	66.5

focus on the feasibility for cell phones, the study in (Zhang & Chen, 2008) is limited as follows:

- The dominating camera motion is translational movement in a plane, the effects of camera zoom and rotations are not considered.
- Illumination is spatially and temporally uniform.
- If there exist some moving objects in the scene, the objects should not be so large that most macro-blocks in the central zone have a moving content.

The system has first been developed in a MatLAB environment. After the simulation, a camera phone implementation is prototyped (Zhang & Chen, 2008). This mobile platform consists of a power supply, mobile phone, a USB cable connecting the phone with a PC, and DebugMux, test software installed in the PC to operate the mobile phone and test processor load. The main idea of the measurements is to track the CPU cycle consumption for Global Motion Estimation (GME) during the execution of exemplary demo software on a mobile camera phone. Both free-hand video and gaming applications have been used for the demo. After every GME for a frame, the current frame in the camera buffer is stored into another buffer, created at the initialization for GME, serving as previous frame (reference frame) in the next iteration. The load is calculated as CPU speed (@ 208 MHz) minus the idle time.

The loads of GME and demo are measured while running in GraphicsServer_CB_Process and GviIDBG, respectively. For instance, with Search Window Size (SWS) set as 14 and Number of Selected Macro Blocks (NSMBs) as 5, the CPU load for GME is 42.2 MCPS in average, which is 20% of the total CPU consumption (Table 2), while the core processes take 5.11 MCPS and 2%. There is a significant load increase when the NSMB count is changed from 5 to 15, which illustrates that this factor plays a very important role in the CPU consumption for GME.

Now let us go further to find out which part of GME as shown in Figure 4 costs most. Obviously Feature Filtering, Block Matching (DS) and Global Motion Calculation are most load consuming. Suppose the size of a macro block is 16*16 and the number of selected ones is 15. For each frame we count the number of additions as follows. In the Feature Filtering computation, (16*15*2+1) = 7215 additions are needed. Since the number of additions for GM calculation depends on the motion vectors of selected macro blocks, we cannot get an accurate value here. But it must not be larger than what is needed for the Feature Filtering. However, for Diamond Search, we need at least 9*[2*(16*16)-1]+8 = 4607 additions for Large Search and 5*[2*16*16-1]=2555 for Small Search. For each frame, the computation has a number of Large Searches. Moreover, if we count the additions for the various conditional operations in DS into this calculation, the number of total additions is much larger than for the other two modules (i.e. Feature Filtering and Global Motion Calculation).

To verify the above argument, GME is tested with the DS function disabled. Of course, in this situation we do not get a correct functionality of the demo, since without DS all the outgoing GMVs are zero, but we find out how much load DS takes. The test result, where the Average MCPS is 13.2 (6% in total) agrees with the above calculation.

Figure 4. Motion extraction

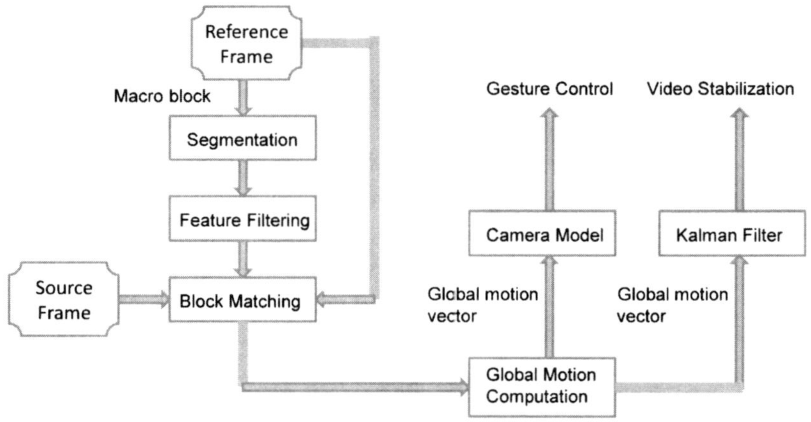

Compared to the DS results listed in Table 2, we conclude now that computation for DS is the main factor for GME load. In other words, the approach of motion estimation should be highly concentrated on block matching techniques for further improvement, which underlines the potential need for hardware acceleration.

In short. CPU usage is crucial for mobile phone applications. Therefore the question is addressed here whether movement detection can be implemented on a mobile platform, well known for the limited execution time and power consumption. It is established that the search pattern is the part that consumes by far most of the time. As we have before selected the DS algorithm for image stabilization, we will look further in that direction.

Tracking the Move

So far, the discussion illustrates that Global Motion Estimation takes a toll. The question therefore arises whether other approaches are viable. We will shortly treat two approaches. The first one is based on an additional accelerator; the second one is based on the use of additional markers.

Movement between two subsequent frames will become apparent when checking corresponding pixels one-by-one. In (Malki & Deepak, 2006), this is performed by pixel subtraction. A typical background object will then disappear, while a foreground object will leave parts that are corresponding to the velocity. In Figure 5, the typical actions using a Cellular Neural Network are depicted. First an averaging template is applied for noise suppression, then the frames are subtracted and the object boundaries are sharpened while removing spurious isolated pixels. Finally a spatial threshold is applied to set a boundary box around the remaining part of the moved object. In (Diaz, 2007), a similar approach is used on the IC3D engine. Both approaches leave a small computational complexity for the host processor, but require that moving objects are wide enough apart that labelling is still feasible.

Another way is by the use of invisible markers (Koch & Zivkovic, 2008). Of rising popularity are Light-Emitting Diodes that flash at a frequency outside the visual spectrum. This is not a necessary requirement for a digital implementation. In a typical system, the image stream is quickly sampled to reveal macro blocks that show an active LED. Macro blocks without LED activity are stored to compose the LED-free image. Separately the macro blocks with LED activity are administrated to find the light-emitting frequency that distinguishes the potentially moving objects. Checking on their location tells whether move-

Figure 5. Template flow diagram in velocity measurement approach

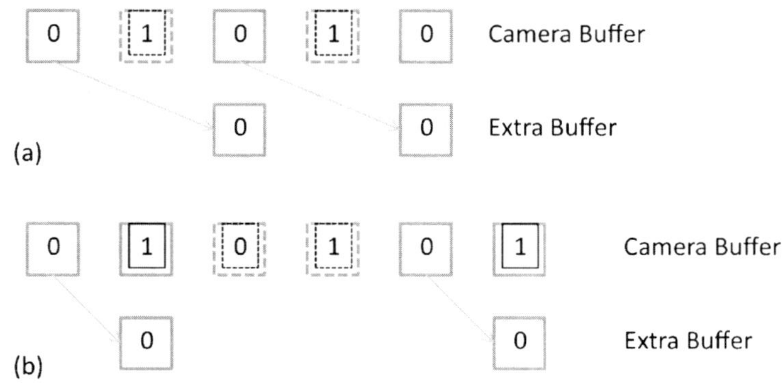

ment has taken place. The beauty of this concept is that it eliminates the implementation problems with H.264 compaction; the drawback is the need to mark the objects in the scene.

Clearly these approaches are interesting but of limited applicability. For the camera phone application, we have looked in a different direction. Although we have got an acceptable real-time performance of GME with NSMB at 5, which is only less than 20% CPU load consumption, the effort for decreasing the load is highly encouraged in that it enables computational resources for running other applications in parallel. This leaves the question whether the implementation of the GME algorithm can be improved further. So far, after every GME for a frame, the current frame in the camera buffer is stored into another buffer. This mechanism is essential for video stabilization, but for other applications such as gaming or GUI control it is redundant. Considering the fact that people's intentional movements for gaming or GUI control are not fast compared with the at least 25 frames available per second, we can skip one frame, or more, in between every two.

This promises to decrease the CPU load for GME by at least half. In (Zhang & Chen, 2008) there are two approaches tested. One is skipping one frame in between every two (Figure 6a), the other is skipping two frames after every two (Figure 6b). The second gets the same GMVs as GME without frame skipping, while the first one gets two times larger GMVs that can be compensated by a simple weight. In order to observe such improvements on GME, it is tested with different NSMBs and Number of Skipped Frame(s) and the result of load consumption agrees with the above expectations. The CPU load is almost halved by skipping 1 frame when NSMBs is set as 15 and SWS as 14, whereas skipping 2 frames does not decreases the load much more, but still helps. In other words, the resolution and sensitivity for global movement are not influenced by frame skipping with the respect of gaming and GUI control.

It seems that DS is the main CPU load consuming process in GME. Since smaller SWS does not help to decrease the load as much as NSMBs but narrows the scope of GMV, SWS at 14 is recommended. Moreover, the macro blocks selected for DS are the ones with the largest activation values in that they contain the most features among others. Considering that we only take 15 macro blocks in the centre of a frame into account for the activation computation and only their NSMBs into motion estimation afterwards, a small value for NSMBs, such as around 5, which is robust enough for DS, is advised to limit the load for GME.

In short. We have experimentally established the existence of an acceptable minimal load that allows suitable global motion estimation in real-time on a platform with scarce resources, such as a mobile camera phone. This leaves the question

Figure 6. Skipping odd (a) and even (b) frames

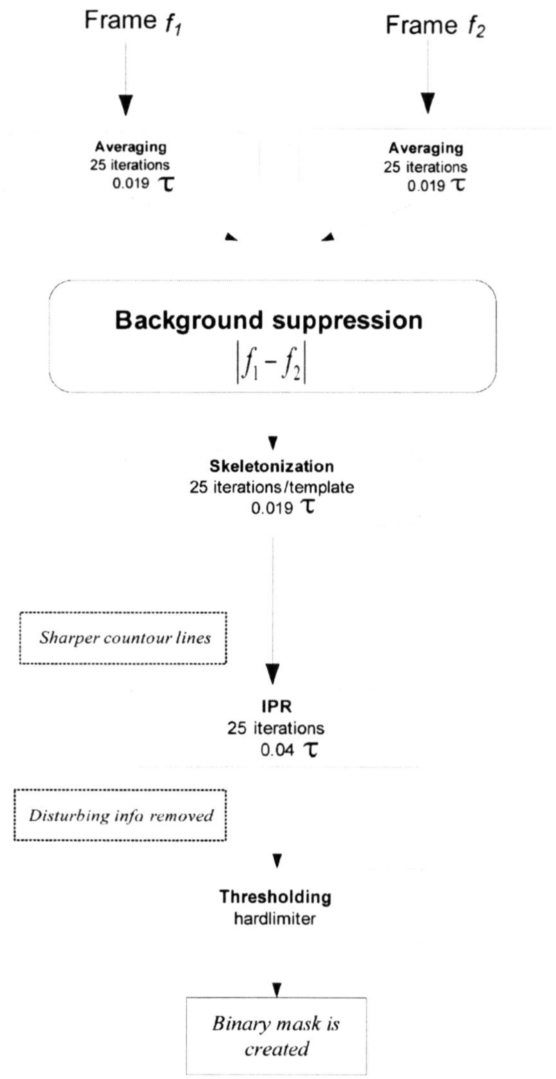

whether the DS algorithm can be implemented so efficiently that camera movements can be followed in real time.

Networked Vision

Though initially movement was extracted to stabilize the image capture, it has later become part of the user interface. In an attempt to simplify the 'look and feel', the iPhone has introduced the touch screen. Navigation on such a screen is supported by shaking the camera. This application requires little accuracy, but still demands high reliability. There are many ghost stories about speech-based navigation (Yoshida, 2009) but errors are not confined to this communication means alone. For instance, a 3-axis accelerometer can react to the non-shaking hand of a person in an elevator. In terms of safety, there seems to be a place for many sensory devices to collectively raise the quality of communication.

The principle objective is to 'extend the hand', which is usually accomplished through the use of gloves (Gershenfeld, 1999). A number of wires are weaved into the glove and connected to rotation and position sensors. These give the information on the hand movements in an arbitrary space (Sarji, 2008). We would like to get away from such special set-ups and devices and simply use the cell-phone. The benefit is that no environmental conditioning is necessary. Our solution starts from the conceptual layering shown in Figure 7. We have to distinguish between copying, signing and waving (Paulson, 2008). These are gestures, but with a different meaning. In copying, the movement should be transferred 1-to-1. In signing and waving, the movement must also be interpreted. For instance, in signing the meaning must be transferred into a language (Westly, 2009).

All these movements can be made small or big, and slow or fast. Even fast movements must allow for the capture of a stable image, as this is the basis for the movement extraction. For small, fast movements the image stabilization problems seem small. A small number of frames may be affected but can easily be bridged. Usually the forward and backward moves are at a different speed. As the forward move carries the essence of the gesture, the recognition is easily realized. For big gestures, life may be harder but they are by nature at a lower speed.

In terms of our communication model, we find that first the gesture needs to be identified by global motion extraction in the observed images (Figure 4). The added problem will be that the images are blurred by the movement. A proper device model

Figure 7. Gesture-based communication

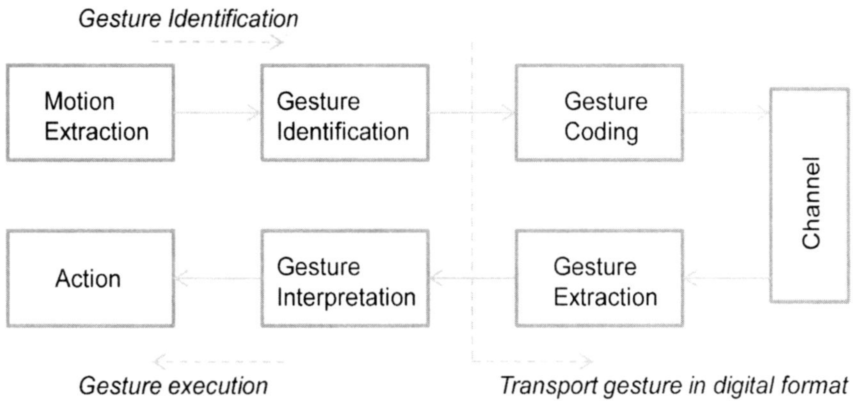

will be needed to separate object motion within the image from intentional and non-intentional device movements. Once motion is identified, a gesture needs to be reconstructed from the movement across an image flow by eliminating superfluous movements, such as returning the hand at the end of a wave.

The communication of gestures over the channel can be in several ways. In a simple command language, the meaning of the gestures can be coded in words that can be artificially spoken or printed. Usually the gestures will be transported in a digital format and this needs not to be confined to signals, though such a special-purpose coding will make the system less maintainable and less fit to work in combination with other telephone brands. After the transport phase through the communication channel, the gesture has to be retrieved from the coded format.

Gestures get their meaning in the application. Therefore the receiving side will again have a model that gives credit to the gesture based on their potential meaning within the application. Though gestures may have been intended at the source side, they get their actual meaning here, at the receiving end. This is therefore where the distinction between copying, signing and waving will be made.

In order to map the gesturing process onto hardware, it is helpful to take guidance in hierarchical layering. A proper skinning of the process from application-oriented global to realization-oriented localized issues allows introducing clear, invariant and certifiable concepts. The underlying idea is a 3-tier architecture, as popularized in many fields of telecommunication with the following meaning:

- The lower (so-called *foundation*) tier expresses the basic technological ingredients. This may introduce facts of common knowledge, abstract fundamental physical parameters to the digital realm, or unify different graphic packages into a single programming interface.
- The middle (so-called *processing*) tier contains the operational functions. It provides the transformation and operations to support a domain of applications in some related technological fields. This may offer a set of classes for the modelling in the envisaged domain and/or a set of algorithms to perform numerical support.
- The final (so-called *application*) tier provides the interface to the application at hand. It personalizes the domain to provide a direct support to the user. The functionality is expressed in terms of the processing

functions. As a consequence, any changes in the application will not induce a major effort as long as they can still be expressed in the available functions.

In the software arena, the 3-tier architecture is directly coupled to concepts of re-use and engineering; for hardware, the tiers image the development phases between concept and product. For intelligent vision systems such as used for gesturing the first layer involves the direct operation on pixels. This will usually involve acceleration where many pixels have to be handled in real-time. The second layer brings features into a model. As there are much less features than pixels, model building can easily be implemented on the typical platform. The third layer uses the model for the desired application. For this, the platform itself can be used, but in the case of a intelligent camera network the overall application will usually run on the camera that serves as momentary network server.

In a typical Wica system, a 8051 controller handles the foundation, while the processing is performed by the IC3D intelligent vision processing chip (Tehrani, 2008). The application layer is available on each camera, but only one of the cameras in the dynamically configured network will function as the cloud server (Ljung & Simmons, 2006). The cameras are coupled over a wireless Zigbee network and communicate observed data instead of images, therefore making an efficient use of the available bandwidth. The Zigbee based operating system allows moving the cloud serving functionality to perform even in the presence of non-functioning cameras.

ACKNOWLEDGMENT

This chapter is based on work performed together with many people. We gratefully acknowledge the contributions of LTH Master students Cheng Wang, Xiao Yang, Erik Ljung, Erik Simmons, Dalong Zhang, Miao Chen, Isael Diaz and Mona Akbarniai Tehrani. Further we thank Rafael Peset Llopis (AXON), Richard Kleihorst (VITO), Peter Meijer (NXP), and Andreas Rossholm (ST-Ericsson) for their fruitful collaboration.

REFERENCES

Adda, O., Cottineau, N., & Kadoura, M. (2003). *A tool for global motion estimation and compensation for video processing (Rpt. project ELEC/COEN 490)*. Concordia University.

Chen, T. C., Chien, S. Y., Huang, Y. W., Tsai, C. H., Chen, C. Y., Chen, T. W., & Chen, L. G. (2006). Analysis and architecture design of an HDTV720p 30 frames/s H.264/AVC encoder. *IEEE Transactions on Circuits and Systems for Video Technology*, *16*(6), 673–688. doi:10.1109/TCSVT.2006.873163

Christensen, C. M. (1997). *The Innovator's Dilemma*. Boston: Harvard Business Press.

Cravotta, R. (2007, 1 September). Recognizing gestures. *EDN Europe*, 22–33.

Diaz Palacios, I. (2007). *A Highly Efficient and Low-Power System for the Detection of Potentially Dangerous Objects*. M.Sc. Thesis, Lund University, Lund, Sweden.

Gershenfeld, N. (1999). *When things start to think*. London: Hodder and Stoughton.

Henning, T. (2008). *State of the Mobile Imaging Industry. Address at 6Sight: The future of imaging*. San Fransisco.

Jacobs, T. R., Chouliaras, V. A., & Mulvaney, D. J. (2006). Thread-Parallel MPEG-2, MPEG-4 and H.264 Video Encoders for SoC Multi-Processor Architectures. *IEEE Transactions on Consumer Electronics*, *52*(1), 269–275. doi:10.1109/TCE.2006.1605057

Kangas, T., Hämäläinen, T. D., & Kuusilinna, K. (2006). Scalable Architecture for SoC Video Encoders. *The Journal of VLSI Signal Processing*, *44*, 79–95. doi:10.1007/s11265-006-5918-x

Knivett, V. (2009, February). MEMS accelerometers: a fast-track to design success? *Electronic Engineering Times Europe*, 32-34.

Koch, M., Zivkovic, Z., Kleihorst, R. P., & Corporaal, H. (2008). Distributed Smart Camera Calibration Using Blinking LED. *Workshop on Advanced Concepts for Intelligent Video Systems* (pp. 242-253). Juan-les-Pins, France.

Ljung, E., & Simmons, E. (2006). *Architecture Development of Personal Healthcare Applications*. M.Sc. Thesis, Lund University, Lund, Sweden.

Malki, S., Deepak, G., Mohanna, V., Ringhofer, M., & Spaanenburg, L. (2006). Velocity Measurement by a Vision Sensor. *IEEE International Conference on Computational Intelligence for Measurement Systems and Applications* (pp. 135-140). La Coruna, Spain.

Or, E. M., & Pundik, O. (2007). Hand Motion and Image Stabilization in Hand-held Devices. *IEEE Transactions on Consumer Electronics*, *53*(4), 1508–1512. doi:10.1109/TCE.2007.4429245

Paulson, L. D. (2008). Software lets a cellphone work like a mouse. [News Briefs]. *IEEE Computer*, *4*(5), 20.

Postma, E., van Dartel, M., & Kortmann, R. (2001). Recognition by fixation. In *Proceedings Belgium/Netherlands Artificial Intelligence Conference, BNAIC* (pp. 425–432). Amsterdam, The Netherlands.

Richardson, I. E. G. (2003). *H.264 and MPEG-4 video compression*. Hoboken, NJ: Wiley. doi:10.1002/0470869615

Rodriguez, A., Gonzalez, A., & Malumbres, M. P. (2006). Hierarchical parallelization of an H.264/AVC video encoder. In Proceedings Parelec (pp. 363-368). Bialystok, Poland.

Rutten, M. J., van Eijndhoven, J. T. J., & Jaspers, E. G. T., vanderWolf, P., Gangwal, O. P., Timmer, A., & Pol, E. J. D. (2002). A Heterogeneous Multiprocessor Architeture for Flexible Media Processing. *IEEE Design & Test of Computers*, *19*(4), 39–50. doi:10.1109/MDT.2002.1018132

Sanchez, V., Nasiopoulos, P., & Abugharbieh, R. (2006). Lossless compression of 4D medical images using H.264/AVC. In *Proceedings ICASSP*, Vol. II (pp. 1116-1119). Toulouse, France.

Sarji, D. K. (2008). HandTalk: Assistive Technology for the Deaf . *IEEE Computer*, *41*(7), 84–86.

Savvas, A. (2008, 17 October). Gartner's top-10 strategic IT technologies for 2009. *Computer Weekly*.

Sibiryakov, A. (2007). Sparse Projections and Motion Estimation in Colour Filter Arrays. In *Proceedings EUSIPCO* (pp. 1814-1818) Poznan, Poland.

Tehrani, M. A. (2008). *Abnormal motion detection and behaviour prediction*. M.Sc. Thesis, Lund University, Lund, Sweden.

Tico, M., & Vehvilainen, M. (2009). Robust Methods of Video Stabilization. In *Proceedings EUSIPCO* (pp. 1819-1822) Poznan, Poland.

Vella, F., Castorina, A., Mancuso, M., & Messina, G. (2002). Digital image stabilization by adaptive block motion vectors filtering. *IEEE Transactions on Consumer Electronics*, *48*(3), 796–801. doi:10.1109/TCE.2002.1037077

Wang, C., & Yang, X. (2007). *H.264 encoding in parallel*. M.Sc. thesis, Lund University, Lund, Sweden.

Westly, E. (2009). Sign Language by Cellphone. *IEEE Spectrum*, (3): 14.

Yoshida, J. (2009, January). Sorry, I didn't mean to change the channel when I sneezed. *Electronic Engineering Times Europe*, 26.

Zhang, D., & Chen, M. (2008). *Global Motion Extraction and Compensation*. M.Sc. Thesis, Lund University, Lund, Sweden.

Chapter 20
Distributed Video Coding for Video Communication on Mobile Devices and Sensors

Peter Lambert
Ghent University, Belgium

Jürgen Slowack
Ghent University, Belgium

Stefaan Mys
Ghent University, Belgium

Rik Van de Walle
Ghent University, Belgium

Jozef Škorupa
Ghent University, Belgium

Christos Grecos
University of the West of Scotland, UK

ABSTRACT

In the context of digital video coding, recent insights have led to a new video coding paradigm called Distributed Video Coding, or DVC, characterized by low-complexity encoding and high-complexity decoding, which is in contrast to traditional video coding schemes. This chapter provides a detailed overview of DVC by explaining the underlying principles and results from information theory and introduces a number of application scenarios. It also discusses the most important practical architectures that are currently available. One of these architectures is analyzed step-by-step to provide further details of the functional building blocks, including an analysis of the coding performance compared to traditional coding schemes. Next to this, it is demonstrated that the computational complexity in a video coding scheme can be shifted dynamically from the encoder to the decoder and vice versa by combining conventional and distributed video coding techniques. Lastly, this chapter discusses some currently important research topics of which it is expected that they can further enhance the performance of DVC, i.e., side information generation, virtual channel noise estimation, and new coding modes.

DOI: 10.4018/978-1-61520-761-9.ch020

INTRODUCTION

In traditional video coding schemes, such as MPEG-2, H.264/AVC, or VC-1, it is the encoder that exploits the statistics of the source signal. As a result, encoding requires significantly more computational resources than decoding, which very well suits traditional application scenarios like broadcasting or video-on-demand, where video is compressed once and decoded many times.

However, emerging applications such as wireless low-power video surveillance, video conferencing with mobile devices, or video communications in sensor networks, require ultra low-complexity encoders, possibly at the expense of a more complex decoder.

Surprisingly, results from information theory established in the 1970s suggest that this should be possible without losing any coding efficiency. In the context of digital video coding, these insights have led to a new video coding paradigm called Distributed Video Coding (DVC), which is based on Distributed Source Coding (DSC), and characterized by low-complexity encoding and high-complexity decoding.

Distributed Source Coding

DSC is a coding paradigm based on two major results from information theory: the Slepian-Wolf theorem and the Wyner-Ziv theorem. Slepian and Wolf (1973) proved that two correlated random sequences generated by repeated independent drawings of a pair of discrete random variables X and Y can be coded as efficiently by two independent coders as by a joint encoder, provided that the resulting bit streams are jointly decoded (Figure 1). In particular, this result states that $R_X+R_Y \geq H(X,Y)$, $R_X+R_Y \geq H(X|Y)$, and $R_Y \geq H(Y|X)$. This means that the sum of the rates of the sources X and Y can indeed achieve the joint entropy, just as for joint encoding (Figure 2).

A special case of DSC is when a decoder makes use of so-called side information. Here, the source sequence X is correlated with some side information Y which is unavailable at the encoder, but available at the decoder (Figure 3). Since conventional encoding techniques can code Y at a rate $R_Y=H(Y)$, the above results indicate that $R_X=H(X|Y)$ is achievable. This case will be the starting point for DVC architectures, as discussed later in this chapter.

Figure 1. Joint source coding (left) vs. distributed source coding (right)

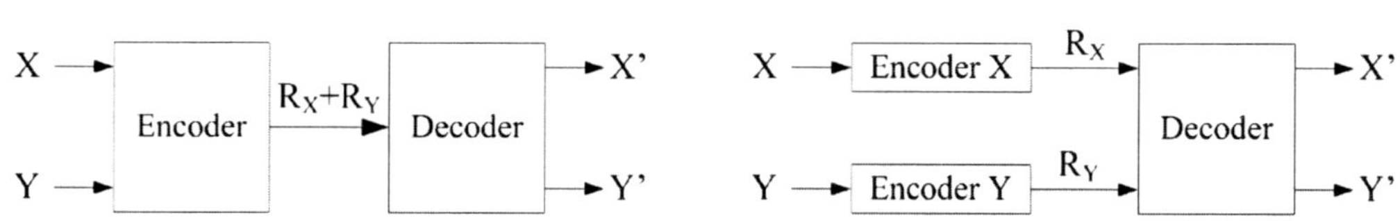

Figure 2. Achievable rate regions for the coding schemes from Figure 1

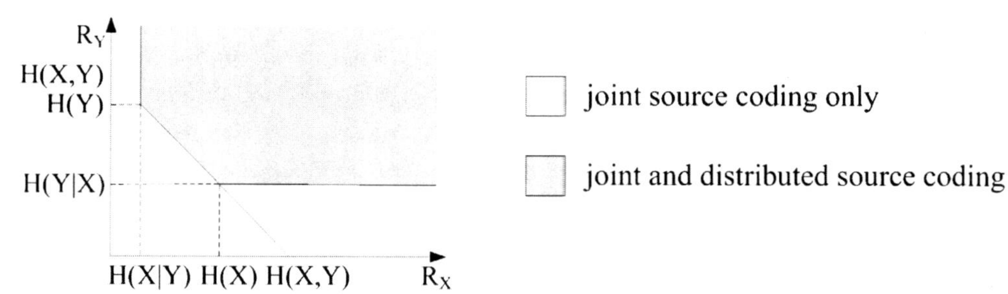

Figure 3. Slepian-Wolf coding with side information at the decoder

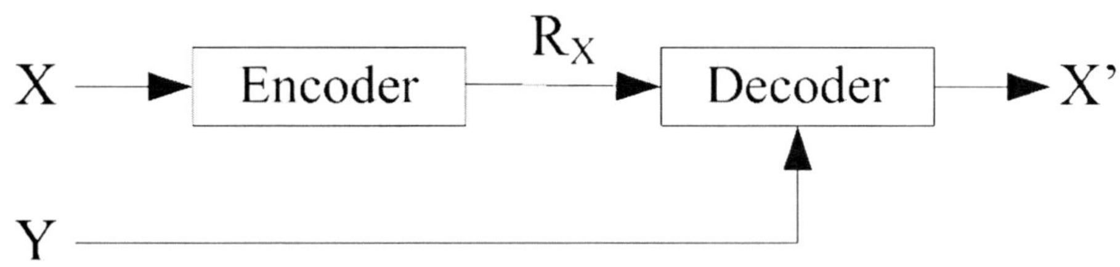

The work of Slepian and Wolf, which involved lossless compression, was extended to lossy compression by Wyner and Ziv (1976). They considered compression with decoder side information. This time, however, a distortion $D=E[d(X,X')]$ between the original signal X and the decoded signal X' is allowed. Let $R^{WZ}_{X|Y}(D)$ be the achievable lower bound for the bit rate given a distortion D, and $R_{X|Y}(D)$ the rate required in case the side information is available at the encoder as well.

Given these notations, Wyner and Ziv (1976) proved that a rate loss $R^{WZ}_{X|Y}(D) - R_{X|Y}(D) \geq 0$ occurs when the encoder does not have access to the side information. More importantly, they also proved that the equality holds in the case of Gaussian memoryless sources and a mean squared error distortion metric d. Later, these results were extended to more general cases, proving that the equality also holds for source sequences X that are the sum of arbitrarily distributed side information Y and independent Gaussian noise N (Pradhan, 2003), and that the rate loss for sources with general statistics and a mean squared error distortion metric d is less than 0.5 bits per sample (Zamir, 1996).

Distributed Video Coding and Its Applications

Distributed video coding tries to exploit the theoretical results from distributed source coding in the context of video compression. In doing so, DVC comes with a number of advantages and accompanying application scenarios. Firstly, using independent encoders and a joint decoder means in practice that the complexity burden is now at the decoder side instead of at the encoder (as in conventional video coding solutions). This in turn could lead to low production cost, low power consumption, and very small encoders. Application scenarios include video capturing with mobile devices or wireless low-power video sensors for surveillance.

Secondly, DVC comes with inherent error robustness. Since there is no prediction loop in the encoder of a distributed video codec, error propagation is not an issue as it is in conventional predictive video codecs like MPEG-2 or H.264/AVC. Furthermore, since DVC is based on error correcting codes (see later), it is straightforward to efficiently extend such a codec with extra protection against transmission errors.

Thirdly, DVC looks promising for the compression of multi-view video sequences. Usually there is a strong correlation between sequences generated by different cameras in a multi-view setup. Using the principles of DVC, it should be possible to efficiently compress these different sequences using separate encoders (i.e., the different cameras/encoders do not need to communicate with each other), provided that the resulting bit streams are decoded using a single joint decoder.

OVERVIEW OF DVC SOLUTIONS

The fundamental theoretical results of Slepian-Wolf and Wyner-Ziv only provide bounds for the rate (and distortion in case of Wyner-Ziv) of a DVC system, but these results do not provide insights in how to build such a system. In this section, we explain how to apply the concepts of DSC to video compression. Starting point is a global point of view and an indication of the main problems that need to be dealt with to build an efficient codec. This knowledge is then used to introduce the two pioneering architectures, i.e., the PRISM system developed at Berkeley by Puri and Ramchandran (2002), and the system proposed by Aaron et al. (2004) at Stanford University. Although many extensions have been proposed by other researchers (including the authors of this chapter), the main architectures have stayed more or less the same.

A General Architecture for DVC

A video sequence consists of a number of frames (temporal axis) and each frame can be divided into non-overlapping blocks, often referred to as macroblocks (spatial axis). If one wants to use distributed source coding principles for video compression of one video sequence, the sequence must be partitioned in a certain way, as to obtain the two correlated sources mentioned in the Slepian-Wolf and Wyner-Ziv theorems (see Figure 1). For example, the sequence can be partitioned into I frames and W frames, using the temporal direction only. The description below assumes this particular partitioning but remains valid for other partitioning strategies. The I frames are coded independently from other frames in the sequence, for example using intra coding techniques available in H.264/AVC. On the other hand, W frames are coded by exploiting correlation between the frames. However, this correlation is exploited at the decoder only, i.e., the encoder codes each frame (W or I) independently from other frames. To achieve this, the decoder generates side information Y using one or more previously decoded frames (I' and/or W').

The side information Y generated at the decoder can be regarded as a noisy version of the original W. After all, the goal of the side information generation module is to estimate the original W as accurately as possible. Therefore, the correlation between W and Y characterizes a virtual channel: it is as if W has been sent to the decoder over a noisy communication channel, so that instead of W, the corrupted version Y is received at the decoder. Hence, for reliable communication, this channel can be protected using channel codes. Y can be considered as the systematic part of this code and since Y is already available at the decoder, only the parity bits need to be sent. Obviously, the amount of error correcting bits that need to be sent depends on the amount of noise on the virtual channel, i.e., it depends on the correlation between W and Y. If Y is a fairly accurate approximation of W, only a small number of parity bits need to be sent and vice versa.

The main problem with the virtual channel is that the virtual noise statistics need to be known by both encoder and decoder: the Wyner-Ziv (WZ) encoder needs to know how many parity bits should be sent to the decoder, and the WZ decoder needs the conditional distribution $P(W|Y)$ for efficient channel decoding (e.g., Viterbi-like decoding). However, the encoder only has access to W while the decoder can only access Y. To solve this problem, two solutions are frequently used in the literature. In the first solution, Y is estimated at the encoder and this information is used to estimate R_W. Information about the conditional distribution $P(W|Y)$ is then sent to the decoder along with the error correcting bits. The disadvantage of this solution is that complexity is added to the encoder which is typically not desired in a DVC context. In the second solution, it is the decoder that estimates $P(W|Y)$ and calculates the number of parity bits that are needed to correct Y reliably. Subsequently, a feedback channel is used to request the amount of parity bits from the encoder. However, the use of a feedback channel is impractical in video storage scenarios, and even in streaming scenarios, the use of a feedback channel should be limited to avoid excessive delays.

Figure 4. General architecture for distributed video coding

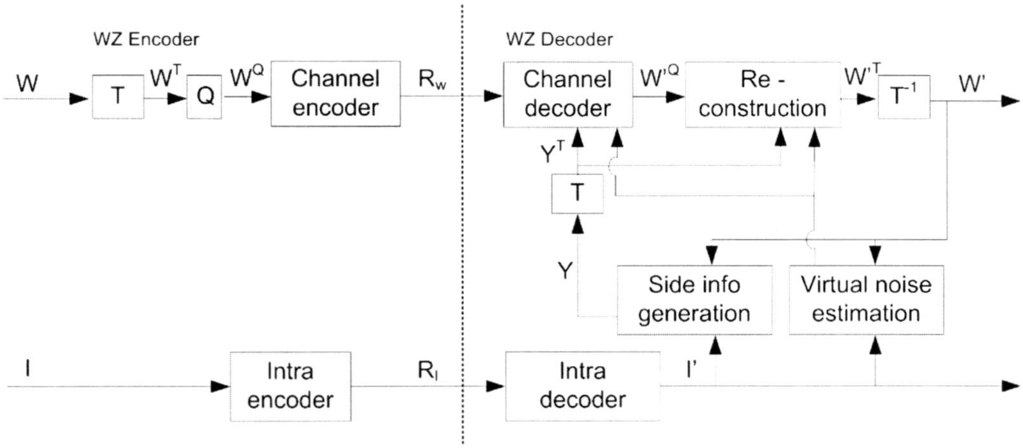

In short, the side information Y plays a crucial role in a DVC system. Estimating W at the decoder using previously decoded data is not straightforward due to complex motion, deformation, occlusion, and so on. In addition, modeling the virtual noise is difficult and results in a trade-off between additional complexity at the encoder and the use of a feedback channel.

To achieve different degrees of compression in a practical system, quantization is performed, usually preceded by a transformation step such as a discrete cosine transform (DCT) or a wavelet transformation. Therefore, a general DVC architecture can be drawn as depicted in Figure 4.

Assume again that the video sequence is partitioned into intra frames I and WZ frames W. As before, I frames are intra coded and sent to the decoder where they can be decoded independently from other frames. At the encoder, each W frame is transformed (T), quantized (Q) and channel-coded. At the decoder, the side information Y is generated and transformed into Y^T. The channel decoder selects the most likely quantization bin for W^Q, denoted as W'^Q, using the side information Y^T and the correlation noise statistics $P(W^T|Y^T)$ provided by the virtual noise estimation module. Next, reconstruction is performed. This step is equivalent to inverse quantization but it also takes Y^T into account: $W'^T=E[W^T|Y^T, W'^Q]$. Finally, the result is inverse transformed.

Although the theorems of Slepian-Wolf and Wyner-Ziv date back to the seventies, it was only quite recently that practical systems have been developed, first by Pradhan and Ramchandran from the University of California at Berkeley, and later on by Aaron and Girod at Stanford University. Many researchers have proposed extensions to both systems in the years that followed, but the basic architecture has stayed more or less the same. In the next subsections we will first discuss the PRISM system proposed by Ramchandran et al., and then we will discuss the Stanford codec.

PRISM

Pradhan and Ramchandran (1999) proposed a technique called DIstributed Source Coding Using Syndromes (DISCUS) and they created a framework for video coding which they called PRISM: Power-efficient, Robust, hIgh-compression, Syndrome-based Multimedia coding (Puri, 2002). PRISM adopts a spatial partitioning approach to obtain the correlated sources of the WZ theorem. At the encoder, each macroblock B is first classified into one of several classes, based on the squared error difference between B and the co-located macroblock in the previous frame. If both macroblocks are strongly correlated, B is not coded and signaled as SKIP. In this case, the decoder can simply take the co-located macroblock in the previous frame without the need for channel coding. On the other hand, if there is very little correlation, B is intra coded because it is likely that the generated side information Y will not be a very accurate prediction.

In all other cases, B is WZ-coded as follows. Each block is first transformed using a DCT transformation and the coefficients are scanned in zig-zag order. The side information Y will be correlated with B for the low frequency coefficients, but the correlation between the higher frequency coefficients is usually low. Therefore, the higher frequency coefficients are intra-coded, i.e., they are quantized and run-length Huffman (entropy) coded. Hence, only the low frequency coefficients are actually WZ coded. In PRISM, syndrome codes are used as channel codes, and both the code and quantization step size are determined by the estimated correlation noise. First, the remaining low frequency coefficients are base quantized and syndrome coded. Next, to achieve the target distortion, the quantization is further refined, and the index of the refinement interval inside the base interval is transmitted to the decoder. In addition, a CRC check is calculated which serves as a signature of the quantized codeword sequence.

At the decoder, all possible (half-pixel accurate) side information blocks are listed (within a given search range). From this set of candidate blocks, the best predictor is selected using the Viterbi algorithm and the received syndrome bits. If this best predictor matches the CRC, syndrome decoding terminates and the block is returned as output. Otherwise, the next best predictor is chosen and so on.

In summary, no conventional motion estimation is performed in PRISM, but instead all candidate side information blocks are listed and tried one by one until the output matches the CRC. Also, since the co-located macroblock in the previous frame is used to perform the initial classification of the macroblock at the encoder, each frame is not coded independently from other frames. Hence, the correlated sources of the Wyner-Ziv theorem are in fact the already classified intra and WZ blocks.

Stanford Codec

Aaron et al. use a frame-based approach and partition the video sequence into I frames (or *key* frames) and W frames. First, they developed a pixel-domain codec (Aaron, 2002) which was later on extended to the transform domain (Aaron, 2004). The overall architecture is the same as the frame-based example described in the general overview above (also see Figure 4).

This architecture is used as a basis for DVC systems by the majority of the research community, including the authors of this chapter. Therefore, this architecture will be used in the next section to explain the functional building blocks of a DVC system in more detail.

Other Systems

Many researchers have proposed new techniques, including techniques to improve the generation of side information by using bidirectional motion refinement and spatial smoothing (Ascenso, 2005), by exploiting both temporal and spatial correlation (Adikari, 2006a; Tagliasacchi, 2006a), and by using multiple side information streams (Adikari, 2006b). Rate-distortion analysis has been provided for motion extrapolation (Li, 2007), motion compensated interpolation (Tagliasacchi, 2007a) and hash-based side information generation (Tagliasacchi, 2007b). The feedback channel has been studied (Pedro, 2007) and practical request stopping criteria have been formulated (Tagliasacchi, 2007c) as well as how to eliminate the feedback channel (Morbée, 2007). Alternative channel codes have been studied such as LDPC codes (Aaron, 2006; Liu, 2006) and overlapped quasi-arithmetic codes (Artigas, 2007a). Also enhanced reconstruction techniques have been proposed (Vatis, 2007).

DVC techniques have also been used in other scenarios, such as traditional video coding with forward error correction using WZ coding (Baccichet, 2006; Bernardini, 2007; Rane, 2004) and multi-view coding (Flierl, 2006; Tosic, 2007; Yang, 2007). An interesting hybrid system has been proposed by Mukherjee (2006), who uses H.264/AVC to code subsampled frames. At the decoder, the decoded subsampled frame is used to generate the side information for the full resolution frame. Subsequently, the frame is decoded using WZ bits.

FUNCTIONAL BLOCKS IN A DVC SYSTEM

To illustrate the internal working of a DVC system, we provide a block-by-block overview of our DVC system, depicted in Figure 4. This system is based on the architecture proposed by Aaron et al. (2004), but it uses different methods for side information generation and virtual noise estimation. In the following we will discuss each building block in some more detail: intra coding, transformation and quantization, bitplane extraction, turbo coding, side information generation, virtual noise estimation, and, finally, reconstruction.

Intra Coding

Frames are intra coded and decoded following the H.264/AVC specification (Wiegand, 2003) for intra-coded pictures. Both coding and decoding can be done completely independently from any other frame. At the decoder, the intra decoded frames are stored in a buffer so they can be used as reference frames later.

Transformation and Quantization

For transformation, the H.264/AVC integer transform (Malvar, 2003) is used as a computationally efficient approximation of the DCT. Each transform coefficient is uniformly quantized in the same way, using one of six available quantization patterns, resulting in quantized coefficients consisting of 8-3 bits.

In particular, in H.264/AVC, the DCT is approximated as a forward transformation T_1 followed by a scaling part S_1 (Figure 5). Analogously, the IDCT is approximated by a backward scaling S_2^{-1} and backward transformation T_2^{-1} The scaling steps S_1 and S_2^{-1} are integrated within the quantization, resulting in Q_1 and Q_2^{-1}, respectively. An important consequence is that, due to this combination of transformation and quantization, the backward transformation T_2^{-1} and the backward quantization Q_2^{-1} are each individually no longer the exact inverse operations of the forward transformation T_1 and quantization Q_1. In the next paragraph, we will explain how to take this into account when implementing H.264/AVC transformation and quantization in the DVC codec.

In the DVC codec, at the encoder, W frames are transformed using T_1 and quantized using Q_1. At the decoder, Y is transformed using T_1 before being used as side info by the turbo decoder. The turbo decoder decodes the frame and returns for each transform coefficient w_k^T the quantization bin q'_k in which it is contained. Using these decoded quantization bins and the transformed side info Y^T, the reconstruction acts as an inverse quantizer and calculates the most likely value w'^T_k for each bit. However, since reconstruction really acts as an inverse quantizer for Q_1, and since S_2^{-1} is not actually the inverse operation of S_1, the decoded values are not properly scaled and need to be rescaled by both S_1 and S_2^{-1} before executing the backward integer transform T_2^{-1}. Naturally, the successive scaling by S_1 and S_2^{-1} can be combined

Figure 5. The H.264/AVC transformation and quantization scheme

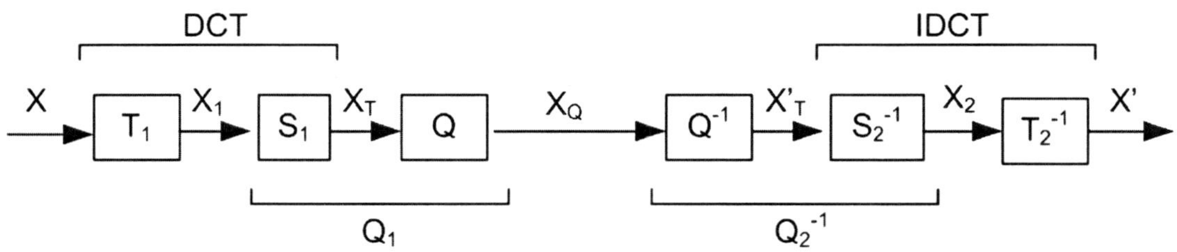

so that in practice only one scaling operation S needs to be performed.

Bitplane Extraction

Bitplane extraction is performed after quantization. First, we shift the AC coefficients so that all quantized values are positive. Next, coefficients are grouped in so-called coefficient bands. For example, all DC coefficients are grouped into one coefficient band. Then, for each band, the bits at corresponding positions are grouped into bitplanes. For example, the most significant bit of the DC coefficients form one bitplane. After this step, each of the bitplanes is fed as a binary input word to the turbo coder, which will encode it.

Bitplane coding comes with two advantages. Firstly, the codewords that are inputted into the turbo encoder are fixed length. More precisely, the length of the codewords corresponds with the number of 4x4 (transformation) blocks per frame. Secondly, decoding can start with the most significant bitplanes and less significant bitplanes could be skipped.

Turbo Coding

Turbo codes are used as channel codes in the described system. In short, turbo coding consists of two fundamental ideas: a code design that produces a code with randomlike properties, and a decoder design that makes use of soft-output values and iterative decoding. In the following, first a brief description of the turbo encoder is given, followed by a short discussion of the turbo decoder. After that, a discussion of the puncturing process follows. For more detailed information on turbo coding, we refer the interested reader to Lin and Costello (2004).

Turbo Encoding

A turbo encoder (Figure 6) consists of two (in most cases identical) systematic convolutional encoders. More precisely, systematic feedback encoders are used. This means that there is a closed loop in the encoder's memory circuit, because of which the parity bits depend not only on a finite number (equal to the number of memory blocks) of previous input bits, but on all previous input bits. Each such systematic feedback encoder normally produces two output bits per input bit: a systematic bit, which equals the input bit, and a parity bit. In DVC however, the systematic bits are discarded, and thus the DVC turbo encoder generates two output (parity) bits $p_k^{(1)}$ and $p_k^{(2)}$ per input bit w_k^Q.

An interleaver π is put in front of the second encoder. This pseudo-random interleaver (which needs to be known at the decoder also) permutes the input sequence before feeding it to the second encoder, ensuring that both encoders generate different parity bits. This introduces some kind of randomness to the code, which highly attributes to the high performance.

Convolutional encoders require very little computational and hardware resources. The presence of the interleaver in the turbo encoder requires

Figure 6. A turbo encoder consists of two (identical) convolutional coders and an interleaver. Turbo decoding is an iterative process involving two soft-input, soft-output (SISO) convolutional decoders passing extrinsic information to each other

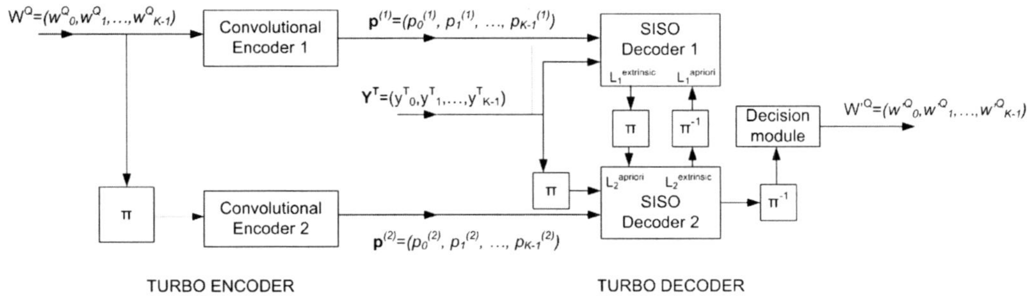

the whole input sequence to be stored in memory, which can be quite long, but this is still nominal compared to, for instance, YUV frames that need to be stored in memory. Thus, the turbo encoder can be implemented very efficiently with small hardware requirements, which perfectly fits the low encoder complexity paradigm of DVC.

Turbo Decoding

At the turbo decoder (Figure 6), optimal decoding of the concatenated code would computationally be too complex. Therefore a suboptimal iterative process is applied involving two soft-input, soft-output (SISO) convolutional decoders passing information to each other. Each SISO decoder calculates the Logarithm of the A Posteriori Probability (LAPP) ratio for each encoded bit w_k^Q. From the LAPP ratios, extrinsic information (i.e., information calculated from data not accessible by the other SISO decoder, namely the parity bits $\mathbf{p}^{(j)}$ corresponding to the current SISO decoder) is extracted and passed as a priori information to the other SISO decoder. Once a given convergence criterion is satisfied, the iterative decoding process stops. The LAPP ratios calculated in the last iteration are fed into the decision module, where a final decision for each bit is made.

So, each SISO decoder has to calculate the LAPP ratio $L(w_k^Q) = ln\left(P(w_k^Q = 1|\mathrm{r}) \,/\, P(w_k^Q = 0|\mathrm{r})\right)$, where **r** represents all information known at the SISO decoder, i.e., both the transformed side information Y^T and the corresponding parity bits $\mathbf{p}^{(1)}$ or $\mathbf{p}^{(2)}$. As can be seen, a positive value of the LAPP ratio denotes a higher probability that the input bit was 1, whilst a negative LAPP ratio indicates an input bit 0.

Based on the structure of the trellis diagram of the code (Figure 7), the a posteriori probabilities $P(w_k^Q|\mathrm{r})$ can be rewritten (Lin and Costello (2004)) so that

$$L(w_k^Q) = ln\left(\frac{\sum_{S^1} p(s_{k-1} = s', s_k = s, \mathrm{r})}{\sum_{S^0} p(s_{k-1} = s', s_k = s, \mathrm{r})}\right)$$

where S^1 (resp. S^0) represents all edges in the trellis corresponding to a state transition $S_{k-1} \rightarrow s_k$ considering an input $w_k^Q = 1$ (resp. $w_k^Q = 0$).

It has been shown (Lin and Costello (2004)) that the probability $p(s_{k-1}=s',s_k=s,\mathbf{r})$ can be written as $\alpha_{k-1}(s')\cdot\gamma_k(s',s)\cdot\beta_k(s)$ with α, β and γ formally defined as $\alpha_{k-1}(s') \triangleq p(s_{k-1} = s', \mathrm{r}_1^{k-1})$, $\beta_k(s) \triangleq p(\mathrm{r}_{k+1}^K | s_k = s)$ and $\gamma_k(s', s) \triangleq p(s_k = s, \mathrm{r}_k | s_{k-1} = s')$.

Intuitively, $\alpha_{k-1}(s')$ is the probability of getting from the start of the trellis to state s' at time k-1. It is often called the forward metric, since it can be computed recursively using γ_k as

Figure 7. Trellis diagram for a simple systematic convolutional code. On the vertical axis, the encoder states are given; on the horizontal axis, the time steps k. A dotted line corresponds with an input bit 0, a full line with an input bit 1. The systematic output bits are identical to the input bits; the corresponding parity output bits are written as labels on the edges

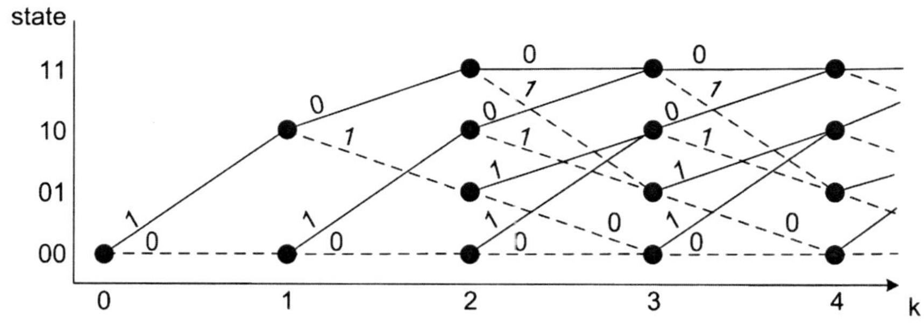

follows: $\alpha_k(s) = \sum_{s' \in \sigma_{k-1}} \alpha_{k-1}(s')\gamma_k(s',s)$. Analogously, $\beta_k(s)$ is the probability of getting from state s at time k to the end of the trellis. It is called the backward metric and can be written as $\beta_{k-1}(s') = \sum_{s \in \sigma_k} \beta_k(s)\gamma_k(s',s)$.

$\gamma_k(s',s)$, often called the branch metric, is the probability of making the state transition from state s' at time k-1 to state s at time k, all given the received values **r**. It is calculated as a function of the a priori probability $L_j^{apriori}$ and the probability $p(\mathbf{r}|s_{k-1}=s', s_k=s)$ which is the conditional probability that $r_k = (y_k^T, p_k^{(j)})$ is received when the state transition $s' \rightarrow s$ has occurred. This probability is calculated at the decoder based on the estimation of the virtual noise.

Puncturing and Feedback Channel

The turbo code as described above does not compress the input sequence. On the contrary, two parity bits are generated for each input bit. However, not all parity bits are needed at the decoder to correct all errors in the side information. Therefore, we need a way to adapt the amount of parity bits that are transmitted to and used at the decoder. This is done by a procedure called puncturing.

In short, at the encoder, instead of directly transmitting all parity bits to the decoder, they are saved in a buffer. Then, the decoder asks for a certain amount of parity bits from the buffer through a feedback channel. If the amount of bits is not big enough to correct the side information, the decoder asks for more bits from the buffer. This way the decoder estimates the smallest number of bits (i.e., minimal rate) required to transmit the current bitplane.

Several online techniques to decide whether or not the decoding was successful have been described in the literature (Tagliasacchi, 2007c; Kubasov, 2007a; Artigas, 2007b). Recently, the authors of this chapter (Škorupa, 2009) showed that better results can be achieved by using the sign-difference ratio as a stopping criterion. This criterion counts the number of sign differences between the a priori and the extrinsic value at the SISO decoder. If that number reaches zero, decoding is considered successful. Otherwise, more turbo decoding iterations are performed. If a predefined maximum number of turbo decoding iterations are performed without triggering the stopping criterion, decoding is considered unsuccessful and more parity bits are requested.

Furthermore, to save the number of requests through the feedback channel, the initial rate is estimated from the error probabilities estimated during virtual noise estimation (see later).

For optimal performance it is desirable that the transmitted parity bits are spread over the whole parity sequences $\mathbf{p}^{(1)}$ and $\mathbf{p}^{(2)}$. At the same time, strong structure in the puncturing pattern seems to degrade the performance of the code. To address these rather contradictory needs the following puncturing procedure is employed. The parity sequences are divided into blocks with a fixed length. When the decoder requests a certain amount of parity bits, the number of bits to be transmitted from each block is calculated, and these bits are chosen in a pseudo-random fashion in each block.

Side Information Generation

The side information generation module is responsible for creating a prediction of the Wyner-Ziv frame to be decoded based on already decoded I or W frames. Since the frame to be predicted is not available at the decoder, traditional motion compensated prediction (using block matching) is not possible. Therefore, interpolation or extrapolation techniques are used. So far, best results are achieved using a technique called motion compensated interpolation (see below). Extrapolation techniques were also proposed in the literature (Borchert, 2007), and are especially useful when long GOP sizes are used.

Motion compensated interpolation is a motion estimation technique that is based only on information contained in the reference frames, and does not use any information from the frame that is being predicted. Therefore, it can be applied in the DVC decoder, where the frame being predicted is not available. Originally, the technique was proposed by Ascenso et al. (2005). Later on, several extensions and improvements were added. Of these improvements, the ones proposed by Ascenso et al. (2006) and Klomp et al. (2006) are incorporated in the described codec. The technique consists of five steps, which are each briefly discussed below.

Mean Filtering of Reference Frames

First of all, both past and future reference frames are filtered using a mean filter with a 5x5 square kernel. This step will help the motion estimation to choose motion vectors that better match the true motion field instead of simply choose the best motion vectors in the rate-distortion sense. This is important because the motion vectors will be interpolated later on.

Forward Motion Estimation Between Reference Frames

After mean filtering the reference frames, a traditional block matching algorithm is used to estimate the motion between future and past reference frames. A modified version of the Mean Absolute Difference (MAD), given below, is used as cost function. In this cost function a penalizing term is introduced which adds an increasing cost when motion vectors go to more extreme positions in the search range.

$$cost(d_x, d_y) = (1 + 0.05\sqrt{d_x^2 + d_y^2}) \cdot MAD(d_x, d_y)$$

The forward motion estimation is conducted using full pel precision. For each of the resulting motion vectors(d_x,d_y), the intersection point with the frame to be predicted is calculated based on the distance from the frame to be predicted to both reference frames.

Obtaining Bidirectional Motion Vectors for the Frame to be Predicted

Next, for each motion block in the frame to be predicted, the motion vector with its intersection point the closest to the middle of the block is selected from the obtained list of motion vectors between the reference frames. The selected motion vector is split up into a forward and a backward motion vector, taking into account the distance

Figure 8. The refinement search range for a motion vector is derived based on the motion vectors of its neighboring motion blocks (Ascenso, 2006)

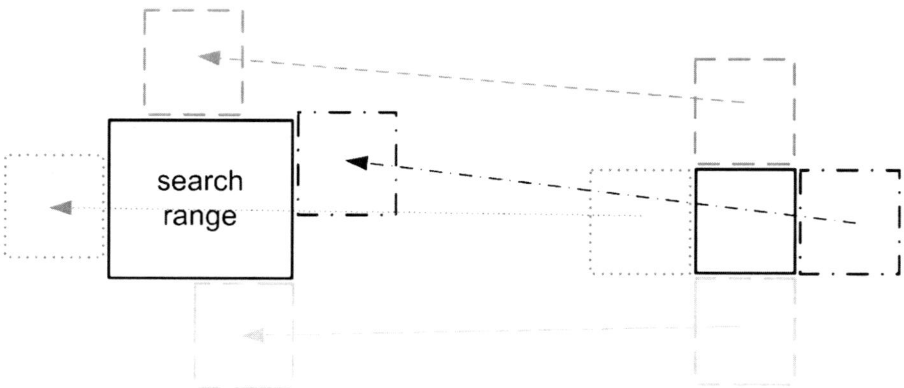

from the frame to be predicted to the past and future reference frames.

From this stage on, motion vectors with half pel precision are used. Therefore, both past and future reference frames are interpolated using the six tap Wiener filter defined in the H.264/AVC specification (Wiegand, 2003), with filter coefficients (1,-5,20,20,-5,1)/32.

Bidirectional Motion Refinement

In this stage, the obtained bidirectional motion vector is further refined assuming a linear trajectory between the future and past reference frames. The refinement vectors for the forward and backward motion vectors are symmetric, and the optimal refinement vector is determined by minimizing the MAD between the past and future reference blocks. The refinement search range is derived based on the motion vectors of its neighboring motion blocks as shown in Figure 8.

A hierarchical coarse-to-fine approach is used: a first iteration uses large block sizes (16x16) to track fast motion reliably. In a second iteration higher precision is achieved using smaller block sizes (8x8).

Spatial Smoothing of the Motion Vectors

The spatial coherence of the motion vectors is then improved using a weighted median filter. For each macroblock MB_i, its motion vector MV_i is replaced with the weighted median of MV_i itself and the motion vectors of the neighboring macroblocks (i.e. up to eight motion vectors MV_k). The weight for each motion vector MV_k is defined as the mean squared error between the past and future reference blocks when MV_k is applied to the macroblock MB_i.

Motion-Compensated Interpolation

Finally, the prediction for each pixel is obtained by applying conventional bidirectional motion compensation using the refined and spatially smooth motion vectors.

Virtual Noise Estimation

The decoder generates the side-information frame Y as a prediction of the original frame X. The knowledge of the correlation between X and Y

has a great impact on the coding performance. Since the difference between X and Y is called virtual noise, the estimation of the correlation properties is often referred to as a *virtual noise estimation*. Because we use a feedback channel, we do not need to estimate the correlation at the encoder. At the decoder, an accurate correlation model is needed for optimal reconstruction and for efficient turbo decoding.

Brites and Pereira (2008) proposed a method to estimate the correlation statistics from the information available at the decoder. In short, the online noise estimation can be described as follows. At the decoder, the side information is generated as an interpolation from two motion compensated reference frames (see before). If the difference between these two frames is small, it is assumed that the motion estimation was accurate and that the generated side information is likely to be very similar to the original signal. On the other hand, if the difference between motion-compensated reference frames is big, it is assumed that the motion estimation failed and the generated side information is not an accurate prediction of the original signal. The authors of this chapter have proposed some refinements to this method (Škorupa, 2008a), and recently also proposed to take quantization noise into account when estimating the virtual noise (Slowack, 2009).

The online noise estimation models the correlation between the corresponding transformed coefficients (or between the corresponding samples in pixel domain) from the side information and the original frame. This correlation is used in the reconstruction function as is. To use it in the turbo decoding, bit-error probabilities for each bit in each particular bitplane can easily be inferred from it.

Reconstruction

The turbo decoder outputs the quantization bins in which each coefficient is located. Given these quantization bins and the side information Y^T, the reconstruction step selects the most likely value for the original coefficient w_k.

Reconstruction is the equivalent of inverse quantization in a traditional video codec. However, in DVC we have calculated side information, which is a non-quantized approximation of the original frame. We can use this extra information to achieve a more accurate reconstruction compared to conventional inverse quantization.

Denote the lower bound of the decoded quantization bin L and its higher bound H. A basic approach for reconstruction, as used in the Stanford codec, could be the following:

$$w'_k = \begin{cases} L & \text{if } y_k \le L \\ y_k & \text{if } L < y_k < H \\ H & \text{if } H \le y_k \end{cases}$$

However, this approach is suboptimal. For an optimal reconstruction (Kubasov, 2007b), w_k should be reconstructed as the expectation $E[W_k|q'_k,y_k]$ of the random variable W given the quantization interval q'_k and the value of the side information y_k.

Since the parameters of the conditional distribution $P(W_k|Y_k{=}y_k)$ have been calculated or estimated to model the virtual noise, $E[W_k|q'_k,y_k]$ is the center of the mass of $P(W_k|Y_k{=}y_k)$ inside the bin q'_k. Hence:

$$w'_k = \frac{\int_L^H w \cdot P(W_k = w \mid Y_k = y_k)dw}{\int_L^H P(W_k = w \mid Y_k = y_k)dw}$$

FLEXIBLE DISTRIBUTION OF COMPUTATIONAL COMPLEXITY THROUGH HYBRID PREDICTIVE-DISTRIBUTED VIDEO CODING

One of the challenges of video streaming scenarios is coping with the bandwidth requirements of the networks and the heterogeneity of the devices. Devices with different characteristics are performing video compression and streaming, ranging from high-end servers to PDAs and mobile phones. In addition, since many systems are multi-tasking systems, available computational complexity is often non static. Furthermore, some devices are battery-constrained, and if computational complexity is decreased, techniques such as dynamic voltage scaling can be used to extend battery lifetime (He, 2005).

As such, besides rate and distortion, available computational complexity is considered an important parameter in a video coding system. This prompted methods for rate-distortion-complexity (RDC) optimization, which have been developed primarily for predictive video coding with encoder-side motion estimation (Kaminsky, 2007; Ates, 2006; Stottrup-Andersen, 2004). However, only encoder complexity is considered, and there is no way for the decoder to take over some of the workload. For DVC systems, some complexity analysis has been provided, e.g. by Liu et al. (2007). However, reducing decoder complexity is not considered.

We can summarize by stating that current solutions for RDC optimization do not allow motion estimation to be shifted between encoder and decoder, but only allow encoder complexity to be decreased. Dynamic aspects such as multi-tasking, variable power-supply and session mobility need systems that adapt better to the changing conditions, by distributing complexity between encoder and decoder according to the amount of resources available at both devices. At one time instance, the encoder device could have more resources available than the decoder device but at a later point in time this could be the other way around.

Therefore, we let the encoder and decoder share the complex task of motion estimation, by developing several modes for coding inter frames. While the basic video coding architecture developed in the previous section remains the same, we introduce a mode-dependent part responsible for motion estimation. In addition, we extend the codec by coding the residual between the original frame W and a prediction Z generated at both encoder and decoder, instead of coding the original frame W. Such a residual approach has proved to be advantageous also for DVC (Aaron, 2006).

The mode-dependent part features several modes of operation for coding inter frames: the predictive mode with all motion estimation at the encoder side, the DVC mode with all motion estimation at the decoder side, and the hybrid modes which share motion estimation between encoder and decoder. Two variants for implementing the hybrid modes are identified. A first approach is to apply a spatial partitioning technique and predict some of the macroblocks in a frame at the encoder while estimating the remaining macroblocks at the decoder. In the second approach, the motion search algorithm is split in two parts, i.e., the encoder calculates coarse motion vectors which are further refined by the decoder.

We provide complexity analysis for the mode-dependent part, by counting the number of pixel read and write operations. These theoretical results will be matched later on by practical measurements, illustrating how complexity is distributed between encoder and decoder in each mode. Finally, since we define the modes on a frame-by-frame basis (i.e., the coding mode for coding a particular frame does not depend on previous coding modes), we can combine different modes per set of frames, and develop a strategy for achieving fine-grain sharing of motion estimation complexity between encoder and decoder.

Description of the Codec

At the encoder, the frame sequence is partitioned into key frames I and Wyner-Ziv (WZ) frames W, as shown in Figure 9. I frames are coded using H.264/AVC intra coding. Decoded intra frames I' are available anyway after intra coding due to mode decision and rate-distortion optimization, hence, I' frames are stored in the *decoded I frame*

Figure 9. Codec architecture consisting of a WZ codec, an intra codec, a mode-dependent part and a mechanism for buffering frames

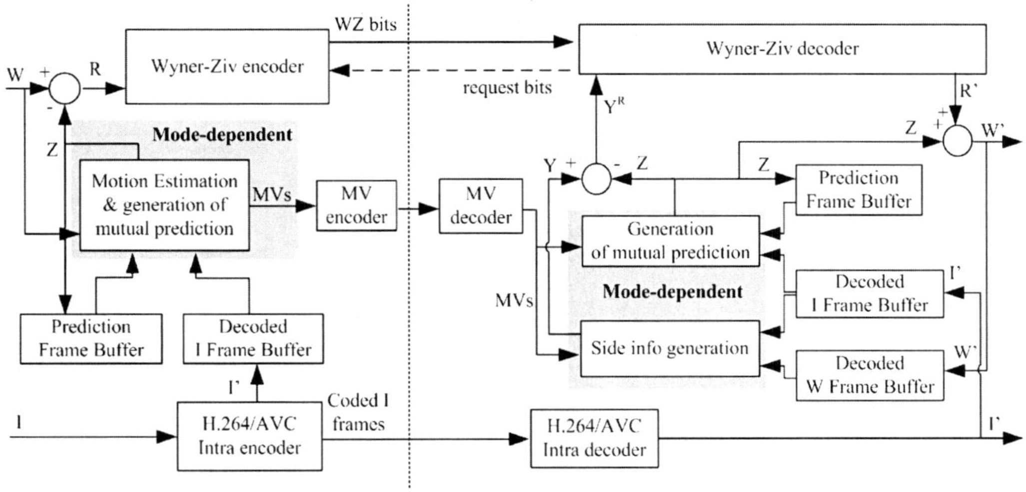

buffer. For each W frame, a certain prediction Z is generated (that will also be generated at the decoder side). The residual R between W and Z is calculated and Z is stored in the *prediction frame buffer*. R is WZ coded using the WZ coder described earlier.

At the decoder, key frames are decoded into I' and stored in the *decoded I frame buffer*. For each W frame, side information Y is generated as well as the mutual prediction Z. The residual Y^R between Y and Z is transformed and used as side information for the WZ decoder. Z is added to the result R' to obtain the decoded frame W'. For future reference, W' is stored in the *decoded W frame buffer*.

Mode-Dependent Part in the Predictive Video Coding Mode

In predictive video coding mode, motion estimation is performed solely at the encoder. Each inter frame W is partitioned into macroblocks of size 8-by-8, and motion vectors are calculated for each block using bidirectional motion estimation. More precisely, the *prediction frame buffer* and *decoded I frame buffer* are consulted and the closest past frame P and future frame F are retrieved. For each macroblock MB_i in W, the best match of full pixel precision in P is determined using a search window of size S, where the size of the search window indicates the number of macroblocks in the search space. Only the luma component is used in motion estimation. The best match is found by minimizing a lagrangian cost function:

$$Cost_i(\vec{v}) = D_i(\vec{v}) + \lambda R_i(\vec{v}),$$

where the distortion metric $D_i(\vec{v})$ is the Sum of Squared Errors (SSE) between MB_i and the macroblock in P defined by the motion vector $\vec{v}$. λ is the Langrange multiplier and $R_i(\vec{v})$ represents the rate to code $\vec{v}$. Motion vectors are coded by first predicting them from their neighbors as in H.264/AVC, and coding the residual between $\vec{v}$ and its prediction using signed exponential Golomb coding, resulting in $R_i(\vec{v})$ bits.

After motion estimation between W and P, for each block in W, the macroblock in P with the lowest cost is interpolated with every candidate block in F, and a similar cost function is used for determining the best match. This results in two motion vectors for each macroblock in W. Since

Table 1. Pixel read and write operations executed at encoder and decoder, per W frame

Mode	Encoder-side			Decoder-side		
Pred.	ME (W, P) ME (W, F) Construct Z	2SHV 3SHV 9HV/2	220.8 331.2 0.5	Construct Z	9HV/2	0.5
	Total	HV (5S + 9/2)	552.4	Total	9HV/2	0.5
DVC				LP filtering ME (F and P) Wiener interpolation Refinement (16x16) Refinement (8x8) Spatial smoothing Bidir. interpol.	2HV(L+1) 2SHV 144HV S^{DVC}_R * 512HV/M S^{DVC}_R * 128HV/M 64HV 18HV	2.0 220.8 14.6 7.3 1.8 6.5 1.8
	Total	-	-	Total	HV(2L+2S+ $10S^{DVC}_R$+231)	261.9
Spat.	ME for S_1 Construct Z	HV(5S + 9/2)/2 9HV/2	276.2 0.5	LP filtering Wiener interpolation Refinement (16x16) Refinement (8x8) Bidir. interpol. Construct Z	2HV(L+1) 144HV S^{SPAT}_R * 256HV/M S^{SPAT}_R * 64HV/M 18HV 9HV/2	2.0 14.6 10.1 2.5 1.8 0.5
	Total	HV(2.5S + 6.75)	276.7	Total	HV(2L+$5S^{SPAT}_R$+ 171.5)	31.6
Subs.	Subs. P and F ME Construct Z	5HV/4 HV/4*(5S/2+9/2) 9HV/2	0.1 34.6 0.5	LP filtering Wiener interpolation Refinement (16x16) Refinement (8x8) Bidir. interpol. Construct Z	2HV(L+1) 144HV S^{SUB1}_R * 512HV/M S^{SUB2}_R * 128HV/M 18HV 9HV/2	2.0 14.6 20.3 1.8 1.8 0.5
	Total	HV(110 + 5S)/16	35.2	Total	HV(2L+$8S^{SUB1}_R$+$2S^{SUB2}_R$+171.5)	261.9

the motion vectors calculated at the encoder will also be available at the decoder, they are used to generate the mutual prediction Z through bidirectional motion compensation, performed on all color components.

At the decoder, Z is constructed using the received motion vectors and reference frames P and F (retrieved from the *prediction frame buffer* and/or *decoded I frame buffer*). No motion estimation is performed by the decoder, therefore, the side information Y is taken equal to Z. As such, the residual Y^R between Y and Z that is used by the turbo decoder contains only zeros.

We will now analyze encoder and decoder computational complexity in terms of the number of pixel read and write operations. An overview of this complexity analysis is provided in Table 1. At the encoder, unidirectional motion search between W and P using a search window of size S requires S comparisons between a certain pixel in W and a certain pixel in P. Frames are represented in the YUV color space, with H(orizontal) by V(ertical) pixels for the Y (i.e. luma) component and H/2 by V/2 pixels for each chroma component U and V. Hence, since motion estimation is only performed on the luma component, 2SHV pixel read operations are performed. Motion estimation between W and F reads one pixel from P, interpolates it with a pixel in F, and compares it with a pixel in W. This requires one additional pixel read operation per pixel, i.e., a total of 3SHV pixel read operations. Constructing Z means interpolating best matches from P and F, requiring two pixel read operations and one pixel write operation, per pixel, for all components. Hence, 9/2HV pixel read/write operations are executed.

At the decoder, the motion vectors are used for constructing Z only, requiring 9/2HV operations.

Mode-Dependent Part in the DVC Mode

In the DVC mode, no motion estimation is performed by the encoder. The mutual prediction Z equals the closest frame, past or future, that can be retrieved from the prediction frame buffer and decoded I frame buffer.

At the decoder, side information Y is generated for each inter frame W using the closest past frame P and closest future frame F, retrieved from the *decoded W frame buffer* and *decoded I frame buffer* only, since decoded frames have better quality than mutual prediction frames. Y is generated using the techniques described earlier in the detailed overview of the DVC architecture. Firstly, for better capturing the true motion field, the luma component of P and F is LP filtered by replacing each pixel by the average of a group of L = 3-by-3 pixels having this pixel as a center. This requires a total of 2HV(L+1) operations. Note that from here on, we will only provide complexity analysis for steps that are not similar to techniques analyzed earlier in this work. We refer to Table 1 for a complete overview.

Next, unidirectional block-based motion estimation is performed between the filtered versions of P and F with $\Delta_{P,F}$ pixel precision using an extended search window of size $\Delta_{P,F}$ S. $\Delta_{P,F}$ denotes the distance between P and F, with $\Delta_{P,F} = 1$ if the frames are adjacent to each other.

After obtaining the motion vectors from F to P, for each macroblock in Y the motion vector intersecting the block closest to the block center is chosen and treated as a bidirectional motion vector, i.e., the intersection point splits the motion vector into a forward vector from W to F and a backward vector W to P.

Next, both the full quality reference frames P and F, and the LP-filtered versions are upsampled to half pixel resolution using a 6-tap Wiener interpolation filter. The number of operations executed by this Wiener filter can be calculated as follows. For each frame, 3/2HV pixels from the low resolution frame need to be copied to the high resolution frame, requiring 9/2HV operations. The remaining 9/2HV pixels are calculated using the 6-tap Wiener filter. Hence, to filter the two original frames and two LP-filtered versions, a total of 144HV operations are required.

Using the half pixel LP-filtered reference frames, the motion vector of each block in Y is refined in two passes: first using reference blocks of size 16x16 and next using reference blocks of size 8x8. At all times, we assume that the motion vector is linear between P and F, and that it goes through the block center. The refinement window S^{DVC}_R for a certain block is defined by the motion vectors of the neighboring blocks, as seen before. Due to this fact, S^{DVC}_R is not constant in size. S^{DVC}_R will be small (on average) if the motion vector field is smooth. On the other hand, S^{DVC}_R will be large if there are a lot of discontinuities in the motion vector field. We obtained a value for S^{DVC}_R to use in our complexity analysis experimentally, using the setup described further in the results section. By averaging over all sequences and rate points, a value of $S^{DVC}_R = 16$ has been obtained. Hence, the complexity of the refinement process is given by S^{DVC}_R * 512HV/M operations in the first pass and S^{DVC}_R * 128HV/M operations in the second pass.

After motion refinement, the motion vectors are spatially smoothed by weighted vector median filtering of the motion vector for MB_i and the motion vectors of neighboring macroblocks applied to MB_i. Concerning computational complexity, for each of the (approximately) eight neighboring motion vectors two macroblocks need to be read, requiring a total of 16HV pixel read/write operations for this step.

Finally, the calculated motion vectors are used for bidirectional motion compensation.

Mode-Dependent Part in the Hybrid Predictive-Distributed Video Coding Modes

In the hybrid modes, motion estimation is shared between encoder and decoder. Two variants for sharing motion estimation can be identified, based on spatial partitioning on the one hand and splitting of the motion estimation algorithm on the other hand.

Hybrid Video Coding Using Spatial Partitioning

Motion estimation can be shared between encoder and decoder by applying a spatial partitioning strategy, and predicting a subset of the macroblocks in an inter frame W at the encoder, while predicting the remaining blocks at the decoder side.

A checkerboard pattern is used to partition the macroblocks of W in two subsets S_1 and S_2. At the encoder, the macroblocks in S_1 are predicted using bidirectional motion estimation, as in the predictive mode. No motion estimation is performed at the encoder for S_2. To construct the mutual prediction frame Z, each block in S_2 is assigned the motion vector of its left neighbor or if it does not exist, the vector of its right neighbor. This technique adds very little complexity to the encoder, but enables us to construct a mutual prediction frame Z through bidirectional motion compensation.

At the decoder, the closest past frame P and the closest future frame F are retrieved from the buffers containing decoded frames. These reference frames are first LP filtered (as in the DVC mode). For each block in S_2, an initial motion vector estimate is generated by using the motion vectors calculated for S_1. First, for a block in S_2, the motion vectors of its neighbors in S_1 are retrieved. Consider for example a non-border block H with a certain neighbor A. Since A is in S_1, it has a backward motion vector (d_x^P, d_y^P) and forward motion vector (d_x^F, d_y^F). This vector has been obtained through rate-distortion optimization at the encoder side, and so it does not necessarily represent the actual motion path. Therefore, we obtain an initial motion vector estimate for H by treating the backward and forward motion vector of A separately. As such, we extend the backward motion vector to F and the forward motion vector to P. This is done for all neighbors A, B, C, and D, resulting into eight linear motion vectors. Each of these motion vectors is applied to H, and the one minimizing the Sum of Squared Errors (SSE) between the reference blocks is chosen.

This initial estimate is then used as a starting point for half-pixel motion refinement, as in the DVC mode. However, in this case a refinement window of fixed size 5-by-5 ($S^{SPAT}{}_R = 25$) showed better results.

Subsequent to half-pixel refinement, bidirectional motion compensation is performed to construct the side info frame Y. No spatial smoothing is performed, since information from neighboring blocks is already taken into account during initialization of the motion vector.

Hybrid Video Coding by Splitting the Motion Search Algorithm

A second way to combine predictive video coding techniques and DVC is to split up the motion search algorithm, i.e., restrict the decoder search space for each macroblock instead of restricting the number of macroblocks for which decoder-side motion estimation needs to be performed. This means that the encoder calculates coarse motion vectors which are further refined by the decoder. Due to the generality of this definition, hybrid modes can be constructed in several ways. In this section we use a subsampling approach.

At the encoder, frames are subsampled by averaging four pixel values at high resolution for calculating one pixel value at low resolution. Next, bidirectional motion search is performed, as in the predictive mode, using a down-scaled search window of size S/4. As such, encoder-side computational complexity is reduced drastically. Subsequently, the low resolution motion vectors are coded and sent to the decoder. To create the mutual prediction frame Z, the motion vectors are upscaled and used for bidirectional motion compensation.

At the decoder, P and F are retrieved from the buffers with decoded frames and LP filtered. The decoded motion vectors are upscaled and used as an initial estimate for motion refinement. Due to subsampling to double pixel resolution at the encoder, we constrain the window for half pixel refinement and set it to a fixed size of 5-by-5 ($S^{SUB1}_R = 25$) in the first pass and 3-by-3 ($S^{SUB2}_R = 9$) in the second pass. As such, the vector is never refined more than three half pixels (which is less than two pixels, i.e., the accuracy of encoder-side motion estimation). No spatial smoothing step is performed afterward, but bidirectional motion compensation follows directly.

Instead of using a subsample approach, other techniques can be used. For example, a heuristic motion search algorithm such as a Three Step Search (TSS) can be split up in executing one or two steps at the encoder side while executing the remaining steps at the decoder.

Results

We present RD results for each mode as well as complexity results by comparing our model of complexity to measurements performed on a test system. Using these results, a method for choosing appropriate coding modes for each frame will be proposed later on.

Rate-Distortion Performance of the System

Tests have been conducted on the Mother and Daughter sequence, but the similar conclusions can be drawn for other sequences as well. An IWWW GOP structure is used, and for each sequence 101 frames are coded (25 GOP's + one closing frame). Inter frames are hierarchically coded, meaning that the sequence I_1 W_1 W_2 W_3 I_2 is coded and decoded in the following order: I_1 I_2 W_2 W_1 W_3. Four different quantization patterns are used, quantizing each coefficient from 3 to 6 bits respectively. Quantization of intra and inter frames is chosen in such a way so that the quality of the decoded frames is constant.

The H.264/AVC reference software (JM 13.2) was used to create two reference RD curves for each sequence: one using a (hierarchical) IBBB GOP structure, and one using only intra-coded frames (IIII).

The results (see Figure 10) show that the compression performance of the different modes is comparable. This is an advantage, since switching between modes (i.e., coding frames using possibly different modes) does not have a significant impact on the performance of the system.

Figure 10. Rate-distortion results for the mother and daughter sequence

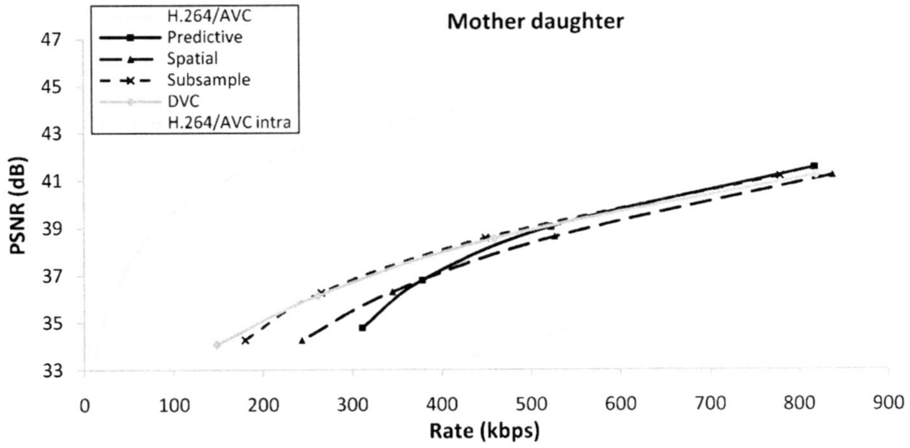

Validation of Complexity Analysis

The complexity for each of the modes has been calculated using the number of pixel read and write operations (see Table 1). How these results should be mapped to the number of CPU cycles or milliseconds spent for coding/decoding each inter frame W depends on hardware details such as calculation speed, cache behavior etc. However, the scaling factor used to convert the number of pixel read/writes to the specific metric desired should be more or less the same for each mode. Hence, theoretical and practical metrics should show the same relationships between encoder and decoder complexity. In other words, if theoretical analysis indicates that the encoder in the predictive mode is twice as complex as the encoder in the spatial mode, then this should be observed in practice also. Therefore, we normalize theoretical complexity and measured complexity to the total complexity in the predictive mode (encoder and decoder) (see Table 2). Practical measurements have been performed on our test system, averaging values for all sequences and rate points. The results clearly show that our model for measuring complexity using the number of pixel read/writes is accurate.

Video Coding with Controllable Complexity

The previous sections illustrated how several modes can be constructed with different distributions of complexity between encoder and decoder, with comparable rate-distortion performance of the modes. Here we will develop a strategy for choosing which coding mode to use for coding each inter frame, in order to meet encoder and decoder complexity constraints. An example is provided to illustrate the different configurations that can be achieved by combining modes.

We assume that techniques are available to estimate or calculate the complexity available at encoder and decoder, and that these values are accurate for coding the following K inter frames W. Denote the total complexity available at the encoder to code these K frames as KC_E and denote the total complexity available at the decoder as KC_D. We will define a method to calculate the optimal linear combination of modes that meets this constraint. The method will be optimal in the sense that the total complexity of the system is minimized.

To code one frame using mode m_i, a complexity budget of M_i^E is needed at the encoder, and a budget of M_i^D is needed at the decoder. From the K inter frames, α_i frames will be coded using mode i. Hence, this optimization problem can be formulated as follows:

Minimize: $\sum_i \alpha_i (M_i^E + M_i^D)$

subject to: $\sum_i \alpha_i = K$, $\sum_i \alpha_i M_i^E \leq K \cdot C_E$,

$\sum_i \alpha_i M_i^D \leq K \cdot C_D$, and $0 \leq \alpha_i \leq K$, $\forall i$.

Remark also that one could favor encoder or decoder complexity decrease, by introducing a weighing factor δ in the cost function: $\sum_i \alpha_i (M_i^E + \delta M_i^D)$.

Table 2. Comparing calculated and measured complexity shows that our model is fairly accurate

	Calculated Encoder Decoder	Measured Encoder Decoder
Predictive mode	> 99.5% < 0.5%	99% 1%
DVC mode	< 0.5% 47%	< 0.5% 48%
Spatial mode	50% 6%	50% 8%
Subsample mode	6% 7%	7% 8%

This problem can be solved, for example, exhaustively or by using integer linear programming techniques. In the following, we will use a graphical exhaustive method to find a solution.

We will illustrate how to choose coding modes by means of an example for a particular set of parameters (K, C_E, C_D, and δ). While this is only one out of many configurations, similar reasoning can be used for other parameters. As an example, we will decide which coding modes to use for coding the following $K = 3$ inter frames. We use the results from our complexity analysis (Tab. 1), and express available complexity in terms of the number of pixel read/write operations that can be performed for coding these K frames. Assume for example that $C_E = 325 * 10^6$, $C_D = 225 * 10^6$, and $\delta = 1$.

Each out of three inter frames can be coded using one out of four modes, resulting into 20 possible ways to code the GOP. Each of these 20 solutions can be described by the tuple (α_0, α_1, α_2, α_3) indicating how the three inter frames are coded: using α_0 times the predictive mode, α_1 times the spatial mode, α_2 times the subsample mode, and α_3 times the DVC mode. Given the complexity of each mode (Tab. 1), each tuple has an associated average encoding complexity per frame $\frac{\sum_i \alpha_i M_i^E}{K}$ and an average decoding complexity of $\frac{\sum_i \alpha_i M_i^D}{K}$. We can use these two values as coordinates for representing the 20 solutions in a plane (see Figure 11). It is easily verified that solutions featuring only two modes i and j lie on a straight line connecting the solutions where i is used exclusively and where j is used exclusively.

From these 20 points, some are suboptimal in the sense that there exists always a better way to code the GOP, i.e., with lower or at most equal encoder and decoder complexity. These points are indicated in gray. The other points (indicated in black) are so-called pareto-optimal. These 12 pareto-optimal points provide the range for distributing complexity between encoder and decoder. We can see that (1,0,0,2) and (2,0,0,1) are not pareto-optimal, which indicates that using the hybrid modes enables more efficient distribution of complexity than combining only the DVC mode and the predictive mode. In addition, since all modes are present in the optimal set, this figure shows that no mode is redundant.

Given the encoder and decoder constraints, illustrated by the gray rectangle, from the pareto-optimal points the point minimizing the cost function is chosen. This means that in this case all three frames should be coded using only the hybrid subsample mode (0,0,3,0). Other solutions satisfying the complexity constraints are suboptimal in this context, since they do not minimize the cost function.

From this example several advantages for using our system can be identified. Firstly, our system provides a solution in case neither encoder nor decoder have enough resources to perform all motion estimation (in our example, neither the predictive mode (3,0,0,0) nor the DVC mode (0,0,0,3) satisfy C_E and C_D). Secondly, due to the fact that there is no dependency between the modes, any frame can be coded using any mode at any time. As a consequence, it is possible to adapt rapidly (with a maximum delay of K frames) to varying complexity constraints that can be imposed by devices with variable power supply, or by systems featuring multi-tasking.

In this example we assumed a very short time during which encoder and decoder complexity constraints remain constant. In practice, however, complexity constraints are likely to be constant over a larger period of time. As K becomes larger, distribution of complexity can be performed more subtle, since the pareto-optimal set will contain more solutions. For example, K = 10 results into 40 pareto-optimal solutions.

Figure 11. Encoder and decoder complexity constraints define a set of possible combinations for coding the GOP (indicated by the gray rectangle). From this set the point minimizing the cost function can be selected

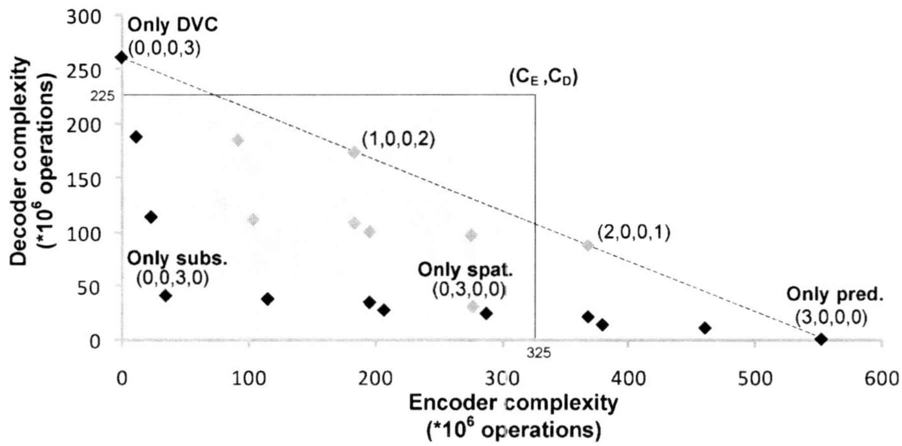

FUTURE RESEARCH DIRECTIONS

Despite the recent advances in the field there is still a substantial gap between the performance of DVC and traditional video coding. This should not be a surprise considering steadily improving traditional video coding in the past 20 years and the first DVC architectures proposed only in 2002. However, it is crucial that DVC soon become competitive to traditional coding in the terms of coding efficiency, because only then it can draw enough attention from the industry and in turn keep its attention from the research community.

In the rest of this section we will discuss several fields of interest in DVC where the biggest improvements to the coding efficiency can be achieved. These areas are subject to a considerable research effort lately, but because the problems at hand which are far from trivial it is difficult to predict when or whether DVC will achieve a performance competitive with traditional video coders.

Side-Information Generation

The generation of side information is the cornerstone of a DVC system. Increasing the accuracy of the side information decreases the number of errors that have to be corrected, and thus effectively reduces the bit rate. At the same time, through the active use of side information in the reconstruction process, its accuracy has a direct impact on the distortion in the decoded video. Therefore it is crucial to generate side information with as high as possible accuracy.

The current state-of-the-art techniques for side-information generation are based on the motion-compensated interpolation. As such these techniques are facing difficulties when dealing with effects like non-linear motion or occlusions. These problems are even more pronounced when longer GOPs are used. Because the efficiency of generating the prediction with the current techniques deteriorate at GOP length as short as 8 (Pereira, 2007), in order to make DVC competitive in the practical scenarios it is important that the problem of long GOP is addressed. This can be done either by improving the interpolation techniques or by developing an effective extrapolation scheme (Borchet, 2007). While there are results indicating that long GOP size should not be feared, it is still not common to use significantly long GOPs in current literature.

Despite the obvious problems of performing the motion estimation at the decoder without knowledge about the original video signal in DVC, this design brings one considerable advantage—no motion vectors have to be signaled in the video stream. This opens the door for techniques which are out of limits of traditional block-based standards like shape-adaptive motion estimation or dense motion fields. Clearly, in DVC it is much easier to use the motion vector for an arbitrary part of the frame. Techniques for estimating the optical flow (Beauchemin, 1995) produce a motion vector for each pixel in the frame, which, in contrast to the block-based approach, is more suitable for adapting to the shape of the moving objects and to certain types of motion like rotation, zooming or deformation. As promising as these techniques might sound, so far, no significant results have been reported in this field.

Another two advantages of DVC are that the prediction signal can be adjusted at any time during the decoding process and that the channel decoder can make use of some knowledge about the accuracy of the prediction. The former allows the decoder to regenerate better side information after the frame is partially decoded and use this more accurate side information through the rest of the decoding process. The latter advantage allows multiple prediction signals or probabilistic models for the motion field (Varodayan, 2008).

To sum up, DVC offers unlimited flexibility in generation and usage of the prediction signal. This stands in strong contrast to traditional video coding, where the usage of the prediction signal is prescribed by standards. It is now the role of the research community to grasp the offered possibilities and use them to improve the coding efficiency of DVC.

Estimating Virtual Channel Noise

One of the assumptions in the Slepian-Wolf and Wyner-Ziv theorems is that the joint probability distribution of the two correlated sources is given. In other words, the rate regions described in the theorems are achievable only if the knowledge about the correlation between the sources is available. That means that in DVC one has to estimate the correlation between the original signal and the side information, i.e., estimate the noise in the virtual channel. Moreover, deviations from the true correlation will decrease the coding performance.

As explained earlier in this chapter, knowledge about the virtual-channel noise is required at the encoder for rate control. A common way to avoid the necessity of noise estimation at the encoder side is using a feedback channel. While not hindering the coding performance, the feedback channel is not practical or even possible in certain application scenarios. On the other hand, doing the noise estimation at the encoder would require generating side information which would countermand the low-complexity encoding. To avoid generating side information at the encoder one can estimate certain properties of the noise from different (motion) characteristics of the video sequence—mostly used are the difference between successive frames, and the difference between the frame and an interpolation of neighboring frames (Martinez, 2008). All these methods sacrifice some RD performance and encoder complexity in order to remove the feedback channel.

The biggest challenge in noise estimation at the decoder side is the lack of the original video signal. Therefore, the estimated correlation between the original signal and the predicted signal will always be a mere approximation of the true correlation. Moreover, accurate correlation estimation has to deal with dynamically changing properties of the virtual noise. Clearly, the accuracy of the prediction depends on the amount and complexity of motion in the video sequence. Because motion can change abruptly in both temporal and spatial directions, so can the properties of the correlation between the original video and the side information. Advances in this field have proved that considerable improvements can be achieved by better modeling of the virtual noise at the decoder (Brites, 2008; Škorupa, 2008a).

Because the noise is determined by the side-information generation, one can expect that as techniques for generating side information are improving, the noise estimation will (have to) evolve as well.

New Coding Modes

As we discussed earlier, statistical properties of a video signal are strongly spatially and temporally varying due to motion in the sequence. Traditional coders deal with these variations by switching between various coding modes on a block basis. Indeed, coding gain achieved between successive generations of coders is contributed greatly by increasing the number of coding modes. In this light it is obvious that the frame-based approach with only two modes (I and W) as described earlier will not perform optimally and the need for different coding modes becomes clear.

While the PRISM codec offered a block-based approach with several coding modes, it—somewhat unexpectedly—lost attention in the scientific community and the codec of Stanford University has become the most important architecture for DVC nowadays. But also in the latter architecture new modes can be introduced. This, however, requires abandoning the frame-based approach or tampering with the channel code. An example of such an approach is to allow intra-coded blocks also in WZ frames (Tagliasacchi, 2006b). A second approach was taken by the authors of this chapter (Mys, 2009). Here, a notion similar to skipped blocks in traditional coding was introduced by adjusting the puncturing and the turbo decoder appropriately. In sequences with low motion, i.e., sequences where skip mode would be used extensively by a traditional coder, this concept can reduce the bit rate up to 52%.

Introducing new coding modes may require significant changes to the codec design, especially for architectures based on the Stanford codec. This, however, may turn into an advantage to DVC, because only constant probing for a change can assure that DVC will keep evolving and improving.

CONCLUSION

In this chapter, we addressed the concept of distributed video coding which is currently emerging as a new video coding paradigm allowing the construction of ultra-low complex video encoder at the expense of a more complex decoder. The theoretical foundations of DVC were discussed briefly after which an overview was given of existing DVC solutions and architectures. One of these architectures was used as reference for a more in-depth discussion of the functional building blocks of a DVC system. As computational complexity plays an important role in the context of DVC, the latter DVC system was extended with a number of coding modes allowing to dynamically shift the complexity between encoder and decoder, facilitating the requirements of emerging video communication applications. Finally, we provided an outlook to some future research directions for which it is believed that advances in these domains will contribute to the overall coding performance of DVC systems.

REFERENCES

Aaron, A., Rane, S., Setton, E., & Girod, B. (2004). Transform-domain Wyner-Ziv codec for video. In . *Proceedings of SPIE Visual Communications and Image Processing*, *5308*, 520–528.

Aaron, A., Varodayan, D., & Girod, B. (2006). Wyner-Ziv residual coding of video. In *Proceedings of the Picture Coding Symposium*.

Aaron, A., Zhang, R., & Girod, B. (2002). Wyner-Ziv coding of motion video. In *Proceedings of the Asilomar Conference on Signals, Systems and Computers: Vol. 1.* (pp.240-244).

Adikari, A. B. B., Fernando, W. A. C., Arachchi, H. K., & Weerakkody, W. A. R. J. (2006a). Wyner-Ziv coding with temporal and spatial correlations for motion video. *Canadian Conference on Electrical and Computer Engineering* (pp. 1188-1191).

Adikari, A. B. B., Fernando, W. A. C., Arachchi, H. K., & Weerakkody, W. A. R. J. (2006b). Multiple side information streams for distributed video coding. *IEEE Electronics Letters*, *42*, 1447–1449. doi:10.1049/el:20062268

Artigas, X., Ascenso, J., Dalai, M., Klomp, S., Kubasov, D., & Ouaret, M. (2007b). The discover codec: architecture, techniques and evaluation. In *Proceedings of the Picture Coding Symposium*.

Artigas, X., Malinowski, S., Guillemot, C., & Torres, L. (2007a). Overlapped quasi-arithmetic codes for distributed video coding. In . *Proceedings of the IEEE International Conference on Image Processing*, *2*, 9–12.

Ascenso, J., Brites, C., & Pereira, F. (2005). Improving frame interpolation with spatial motion smoothing for pixel domain distributed video coding. In *Proceedings of the 5th EURASIP Conference on Speech and Image Processing, Multimedia Communications and Services*.

Ascenso, J., Brites, C., & Pereira, F. (2006). Content adaptive Wyner-Ziv video coding driven by motion activity. In *Proceedings of the IEEE International Conference on Image Processing*. (pp. 605-508).

Ates, H., Kanberoglu, B., & Altunbasak, Y. (2006). Rate-distortion and complexity joint optimization for fast motion estimation in H.264 video coding. In *Proceedings of the IEEE International Conference on Image Processing* (pp. 37-40).

Baccichet, P., Rane, S., & Girod, B. (2006). Systematic lossy error protection based on H.264/AVC redundant slices and flexible macroblock ordering. *Journal of Zheijang University . Scientific American*, *5*, 727–736.

Beauchemin, S. S., & Barron, J. L. (1995). The computation of optical flow . *ACM Computing Surveys*, *27*(3), 433–467. doi:10.1145/212094.212141

Bernardini, R., Fumagalli, M., Naccari, M., Rinaldo, R., Tagliasacchi, M., Tubaro, S., & Zontone, P. (2007). Error concealment using a DVC approach for video streaming applications. In *Proceedings of the EURASIP European Signal Processing Conference*.

Berrou, C., Glavieux, A., & Thitimajshima, P. (1993). Near Shannon limit error correcting coding and decoding: turbo codes. In *Proceedings of the IEEE International Conference on Communications* (pp. 1064-1070).

Borchert, S., Westerlaken, R. P., Klein Gunnewiek, R., & Lagendijk, R. L. (2007). On extrapolating side information in Distributed Video Coding. *Proceedings of the Picture Coding Symposium*.

Brites, C., & Pereira, F. (2008). Correlation noise modeling for efficient pixel and transform domain Wyner-Ziv video coding. *IEEE Transactions on Circuits and Systems for Video Technology*, *18*(9), 1177–1190. doi:10.1109/TCSVT.2008.924107

Flierl, M., & Girod, B. (2006). Coding of multiview image sequences with video sensors. In *Proceedings of the IEEE International Conference on Image Processing*. (pp. 609-612).

He, Z., Liang, Y., Chen, L., Ahmad, I., & Wu, D. (2005). Power-rate-distortion analysis for wireless video communication under energy constraints. *IEEE Transactions on Circuits and Systems for Video Technology*, *15*(5), 645–658. doi:10.1109/TCSVT.2005.846433

Kaminsky, E., Grois, D., & Hadar, O. (2008). Dynamic computational complexity and bit allocation for optimizing H.264/AVC video compression. *Journal of Visual Communication and Image Representation*, *19*(1), 56–74. doi:10.1016/j.jvcir.2007.05.002

Klomp, S., Vatis, Y., & Ostermann, J. (2006). Side information interpolation with subpel motion compensation for Wyner-Ziv decoder. In *Proceedings of the International Conference on Signal Processing and Multimedia Applications*.

Kubasov, D., Lajnef, K., & Guillemot, C. (2007a). A hybrid encoder/decoder rate control for a Wyner–Ziv video codec with a feedback channel. In *Proceedings of the IEEE Multimedia Signal Processing Workshop*. (pp. 251-254).

Kubasov, D., Nayak, J., & Guillemot, C. (2007b). Optimal Reconstruction in Wyner-Ziv Video Coding with Multiple Side Information. In *Proceedings of the International Workshop on Multimedia Signal Processing*. (pp. 183-186).

Li, Z., Liu, L., & Delp, E. J. (2007). Rate distortion analysis of motion side estimation in Wyner-Ziv video coding. *IEEE Transactions on Image Processing, 16*(1), 98–113. doi:10.1109/TIP.2006.884934

Lin, S., & Costello, D. J. (2004). *Error Control Coding* (2nd ed., pp. 563–582). Upper Saddle River, NJ: Pearson Prentice Hall.

Liu, L., & Delp, E. J. (2006). Wyner-Ziv video coding using LDPC codes. In *Proceedings of the IEEE Nordic Signal Processing Symposium*. (pp.258-261).

Liu, L., Li, Z., & Delp, E. (2007). Complexity-rate-distortion analysis of backward channel aware Wyner-Ziv coding. In *Proceedings of the IEEE International Conference on Image Processing* (pp. 25-28).

Malvar, H. S., Hallapuro, A., Karczewicz, M., & Kerofsky, L. (2003). Low-complexity transform and quantization in H.264/AVC. *IEEE Transactions on Circuits and Systems for Video Technology, 13*(7), 598–603. doi:10.1109/TCSVT.2003.814964

Martinez, J. L., Fernandez-Escribano, G., Kalva, H., Weerakkody, W. A. R. J., Fernando, W. A. C., & Garrido, A. (2008). Feedback free DVC architecture using machine learning. In *Proceedings of International Conference on Image Processing* (pp. 1140-1143).

Milani, S., & Calvagno, G. (2007). A distributed video coder based on the H.264/AVC standard. In *Proceedings of the European Signal Processing Conference* (pp. 673-677).

Morbée, M., Prades-Nebot, J., Pizurica, A., & Philips, W. (2007). Rate allocation algorithm for pixel-domain distributed video coding without feedback channel. In *Proceedings of IEEE International Conference on Acoustics, Speech and Signal Processing* (pp. 521-524).

Mukherjee, D. (2006). *A robust reversed-complexity Wyner-Ziv video codec introducing sign-modulated codes* (Tech. Rep. HPL-2006-80), HP Laboratories Palo Alto.

Mys, S., Slowack, J., Škorupa, J., Lambert, P., & Van de Walle, R. (2007). Dynamic complexity coding: Combining predictive and Distributed Video Coding. In *Proceedings of the Picture Coding Symposium.*

Mys, S., Slowack, J., Škorupa, J., Lambert, P., & Van de Walle, R. (2009). (in press). Introducing skip mode in distributed video coding. *Signal Processing Image Communication, 24*, 200–213. doi:10.1016/j.image.2008.12.004

Pedro, J., Brites, C., Ascenso, J., & Pereira, F. (2007). Studying the feedback channel in transform domain Wyner-Ziv video coding. In *Proceedings of the 6th Conference on Telecommunications*.

Pereira, F., Ascenso, J., & Brites, C. (2007). *Studying the GOP size impact on the performance of a feedback-channel based Wyner-Ziv video codec* (pp. 801–815). Advances in Image and Video Technology.

Pradhan, S. S., Chou, J., & Ramchandran, K. (2003). Duality between source coding and channel coding and its extension to the side information case. *IEEE Transactions on Information Theory, 49*(5), 1181–1203. doi:10.1109/TIT.2003.810622

Pradhan, S. S., & Ramchandran, K. (1999). Distributed source coding using syndromes (DISCUS): Design and construction. In *Proceedings of the IEEE Data Compression Conference* (pp.158-167).

Puri, R., & Ramchandran, K. (2002). PRISM: A new robust video coding architecture based on distributed compression principles. In *Proceedings of the Allerton Conference on Communication, Control and Computing*.

Rane, S., & Girod, B. (2004). Analysis of error-resilient video transmission based on systematic source-channel coding. In *Proceedings of the Picture Coding Symposium*.

Škorupa, J., Mys, S., Slowack, J., Lambert, P., & Van de Walle, R. (2008b). Heuristic dynamic complexity coding. In *Proceedings of SPIE, Optical and Digital Image Processing* (pp. 1-8).

Škorupa, J., Slowack, J., Mys, S., Lambert, P., & Van de Walle, R. (2008a). Accurate Correlation Modeling for Transform-Domain Wyner-Ziv Video Coding. In *Proceedings of the Pacific-Rim Conference on Multimedia* (pp. 1-10).

Škorupa, J., Slowack, J., Mys, S., Lambert, P., Van de Walle, R., & Grecos, C. (2009). Stopping criterions for turbo coding in a Wyner-Ziv video codec. In *Proceedings of the Picture Coding Symposium*.

Slepian, D., & Wolf, J. K. (1973). Noiseless coding of correlated information sources. *IEEE Transactions on Information Theory, 19*(4), 471–480. doi:10.1109/TIT.1973.1055037

Slowack, J., Mys, S., Škorupa, J., Lambert, P., Van de Walle, R., & Grecos, C. (2009). Accounting for quantization noise in online correlation noise estimation for distributed video coding. In *Proceedings of the Picture Coding Symposium*.

Stottrup-Andersen, J., Forchhammer, S., & Aghito, S. (2004). Rate-distortion-complexity optimization of fast motion estimation in H.264/MPEG-4 AVC. In *Proceedings of the IEEE International Conference on Image Processing* (pp. 111-114).

Tagliasacchi, M., Frigerio, L., & Tubaro, S. (2007a). Rate-distortion analysis of motion-compensated interpolation at the decoder in distributed video coding. *IEEE Signal Processing Letters, 14*(9), 625–628. doi:10.1109/LSP.2007.896187

Tagliasacchi, M., Pedro, J., Pereira, F., & Tubaro, S. (2007c). An efficient request stopping method at the turbo decoder in distributed video coding. In *Proceedings of the EURASIP European Signal Processing Conference*.

Tagliasacchi, M., Trapanese, A., Tubaro, S., Ascenso, J., Brites, C., & Pereira, F. (2006a). Exploiting spatial redundancy in pixel domain Wyner-Ziv video coding. In *Proceedings of the IEEE International Conference on Image Processing*. (pp. 253-256).

Tagliasacchi, M., Trapanese, A., Tubaro, S., Ascenso, J., Brites, C., & Pereira, F. (2006b). Intra mode decision based on spatio-temporal cues in pixel domain Wyner-Ziv video coding, In . *Proceedings of International Conference on Acoustics, Speech, and Signal Processing, 2*, 57–60.

Tagliasacchi, M., & Tubaro, S. (2007b). Hash-based motion modeling in Wyner-Ziv video coding. In . *Proceedings of the IEEE International Conference on Acoustics Speech and Signal Processing, 1*, 509–512.

Tosic, I., & Frossard, P. (2007). Wyner-Ziv coding of multi-view omnidirectional images with overcomplete decompositions. In . *Proceedings of the IEEE International Conference on Image Processing, 3*, 17–20.

Varodayan, D., Chen, D., Flierl, M., & Girod, B. (2008). Wyner-Ziv coding of video with unsupervised motion vector learning. *Signal Processing Image Communication*, *23*(5), 369–378. doi:10.1016/j.image.2008.04.009

Vatis, Y., Klomp, S., & Ostermann, J. (2007). Enhanced reconstruction of the quantised transform coefficients for Wyner-Ziv coding. In *Proceedings of the IEEE International Conference on Multimedia & Expo.* (pp. 172-175).

Wiegand, T., Sullivan, G. J., Bjøntegaard, G., & Luthra, A. (2003). Overview of the H.264/AVC video coding standard. *IEEE Transactions on Circuits and Systems for Video Technology*, *13*(7), 560–576. doi:10.1109/TCSVT.2003.815165

Wyner, A. D., & Ziv, J. (1976). The rate-distortion function for source coding with side information at the decoder. *IEEE Transactions on Information Theory*, *22*(1), 1–10. doi:10.1109/TIT.1976.1055508

Yang, F., Dai, Q., & Ding, G. (2007). Multi-view images coding based on multiterminal source coding". In*Proceedings of the IEEE International Conference on Acoustics, Speech and Signal Processing* (pp. 1037-1040).

Zamir, R. (1996). The rate loss in the Wyner-Ziv problem. *IEEE Transactions on Information Theory*, *42*(6), 2073–2084. doi:10.1109/18.556597

Chapter 21
Fast Mode Decision in H.264/AVC

Peter Lambert
Ghent University, Belgium

Stefaan Mys
Ghent University, Belgium

Jozef Škorupa
Ghent University, Belgium

Jürgen Slowack
Ghent University, Belgium

Rik Van de Walle
Ghent University, Belgium

Ming Yuan Yang
University of the West of Scotland, UK

Christos Grecos
University of the West of Scotland, UK

Vassilios Argiriou
University of East London, UK

ABSTRACT

The latest video coding standard (Wiegand, 2003), H.264/AVC, uses variable block sizes ranging from 16x16 to 4x4 to perform motion estimation in inter-frame coding and a rich set of prediction patterns for intra-frame coding. Then a robust RDO (Rate Distortion Optimization) technique is employed to select the best coding mode and reference frame for each macroblock. As a result, H.264/AVC exhibits high coding efficiency compared to older video coding standards [2, 3] and shows significant future promise in the fields of video broadcasting and communication. However, high coding efficiency also carries high computational complexity. Fast mode decision is one of the key techniques to significantly reducing computational complexity for a similar RD (Rate Distortion) performance. This chapter provides an up-to-date critical survey of fast mode decision techniques for the H.264/AVC standard. The motivation for this chapter is twofold: Firstly to provide an up-to-data review of the existing techniques and secondly to offer some insights into the studies of fast mode decision techniques.

DOI: 10.4018/978-1-61520-761-9.ch021

INTRODUCTION

The H.264/AVC video coding standard is the newest video coding standard which is proposed by JVT (Joint Video Team). A number of new design features are adopted in this standard which significantly improve the rate distortion performance as compared to other standards. These features include variable block size and quarter sample accurate motion compensation with motion vectors even outside picture boundaries, multiple reference frames selection, decoupling of referencing from display order for flexibility and removal of extra delay associated with bi-predictive coding, bi-predictive pictures to be used as references for better motion compensation, weighted offsetting of prediction signals for coding efficiency in scenes including fades etc, improved "skipped" and "direct" mode inference for better RD performance in video sequences containing neighboring macroblocks (of the same scene object) moving in a common direction etc. H.264/AVC further allows directional edge extrapolation in intra coded areas for improving the quality of the prediction signal and allowing prediction from neighboring areas that are inter coded, in-loop deblocking filter for removing compression artifacts as well as providing better quality reconstructed signals for subsequent motion compensation, and hierarchical block size transforms that enable signals with sufficient correlation to use longer basis functions than 4x4 transforms. There are also provisions for embedded processors such as exact match inverse transforms for "drift free" decoded representations and finally the standard provides advanced entropy coding techniques such as CAVLC (Context Adaptive Variable Length Coding) and CABAC (Context Adaptive Binary Arithmetic Coding) which are also present in the H.263 and JPEG2000 standards.

However, the improvements in the RD performance come with significant complexity increases. These new features not only increase the complexity of H.264/AVC encoders but also of the corresponding decoders. Variable block size motion estimation and compensation, Hadamard transform, RDO mode decision, displacement vector resolution and multiple reference frames are the main H.264/AVC encoding tools which increase the complexity of H.264/AVC encoders. In (Ostermann, 2004) an analysis of the complexity increase in the H.264/AVC video coding standard is presented and compared with previous standards. The significant computational complexity makes it very difficult to use the standard as it is in real-time applications. Reducing the complexity without degrading RD performance thus becomes a critical problem.

In order to understand the complexity of H.264/AVC more clearly, an experiment of complexity analysis is performed here. The Intel® VTune™ Performance Analyzer7.0 is used in this work as the evaluation tool to evaluate the software performance and obtain the complexity profile of an H.264/AVC encoder. In this experiment, the Foreman sequence (100 frames, QCIF (Quarter Common Intermediate Format) format, Baseline profile) is encoded on an Intel Pentium-4 3.09GHz PC with 768 MB memory and using the Microsoft Windows XP operating system. Figure. 1 shows the complexity proportion of different encoding modules in the H.264 JM8.1 [21] reference encoder.

According to Figure 1, the most time-consuming modules of the H.264/AVC encoder are Motion Estimation, Interpolation, SATD (Sum of Absolute Transformed Differences), and DCT (Discrete cosine transform) which are all related to the RDO based motion estimation and mode decision. Because mode decision covers all these four aspects, a good fast mode prediction algorithm for H.264/AVC is a promising way to reduce the complexity of video encoders.

The relation between complexity reduction and seamless video communication can be best described by using Figure 2 (Hsu 1997) below:

The total delay of the system is the sum of the delays in the encoder, decoder buffers and

Figure 1. Complexity proportion of different encoding modules in H.264/AVC encoder by Intel® VTune™

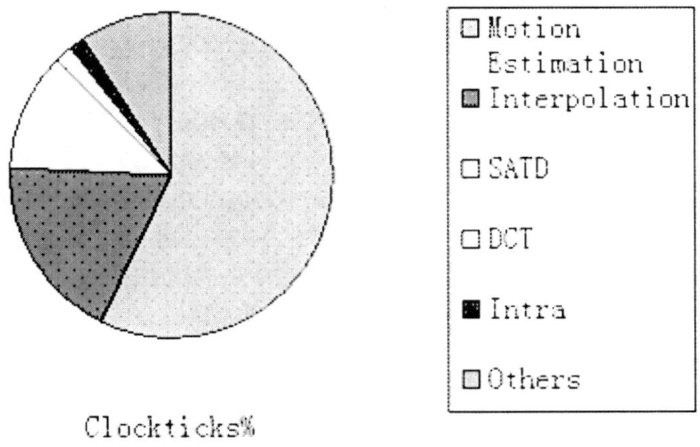

the channel delay. Evidently, there needs to be a balance among the rates of filling the encoder buffer with bits, emptying it to the channel, filling the decoder buffer with bits and emptying it for playing a specified number of video frames/sec. A computationally intensive encoding process will result in disturbing this balance since it will reduce (drastically) the rate of filling up the encoder buffer. This in turn will result in reducing the frames/sec at the decoder end, thus producing a non smooth visual experience (video will appear jerky or frozen at periods). It can be easily deduced that techniques to reduce the computational load at the encoder would restore the rate balance between different components of a communication system and thus enhance the end user experience. Provided that the relationships between the buffer sizes and the rates in different parts of the system are upheld, reduced computation techniques may also imply more frames/sec, thus smoother visual

Figure 2. Time delay diagram of a video communication system

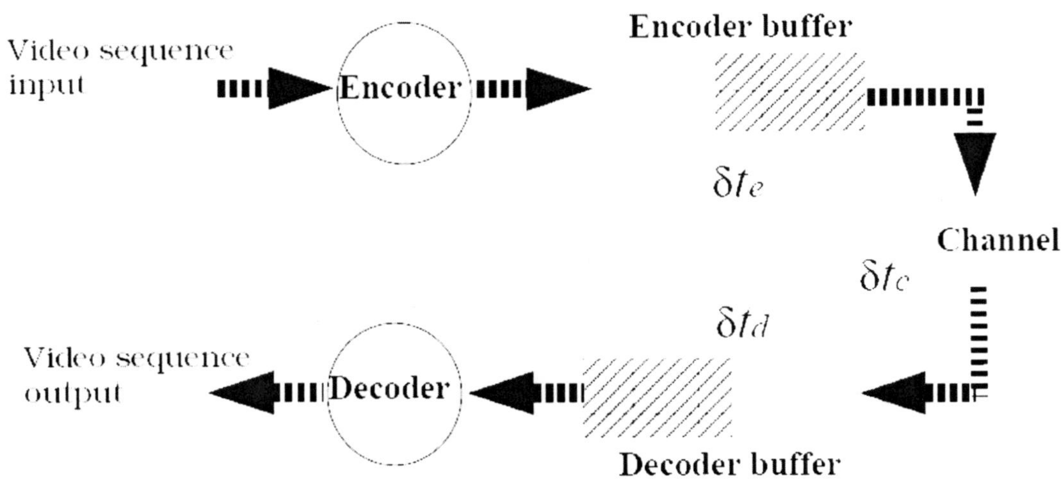

experience at the decoder end. In the context of H264 AVC, most of these techniques are mode decision techniques which are described and analyzed in this chapter.

The necessity of computational complexity reduction for using video on mobile handsets can be evidenced from the fact that consumers increasingly want, and expect their mobile devices to capture and playback HD video without adversely impacting battery life. Handset developers have a range of video codec options to satisfy this consumer demand such as fully customized hardware blocks integrated into system-on-chip (SoC) designs, optimized software codecs running on enhanced-instruction set RISC or DSP processors, or software running on standard processor cores, like ARM 9 or ARM11. In order to be able to bring high-definition video to the hands of users, mobile device designers must optimize the power consumption of all components. Therefore, minimizing the maximum power consumption figure of a chip often concentrates on finding the most power-efficient way to implement video processing algorithms. Again in the context of H264 AVC, most of these power efficient schemes are hardware solutions combined with the software mode decision techniques which are described and analyzed in this chapter. This combination is a classical case of hardware/software co-design.

This chapter is organized as follows: Section 2 gives a brief overview of the mode decision mechanism of H.264/AVC and focuses both on RDO based and on low complexity mode decision preliminaries. Section 3 focuses on speeding up RDO based mode decision techniques. It presents, analyses and compares skip/direct, inter, intra and selective inter/intra schemes. Section 4 concludes the paper and presents future research directions.

MODE DECISION IN H.264/AVC

Macroblocks in H.264/AVC standard have many mode candidates due to variable block size motion estimation and directional intra modes. Consequently, some criteria need to be used to decide which candidate mode is the best one for the current macroblock. The H.264/AVC standard suggests two mode decision schemes for encoder: a high-complexity mode decision, also known as Rate Distortion Optimized (RDO) based mode decision, and a low-complexity mode decision. Compared to the low-complexity mode decision, RDO-based motion estimation and mode decision improves PSNR (Peak Signal-to-Noise Ratio) (up to 0.35 dB) and bit rate (up to 9% bit savings) but also comes at the cost of significant computational complexity for common test video sequences. In the rest of this section we will briefly introduce these two mode decision schemes.

RDO Based Mode Decision for H.264/AVC

In the high complexity mode of the H.264/AVC standard, the macroblock mode is chosen by minimizing the Lagrangian function:

$$J(s,c,MODE|QP,\lambda_{MODE})=SSD(s,c,MODE|QP) + \lambda_{MODE}*R(s,c,MODE|QP) \quad (1)$$

In the above equation, J denotes the cost function and is dependent on s (the original signal macroblock), c (the reconstructed signal macroblock) and MODE (selected from a set of modes as explained later). J is found for a given QP (Quantization Parameter) and λ_{MODE} (the Lagrange multiplier for mode decision). SSD is the sum of the squared differences between the original macroblock and its reconstruction with QP and it also depends on the original and reconstructed macroblock, as well as the mode decision (MODE).

Finally, the rate $R(s,c,MODE|QP)$ depends on the original and reconstructed macroblock with QP, as well as the chosen MODE, and reflects the number of bits produced for header(s) (including MODE indicators), motion vector(s) and coefficients. It is worth mentioning that an encoder is free to calculate this rate by either measuring or by estimating it. In RDO mode, the reference encoder will actually measure it, in other words, it will code the current macroblock up to and including entropy encoding

In equation (1), MODE is chosen from the set of potential prediction modes as follows:

For Intra slices:

$$MODE \in \{INTRA4*4, INTRA16*16\} \quad (2)$$

For P slices: {single reference forward or backward prediction}

$$MODE \in \{INTRA4*4, INTRA16*16, SKIP, MODE\_16*16, MODE\_16*8, MODE\_8*16, MODE\_8*8\} \quad (3)$$

For B slices: {bi-directionally predicted slices}

$$MODE \in \{INTRA4*4, INTRA16*16, DIRECT, MODE\_16*16, MODE\_16*8, MODE\_8*16, MODE\_8*8\} \quad (4)$$

DIRECT mode is particular to the bi-directionally predicted macroblocks in B slices, while SKIP mode implies that no motion or residual information will be encoded (only the MODE indicator is actually transmitted).

In the above mode sets, any mode with the prefix INTRA will result in encoding the spatially predicted signal rather than its temporal residual, while any mode with the prefix MODE_ refers to inter modes. Furthermore, when MODE is equal to INTRA4*4 or INTRA16*16, the best intra mode for each case is chosen through evaluation of the functional of equation 3-12 with mode choices from the following sets:

$$INTRA4*4 \in \{DC, HORIZONTAL, VERTICAL, DIAGONAL\_DOWN\_RIGHT, DIAGONAL\_DOWN\_LEFT, VERTICAL\_LEFT, VERTICAL\_RIGHT, HORIZONTAL\_UP, HORIZONTAL\_DOWN\} \quad (5)$$

$$INTRA16*16 \in \{DC, HORIZONTAL, VERTICAL, PLANE\} \quad (6)$$

A similar functional minimization results in the choice of the best 8*8 mode for P and B slices from the following set:

$$MODE\_8*8 \in \{INTER\_8*8, INTER\_8*4, INTER\_4*8, INTER\_4*4\} \quad (7)$$

Any mode with the prefix MODE_ in equations (3,4) and INTER_ in equation (7) assumes that the best motion vector is known for this mode and implies functional minimizations for each candidate motion vector $m = (m_x, m_y)$ inside the search window of the form:

$$J(m, \lambda_{MOTION}) = DD_{FD}(s, c(m)) + \lambda_{MOTION} * R(m-p) \quad (8)$$

where λ_{MOTION} is the Lagrange multiplier for motion estimation, p=(px,py) is a predicted motion vector and R$(m-p)$ is the number of bits for encoding motion residuals only. The rate term is computed from a look-up table, while the distortion term $DD_{FD}(s, c(m))$ depends on the original signal s and the reconstructed best match c that in turn depends on the candidate motion vector m. It has to be noted that the choice of λ_{MOTION} in equation (8) is affected by the choice of the distortion metric.

Once the best 8*8, INTRA4*4, INTRA16*16 modes are found, the minimal cost for the macroblock is evaluated by looping through the different mode possibilities (equations 3, 4).

A straightforward mode decision complexity assessment for a 16*16 macroblock (luma component only) reveals that we need 144 cost evaluations for the best INTRA4*4 mode. Adding 4 more evaluations for the INTRA16*16 case, 16 more for the best 8*8 inter mode and 7 more for selecting the minimal cost among all modes results in 148 evaluations for macroblocks in Intra

Figure 3. Flow chart of low complexity mode decision

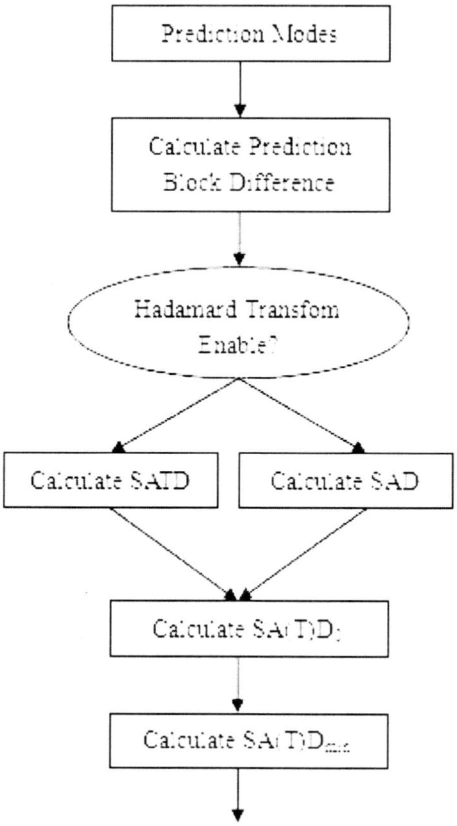

slices and 171 evaluations for macroblocks in P or B slices. Coupled with similar cost evaluations for the chroma components and the fact that our complexity assessment did not consider evaluations for the best motion vector that depend on the size of the search window and on the sub-pixel accuracy, clearly shows that the mode decision process is computationally intensive.

Low-Complexity Mode Decision

In the low-complexity mode decision scheme, the block difference between the prediction for each candidate mode and the block to be encoded is calculated in the first step. In the second step, this block difference is either transformed to a single value using the Sum Absolute Transformed Difference (SA(T)D) or the absolute values are calculated and summed for each block location using the Sum of Absolute Difference (SAD) metric. The transform used for SA(T)D is the Hadamard transform. In any case, a set of predicted initial values for $SA(T)D_0$ is then calculated in the third step by using quantization parameters and bit usage estimates for motion vectors (magnitude + labels). Finally, the minimum $SA(T)D_{min}$ is chosen as shown in the flow-chart of Figure 3.

Low complexity mode decision uses SA(T)D which includes only subtraction and a simple convolution to represent the distortion term. It also uses $SA(T)D_0$ which only includes the table look up computation to find the bit rate term. No DCT, IDCT (Inverse Discrete Cosine Transform) and entropy coding are included in the low complexity mode decision scheme, which implies much lower computation as compared with the high complexity scheme. The average execution time of low-complexity mode decision is only 7% of that of high-complexity mode decision. However, low-complexity mode decision loses an average of 0.48dB in PSNR compared to the RDO based mode decision. In the following sections, unless we mention low-complexity mode decision explicitly, mode decision means RDO based mode decision.

SPEEDING UP RDO BASED MODE DECISION IN H.264/AVC

As explained before, the RDO based mode decision is a very computationally intensive process due to the multiple motion estimations involved. This fact inspired a variety of fast mode decision techniques to be developed for reducing the computational complexity, while retaining similar rate distortion performance. In this section, the most important four categories of fast mode decision techniques — fast skip/direct mode, fast inter mode, fast intra mode and fast intra/inter mode—

will be discussed. Inside each category, different mode decision approaches will be presented and compared. The comparisons will enable us to better understand the assumptions, advantages and limitations of these approaches.

Fast Skip/Direct Mode Decision Techniques

In some prior standards, for example H.263+, a skipped macroblock is defined as a macroblock with zero Motion Vector and zero Quantized Transform Coefficients. A significant proportion of macroblocks are skipped (not coded), particularly in low-motion sequences and/or at higher quantizer step sizes (and hence lower bitrates). The difference between prior standards and H.264/AVC is that in prior standards, a skipped area of a predicted slice was assumed stationary, while in H.264/AVC a predicted motion vector derived directly from previously encoded information is used in skipped areas.

C. Grecos and M.Y Yang (2005) presented a fast skip mode decision technique based on spatio-temporal neighborhood information. The idea is based on the observation (which is particularly evident in slow moving sequences) that the areas of a slice consisting of macroblocks in SKIP mode, slowly change over time. If the macroblocks on the top and left of the one to be encoded in the current frame and the macroblocks on the right and bottom of the co-located macroblock in the previous frame are all in skip mode, then this mode pattern can be a good indication that the current macroblock can be skipped. To strengthen the accuracy of the skip mode prediction, the authors also add an extra condition that the SAD between the current macroblock and its co-located one should be less than the average SAD among the skipped macroblocks in the reference picture and their co-located predictors. The proposed scheme predicts SKIP modes without any motion estimation which can significantly reduce the computational complexity.

A.C.W. Yu, G. R. Martin and H. Park (2008) proposed a novel skip mode detection technique based on three layers. Considering that skip macroblocks tend to occur in clusters (in a similar idea to (Grecos, 2005), spatial and temporal skip mode information is used in the first layer. If the co-located macroblock in the reference frame or at least one of the upper or the left macroblock of the current macroblock in the current frame is in skip mode, the current macroblock passes through the first layer. Otherwise the current macroblock cannot be predicted as SKIP mode directly. In the second layer, if the SAD between the current macroblock and its co-located macroblock in the reference frame is smaller than an adaptive threshold (the average SAD of the available skipped neighbors and their collocated ones in the first layer), the current macroblock can be considered as a potential SKIP mode macroblock. Otherwise the current macroblock cannot be predicted as SKIP mode. A fast transform-quantized implementation based on (Malvar, 2003) is used in the third layer to detect whether all quantized coefficients of SATD between the current and co-located macroblocks are zeros. Finally, the current macroblock is in SKIP mode if it passes through all the three layers. The authors include the quantization parameter and the integer transform in the mode decision process which makes their work robust, however they do not consider motion information.

C.S. Kannangara et. al. (2006) also proposed a low-complexity SKIP mode decision scheme. The SKIP mode RD cost of the current macroblock can be calculated based on predicted motion vectors without performing any motion estimation. The best inter mode RD cost can be predicted through a model based on local sequence statistics and for a given Lagrange multiplier parameter. This model requires no motion estimation either. The early SKIP mode detection is made by comparing the two RD costs. The achievable computational savings for typical video sequences are in the range of 19%-67% in the baseline profile without significant loss of rate-distortion performance as compared to the standard. The experiments show that the technique achieves very similar Rate Distortion performance with the low complexity mode of H.264/AVC

A Bayesian framework for fast SKIP mode decision is proposed by M. Bystrom, I. Richardson and Y. Zhao (2008). The RD cost difference, which is the difference between the RD costs of the SKIP mode and the best other mode, is used as a discriminator. This difference depends on the QP and the frame content activity and its modeling is probabilistic. Firstly, the conditional probability density functions (PDF) of the RD cost difference are modeled for a set of training sequences at different bit rates. These PDF are measured for the SKIP mode and the other modes. The model of RD cost difference for the SKIP mode is simulated by a Gaussian PDF and the model for other modes is simulated by a Rayleigh PDF. The final RD cost difference can be calculated using Maximum Likelihood (ML) and Maximum a Posteriori (MAP) algorithms based on these conditional PDFs and the priori probabilities of the SKIP and other modes. The a priori probabilities of SKIP and other modes can be calculated as the average frequency of SKIP and other modes in previous frames. Alternatively a look-up table, from where the a priori probabilities can be indexed, can be built based on the video content activity factor and QP. Compared with (Jeon, 2003), this work can achieve 12%-58% more time savings with average 0.04dB PSNR decrease and 0.15% bit rate increase. However, the performance is dependent on the range of content for the training set, the accuracy of modeling and even the sequence resolution. For example, comparing with (Jeon, 2003) in the particular case of mobile video sequences, 13% less time saving is attained with 0.15dB PSNR decrease and 2.75% bit rate increasing.

A fast SKIP mode decision technique which resulted in contributions to the standard, is the work of Jeon et al. (Jeon 2003). According to this work, a macroblock can have SKIP mode in the baseline profile, when the following set of four conditions is satisfied:

1. The best motion compensation block size for this macroblock is 16x16 (MODE_16*16)
2. The best reference slice is the previous slice
3. The best motion vector is the predicted motion vector (regardless of this being a zero motion vector or a non-zero one)
4. The transform coefficients of the 16x16 block size are all quantized to zero.

This set of four conditions is non sufficient due to the assumption that the mode with the lowest RD cost is the inter MODE_16*16 (condition 1), which may be true or not. If it is true, then we can safely say that the macroblock can be skipped since $J_{SKIP} < J_{MODE\_16*16}$ for the same motion vectors as condition 3, thus making the SKIP mode the chosen one due to its lowest cost. If the first condition above is not true though, the algorithm will miss-predict macroblocks as skipped and the RD performance will suffer. The important point here is that although condition 1 makes the set of the above conditions non sufficient for SKIP mode decision, it is "good enough" (dependent on the video content of course). So in order to predict SKIP mode, the approach described in (Jeon, 2003) only needs to perform motion estimation for the 16x16 mode and thus motion estimation for the remaining mode types can be saved if the above conditions are satisfied. Experimental results show that the proposed method results in time savings of 15% on the average without any noticeable RD performance loss as compared to the standard. The proposed technique cannot achieve very significant time savings since motion estimation is still performed for MODE_16x16, however good RD performance is retained due to the use of sufficiently accurate temporal information from the motion estimation step.

J.Lee, B Jeon et al (2004) also presented similar

ideas to predict direct mode for B slices. A macroblock is predicted as having SKIP mode when the following set of two conditions is satisfied:

1'. The reference slices and the motion vectors are the same as the ones decided under the DIRECT mode.
2'. The transform coefficients of the 8x8 sub-blocks of this macroblock are all quantized to zero.

With the same reasoning as above, we can observe that the set of conditions for SKIP mode decision in the B slices of the main profile is non sufficient either. This is due to the fact that no motion compensation takes place as can be seen in conditions 1' and 2' and it is implicitly assumed that the DIRECT_16*16 mode is the one with the minimal cost. If true, the RD performance is not hampered, otherwise it is. Condition 1' in fact enforces the mode of the sub-blocks to be DIRECT_8*8 and the mode of the whole macroblock to be DIRECT_16*16. Furthermore, $J_{SKIP} < J_{MODE\_16*16}$ for the same motion vectors and reference slices from condition 1', clearly implying that the macroblock can be safely skipped. The reader should note that in contrast to the baseline profile, there is actually no assumption in terms of the best reference slice for the SKIP mode detection in the B slices of the main profile. From the design of the standard, the DIRECT and SKIP modes both have the same reference slices and motion vectors, while they are different only in the fact that the SKIP mode needs to have all quantized coefficients in the 8x8 sub-blocks equal to zero, while some coefficients are not zero for the DIRECT mode (condition 2'). Theoretically, condition 2' is a relatively strong one in terms of skipping, since the likelihood of having non zero coefficients in a larger block size (16x16) is very small. Time savings of 52% on the average can be achieved with an average PSNR decrease of 0.01db and average bit rate increase of 0.26% as compared to the standard.

In summary, the above analysis shows that SKIP/DIRECT modes are especially useful for low bit rate coding. Early detection of these modes can lower the encoder complexity significantly (roughly in the order of 20% - 60%) with only small losses in quality. Most SKIP/DIRECT mode decision techniques exploit temporal and spatial neighbourhood information in combination with adaptive thresholds. However, to reduce the impact on the RD performance, such methods should incorporate the QP and the fact that SKIP/DIRECT modes have an inferred motion vector in H.264/AVC, something that is often omitted.

Fast Inter Mode Decision Techniques

X. Jing and L. Chau (2004) propose a fast inter mode decision method which only depends on the pixel based Mean Absolute Difference (MAD) metric between the current and previous frames. The main idea is to use large blocks for smooth areas and small blocks for areas containing complex motions. If the MAD between the current and co-located macroblocks is smaller than the weighted Mean Absolute Frame Difference (MAFD) between the current and previous frames, large mode types {MODE_16*16, MODE_16*8, MODE_8*16} are chosen. Otherwise all mode types are examined. The disadvantage of this algorithm is that the weighting used in the MAFD metric is based on the Quantization Parameter (QP). This implies that initial offline training is needed for relating the weights to QPs for every sequence. Furthermore generalizing the use of these weights to arbitrary sequences is likely not to perform optimally due to differences in motion characteristics, thus making this algorithm not very practical in real time applications. The proposed algorithm can obtain up to 48% computational savings with similar rate distortion performance to the standard for a variety of test sequences.

A. Chang (2003) proposes another algorithm which uses pixel based Sum of Absolute Difference (SAD) information to predict inter modes. If the texture undergoes an integer-pixel translational motion, the texture will look exactly the same in the two consecutive frames and it can be predicted perfectly by integer-pixel motion estimation. If the edges of the texture have a half-pixel or quarter-pixel offset, they may be blurred thus sub-pixel motion estimation will be important. The SAD of the MODE_16*16 is calculated after integer-pixel motion estimation. If this SAD is smaller than an adaptive threshold based on the average SAD of the macroblocks having MODE_16*16 as their best mode, the current macroblock is assumed to have integer-pixel translational motion and MODE_16*16 is chosen as the best mode. Otherwise, the current macroblock may have a half-pixel or quarter-pixel offset. If MODE_16x16 is the best mode, texture analysis and segmentation will be subsequently performed on the best match to potentially refine the best mode to MODE_16*8 or MODE_8*16. However, (Chang, 2003) only considers three coding modes namely MODE_16*16, MODE_16*8 and MODE_8*16 which may be limiting in cases where more refined mode decision is required for improved RD performance due to motion characteristics of sequences. The improvements in RD come of course to the expense of CPU/ run time savings. By considering only three modes, 40.73% run time savings can be achieved with 0.04dB PSNR degradation and 0.92% bit rate increase as compared to the standard.

Motion vector information is used by (Ahmad, 2004) to predict inter modes. This algorithm is based on the principle of 3D recursive search algorithm (Haan, 1993) which provides a fast, convergent and highly accurate motion vector prediction, taking into account the total cost of some modes in the previous and current frames. The total cost of a mode is defined as the cost of the mode itself plus the motion vector cost for that mode. The motion vector cost in turn is calculated using Lagrange multipliers and motion vector magnitudes. Candidate modes for total cost calculations are chosen from the modes of the left, top, and top-left macroblocks of the current one and the macroblock which is two rows down and one column right of the collocated macroblock in the previous frame. The mode with the minimal total cost is chosen as the best mode for the current macroblock. Using this algorithm, a maximum increase of about 15% in bitrate is achieved at the same quality compared with standard. Evidently, the increase in bitrate is significant at the same quality and thus affects negatively the RD performance.

K.P.Lim et. al (2003) propose a new algorithm called "homogeneous regions detection" to classify the inter modes into groups. It is observed that in non-deforming, smoothly moving video sequences, the smooth regions of video objects move together. One of the main reasons for using variable block sizes in H.264/AVC is to represent motion of video objects more accurately. Since homogeneous regions tend to move together, homogeneous blocks in the frame should have similar motion and should not be further split into smaller blocks. A region is homogeneous if the texture in the region has very similar pixel values. So some macroblocks which are homogeneous could belong to a specific subset {MODE_16*16, MODE_16*8, MODE_8*16} of modes, and do not need to be motion estimated for the rest of the modes. The authors use the edge map computed in the fast intra mode decision technique of (Pan, 2003) to decide which macroblocks belong to homogeneous regions. If the current macroblock is homogeneous, its mode belongs to the set {MODE_16*16, MODE_16*8, MODE_8*16}, otherwise all modes are tested. The results show a speed up of 30% in run times with a maximum of 0.08dB PSNR decrease and a maximum of 1.44% bit rate increase as compared to the standard. However, this algorithm needs a fixed threshold to decide which macroblock is homogeneous and thus will not be very suitable for different video sequences.

D. Zhu (2004) uses a 7-tap filter on horizontal

and vertical directions of the original and reference images respectively to get down-sampled half resolution small images. The mode selection method of (Lim, 2003) is used to get a set of prediction mode candidates in small images. This set of mode candidates is then mapped to another set of mode candidates for the current macroblock in the original image. Because motion estimation is performed in small images, a small set of mode candidates is chosen and thus a lot of time savings can be achieved. The experimental result shows that this algorithm can reduce by nearly 50% the encoding time with PSNR reduction of about 0.2dB as compared to the standard.

C. Grecos and M. Y. Yang (2007) extend (Lim, 2003) into the error domain. The basic idea can be summarized as follows: Video objects in consecutive frames are not always deformed or divided. They may be still or just change location translationally, especially for slow motion video sequences or for the slow motion frame parts of fast video sequences. In terms of computational speed ups, there is potential mis-prediction of modes in macroblocks of those areas from spatial only mode decision techniques such as (Lim, 2003). This occurs in the cases where these macroblocks have high spatial detail and as such will be assigned smaller size modes, whereas by examining error characteristics these macroblocks can be assigned larger size modes. Similarly with (Lim, 2003), the authors check homogeneity but in the error domain. A novel concept of the "moving average sum of amplitudes of edge error vectors" is exploited for designing adaptive thresholds, which makes (Grecos, 2007) suitable not only for the slow motion video sequences but also for the fast motion video sequences. BDPSNR (Bjontegaard Delta Peak Signal-to-Noise Ratio) and BDBR (Bjontegaard Delta Bit Rate) (Bjontegaard, 2001) recommended by JVT are used as metrics to measure the performance difference between methods. Experimental results show that compared with (Lim, 2003), the algorithm gains an average of 12% time savings for the baseline profile with BDPSNR average reduction of 0.04dB and BDBR average increase of 0.43% for the simple profile. For the main profile, a BDPSNR average reduction of 0.03dB and BDBR average increase of 0.07% are observed depending on motion characteristics.

Cost comparisons based on the SATD metric have been used in (Kim, 2004; Tanizawa, 2004) for fast mode decision. The basic idea comes from experiments showing that there is a strong relationship between the costs of low complexity mode decision and the costs of RD based mode decision. In the above algorithms, three most probable modes with lowest costs in low-complexity mode decision are chosen for high complexity mode decision. The disadvantage of these schemes is that mode candidates are known only after all motion estimation has been performed for the low complexity case, thus only part of the mode decision process can be saved time-wise. About 80% of execution time of RD based mode selection can be saved for an average PSNR loss is 0.07dB as compared to the standard.

A fast inter mode decision algorithm is proposed in (Zhou, 2004) by exploiting the correlation of J costs. The basic idea of this algorithm is that if the cost of larger block-size modes is higher than the cost of the current block-size mode, then the best mode of current macroblock cannot be of a larger block-size. Meanwhile, if the cost of a smaller block-size mode is higher than that of current block-size mode, then best mode of current macroblock cannot be of a smaller block-size. A similar idea has been used in (Yin, 2003) which is based on the monotonicity of the error surface as another way to group the modes. The error surface is built initially by 3 modes: MODE_16*16, INTER_8*8 (the entire macroblock is examined using only 8x8 partitions which means four 8x8 sub-blocks for this macroblock), INTER_4*4 (the entire macroblock is examined using only 4x4 partitions which means the entire macroblock is partitioned equally into sixteen 4x4 sub-blocks). If the error surface is monotonic, that is if J(16x16)

< J(8x8) < J(4x4) or J(16x16) > J(8x8) > J(4x4) where J denotes a cost function, only modes (block sizes) between the best two modes are tested. If not, all other modes need to be tested. The order of motion estimation suggested by the H.264/AVC standard is MODE_16*16, MODE_16*8, MODE_8*16 and MODE_8*8. And then in each 8x8 sub-block, motion estimations of INTER_8*8, INTER_8*4, INTER_4*8 and 4x4 are performed one by one. In this structure, best motion vectors of neighbor sub-blocks can be utilized to predict the predicted motion vector (the initial position for motion estimation) of current sub-block. However in (Yin, 2003) the order of motion estimation for different partition block size is changed which will introduce complexity and will affect negatively the RD performance.

B. G. Kim (2008) proposed a fast inter mode decision algorithm based on the temporal correlation of mode information for P slices. In the slow motion video sequences, the mode information in the previous reference slice is highly correlated with the mode decision in the current slice. Experimental test shows that if the co-located macroblock is in skip mode, the probability that the current macroblock will be in skip or MODE_16*16 is greater than 91%. When the co-located macroblock is in MODE_16*16, the probability that the current macroblock will be in skip, MODE_16*16, MODE_16*8 or MODE_8*16 modes is greater than 80%. Even when the co-located macroblock is in MODE_16*8 or MODE_8*16, the probability that the current macroblock will be in skip, MODE_16*16, MODE_16*8 or MODE_8*16 is still greater than 70%. Due to this observation, a simple macroblock mode tracking strategy is devised. Initially, motion estimation for the MODE_16*16 is performed and the best match will intersect at most four macroblocks in the reference slice. The most correlated macroblock of the current one is found in the reference slice and based on its mode type, a sub set of candidate modes is checked initially. If the minimum J cost of candidate modes is less than the cost of the tracked macroblock (the most correlated macroblock in the reference slice), the mode of the current macroblock is the one that has the minimum J cost. Otherwise, the sub set of candidate modes is enlarged with the co-located macroblock mode type. Image intensity analysis is also used to refine the choice of candidate modes. The author compared his algorithm with other three fast mode decision algorithms (Jeon, 2003; Jing, 2004; Salagdo, 2006) and showed that he can achieve a good balance between time savings and RD performance. A speed-up factor of 57% on the average was shown, with a bit rate increment of 0.07% and a loss of 0.05dB as compared to the standard.

C Grecos (2005) presented a layered inter mode prediction scheme for P slices. In the first stage, an enhanced fast skip mode decision technique is used. After a percentage of macroblocks in characterized as skipped, the conditions of (Jeon, 2003) are used to identify even more skipped macroblocks in the second stage. For the remaining macroblocks (which also include a percentage of skipped macroblocks that were not classified with the first two stages), (Jeon, 2003) proposed a set of three smoothness conditions. Firstly, the J_{MODE_16*16} cost of the current macroblock should be less than the average J_{MODE_16*16} cost of the macroblocks in this mode in the reference frame. Secondly the co-located macroblock in the reference frame should be of skip mode or of MODE_16*16. Thirdly, the SAD between the current and the co-located macroblock should be less than the average SAD among the skip mode macroblocks in the previous frame and the collocated macroblocks in the current frame. For RD performance very close to the reference encoder, (Jeon, 2003) achieves 35-58% reduction in run times and 33-55% reduction in CPU cycles for both rate-controlled and non-rate-controlled encoding. Compared to (Jeon, 2003), gains of 9-23% in run times and 7-22% in CPU cycles are reported from the scheme of (Grecos 2005). In order to increase the time savings, C. Grecos and M. Y. Yang proposed another algorithm in (Grecos, 2006) which could be considered as an extension of (Jeon, 2003). Their algorithm devised three heuristics instead of three smoothness conditions to be used for predicting subsets of decidable modes. Firstly, the J_{MODE_16*16} cost of the current macroblock should be less than the average J_{MODE_16*16} cost of the macroblocks in this mode in the previous slice of the same slice type. Secondly, for all macroblocks not satisfying the first heuristic, the J_{INTER_8*8} cost of the current macroblock should be less than the average J_{INTER_8*8} cost of the macroblocks in the previous slice that are neither skipped nor were satisfied with the first heuristic. Thirdly, the J_{MODE_16*16} cost should be less than J_{INTER_8*8} for the current macroblock. If any of these three heuristics is satisfied, a subset of candidate modes {SKIP, MODE_16*16, MODE_16*8, MODE_8*16} is assigned to the current macroblock. For the macroblocks that are neither skipped nor belonging to the above subset of modes, the monotonicity property (Zhou, 2004) is used to predict even more macroblocks. For very similar RD performance, (33%-90%) reduction in run times can be achieved as compared to the standard. Compared to (Jeon, 2003; Lee, 2004) that were used as input to the standard, (Grecos, 2006) is faster by 9-23% for very similar RD performance. The ideas in (Grecos, 2006) can be implemented in both the simple and main profiles.

In (Seok, 2008), the authors propose a fast mode decision algorithm using a filter bank of Kalman filters for the H.264/AVC. The basic idea is similar with (Jeon, 2003; Grecos, 2006). A simplified Kalman filter that evaluates an expected RD cost for current macroblock can be built based on the estimated RD cost for the previous macroblock mode, the real RD cost of previous macroblock mode, and the adaptation gain. In order to classify each category, the authors employ three Kalman filters: EJa (the expected RD cost for all macroblock modes together), EJp (the expected RD cost for all inter macroblock modes) and EJi (the expected RD cost for all intra macroblock modes). Macroblock modes can be categorized into four classes based on these three filters and each class contains a subset of all modes. The algorithm has two steps. Firstly the current macroblock is encoded using some candidate mode types of each class. In the second step, if the minimal RD cost of the current macroblock is less than the average RD cost of the mode types of all previously encoded macroblocks in the current frame, the candidate mode set for the current macroblock is reduced. Otherwise, the candidate mode set for the current macroblock is increased. The authors claimed encoding speed ups of about 30% with small degradation of video quality as compared to the standard. Since this algorithm is designed especially for high definition video encoding, it cannot be used for low bit rate and/or low resolution video sequences for two reasons. Firstly the MODE_8*8 is disabled in the technique due to the insignificant effect of the MODE_8*8 in high definition video in high bit rates. But without using the MODE_8*8, the RD performance will be degraded significantly for the low bit rate cases. Secondly, an initial phase of collecting at least 15 macroblock modes is needed before the estimation can be performed by using Kalman filters. For low resolution video sequences which are typical for video on mobile devices, the collection time for filter evaluation is prohibitive thus both reducing overall time savings but more importantly potentially violating real time constraints for bidirectional communications.

The works in (Choi, 2006; Kuo, 2006; Wang, 2007) propose fast inter mode decisions based on the useful information extracted from the motion estimation step. B. D. Choi (2006) contributed a scheme to jointly optimize inter mode selection and ME using multi-resolution analysis. The multi-resolution motion estimation based on the discrete wavelet transform and modified integer transform is employed in the 4x4 sub macroblock level. Different search patterns are chosen in different bands. Subsequently, edge intensity is calculated in each band for each 4x4 sub macroblock. Homogeneity properties are found using linear or quadrature discriminant functions for both the 8x8 sub-macroblock and 16x16 macroblock levels. Candidate modes for the current macroblock are assigned based on this homogeneity information. Experiments are performed in slow and average motion video sequences. The encoding time is reduced by 60% on average with up to 0.15 dB PSNR decrease as compared to the standard. However, no bit rate information is provided. Instead of using multi-resolution motion estimation, (Kuo, 2006) simply used a diamond search (DS) motion estimation algorithm for each 4x4 block of the current macroblock. Sixteen motion vectors called the seed motion field are collected after DS. The Bhattacharyya distance is calculated to measure the separability of motion field classification. If the seed motion field is separable, maximum likelihood classification based on the motion vectors distribution is employed to find which mode is most likely from the set {MODE_16*16, MODE_16*8, MODE_8*16, MODE_8*8}. If the motion field not separable, the predicted RD costs are calculated using the seed motion vectors and the seed motion vector which gives the minimal cost is set as the search center for motion refinement. If MODE_8*8 is selected as the best mode, the motion estimation of block types in the set {INTER_8*8, INTER_8*4, INTER_4*8, INTER_4*4} can be sped up by using a predicted initial search position and an adaptive search range based on the seed motion vectors information. Compared with (Yin, 2003; Kuo, 2006), this algorithm achieved more time savings with similar RD performance. In (Wang, 2007), motion estimation is implemented in the conventional way from 16x16 to 4x4 block sizes. An all-zero coefficient blocks detection technique similar to (Yu, 2008) is adopted during the motion estimation. If all the 4x4 blocks within the current block are determined as all-zero coefficient blocks, the motion estimation is terminated and the rest of modes are skipped. If there are some none zero coefficients in the current macroblock, spatial and temporal homogeneity information with the help of the fixed thresholds is employed to predict the candidate modes. About 52% (Simple Profile) and 44% (Main Profile) of time on the average can be saved with 0.06 PSNR reduction and 0.80% bit rate increase as compared to the standard.

In (Yu, 2008), the authors proposed a hierarchical structure comprising of three layers for fast inter mode decision. The first layer is a fast skip mode detection algorithm using all-zero coefficient blocks as well as the spatial and temporal skip mode information. The central idea in the second layer is homogeneous contents analysis in the DCT domain. High AC components of the current macroblock after the DCT transform indicate a homogeneous macroblock. So the total energy of the AC coefficients of a macroblock is used for classifying it into two categories (low spatial complexity and high spatial complexity) with the help of a fixed threshold after empirical evaluation. Low spatial complexity macroblocks will be assigned the candidate mode set {SKIP, MODE_16*16}, whereas high complexity macroblocks the set {MODE_16*8, MODE_8*16}. From the above two candidate mode sets, if the lowest RD cost mode is in the low spatial complexity set, MODE_8*8 and its sub-modes will not be examined. Otherwise, the algorithm will go to the third layer. In the third layer, motion estimation will be performed in the partition size of the 8x8 block. If the RD cost of the current 8x8 block is bigger than a quarter of the RD cost of the best mode in the previous layers, modes of smaller partition blocks are ignored. The simulations show that up to 75% time savings is achievable with very similar RD performance as compared to the standard.

In summary, due to the extensive use of inter prediction in coded video sequences and the multitude of INTER coding modes in H.264/AVC, a lot of research efforts target fast INTER mode decision algorithms. Despite the very high diversity in techniques that are applied to achieve a fast mode decision, a number of general classes can be identified: techniques that predict the current mode based on spatio-temporal information (e.g., based on SA(T)D, MAD), techniques that use a prediction model based on mode statistics, probabilities, or mode regions, techniques that also incorporate (fast) motion estimation in the mode decision process, and techniques that efficiently predict the bit rate of certain modes (thus eliminating multiple redundant encoding steps). The reported results indicate that overall, gains in the range of 30% - 80% can be achieved by applying fast INTER mode decision. Due to the high diversity in techniques and the different comparison points for complexity and RD performance, it is impossible to state that one class of techniques performs better than another. However, mode decision techniques that use a statistical model to predict modes tend to report slightly higher gains. Also combining motion estimation and mode decision looks very promising.

Fast Intra Mode Decision

The total number of candidate intra modes for a macroblock is five hundred and ninety two, thus imposing a high computation load of the encoder and triggering of course a flurry of research efforts for the reduction of this load. In (Meng, 2003; Zhang, 2004; Pan, 2003; Fu, 2004; Tsai, 2008; Yang, 2004; Cheng, 2005) fast intra mode decision is based on pixel domain analysis. B. Meng (2003) proposed a fast intra-prediction algorithm for INTRA4*4 modes. According to this work, pixels in 4x4 blocks are categorized into 4 groups and each group is a "down-sampled" version of the original block. The best prediction mode is chosen in a computationally efficient manner by using both SAD and the quantisation parameter to check some of these groups of blocks. Due to the correlation of intra modes directions, the best prediction mode's two neighbouring directional modes are also chosen as candidate modes, so finally the candidate mode set has cardinality of three. For improving the speed of intra mode prediction, thresholds for early termination are also used. Computation can be saved by not only examining a small set of INTRA4*4 mode candidates but also by using fewer pixels due to down-sampling of the original block. In order to achieve significant time savings and good RD performance, different thresholds for different video sequences based on different quantization parameters need to be set before encoding. Evidently, the setting of thresholds can be problematic in one pass encoding schemes with video of unknown contents.

Y. Zhang's (Zhang, 2004) algorithm is based on two observations. One is that the best INTRA4*4 mode in 4x4 blocks is highly likely to be in the dominant direction of the local edges. The other is that the DC mode has higher probability to be the best mode compared to other intra 4x4 modes. According to the algorithm, a 4x4 block is initially divided into four 2x2 blocks and feature analysis is performed to find the local dominant edge direction. The local dominant edge direction can be classified into 7 types namely, no obvious edge, vertical edge, horizontal edge, diagonal down/left edge, diagonal down/right edge, vertical-dominant edge and horizontal-dominant edge. A set of modes based on local edge direction information plus the DC mode are chosen to be the set of candidate modes. However there are two problems in the above algorithm which result in increase of bitrates and loss of PSNR as compared to the standard. Firstly the assumption of the best mode being in the dominant edge direction is not always true and secondly analysis of local edge information extraction based on the 2x2 blocks' intensity cannot be very accurate. According to the authors, 40% to 70% of computational complexity can be saved with less than 5.5% of bit rate increase and not more than 0.05dB PSNR degradation as compared to the standard.

Pan's algorithm (Pan et. al 2003) is based on local edge directional information in order to reduce the amount of calculations in intra prediction. Firstly, the Sobel edge operators (Gonzalez, 2002) are applied to the current frame to generate the edge map. Then an edge direction histogram is calculated from all the pixels in the block by summing up the amplitudes of those pixels with similar directions in the block. The histogram cell with the maximum amplitude indicates that there is a strong edge presence in that direction, and thus it is the direction of the best prediction mode. For increased accuracy, a small number of the most likely intra prediction modes are also chosen for RDO calculation. The drawbacks are similar with (Zhang, 2004). This method shows average gains of 60% in time savings which can be achieved with average increase of 5.9% in bit-rate and induces negligible loss of PSNR as compared to the standard.

In Pan's algorithm (2003), an edge map needs to be produced for the whole picture and this needs some computation. F. Fu (2004) proposed a faster algorithm which performs only partial edge detection since he observed that mode decision normally depends on the edge information between the left, top blocks and current block. The set of candidate modes is chosen based on Pan's intra candidate mode selection but using partial edge information. A most probable intra mode type for current macroblock can also be obtained based on the intra mode types and costs of the left and top macroblocks of the current macroblock, and this most probable intra mode type can be used for an early termination criterion to disable INTRA4*4 or INTRA16*16 mode decision. If the most probable intra mode type is INTRA4*4 and the Rate distortion cost of the best INTRA4*4 mode is significant, the INTRA16*16 will be disabled. If the most probable intra mode type is INTRA16*16 and the Rate distortion cost of the best INTRA16*16 is negligible, the INTRA4*4 modes will not be examined. Compared with Pan (2003) this work reduces computation time by a factor of 2 to the expense of very small RD performance degradation.

In (Tsai, 2008) another intensity gradient filter (Ma, 1998) called texture edge flow is adopted instead of the Sobel edge detector. The difference between these two gradient filters is that the Sobel detector only uses two directional gradients to extract edge information but the texture edge flow uses four. This difference makes texture edge flow more suitable for the H.264/AVC than the Sobel detector since the energy information it extracts from different directions is more accurate. By ordering the energy data of different intra modes, a sub-set of candidate intra modes can be selected. A good balance can be achieved between computation complexity and RD performance by selecting an appropriate number of candidate modes. Based on experiments, four modes for INTRA4*4 and 2 modes for INTRA16*16 were found to be a good option. Compared with the standard, the algorithm can achieve around 76% time savings with an average of 0.17 dB PSNR decrease and 2.83% bit rate increase. Authors also claim that their scheme outperforms the work in (Pan, 2003) in terms of PSNR (0.05dB increase), bit-rate (0.66% decrease) and encoding time savings (around 43% savings).

A fast INTRA16*16/INTRA4*4 mode selection algorithm is proposed in (Yang, 2004) which uses macroblock properties. The main idea is that INTRA16*16 modes are more suitable for predicting smooth areas and INTRA4*4 modes can achieve good prediction in regions with significant detail. Based on this idea, pixel level analysis based on thresholds is performed to classify macroblocks into smooth and non-smooth ones and assign to them the aforementioned mode groups. The proposed algorithm can achieve 10%-40% computation reduction with similar PSNR and bitrate performance compared to the standard.

C. Cheng (2005) assumed that the J costs of INTRA4*4 modes are monotonic and proposed a three-step fast intra mode prediction algorithm at the end of which only six INTRA4*4 modes need to be examined instead of nine in the standard. Experiments show that this algorithm can obtain about 31% time savings on the average for intra mode decision, with similar PSNR and about 1%

of bit rate increase compared to the standard.

Instead of using spatial (pixel) domain analysis, the algorithms in (Sarwer, 2008; Wang, 2007; Yu, 2005) develop fast intra mode decision algorithms using transformed domain analysis. The work in (Sarwer, 2008) used the SATD metric for some of the INTRA4*4 modes. Due to the high spatial correlation information in the natural video sequences, the probability that the best mode for the current macroblock is the same as the mode of the upper or left macroblocks is very high. The proposed method reduces the number of candidate modes from nine to one based on the combination of the rank of the SATD values for the subset of the examined INTRA4*4 modes and the most probable mode i.e the one of the top or left macroblocks. Thresholds are also needed in this algorithm. Experimental results show that this scheme saves about 70% of the encoding time with 0.06dB loss and 2.24% bit rate increase as compared to the standard. H. M. Wang et al (Wang 2007) proposed a fast intra 4x4 mode decision algorithm based on a fast SATD computation scheme which reduces the full computation of the metric by about 50% The INTRA 4*4 mode decision is further sped up by a two stage simplified scheme. In the first step, only five out of nine possible modes are examined and an extra mode based on the SAD criterion is further examined for the current block in the second step. The achievable time savings are 70% on the average, with 0.05dB quality loss and 0.51% bit rate increase as compared to the standard. In (Yu, 2005), a fast intra mode prediction algorithm is proposed based on a fast partial DCT transform scheme. The DC coefficient and the low-frequency AC coefficients which contain more energy than the high-frequency ones, are calculated using this fast transform scheme for all intra modes. Between one and four modes with the smallest energy plus the most probable mode are selected as the candidate modes for the current block. Compared with Pan (2003) the average time savings are increased and the RD performance is improved.

Changsung Kim (2004; 2006) proposed a new fast intra prediction mode scheme which is a combination of spatial (pixel) and transformed domain analysis. The proposed algorithm adopts a multi-stage mode decision process which uses a spatial domain feature (SAD) and a transform domain feature (SATD) together to remove unlikely intra modes. In the final step of this scheme, a new Rate Distortion model is used to find the best intra mode from the set of intra modes when the QP is larger than a threshold (sixteen in this case). Since the RD model predicts the rate and distortion instead of actually measuring them, some computation can be further saved in the mode decision step due to avoiding macroblock reconstructions. When the QP is smaller than the threshold, full RD-based mode decision (including reconstructions) will be used. Experiments show reduction of the computational complexity of intra mode decision of up to 90%, with little PSNR degradation and at very similar bitrates compared with the standard.

In summary, there are two main approaches for intra mode decision techniques: based on pixel information (e.g., edge detection) or based on transformed domain analysis (e.g., SATD). Because pixel-domain processing also implies some complexity, the reported gains in complexity (for the same RD performance) for pixel-domain techniques (30% - 70%) tend to be lower than those for the transform-domain techniques (50% - 90%). Of course, these complexity gains only apply for intra-only coding and the impact thereof on the overall mode decision process will be much smaller (since the majority of execution time is spent on inter coding).

Fast Mode Selection Between Intra and Inter Sets of Modes

The selective intra/inter mode decision technique of (Jeon, 2003) contains a spatial (pixel) domain analysis which aids the decision of whether the intra modes for the current macroblock should be checked or not. This technique uses the average boundary error (ABE) between the pixels on the boundary of the current and its adjacent encoded blocks under the best inter mode as an indicator of the degree of spatial correlation and the average rate (AR), i.e., the average number of bits consumed to encode the motion-compensated residual data under the best inter mode as an indicator of the degree of temporal correlation. Subsequently, the average rate for the best inter mode and the average boundary error for the current block are compared. If AR <k*ABE which k is a certain user-defined positive number, there is no need to consider any intra modes for the current macroblock so only inter mode decisions are performed. Otherwise, both inter and intra modes should be considered. Experiments show that about 20% of time savings on the average can be achieved, at the expense of 0.036dB PSNR reduction and 0.7% bitrates increase as compared to the standard.

A feature-based fast intra/inter mode decision method is proposed by Changsung Kim (Kim 2007) to reduce the encoder complexity of the H.264 video coding standard. Three extracted features are considered: the spatial domain correlation found by applying SATD on the intra prediction residual, the temporal domain correlation found by applying SATD on the difference between the current macroblock and the best matched one in the reference slice, and finally the motion vector length. These three features are extracted from the current macroblock to form a 3D (three-Dimension) feature vector. Subsequently, based on the location of feature vector in the feature space, the video slices are partitioned into three regions: risk-free, risk-tolerable and risk-intolerable regions. Depending on the small cells into which the risk region is quantized, the risk minimizing mode can be found. Finally, if the feature vector lies in the risk-free region, the mode decision is made based on simple feature comparison. If the feature vector lies in the risk-tolerable region, the risk minimizing mode is selected. If the feature vector is in the risk-intolerable region, a full RD based mode decision process is performed. Experiments show reductions of about 19-25% of the total encoding time at the expense of 4.1% average rate increase and 0.27% average PSNR loss as compared to the H.264/AVC standard.

CONCLUSION AND FUTURE RESEARCH DIRECTIONS

The H.264/AVC standard has shown significant Rate Distortion improvements as compared to other standards for video compression. Special design of the network abstraction layer also makes it more flexible for application to a wide variety of network environments. Considering that transmission bandwidth is still a valuable commodity, H.264/AVC becomes a very promising standard in applications ranging from television broadcast to video for mobile devices. However, high coding performance comes with high computation and power consuming cost, which makes this standard problematic in its use for low delay applications such as video conferencing. In this context, fast mode decision techniques are very important since they enable meeting these low delay requirements. This chapter gives an overview of the state of the art in fast mode decision algorithms. Some of our most important findings are firstly that motion estimation and mode decision are interwined and the majority of time savings in mode decision occurs due to the avoidance of motion estimations, secondly that reported gains in speed are often hard to interpret since they depend on the experimental conditions/content and thirdly that the literature rarely compares mode decision methods and whenever this occurs there is not a clear winner,

We would like to conclude by stressing two facts. Firstly, in this chapter we concentrated on algorithmic rather than micro architectural optimizations and added speed-ups and power savings are possible if we take the latter class of optimizations into account. Classic techniques for optimizations at the micro level include module and functional unit level parallelism and clock gating. In module and functional unit level parallelism, different parts of an algorithm and different operations in each module are executed concurrently in a pipelined manner, thus improving the speed significantly. In clock gating, clocks can be deactivated for functions when they are not required, thereby reducing the chip power consumption. Secondly that fast mode decision is still an on-going research area and as such there is a lot of room for improvement but also for extensions in the recent H264 derivatives. A typical example is the lack of much research work in the contexts of H264/SVC (Schwarz, 2007) and H264/MVC (Vetro, 2006), the scalable and multi-view extensions of the H.264/AVC. We hope that this chapter will ignite more research efforts in these directions.

REFERENCES

Ahmad, A., Khan, N., Masud, S., & Maud, M. A. (2004, January). Efficient Block Size Selection in H.264 Video Coding Standard. *Electronics Letters, 40*(1). doi:10.1049/el:20040068

Bjontegaard, G. (Apri, l2001), *Calculation of Average PSNR Differences Between RD-Curves*, ITU-T Q6/SG16, Doc. VCEG-M33.

Bystrom, M., Richardson, I., & Zhao, Y. (2008, February). Efficient Mode Selection for H.264 Complexity Reduction in a Bayesian Framework. *Signal Processing Image Communication, 23*(2), 71–86. doi:10.1016/j.image.2007.11.001

Chang, A., Au, O. C., & Yeung, Y. M. (July, 2003) A Novel Approach to Fast Multi-block Motion Estimation for H.264 Video Coding. *International Conference on Multimedia and Expo, 2003. ICME'03, 1*, 6-9.

Cheng, C., & Chang, T. (May, 2005) Fast Three Step Intra Prediction Algorithm for 4x4 Blocks in H.264. *IEEE International Symposium on Circuits and System, 2005*, ISCAS 2005.

Chia A., Woo, Y., Martin, G. R., & Park, H. (February, 2008,). Fast Inter-Mode Selection in the H.264/AVC Standard Using a Hierarchical Decision Process. *IEEE Transaction on Circuits and System for Video Technology, 18*(2).

Choi, B. D., Nam, J. H., Hwang, M. C., & Ko, S. J. (2006). Fast motion estimation and intermode selection for H.264. *EURASIP Journal on Applied Signal Processing, 2006*, 1–8. doi:10.1155/ASP/2006/71643

Fu, F., Lin, X., & Xu, L. (August, 2004,). Fast Intra Prediction Algorithm in H.264/AVC. In *Proceedings of7th International Conference On Signal Processing*, ICSP'04, Vol 2, 31Aug-4.

Gonzalez, R. C., & Woods, R. E. (2002). *Digital Image Processing*. Upper Saddle River, NJ: Prentice Hall.

Grecos, C. & Yang, M. (June, 2005). Fast inter mode prediction for P slices in the H264 video coding standard. *IEEE Transaction on Broadcasting, 51*(2).

Grecos, C., & Yang, M. (2006, December). Fast mode prediction for the Baseline and Main profiles in the H.264 video coding standard. *IEEE Transactions on Multimedia, 8*(6). doi:10.1109/TMM.2006.884631

Grecos, C., & Yang, M. Y. (2007). Coding of Audio Visual Objects Part 2: Visual, ISO/IEC JTC1, ISO/IEC 14496-2 (MPEG-4 Visual version 1). *Digital Signal Processing, 17*(3), 652–664. doi:10.1016/j.dsp.2005.11.005

Grecos, C., & Yang, M. Y. (2007). Exploiting temporal information and adaptive thresholding for fast mode decision in H264 video coding standard . *Multidimensional Systems and Signal Processing, 18*, 309–316. doi:10.1007/s11045-006-0006-8

Haan, D., Biezen, G.,, P. W. A. C., Ojo, O. A., & Huijgen, H. (1993). True motion estimation with 3-D recursive search block matching. *IEEE Trans. Circuits Syst. Video Technol., 3*,368–379, 388

Hsu, C. Y., Ortega, A., & Reibman, A. R. (1997). Joint Selection of Source and Channel Rate for VBR Video Transmission under ATM policing constraints. *IEEE Journal on Sel. Areas in Communications . Special Issue on Real-Time Video Services in Multimedia Networks, 15*(6), 1016–1028.

Jeon, B. & J. Lee, (December, 2003). *Fast Mode Decision for H264,* ISO/IEC JTC1/SC29/WG11 and ITU-T SG16, Input Document JVT-J033.

Jing, X., & Chau, L. P. (2004, August). Fast Approach for H.264 Inter Mode Decision. *Electronics Letters, 40*(17). doi:10.1049/el:20045243

Jing, X. & Chau, L. P. (September, 2004). Fast approach for H.264 inter-mode decision. *Electron. Lett., 40*17),1050-1052.

Kannangara, C. S., Richardson, I. E. G., Bystrom, M., Solera, J., Zhao, Y., MacLennan, A., & Cooney, R. (2006, February). Low Complexity Skip Prediction for H.264 Through Lagrangian Cost Estimation. *IEEE Transactions on Circuits and Systems for Video Technology, 16*(2), 202–208. doi:10.1109/TCSVT.2005.859026

Kim, B. G. (2008, February). Novel Inter-Mode Decision Algorithm Based on Macroblock Tracking for the P-Slice in H.264/AVC Video Coding. *IEEE Transactions on Circuits and Systems for Video Technology, 18*(2). doi:10.1109/TCSVT.2008.918121

Kim, C., Hsuan-Huei, S., & Kuo, C. C. J. (January, 2004). *Multistage Mode Decision for Intra Prediction in H.264 Codec.* IS&T/SPIE 16th Annual Symposium EI, Visual Communications and Image Processing, Orlando, FL.

Kim, C., Hsuan-Huei, S., & Kuo, C. C. J. (October, 2004). Feature-Based Intra- Prediction Mode Decision for H.264. In *IEEE Proceedings of International Conference Image Processing,* submitted, Singapole.

Kim, C., Hsuan-Huei, S., & Kuo, C. C. J. (April, 2006). Fast H.264 Intra-Prediction Mode Selection Using Joint Spatial and Transform Domain Features. *Journal of Visual Communication and Image Representation,* ELSEVIER.

Kim, C., & Kuo, C. C. J. (2007, April). Feature-Based Intra/Inter Coding Mode Selection for H.264/AVC. *IEEE Transactions on Circuits and Systems for Video Technology, 17*(4), 441–453. doi:10.1109/TCSVT.2006.888829

Kim, H., & Altunbasak, Y. (October, 2004). Low-complexity Macroblock Mode Selection for H.264/AVC Encoders. *IEEE International Conference on Image Processing, 2004, ICIP '04, 2*, 24-27.

Kuo, T. Y., & Chan, C. H. (2006, October). Fast Variable Block Size Motion Estimation for H.264 Using Likelihood and Correlation of Motion Field. *IEEE Transactions on Circuits and Systems for Video Technology, 16*(10). doi:10.1109/TCSVT.2006.883512

Lee, J., Choi, I. Choi, W., & Jeon, B. (March, 2004). *Fast Mode Decision for B slice,* ISO/IEC JTC1/SC29/WG11 and ITU-T SG16, Input Document JVT-K021.

Lim, K. P., Wu, S., Wu, D. J., Rahardja, S., Lin, X., Pan, F., & Li, Z. G. (September, 2003). *Fast Inter Mode Selection,* ISO/IEC JTC1/SC29/WG11 and ITU-T SG16, Input Document JVT-I020.

Ma, W. Y., & Manjunath, B. S. (1998, May). A Texture Thesaurus for Browsing Large Aerial Photographs. *Journal of the American Society for Information Science American Society for Information Science, 49*(7), 633–648. doi:10.1002/(SICI)1097-4571(19980515)49:7<633::AID-ASI5>3.0.CO;2-N

Malvar, H., Karczewicz, M., & Kerofsky, L. (July, 2003). Low complexity transform and quantization in H.264/AVC. *IEEE Trans. Circuits Syst. Video Technol., 13*(7).

Meng, B., Au, O. C., Wong, C., & Lam, H. (July, 2003). *Efficient intra-prediction mode selection for 4x4 blocks in H.264*, 2003 IEEE lot. Conf. Multimedia &Expo (ICME2003). Baltimore, MD

Ostermann, J., Bormans, J., List, P., Marpe, D., Narroschke, M., Pereira, F., Stockammer, T., & Wedi, T. (2004). Video Coding with H.264/AVC: Tools, Performance, and Complexity. *IEEE Circuits and Systems, 4*(1).

Pan, F., Lin, X., Susanto, R., Lim, K. P., Li, Z. G., Feng, G. N., Wu, D. J., & Wu, S. (n.d.). *Fast Mode Decision for Intra Prediction*, ISO/IEC JTC1/SC29/WG11 and ITU-T SG16, Input Document JVT-G013, March 2003.

Salagdo, L., & Nieto, M. (October, 2006). Sequence independent very fast mode decision algorithm on H.264/AVC baseline profile. In *Proceedings of Int. Conf. Image Process.*, Atlanta, GA, USA, pp. 41-44.

Sarwer, M. G., Po, L. M., & Wu, Q. M. (2008). Fast sum of absolute transformed difference based 4x4 intra-mode decision of H.264/AVC video coding standard . *Signal Processing Image Communication, 23*, 571–580. doi:10.1016/j.image.2008.05.002

Schwarz, H., Marpe, D., & Wiegand, T. (2007, September). Overview of the Scalable Video Coding Extension of the H.264/AVC Standard. *IEEE Transactions on Circuits and Systems for Video Technology, 17*(9). doi:10.1109/TCSVT.2007.905532

Seok, J., Lee, J. W., & Cho, C. S. (2008, February). Fast Block Mode Decision Algorithm in H.264/AVC using a Filter Bank of Kalman Filters for High Definition Encoding. *Multimedia Systems, 13*(5-6). doi:10.1007/s00530-007-0100-2

Source code link: http://iphome.hhi.de/suehring/tml/download/old_jm/ jm81a.zip

Tanizawa, A., Koto, S., Chujoh, T., & Kikuchi, Y. (October, 2004) A Study on Fast Rate-distortion Optimized Coding Mode Decision for H.264. *IEEE International Conference on Image Processing, 2004, ICIP '04., 2*, 24-27.

Tsai, A. C., Paul, A. P., & Wang, J. C. (May, 2008). Intensity Gradient Technique for Efficient Intra-Prediction in H.264/AVC. *IEEE Transactions on Circuits and Systems for Video Technique, 18*, (5).

Vetro, A., Su, Y., Kimata, H., & Smolic, A. (October, 2006). *Joint Draft 1.0 on Multiview Video Coding*, Doc. JVT-U209 Joint Video Team, Hangzhou, China.

Wang, H., Kwong, S., & Kok, C. W. (2007, June). An Efficient Mode Decision Algorithm for H.264/AVC Encoding Optimization. *IEEE Transactions on Multimedia, 9*(4). doi:10.1109/TMM.2007.893345

Wang, H. M., Tseng, C. H., & Yang, J. F. (2007, September). Computation Reduction for Intra 4x4 Mode Decision with SATD Criterion in H.264/AVC. *Signal Processing, IET, 1*(3), 121–127. doi:10.1049/iet-spr:20065007

Wiegan, T., & Girod, B. (September, 2001), *Lagrange Multiplier Selection in Hybrid Video Coder Control,* IEEE International Conference on Image Processing (ICIP'01), Thessaloniki, Greece.

Wiegand, T., Sullivan, G. J., Bjntegaard, G., & Luthra, A. (2003, July). Overview of the H.264/ AVC video coding standard. *IEEE Transactions on Circuits and Systems for Video Technology, 13*(7). doi:10.1109/TCSVT.2003.815165

Yang, C. PO, L., & Lam, W. (October, 2004). A Fast H.264 Intra Prediction Algorithm using Macroblock Properties. *International Conference on Image Processing, ICIP'04, 1*, 24-27.

Yin, P., Cheong, H. Y., Tourapis, A., & Boyce, J. (2003). Fast *Mode Decision and Motion Estimation for JVT/H.264*, IEEE International Conference in Image Processing.

Yu, A. C., Martin, G., & Park, H. (September, 2005).A Frequency Domain Approach to Intra Mode Selection in H.264/AVC. In *Proceedings of 13th European Signal Processing Conference (EUSIPCO) 05, 4PP.*, Antalya, Turkey.

Zhang, Y., Dai, F., & Lin, S. (June, 2004). Fast 4x4 Intra-prediction Mode Selection for H.264. *2004 IEEE International Conference on Multimedia and Expo (ICME2004), 2*, 27-30.

Zhou, Z., & Sun, M. T. (2004, October) "Fast Macroblock Inter Mode Decision and Motion Estimation for H.264/MPEG-4 AVC," IEEE International Conference on Image Processing, ICIP'04. vol 2, 24-27.

Zhu, D., Dai, Q., & Ding, R. (June 2004). Fast Inter Prediction Mode Decision for H.264. *IEEE International Conference on Multimedia and Expo, ICME '04, 2*, 27-30.

Chapter 22
Mobile Video Streaming

Chung-wei Lee
University of Illinois at Springfield, USA

Joshua L. Smith
University of Illinois at Springfield, USA

ABSTRACT

Mobile video streaming is a natural augmentation to today's thriving Internet video streaming service. With the rapid growth of the capability of mobile handheld devices and abundant bandwidth from high-speed wireless networks, it is expected that mobile video streaming service will soon become a lucrative business section and a thrust for technological advancement on computer and telecommunication industries. In this chapter, essential technical components for constructing mobile video streaming systems are introduced. They include the latest development on broadband wireless technology and video-capable mobile handheld devices. As many modern technologies are often driven by consumer demand, user experience and expectation are discussed from the perspective of mobile video streaming. At the end, several cutting-edge research and development breakthroughs are presented as they may change the future of mobile video streaming systems.

INTRODUCTION

Mobile video streaming not only is a rising mobile commerce model but also facilitates other mobile commerce businesses. Within the past several years, digital video streaming has become one of the key Internet applications that have profound impact on our daily life. This success changes our everyday entertainment activities as well as many business operations around the world. As "YouTube" (YouTube, 2009) grows to be one of the most popular online video streaming websites, the demand for mobile video streaming service has gained tremendous momentum to becoming the next big wave in the next-generation wireless Internet. The main driving forces behind this trend are the popularity of modern mobile handheld devices which are capable of processing digital video, and the fast development and deployment of broadband wireless networks.

DOI: 10.4018/978-1-61520-761-9.ch022

In the early age of Internet there was no video streaming because of the insufficient bandwidth on computer networks. Compared with typical Internet data services such as emails and file transfers, digital video streaming requires much more stringent requirements on the end-to-end packet delivery latency and considerable amount of network bandwidth consumption. Only after broadband Internet access became commonly available to general public though massive deployment of DSL (digital subscriber line) and high-speed cable modem, online video streaming service began to blossom.

A typical network-based digital video system involves the following three processes: video content creation, video storage, and video distribution. Apparently the first step is to create video contents. While it was once a costly process requiring expensive professional camera equipments, today almost everybody can shoot video with inexpensive digital cameras/camcorders. This wide availability of video content production makes everyone a show producer, and therefore furnishes the enormous video collection in social networking websites such as YouTube. Since recorded videos require significant memory space for storage (and further distribution), it is almost necessary to apply state-of-the-art video compression algorithms/standards before they are ready for streaming over the network. Currently the most popular digital video standards include MPEG-2. H.263, and H.264/MPEG-4 AVC. The last, but definitely not the least, step is to distribute the video contents from one place to the other(s) though wired or wireless networks. This process is also known as "streaming" because video content is delivered packet by packet (like continuous water flow) from the source to the destination. The source and destination of video stream can be powerful computer servers (such as YouTube web server), desktop computers, or even small handheld computing devices.

The focus in this chapter is to provide readers with various technical perspectives surrounding the mobile video streaming technologies. First, different types and configurations of mobile video streaming systems are classified into three general categories. Then, the key components in such systems will be introduced and compared (when applicable). This includes wireless networks and video-capable mobile handheld devices. Also, in this consumer-centered commerce, user experience and expectation are discussed. Finally, some thoughts on future research directions and conclusions are presented.

MOBILE VIDEO STREAMING SYSTEM CLASSIFICATION

When it comes to the business life today everyone is always on the move and rarely able to take the time to watch their favorite shows or sports games. Because of the lack of time and the demand for entertainment the market for video streaming handheld devices is one of the only markets in America that is constantly rising. Mobile video streaming systems can be categorized into three different types. There is broadcasting, which is mobile streaming television directly from the source as it appears, such as regular television. There is also the client-server approach for mobile video which is often referred to as mobile video on demand. This is when the user pulls a specific video from a host at any given time, such as YouTube. Lastly there is peer to peer sharing which is when every mobile device acts as a server and a client both hosting and receiving information. Regardless of the type of mobile video streaming system the demand for such service is at an all time high and experts are saying that the demand for such systems will only grow in the future.

Broadcasting

Broadcasting in the paradigm of mobile video streaming means that television program videos are directly sent to mobile handheld devices from the source as it appears, such as regular television stations and cable TV service providers. One of the first broadcasting devices on the U.S. market was running MobiTV (MobiTV, 2009) and was offered by Sprint. Currently AT&T also uses MobiTV for broadcasting TV with selected handheld devices. Because of the abundant network systems involved this is one of the easiest ways to stream video to handheld devices. The images are sent from a single source to all devices currently viewing that video. This one-to-many feature drastically cuts down on network bandwidth usage. Therefore, for example, AT&T currently doesn't allow much video streaming to occur concurrently because of the bandwidth that is required to run such network application. There are stipulations that will allow certain broadcasting to mobile handheld devices. Applications like Sling, which takes an image directly from a television set and sends it to a mobile handheld device, are not allowed by the terms of service associated with AT&T. This one-to-one type of broadcasting is said to take too much bandwidth and would end up crippling the network. Some more popular things that are approved by AT&T include the major league baseball broadcasting. This broadcasting is achieved because it is a one-to-many type of broadcasting which takes significantly less accumulated bandwidth and relieves some of the strain that would typically be put on the network.

MobiTV has been one of the leading broadcasting applications on the market today and to prove this a third carrier tried to pick up the ability to use MobiTV on all of their handheld devices. (Krakow, 2004) states that "Sprint's basic MobiTV service allows you to sample content from a number of providers including NBC News. NBC Mobile is constantly updating newscasts specially tailored for MobiTV's small screen as well providing longer interviews and stories from NBC News shows and MSNBC".

By the end of the fiscal year 2009, more than 15 television networks are planning to be able to broadcast to mobile handheld devices. This jump in commitment does not come lightly in these difficult economic times. According to the Datamonitor Research store, "By 2009, 69 million people worldwide are expected to subscribe to mobile television service, generating total expected revenue of $5.5 billion" (Schatz, Wagner, Jordan, 2007).

Client-Server

Client-server is often referred to as mobile video on demand. This is possibly the most desired feature for the majority of customers when purchasing a handheld device. This allows the user to browse videos and watch exactly what they want to watch when they want to watch it. YouTube is a prime example of this and because they have changed their website to allow handheld devices to browse it they have increased their viewing range. Instead of only getting to users sitting at a computer because of this people can now browse YouTube while on a bus, a plane or even in a car (however, driving safety should be seriously taken into consideration). Another form of this mobile video on demand is the new sensation "Hulu" (Hulu, 2009). Television networks can upload their videos to the servers at Hulu, or use their own servers, so that users can go to a centralized place to receive these videos. Due to the easy installation of FlashLite on most handheld devices these websites are available to view within seconds.

According to Brough Turner, co-founder, senior vice president and CTO of NMS Communication, mobile video on demand is the only choice that makes sense when it comes to the type of video based systems. He claims that broadcasting television is only holding back the revenue growth in the market of video-based systems and that mobile video on demand is the only "forward looking" option that we have.

Search Toppers (Search Toppers, 2009) is another company focused on hosting television shows and video clips for the use of user viewing. The only difference is that this company specializes in the use of mobile handheld devices unlike YouTube and Hulu where their main focus is for the user to use an entire computer.

Peer-to-Peer (P2P)

Peer-to-peer (P2P) is when every mobile device acts as a server and a client both hosting and receiving information. Many operating systems on handheld devices are capable of taking advantage of this. The development of mobile P2P technology has been deeply affected by the Internet-based file sharing network systems such as Gnuttela and BitTorrent. For example, Symella (Symella, 2009) is a Gnuttela client system for Nokia Series 60 mobile devices running Symbian OS platform, and SymTorrent (SymTorrent, 2009) is a BitTorrent client for Symbian S60 smartphones.

With increasing interest of mobile video streaming from general public, newer mobile P2P systems that are capable of video streaming has begun to emerge. Constrained by the limited capabilities (e.g., processor, memory, and battery), a hybrid approach for streaming low-resolution video to mobile handheld devices is demonstrated in the work of a Nokia research technical report (Karonen & Lahtinen, 2007). The feasibility of full-fledged P2P video streaming has been tested in a project utilizing Symbian-based high-end Nokia devices together with Wi-Fi networks (Xie, Li, Keung, 2008). As experiments showed, radio interference caused by multiple Wi-Fi devices operating in close vicinity and relatively low computation power may degrade the mobile video streaming performance.

Peer to peer sharing is the most legally controversial of the video based systems because of copyright infringements. CloudTrade is one of the leading competitors in the peer to peer networking game for the mobile handheld devices. "Our mission is to offer FREE anytime, anywhere access to your personal content through a platform that easily and LEGALLY lets you share with your friends" (CloudTrade, 2009). Through this application a user in China can upload family photos or a song that was recently obtained and pass these along to a friend in Europe, South America or anywhere else that can receive a signal. This system doubles as a recovery system so that if a user was to lose their handheld device they would be able to access all of the data that they recently uploaded to CloudTrade.

These three different types of mobile video streaming systems are illustrated in Figure 1. In the case of broadcasting, all receivers receive the same video from the source (e.g., MobiTV website). In the client-server structure, users can request different videos from the same server concurrently (i.e., indicated by different line shapes). In the peer-to-peer paradigm, each mobile station can serve as a video provider or receiver to any other station.

WIRELESS NETWORK INFRASTRUCTURE

Wireless network infrastructure provides essential digital video communication capability for mobile video streaming consumers and service suppliers. From the point of view of video-based mobile commerce, sometimes it is necessary for a wired network infrastructure, such as the Internet, to be augmented by wireless networks that support mobility for end users. Since Internet-based wired networks are more mature than their wireless counterparts, this section focuses on the wireless network infrastructure.

Wireless Local Area Network (WLAN)

Mobile handheld devices used in wireless networks are usually light-weight, easy to carry, and flexible in network configuration. Therefore, they are suitable for dynamic WLAN environments such as office networks, home networks, personal area networks (PANs), and ad hoc networks. In a one-hop WLAN, where an access point (AP)

Figure 1. Three different types of mobile video streaming systems

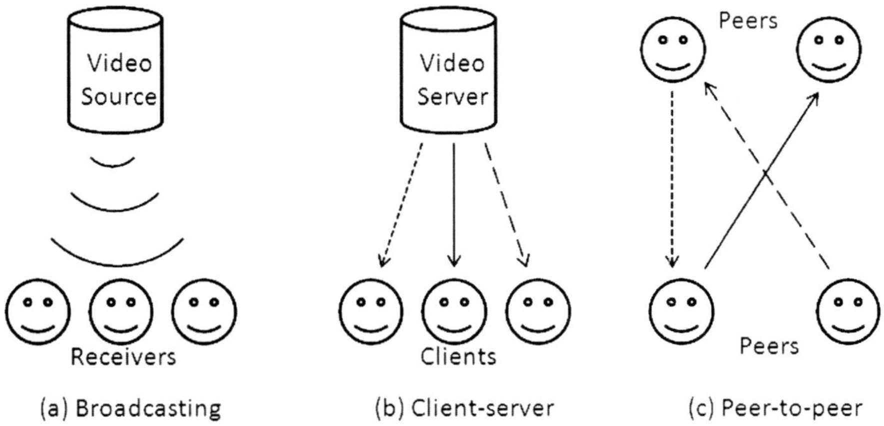

acting as a router or switch is a part of a wired network, mobile devices connect directly to the AP through radio channels. Data (including digital video) packets are relayed by the AP to the other end of a network connection. If no APs are available, mobile devices can form a wireless ad hoc network among themselves and exchange data packets or perform business transactions as necessary.

Currently the most popular WLAN technology is Wi-Fi, which is based on the IEEE 802.11 standard (IEEE 802.11, 2007). As the Wi-Fi evolves, its maximum transmission data rate continues to rise. For example, the original 802.11 standard could support up to 2 Mbps (mega-bit-per-second). Then 802.11b could reach 11 Mbps, 802.11g could achieve 54 Mbps. Recent 802.11n even claims the maximum rate to be 600 Mbps. With such a high-speed wireless communication channel, mobile video streaming becomes a feasible application that attracts millions of people around the world.

However, high-speed transmission channel does not guarantee high-performance for an end-to-end video streaming application because the underlying medium access control (MAC) protocol defined by IEEE 802.11 standards. The default MAC protocol in 802.11-based products is called carrier sense multiple access with collision avoidance (CSMA/CA). While pure CSMA/CA provides fair channel access to all stations (mobile devices), it lacks the capability of supporting functions for real-time continuous data traffic such as digital video. As the result, mobile video streaming performance can be bad in an uncontrolled or congested WLAN environment.

IEEE 802.11e is an approved amendment to the original IEEE 802.11 standard. IEEE 802.11e defines functions and mechanisms that can be used to support real-time traffic requirements. Collectively, they can be called as quality of service (QoS) enhancements. While 802.11e strives to be compatible with the original CSMA/CA, the QoS enhancements mainly come from EDCA (enhanced distributed channel access) and HCCA (HCF (hybrid coordinator function) controlled channel access). In EDCA, traffic data are assigned different levels of priority, with high-priority traffic being delivered faster statistically. HCCA is a very sophisticated coordination function which allows the coordinator (usually the AP) to perform precise QoS configuration so that detailed station and packet scheduling can be facilitated to improve multimedia (video and audio) streaming performance.

Wireless Cellular Network

There's a long history regarding the evolution of wireless cellular networks. Originally designed for voice-only communication, wireless cellular systems have been evolving from analog to digital, from circuit-switching to packet-switching, in order to accommodate data-oriented modern telecommunication applications. The first generation (1G) system such as the advanced mobile phone system (AMPS) and total access control system (TACS) had become obsolete, and thus does not play a significant role in today's wireless systems. The second generation (2G, 2.5G) such as the global system for mobile communications (GSM) and its enhancement general packet radio service (GPRS) can support data rate of only about 100 kbps. Its upgraded version, named enhanced data for global evolution (EDGE), is capable of supporting 384 kbps. In North America, most wireless system operators use time division multiple access (TDMA) and code division multiple access (CDMA) technologies in their cellular networks.

Currently, most of the cellular wireless networks in the world follow 2G, 2.5G, and 3G standards. The 3G systems with quality-of-service (QoS) capability are expected to dominate wireless cellular services in the near future. The two main technologies for 3G are Wideband CDMA (WCDMA), proposed by Ericsson, and CDMA2000, proposed by Qualcomm. Both are based on direct sequence spread spectrum (DSSS) technology. Technical differences between them include different chip rates, frame times, spectrum used, and time synchronization mechanisms. The WCDMA system can internetwork with GSM networks and has been strongly supported by the European Union, which calls it the Universal Mobile Telecommunications System (UMTS). CDMA2000 is backward-compatible with IS-95, which is widely deployed in North America. Generally services provided by 3G operators go beyond traditional voice-based communication. Many 3G subscribers use applications such as web browsing, emailing, and video/music downloading on daily basis, which is an integral part or our everyday life (Hu, Lee, Yeh, 2004).

While many wireless carriers are still in the process of deploying their 3G network systems, some operators have already advertised their 4G plans. In general, 4G is the system with capabilities beyond existing 3G services. It is expected for 4G to integrate all telecommunication services so that voice, data, music, and video can all be accessed transparently anytime, anywhere. To achieve this goal, an all-IP-based solution seems to be the most likely approach. When 4G systems are fully deployed, with its transmission rate at 15 – 100 Mbps, streaming high-definition television programs and movies to mobile handheld devices can become a reality.

Mobile WiMAX

A new wireless technology that has been gaining a tremendous momentum is called "Mobile WiMAX". It is based on the standard of IEEE 802.16m. According to a recent report on its standardization development and market growth (Kim, 2009), Mobile WiMAX is a strong competitor to be considered as a core technology for the next generation IMT-Advanced standard. At the medium access control level, WiMAX uses a scheduling process for each subscriber station to obtain a time slot for data transmission. This process is done only once for the whole duration of a connection. In contrast, Wi-Fi's CSMA/CA requires channel contention to take place for each packet, which may result in more bandwidth waste.

The current wireless technologies that are most suitable for mobile video streaming are summarized in Table 1. As shown in the table, generally there is a tradeoff between data throughput and radio transmission range among these competing wireless technologies. This implies that each technology may have its own unique market share depending on communication needs from a variety of mobile applications and consumers.

Table 1. Summary of high-throughput wireless technologies

Technology	Standard source/basis	Maximum data throughput (Mbits/sec)	Relative radio transmission range
Wi-Fi	IEEE 802.11g	22	Short
	IEEE 802.11n	144	Short
HSPA	WCDMA	14.5 (download) / 7.56 (upload)	Long
CDMA EV-DO Rev. A	CDMA2000	3.1 (download) / 1.8 (upload)	Long
Mobile WiMAX	IEEE 802.16m	37 (download) / 10 (upload)	Medium

VIDEO-CAPABLE MOBILE HANDHELD DEVICES

There are many choices regarding video-streaming-capable mobile handheld devices. Some of them are an integrated part of modern smartphobnes, others are embedded in PDAs (personal digital assistants). The operating systems on those devices are typically tied to their hardware. Therefore, application developers may face incompatibility issues. In this section, main features of some popular smartphones (Cha, July 2009), PDAs (Cha, May 2009), and operating systems are discussed.

Smartphones

RIM Blackberry Curve 8900: The Blackberry Curve's screen is one of the smallest out of all the devices with it only being 2.4 inches. This limits the size of the video the user can view. Internet access is limited to only areas that have Wi-Fi and video streaming is limited because of the lack of support for Flash. This smartphone is the only one compared that does not have a touch screen.

Nokia E71x: The Nokia E71x has a screen size that matches that of the Blackberry Curve at 2.4 inches. The E71x has the ability to connect to the internet through Wi-Fi and 3G making video streaming possible is most areas. Flash lite comes preinstalled on the E71x which makes streaming videos from certain websites a breeze. Most other websites using Flash can be viewed after downloading a readily available application.

Samsung Omnia: Samsung Omnia has a 3.2 inch screen making it slightly above average. Third party applications for this device are hard to come by but are slowly being produced. The Samsung Omnia currently comes with Flash lite installed making streaming video from certain websites possible. There is no full Flash player available for this phone making the video streaming process limited. The Samsung Omnia is further limited by only being able to access the internet through Wi-Fi.

Palm Pre: The Palm Pre has a 3.1 inch screen which is considered average in the smartphone market. The Palm Pre connects to the internet through both Wi-Fi and 3G depending on availability. The Pre is currently loaded with Flash lite but by the end of the year a patch will be implemented incorporating a full Flash player making it stream video from any website.

Apple iPhone 3G S: The iPhone has a 3.5 inch screen that makes viewing videos quite ideal considering the competition. This device connects to the internet via Wi-Fi and 3G. To stream videos from most websites third party software is required because the iPhone does not currently come with Flash player.

PDAs

HP iPaq hx2750: The iPaq hx2750 has a 3.5 inch screen. This device supports Wi-Fi connectivity and has a full Flash player installed making it ideal for streaming video from an office or home.

HP iPaq rx5900: Matching the iPaq hx2750 the iPaq rx5900 is showing a 3.5 inch screen. This device supports Wi-Fi connectivity and has a full Flash player installed making it ideal for streaming video from an office or home. The battery life of this device is less than impressive according to the CNET reviews.

Palm TX: The palm TX comes to the table boasting a massive 3.8 inch screen. This device is loaded with Flash player and connects through Wi-Fi keeping the standards with the PDA market.

Pharos Traveler GPS 525: The Pharos 525 is equipped with a 2.9 inch screen making it one of the smaller screens examined in this article. This device is loaded with Flash player and connects through Wi-Fi also keeping the standards with the PDA market.

Operating Systems

Windows Mobile devices typically have more versatility because of the easy attainment of applications, which are third party programs. "Windows Mobile devices also seamlessly integrate with Microsoft Exchange and it makes viewing Word, Excel, PowerPoint and Windows Media files easy as well" (eBay, 2009). This makes Windows Mobile a very good choice for users that are interested in more of a business phone.

Palm OS devices are typically limited on the number of third party applications. "Although program variety can be limited, many programs are award-winners" (eBay, 2009). The Palm OS is typically regarded as the fastest operating system because it was designed to run on devices with low processor speeds. Because of the simplicity of the Palm OS it is very stable and users rarely encounter problems.

Symbian devices can count on experiencing the implementation of the three main design principles incorporated with this world-famous operating system, which are: data integrity and security, user time and system resourcefulness. Symbian has made various attempts to compete with Windows Mobile in the United States market and one of the ways of doing so is to implement email and Microsoft Office file viewers.

Devices that use the **Apple Mac OS X** are able to find a wide array of applications to download to bring the smartphone user experience to a whole new level. Seamless integration of the email and Microsoft Office file viewers has made this operating system one of the top competitors in the United States. This operating system is not aimed towards any specific group of people but instead people as a whole. Multitasking is still not available with this operating system but it is planned to be implemented in the near future. This upgrade may help this operating system become competitive on the global level.

END-USER EXPERIENCE AND EXPECTATION

Since mobile video streaming technologies are still in their early stage, much of the future development depends on the consumer demand and market trend. In this section, crucial end-user experience and expectation regarding mobile video streaming are discussed, which may shape the technological advances in the future.

Network

The first thing that most users look for when buying a smart phone or PDA is the type of network connections it has. The Internet now plays a huge roll in most smartphone and PDA user's lives and therefore has become a necessity. There are three main types of networks advertised today and they include the following; 3G, 4G and Wi-Fi. Wi-Fi tends to be the quickest but is only available in limited areas so this is nice for people that have an access point near them at all times. 3G is the slowest of these three connections but is available where cell phones can get service. 4G is quicker than 3G but still slower than Wi-Fi but is available where cell phone service is attainable. Most end-users now demand at least 3G for their hand held devices and should honestly settle for nothing less.

CPU Speed

CPU speed varies greatly between handheld devices but is considered to be one of the most important features. CPU speed effects loading time of programs and contact lists. Currently the average CPU speed is about 624 MHz with a handful of phones sitting in this bracket, such as: iPhone, BlackBerry storm, and Samsun Omnia.

Video Rate

Currently the standard video streaming rate is 24 frames (or pictures) per second on a handheld device and nearly every main stream device fits into this category. There are a few exceptions with some being a bit higher and some being a bit lower.

User Interface

The user interface of each handheld device varies greatly and is usually one of the large determining factors of decision making. Because of the vast difference in user interfaces many users will have different preferences and although every handheld device maker is trying to be the best this is simply impossible. Some people prefer a large screen and others prefer a small one, some users prefer using a stylus while others like a touch screen or a device that is controlled through hardware.

Keyboard Layout

Many people now use their handheld devices as portable computers, and because of this they do a lot of typing. These people typically prefer a QWERTY keyboard while the more traditional users prefer the standard 12 button layout. This is then taken further when the idea of a touch screen is incorporated. This provides the possibility of the keyboard being physical or on the screen. The size of the keyboard is another factor that users consider when looking at a handheld device.

Application Programs

When looking to purchase a handheld device many users look at the expandability of that device. Expandability involves the number and type of third party programs that can be used to customize the handheld device to make a more correct fit to the user's direct needs. Many users want to purchase a phone with everything that they need already installed but this would require each user wanting the same thing as the previous user. Some companies have tried to load up the phone with as many programs as they could fit on there but many users complained that many of the programs weren't needed and caused more of a problem than a solution. The easiest way to alleviate this problem is to give each user an easy way to modify the device to fit their exact needs.

Memory Size

RAM is used by every handheld device on the market now and this affects the speed of the handheld device. Typically more RAM is a better thing but the manufacturers do not want to put needless amounts of RAM in each handheld device because this would increase the cost of the handheld device without improving performance. Most handheld devices have either 64MB or 128MB of RAM although a few do reign supreme with 256, 288 or the massive 384MB, which is currently only in the Sony Ericsson XPERIA X1.

Internal Storage

Internal storage plays a major factor when users are deciding to purchase a handheld device. To expand a handheld device additional storage is needed to hold the third party programs. Internal storage is also needed if the user wishes to record videos, take pictures or upload music. The amount of storage needed depends on users' desire to carry data with them. Many users try to eliminate their need to carry multiple devices so they purchase a handheld device with the goal to multitask. Most handheld devices are now all considered MP3 players and personal computers so they need not only hold and play lots of music but also maintain a lot of information. The average handheld device now has 8 GB of storage although some are pushing the envelope at 32GB of internal storage. Many devices have left this up to the end user and allowed for additional storage to be added to the device at the user's discretion.

FUTURE RESEARCH DIRECTIONS

Computer/telecommunication hardware manufacturers and software companies have been working vigorously to develop next-generation mobile video streaming systems. There are three major technological breakthroughs that are planned to be released to the public in the near future that may change the way that mobile handheld devices are used. One of the first major breakthroughs is an upgraded network able to handle the vast amounts of real-time streaming data. The second breakthrough is a hardware advancement that is going to increase the quality and speed of the video. The last major breakthrough planned is a software upgrade that is going to allow a third party program to stream video on the current networks.

Upgraded and Integrated Network

With the 3G network over halfway through its life expectancy something new is bound to happen. Multiple carriers have advertised that they are currently working on a 4G network that is going to be able to stream video, voice and data more securely and quickly (Higginbotham, 2008). One of the key features that is planned to be implemented is the ability to seamlessly switch between networks when a signal is fading. This would create a virtually global network that would involve many carriers relying on each other and cooperating together to give the users what they desire.

Graphics Accelerator

Current mobile handheld devices stream video, when capable, around 12 frames per second. According to many studies that human eyes can detect delays in video when it falls below 30 frames per second. Intel has been developing a graphics accelerator that is made specifically for mobile handheld devices. Not only does this graphics accelerator make the rendering of graphics a lot quicker and smoother, it uses 10 times less energy than previous graphics accelerators (Shilov, 2009). It has been proposed that with a graphics accelerator this fast it would be possible to even play a high end video game on a mobile handheld device proving that this advancement is one of epic proportions.

Adaptive Software

The performance of mobile video streaming depends on the network (wired and wireless) conditions. Therefore it is desirable to make the streaming software adaptive to the fast changing network environment. Also, supporting system compatibility is the key to universal seamless streaming experience since various hardware/software providers are involved in the end-to-end streaming process. For example, researchers at the University of Mannheim have developed a prototype program that is designed with the idea of video-streaming mobile handheld devices in mind. (Baumgart, Knapp, Schader, Mill, 2005) claim that "we provide lightweight client support for disruption tolerant end-to-end optimized video streaming, which is realized as a prototype in a platform independent J2ME environment". This prototype is meant to be usable on any operating system that is able to run Java programs. This breakthrough will be a huge one because of its ability to run on devices with different computing powers and deliver an exceptional quality of service. This prototype program contains many layers and sophisticated algorithms that in the end are expected to provide high-performance video streaming.

CONCLUSION

Although the market for mobile video streaming is a fairly new one, it offers great potential and promise in the near future. With several different categories of such systems available, there is vast amount of room for competition and growth. Broadcasting, which is mobile video streaming directly from the television program source, provides diversity for end-user handheld devices. The client-server approach for mobile video is possibly the area with the most growth potential. This is often referred to as mobile video on demand. Peer to peer sharing is when every mobile device acts as a server and a client both hosting and receiving information. Companies like Idetic, creators of MobiTV, CloudShare and Search Toppers have already claimed their stake in the mobile video streaming market. Regardless of the type these systems, the demand for such service is at an all time high and experts are saying that the demand for such systems will only grow in the future.

While there are many outstanding smartphones and PDAs on the market, not all of them are capable of sustaining the stringent mobile video streaming requirements. When these devices are compared on the sole basis of video streaming capabilities, so far the iPhone 3G S has the best overall performance. It offers fast internet through Wi-Fi and 3G and a large screen for viewing videos. The few minor errors with the smartphone are being addressed with fixes coming in the near future. When it comes to the users specific needs only they can make the judgment call on the device that suits them best.

User experience and expectation are of great importance to service providers and equipment manufactures. When it comes to handheld devices today there are many different specifications that users look at to determine which to purchase. CPU speed needs to be adequate, which basically means not noticeably slow. Video rate should seem smooth and avoid choppiness. There is no correct way to make a user interface as it is completely determinate to the user. Keyboard preference is also determinate to the user's needs. Application programs are needed to customize the handheld device to make each device seem personalized and make each user feel as if the device was designed with them in mind. RAM is needed to assist the speed of the device which makes the experience of using a handheld device a more pleasing one. Internal storage is used by users in the attempt to make their handheld device a personal computer that fits in their hand, and the size of this internal storage is based on the individual user's needs. At the time a user is going to purchase a handheld device many things to come into play that may persuade the user to pick one device over another. When it all gets boiled down it comes to the user's needs and the availability of a device to fit their current needs.

As mobile video streaming is still in its infant stage, great technology breakthrough news are constantly expected by everyone who owns a mobile handheld device. So far, technical advances made by R&D teams around the world are very promising in the near future. Within a year wireless carriers plan to have an upgraded network in place that is capable of handling the vast amount of data required for video streaming. The development of Intel SIMD graphics accelerator can take mobile video streaming to a higher level with better quality and lower energy consumption. When multiple vendors compete for the share of mobile handheld device market, system compatibility issue becomes critical for smooth video streaming experience that users expected. Platform-independent adaptive client software aims to overcome this obstacle.

REFERENCES

Baumgart, A. S., Knapp, H., Schader, M., & Mill, S. (2005, September). *A Platform-Independent Adaptive Video Streaming Client for Mobile Devices*. The 7th IFIP International Conference on Mobile and Wireless Communications Networks (MWCN 2005), Marrakech, Morocco.

Cha, B. (2009, May). *Best 5 PDAs*. Retrieved July 10, 2009, from http://reviews.cnet.com/best-pdas/

Cha, B. (2009, July). *Best Smartphones*. Retrieved July 10, 2009, from http://reviews.cnet.com/best-smartphones/

CloudTrade. LLC. (2009). *CloudTrade* Retrieved July 21, 2009, from http://cloudtrade.com/press.htm?TitleType=about&actionTaken=abouttwo

eBay Inc. (2009). *Smartphones Buying Guide*. Retrieved July 8, 2009, from http://pages.ebay.com/buy/guides/smart-phones-buying-guide/

Higginbotham, S. (2008). *Countdown to 4G: Who's Doing What, When. GigaOM*. Retrieved July 29, 2009, from http://gigaom.com/2008/08/13/countdown-to-4g-whos-doing-what-when/

Hu, W. C., Lee, C. W., & Yeh, J. H. (2004). Mobile Commerce Systems . In Shi, N. (Ed.), *Mobile Commerce Applications* (pp. 1–23). Hershey, PA: Idea Group Publishing.

Hulu, L. L. C. (2009). *Hulu*. Retrieved July 21, 2009, from http://www.hulu.com/

IEEE. 802.11. (2007). IEEE Standard for Information technology-Telecommunications and information exchange between systems-Local and metropolitan area networks-Specific requirements - Part 11: Wireless LAN Medium Access Control (MAC) and Physical Layer (PHY) Specifications. Piscataway, NJ: IEEE.

Karonen, O., & Lahtinen, P. (2007). *Video Sharing between Handheld Devices: Combining Live Streams with File Downloads (Technical Report)*. Nokia Research.

Kim, W. (2009, June). Mobile WiMAX, the Leder of the Mobile Internet Era. *IEEE Communications Magazine*, 10–12. doi:10.1109/MCOM.2009.5116792

Krakow, G. (2004). *Watching TV on a cell phone. MSNBC*. Retrieved July 20, 2009, from http://www.msnbc.msn.com/id/6305929/

Mobi, T. V. Inc. (2009). *MobiTV*. Retrieved July 20, 2009, from http://www.mobitv.com/technology/

Schatz, R., Wagner, S., & Jordan, N. (2007, July). *Mobile Social TV: Extending DVB-H Services with P2P-Interaction*. The Second International Conference on Digital Telecommunications (ICDT 2007), Silicon Valley, USA (pp. 14-14).

Search Toppers, L. L. C. (2009). *Search Toppers*. Retrieved July 21, 2009, from http://www.searchtoppers.com/

Shilov, A. (2009). *Intel Develops Breakthrough Graphics Accelerator for Small Mobile Devices. XBit Labs*. Retrieved July 29, 2009, from http://www.xbitlabs.com/news/video/display/20090317231622_Intel_Develops_Breakthrough_Graphics_Accelerator_for_Small_Mobile_Devices.html

Symella. (2009). *Symella*. Retrieved July 20, 2009, from http://amorg.aut.bme.hu/projects/symella

SymTorrent. (2009). *SymTorrent*. Retrieved July 20, 2009, from http://amorg.aut.bme.hu/projects/symtorrent

UTRAN overall description. (1999). 3GPP. *TS 25.401 v3.3.0*, R-99, RAN WG3, 1999.

Vriendt, J. D., Lainé, P., Lerouge, C., & Xu, X. (2002). Mobile network evolution: A revolution on the move. *IEEE Communications Magazine*, *40*(4), 104–111. doi:10.1109/35.995858

Xie, S., Li, B., & Keung, G. Y. (2008, January). *The Peer-to-Peer Live Video Streaming for Handheld Devices*. The Fifth IEEE Consumer Communications & Networking Conference (CCNC 2008), Las Vegas, NV

YouTube. LLC. (2009). *YouTube*. Retrieved June 15, 2009, from http://www.youtube.com/

ADDITIONAL READING

Ahmadi, S. (2009, June). An Overview of Next-Generation Mobile WiMAX Technology. *IEEE Communications Magazine*, 84–98. doi:10.1109/MCOM.2009.5116805

Davies, M., Dantcheva, A., & Fröhlich, P. (2008, April). Comparing Access Methods and Quality of 3G Mobile Video Streaming Services. In *Proceedings of CHI 2008*, Florence, Italy.

Etemad, K. (2008, October). Overview of Mobile WiMAX Technology and Evolution. *IEEE Communications Magazine*, 31–40. doi:10.1109/MCOM.2008.4644117

Gualdi, G., Prati, A., & Cucchiara, R. (2008, October). Video Streaming for Mobile Video Surveillance. *IEEE Transactions on Multimedia*, *10*(6), 1142–1154. doi:10.1109/TMM.2008.2001378

Mohapatra, S., Cornea, R., Dutt, N., Nicolau, A., & Venkatasubramanian, N. (2003, November). *Integrated Power Management for Video Streaming to Mobile Handheld Devices*. ACM MM 2003, Berkeley, CA.

Pyun, J. Y. (2008, November). Error concealment aware streaming video system over packet-based mobile networks. *IEEE Transactions on Consumer Electronics*, *54*(4), 1705–1713. doi:10.1109/TCE.2008.4711224

Qu, Q., Appuswamy, P., & Chan, Y. S. (2006, October). *QoS Guarantee and Provisioning for Real-Time Digital Video over Mobile Ad Hoc CDMA Networks with Cross-Layer Design*. IEEE Wireless Communications.

Quan, H. T., & Ghanbari, M. (2008, September). Temporal Aspect of Perceived Quality in Mobile Video Broadcasting. *IEEE Transactions on Broadcasting*, *54*(3), 641–651. doi:10.1109/TBC.2008.2001246

Schierl, T., Ganger, K., Hellge, C., & Weigand, T. (2006, October). *SVC-Based Multisource Streaming for Robust Video Transmission in Mobile Ad Hoc Networks*. IEEE Wireless Communications.

Shen, B., Tan, W. T., & Huve, F. (2008). Dynamic Video Transcoding in Mobile Environments. *IEEE MultiMedia*, *15*(1), 42–51. doi:10.1109/MMUL.2008.5

Song, W. J., Chung, J. M., Lee, D., Lim, C., Choi, S., & Yeoum, T. (2009, April). Improvements to Seamless Vertical Handover between Mobile WiMAX and 3GPP UTRAN through the Evolved Packet Core. *IEEE Communications Magazine*, 66–73. doi:10.1109/MCOM.2009.4907409

Sun, Q. (2008). Video Browsing on Handheld Devices – Interface Designs for the Next Generation of Mobile Video Players. *IEEE MultiMedia*, *15*(3), 76–83. doi:10.1109/MMUL.2008.66

Tu, W., & Steinbach, E. (2009, February). Proxy-Based Reference Picture Selection for Error Resilient Conversational Video in Mobile Networks. *IEEE Transactions on Circuits and Systems for Video Technology*, *19*(2), 151–164. doi:10.1109/TCSVT.2008.2009240

Wang, Y. K., Bouazizi, I., Hannuksela, M. M., & Curcio, I. D. D. (2007, September). *Mobile Video Applications and Standards*. MV 2007, Augsburg, Germany.

Zhu, J., & Yin, H. (2009, June). Enabling Collocated Coexistence in IEEE 802.16 Networks via Perceived Concurrency. *IEEE Communications Magazine*, 108–114.

Compilation of References

Aaron, A., Rane, S., Setton, E., & Girod, B. (2004). Transform-domain Wyner-Ziv codec for video. In . *Proceedings of SPIE Visual Communications and Image Processing, 5308*, 520–528.

Aaron, A., Varodayan, D., & Girod, B. (2006). Wyner-Ziv residual coding of video. In *Proceedings of the Picture Coding Symposium.*

Aaron, A., Zhang, R., & Girod, B. (2002).Wyner-Ziv coding of motion video. In *Proceedings of the Asilomar Conference on Signals, Systems and Computers: Vol. 1.* (pp.240-244).

Abraham, R. (2007). Mobile phones and economic development: Evidence from the fishing industry in India. *MIT Press Journal, 4*(1), 5–17.

Access Economics Report. (2005). *The shifting burden of cardiovascular disease in Australia, A report of Heart foundation.* Retrieved March 20, 2009, from http://www.heartfoundation.com.au/ media/nhfa_shifting_burden_cvd_0505.pdf

ACCESS. (n.d.). *Small-Fit Rendering.* Retrieved June 14, 2008, from http://www.access-company.com/products/netfrontmobile/contentviewer/mcv_tips.html#Anchor-Smar-45765

ACMA. 2009. *Australian Communications and Media Authority Report (2009): Convergence and Communications.* Retrieved March 20, 2009, from http://www.acma.gov.au/webwr/_assets/main/lib100068/convergence_%20comms_rep-1_household_consumers.doc

Adamic, L. A., Lukose, R. M., Puniyani, A. R., & Huberman, B. A. (2001). Search in power-law networks. *Physical Review E: Statistical, Nonlinear, and Soft Matter Physics, 64*(4), 1842–1845. doi:10.1103/PhysRevE.64.046135

Adda, O., Cottineau, N., & Kadoura, M. (2003). *A tool for global motion estimation and compensation for video processing (Rpt. project ELEC/COEN 490).* Concordia University.

Adeya, C. N. (2005). *Wireless technologies and development in Africa.* Unpublished report. Retrieved June 15,2009, Fromhttp://arnic.info/workshop05/Adeya_WirelessDev_Sep05.pdf

Adikari, A. B. B., Fernando, W. A. C., Arachchi, H. K., & Weerakkody, W. A. R. J. (2006a). Wyner-Ziv coding with temporal and spatial correlations for motion video. *Canadian Conference on Electrical and Computer Engineering* (pp. 1188-1191).

Adikari, A. B. B., Fernando, W. A. C., Arachchi, H. K., & Weerakkody, W. A. R. J. (2006b). Multiple side information streams for distributed video coding. *IEEE Electronics Letters, 42*, 1447–1449. doi:10.1049/el:20062268

Adipat, B., & Zhang, D. (2005). Adaptive and personalized interfaces for mobile Web. In *Proceedings of the 15th Annual Workshop on Information Technolgies & Systems (WITS)*, Las Vegas, NV.

AfriGadget. (2009). *Harnessing Personal Movement for Power in Rural Africa.* Retrieved June 15, 2009, fromhttp://www.afrigadget.com/2009/02/12/harnessing-personal-movement-for-power-in-rural-africa/

Agha, G. (1986). *Actors: a Model of Concurrent Computation in Distributed Systems.* Cambridge, MA: MIT Press.

Ahmad, A., Khan, N., Masud, S., & Maud, M. A. (2004, January). Efficient Block Size Selection in H.264 Video Coding Standard. *Electronics Letters, 40*(1). doi:10.1049/el:20040068

Aigner, M., Dominikus, S., & Feldhofer, M. (2007). A System of Secure Virtual Coupons Using NFC Technology. In *Proceedings of the 5th Ann. IEEE Int'l Conf. Pervasive Computing and Communications Workshops (PerComW 07)*, (pp. 362-366). IEEE CS Press.

AIHW. (2004a). *Indigenous Australians carrying heaviest burden of cardiovascular disease*. Retrieved March 20, 2009, from http://www.aihw.gov.au/mediacentre/2004/mr20040505.cfm

AIHW. (2004b). Heart, stroke and vascular diseases—Australian Facts 2004. *AIHW Cat. No. CVD 27. Canberra: AIHW and National Heart Foundation of Australia (Cardiovascular Disease Series No. 22).*

Akay, M. (1995). Wavelets in biomedical engineering. *Annals of Biomedical Engineering, 23*, 531–542. doi:10.1007/BF02584453

Ala-Laurila, J., Mikkonen, J., & Rinnemaa, J. (2001). Wireless LAN access network architecture for mobile operators. *IEEE Communications Magazine, 39*(11), 82–89. doi:10.1109/35.965363

Alam, H., & Rahman, F. (2003). Web document manipulation for small screen devices: a review.In *Proceedings of the 2nd International Workshop on Web Document Analysis (WDA2003)*, Edinburgh, UK.

Alam, M., & Prasad, N. (2008). Convergence transforms digital home: Techno-economic impact. *Wireless Personal Communications, 44*(1), 75–93. doi:10.1007/s11277-007-9380-2

Albon, R., & York, R. (2008). Should mobile subscription be subsidised in mature markets? *Telecommunications Policy, 32*(5), 294–306. doi:10.1016/j.telpol.2008.02.003

Alliez, P., Desbraun, M., (2001). Compression for Lossless Transmission of Triangle Meshes. In Proceeding of SIGGRAPH 2001 (pp. 195–202). Progressive.

Al-Regib, G., Altunbasak, Y., & Rossignac, J. (2005). An unequal error protection method for progressively transmitted 3D models . *IEEE Transactions on Multimedia, 7*(4), 766–776. doi:10.1109/TMM.2005.850981

Amitay, E., & Paris, C. (2000). Automatically summarizing Web sites - Is there a way around it? In *Conference on Information and Knowledge Management* (pp. 173-179).

Amoretti, M. (2009B). A Framework for Evolutionary Peer-to-Peer Overlay Schemes. In *European Workshops on the Applications of Evolutionary Computation*, Tubingen, Germany.

Amoretti, M., Agosti, M., & Zanichelli, F. (2009A). DEUS: a Discrete Event Universal Simulator. *In 2nd ICST/ACM International Conference on Simulation Tools and Techniques (SIMUTools 2009)*, Roma, Italy.

Amoretti, M., Bisi, M., Zanichelli, F., & Conte, G. (2005). Introducing Secure Peergroups in SP2A. In *2nd IEEE International Workshop on Hot Topics in Peer-to-Peer Systems, co-located with Mobiquitous*, 2005, San Diego, California.

Amoretti, M., Bisi, M., Zanichelli, F., & Conte, G. (2008). *Enabling Peer-to-Peer Web Service Architectures with JXTA-SOAP*. In IADIS International Conference e-Society 2008, Algarve, Portugal.

Andace, D. (2004). ***E-Commerce** and **M-commerce** technologies*. London: IRM press.

Android(2009). *Android*. Retrieved March 26, 2009, from http://www.android.com

Anick, P. G., & Tipirneni, S. (1999). The paraphrase search assistant: Terminological feedback for iterative information seeking. In *ACM SIGIR Conference* (pp. 153-159).

Appfrica (2008). *The current state of Internet penetration in Africa*. Retrieved June 15, 2009, from http://appfrica.net/blog/archives/248

Arase, Y., Hara, T., Uemukai, T., & Nishio, S. (2007). OPA Browser: A Web browser for cellular phone users.

In *ACM Symposium on User Interface Software and Technology* (pp. 71-80).

Arase, Y., Maekawa, T., Hara, T., Uemukai, T., & Nishio, S. (2006). A Web browsing system based on adaptive presentation of Web contents for cellular phones. In *Proceedings of the 2006 International Cross-Disciplinary Workshop on Web Accessibility (W4A)*, (pp. 86-89). Edinburgh, U.K.

Arase, Y., Maekawa, T., Hara, T., Uemukai, T., & Nishio, S. (2007). A Web browsing system for cellular phone users based on adaptive presentation. *Universal Access in the Information Society*, *6*(3), 259–271. doi:10.1007/s10209-007-0088-6

Artigas, X., Ascenso, J., Dalai, M., Klomp, S., Kubasov, D., & Ouaret, M. (2007b). The discover codec: architecture, techniques and evaluation. In *Proceedings of the Picture Coding Symposium.*

Artigas, X., Malinowski, S., Guillemot, C., & Torres, L. (2007a). Overlapped quasi-arithmetic codes for distributed video coding. In . *Proceedings of the IEEE International Conference on Image Processing*, *2*, 9–12.

Ascenso, J., Brites, C., & Pereira, F. (2005). Improving frame interpolation with spatial motion smoothing for pixel domain distributed video coding. In *Proceedings of the 5th EURASIP Conference on Speech and Image Processing, Multimedia Communications and Services.*

Ascenso, J., Brites, C., & Pereira, F. (2006). Content adaptive Wyner-Ziv video coding driven by motion activity. In *Proceedings of the IEEE International Conference on Image Processing.* (pp. 605-508).

Aspert, N., Santa-cruz, D., & Ebrahimi, T. (2002). Mesh:measuring errors between surfaces using the hausdoff distance. In *Proceeding of IEEE Int'l Conf. on Multimedia and Expo* (pp. 705–708)

Ates, H., Kanberoglu, B., & Altunbasak, Y. (2006). Rate-distortion and complexity joint optimization for fast motion estimation in H.264 video coding. In *Proceedings of the IEEE International Conference on Image Processing* (pp. 37-40).

Avoine, G. (2006). *Security and Privacy in RFID Systems.* Retrieved February 1, 2009, from http://lasecwww.epfl.ch/~gavoine/rfid

Axmor (2009). *Axmor's Symbian Security Applications.* Retrieved June 10, 2009, from http://www.axmor.com/symbian-development/phone-confidentiality.aspx

Baccichet, P., Rane, S., & Girod, B. (2006). Systematic lossy error protection based on H.264/AVC redundant slices and flexible macroblock ordering. *Journal of Zhejiang University . Scientific American*, *5*, 727–736.

Bahl, P., & Padmanabhan, V. N. (2000). An . In *Building RF-based User Location and Tracking System, in IEEE INFOCOM 2000* (*Vol. 2*, pp. 775–784). Tel-Aviv, Israel: RADAR.

Bajaj, C., & Schikore, D. (1996). Error-bounded reduction of trianges meshes with multivariate date. *SPIE*, *2656*, 34–45.

Bajaj, C., Cutchin, S., Pascucci, V., Zhuang, G., (1998). *Error Resilient Transmission of Compressed VRML* (Tech. Rep.). Austin, TX: TICAM, The University of Texas at Austin.

Bal, H. E., Steiner, J. G., & Tanenbaum, A. S. (1989). Programming Languages for Distributed Computing Systems. *ACM Computing Surveys*, *21*(3), 261–322. doi:10.1145/72551.72552

Balan, E. (2007). *8,000**iPhones** Sold in the UK on First Day.* Retrieved April 15, 2008, from http://news.softpedia.com/news/8-000-**iPhone**s-Sold-in-the-UK-on-First-Day-70696.shtml

Baluja, S. (2006). Browsing on small screens: Recasting web-page segmentation into an efficient machine learning framework. In *International World Wide Web Conference* (pp. 33-42).

Banerjee, K., Wu, F., & Agu, E. (2005). Estimating Mobile Memory Requirements and Rendering Time for Remote Execution of the Graphics Pipeline. In*Proceeding of Eurographics 2005.*

Bardram, J. E. (2004, March 14 - 17). Applications of context-aware computing in hospital work: examples and design principles. In *Proceedings of the 2004 ACM symposium on Applied computing,* Nicosia, Cyprus.

BarkaiD. (2002). Peer-to-Peer Computing: Technologies for Sharing and Collaborating on the Net. Santa Clara, CA: Intel Press.

Baset, S. A., & Schulzrinne, H. G. (2006). An Analysis of the Skype Peer-to-Peer Internet Telephony Protocol. In *25th IEEE International Conference on Computer Communications (INFOCOM 2006),* Barcelona, Spain.

Baudisch, P., Xie, X., Wang, C., & Ma, W. Y. (2004). Collapse-to-zoom: Viewing web pages on small screen devices by interactively removing irrelevant content. In *ACM Symposium on User Interface Software and Technology* (pp. 91-94).

Baumgart, A. S., Knapp, H., Schader, M., & Mill, S. (2005, September). *A Platform-Independent Adaptive Video Streaming Client for Mobile Devices.* The 7th IFIP International Conference on Mobile and Wireless Communications Networks (MWCN 2005), Marrakech, Morocco.

Beale, R. (2005). Supporting Social Interaction with Smart Phones. *IEEE Pervasive Computing / IEEE Computer Society [and] IEEE Communications Society, 4*(2), 35–41. doi:10.1109/MPRV.2005.38

Beauchemin, S. S., & Barron, J. L. (1995). The computation of optical flow . *ACM Computing Surveys, 27*(3), 433–467. doi:10.1145/212094.212141

Becker, D. (2006). Fundamentals of electrocardiography interpretation. *Anesthesia Progress, 53*(2), 53–64. doi:10.2344/0003-3006(2006)53[53:FOEI]2.0.CO;2

Bell, A. (2007). *The latest apple* ***iPhone.*** Retrieved February 10, 2008, from http://www.ianbell.com/2007/09/26/**iPhone**-mania-persists-despite-apples-cold-shoulder

Ben Mokhtar, S., & Capra, L. (2009). From Pervasive To Social Computing: Algorithms and Deployments. *To Appear in the ACM International Conference on Pervasive Services (ICPS '09).*

Berger, S., McFaddin, S., Narayaswami, C., & Raghunath, M. (2003). Web Services on Mobile Devices - Implementation and Experience. In *5th IEEE Workshop on Mobile Computing Systems & Applications*, Monterey, CA.

Bernard, J. J., & Tracy, M. (2008). Sponsored search: an overview of the concept, history, and technology. *International Journal of Electronic Business, 6*(2), 114–131. doi:10.1504/IJEB.2008.018068

Bernardini, R., Fumagalli, M., Naccari, M., Rinaldo, R., Tagliasacchi, M., Tubaro, S., & Zontone, P. (2007). Error concealment using a DVC approach for video streaming applications. In *Proceedings of the EURASIP European Signal Processing Conference.*

Berrou, C., Glavieux, A., & Thitimajshima, P. (1993). Near Shannon limit error correcting coding and decoding: turbo codes. In *Proceedings of the IEEE International Conference on Communications* (pp. 1064-1070).

Bhaskara, G., Helmy, A., & Gupta, S. (2003). Micromobility protocol design and evaluation: a parameterized building block approach. *IEEE 58th Vehicular Technology Conference* (pp. 2019- 2024.)

Bhatti, N., Bouch, A., & Kuchinsky, A. (2000). Integrating user-perceived quality into Web server design. *Computer Networks, 33,* 1–16. doi:10.1016/S1389-1286(00)00087-6

Bhimani, A. (1996). Securing The Commercial Internet. *Communications of the ACM, 39*(6), 29–35. doi:10.1145/228503.228509

BidgoliH. (2002). Electronic Commerce Principles and Practice. London: Academic Press.

Biegel, G., & Cahill, V. (2004). A framework for developing mobile, context-aware applications.In *Proc, Second IEEE Annual Conference on Pervasive Computing and Communications,* PERCOM, 2004

Bischoff, S., & Kobbelt, L. (2002). Toward robust broadcasting of geometry data. *Computer Graphics, 26*(5), 665–675. doi:10.1016/S0097-8493(02)00122-X

Bjontegaard, G. (Apri, l2001), *Calculation of Average PSNR Differences Between RD-Curves,* ITU-T Q6/SG16, Doc. VCEG-M33.

BlackBerry. (2009). *BlackBerry Enterprise Solution for Mobile Security*. Retrieved June 10, 2009, from http://na.blackberry.com/eng/ataglance/security/features.jsp

Blekas, A., Garofalakis, J., & Stefanis, V. (2006). Use of RSS feeds for content adaptation in mobile Web browsing. In *Proceedings of the 2006 International Cross-Disciplinary Workshop on Web Accessibility (W4A)* (pp. 79-85). Edinburgh, U.K.

Bonn, U. (2006). *Bidirectional Texture Function Database Bonn. Retrieved* 2006, from http://btf.cs.uni-bonn.de/index.html

Borchert, S., Westerlaken, R. P., Klein Gunnewiek, R., & Lagendijk, R. L. (2007). On extrapolating side information in Distributed Video Coding. *Proceedings of the Picture Coding Symposium.*

Borodin, Y., Mahmud, J., & Ramakrishnan, I. V. (2007). Context browsing with mobiles—when less is more. In *Proceedings of the 5th International Conference on Mobile Systems, Applications, and Services (MobiSys'07)* (pp. 3-15). San Juan, PR

Bos, B., Celik, T., Hickson, I., & Håkon, W. L. (2009). Cascading Style Sheets (CSS 2.1). *W3C working note*. Retrieved, from http://www.w3.org/TR/CSS21/, 2009

Bouwman, H., De Vos, H., Haaker, T., (Eds.). (2008). Mobile Service Innovation and Business Models. New York: Springer. 10.1007/978-3-540-79238-3

Bradley, J. N., & Brislawn, C. M. (1994). The wavelet/scalar quantization compression standard for digital fingerprint images. In *Proceeding of IEEE International Symposium on Circuits and Systems(ISCAS).*

Briot, J.-P., Guerraoui, R., & Lohr, K.-P. (1998). Concurrency and Distribution in Object-Oriented Programming. *ACM Computing Surveys, 30*(3), 291–329. doi:10.1145/292469.292470

Brites, C., & Pereira, F. (2008). Correlation noise modeling for efficient pixel and transform domain Wyner-Ziv video coding. *IEEE Transactions on Circuits and Systems for Video Technology, 18*(9), 1177–1190. doi:10.1109/TCSVT.2008.924107

Broder, A., Fontoura, M., Josifovski, V., & Riedel, L. (2007). A semantic approach to contextual advertising, In *SIGIR '07: Proceedings of the 30th annual international ACM SIGIR conference on Research and development in information retrieval* (pp. 559-566). New York: ACM.

Brown, G. N. (2007). Linux: a platform for innovation in converged mobile handsets. *BT Technology Journal, 25*(2), 126–132. doi:10.1007/s10550-007-0036-2

Bruijin, O., Spence, R., & Chong, M. Y. (2002). RSVP browser: Web browsing on small screen devices. *Personal and Ubiquitous Computing, 6*(4), 245–252. doi:10.1007/s007790200024

Bulterman, D. (2005). Synchronized Multimedia Integration Language (SMIL2.1). *W3C recommendation.* Retrieved December 2005, fromhttp://www.w3.org/TR/2005/REC-SMIL2-20051213/

Burigat, S., Chittaro, L., & Gabrielli, S. (2008). Navigation techniques for small-screen devices: An evaluation on maps and web pages. *International Journal of Human-Computer Studies, 66*(2), 78–97. doi:10.1016/j.ijhcs.2007.08.006

Buyukkokten, O., Garcia-Molina, H., & Paepcke, A. (2001). Seeing the whole in parts: Text summarization for web browsing on handheld devices. In *International World Wide Web Conference* (pp. 652-662).

Buyukkokten, O., Garcia-Molina, H., & Paepcke, A. (2000). Efficient web browsing for PDAs. In CHI conference (pp. 430–437). Power Browser.

Bystrom, M., Richardson, I., & Zhao, Y. (2008, February). Efficient Mode Selection for H.264 Complexity Reduction in a Bayesian Framework. *Signal Processing Image Communication, 23*(2), 71–86. doi:10.1016/j.image.2007.11.001

Callsen, C. J., & Agha, G. (1994). Open Heterogeneous Computing in ActorSpace. *Journal of Parallel and Distributed Computing, 21*(3), 289–300. doi:10.1006/jpdc.1994.1060

Campbell, A. T., & Gomez-Castellanos, J. (2000). IP micro-mobility protocols. *ACM SIGMOBILE Mobile*

Computing and Communications Review, *4*(4), 45–53. doi:10.1145/380516.380537

Campbell, A. T., Gomez, J., Wan, K. C. y., Zoltán, R. T., & Valko, A. G. (2002). Comparison of IP micro mobility protocols. *IEEE Wireless Communications*, *9*, 72–82. doi:10.1109/MWC.2002.986462

Campbell, A., Aurrecoechea, C., & Hauw, H. (1996). *A survey of qos architectures.* New York: Multimedia Systems.

Cardelli, L. (1995). A Language with Distributed Scope. In *Proceedings of the 22nd ACM SIGPLAN-SIGACT Symposium on Principles of Programming Languages*, 286-297.

Carli, M., Cappabianca, F., Tenca, A., & Neri, A. (2004). Mobility Management for the Next Generation of Mobile Cellular Systems . In *Lecture notes of Telecommunications and Networking* (pp. 991–996). Berlin, Heidelberg: Springer.

Castells, F., Cebrián, A., & Millet, J. (2007). The role of independent component analysis in the signal processing of ECG recordings. *Biomedizinische Technik. Biomedical Engineering*, *52*(1), 18–24. doi:10.1515/BMT.2007.005

Castelluccia, C., & Avoine, G. (2006, April 19-21). Noisy Tags: A Pretty Good Key Exchange Protocol for RFID Tags. *7th IFIP WG 8.8/11.2 International Conference, Smart Card Research and Advanced Applications*, Tarragona, Spain.

Central Intelligence Agency. (2009). *The World Factbook.* Retrieved June 15, 2009, from https://www.cia.gov/library/publications/the-world-factbook/fields/2103.html.

CEVA. (2009). *Glossary of Terms.* Retrieved June 10, 2009, from http://ceva-dsp.mediaroom.com/index.php?s=glossary

Cha, B. (2009, July). *Best Smartphones.* Retrieved July 10, 2009, from http://reviews.cnet.com/best-smartphones/

Cha, B. (2009, May). *Best 5 PDAs.* Retrieved July 10, 2009, from http://reviews.cnet.com/best-pdas/

Cha, K., Jun, D., Wilson, A., & Park, Y. (2008). Managing and modeling the price reduction effect in mobile telecommunications traffic. *Telecommunications Policy*, *32*(7), 468–479. doi:10.1016/j.telpol.2008.04.005

Chan, T.W., Roschelle, J., Hsi, S., Kinshuk, Sharples, M., Brown, T., Patton, C., Cherniavsky, J., Pea, R., Norris, C., Soloway, E., Balacheff, N., Scardamalia, M., Dillenbourg, P., Looi, C. K., Milrad, M., & Hoope, U. (2006). One-to-one technology enhanced learning: an opportunity for global research collaboration. *Research and Practice in Technology Enhanced Learning* 1, *3*(29).

Chandrasekhar, V., Andrews, J., & Gatherer, A. (2008). Femtocell networks: A survey. *IEEE Communications Magazine*, *46*(9), 59–67. doi:10.1109/MCOM.2008.4623708

Chang, A., Au, O. C., & Yeung, Y. M. (July, 2003) A Novel Approach to Fast Multi-block Motion Estimation for H.264 Video Coding. *International Conference on Multimedia and Expo, 2003. ICME'03, 1*, 6-9.

ChanY. T. (1995). Wavelets basics. Boston: Klumer Academic Publishers.

Charlesworth, A. (2009). The ascent of smartphone. *Engineering & Technology*, *4*(3), 32–33. doi:10.1049/et.2009.0306

Chatterjee, P., Hoffman, D. L., & Novak, T. P. (2006). Modeling the clickstream: Implications for web-based advertising efforts . *Marketing Science*, *22*(4), 520–541. doi:10.1287/mksc.22.4.520.24906

Chen, B. Y., & Nishita, T. (2002). Multiresolution Streaming Mesh with Shape Preserving and QoS-like Controlling. In *Proceeding of Web3D 2002* (pp. 35-42)

Chen, B., & Nguyen, M. (2001). Pop: A Hybrid Point and Polygon Rendering System for Large Data, In *Proceeding of IEEE Visualization 2001.*

Chen, R. (2005). *Modeling of User Acceptance of Customer E-Commerce Website.* Paper presented at WISE 2005, 454-462.

Chen, S. W. (2007, Nov-Dec). A nonlinear trimmed moving averaging-based system with its application to real-time QRS beat classification. *Journal of Medical Engineering & Technology*, *31*(6), 443–449. doi:10.1080/03091900701234267

Chen, T. C., Chien, S. Y., Huang, Y. W., Tsai, C. H., Chen, C. Y., Chen, T. W., & Chen, L. G. (2006). Analysis and architecture design of an HDTV720p 30 frames/s H.264/AVC encoder. *IEEE Transactions on Circuits and Systems for Video Technology*, *16*(6), 673–688. doi:10.1109/TCSVT.2006.873163

Chen, Y., Ma, W. Y., & Zhang, H. J. (2003). Detecting Web page structure for adaptive viewing on small form factor devices. In *Proceedings of the 12th International Conference on World Wide Web* (pp. 225-233).Budapest, Hungary.

Cheng, C., & Chang, T. (May, 2005) Fast Three Step Intra Prediction Algorithm for 4x4 Blocks in H.264. *IEEE International Symposium on Circuits and System, 2005*, ISCAS 2005.

Chia A., Woo, Y., Martin, G. R., & Park, H. (February, 2008,). Fast Inter-Mode Selection in the H.264/AVC Standard Using a Hierarchical Decision Process. *IEEE Transaction on Circuits and System for Video Technology, 18*(2).

Choi, B. D., Nam, J. H., Hwang, M. C., & Ko, S. J. (2006). Fast motion estimation and intermode selection for H.264. *EURASIP Journal on Applied Signal Processing, 2006*, 1–8. doi:10.1155/ASP/2006/71643

Chow, C. Y., Mokbel, M. F., & Liu, X. (2006). A peer-to-peer spatial cloaking algorithm for anonymous location-based services. In *Proc. of ACM GIS* (pp. 171-178).

Chow, M. (1997). Optimized geometry compression for real-time rendering. In *Proceeding of IEEE Visualization 97* (pp. 347–354)

Christensen, C. M. (1997). *The Innovator's Dilemma*. Boston: Harvard Business Press.

Christopoulos, C., Skodras, A., & Ebrahimi, T. (2000). The JPEG2000 still image coding system: an overview. In *Proceeding of IEEE Trans. on Consumer Electronics* (pp. 1103–1127), Vol. 46, Issue 4, Cignini, P., Rocchini, C., & Scopigno, R. (1998). *Metro: measuring error on simplified surfaces*. Computer Graphics Forum, pp. (167–174).

Chu, K. M., Leung, K. R. P. H., Ng, J. K., & Li, C. H. (2004), Locating Mobile Stations with Statistical Directional Propagation Model, in *Proc. of the 18th Intl. Conf. on Adv. Info. Networking and Applications (AINA 2004)*, Fukuoka, Japan, pp. 230-235.

Chuah, M., Fu, F., (2007). ECG *Anomaly detection via time series analysis*. Lecture Notes in Computer Science: Frontiers of High Performance Computing and Networking ISPA 2007 Workshops, 2007 (pp. 123–135). Springer.

Chui, C. K., (1992). An introduction to wavelets. New York: Academic.

Clarberg, P., Jarosz, W., Akenine-Moller, T., Jensen, H. W., (2005). Efficiently Evaluating Products of Complex Functions. In Proceeding of ACM SIGGRAPH 2005 (pp. 1166–1175). Wavelet Importance Sampling.

Clark, H. H., & Brennan, S. E. (1991). Grounding in Communication. In L. Resnick, J. Levine & S. Teasley (Eds.), *Perspectives on Socially Shared Cognition* (127-149). Hyattsville, MD: American Psychological Association.

Clarke, I. (2001). Emerging value propositions for m-commerce. *The Journal of Business Strategy*, *18*(2), 133–148.

CloudTrade. LLC. (2009). *CloudTrade* Retrieved July 21, 2009, from http://cloudtrade.com/press.htm?TitleType=about&actionTaken=abouttwo

Cohen, E., & Shenker, S. (2002). Replication Strategies in Unstructured Peer-to-Peer Networks. In *ACM SIGCOMM '02*, Pittsburgh, PA.

Cohen, J., Olano, M., & Manocha, D. (1998). Appearance Preserving Simplification. In *Proceeding of ACM SIGGRAPH 1998* (pp. 115-122).

Cohen, J., Varshney, A., Manocha, D., Turk, G., Weber, H., Agarwal, P., et al. (1996) Simplification Envelopes. In *Proc. of ACM SIGGRAPH 1996.*

Collins, J., & Vile, D. (2007, May). Mobile Security A primer on the security of mobile devices, and the implications for enterprise IT, *Freeform Dynamics Ltd.*

Cosman, P. C., Rogers, J. K., Sherwood, P. G., & Zeger, K. (2000). Combined Forward Error Control and Packetized Zero TreeWavelet Encoding for Transmission of Images over Time-Varying Channels. *IEEE Transactions on Image Processing, 9*(6), 982–993. doi:10.1109/83.846241

Cote, M., Suryn, W., Laporte, C., & Martin, R. (2005). The Evolution Path for Industrial Software Quality Evaluation Methods Applying ISO/IEC 9126:2001 Quality Model: Example of MITRE's SQAE Method. *Software Quality Journal, 13*(1), 17–30. doi:10.1007/s11219-004-5259-6

Couder, P., & Kermarrec, A. M. (1999). *Improving Level of Service of Mobile User Using Context-Awareness.* Paper presented at the 18th IEEE Symposium on Reliable Distributed System, Lausanne, Switzerland.

Coursaris, C., & Hassanein, K. (2002). Understanding m-commerce. *Quarterly Journal of Electronic Commerce, 3*(3), 247–271.

Cravotta, R. (2007, 1 September). Recognizing gestures. *EDN Europe*, 22–33.

Curtis, M., Luchini, K., Bobrowsky, W., Quintana, C., & Soloway, E. (2002) Handheld use in K-12: a descriptive account. In *Proceedings of IEEE International Workshop on Wireless and Mobile Technologies in Education (WMTE)*, pp. 23–30. Los Alamitos, CA: IEEE Computer Society Press

Dahlman, E., Gudmundson, B., Nilsson, M., & Skold, J. (1998). UMTS/IMT-2000 based on wideband CDMA. *IEEE Communications Magazine, 36*(9), 70–80. doi:10.1109/35.714620

Dana, P. H. (1998), *Global Positioning System Overview*, The University of Texas, Retreived (n.d.)., from http://www.colorado.Edu/geography/gcraft/notes/gps/gps.html

Danis, C. Bailey, M., Christensen, J., Ellis, J., Erickson, T., Farrell, R., & Kellogg, W. A. (2009) Social Computing Applications for the Next Billion Users. In *Designing Future Mobile Software for Underserved Users Workshop* at CSCW 2008.

Data, R. (2006) *Cornell University, Program of Computer Graphics*, Re trieved 2006, from http://www.graphics.cornell.edu/online/measurements/reflectance/index.html.

Daubechies, I., (1992). Ten lectures on wavelets. Philadelphia: Society for Industrial and Applied Mathematics.

Davies, N., Friday, A., Wade, S. P., & Blair, G. S. (1998). L2imbo: a distributed systems platform for mobile computing. *Mobile Networks and Applications, 3*(2), 143–156. doi:10.1023/A:1019116530113

Davison, B. D. (2001). A Web caching primer. *IEEE Internet Computing, 5*(4), 38–45. doi:10.1109/4236.939449

de Berg, M., Cheong, O., van Kreveld, M., & Overmars, M. (2008). *Computational Geometry: Algorithms and Applications.* Springer, 3rd edition.

De Capual, C., De Falco, S., & Morellol, R. (2006). *A Soft Computing-Based Measurement System for Medical Applications in Diagnosis of Cardiac Arrhythmias by ECG Signals Analysis.* 2006 IEEE International Conference on Computational Intelligence for Measurement Systems and Applications. pp: 2-7.

Dean, J., & Ghemawat, S. (2004). Mapreduce: simplified data processing on large clusters. In *Sixth Symposium on Operating System Design and Implementation*, pages 137–150.

Debevec, P. (2006). *Paul Debevec's Light Probe Image Gallery.* Retrieved (n.d.)., from http://www.debevec.org/Probes/

Dedecker, J., Van Cutsem, T., Mostinckx, S., D'Hondt, T., & De Meuter, W. (2006). Ambient-oriented Programming in AmbientTalk. In *Proceedings of the 20th European Conference on Object-oriented Programming (ECOOP), 4067*, 230-254.

Deloitte (2005, January), *Worldwide Mobile Phone Subscriber Research*, New York: Deloitte & Touche.

Derose, T., Lounsbery, M., & Warren, J. (1997). Multiresolution analysis for surfaces of arbitrary topological type. *ACM Transactions on Graphics*, *16*, 34–73. doi:10.1145/237748.237750

Deutsch, M., Willis, R., (1988). Software Quality Engineering: A Total Technical and Management Approach. Englewood Cliffs, NJ: Prentice-Hall.

Dey, A. (2001, February). Understanding and using context. *Personal and Ubiquitous Computing*, *5*(1), 4–7. doi:10.1007/s007790170019

Diaz Palacios, I. (2007). *A Highly Efficient and Low-Power System for the Detection of Potentially Dangerous Objects.* M.Sc. Thesis, Lund University, Lund, Sweden.

Dimitriou, T. (2008, January 10-12). Proxy Framework for Enhanced RFID Security and Privacy. *Consumer Communications and Networking Conference*, Las Vegas, NV.

Divakaran, D. (2002). *RTLinux HOWTO. Internet FAQ Archives Online Education.* Retrieved August 8th, 2002, from http://www.faqs.org/docs/Linux-HOWTO/RTLinux-HOWTO.html

Dixon, J., & Jakl, M. (2005). *Symbian OS v9 Security Architecture.* Retrieved June 10, 2009, from http://developer.symbian.com/main/documentation/sdl/symbian94/sdk/doc_source/guide/platsecsdk/SGL.SM0007.013_Rev2.0_Symbian_OS_Security_Architecture.doc.html

Djuknic, G. M., & Richton, R. E. (2001). Geolocation and Assisted GPS. *Computer*, *34*(2), 123–125. doi:10.1109/2.901174

Donegan, M. (2005). The business case: Can convergence really pay off? Total Telecom, (APR.), 35. Engel, C. (2007). Competition in a pure world of Internet telephony. *Telecommunications Policy*, *31*(8-9), 530–540.

Du, W. (2001). *A study of several specific secure two-party computation problems.* Ph.D. dissertation, Purdue University.

Duguet, F., & Drettakis, G. (2004). Flexible Point-Based Rendering on Mobile Devices. *IEEE Computer Graphics and Applications*, (July/August): 57–63. doi:10.1109/MCG.2004.5

Duhl, J. (2003). *White Paper: Rich Internet Applications.* Retrieved March 27, 2009, from http://www.adobe.com/platform/whitepapers/idc_impact_of_rias.pdf.

Dyn, N., Levin, D., & Gregory, J. A. (1990). *A* butterfly subdivision scheme for surface interpolation with tension control. *ACM Transactions on Graphics*, pp160–pp169.

eBay Inc. (2009). *Smartphones Buying Guide.* Retrieved July 8, 2009, from http://pages.ebay.com/buy/guides/smart-phones-buying-guide/

Elfriede, D., Rashka, J., (2001). Quality Web Systems, Performance, Security, and Usability. Reading, MA: Addison Wesley.

Ellison, R., Fisher, D., Linger, R., & Lipson, H. (1997). *Survivable Network Systems: An Emerging Discipline.* (Technical Report CMU/SEI-97-TR-013), Software Engineering Institute, Carnegie Mellon University.

Embey, D. W., Jiang, Y., & Ng, Y. K. (1999). Record-boundary discovery in web documents. In *ACM SIGMOD Conference* (pp. 467-478).

EnglishEnglish.com. (2003). *What percentage of the internet is in English?* Retrieved June 15, 2009, from http://www.englishenglish.com/english_facts_8.htm

Ercelebi, E. (2004). Electrocardiogram signals de-noising using lifting-based discrete wavelet transform. *Computers in Biology and Medicine*, *34*, 479–493. doi:10.1016/S0010-4825(03)00090-8

Erickson, T., Kellogg, W. A., Laff, M., Sussman, J., Wolf, T. V., Halverson, C. A., & Edwards, D. (2006). A persistent chat space for work groups: the design, evaluation and deployment of loops. In *Proceedings of the 6th Conference on Designing Interactive Systems 06* (pp. 331-340) New York: ACM Press.

Ethnologue (2006). *Ethnologue, Languages of the World.* Retrieved June 15, 2009, from http://www.ethnologue.com

Etsion, Y., Tsafrir, D., & Feitelson, D. G. (2003). Effects of clock resolution on the scheduling of interactive and soft real-time processes. In Joint International Conference on

Measurement and Modeling of Computer Systems: *Proceedings of the 2003 ACM SIGMETRICS international conference on Measurement and modeling of computer systems: Operating systems* (pp. 172 - 183). New York: Association for Computing Machinery.

Eugster, P., Felber, P., Guerraoui, R., & Kermarrec, A. (2003). The many faces of publish/subscribe. *ACM Computing Surveys*, *35*(2), 114–131. doi:10.1145/857076.857078

Eugster, P., Garbinato, B., & Holzer, A. (2005). Location-based Publish/Subscribe. *Fourth IEEE International Symposium on Network Computing and Applications*, 279-282.

European Space Agency (ESA). (n.d.). *Galileo*. Retrieved (n.d.)., from http://www.esa.int/esaNA/galileo.html

Everson, E. (2007). *Holiday shopping scams targetingMobile*. Retrieved March 18, 2008, from http://community.zdnet.co.uk/blog/0,1000000567,10006581o-2000440756b,00.htm

Fabrizi, S., & Wertlen, B. (2008). Roaming in the Mobile Internet. *Telecommunications Policy*, *32*(1), 50–61. doi:10.1016/j.telpol.2007.11.003

Fain, D. C., & Pedersen, J. O. (2006). Sponsored search: A brief history. *Bulletin of the American Society for Information Science and Technology*, *32*(2), 12–13. doi:10.1002/bult.1720320206

Feldhofer, M., Dominikus, S., & Wokerstorfer, J. (2004, August 11-13). Strong Authentication for RFID Systems Using the AES Algorithm. *Cryptographic Hardware and Embedded Systems – CHES 2004, 6th International Workshop*, p. 357-370, Cambridge, MA.

Fensli, R., Gunnarson, E., & Gundersen, T. (2005). A Wearable ECG-recording System for Continuous Arrhythmia Monitoring in a Wireless Tele-Home-Care Situation. In *Proceedings of the 18th IEEE Symposium on Computer-Based Medical Systems (CBMS'05).*

Fettig, A. (2005). Twisted Network Programming Essentials. Cambridge, MA: O'Reilly Media, Inc.

Fielding, R. T., & Taylor, R. N. (2000). Principled design of the modern Web architecture. In *Proceedings of the 22nd international Conference on Software Engineering (Limerick, Ireland, June 04 - 11, 2000)* (407-416), ICSE '00. New York:ACM

Fitchard, K. (2004). *Qualcomm re-imagines mobile media. Telephony, 245(22), 6-7.* Retrieved (n.d.)., from http://www.scopus.com/scopus/inward/record.url?eid=2-s2.0-9744254533&partnerID=40

Flierl, M., & Girod, B. (2006). Coding of multi-view image sequences with video sensors. In *Proceedings of the IEEE International Conference on Image Processing.* (pp. 609-612).

Flinn, J., deLara, E., Satyanarayanan, M., Wallach, D., & Zwaenepoel, W. (2001). Reducing the energy usage of office applications. In *Proceeding. of Middleware'01.*

Fogel, E., Cohen, D., Revital, I., & Zvi, T. (2001). A Web Architecture for Progressive Delivery of 3D Content. *In Proceeding of ACM Web3D 2001* (pp. 35-41)

Forin, A., Forin, R., Raffman, A., Raffman, A., & Aken, J. V. (1998). *Asymmetric Real Time Scheduling on a Multimedia Processor.* (Technical Report MSR-TR-98-09). Redmond, WA: Microsoft Research.

Forum, W. (2006). 'Mobile WiMAX Part I: A Technical Overview and Performance Evaluation'. *Mobile WiMAX - Part I: A Technical Overview and Performance Evaluation.*

Francis, J. (2007). Techno-economic analysis of the open broadband access network wholesale business case. In 2007 16th IST Mobile and Wireless Communications Summit, 2007 16th IST Mobile and Wireless Communications Summit. Budapest.

Franke, M. (2007). *Seminar Paper: A Quantitative Comparison of Realtime Linux Solutions.* Chemnitz, Germany: Chemnitz University of Technology, Department of Computer Science.

Frohlich, D. M., Rachovides, D., Riga, K., Bhat, R., Frank, M., Edirisinghe, E., et al. (2009). StoryBank: mobile digital storytelling in a development context. In *Proceedings of CHI 2009* (1761-1770), New York: ACM.

Fu, F., Lin, X., & Xu, L. (August, 2004,). Fast Intra Prediction Algorithm in H.264/AVC. In *Proceedings of7th International Conference On Signal Processing*, ICSP'04, Vol 2, 31Aug-4.

Funkhouser, T., & Sequin, C. (1993). Adaptive display algorithm for interactive frame rates during visualization of complex virtual environments. In *Proceeding of ACM SIGGRAPH'93* (pp. 247–254)

Gaber, J. (2007). GLOBECOM Workshop 07. Washington, DC: Spontaneous Emergence Model for Pervasive Environments. In IEEE.

Games, M. (2005). *Mobile Games Industry Worth US $11.2 Billion by 2010*. Retrieved 2005 from http://www.3g.co.uk/PR/May2005/1459.htm

Gamma, E., Helm, R., Johnson, R., Vlissides, J. (1995). Design Patterns. Reading: Addison-Wesley.

Gao, F., & Hope, M. (2008). Collaborative middleware on Symbian OS via Bluetooth MANET. *WSEAS TRANSACTIONS on COMMUNICATIONS*, *7*(4), 300–310.

Garces-Erice, L., Biersack, E. M., Felber, P. A., Ross, K. W., & Urvoy-Keller, G. (2003). Hierarchical Peer-to-Peer Systems. In *International Conference on Parallel and Distributed Computing (Euro-Par 2003)*, Klagenfurt, Austria.

Garland, M., & Heckbert, P. (1997). Surface Simplification using Quadric Error Metrics. In *Proceeding of ACM SIGGRAPH 1997* (pp. 209-216)

Garofalakis, J., Stefani, A., Stefanis, V., & Xenos, M. (2007). *Quality attributes of consumer-based m-commerce systems.* Paper presented at the 2007 ICETE-Business Conference, 130-136.

Gartner (2004a). *2004 mobile security research reports*. Stamford, MA: Gartner Inc

Gartner (2004b). Q1 *2004 research report on wireless mobile security and hackers*. Stamford, MA: Gartner Inc.

Gartner Group. (2009). *Gartner Says Worldwide Mobile Phone Sales Declined 8.6 Per Cent and Smartphones Grew 12.7 Per Cent in First Quarter of 2009.* Press released dated May 20, 2009. Retrieved June 15, 2009, from http://www.gartner.com/it/page.jsp?id=985912

Gartner, Inc. (2009a). *Gartner Says in the Fourth Quarter of 2008 the PC Industry Suffered Its Worst Shipment Growth Rate Since 2002*. Retrieved March 15, 2009, from http://www.gartner.com/it/page.jsp?id=856712

Gartner, Inc. (2009b). *Gartner Says Worldwide Smartphone Sales Reached Its Lowest Growth Rate With 3.7 Per Cent Increase in Fourth Quarter of 2008*. Retrieved March 18, 2009, from http://www.gartner.com/it/page.jsp?id=910112

Gartner, Inc. (2009c). *Gartner Says Worldwide Mobile Phone Sales Grew 6 Per Cent in 2008, But Sales Declined 5 Per Cent in the Fourth Quarter*. Retrieved March 19, 2009, from http://www.gartner.com/it/page.jsp?id=904729

Gauthier, Dickey., C., & Grothoff, C, (2008). Bootstrapping of Peer-to-Peer Networks. In *International Workshop on Dependable and Sustainable Peer-to-Peer Systems*, Turku, Finland.

Gedik, B., & Liu, L. (2005). Location privacy in mobile systems: A personalized anonymization model. In *Proc. of ICDCS*, 2005 (pp. 620-629).

Gelernter, D. (1985). Generative communication in Linda. *ACM Transactions on Programming Languages and Systems*, *7*(1), 80–112. doi:10.1145/2363.2433

Gershenfeld, N. (1999). *When things start to think*. London: Hodder and Stoughton.

Ghinea, G., & Angelides, M. C. (2004). A User Perspective of Quality of Service in m-Commerce. *Multimedia Tools and Applications*, *22*(2), 187–206. doi:10.1023/B:MTAP.0000011934.59111.b5

Ghinita, G., Kalnis, P., & Skiadopoulos, S. (2007). Prive: Anonymous location-based queries in distributed mobile systems. In *Proc. of WWW 2007* (pp. 371-380).

Glover, B., & Bhatt, H. (2006). *RFID Essentials*. Norfolk, UK: O'Reilly Publisher.

Gobbeti, E., & Bouvier, E. (1999). Time-Critical Multiresolution Scene Rendering. In *Proceeding of IEEE Visualizatoin* (pp. 123–130).

Goh, K., Lavanya, J., Kim, Y., Tan, E., & Soh, C. (2005, September 1-4). A PDA-based ECG Beat Detector for Home Cardiac Care. In *Proceedings of the 2005 IEEE Engineering in Medicine and Biology 27th Annual Conference*(pp: 375-378). Shanghai, China.

Golatowski, F., Hildebrandt, J., Blumenthal, J., & Timmermann, D. (2002). Framework for Validation, Test and Analysis of Real-Time Scheduling Algorithms and Scheduler Implementations. In RSP, *Proceedings of the 13th IEEE Intl. Workshop on Rapid System Prototyping* (RSP'02), (pp. 146).

Goldwasser, S. (1997). Multi party computations: past and present. In *Annual ACM Symposium on Principles of Distributed Computing*, pp.1-6.

Goldwasser, S., Micali, S., & Rivest, R. (1988). A digital signature scheme secure against adaptive chosen-message attacks . *SIAM Journal on Computing, 17*(2), 281–308. doi:10.1137/0217017

Gonzalez, R. C., & Woods, R. E. (2002). *Digital Image Processing.* Upper Saddle River, NJ: Prentice Hall.

Gopakumar, K. (2006). E-governance services through telecentres: Role of human intermediary and issues of trust. *Information Technologies and Development, 4*(1), 19–35.

Govoni, D., & Soto, J. C. (2002). JXTA and security. In *JXTA: Java P2P Programming.* Indianapolis, IN: Sams Publishing.

Graf, A., Dabrunz,O,, Assmann, S. (2009). *Interrupt Handling on x86 (RT) and Boot Interrupt Quirks.* Nürnberg, Germany: Maxfeldstr

Graham-Rowe, D. (2004*). Camera phones will be high-precision scanners, NewScientist.com news service.* Retrieved Oct 10, 2008, from http://www.newscientist.com/article.ns?id=dn7998.

Graps, A. (1995). A friendly guide to wavelets. *IEEE Computational Science & Engineering, 2*(2).

Gray, C., & Cheriton, D. (1989). Leases: an efficient fault-tolerant mechanism for distributed file cache consistency. *SOSP '89: Proceedings of the twelfth ACM symposium on Operating systems principles*, 202-210.

Grecos, C. & Yang, M. (June, 2005). Fast inter mode prediction for P slices in the H264 video coding standard. *IEEE Transaction on Broadcasting, 51*(2).

Grecos, C., & Yang, M. (2006, December). Fast mode prediction for the Baseline and Main profiles in the H.264 video coding standard. *IEEE Transactions on Multimedia, 8*(6). doi:10.1109/TMM.2006.884631

Grecos, C., & Yang, M. Y. (2007). Coding of Audio Visual Objects Part 2: Visual, ISO/IEC JTC1, ISO/IEC 14496-2 (MPEG-4 Visual version 1). *Digital Signal Processing, 17*(3), 652–664. doi:10.1016/j.dsp.2005.11.005

Grecos, C., & Yang, M. Y. (2007). Exploiting temporal information and adaptive thresholding for fast mode decision in H264 video coding standard . *Multidimensional Systems and Signal Processing, 18*, 309–316. doi:10.1007/s11045-006-0006-8

Gruteser, M., & Grunwald, D. (2003). *Anonymous usage of location-based services through spatial and temporal cloaking* (pp. 31–42). In Proc. of MobiSys.

Gu, T., et al. (2004). An ontology-based context model in intelligent environments. In *Proc, Communication Networks and Distributed Systems Modeling and Simulation Conf., Soc, for modeling and simulation intl's, 2004.*

Gueziec, A. (1999). Locally toleranced surface simplification. *IEEE Transactions on Visualization and Computer Graphics, 5*(2), 168–189. doi:10.1109/2945.773810

Gumbold, S., & Straber, W. (1998). Real Time Compression of Triangle Mesh Connectivity. In *Proceeding of ACM SIGGRAPH 1998* (pp. 133-140).

Gunasekaran, V., & Harmantzis, F. (2008). Towards a Wi-Fi ecosystem: Technology integration and emerging service models. *Telecommunications Policy, 32*(3-4), 163–181. doi:10.1016/j.telpol.2008.01.002

Gupta, A., & Kumar, A. Mayank, Tripathi, V. N., & Tapaswi, S. (2007). Mobile Web: Web manipulation for small displays using multi-level hierarchy page segmentation. In *Proceedings of the 4th International Conference on Mobile Technology, Applications, and Systems* (pp. 599-606). Singapore.

Haan, D., Biezen, G.,, P. W. A. C., Ojo, O. A., & Huijgen, H. (1993). True motion estimation with 3-D recursive search block matching. *IEEE Trans. Circuits Syst. Video Technol., 3*,368–379, 388

Haddow, G. D., Bullock, J. A. (2004). Introduction to Emergency Management. Amsterdam: Butterworth-Heinemann.

Haghighi, P. D., Zaslavsky, A., Krishnaswamy, S., & Gaber, M. M. (2009). *Mobile Data Miningfor Intelligent Healthcare Support.* 42nd Hawaii International Conference on System Sciences, 2009, pp: 1-10.

Hamming, R. W. (1950). Error Detecting and Error Correcting Codes. *The Bell System Technical Journal, 29*, 147–160.

Harrison, R., & Shackman, M. (2007). *Symbian OS C++ for Mobile Phones.* Hoboken, NJ: Wiley Publishing.

Haselsteiner, E. & Breitfuß, K. (2006). Security in near field communication (NFC), *Printed handout of Workshop on RFID Security, 6.* Amsterdam: Philips Semiconductors.

Hashim, A. H. A., & Anwar, F. Mohd. S., & Liyakthalikh, H. (2005). Mobility Issues in Hierarchical Mobile IP. *3rd IEEE International Conference: Science of Electronic, Technologies of Information and Telecommunications*, pages.

Hattori, G., Hoashi, K., Matsumoto, K., & Sugaya, F. (2007). Robust Web page segmentation for mobile terminal using content-distances and page layout information. In *Proceedings of the 16th International Conference on World Wide Web* (pp. 361-370). Banff, Canada

He, Z., Liang, Y., Chen, L., Ahmad, I., & Wu, D. (2005). Power-rate-distortion analysis for wireless video communication under energy constraints. *IEEE Transactions on Circuits and Systems for Video Technology, 15*(5), 645–658. doi:10.1109/TCSVT.2005.846433

Heine, G., & Horrer, M. (1999). *GSM Networks: Protocols, Terminology and Implementation.* Norwood, MA: Artech House.

Helfenbein, E., Zhou, S., Lindauer, J., Field, D., Gregg, R., & Wang, J. (2006). An algorithm for continuous real-time QT interval monitoring. *Journal of Electrocardiology, 39*, 123–S127. doi:10.1016/j.jelectrocard.2006.05.018

Henning, T. (2008). *State of the Mobile Imaging Industry. Address at 6Sight: The future of imaging.* San Fransisco.

Henrici, D., & Muller, P. (2004, March 14-17). Hash-based Enhancement of Location Privacy for Radio-frequency Identification Devices using Varying Identifiers. *Second IEEE Annual Conference on Pervasive Computing and Communications Workshops*, Orlando, FL.

Herness newsletter. (2001). ***Mobile commerceand*** *its future.* Retrieved February 21, 2008, from http://www.netmode.ntua.gr/courses/postgraduate/edi/material/11th_Hermes_Newsletter(Mobicom).pdf

Herring, S. C., Scheidt, L. A., Kouper, I., Wright, E. (2006). A longitudinal content analysis of weblogs: 2003-2004. In TremayneM. (Ed.), Blogging, Citizenship, and the Future of Media (pp. 3–20). London: Routledge.

Hewitt, C. (2008). ORGs for Scalable, Robust, Privacy-Friendly Client Cloud Computing . *Internet Computing, 12*(5), 96–99. doi:10.1109/MIC.2008.107

Higel, S. (2003). *Towards an Intuitive Interface for Tailored Service Compositions*, Compositions, - DAIS 2003 . *Lecture Notes in Computer Science, 2893*, 17–21.

Higginbotham, S. (2008). *Countdown to 4G: Who's Doing What, When. GigaOM.* Retrieved July 29, 2009, from http://gigaom.com/2008/08/13/countdown-to-4g-whos-doing-what-when/

Hiltunen, M., Schlichting, R., Ugarte, C., & Wong, G. (2000). Survivability through Customization and Adaptability: The Cactus Approach. *DARPA Information Survivability Conference and Exposition* (pp. 294-307).

Hollan, J., Hutchins, E., & Kirsh, D. (2000). Distributed cognition: toward a new foundation for human-computer interaction research. *ACM Transactions on Computer-Human Interaction*, *7*(2), 174–196. doi:10.1145/353485.353487

Holsapple, C., & Sasidharan, S. (2005). *The dynamics of trust in B2C e-commerce: a research model and agenda.* Paper presented at ISeB, 377-403.

Holzinger, A. (2005). Usability Engineering Methods for Software Developers. *Communications of the ACM*, *48*(1), 71–74. doi:10.1145/1039539.1039541

Hong, S. J., & Lerch, F. J. (2002). A Laboratory study of Customers' preferences and purchasing behavior with regards to software components. *The Data Base for Advances in Information Systems*, *33*(3), 23–37.

Hong, S., Thong, J. Y., Moon, J., & Tam, K. (2008). Understanding the behavior of mobile data services consumers. *Information Systems Frontiers*, *10*(4), 431–445. doi:10.1007/s10796-008-9096-1

Hong, X., Zhang, L., & Jin-Long, Hu. (2006). New Scheme of Implementing Real-Time Linux. In icsea, *(pp.67), International Conference on Software Engineering Advances* (ICSEA'06).

Hoppe, H. (1996) Progressive meshes. In *Proceeding of ACM SIGGRAPH* (pp. 99–108).

Hoppe, H. (1998). Efficient Implementation of Progressive Meshes. *Computers & Graphics*, *22*(1), 27–36. doi:10.1016/S0097-8493(97)00081-2

Hsiao, J. L., Hung, H. P., & Chen, M. S. (2008). Versatile transcoding proxy for Internet content adaptation. *IEEE Transactions on Multimedia*, *10*(4), 646–658. doi:10.1109/TMM.2008.921852

Hsu, C. Y., Ortega, A., & Reibman, A. R. (1997). Joint Selection of Source and Channel Rate for VBR Video Transmission under ATM policing constraints. *IEEE Journal on Sel. Areas in Communications. Special Issue on Real-Time Video Services in Multimedia Networks*, *15*(6), 1016–1028.

Hu, F., Jiang, M., Celentano, L., & Xiao, Y. (2008). Robust medical ad hoc sensor networks (MASN) with wavelet-based ECG data mining. *Ad Hoc Networks*, *6*, 986–1012. doi:10.1016/j.adhoc.2007.09.002

Hu, H., & Lee, D. (2006). Range nearest neighbor query. *IEEE Transactions on Knowledge and Data Engineering*, *18*(1), 78–91. doi:10.1109/TKDE.2006.15

Hu, H., & Xu, J. (2009). Non-exposure location anonymity. *IEEE International Conference on Data Engineering*, 2009, to appear.

Hu, W. C., Lee, C. W., & Yeh, J. H. (2004). Mobile Commerce Systems . In Shi, N. (Ed.), *Mobile Commerce Applications* (pp. 1–23). Hershey, PA: Idea Group Publishing.

Hu, W. C., Yeh, J. H., Chu, H. J., & Lee, C. W. (2005). Internet-enabled mobile handheld devices for mobile commerce. *Contemporary Management Research*, *1*(1), 13–34.

Hu, W. C., Zuo, Y., Chen, L., & Yang, C. H. (2008). Adaptive mobile Web browsing using Web mining technologies. Memmola M. & Al-Hakim, L., editors, *Business Web Strategy: Aligning the Internet with Corporate Design*, Hershey, PA: Information Science Reference.

Hu, W., Zuo, Y., Wiggen, T., & Krishna, V. (2008, May 18-20). Handheld Data Protection Using Handheld Usage Pattern Identification. 2008 IEEE International Conference on Electro/Information Technology, p. 237-240, Ames, IA

Hua, Z., Xie, X., Liu, H., Lu, H., & Ma, W. Y. (2006). Design and performance studies of an adaptive scheme for serving dynamic Web content in a mobile computing environment. *IEEE Transactions on Mobile Computing*, *5*(12), 1650–1662. doi:10.1109/TMC.2006.182

Huang, W. W., Wang, Y., Day, J., (2007). Global Mobile Commerce: Strategies, Implementation and Case Studies. Hershey, PA: Idea Group Reference.

Hudson, J. H., Christensen, J., Kellogg, W. A., & Erickson, T. (2002). I'd be overwhelmed, but it's just one more thing to do. In *Proceedings of the SIGCHI conference on Human Factors in Computing Systems*. (pp. 97-104) New York: ACM Press.

Hulu, L. L. C. (2009). *Hulu*. Retrieved July 21, 2009, from http://www.hulu.com/

Hwang, Y., Kim, J., & Seo, E. (2003). Structure-aware Web transcoding for mobile devices. *IEEE Internet Computing*, *7*(5), 14–21. doi:10.1109/MIC.2003.1232513

Iannello, G., Pescapè, A., Ventre, G., & Vollero, L. (2004). *Experimental analysis of heterogeneous wireless networks. WWIC 2004, Wired/Wireless Internet Communications 2004*. LNCS.

IDC. (2004). *Worldwide Mobile Phone Shipment Research*. International Data Corp, Press Release 2003 and 2004

IEEE 802.11 Wireless LAN Standard, 2001.

IEEE. 802.11. (2007). IEEE Standard for Information technology-Telecommunications and information exchange between systems-Local and metropolitan area networks-Specific requirements - Part 11: Wireless LAN Medium Access Control (MAC) and Physical Layer (PHY) Specifications. Piscataway, NJ: IEEE.

IETF. (2009). *hybi: Bidirectional communication for hypertext*. Retrieved April 11, 2009, from http://trac.tools.ietf.org/bof/trac/wiki/HyBi

Iftode, L., Borcea, C., Ravi, N., Kang, P., & Zhou, P. (2004). Smart phone: An embedded system for universal interactions. In *Proceedings of the tenth International Workshop on Future Trends in Distributed Computing Systems* (pp. 88-94).

Infotrend/CAP Ventures. (2004). W*orldwide Camera Phone and Photo Messaging Forecast: 2004-2009.*London: Kluwer. Retrieved June 15, 2009, from http://store.infotrendsresearch.com/PhotoGallery.asp?ProductCode=MobileImagingStudy10106

INSTAT. (2007). *Size and Growth of Smartphone Market Will Exceed Laptop Market for Next Five Years*. Retrieved June 15, 2009, from http://www.instat.com/press.asp?ID=2148&sku=IN0703823WH.

Internet World Stats. (2008). *Internet usage in Asia*. Retrieved June 15, 2009, fromhttp://www.internetworldstats.com/stats3.htm

ISO/IEC 9126 (2004). *Software Product Evaluation –Quality Characteristics and Guidelines for the User*. Geneva, Switzerland: International Organization for Standardization.

ITU. (2009). *New ITU ICT Development Index compares 154 countries*. (press release dated March 2, 2009).http://www.itu.int/newsroom/press_releases/2009/07.html

Jacobs, T. R., Chouliaras, V. A., & Mulvaney, D. J. (2006). Thread-Parallel MPEG-2, MPEG-4 and H.264 Video Encoders for SoC Multi-Processor Architectures. *IEEE Transactions on Consumer Electronics*, *52*(1), 269–275. doi:10.1109/TCE.2006.1605057

Jakkula, S. (June, 2002). *Proactive Micro Mobility Protocol Design*. Unpublished Master's dissertation, IIIT-Allahabad, India.

Jaokar, A., & Fish, T. (2006). *Mobile Web 2.0 -- The innovator guide to developing and marketing next generation wireless/mobile applications*, London: Futuretext. Retrieved March 22, 2009, from http://mobileweb20.futuretext.com

Jarno (2008). *Symbian Jailbreak by Spanish modder*. Retrieved June 10, 2009, from http://www.f-secure.com/weblog/archives/00001451.html

Jasemian, Y., & Arendt-Nielsen, L. (2005). Evaluation of a realtime, remote monitoring telemedicine system using the Bluetooth protocol and a mobile phone network. *Journal of Telemedicine and Telecare*, *11*(5), 256–160. doi:10.1258/1357633054471911

Jeng, I. H., & Wang, Y. R. (2008). *Gosport Video*. Retrieved March 30, 2009, from http://faculty.pccu.edu.tw/~zyh2/gosport/Gosport.AVI.

Jeng, I. H., Chang, A. Y., & Wang, Y. R. (2008). Plug into the online database and play Mobile Web 2.0. *IT Professional*, *10*(5), 34–38. doi:10.1109/MITP.2008.107

Jeng, I. H., Lee, C. J., Wang, Y. R., & Cheng, C. K. (2009). Secure Information Retrieval and Reveal for Mobile Apparatus Based on 2D Barcode Digital Signature. In*Proc. 13rd Ann. IEEE Int'l Symposium on Consumer Electronics (ISCE 09)*, (pp. 683-686). IEEE Press.

Jeon, B. & J. Lee, (December, 2003). *Fast Mode Decision for H264,* ISO/IEC JTC1/SC29/WG11 and ITU-T SG16, Input Document JVT-J033.

Jiang, T., Xiang, W., Chen, H., & Ni, Q. (2007). Multicast broadcast services support in OFDMA-based WiMAX systems. *IEEE Communications Magazine, 45*(8), 78-86. Retrieved (n.d.), from http://www.scopus.com/scopus/inward/record.url?eid=2-s2.0-34548642639&partnerID=40.

Jiang, W., & Kong, S. G. (2007, Nov). Block-based neural networks for personalized ECG signal classification. *IEEE Transactions on Neural Networks, 18*(6), 1750–1761. doi:10.1109/TNN.2007.900239

Jindal, A., Crutchfield, C., Goel, S., Kolluri, R., & Jain, R. (2008). The mobile Web is structurally different. In *Proceedings of the 27th Conference on Computer Communications* (pp. 1-6). Phoenix, AZ.

Jing, X., & Chau, L. P. (2004, August). Fast Approach for H.264 Inter Mode Decision. *Electronics Letters, 40*(17). doi:10.1049/el:20045243

Joseph, A. D., deLespinasse, A. F., Tauber, J. A., Gifford, D. K., & Kaashoek, M. F. (1995). Rover: a toolkit for mobile information access. In *Proceedings of the 15th ACM Symposium on Operating Systems Principles (SOSP '95)*, 156-171.

Joshi, A., Welankar, N., Bl, N., Kanitkar, K., & Sheikh, R. (2008, September 2-5). Rangoli: A Visual Phonebook for Low-literate Users. *MobileHCI 2008*, Amsterdam, The Netherlands.

Juels, A. (2004, September 8-10). Minimalist Cryptography for Low-cost RFID Tags. *The Fourth Conference on Security in Communication Networks* (pp. 149-153), Amalfi, Italy.

Juels, A. (2006, February). RFID Security and Privacy: A Research Survey (2006). *IEEE Journal on Selected Areas in Communication, 24*(2).

Juels, A., Rivest, R., & Szydlo, M. (2003). The Blocker Tag: Selective Blocking of RFID Tags for Consumer Privacy. *ACM Conference on Computer and Communication Security*, (pp. 103-111).

Jul, E., Levy, H., Hutchinson, N., & Black, A. (1988). Fine-Grained Mobility in the Emerald System. *ACM Transactions on Computer Systems, 6*(1), 109–133. doi:10.1145/35037.42182

Junglas, I. (2007). On the usefulness and ease of use of location-based services: insights into the information system innovator's dilemma. *International Journal of Mobile Communications, 5*(4), 389–408. doi:10.1504/IJMC.2007.012787

Justin, C., & Rajive, B. (2008). Programming in Mobile Ad Hoc Networks. *The Fourth International Wireless Internet Conference (WICON).* 10.4108/ICST.WICON2008.4932

Kagami, S. (2001). *Humanoid robot h7 for autonomous and intelligent software research.* In Real Time Linux Workshop, Milan, Italy, 2001. ftp://ftp.realtimelinuxfoundation.org/pub/events/rtlws-2001/proc/k02-kagami.pdf

Kaiduan, X., Wong, V. W. S., & Leung, V. C. M. (2004, September). Support of Micro-Mobility in MPLS-Based Wireless Access Networks. *Oxford Journals IEICE-Transactions on Communications, 88*(7), 2735–2742.

Kail, E., Khoor, S., & Nieberl, J. (2005). *Ambulatory Wireless Internet Electrocardiography: New concepts & Maths.* 2nd International Conference on Broadband Networks (pp: 1001-1006).

Kalaspur, S., Kumar, M., & Shirazi, B. A. (2007). Dynamic Service Composition in Pervasive Computing. *IEEE Transactions on Parallel and Distributed Systems, 18*(7), 907–917. doi:10.1109/TPDS.2007.1039

Kalofonos, D. N., Antoniou, Z., Reynolds, F. D., Van-Kleek, M., Strauss, J., & Wisner, P. (2008). MyNet: a Platform for Secure P2P Personal and Social Networking Services. *Sixth Annual IEEE International Conference on Pervasive Computing and Communications (PerCom)*, 135-146.

Kaminsky, E., Grois, D., & Hadar, O. (2008). Dynamic computational complexity and bit allocation for optimizing H.264/AVC video compression. *Journal of Visual Communication and Image Representation, 19*(1), 56–74. doi:10.1016/j.jvcir.2007.05.002

Kan, K. K. H., Chan, S. K. C., & Ng, J. K. (2003). A Dual-Channel Location Estimation System for providing Location Services based on the GPS and GSM Networks, in *Proceedings of The 17th International Conference on Advanced Information Networking and Applications(AINA 2003)*, Xi'an, China, pp. 7-12.

Kangas, T., Hämäläinen, T. D., & Kuusilinna, K. (2006). Scalable Architecture for SoC Video Encoders. *The Journal of VLSI Signal Processing, 44*, 79–95. doi:10.1007/s11265-006-5918-x

Kannangara, C. S., Richardson, I. E. G., Bystrom, M., Solera, J., Zhao, Y., MacLennan, A., & Cooney, R. (2006, February). Low Complexity Skip Prediction for H.264 Through Lagrangian Cost Estimation. *IEEE Transactions on Circuits and Systems for Video Technology, 16*(2), 202–208. doi:10.1109/TCSVT.2005.859026

Kappel, G., Retschitzegger, W., Kimmerstorfer, E., Pröll, B., Schwinger, W., & Hofer, T. (2002, June 10). *Towards a Generic Customisation Model for Ubiquitous Web Applications.* 2nd International, Workshop on Web Oriented Software Technology, IWWOST'2002; Málaga, Spain.

Karonen, O., & Lahtinen, P. (2007). *Video Sharing between Handheld Devices: Combining Live Streams with File Downloads (Technical Report).* Nokia Research.

Keggan, D. & Fern Univ., H. (2002, September 30). *The future of learning: From e-learning to m-learning. ERIC Document Reproduction* (Service No. ED472435)

Kelli, B., & Vidgen, R. (2005). *A quality framework for web site quality: user satisfaction and quality assurance.* Paper presented at the WWW 2005, 930-931.

Kesteren, A. V. (2008). *The XMLHttpRequest Object.* Retrieved April 11, 2009, from http://www.w3.org/TR/2006/WD-XMLHttpRequest-20060405/

Khedr, M., & Karmouch, A. (2005). ACAI: Agent-based contextaware infrastructure for spontaneous applications. *Journal of Network and Computer Applications*, 19–44. doi:10.1016/j.jnca.2004.04.002

Khoakovsky, A., Schroder, P., & Sweldens, W. (2000). Progressive geometry compression. In *Proceeding of SIGGRAPH 2000* (pp. 271–278)

Kim, B. G. (2008, February). Novel Inter-Mode Decision Algorithm Based on Macroblock Tracking for the P-Slice in H.264/AVC Video Coding. *IEEE Transactions on Circuits and Systems for Video Technology, 18*(2). doi:10.1109/TCSVT.2008.918121

Kim, C., & Kuo, C. C. J. (2007, April). Feature-Based Intra/Inter Coding Mode Selection for H.264/AVC. *IEEE Transactions on Circuits and Systems for Video Technology, 17*(4), 441–453. doi:10.1109/TCSVT.2006.888829

Kim, C., Avoine, G., Koeune, F., Standaert, F., & Pereira, O. (2009, December 2-4). The Swiss-Knife RFID Distance Bounding Protocol. *The 12th International Conference on Information Security and Cryptology*, Seoul, Korea.Knight, J., Strunk, E., and Sullivan, K. (2003). Towards a Rigorous Definition of Information System Survivability, *DARPA Information Survivability Conference and Exposition*, Washington, DC.

Kim, C., Hsuan-Huei, S., & Kuo, C. C. J. (April, 2006). Fast H.264 Intra-Prediction Mode Selection Using Joint Spatial and Transform Domain Features. *Journal of Visual Communication and Image Representation*, ELSEVIER.

Kim, C., Hsuan-Huei, S., & Kuo, C. C. J. (January, 2004). *Multistage Mode Decision for Intra Prediction in H.264 Codec.* IS&T/SPIE 16th Annual Symposium EI, Visual Communications and Image Processing, Orlando, FL.

Kim, C., Hsuan-Huei, S., & Kuo, C. C. J. (October, 2004). Feature-Based Intra-Prediction Mode Decision for H.264. In *IEEE Proceedings of International Conference Image Processing*, submitted, Singapole.

Kim, H., & Altunbasak, Y. (October, 2004). Low-complexity Macroblock Mode Selection for H.264/AVC Encoders. *IEEE International Conference on Image Processing, 2004, ICIP'04, 2*, 24-27.

Kim, J., Lee, S., & Kobbelt, L. (2004). *View*-dependent streaming of progressive meshes. In *IEEE Trans Circuits and Systems for Video Technology* (2004).

Kim, S., Song, S., & Jung, H. (2007). WiBro-based mobile RFID service development. In *IEEE Wireless Communications and Networking Conference, WCNC*, 2007 IEEE Wireless Communications and Networking Conference, WCNC 2007. (pp. 2880-2884). Kowloon.

Kim, W. (2009, June). Mobile WiMAX, the Leder of the Mobile Internet Era. *IEEE Communications Magazine*, 10–12. doi:10.1109/MCOM.2009.5116792

Kirste, T. (1995). An infrastructure for mobile information systems based on a fragmented object model. *Distributed Systems Engineering Journal, 2*, 161–170. doi:10.1088/0967-1846/2/3/004

Kleis, M., Lua, E. K., & Zhou, X. (2005). Hierarchical Peer-to-Peer Networks using Lightweight SuperPeer Topologies. In *10th IEEE Symposium on Computers and Communication (ICSS05)*, La Manga del Mar Menor, Cartagena, Spain.

Klomp, S., Vatis, Y., & Ostermann, J. (2006). Side information interpolation with subpel motion compensation for Wyner-Ziv decoder. In *Proceedings of the International Conference on Signal Processing and Multimedia Applications.*

Knivett, V. (2009, February). MEMS accelerometers: a fast-track to design success? *Electronic Engineering Times Europe*, 32-34.

Koch, M., Zivkovic, Z., Kleihorst, R. P., & Corporaal, H. (2008). Distributed Smart Camera Calibration Using Blinking LED. *Workshop on Advanced Concepts for Intelligent Video Systems* (pp. 242-253). Juan-les-Pins, France.

Kong, J., Ates, K. L., Zhang, K., & Gu, Y. (2008). Adaptive mobile interfaces through grammar induction. *Proceedings of the 20th IEEE International Conference on Tools with Artificial Intelligence* (pp.133-140). Dayton, OH.

KorhonenJ. (2004). *Introduction to 3G**mobile**communications*. Norwood, MA: Artech house.

Krakow, G. (2004). *Watching TV on a cellphone. MSNBC.* Retrieved July 20, 2009, from http://www.msnbc.msn.com/id/6305929/

Kranen, P., Kensche, D., Kim, S., Zimmermann, N., Muller, E., Quix, C., et al. (2008). *Mobile Mining and Information Management in HealthNet Scenarios*. 9th International Conference on Mobile Data Management (pp: 215-216).

Kubasov, D., Lajnef, K., & Guillemot, C. (2007a). A hybrid encoder/decoder rate control for a Wyner–Ziv video codec with a feedback channel. In *Proceedings of the IEEE Multimedia Signal Processing Workshop.* (pp. 251-254).

Kubasov, D., Nayak, J., & Guillemot, C. (2007b). Optimal Reconstruction in Wyner-Ziv Video Coding with Multiple Side Information. In *Proceedings of the International Workshop on Multimedia Signal Processing.* (pp. 183-186).

Kulkarni, N. Kumar, S. Mani, K. & Padmanabhuni, S. (2005). *Web Services: E-Commerce Partner Integration.*, IT-Pro, 23-29.

Kumar, A., Rajput, N., Agarwal, S., Chakraborty, D., & Nanavati, A. A. (2008). Organizing the unorganized: Employing IT to empower the under-privileged. In *Proceedings of the International World-Wide Web Conference* (pp. 935-944), Beijing, China: ACM Press.

Kuo, T. Y., & Chan, C. H. (2006, October). Fast Variable Block Size Motion Estimation for H.264 Using Likelihood and Correlation of Motion Field. *IEEE Transactions on Circuits and Systems for Video Technology, 16*(10). doi:10.1109/TCSVT.2006.883512

Kwon, O. B., & Sadeh, N. (2004). Applying case-based reasoning and multi-agent intelligent system to context-aware comparative shopping. *Decision Support Systems, 37*(2), 199–213.

Kynäslahti, H. (2003). In search of elements of mobility in the context of education . In Kynäslahti, H., & Seppälä, P. (Eds.), *Proceedings of Mobile Learning* (pp. 41–48). Helsinki, Finland: IT Press.

Lacoste, G., et al. (Eds.). (2000). *The Commerce Layer: A Framework for Commercial Transactions.* LNCS 1854, pp. 121–153.

Lai, C., Yang, J., Chen, F., & Chan, T. (2007). (n.d.). Affordances of mobile technologies for experiential learning: the interplay of technology and pedagogical practices. *Journal of Computer Assisted Learning, 23*, 326–337. doi:10.1111/j.1365-2729.2007.00237.x

Laitinen, H., Ahonen, S., Kyriazakos, S., Lahteenmaki, J., Menolascino, R., & Parkkila, S. (2001). Cellular location technology (Tech. Rep. 007), *VTT Information Technology*.

Lalonde, P. (1997). *Representations and uses of light distribution functions*, PhD Dissertation, Vancouver, BC: The University of British Columbia.

Lamberti, F., Zunino, C., Sanna, A., Fiume, A., & Maniezzo, M. (2003). An Accelerated Remote Graphics Architecture for PDAs. In*Proceeding of ACM Web3D 2003* (pp. 55-61).

Lamparter, B., & Westhoff, D. (2002). *Security Challenges in the future mobile Internet*. PAMPAS'02 Workshop on Requirements for Mobile Privacy & Security. Heidelberg, Germany: NEC Network Laboratories

Lampe, C., Ellison, N., & Steinfield, C. (2006). A face(book) in the crowd: social searching vs. social browsing. In *Proceedings of the Conference on Computer-supported Cooperative Work* (pp. 167-170) New York: ACM Press.

Lawton, G. (2001). Browsing the mobile Internet. *IEEE Computer*, *35*(12), 18–21.

Lazaro, O., Gonzalez, A., Aginako, L., Hof, T., Filali, F., & Atkinson, R. (2007). *Enabler for next generation pervasive wireless services. In 2007 16th IST Mobile and Wireless Communications Summit, 2007 16th IST Mobile and Wireless Communications Summit*. Budapest: MULTINET.

Ledvinam, B., Mota, F., & Kintner, P. M. (2000). A coming of age for gps: A rtlinux based gps receiver. In *Proceedings of the Workshop on Real Time Operating Systems and Applications and Second Real Time Linux Workshop (in conjunction with IEEE RTSS 2000)*, Orlando, Florida, 2000.

Lee, J., Choi, I. Choi, W., & Jeon, B. (March, 2004). *Fast Mode Decision for B slice*, ISO/IEC JTC1/SC29/WG11 and ITU-T SG16, Input Document JVT-K021.

Lee, K. (2007). Technology leaders forum - Create the future with mobile WIMAX. *IEEE Communications Magazine*, *45*(5). Retrieved (n.d.), from http://www.scopus.com/scopus/inward/record.url?eid=2-s2.0-34249097660&partnerID=40

Lehner, F., & Nosekabel, H. (2002). The role of mobile devices in e-learning – first experience with a e-learning environment. In *IEEE International Workshop on Wireless and Mobile Technologies in Education* (eds M. Milrad, H.U. Hoppe & Kinshuk), pp. 103–106.Los Alamitos, CA: IEEE Computer Society Press.

Lemarie, P., & Meyer, Y. (1986). Ondelettes et bases hilbertiennes. *Rev. Mat. Iberoamericana*, *2*, 1–18.

Lentczner, M. (2009). *Reverse HTTP*. Retrieved March 27, 2009, from http://tools.ietf.org/html/draft-lentczner-rhttp-00

Levoy, M. (2006). *The Digital Michelangelo Project Archive*. Retrieved (n.d.)., from http://graphics.stanford.edu/data/mich/

Lext, J., Assarsson, U., & Moller, T. (2001). *A* Benchmark for Animated Ray Tracing. *IEEE Computer Graphics and Applications*, *21*(2), 22–31. doi:10.1109/38.909012

Li, Q., & Zhang, X. (2004). *Three Dimensional Model: An Analyzing Sketch for E-commerce Theories and Applications*. Paper presented at the Sixth International Conference on Electronic Commerce, 207-212.

Li, Y., & Ding, X. (2007, March 20-22). Protecting RFID Communications in Supply Chains. *ACM Symposium on InformAtion, Computer and Communications Security*, Singapore.

Li, Z., Liu, L., & Delp, E. J. (2007). Rate distortion analysis of motion side estimation in Wyner-Ziv video coding. *IEEE Transactions on Image Processing*, *16*(1), 98–113. doi:10.1109/TIP.2006.884934

Lieberman, H. (1986). Using prototypical objects to implement shared behavior in object-oriented systems. *Conference proceedings on Object-oriented Programming Systems, Languages and Applications*, 214-223.

Lim, C., & Kwon, T. (2006, December 4-7). Strong and Robust RFID Authentication Enabling Perfect Ownership Transfer. *The 8th Conference on Information and Communications Security*, Raleigh, NC.

Lim, K. P., Wu, S., Wu, D. J., Rahardja, S., Lin, X., Pan, F., & Li, Z. G. (September, 2003). *Fast Inter Mode Selection*, ISO/IEC JTC1/SC29/WG11 and ITU-T SG16, Input Document JVT-I020.

Lin, S., & Costello, D. J. (2004). *Error Control Coding* (2nd ed., pp. 563–582). Upper Saddle River, NJ: Pearson Prentice Hall.

Lindsay, C., & Agu, E. (2005). Wavelength dependent Rendering Using Spherical Harmonics. In *Proceeding of Eurographics* (2005).

Lindstrom, P., & Turk, G. (2000). Image-driven simplification. *ACM Transactions on Graphics, 19*(3), 204–241. doi:10.1145/353981.353995

Linner, D., Krüssel, S., & Steglich, S. (2008). CAPgets: *Mobile Web Runtime Environment for Community Applications.* 1st International Workshop on Next Generation Networks: Open Platforms & Services (NGNOPS 2008), September 2008, Wales, UK.

Liskov, B. (1988). Distributed programming in Argus. *Communications of the ACM, 31*(3), 300–312. doi:10.1145/42392.42399

Liu, L., & Delp, E. J. (2006). Wyner-Ziv video coding using LDPC codes. In *Proceedings of the IEEE Nordic Signal Processing Symposium.* (pp.258-261).

Liu, L., Li, Z., & Delp, E. (2007). Complexity-rate-distortion analysis of backward channel aware Wyner-Ziv coding. In *Proceedings of the IEEE International Conference on Image Processing* (pp. 25-28).

Liu, X., Shenoy, P., Corner, M., (2005). Application level power management with performance isolation. In Proceeding. of ACM MM'05. Chameleon.

Ljung, E., & Simmons, E. (2006). *Architecture Development of Personal Healthcare Applications.* M.Sc. Thesis, Lund University, Lund, Sweden.

Loop, C. (1987). *Smooth subdivision surfaces based on triangles. Master's thesis.* Salt Lake City, UT: Department of Mathematical, University of Utah.

Losavio, F., Chirinos, L., Matteo, A., Levy, N., & Ramdane, A. (2004). ISO quality standards for measuring architectures. *Journal of Systems and Software, 72*, 209–223. doi:10.1016/S0164-1212(03)00114-6

Lounsbery, M. (1994). *Multiresolution analysis for surfaces of arbitrary topological type.* PhD Dissertation, Seattle, WA: University of Washington.

Luebke, D., & Hallen, B. (2001). Perceptually driven simplification. for interactive rendering. In *Proceeding of Eurographics Rendering Workshop* (pp. 7–18)

Lv, Q., Cao, P., Cohen, E., Li, K., & Shenker, S. (2002). Search and Replication in Unstructured Peer-to-Peer Networks. In *ACM International Conference on Supercomputing (ICS '02)*, New York.

Lynch, I. (2000). ***Mobile commerce***- *big or REALLY big?* Retrieved February 2, 2008, from http://www.vnunet.com/vnunet/news/2113522/**mobile**-commerce-big-really-big

Ma, W. Y., & Manjunath, B. S. (1998, May). A Texture Thesaurus for Browsing Large Aerial Photographs. *Journal of the American Society for Information Science American Society for Information Science, 49*(7), 633–648. doi:10.1002/(SICI)1097-4571(19980515)49:7<633::AID-ASI5>3.0.CO;2-N

Macintyre, B., & Feiner, S. (1998). A Distributed 3D Library. In . *Proceedings of SIGGRAPH, 2005*, 361–370.

Maekawa, T., Hara, T., & Nishio, S. (2006). Image classification for mobile Web browsing. In *Proceedings of the 15th International Conference on World Wide Web* (pp. 43-52). Edinburgh, Scotland.

Maekawa, T., Hara, T., & Nishio, S. (2006a). A Collaborative Web Browsing System for Multiple Mobile Users. In *IEEE International Conference on Pervasive Computing and Communications* (pp. 22-33).

Maekawa, T., Hara, T., & Nishio, S. (2006b). Two approaches to browse large web pages using mobile devices. In *International conference on Mobile Data Management.*

Magret, V., & Choyi, V. K. (January, 2001). Multicast Micro-mobility Management. *Lecture Notes in Computer Science*, (pages 260-268).Berlin / Heidelberg: Springer.

Mäkeläinen, R., Di Flora, C., & Mikkonen, T. (2008). Enhanced integration of Java to symbian OS using smart pointers. In *ACM International Conference Proceeding Series; Vol. 343. Proceedings of the 6th international workshop on Java technologies for real-time and embedded systems.* Real-Time JVM implementation issues (pp. 38-47).

Malki, S., Deepak, G., Mohanna, V., Ringhofer, M., & Spaanenburg, L. (2006). Velocity Measurement by a Vision Sensor. *IEEE International Conference on Computational Intelligence for Measurement Systems and Applications* (pp. 135-140). La Coruna, Spain.

Mallat, S. (1989). A theory for multiresolution signal decomposition: The wavelet representation. *IEEE Transactions on Pattern Analysis and Machine Intelligence, 11*, 674–693. doi:10.1109/34.192463

Malvar, H. S., Hallapuro, A., Karczewicz, M., & Kerofsky, L. (2003). Low-complexity transform and quantization in H.264/AVC. *IEEE Transactions on Circuits and Systems for Video Technology, 13*(7), 598–603. doi:10.1109/TCSVT.2003.814964

Malvar, H., Karczewicz, M., & Kerofsky, L. (July, 2003). Low complexity transform and quantization in H.264/AVC. *IEEE Trans. Circuits Syst. Video Technol., 13*(7).

Mamei, M., & Zambonelli, F. (2004). Programming Pervasive and Mobile Computing Applications with the TOTA Middleware. *PERCOM '04: Proceedings of the Second IEEE International Conference on Pervasive Computing and Communications*, 263-276.

Mannila, H., Tikanmki, J., Himberg, J., Korpiaho, K., & Toivonen, H. (2001) Time series segmentation for context recognition in mobile devices. In *First IEEE international conference on data mining* (pp:203–210).

Mantegazza, P., Bianchi, E., Dozio, L., & Papacharalambous, S. (2000). Rtai: Real time application interface. *Linux Journal, 72*, 1-1. Retrieved April 2000, from http://noframes.linuxjournal.com/lj-issues/issue72/3838.html

Marca, D., & Perdue, B. (2000).*A Software Engineering Approach and Tool Set for Developing Internet Applications.* Paper presented at ICSE 2000, Limerick, Ireland, 738-741.

Martin, I. M. (2000). ARTE: An Adaptive Rendering and Transmission Environment for 3D Graphics. In *Proceeding of 8th ACM International Conference on Multimedia* (pp. 413-415).

Martinez, J. L., Fernandez-Escribano, G., Kalva, H., Weerakkody, W. A. R. J., Fernando, W. A. C., & Garrido, A. (2008). Feedback free DVC architecture using machine learning. In *Proceedings of International Conference on Image Processing* (pp. 1140-1143).

Mascolo, C., Capra, L., & Emmerich, W. (2002). Mobile Computing Middleware . In *Advanced lectures on networking* (pp. 20–58). New York: Springer-Verlag New York, Inc.doi:10.1007/3-540-36162-6_2

McAfee-NTT-. *DoCoMo (2007). The Future of Mobile Security – Here Today. McAfee & NTT DoCoMo.* Retrieved June 10, 2009, from http://www.mcafee.com/us/local_content/case_studies/cs_future_mobile_security.pdf

McGuire, M., Plataniotis, K., & Venetsanopoulos, A. (2005). Data Fusion of Power and Time Measurements for Mobile Terminal Location . *IEEE Transactions on Mobile Computing, 4*(2), 58–66. doi:10.1109/TMC.2005.24

M-commerce. (2006). *Separating**Mobile commerce**from Electronic Commerce.* Retrieved March 20, 2008, from http://www.**mobile**info.com/Mcommerce/differences.htm

Meier, R., Cahill, V., Nedos, A., & Clarke, S. (2005). *Proximity-Based Service Discovery in Mobile Ad Hoc Networks* (pp. 115–129). Distributed Applications and Interoperable Systems.

Meng, B., Au, O. C., Wong, C., & Lam, H. (July, 2003). *Efficient intra-prediction mode selection for 4x4 blocks in H.264*, 2003 IEEE lot. Conf. Multimedia &Expo (ICME2003). Baltimore, MD

Micheal, J. (2007). *Smart Phone Operating System Concepts with Symbian OS. West Sussex PO19 8SQ*. England: John Wiley & Sons Ltd.

Milani, S., & Calvagno, G. (2007). A distributed video coder based on the H.264/AVC standard. In *Proceedings of the European Signal Processing Conference* (pp. 673-677).

Miller, M., Tribble, E. D., & Shapiro, J. (2005). Concurrency among strangers: Programming in E as plan coordination. *Symposium on Trustworthy Global Computing, 3705*, 195-229.

Mitchell, C. (2004). Security for mobility. London: IET.

Mobi, T. V. Inc. (2009). *MobiTV*. Retrieved July 20, 2009, from http://www.mobitv.com/technology/

Mobile Marketing Association. (2008). (n.d.). Mobile Advertising Guidelines. [fromhttp://mmaglobal.com/mobileadvertising.pdf]. *Retreived.*

Mohanty, S. (2006). A new architecture for 3G and WLAN integration and inter-system handover management. *Wireless Networks, 12*(6), 733–745. doi:10.1007/s11276-006-6055-y

Mohr, A., Riskin, E., & Ladner, R. (2000). Unequal loss protection: Graceful degradation of image quality over packet erasure channels through forward error correction. *IEEE Journal on Selected Areas in Communications, 18*(6), 819–828. doi:10.1109/49.848236

Mohr, W. (2006). Strategic steps to be taken for future mobile and wireless communications. *Wireless Personal Communications, 38*(1), 143–160. doi:10.1007/s11277-006-9022-0

Mohr, W. (2008). Vision for 2020? *Wireless Personal Communications, 44*(1), 27-49. Retrieved (n.d.), from http://www.scopus.com/scopus/inward/record.url?eid=2-s2.0-36949011635&partnerID=40.

Mokbel, M. F., Chow, C. Y., & Aref, W. G. (2006). The new casper: Query processing for location services without compromising privacy. In *Proc. of VLDB*, 2006 (pp. 763-774).

Momtchev, M., & Marquet, P. (2002). An Asymmetric Real-Time Scheduling for Linux. In *ipdps, vol. 2, (pp.0096), Intl. Parallel and Distributed Processing Symposium: IPDPS 2002 Workshops.*

Moody, S. (2007). *Biometrics in the Here and Now*. Retrieved May 25, 2008, from http://www.technewsworld.com/story/59728.html

Moores, T. (2005). Do customers understand the role pf privacy in E-commerce. *Communications of the ACM, 48*(3), 86–91. doi:10.1145/1047671.1047674

Morbée, M., Prades-Nebot, J., Pizurica, A., & Philips, W. (2007). Rate allocation algorithm for pixel-domain distributed video coding without feedback channel. In *Proceedings of IEEE International Conference on Acoustics, Speech and Signal Processing* (pp. 521-524).

Morris, B. (2006). *Symbian OS Architecture Sourcebook*. Hoboken, NJ: John Wiley & Sons.

Morris, B. (2007). *The Symbian OS Architecture Sourcebook. West Sussex PO19 8SQ*. England: John Wiley & Sons Ltd.

Morris, B. (2008). *A guide to Symbian Signed* (3rd ed.). London: Symbian Software Ltd.

MorrisT. (2009). Multimedia Systems. New York: Springer.

M-Shield. (2009). *Texas Instruments' M-Shield Mobile Security Technology Solution*. Retrieved June 10, 2009, from http://focus.ti.com/general/docs/wtbu/wtbugencontent.tsp?templateId=6123&navigationId=12316&contentId=4629

Mukherjee, D. (2006). *A robust reversed-complexity Wyner-Ziv video codec introducing sign-modulated codes* (Tech. Rep. HPL-2006-80), HP Laboratories Palo Alto.

Muller, M., & Daniels, J. (1990). Toward a definition of voice documents. In Conference on Supporting Group Work. In *Proceedings of the ACM SIGOIS and IEEE CS TC-OA Conference on Office Information Systems* (pp. 174 – 183), New York: ACM Press.

Muller, M. J., Farrell, R., Cebulka, K. D., Smith, J. G. (1992). Issues in the usability of time-varying multimedia. In Blattner, M. M., Dannenberg, R. B. (Eds.), Multimedia interface design (pp. 7–38). New York: ACM Press.

Murphy, A., Picco, G., & Roman, G. C. (2001). LIME: A Middleware for Physical and Logical Mobility. In *Proceedings of the The 21st International Conference on Distributed Computing Systems*, 524-536.

Mys, S., Slowack, J., Škorupa, J., Lambert, P., & Van de Walle, R. (2007). Dynamic complexity coding: Combining predictive and Distributed Video Coding. In *Proceedings of the Picture Coding Symposium*.

Mys, S., Slowack, J., Škorupa, J., Lambert, P., & Van de Walle, R. (2009). (in press). Introducing skip mode in distributed video coding. *Signal Processing Image Communication*, *24*, 200–213. doi:10.1016/j.image.2008.12.004

Naismith, L., Lonsdale, P., Vavoula, G., & Sharples, M. (2004). *Literature Review in Mobile Technologies and Learning REPORT 11: FUTURELAB SERIES.* Retrieved (n.d.), from http://www.google.com/search?q=Literature+Review+in+Mobile&rls=com.microsoft:*&ie=UTF-8&oe=UTF-8&startIndex=&startPage=1 Accessed June 2009

NCC. (2008). ***M-commerce**: more money less hype*. Retrieved May 20, 2008, from http://www.nccmembership.co.uk/pooled/articles/BF_WEBART/view.asp?Q=BF_WEBART_113234

Needle, D. (2005). *Smartphones Take Center Stage*. Retrieved June 10, 2009, from http://www.wi-fiplanet.com/news/article.php/3551686

NetFront. (n.d.). *NetFront*Retrieved (n.d.), from http://www.access-netfront.com/

Neto, B., Cristo, M., Golgher, P., & deMoura, E. (2005). Impedance coupling in content-targeted advertising. In*Proc. SIGIR, 2005.* eMarketer (2007). *eMarketer*Retrieved (n.d.)., from http://www.emarketer.com/Article.aspx?id=1004635.

Ng, J. K., Chan, S. K., & Kan, K. K. (2002). Location Estimation Algorithms for Providing Location Services within a Metropolitan area based on a Mobile Phone Network, in *Proceedings of The 5th International Workshop on Mobility Databases and Distributed Systems (MDDS 2002)*, Aix-en-Provence, France, pp. 710-715.

Ngai, E. W. T., & Gunasekaran, A. (2007). A review for mobile commerce research and applications. *Decision Support Systems*, *43*, 3–15. doi:10.1016/j.dss.2005.05.003

Nielsen, J., & Molich, R. (1990). *Heuristic Evaluation of Users Interfaces*. Paper presented at CHI90, 249–256.

Nielsen, J., Molish, R., Snyder, C., Farreli, S. (2001). E – Commerce User Experience. Boston: Nielsen Norman Group.

Nokia-N97. (2009). *Nokia N97 Tech Specs*. Retrieved June 10, 2009, from http://www.nokiausa.com/find-products/phones/nokia-n97/specifications

Ogata, H., & Yano, Y. (2004) Context-aware support for computer-supported ubiquitous learning. In *Proceedings of IEEE International Workshop on Wireless and Mobile Technologies in Education (WMTE)*, pp. 27–34. Los Alamitos, CA: IEEE Computer Society Press.

OLPC. (2009). *One Laptop Per Child.* Retrieved June 15, 2009, from http://laptop.org

Olsina, L. Lafuente, G. & Rossi, G. (2000). *E-commerce Site Evaluation: a Case Study*. Paper presented at EC-Web 2000, 239-252.

Olson, G. M., & Olson, J. S. (2000). Distance matters. *Human-Computer Interaction*, *15*(2), 139–178. doi:10.1207/S15327051HCI1523_4

OMTP BONDI. (2009). *Home - BONDI*. Retrieved April 11, 2009, from http://bondi.omtp.org/default.aspx

OpenAjax Alliance. (2009). *OpenAjax Alliance*. Retrieved April 11, 2009, from http://www.openajax.org/index.php

Opera for Mobile. (n.d.) *Opera for Mobile*. Retrieved (n.d.), from http://www.opera.com/products/mobile/

Opera Software ASA. (n.d.). *Opera's Small-Screen Rendering*. Retrieved June 23, 2008, from http://www.opera.com/products/mobile/smallscreen/

Or, E. M., & Pundik, O. (2007). Hand Motion and Image Stabilization in Hand-held Devices. *IEEE Transactions on Consumer Electronics, 53*(4), 1508–1512. doi:10.1109/TCE.2007.4429245

OSISA. (2009). *Electricity for Africa?* Retrieved June 15, 2009, from http://www.osisa.org/node/4164

Ostermann, J., Bormans, J., List, P., Marpe, D., Narroschke, M., Pereira, F., Stockammer, T., & Wedi, T. (2004). Video Coding with H.264/AVC: Tools, Performance, and Complexity. *IEEE Circuits and Systems, 4*(1).

Otterbacher, J., Radev, D., & Kareem, O. (2006). News to go: hierarchical text summarization for mobile devices. In*Proceedings of the 29th Annual International ACM SIGIR Conference on Research and Development in Information Retrieval* (pp. 589-596). Seattle, WA.

Pagonis, J., & Sinclair, M. C. (1999). Initial Considerations . In *IEE Colloquium on Lost in the Web: Navigation on the Internet, Digest No. 1999/169.* Evolving Personal Agent Environments to Reduce Internet Information Overload.

Pahlavan, K., & Krishnamurthy, P. (2002). *Principles of Wireless Networks a Unified Approach.* Upper Saddle River, NJ: Pearson Education, Inc.

Pajarola, R., & Rossignac, J. (2000). Compressed pregressive meshes . *IEEE Transactions on Visualization and Computer Graphics, 6*(1), 79–93. doi:10.1109/2945.841122

Pal, P., Loyall, J., Schantz, R., Zinky, J., & Webber, F. (2000). Open Implementation Toolkit for Building Survivable Applications. *DARPA Information Survivability Conference and Exposition, 2*, 197-200.

Pan, F., Lin, X., Susanto, R., Lim, K. P., Li, Z. G., Feng, G. N., Wu, D. J., & Wu, S. (n.d.). *Fast Mode Decision for Intra Prediction*, ISO/IEC JTC1/SC29/WG11 and ITU-T SG16, Input Document JVT-G013, March 2003.

Panken, F., Hoekstra, G., Barankanira, D., Francis, C., & Schwendener, R., Gr°ndalen, O., et al. (2007). Extending 3G/WiMAX networks and services through residential access capacity [Wireless broadband access]. *IEEE Communications Magazine, 45*(12), 62–69. doi:10.1109/MCOM.2007.4395367

Papazoglou, M. (2001). Agent oriented support in e-business technology. *Communications of the ACM, 44*(4), 71–77. doi:10.1145/367211.367268

Park, L., Ramamohanarao, K., & Palaniswami, M. (2005). A novel document retrieval method using the discrete wavelet transform. *Trans. on Graphics (TOG), 23*(3).

Pascoe, J. (1998). Adding generic contextual capabilities to wearable computers. In *Proceedings of 2nd International Symposium on Wearable Computers*(pp. 92-99)

Pashtan, A., Kollipara, S., & Pearce, M. (2003). Adapting content for wireless Web services. *IEEE Internet Computing, 7*(5), 79–85. doi:10.1109/MIC.2003.1232522

Passini, L., & Trassati, A. (2009). *Wireless Universal Resource File (WURFL).* Retrieved 2009, from http://wurfl.sourceforge.net/

Paulson, L. D. (2008). Software lets a cellphone work like a mouse. [News Briefs]. *IEEE Computer, 4*(5), 20.

Payment Processing Expert. (2007). *PayPal**M-commerce**.* Retrieved May 13, 2008, from http://paymentprocessingnews.blogspot.com/2007/10/paypal-**m-commerce**.html

Pedro, J., Brites, C., Ascenso, J., & Pereira, F. (2007). Studying the feedback channel in transform domain Wyner-Ziv video coding. In *Proceedings of the 6th Conference on Telecommunications.*

Peng, Z., Duan, Z., Qi, J. J., Cao, Y., & Lv, E. (2007). HP2P: A Hybrid Hierarchical P2P Network. In *1st International Conference on the Digital Society*, Gaudeloupe.

Pereira, F., Ascenso, J., & Brites, C. (2007). *Studying the GOP size impact on the performance of a feedback-channel based Wyner-Ziv video codec* (pp. 801–815). Advances in Image and Video Technology.

Peris-Lopez, P., Hernandez-Castro, C., Estevez-Tapiador, J., & Ribagorda, A. (2006, July 12-14). LMAP: A Real Lightweight Mutual Authentication Protocol for Low-cost RFID Tags. *2nd Workshop on RFID Security*, Graz, Austria.

Peris-Lopez, P., Hernandez-Castro, C., Estevez-Tapiador, J., & Ribagorda, A. (2006, September 3-6). M^{2}AP: A Minimalist Mutual-authentication Protocol for Low-cost RFID Tags. *International Conference on Ubiquitous Intelligence and Computing* (pp. 912-923), Wuhan, China.

Phones review. (2008). *BT launches BT total broadband anywhere with free smart phone*. Retrieved May 21, 2008, from http://www.phonesreview.co.uk/2008/05/20/bt-launches-bt-total-broadband-anywhere-with-free-**Smartphone**

Pimentel, C., & Blake, I. (1998). Modeling burst channels using partitioned fritchman's markov models. *IEEE Trans Veh Tech*, *47*(3), 885–899. doi:10.1109/25.704842

Plauché, M., & Nallasamy, U. (2008). Speech interfaces for equitable access to information technology. *Information Technologies and International Development, The MIT Press*, *4*(1), 69–86. doi:10.1162/itid.2007.4.1.69

Pomiers, P., & Noel, T. (2000). *SynDEx Communications Under Linux*. INRIA Rocquencourt.

Portio Research (2009). *Mobile Messaging Futures 2009-2013.* Retrieved June 15, 2009, fromhttp://wwww.portioresearch.com/MMF09-13.html

Postma, E., van Dartel, M., & Kortmann, R. (2001). Recognition by fixation. In *Proceedings Belgium/Netherlands Artificial Intelligence Conference, BNAIC* (pp. 425–432). Amsterdam, The Netherlands.

Pradhan, S. S., & Ramchandran, K. (1999). Distributed source coding using syndromes (DISCUS): Design and construction. In *Proceedings of the IEEE Data Compression Conference* (pp.158-167).

Pradhan, S. S., Chou, J., & Ramchandran, K. (2003). Duality between source coding and channel coding and its extension to the side information case. *IEEE Transactions on Information Theory*, *49*(5), 1181–1203. doi:10.1109/TIT.2003.810622

Prahalad, C. K. (2004). The Fortune at the Bottom of the Pyramid: Eradicating Poverty through Profits Pearson Education Inc. Upper Saddle River, NJ: Wharton School Publishing.

Prasithsangaree, P., Krishnamurthy, P., & Chrysanthis, P. K. (2002). On Indoor Position Location with Wireless LANs, in *The 13th IEEE International Symposium on Personal,Indoor, and Mobile Radio Communications (PIMRC 2002)*, Lisboa, Portugal, pp.720-724.

Prekop, P., & Burnett, M. (2003). Activities, context and ubiquitous computing. *Special Issue on Ubiquitous Computing Computer Communications*, *26*(11), 1168–1176.

Proctor, F. M., & Shackleford, W. P. (2001). Timing studies of real-time Linux for control. In *Proceedings of DETC 01 ASME 2001 Design Engineering Technical Conferences & Information in Engineering Conference*, Pittsburgh, PA, September 9-12 2001. ASME.

Proctor, F. M., & Shackleford, W. P. (2002). Embedded real-time Linux for cable robot control. In *Proceedings of DETC'02 ASME 2002 Design Engineering Technical Conf. & Computers & Information in Engineering Conference*, Montreal, Canada, September 29-October 2 2002.

Proctor, F. M., Damazo, B., Yang, C., & Frechette, S. (1993). *Open architectures for machine control.* Technical report, National Institute of Standards and Technology, Gaithersburg . *MD Medical Newsmagazine*, (December): 1993.

Proctor, R., Vu, K., Najjar, L., Vaughan, M., & Salvendy, G. (2003). Content Preparation and Management for E-Commerce Web Sites. *Communications of the ACM*, *46*(12), 289–299. doi:10.1145/953460.953513

Public technology. (2006). *Cheltenham launches coin free parking with new**mobile**phone payment system.* Retrieved April 22, 2008, from http://www.publictechnology.net/modules.php?op=modload&name=News&file=article&sid=5185

Puri, R., & Ramchandran, K. (2002). PRISM: A new robust video coding architecture based on distributed compression principles. In *Proceedings of the Allerton Conference on Communication, Control and Computing.*

Rabin, J., & Nevile, C. (Eds.). (2008). *Mobile Web Best Practices 1.0 – Basic Guidelines – W3C Recommenda-*

tion 29 July 2008. Retrieved March 27, 2009, from http://www.w3.org/TR/mobile-bp/

Rajapakse, D. C. (2008). *Fragmentation of Mobile Applications*. Retrieved April 11, 2009, from http://www.comp.nus.edu.sg/~damithch/df/device-fragmentation.htm

Ramamoorhthi, R., & Hanrahan, P. (2002). Frequency Space Environment Map Rendering. In *Proceeding of ACM SIGGRAPH 2002* (pp. 517-526).

Ramos, C., Augusto, J. C., & Shapiro, D. (2008). Ambient Intelligence - the Next Step for Artificial Intelligence . *IEEE Intelligent Systems*, *23*(2), 15–18. doi:10.1109/MIS.2008.19

Rane, S., & Girod, B. (2004). Analysis of error-resilient video transmission based on systematic source-channel coding. In *Proceedings of the Picture Coding Symposium*.

Reddy, M. (1997). *Perceptually modulated level of detail for virtual environments*. PhD Dissertation, Edinburgh, UK: University of Edinburgh, UK.

Reddy, M. (2001). Perceptually optimized 3D graphics. *IEEE Computer Graphics and Applications*, *21*(5), 68–75. doi:10.1109/38.946633

Reed, I.S., & Solomon, G. (1960). Polynomial Codes over Certain Finite Fields, *Journal of the Society for Industrial and Applied Mathematics*.

Regan, K. (2008). *Amazon Aims to Light **M-commerce** Fire with TextBuyIt*. Retrieved May 21, 2008, from http://www.ecommercetimes.com/story/62417.html

Regelson, M., & Fain, D. (2006). Predicting click-through rate using keyword clusters. *In Proc. of the Second Workshop on Sponsored Search Auctions*, 2006.

Rengnier, P., Lima, G., & Barreto, L. (2008). Evaluation of interrupt handling timeliness in real-time Linux operating systems. *ACM SIGOPS Operating Systems Review*, *42*(6), 52–63. doi:10.1145/1453775.1453787

Retrieved July 30th, 2000, from http://www-rocq.inria.fr/syndex/doc/U/SynDExCommsLinux.html

Rheingold, H. (1993). The virtual community: Homesteading on the electronic frontier. Reading, MA: Addison Wes.

Ribeiro-Neto, B., Cristo, M., Golgher, P. B., & de Moura, E. S. (2005). Impedance coupling in content-targeted advertising. In *SIGIR '05: Proc. of the 28th annual intl. ACM SIGIR conf.*,pages 496–503, New York: ACM.

Richardson, I. E. G. (2003). *H.264 and MPEG-4 video compression*. Hoboken, NJ: Wiley. doi:10.1002/0470869615

Rieback, M., Crispo, B., & Tanenbaum, A. (2005, July). RFID Guardian: A Battery-powered Mobile Device for RFID Privacy Management. *Australian Conference on Information Security and Privacy, 3574*, 184-194

Rieger, R., & Gay, G. (1997). Using mobile computing to enhance field study. In . *Proceedings of the Computer-Supported Collaborative Learning Conference: CSCL, 97*, 215–223.

Rijnbeek, P., Kors, J., & Witsenburg, M. (2001). Minimum Bandwidth Requirements for Recording of Pediatric Electrocardiograms. *Circulation, 104*, 3087–3090. doi:10.1161/hc5001.101063

Rivest, R., Shamir, A., & Adleman, L. (1978). A Method For Obtaining Digital Signatures and Public-Key Cryptosystems . *Communications of the ACM, 21*(2), 120–126. doi:10.1145/359340.359342

Roberts, C. M. (2006). Radio Frequency Identification (RFID). *Computers & Security, 25*(1), 1, 18–26. doi:10.1016/j.cose.2005.12.003

Roberts, R. (2006). *Use of Remote Monitoring Devices Increases, Telemedicine Information Exchange, (Original Source: Wall Street Journal, April 18, 2006)*. Retrieved Feb 20, 2008, from http://tie.telemed.org/legal/news.asp

Rodriguez, A., Gonzalez, A., & Malumbres, M. P. (2006). Hierarchical parallelization of an H.264/AVC video encoder. In Proceedings Parelec (pp. 363-368). Bialystok, Poland.

Rodríguez, J., Goñi, A., & Illarramendi, A. (2005). Real-Time Classification of ECGs on a PDA. *IEEE*

Transactions on Information Technology in Biomedicine, 9(1), 23:34.

Roe, P., & Chan, S. Y. (1999). *I/O in the gardens non-dedicated cluster computing environment.* IEEE Press.

Rohlf, J., Helman, J., (1994). A High Performance Multiprocessing Toolkit for Real- Time 3D Graphics. In Proceeding of ACM SIGGRAPH'94 (pp. 381–395). IRIS Perfromer.

Rohs, M. (2005). Camera Phones with Pen Input as Annotation Devices. In *Proceedings of the Workshop PERMID* (pp. 23-26).

Ronfard, R., & Rossignac, J. (1996). Full-range approximation of triangulated polyhedral. *Computer Graphics Forum, 15*(3), 67–76. doi:10.1111/1467-8659.1530067

RoR. (2009). *Ruby on Rails.* Retrieved June 15, 2009, from http://rubyonrails.org

Roschelle, J. (2003). Unlocking the learning value of wireless mobile devices. *Journal of Computer Assisted Learning, 9*, 260–272. doi:10.1046/j.0266-4909.2003.00028.x

Rossignac, J. (1999). Edge breaker: connectivity compression for triangle meshes. *IEEE Transactions on Visualization and Computer Graphics, 5*(1), 47–61. doi:10.1109/2945.764870

Roto, V., Popescu, A., Koivisto, A., Vartiainen, E. (2006). A Web page visualization method for mobile phones. In CHI conference (pp. 35–45). Minimap.

Rusinkiewicz, S., Levoy, M., (2000). A Multiresolution Point Rendering System for Large Meshes. In Proceeding of ACM SIGGRAPH 2000 (pp. 343–352). Qsplat.

Russel, A. (2006). *Comet: Low Latency Data for the Browser.* Retrieved April 11, 2009, from http://alex.dojotoolkit.org/2006/03/comet-low-latency-data-for-the-browser/

Russell, A., Wilkins, G., Davis, D., & Nesbitt, M. (2007). *Bayeux Protocol -- Bayeux 1.0draft1.* Retrieved March 27, 2009 from http://svn.cometd.org/trunk/bayeux/bayeux.html

Russian Space Agency. (n.d.). *Global navigation satellite system (glonass).* Retrieved (n.d.)., from http://www.glonass-ianc.rsa.ru/

Rutten, M. J., van Eijndhoven, J. T. J., & Jaspers, E. G. T., vanderWolf, P., Gangwal, O. P., Timmer, A., & Pol, E. J. D. (2002). A Heterogeneous Multiprocessor Architeture for Flexible Media Processing. *IEEE Design & Test of Computers, 19*(4), 39–50. doi:10.1109/MDT.2002.1018132

S60-Hacked. (2008). *Symbian S60 Hacked.* Retrieved June 10, 2009, from http://www.symbianfreak.com/news/008/03/s60_3rd_ed_feature_pack_1_has_been_hacked.htm

SadehS. (2002). ***M-commerce****: Technologies, Services and Business Models.* New York: John Wiley and Sons.

Safenet (2009). *Two Factor Authentication.* Retrieved June 10, 2009, from http://www.safenet-inc.com/products/tokens/index.asp

Safe-Pro. (2009). *Handy Safe Pro for Symbian.* Retrieved June 10, 2009, from http://www.software.com/downloads/business-applications/review-Handy-Safe-Pro-for-Nokia-9500-9300-521060.html

Salagdo, L., & Nieto, M. (October, 2006). Sequence independent very fast mode decision algorithm on H.264/AVC baseline profile. In *Proceedings of Int. Conf. Image Process.*, Atlanta, GA, USA, pp. 41-44.

Salkintzis, A., Fors, C., & Pazhyannur, R. (2002). WIAN-GPRS integration for next-generation mobile data networks. *IEEE Wireless Communications, 9*(5), 112–124. doi:10.1109/MWC.2002.1043861

Samuel, J. (2005). Mobile communications in South Africa, Tanzania and Egypt: Results from Community and Business Surveys. *Africa: The Impact of Mobile Phones . The Vodafone Policy Paper Series, 2*(March), 44–52.

Sanchez, V., Nasiopoulos, P., & Abugharbieh, R. (2006). Lossless compression of 4D medical images using H.264/AVC. In *Proceedings ICASSP*, Vol. II (pp. 1116-1119). Toulouse, France.

Sarji, D. K. (2008). HandTalk: Assistive Technology for the Deaf . *IEEE Computer, 41*(7), 84–86.

Sarwer, M. G., Po, L. M., & Wu, Q. M. (2008). Fast sum of absolute transformed difference based 4x4 intra-mode decision of H.264/AVC video coding standard . *Signal Processing Image Communication, 23*, 571–580. doi:10.1016/j.image.2008.05.002

Saunders, S., Ross, M., Staples, G., & Wellington, S. (2006). The software quality challenges of service oriented architectures in e-commerce. *Software Quality Journal, 14*, 65–75. doi:10.1007/s11219-006-6002-2

Savvas, A. (2008, 17 October). Gartner's top-10 strategic IT technologies for 2009. *Computer Weekly.*

Schatz, R., & Egger, S. (2008). Social interaction features for mobile TV services. In *IEEE International Symposium on Broadband Multimedia Systems and Broadcasting 2008, Broadband Multimedia Symposium 2008, BMSB*, IEEE International Symposium on Broadband Multimedia Systems and Broadcasting 2008, Broadband Multimedia Symposium 2008, BMSB. Las Vegas, NV.

Schatz, R., Wagner, S., & Jordan, N. (2007, July). *Mobile Social TV: Extending DVB-H Services with P2P-Interaction.* The Second International Conference on Digital Telecommunications (ICDT 2007), Silicon Valley, USA (pp. 14-14).

Schilit, B., & Theimer, M. (1994). Disseminating Active Map Information to Mobile Hosts. *IEEE Network, 8*(5), 22–32. doi:10.1109/65.313011

Schilit, B., Adams, N., & Want, R. (1994, December). Context aware computing applications. In *Proceedings of IEEE Workshop on Mobile Computing Systems and Applications* (pp85-90). Santa Cruz, CA

Schmalstieg, D. (1997). *The Remote Rendering Pipeline*, PhD Dissertation, Vienna University of Technology, Austria.

Schmidt, A., Aidoo, K. A., Takaluoma, A., Tuomela, U., Laerhoven, K. V., & de Velde, W. V. (1999, September). Advanced interaction in context. In *Proceedings of First International Symposium on Handheld and Ubiquitous Computing* (pp.89-101), Karlsruhe, Germany.

Schreier, P. G. (2001). *Interfacing DA Hardware To Linux (Technical report).* United Electronic Industries.

Schroder, P., & Sweldens, W. (1995). Spherical wavelets: Efficiently representing functions on the sphere. In *Proceeding of ACM SIGGRAPH . Computer Graphics, 29*, 161–172.

Schroeder, W. (1992). Decimation of triangle meshes. In *Proceeding of ACM SIGGRAPH* (pp. 65–70).

Schwarz, H., Marpe, D., & Wiegand, T. (2007, September). Overview of the Scalable Video Coding Extension of the H.264/AVC Standard. *IEEE Transactions on Circuits and Systems for Video Technology, 17*(9). doi:10.1109/TCSVT.2007.905532

Search Toppers, L. L. C. (2009). *Search Toppers.* Retrieved July 21, 2009, from http://www.searchtoppers.com/

Seok, J., Lee, J. W., & Cho, C. S. (2008, February). Fast Block Mode Decision Algorithm in H.264/AVC using a Filter Bank of Kalman Filters for High Definition Encoding. *Multimedia Systems, 13*(5-6). doi:10.1007/s00530-007-0100-2

Seppälä, P., & Alamäki, H. (2003). Mobile learning in teacher training. *Journal of Computer Assisted Learning, 19*, 330–335. doi:10.1046/j.0266-4909.2003.00034.x

Seshagiri, S., Sagar, A., & Joshi, D. (2007). Connecting the 'Bottom of the Pyramid' – An Exploratory Case Study of India's Rural Communication Environment. *WWW 2007, May 8-12, 2007*, Alberta, Canada.

Shen, D., Chen, Z., Yang, Q., Zeng, H. J., Zhang, B., Lu, Y., & Ma, W. Y. (2004). Web-page classification through summarization. In *ACM SIGIR Conference* (pp. 242-249).

Shilov, A. (2009). *Intel Develops Breakthrough Graphics Accelerator for Small Mobile Devices. XBit Labs.* Retrieved July 29, 2009, from http://www.xbitlabs.com/news/video/display/20090317231622_Intel_Develops_Breakthrough_Graphics_Accelerator_for_Small_Mobile_Devices.html

Shim, Y. C., Kim, H. A., & Lee, J. I. (2005). Design and Evaluation of a New Micro-mobility Protocol in Large

Mobile and Wireless Networks . In *Lecture Computational Science and Its Applications* (pp. 9–12). Berlin, Heidelberg: Springer.

Sibiryakov, A. (2007). Sparse Projections and Motion Estimation in Colour Filter Arrays. In *Proceedings EUSIPCO* (pp. 1814-1818) Poznan, Poland.

Sidnal, N. S., & Manvi, S. S. (2006). Context aware mobile commerce using agent technology. In *Ad Hoc and Ubiquitous Computing, 2006. ISAUHC '06. International Symposium,* (pp. 163-168).

Škorupa, J., Mys, S., Slowack, J., Lambert, P., & Van de Walle, R. (2008b). Heuristic dynamic complexity coding. In *Proceedings of SPIE, Optical and Digital Image Processing* (pp. 1-8).

Škorupa, J., Slowack, J., Mys, S., Lambert, P., & Van de Walle, R. (2008a). Accurate Correlation Modeling for Transform-Domain Wyner-Ziv Video Coding. In *Proceedings of the Pacific-Rim Conference on Multimedia* (pp. 1-10).

Škorupa, J., Slowack, J., Mys, S., Lambert, P., Van de Walle, R., & Grecos, C. (2009). Stopping criterions for turbo coding in a Wyner-Ziv video codec. In *Proceedings of the Picture Coding Symposium.*

Slepian, D., & Wolf, J. K. (1973). Noiseless coding of correlated information sources. *IEEE Transactions on Information Theory, 19*(4), 471–480. doi:10.1109/TIT.1973.1055037

Slowack, J., Mys, S., Škorupa, J., Lambert, P., Van de Walle, R., & Grecos, C. (2009). Accounting for quantization noise in online correlation noise estimation for distributed video coding. In *Proceedings of the Picture Coding Symposium.*

Sohn, K., Lee, C., Ryou, J., & Jang, W. (2001*). Error-resilient Zerotree Wavelet Video Coding. SPIE Journal of Optical Engineering,* (pp. 2480-2488).

Song, B. (2008, July 9-11). *RFID Tag Ownership Transfer.* The 4th Workshop on RFID Security, Budapest, Hungary.

Source code link: http://iphome.hhi.de/suehring/tml/download/old_jm/ jm81a.zip

Srinivasan, J. (2007). The role of trustworthiness in information service usage: The case of Parry information kiosks in Tamil Nadu, India. In *Proceedings of the International Conference on Information and Communication Technologies for Development (ICTD),* (pp. 345-352) New York: ACM Press.

Srirama, S. N., Jarke, M., & Prinz, W. (2006). *A* Mediation Framework for Mobile Web Service Provisioning. In *10th IEEE International Enterprise Distributed Object Computing Conference Workshops (EDOCW 2006),* Hong Kong, China.

Srirama, S. N., Jarke, M., & Prinz, W. (2008). MWSMF: a Mediation Framework Realizing Scalable Mobile Web Service. In *Mobilware 2008,* Innsbruck, Austria.

Stajano, F., & Anderson, R. (1999). The Resurrecting Duckling: Security Issues for Ad-hoc Wireless Networks. *7th International Workshop on Security Protocols, 1796,* Lecture Notes in Computer Science, p. 172-194, New York: Springer-Verlag.

Stanford 3D Scanning (2006). Stanford 3D Scanning Repository. Retrieved 2006, from http://graphics.stanford.edu/data/3Dscanrep/

Steels, L., & Tisselli, E. (2008), Social Tagging in Community Memories. In *Proceedings of the 2008 AAAI Spring Symposium: Social Information Processing. Stanford University, ed.,* Menlo Park, CA: AAAI Press.

Stefani, A., & Xenos, M. (2008). E-commerce system quality assessment using a model based on ISO 9126 and Belief Networks. *Software Quality Control, 16*(1), 107–129.

Stottrup-Andersen, J., Forchhammer, S., & Aghito, S. (2004). Rate-distortion-complexity optimization of fast motion estimation in H.264/MPEG-4 AVC. In *Proceedings of the IEEE International Conference on Image Processing* (pp. 111-114).

Strang, G., Nguyen, T. (1996). Wavelets and filter banks. Wellesley, MA: Wellesley-Cambridge Press.

Sufi, F., Fang, Q., Khalil, I., & Mahmoud, S. (2009). Novel Methods of Faster Cardiovascular Diagnosis in Wireless Telecardiology. *IEEE Journal on Selected Areas in Communications, 27*(4), 537–553. doi:10.1109/JSAC.2009.090515

Sufi, F., Fang, Q., Mahmoud, S., & Cosic, I. (2006). A mobile phone based intelligent telemonitoring platform. In *The Proceedings of 3rd IEEE EMBS International Summer School on Medical Devices and Biosensors*(pp: 101–104).

Sweldens, W. (1996). The lifting scheme: A custom-design construction of biothogonal wavelets. *Applied and Computational Harmonic Analysis, 3*, 186–200. doi:10.1006/acha.1996.0015

Symantec_MAWM (2009). *Symantec Mobile AntiVirus for Windows Mobile*. Retrieved June 10, 2009, from http://www.symantec.com/business/mobile-antivirus-for-windows-mobile

Symantec_MSS (2009). *Symantec Mobile Security for Symbian*, Threat protection for Symbian OS Series 60 and UIQ through integrated antivirus and firewall technologies. California: Symantec.

Symella. (2009). *Symella*. Retrieved July 20, 2009, from http://amorg.aut.bme.hu/projects/symella

SymTorrent. (2009). *SymTorrent*. Retrieved July 20, 2009, from http://amorg.aut.bme.hu/projects/symtorrent

Tack, N., Lafruit, G., Catthoor, F., & Lauwereins, R. (2005). Pareto based optimization of multi-resolution geometry for real time rendering. In *Proceeding of ACM Web 3D* (pp. 19–27).

Tack, N., Moran, F., Lafruit, G., Lauwereins, R., (2004). 3D Rendering Time Modeling and Control for Mobile Terminals. In Proceeding of ACM Web3D (pp. 109–117). Synposium.

Tagliasacchi, M., & Tubaro, S. (2007b). Hash-based motion modeling in Wyner-Ziv video coding. In . *Proceedings of the IEEE International Conference on Acoustics Speech and Signal Processing, 1*, 509–512.

Tagliasacchi, M., Frigerio, L., & Tubaro, S. (2007a). Rate-distortion analysis of motion-compensated interpolation at the decoder in distributed video coding. *IEEE Signal Processing Letters, 14*(9), 625–628. doi:10.1109/LSP.2007.896187

Tagliasacchi, M., Pedro, J., Pereira, F., & Tubaro, S. (2007c). An efficient request stopping method at the turbo decoder in distributed video coding. In *Proceedings of the EURASIP European Signal Processing Conference*.

Tagliasacchi, M., Trapanese, A., Tubaro, S., Ascenso, J., Brites, C., & Pereira, F. (2006a). Exploiting spatial redundancy in pixel domain Wyner-Ziv video coding. In *Proceedings of the IEEE International Conference on Image Processing*. (pp. 253-256).

Tagliasacchi, M., Trapanese, A., Tubaro, S., Ascenso, J., Brites, C., & Pereira, F. (2006b). Intra mode decision based on spatio-temporal cues in pixel domain Wyner-Ziv video coding, In . *Proceedings of International Conference on Acoustics, Speech, and Signal Processing, 2*, 57–60.

Tamai, M., Sun, T., Yasumoto, K., Shibata, N., & Ito, M. (2004). Energy-aware video streaming with QoS control for portable computing devices. In *Proceeding of ACM NOSSDAV'04* (pp. 68–73).

Tan, C., Sheng, B., & Li, Q. (2008, March). Secure and Serverless RFID Authentication and Search Protocol. *IEEE Transactions on Wireless Communications, 7*(3).

Tang, X., Xu, J., & Lee, W. C. (2008). Analysis of TTL-Based Consistency in Unstructured Peer-to-Peer Networks. *IEEE Transactions on Parallel and Distributed Systems, 19*(12), 1683–1694. doi:10.1109/TPDS.2008.44

Tanizawa, A., Koto, S., Chujoh, T., & Kikuchi, Y. (October, 2004) A Study on Fast Rate-distortion Optimized Coding Mode Decision for H.264. *IEEE International Conference on Image Processing, 2004, ICIP'04., 2*, 24-27.

Tan, P. N., Steinbach, M., Kumar, V. (2006). Introduction to Data Mining. Boston: Pearson Education, Inc.

Tarvainen, P. (2004, November 4-5). Survey of the Survivability of IT Systems. *The 9th Nordic Workshop on Secure IT-systems*, Helsinki, Finland.

Tehrani, M. A. (2008). *Abnormal motion detection and behaviour prediction.* M.Sc. Thesis, Lund University, Lund, Sweden.

Teller, S. (1992). *Visibility Computations in Densely Occluded Polyhedral Environments.* PhD Dissertation.

Terrasa, A., & García-Fornes, A. (1999). Real-Time Synchronization Between Hard and Soft Tasks in RT-Linux. In rtcsa, pp.434, *Sixth International Conference on Real-Time Computing Systems and Applications* (RTCSA'99).

Terziyan, V. (2001). *Architecture for Mobile P-Commerce: Multilevel Profiling Framework.* In Workshop Notes for the IJCAI01, Workshop on E-business & the Intelligent.

Tewari, S., & Kleinrock, L. (2006), Proportional Replication in Peer-to-Peer Networks. In *25th Annual Joint Conference of the IEEE Computer and Communications Societies (INFOCOM)*, Barcelona, Spain.

Thakor, N.V., Webster, J.G., & Tompkins, W.J. (1984). Estimation of QRS complex power spectra for design of a QRS filter. *IEEE Transactions on Biomedical Engineering. 31*(11), 702:706.

The Pew Research Center for the People and the Press. (2007). *The Pew Global Attitudes Project.* Retrieved October 4, 2007, from http://www.pewglobal.org

Thomas, J., Rose, C., & Charpillet, F. (2007). A support system for ECG segmentation based on Hidden Markov Models. In . *Proceedings of Annual Conference of IEEE Eng Med Biol Soc., 2007*, 3228–3231.

Tico, M., & Vehvilainen, M. (2009). Robust Methods of Video Stabilization. In *Proceedings EUSIPCO* (pp. 1819-1822) Poznan, Poland.

Times, E. E. (2008). *India's wireless network base will soon be the world's second largest.* Article by K.C. Krishnadas on March 24, 2008. RetrievedJune 15, 2009, from http://www.eetimes.com/news/latest/showArticle.jhtml?articleID=206905386

Tisal, J. (May 2001). *The GSM Network: The GPRS Evolution:One Step Towards UMTS* Wiley, Forth Worth, TX: John & Sons.

Tobagi, F. A., Binder, R., Leiner, B., (1984). Packet Radio and Satellite Networks (pp. 24–40). IEEE Communications.

Toorani, M., & Shirazi, A. A. B. (2008). SSMS - A Secure SMS Messaging Protocol for the M-Payment Systems, In *Proceedings of the 13th IEEE Symposium on Computers and Communications (ISCC'08)*, (pp. 700-705)., Marrakesh, Morocco; IEEE ComSoc.

Tosic, I., & Frossard, P. (2007). Wyner-Ziv coding of multi-view omnidirectional images with overcomplete decompositions. In . *Proceedings of the IEEE International Conference on Image Processing, 3*, 17–20.

Touma, C., & Gotsman, C. (1998). Triangle Mesh Compression. In *Proceeding of Graphics Interface* (pp. 26-34).

Transmission Control Protocol (TCP), Request For Comment (RFC) 793, Internet Engineering Task Force (IETF), September 1981.

Traversat, B., Arora, A., Abdelaziz, M., Duigou, M., Haywood, C., Hugly, J.-C. (2003). *Project JXTA 2.0 Super-Peer Virtual Network* (Tech. Rep.). Sun Micro-Systems.

Traxler, J. (2007). Defining, Discussing, and Evaluating Mobile Learning: The moving finger writes and having writ. *International Review of Research in Open and Distance Learning, 8*(2), 12.

Trkman, P., Jerman Blazic, B., & Turk, T. (2008). Factors of broadband development and the design of a strategic policy framework. *Telecommunications Policy, 32*(2), 101–115. doi:10.1016/j.telpol.2007.11.001

Tsai, A. C., Paul, A. P., & Wang, J. C. (May, 2008). Intensity Gradient Technique for Efficient Intra-Prediction in H.264/AVC. *IEEE Transactions on Circuits and Systems for Video Technique, 18*, (5).

Tsudik, G. (2006, March 13-17). YA-TRAP:Yet Another Trivial RFID Authentication Protocol. *The 4th Annual IEEE International Conference on Pervasive Computing and Communications*, Pisa, Italy.

Tuyls, P., & Batina, L. (2006, February 13-17). RFID-Tags for Anti-Counterfeiting. *The Cryptographer's Track at the RSA Conference*, San Jose, CA.

UNESCO. (2008). *UNESCO WebWorld News | Point of View*. Retrieved June 15, 2009, from http://www.unesco.org/webworld/points_of_views/tawfik_2.shtml

UNESCO. (2009). *UNESCO Institute for Statistics*. Retrieved June 15, 2009, fromhttp://www.uls.unesco.org

User Agent Specification, W. A. P. (1999). Received (n.d.)., from http://www.wapforum.org/what/technical.htm, 1999

Ushahidi.com. (2009). *Crowdsourcing Crisis Information (FOSS)*. Retrieved June 15, 2009, from http://www.ushahidi.com/

UTRAN overall description. (1999). 3GPP. *TS 25.401 v3.3.0*, R-99, RAN WG3, 1999.

Vajda, I., & Buttyan, L. (2003, October 12). Lightweight Authentication Protocols for Low-cost RFID Tags. *Second Workshop on Security in Ubiquitous Computing*, Seattle, WA.

Valette, S., & Prost, R. (2003). Wavelet-based progressive compression scheme for triangle meshes: Wavemesh. *IEEE Transactions on Visualization and Computer Graphics*, *10*(2).

Valette, S., & Prost, R. (2004). Multiresolution analysis of irregular surface meshes. *IEEE Transactions on Visualization and Computer Graphics*, *10*, 113–122. doi:10.1109/TVCG.2004.1260763

Van Cutsem, T., Mostinckx, S., & De Meuter, W. (2008). Linguistic Symbiosis between Actors and Threads. *Computer Languages, Systems & Structures*, *1*(35).

Van Cutsem, T., Mostinckx, S., Gonzalez Boix, E., Dedecker, J., & De Meuter, W. (2007). AmbientTalk: object-oriented event-driven programming in Mobile Ad hoc Networks. In *Proceedings of the XXVI International Conference of the Chilean Computer Science Society (SCCC 2007)*, 3-12.

van Engelen, R. A., & Gallivan, K. (2002). The gSOAP Toolkit for Web Services and Peer-To-Peer Computing Networks. In *2nd IEEE International Symposium on Cluster Computing and the Grid (CCGrid 2002)*, Berlin, Germany.

Varela, C., & Agha, G. (2001). Programming dynamically reconfigurable open systems with SALSA. *SIGPLAN Not.*, *36*(12), 20–34. doi:10.1145/583960.583964

Varodayan, D., Chen, D., Flierl, M., & Girod, B. (2008). Wyner-Ziv coding of video with unsupervised motion vector learning. *Signal Processing Image Communication*, *23*(5), 369–378. doi:10.1016/j.image.2008.04.009

Vatis, Y., Klomp, S., & Ostermann, J. (2007). Enhanced reconstruction of the quantised transform coefficients for Wyner-Ziv coding. In *Proceedings of the IEEE International Conference on Multimedia & Expo*. (pp. 172-175).

Vella, F., Castorina, A., Mancuso, M., & Messina, G. (2002). Digital image stabilization by adaptive block motion vectors filtering. *IEEE Transactions on Consumer Electronics*, *48*(3), 796–801. doi:10.1109/TCE.2002.1037077

Vetro, A., Su, Y., Kimata, H., & Smolic, A. (October, 2006). *Joint Draft 1.0 on Multiview Video Coding*, Doc. JVT-U209 Joint Video Team, Hangzhou, China.

Vicente, C. G., Pablo, G. E., & Vincent, P. (2004). Evaluation of cellular IP mobility Tracking procedures. *The International Journal of Computer and Telecommunication Networking*, *45*(3), 261–279.

Vriendt, J. D., Lainé, P., Lerouge, C., & Xu, X. (2002). Mobile network evolution: A revolution on the move. *IEEE Communications Magazine*, *40*(4), 104–111. doi:10.1109/35.995858

W3C (2007). *Mobile Web Best Practices 1.0, W3C Proposed Recommendation*, 2007 Retrieved February 22, 2009, from http://www.w3.org/TR/mobile-bp/

W3C (2008). *Mobile Web Application Best Practices Working Draft*. 22 December 2008. Retrieved March 1, 2009, from http://www.w3.org/TR/2008/WD-mwabp-20081222/

W3C (2008). *W3C WebApps Working Group*. Retrieved April 11, 2009, from http://www.w3.org/2008/webapps/

W3C (2009). *About W3C*. Retrieved April 11, 2009, from http://www.w3.org/Consortium/

W3C (2009). *HTML5*. Retrieved April 11, 2009, from http://www.w3.org/TR/html5/

W3C (2009). *Mobile Web Initiative*. Retrieved April 11, 2009, from http://www.w3.org/Mobile/

W3C (2009). *Web Sockets API*. Retrieved March 27, 2009, from http://dev.w3.org/html5/websockets/

W3C (2009). *Widgets 1.0: Packaging and Configuration*. Retrieved April 11, 2009, from http://www.w3.org/TR/widgets/

Waldo, J. (2001). Constructing Ad Hoc Networks. *IEEE International Symposium on Network Computing and Applications (NCA'01), 9*.

Wang, C., & Yang, X. (2007). *H.264 encoding in parallel*. M.Sc. thesis, Lund University, Lund, Sweden.

Wang, C., Zhang, P., Choi, R., & Eredita, M. (2002). Understanding consumers attitude toward advertising. In *Eighth Americas conf. on Information System* (pages 1143–1148)

Wang, H. M., Tseng, C. H., & Yang, J. F. (2007, September). Computation Reduction for Intra 4x4 Mode Decision with SATD Criterion in H.264/AVC. *Signal Processing, IET, 1*(3), 121–127. doi:10.1049/iet-spr:20065007

Wang, H., Kwong, S., & Kok, C. W. (2007, June). An Efficient Mode Decision Algorithm for H.264/AVC Encoding Optimization. *IEEE Transactions on Multimedia, 9*(4). doi:10.1109/TMM.2007.893345

Wang, X., Silva, F., & Heidemann, (2004). J. Demo abstract: Follow-me application—active visitor guidance system. In *Proceedings of the 2nd ACM SenSys Conference*.

Wang, Y. C., & Lin, K. J. (1998). Enhancing the Real-Time Capability of the Linux Kernel, In rtcsa, (pp.11), *Fifth Intl Conference on Real-Time Computing Systems and Applications* (RTCSA'98).

Weis, S., Sarma, S., Rivest, R., & Engels, D. (2003, March 12-14). Security and Privacy Aspects of Low-cost Radio Frequency Identification Systems. *The 1st International Conference on Security in Pervasive Computing*, Boppard, Germany.

Weiser, M. (1991, September). The computer for the 21st century. *Scientific American*, 94–104.

Weiser, M. (1993, July). Some computer science issues in ubiquitous computing. *Communications of the ACM, 36*(7), 75–84. doi:10.1145/159544.159617

Wennlund, A. (April 2003). *Context-aware Wearable Device for Reconfigurable Application Networks*, Department of Microelectronics and Information Technology(IMIT) WURFL (2008) Retrieved April, 2003, from http://wurfl.sourceforge.net/

Westly, E. (2009). Sign Language by Cellphone. *IEEE Spectrum*, (3): 14.

WHATWG. (2009). *HTML5 - Draft Standard*. Retrieved March 27, 2009, from http://www.whatwg.org/specs/web-apps/current-work/.

WHATWG. (2009). *Web Hypertext Application Technology Working Group*. Retrieved April 11, 2009, from http://www.whatwg.org/

Wiegan, T., & Girod, B. (September, 2001), *Lagrange Multiplier Selection in Hybrid Video Coder Control*, IEEE International Conference on Image Processing (ICIP'01), Thessaloniki, Greece.

Wiegand, T., Sullivan, G. J., Bjntegaard, G., & Luthra, A. (2003, July). Overview of the H.264/AVC video coding standard. *IEEE Transactions on Circuits and Systems for Video Technology, 13*(7). doi:10.1109/TCSVT.2003.815165

Wiki_LS (2009). *Layered Security*. Retrieved June 10, 2009, from http://en.wikipedia.org/wiki/Layered_security

Wilce, M. (2001). *High Level Requirements for Release 7.0 of the Symbian Platform v0.04*. London: Symbian.

Williams, N., Luebke, D., Cohen, J., Kelley, M., Schubert, B., (2003). Perceptually guided simplification of lit, textured meshes. In Proceeding of Interactive 3D (pp. 113–121). Graphics.

Winmmer, M., & Wonka, P. (2003). Rendering time estimation for Real-Time Rendering. In *Proceeding of the Eurographics Symposium on Rendering* (pp. 118–129).

WM-Security. (2008). *Security Mobdel for Windows Mobile 5.0 and Windows Mobile 6*. Seattle, WA: Microsoft Corporation.

Wobbrock, J., Forlizzi, J., Hudson, S., & Myers, B. (2002). WebThumb: Interaction techniques for small-screen browsers. In *ACM Symposium on User Interface Software and Technology* (pp. 205-208).

Wright, A. (2009). Get Smart. *Communications of the ACM*, *52*(1), 15–16. doi:10.1145/1435417.1435423

Wu, B., Zhuo, Y., Zhu, X., Yan, Q., Zhu, L., & Li, G. (2005). A Novel Mobile ECG Telemonitoring System. In *Proceedings of 27th Annual International Conference of the IEEE Engineering in Medicine and Biology Society* (pp: 3818 – 3821).

Wu, F., Agu, E., & Lindsay, C. (2007). Pareto-Based Perceptual Metric for Imperceptible simplification on mobile displays. In *Proceeding of Eurographics 2007*, Prague, Czech Republic.

Wu, F., Agu, E., & Lindsay, C. (2008), Adaptive CPU Scheduling to Conserve Energy in Real-Time Mobile Graphics Applications. In *Proceeding of ISVC 2008*, Las Vegas, NV.

Wu, F., Agu, E., & Ward, M. (2006). Multiresolution Graphics on Ubiquitous Displays using Wavelets. *International Journal of Virtual Reality*, *5*(3).

Wu, F., Agu, E., (2006). Unequal Error Protection for Wavelet-Based Wireless Mesh Transmission. Boston, MA: ACM SIGGRAPH.

Wyner, A. D., & Ziv, J. (1976). The rate-distortion function for source coding with side information at the decoder. *IEEE Transactions on Information Theory*, *22*(1), 1–10. doi:10.1109/TIT.1976.1055508

Xie, K., Wong, W. S. V., & Leung, C. M. V. (2005). Support of Micro-Mobility in MPLS-Based Wireless Access Networks. Oxford Journal in IEICE Transactions on Communications. E88(B), 2735-2742.

Xie, S., Li, B., & Keung, G. Y. (2008, January). *The Peer-to-Peer Live Video Streaming for Handheld Devices*. The Fifth IEEE Consumer Communications & Networking Conference (CCNC 2008), Las Vegas, NV

Xie, X., Miao, G., Song, R., Wen, J. R., & Ma, W. Y. (2005). Efficient browsing of Web search results on mobile devices based on block importance model. In *Proceedings of the 3rd IEEE International Conference on Pervasive Computing and Communications (PerCom 2005)* (pp. 17-26). Kauai, HI.

Xiea, B., Kumara, A., Agrawal, D. P., & Srinivasan, S. (2006). Secured macro/micro-mobility protocol for multi-hop cellular IP. *Journal Security in Wireless Mobile Computing Systems*, *2*(2), 111–136.

XMPP.org. (2009). XEP-0124: Bidirectional-*streams Over Synchronous HTTP (BOSH)*. Retrieved March 27, 2009, from http://xmpp.org/extensions/xep-0124.html

Yair, A., Claudiu, D., & Hilsdale, M. (2006). Fast handoff for seamless wireless mesh network, *ACM International Conference On Mobile Systems, Applications And Service*, (pp. 83-95).

Yamakami, T. (2006). Lessons in business model development from early mobile Internet services in Japan. In *International Conference on Mobile Business, ICMB 2006*, International Conference on Mobile Business, ICMB 2006. Copenhagen.

Yan, Z., & Hee, S. B. (2004). Counting in Hierarchical Cellular System with overflow scheme "Handoff Counting in Hierarchical Cellular System with overflow scheme. *The International Journal of Computer and Telecommunications Networking*, *46*(4), 541–554.

Yan, Z., Kumar, S., & Kuo, C. (2001). Error resilient coding of 3-D graphic models via adaptive mesh segmentation.

IEEE Transactions on Circuits and Systems for Video Technology, 11(7), 860–873. doi:10.1109/76.931112

Yang, C. C., & Wang, F. L. (2003). Fractal summarization for mobile devices to access large documents on the Web. In *Proceedings of the 12th International Conference on World Wide Web* (pp. 215-224). Budapest, Hungary.

Yang, C. K., & Chiueh, T. (2005). An Integrated Pipeline of Decompression, Simplification and Rendering for Irregular Volume Data. In *Proceeding of 4th International Workshop on Volume Graphics* (pp 147-237)

Yang, C. PO, L., & Lam, W. (October, 2004). A Fast H.264 Intra Prediction Algorithm using Macroblock Properties. *International Conference on Image Processing, ICIP'04, 1*, 24-27.

Yang, F., Dai, Q., & Ding, G. (2007). Multi-view images coding based on multiterminal source coding". In*Proceedings of the IEEE International Conference on Acoustics, Speech and Signal Processing* (pp. 1037-1040).

Yang, G., Tan, W., Mukherjee, S., Ramakrishnan, I. V., & Davulcu, H. (2003) On the power of semantic partitioning of web documents. In *Information Integration on the Web* (pp. 39-46).

Yang, N., Ulrich, K., Adrian, S., & Liviu, I. (2005). Programming ad-hoc networks of mobile and resource-constrained devices. *SIGPLAN Not., 40*(6), 249–260. doi:10.1145/1064978.1065040

Yang, S., Kim, C., Kuo, C., (2004). A progressive view-dependent technique for interactive 3D mesh transmission. In IEEE Trans. Circuits and Systems for Video Technology.

Yankelovich, W. W., Roberts, P., Wessler, M., Kaplan, J., & Provino, J. (2004). Meeting Central: Making distributed meetings more effective. In *Proceedings of the Conference on Computer-supported Cooperative Work (pp.* 419-428) New York: ACM Press.

Yavuz, M., Diaz, S., Kapoor, R., Grob, M., Black, P., & Tokgoz, Y. (2006). VoIP over cdma2000 1xEV-DO Revision A. *IEEE Communications Magazine, 44*(2), 88–95. doi:10.1109/MCOM.2006.1593550

Yih, W., Goodman, J., & Carvalho, V. R. (2006). Finding advertising keywords on web pages.In *WWW '06: Proc. of the 15th intl. conf. on World Wide Web* (pages 213–222), New York: ACM.

Yin, P., Cheong, H. Y., Tourapis, A., & Boyce, J. (2003). Fast *Mode Decision and Motion Estimation for JVT/H.264*, IEEE International Conference in Image Processing.

Yin, X., & Lee, W. S. (2004). Using link analysis to improve layout on mobile devices. In *Proceedings of the 13th International Conference on World Wide Web* (pp. 338-344). New York.

Yodaiken, V., Cloutier, P., Schleef, D., Daly, P. N., Rajkumar, R., & Kuhnm, B. (2000, November 27-30). Development of RTOSes and the position of Linux in the RTOS and embedded market. In *Proceedings of the 21st Symposium on Real-Time Systems (RSS-00)*, (pp. 8-8), Los Alamitos, CA: IEEE Computer Society.

Yonezawa, A., Briot, J. P., & Shibayama, E. (1986). Object-oriented concurrent programming in ABCL/1. *Conference proceedings on Object-oriented programming systems, languages and applications*, 258-268.

Yoshida, J. (2009, January). Sorry, I didn't mean to change the channel when I sneezed. *Electronic Engineering Times Europe*, 26.

YouTube. LLC. (2009). *YouTube*. Retrieved June 15, 2009, from http://www.youtube.com/

Yu, A. C., Martin, G., & Park, H. (September, 2005).A Frequency Domain Approach to Intra Mode Selection in H.264/AVC. In *Proceedings of 13th European Signal Processing Conference (EUSIPCO) 05, 4PP.*, Antalya, Turkey.

Yuan, W., & Nahrstedt, K. (2004). Practical voltage scaling for mobile multimedia device. *In Proceeding Of ACM MM'04* (pp.924–931).

Zamir, R. (1996). The rate loss in the Wyner-Ziv problem. *IEEE Transactions on Information Theory, 42*(6), 2073–2084. doi:10.1109/18.556597

Zhang, D. (2007). Web content adaptation for mobile handheld devices. *Communications of the ACM, 50*(2), 75–79. doi:10.1145/1216016.1216024

Zhang, D., & Chen, M. (2008). *Global Motion Extraction and Compensation*. M.Sc. Thesis, Lund University, Lund, Sweden.

Zhang, H. (2004). The optimality of naive bayes. In Barr, V., Markov, Z., Barr, V., and Markov, Z., editors, *FLAIRS Conference*. AAAI Press.Zhou, H., Hou, K., Ponsonnaille, J., Gineste, L., & De Vaulx, C. (2005). *A Real-Time Continuous Cardiac Arrhythmias Detection System: RECAD* (pp: 875 – 881).

Zhang, H., & Arora, A. (2004). All-IP wireless networks. *IEEE Journal on Selected Areas in Communications, 2*, 613–616.

Zhang, H., Zhang, L., Shan, X., & Li, V. O. K. (2007). Probabilistic Search in P2P Networks with High Node Degree Variation. In *IEEE International Conference on Communications (ICC 2007)*, Glasgow, Scotland.

Zhang, J., Chen, X., Yang, J., & Waibel, A. 2002. A PDA-based sign translator. In Proc. the 4th IEEE Int. Conf. on Multimodal Interfaces.

Zhang, Y., Dai, F., & Lin, S. (June, 2004). Fast 4x4 Intra-prediction Mode Selection for H.264. *2004 IEEE International Conference on Multimedia and Expo (ICME2004), 2*, 27-30.

Zhou, J., Chu, K. M.-K., & Ng, J. K.-Y. (2005), An Improved Ellipse Propagation Model for Location Estimation in facilitating Ubiquitous Computing, in *Proceedings of the 11th IEEE International Conference on Embedded and Real-Time Computing Systems and Applications (RTCSA 2005)*, Hong Kong, pp. 463-466.

Zhou, J., Chu, K. M.-K., & Ng, J. K.-Y. (2005). Providing Location Services within a Radio Cellular Network using Ellipse Propagation Model, in *Proceedings of the 19th International Conference on Advanced Information Networking and Applications (AINA2005)*, Taipei, Taiwan, pp. 559-564.

Zhou, Z., & Sun, M. T. (2004, October) "Fast Macroblock Inter Mode Decision and Motion Estimation for H.264/ MPEG-4 AVC," IEEE International Conference on Image Processing, ICIP'04. vol 2, 24-27.

Zhu, D., Dai, Q., & Ding, R. (June 2004). Fast Inter Prediction Mode Decision for H.264. *IEEE International Conference on Multimedia and Expo, ICME '04, 2*, 27-30.

Zunino, C., Lamberti, F., Sanna, A., Montrucchio, B., (2002). A Wireless Architecture for Performance Monitoring and Visualization on PDA Device*s*. In Proceeding of SCI 02 (Vol. XV, pp. 143–148). Proceedings.

Zuo, Y., Search and Private Tag Search Protocols (2009). *Special Issue on Advanced RFID Technologies, Information Systems Frontier - A Journal of Research and Innovation*, to appear.

Zwass, V. (1996). Electronic Commerce: Structures and Issues. *International Journal of Electronic Commerce, 1*(1), 3–13.

About the Contributors

Wen-Chen Hu received a BE, an ME, an MS, and a PhD, all in Computer Science, from Tamkang University, Taiwan, the National Central University, Taiwan, the University of Iowa, Iowa City, and the University of Florida, Gainesville, in 1984, 1986, 1993, and 1998, respectively. He is currently an associate professor in the Department of Computer Science of the University of North Dakota, Grand Forks. He is the Editor-in-Chief of the International Journal of Handheld Computing Research (IJHCR) and has served as editor and editorial advisory/review board members for over 20 international journals/books and chaired more than 10 tracks/sessions and program committees for international conferences. Dr. Hu has been teaching more than 10 years at the US universities and over 10 different computer/IT-related courses, and advising more than 50 graduate students. He has published over 80 articles in refereed journals, conference proceedings, books, and encyclopedias, edited four books and proceedings, and solely authored a book. His current research interests include handheld computing, electronic and mobile commerce systems, Web technologies, and databases.

Yanjun Zuo is an assistant professor at the University of North Dakota, Grand Forks, USA. He received a Ph.D. in Computer Science from the University of Arkansas, Fayetteville, USA in 2005. He also holds two master's degrees in Computer Science and Business Administration from the University of Arkansas and the University of North Dakota, Grand Forks, USA, respectively. His research interests include information and computer security, trustworthy computing, survivable and self-healing systems, and information privacy protection. He has published numerous articles in referred journals and conference proceedings in those fields.

* * *

Emmanuel Agu received his Masters and PhD degrees in Electrical and Computer Engineering from the University of Massachusetts, Amherst. He is now an Associate Professor in the Department of Computer Science at Worcester Polytechnic Institute in Worcester, MA. His research interests are in scalable transmission of graphics content, energy-efficient rendering on mobile devices, graphics rendering, general purpose computation on graphics processors and wireless networking protocols. His work has been funded by the National Science Foundation.

Ashraf Ahmad obtained his PhD degree in Computer Science and Engineering from National Chiao Tung University (NCTU) in Taiwan. He obtained his B.Sc. degree from Princess Sumya University for Technology (PSUT) in Jordan. Dr. Ahmad is currently an assistant professor at the department of computer since in PSUT, Jordan. His interest area includes multimedia semantic features extraction, and analysis, multimedia retrieval, and multimedia communication. Prof. Ahmad has authored over 40 scientific publications including journal papers, conference papers and book chapters. In addition, Dr. Ahmad has several US and international patents in his field of expertise. He serves in program committee for several international conferences. He is also a reviewer and referee for several conferences and journals. His work has been published and presented at various international conferences. Dr. Ahmad has been listed in Who's Who in the World for the year 2006 and Who's Who in Asia for the year 2007. In addition, He has been elected as one of the 2000 Outstanding Intellectuals of the 21st Century for the year 2006 for his outstanding contribution in field of Video Processing and Communications. Ashraf Ahmad has been chosen as one of the recipients of Leading Scientist award in the year 2006.

Michele Amoretti received the Dr. Ing. degree in Electronic Engineering (2002) and the Ph.D. degree in Information Technologies (2006) from the University of Parma. Currently he is a Research Assistant at the same University, and a member of Distributed Systems Group. His research activity focuses on the formal and simulative characterization of distributed autonomic systems, in particular those based on the peer-to-peer paradigm. He contributed to the design and development of service-oriented architectures in the context of several EU projects related to healthcare and ambient intelligence.

Yuki Arase received her B.E. and M.I.S. from Osaka University, Japan, in 2006 and 2007. She is currently a Ph.D. candidate at the Department of Multimedia Engineering of Osaka University. Her research interests include mobile computing, HCI, and spatial data mining on large scale Web data. She is a member of ACM, IPSJ, and DBSJ.

Vassilios Argiriou received the B.Sc. degree in computer science from Aristotle University of Thessaloniki, Greece, in 2001 and the M.Sc. and Ph.D. degrees from the University of Surrey, Surrey, U.K., in 2003 and 2006, respectively, both in electrical engineering. From 2001 to 2002, he held a research position at the AIIA Lab, Aristotle University, working on image and video watermarking. Between 2002 and 2006, he participated in many European projects for archive file restoration (Presto Space) and sub-pixel motion estimation collaborating with Snell & Wilcox. He joined the Communications and Signal Processing (CSP) Department, Imperial College, London, U.K., in 2007 where he held a Research Fellow position working on 3-D image reconstruction from photometric stereo. Currently he is a Senior Lecturer at the University of East London working on Photometric Stereo and AI for Computer Games.

Mark Bailey's background is in human computer interaction and the psychological side of human performance. Since joining IBM in 2007 his research has focused on designing for groups that are usually marginalized, such as people with disabilities, older adults, low-literacy and illiterate users, and people in developing countries. Mark's current work includes creating applications for low-end phones—devices that are now in the hands of billions of people. These users are among the "next billion" users of IT, whose numbers are growing faster than any other user group and who are driving tremendous innovation in the IT space. Mark earned his Master's degree in Computer Science from the University of Oregon.

Mariam M. Biltawi is a Trainer and Computer Lab Supervisor in Princess Sumaya University for Technology. Mariam M. Biltawi holds the bachelor degree in Computer Science from Princess Sumaya

University for Technology, and Mrs. Biltawi is preparing Master degree in Computer Science from Al-Balqa' Applied University, Jordan. She has research interest in Mobile Operating Systems.

Elisa Gonzalez Boix is a PhD student at the Programming Technology Laboratory of the Vrije Universiteit Brussel, Belgium. Her research interests are communication abstractions, context-awareness, parameter passing and memory management in mobile ad hoc networks. In her PhD research she investigates language constructs to deal with the effects engendered by partial failures in mobile ad hoc networks. She has explored the use of leasing to detect and react to partial failures and to manage lifetime of remote objects in mobile ad hoc networks. Her coordinates can be found at http://prog.vub.ac.be/~egonzale.

Andoni Lombide Carreton is a PhD student at the Programming Technology Laboratory of the Vrije Universiteit Brussel, Belgium. His research interests are context-awareness, event-driven systems and event processing, control flow management and coordination in mobile ad hoc networks. He has been involved in applying the ambient oriented programming paradigm to RFID systems. He is currently investigating new programming language constructs and tools to process and react to distributed events and to manage the control flow and coordination of distributed applications in mobile ad hoc networks, and concretely applying this research to systems containing large amounts of volatile data such as RFID systems. His coordinates can be found at http://prog.vub.ac.be/~alombide.

Chung-Han Chen received the Ph.D. degree in Computer Engineering from the University of Louisiana at Lafayette (renamed to University of Louisiana at Lafayette) in 1993. He is currently an Associate Professor of the Department of Computer Science at Tuskegee University, Alabama. His research interests are in computer networks, information and network security, operating systems, computer architecture, and VLSI design. He has been a member of IEEE, computer society since 1985. He served as the workshop chair of the ACM SE conference 2008.

Lei Chen received his B.Eng. degree in Computer Science from Nanjing University of Technology, China, in 2000, and Ph.D. degree in Computer Science from Auburn University, USA, in Aug. 2007. He has been with Sam Houston State University as an Assistant Professor in Computer Science since 2007. Dr. Chen has been actively working in research in computer networks, network security, and wireless and multimedia networking. He has authored about 20 articles and book chapters in major journals, conference proceedings and books. He also serves in the editorial advisory/review board and the technical program committee of a number of books, journals and conferences.

James E. Christensen is a Senior Technical Staff member at IBM's T.J. Watson Research Center in Hawthorne New York. He has held both management and technical positions in IBM, and worked on a broad range of projects including applications for voice, image and text, on tools for program development and debugging, and at all levels of systems infrastructure. Jim graduated from the University of Illinois in 1977 with a Masters degree in Computer Science.

Gianni Conte received the Dr.Ing. degree in Electronic Engineering (1970) at the Politecnico of Torino. In 1989 he moved to the University of Parma where he presently teaches computer architectures, parallel processing and information systems, and leads the Distributed Systems Group. From 1996 to 2002 he acted as Director of the Department of Information Engineering. He his author of over 90 technical papers in refereed journals and conferences and of three books on parallel computer architectures and performance modeling. His research interests include parallel processing architecture, performance evaluation of distributed systems, stochastic Petri net modeling, and VLSI system architectures and distributed information systems. In 2002 he received the "Faculty Award" from IBM Corporation. Since 2005 is vice-president of the GII (Gruppo Ingegneria Informatica), the Association of Italian Computer Engineers University Professors.

Tom Van Cutsem is a post-doctoral researcher at the Programming Technology Laboratory of the Vrije Universiteit Brussel, Belgium. He finished his PhD dissertation in 2008 on object designation in mobile ad hoc networks and co-designed the AmbientTalk programming language. His broad research interests include programming language design and implementation, distributed programming and reflective architectures. He is a post-doctoral fellow of the Research Foundation, Flanders (FWO). His coordinates can be found at http://prog.vub.ac.be/~tvcutsem.

Catalina M. Danis is a Research Staff Member in the Social Computing Group at IBM's T. J. Watson Research Center in Hawthorne, NY. Her research focus is on understanding how people work, individually and as members of groups and organizations. She then works with designers and application developers to incorporate her research findings into new software systems and follows up with studies of the system's impact on users. Since joining IBM in 1987, Catalina has worked in the areas of automatic speech recognition, collaboration technologies and high performance computing. She publishes in the CHI, CSCW and DIS communities. Catalina holds a Ph.D. in Cognition and Communication from the University of Chicago.

Jason B. Ellis is a researcher in the Social Computing Group at the IBM T.J. Watson Research Center in New York. His research focuses on the design, implementation, and analysis of social computing applications that facilitate collaboration among diverse user populations. Examples include online gaming communities, inter-generational communication, and the grassroots teams in open source. His current work explores using social computing to empower the "next billion" users of IT—those in developing countries who do not yet benefit from computing technology. He has published in key conferences such as ACM CHI, CSCW, DIS and learning sciences conferences CSCL and ICLS. Jason earned his Ph.D. in Computer Science at Georgia Tech.

Thomas D. Erickson is an interaction designer and researcher at IBM Research in New York, to which he telecommutes from his home in Minneapolis. His primary interest is in studying and designing systems that enable distributed groups to interact productively over networks; his approach involves using minimalist visualizations called social proxies to provide cues about activities of participants. More generally, Erickson's approach to systems design is shaped by methods developed in HCI, theories and representational techniques from architecture and urban design, and theoretical and analytical approaches from rhetoric and sociology. Over the last two decades Erickson has published about fifty peer reviewed articles and chapters, and been involved in the design of over a dozen systems ranging from research prototypes to commercial products. He is co-editor of *HCI Remixed*, a book of essays on works that have influenced the HCI community that was published by MIT Press in 2008.

John Q. Fang received his B.S. degree in applied physics from Tsinghua University, China in 1991 and his Ph.D. degree in Biomedical Engineering from Monash University, Australia in 1999 respectively. He has several years of experience in IT industry before he joined RMIT University, Australia, as an academic in 2002 and is currently a senior lecturer in School of Electrical & Computer Engineering, RMIT. His major research interests include intelligent and miniaturised medical instrumentation, wearable and implantable body sensor networks and pervasive computing technologies applicable to healthcare delivery. Dr. Fang leads the Biomedical Engineering Bachelor Program and the Bioinstrumentation Research Group in RMIT.

Robert G. Farrell is a Research Staff Member in the Social Computing Group at IBM T. J. Watson Research Center in Hawthorne, NY. His research focuses on information organization, human learning, and team collaboration. He is currently developing an interactive audiovisual user interface for mobile touch screen phones as part of the group's focus on mobile technology for the "next billion" IT users. Since joining IBM in 1988, Rob has been a group manager, principal investigator, and technical lead. He holds Master of Science and Master of Philosophy degrees in Computer Science from Yale University. Prior to IBM, he worked as a Member of Technical Staff at Bell Communications Research in Morristown, NY. He is the author of over fifty publications in the fields of human-computer interaction, information science, and artificial intelligence.

John Garofalakis (http://athos.cti.gr/garofalakis/) is an Associate Professor at the Department of Computer Engineering and Informatics, University of Patras, Greece, and the director of the applied research department "Telematics Center" of the Research Academic Computer Technology Institute (RACTI). He is responsible and scientific coordinator of several European and national IT and Telematics Projects (ICT, INTERREG, etc.). His publications include more than 110 articles in refereed International Journals and Conferences. His research interests include Web and Mobile Technologies, Performance Analysis of Computer Systems, Computer Networks and Telematics, Distributed Computer Systems, Queuing Theory.

Christos Grecos received his M.Sc. degree in Computer Science from Heriott Watt University, Scotland, UK in 1995 and his Ph.D. degree on the same discipline from Glamorgan University, Wales, UK in 2001. After working as an assistant and associate professor at Robert Gordon, Loughborough and Central Lancashire Universities in UK, he became a full professor in Visual Communications and Head of School of Computing at the University of West of Scotland, UK. His current research interests include design and implementation of image and video standard compliant compression and transmission systems, multimedia visual quality assessment, content description of multimedia data, complexity management, and software engineering for multimedia applications.

Takahiro Hara received his B.E., M.E., and D.E. in Information Systems Engineering from Osaka University in Japan, in 1995, 1997, and 2000. He is currently an Associate Professor at the Department of Multimedia Engineering of Osaka University. His research interests include distributed database systems in advanced computer networks, such as high-speed networks and mobile computing environments. Dr. Hara is a member of ACM, IEEE, IEICE, IPSJ, and DBSJ.

Haibo Hu is an Assistant Professor in the Department of Computer Science, Hong Kong Baptist University. Prior to this, he held several research and teaching posts at HKUST and HKBU. He received his PhD degree in Computer Science from the Hong Kong University of Science and Technology in 2005. His research interests include mobile and wireless data management, location-based services, and privacy-aware computing. He has published over 20 research papers in international conferences, journals and book chapters. He is also the recipient of many awards, including ACM Best PhD Paper Award and Microsoft Imagine Cup.

Weihong Hu received an MS degree in Computer Science and Software Engineering at the Auburn University, Auburn, Alabama in 2003. She is a PhD student at the same department now. At the same time, she is an associate professor at the Computer Information Engineering Center at the Shandong Sport University, China. Her current research interests include human centered computing and human computer interface.

Xiaoyun Huang received the BSc. in computer software from Changchun University of Science and Technology, Changchun, China in 1993. He has more than 10 years of commercial experience in various real-world business applications primarily using Delphi and Java with particular expertise and experience in the design and implementation of dynamic Web-based mobile applications using J2ME and XML technologies. Xiaoyun worked as a research associate in School of Electrical and Computer Engineering, RMIT University, Australia from July 2008 to Dec 2008. Currently, he is the deputy chief of the Computer Management Section of the Sichuan Provincial Committee of College Exam and Admission Office, Sichuan, China.

Ziad Hunaiti is a senior lecturer at Anglia Ruskin University. His research interest includes fourth generation wireless technologies, information systems and location based services (LBS) and technologies. He is also highly involved in mobile learning and mobile commerce research.

I-Horng Jeng received the B.S. degree in Computer Science from Tamkang University in 1992, and M.S. and Ph.D. degrees in Electrical Engineering from National Taiwan University in 1994 and 2001, respectively. He is an assistant professor in the Department of Computer Science at Chinese Culture University, Taiwan. Dr. Jeng is currently both an associated editor for the Far East Journal of Experimental and Theoretical Artificial Intelligence (FEJETAI) and a member of the editorial review board for the International Journal of Handheld Computing Research (IJHCR). Dr. Jeng is included in "Who's Who in the World" during 2008 to 2010. His current research interests are embedded system software, mobile commerce, and human-computer interaction.

Yiming Ji received his Ph.D. degree in Computer Science in 2006 from Auburn University. He is currently an Assistant Professor at the University of South Carolina, Beaufort (USCB). His research is mainly in wireless networks, and the focus is on wireless location determination (indoor, outdoor, and sensor networks). His research also includes embedded systems, modeling and simulation, digital image/video processing and analysis, and scientific computation. He has authored over 25 journal and conference publications and one book. He has served on the program committees of several international meetings including IEEE IC3N, IEEE ISCC, and World Congress on Engineering. He is also an adjunct professor at the University of South Carolina, Columbia and the Professor in residence at the Center of Excellence in Collaborative Learning at USCB.

Nan Jing is currently a Senior R&D engineer in Antenna Software Inc., where he is taking the technical efforts on mobility platform for business solutions including customer relationship management and inter-enterprise workflow modeling. Prior to Antenna Software, Nan worked as mobile specialist at Industrial Scientific where he helped in designing their behavior-based safety management applications on Blackberry and Windows Mobile. Before that he was the first engineer in the mobile player team at DivX and led the development of the first few versions of DivX Mobile Player. He has published numerous papers and one book in software design, web service, decision science and business process management. Nan is a frequent reviewer for conferences and journals in information system and mobile technologies, and he is presently on the editorial review board of International Journal of Handheld Computing Research. Nan has earned a Ph.D. degree and a Master degree from University of Southern California and a Bachelor Degree from Peking University, China, all in Computer Science.

Naima Kaabouch received a BS and an MS from the University of Paris 11, France in 1982 and 1985, respectively, and a PhD from the University of Paris 6, France in 1990. She is currently an assistant professor and graduate director in the Department of Electrical Engineering at the University of North Dakota. Her research interests include signal/image processing biomedical engineering, bioinformatics, embedded systems, digital communications, and nondestructive techniques.

Wendy A. Kellogg manages the Social Computing Group at IBM's T. J. Watson Research Center. Topics addressed by the group have included social translucence, computer-mediated communication, and recently social computing for green IT and the next billion users. Kellogg holds a Ph.D. in Cognitive Psychology and publishes in HCI and CSCW. She serves on the editorial board of ACM's Transactions on Computer-Human Interaction (ToCHI) and has chaired a variety of key conferences and technical programs and venues (CHI, DIS, CSCW). Wendy served on the National Academy of Science's Computer Science and Telecommunications Board (CSTB). She was named ACM Fellow in 2002 for contributions to social computing and human-computer interaction, and elected to the CHI Academy in 2008.

Anna Kress studied Computer Science at the Eberhard Karls Universität Tübingen, the Saint-Petersburg State University (Russian Federation) and at the Freie Universität Berlin. She received her degree in Computer Science in 2006. Since May 2007 Anna has been a research associate at the Fraunhofer Institute FOKUS, where she has worked in national and international research projects. Her research areas are Future Mobile Web Applications and Platforms and Location Based Services. Furthermore, she also teaches at the TU Berlin, chair for Open Communications Systems (OKS).

Maria Chiara Laghi received the Dr. Ing. degree in Electronic Engineering (2004) from the University of Parma. Currently she is a Ph.D. student in Information Technologies at the same University, and a member of Distributed Systems Group. Her research activity focuses on service-oriented architectures and peer-to-peer systems with particular interest for mobile device applications and security aspects.

Peter Lambert received his M.Sc. degrees in Mathematics and Applied Informatics from Ghent University in 2001 and 2002, respectively. He obtained the Ph.D. degree in computer science in 2007 at the same university. In 2007 he became a post-doctoral research fellow at the Multimedia Lab of the Department of Electronics and Information Systems of Ghent University – IBBT (Belgium) where he currently holds a position as Technology Developer. His research interests include multimedia applications, (scalable) video coding technologies, multimedia content adaptation, and error robustness of digital video.

Chung-wei Lee is an assistant professor in the Department of Computer Science, University of Illinois at Springfield. He is interested in wireless networking, mobile computing, multimedia streaming, network quality of service (QoS), and computer/network security. He received his Ph.D. in Computer and Information Science and Engineering from University of Florida, Gainesville, Florida, USA in August 2001. Dr. Lee has taught courses related to his research, including wireless and mobile networks, networked multimedia systems, and advanced computer networks. His achievement in research and teaching has resulted in grants/awards from US federal agencies such as National Science Foundation (NSF) and Department of Education.

Hoon-Jae Lee received his BS, MS, and PhD degrees in electronic engineering from Kyungpook National University, Daegu, Korea in 1985, 1987, and 1998, respectively. He is currently an associate professor in the School of Computer & Information Engineering at Dongseo University. From 1987 to till date, he was a research associate at the Agency for Defense Developmemt (ADD). He has more than 150 national/international technical publications as well as about 50 patents. His current research interests include developing secure communication system, secure Wireless Sensor Network, and Side-Channel Attack.

Shunn-Yuh Lee was born in Taichung, Taiwan, in 1966. He received the B.S. degree from the National Taiwan Ocean University, Chilung, Taiwan, in 1988, and the M.S. and Ph.D. degree from National Cheng Kung University, Tainan, Taiwan, in 1994 and 1999, respectively. Since 2002 and 2006, he has been an Assistant Professor and Associate Professor, respectively, at the Institute of Electrical Engineering, National Chung Cheng University, Chia-Yi, Taiwan. Currently, he the chairman of Heterogeneous Integration Consortium (HIC) sponsored by Ministry of Education, Taiwan. His present research activities involve the design of analog and mixed-signal integrated circuits including filter, high-speed ADC/DAC, and sigma-delta ADC/DAC, biomedical circuits and systems, low-power and low-voltage analog circuits, and RF front-end integrated circuits for wireless communications. Dr. Lee now is a member of Circuits and Systems (CAS) Society, Solid-State Circuits Society, Medicine and Biology Society (EMBS), and Communication Society of IEEE.

Clifford Lindsay is a fourth year Ph.D. student in the Computer Science department at Worcester Polytechnic Institute in Worcester, Massachusetts. He received his Bachelors of Science degree in Computer Science from the University of California, San Diego. He specializes in Computer Graphics and Computational Photography and has written several publications related to these topics. Currently he is working on designing and developing programmable camera framework for 3D capture, image processing and rendering.

David Linner has been a research scientist at the Fraunhofer Institute for Open Communication Systems FOKUS and at the department for Open Communication Systems at the Technical University of Berlin (TUB) since 2006. His interest in research developed over several years spent as a programmer and software developer at a young IT company. David Linner's connection with the Fraunhofer Institute

FOKUS goes back to his undergraduate days. His research fields include the development and implementation of mobile software platforms and innovative applications in the World Wide Web (WWW). In this capacity he has been head of development for projects in the field of Web 2.0, Mixed-Reality Gaming and Telco/Web convergence. He currently heads several research projects for the realization of components, platforms and protocols for a Web of the future.

Suleyman Malki studied at Lund University (Sweden) for his M.Sc degree. In 2003 he began to research Cellular Neural Network (CNN) implementations on Field-Programmable Gate-Arrays, leading to his Ph.D. Degree in 2008. During this period he also developed a number of vision-based applications, notably in motion measurement and person authentication. Currently he is connected to the BroadCom group of the Electrical and Information Technology Department of LTH, where he is investigating wireless communication technologies. Suleyman has co-authored around 20 reviewed publications.

Wolfgang De Meuter is a professor at the Vrije Universiteit Brussel. He has been active in the field of object-orientation since the early nineties. His research interests include programming languages and their evaluators, aspect-oriented programming, meta-programming and more recently also language constructs for ambient-oriented systems. He has organized numerous successful workshops at previous ECOOP's and OOPSLA's. In 2008 he was awarded the Dahl-Nygaard Prize for his contribution to object-oriented programming of ambient systems. His coordinates can be found at http://prog.vub.ac.be/~wdmeuter/WolfHome.

Stijn Mostinckx is a PhD student at the Programming Technology Laboratory of the Vrije Universiteit Brussel, Belgium. In the past five years, he has been involved in the design of AmbientTalk, an object-oriented programming language tailored for mobile ad hoc networks, and the Fact Space model which allows coordinating mobile applications using logic rules and a dynamically shared knowledge base. His current research focuses on developing a bridge between both models to allow applications to dynamically respond to the (dis)appearance of physical objects tagged with RFID tags. His coordinates can be found at http://prog.vub.ac.be/~smostinc.

Stefaan Mys received his M.Sc. degree in Informatics from Ghent University, Belgium in 2005. Since his graduation he has been working as a Ph.D. student at the Multimedia Lab of the Department of Electronics and Information Systems of Ghent University – IBBT (Belgium). His main research interest currently is distributed video coding. Previously, it also included error resilient video coding.

Joseph Kee-Yin Ng received a Ph.D. in Computer Science from the University of Illinois at Urbana-Champaign in 1993. Prof. Ng joined Hong Kong Baptist University in 1993, and is a Professor in the Department of Computer Science. His current research interests include Real-Time Networks, Multimedia Communications, Ubiquitous/Pervasive Computing, and Distributed Computing. Prof. Ng is the Chair of the Steering Committee of the International Conference series on Embedded and Real-Time Computing Systems and Applications (RTCSA) and he also served as Program Chairs or General Chairs for numerous International Conferences which include TENCON'06, RTCSA'05, ICPADS'05, AINA'05, AINA'04, ICSC'01, ICC'01, RTCSA'99, ICSC'99 and ICC'99. Prof. Ng has obtained 2 Patents and published over 120 technical papers in journals, conference proceedings, book chapters, and technical reports. Prof. Ng is a member of the Editorial Board of the Service Oriented Computing and Applications Journal, Journal of Pervasive Computing and Communications, Journal of Ubiquitous Computing and Intelligence, and Journal of Embedded Computing. He had been an Associate Editor of Real-Time Systems Journal, and currently Associate Editors for Journal of Systems Architecture, HKIE Transactions, and Journal of Mobile Multimedia.

Shojiro Nishio received his B.E., M.E., and Ph.D. degrees from Kyoto University, Japan, in 1975, 1977, and 1980, respectively. He has been a full professor at Osaka University since August 1992. At Osaka University, he served as the founding director of the Cybermedia Center from April 2000 to August 2003, and as the Dean of the Graduate School of Information Science and Technology from August 2003 to August 2007. Since August 2007, he has been serving as a Trustee and Vice President of Osaka University. His current research interests include database systems and multimedia systems. Dr. Nishio has served on the Editorial Boards of IEEE Transactions on Knowledge and Data Engineering and ACM Transactions on Internet Technology, and is currently involved with the editorial board of Data and Knowledge Engineering. Dr. Nishio is a fellow of IEICE and IPSJ, and is a member of six learned societies, including ACM and IEEE.

Phillip Olla is the endowed Phillips Chair of Management and Professor of MIS at the school of business at Madonna University in Michigan USA. His research interests include knowledge management, mobile telecommunication, and health informatics. In addition to University level teaching, Dr. Phillip Olla is also a Chartered Engineer and has over 10 years experience as an independent Consultant and has worked in the telecommunications, space, financial and healthcare sectors. He was contracted to perform a variety of roles including Chief Technical Architect, Program Manager, and Director. Dr. Olla is the Associate Editor for the Journal of Information Technology Research and the Software / Book Review Editor for the International Journal of Healthcare Information Systems and Informatics, and is also a member of the Editorial Advisory & Review Board for the **Journal of Knowledge Management Practice.** Dr. Phillip Olla has a PhD in Mobile Telecommunications from Brunel University in the UK, he is an accredited Press member of the British Association of Journalism, Chartered IT Professional with the British Computing Society and a member of the IEEE society.

Yanbo Ru received his B.S. degree from Huazhong University of Science and Technology, and his M.S. and Ph.D. degrees from University of Southern California, all in computer science. Dr. Ru's areas of expertise include web technology, information retrieval, online advertising, parallel processing, and cloud computing. He applies his research to problems of building business search engine and directory and pay-per-click advertising network, focusing on information extraction and classification and list generation. Dr. Ru is also a core developer of CloudBase project - an open source data warehouse system built on top of Map-Reduce architecture. Dr. Ru currently works at business.com Inc.

Christophe Scholliers is a PhD student at the Programming Technology Laboratory of the Vrije Universiteit Brussel, Belgium. His research interests are coordination, context awareness, concurrency and code mobility. He has been involved in the design and implementation of the context aware programming paradigm dubbed the fact-space model. He is currently working on language abstractions to deal with multi-hop communication in uncertain networks. His coordinates can be found at http://prog.vub.ac.be/~cfscholl.

Eliamani Sedoyeka is a PhD student at Anglia Ruskin University. His has research interests in new generation broadband wireless technologies and the way they can be utilized to improve business efficiency and quality of life. He is also interested in QoS and SLA issues as well as data technologies. He is also involved in location based services (LBS) and mobile learning researches.

Dhananjay Singh received his B.Tech. degree in Computer Science and Engineering from Purvanchal University, Jaunpur, 2003; and M. Tech. (IT) degree with specialization in Wireless Communication and Computing from Indian Institute of Information Technology, Allahabad, 2006, Uttar Pradesh, India. At present, he is pursuing his Ph.D degree in the Dept. of Ubiquitous IT, at Graduate School of Design & IT, Dongseo University, Busan, South Korea. He has more than 20 national/international technical publications as well as 3 international patents. His areas of interest are Ubiquitous Sensor Networks, IP-USN, Mobile Computing, MANETs and Signal Processing.

Jozef Škorupa received his M.Sc. degree in Mathematics from Comenius University, Slovakia, in 2004. In 2006 he joined the Multimedia Lab of the Department of Electronics and Information Systems of Ghent University – IBBT (Belgium) where he is currently working towards the Ph.D. degree. His research interests include distributed video coding and signal processing.

Jürgen Slowack received his M.Sc. degree in Engineering (Computer Science) from Ghent University, Belgium, in 2006. From then on, he has been working towards a Ph.D. in computer science at the Multimedia Lab of the Department of Electronics and Information Systems of Ghent University – IBBT (Belgium). His research interests include video coding with a special focus on Distributed Video Coding.

Joshua L. Smith received his Master's degree in computer science from the University of Illinois at Springfield, where he specialized in computer programming, in 2009. He was graduate assistant to the computer science department for the duration of his tenure at the University of Illinois at Springfield. Since graduation he has been accepted on at the University of Illinois at Springfield to teach as an adjunct professor teaching computer programming. While still attending college he was able to become president of the computer science club at the University of Illinois at Springfield. During this time his main area of research was in the development of supercomputers. After graduation his research area shifted towards video streaming mobile hand held devices. He is currently continuing to further his education in computer programming and software development by working as an independent consultant.

Lambert Spaanenburg is currently Professor in Silicon Systems at Lund University (Sweden). From 1988 to 1992 he was heading the Signal Processing Department of the Institute for Microelectronic in Stuttgart (Germany), where his group prototyped neural vehicle control for the Daimler Optically Steered Car OSCAR. In 2002 he moved to Groningen University in Groningen (The Netherlands) to found the Technical Computing Science. In 2002 his group split off Dacolian, a company in optical license-plate recognition, while he went to Sweden. Lambert has co-authored more than 200 reviewed publications and owns 6 patents. His research interest is in intelligent wireless systems and in mixed-signal architectures for fault-tolerant distributed systems in hardware/software co-implementations.

Antonia Stefani graduated from the University of Patras, Department of Mathematics in 1999. She received her MSc degree in Computer Science, "Mathematics on Computers and Decision Making", from the department of Mathematics and the department of Computer Engineering and Informatics of the University of Patras, in 2001. Her scientific area was "Foundations of Computer Science and its Applications in Automated Decision Making". She received the PhD degree in 2008 at Hellenic Open University, School of Science and Technology. Her research interests include software quality, software metrics, e-commerce and m-commerce systems.

Vassilios Stefanis (http://www.stefanis.net) received his diploma from the University of Patras, Computer Engineering and Informatics Department in 2005. He also received a MSc degree in "Computer Science and Engineering" from the same department in 2008 and now he is a PhD candidate. From 2005 he works as a researcher and software engineer to the research academic computer technology institute (RA-CTI) in Patras. Also, from 2008 he teaches as a Lecturer at TEI of Messolonghi, Department Of Applied Informatics in Management & Finance. His research interests include internet technologies, mobile web, mobile applications and p2p technologies.

Stephan Steglich is Head of the department "Future Applications and Media" (FAME) at Fraunhofer FOKUS. He received his M.Sc. (in 1998) and PhD (in 2003) in Computer Science from the TU Berlin. His fields of interest include context-awareness, user-interaction, and service front-ends. Currently Stephan is working in the area of next Generation Web platforms (WebX.0) and Web-Telco convergence with focus on converged media. Stephan is managing international and national level research activities and has been an organizer and a member of program committees of several international conferences. He has actively participated in standardization activities in these research areas and gives lectures in "Mobile Telecommunication Systems" and "Advanced Communication Systems" at the TU Berlin, chair for Open Communications Systems (OKS).

Daniel Tairo is a research student at the University of Greenwich. He has a masters degree in Data Warehousing and Data Mining. His research interests are in the integration of human activities into computer supported business processes using Web Services Business Process Execution Language (WS-BPEL). He is interested in the automation of business processes lifecycle from modeling to deployment and execution.

Rik Van de Walle received his M.Sc. and Ph.D. degrees in Engineering from Ghent University, Belgium in 1994 and 1998 respectively. After a visiting scholarship at the University of Arizona (Tucson, USA), he returned to Ghent University, where he became professor of multimedia systems and applications, and head of the Multimedia Lab of the Department of Electronics and Information Systems of Ghent University – IBBT (Belgium). His current research interests include multimedia content delivery, presentation and archiving, coding and description of multimedia data, content adaptation, and interactive (mobile) multimedia applications.

Fan Wu is currently an Assistant Professor in Computer Science Department at Tuskegee University in Tuskegee, AL. He received his B.S. and M.S. degrees in Computer Science from Nanjing University of Posts and Telecommunications, Nanjing, China, in 2000 and 2003 respectively and his Ph.D. degree in Computer Science from Worcester Polytechnic Institute in Worcester, MA in 2008. His research interests are in mobile graphics, ubiquitous computing, computer graphics and general purpose computation on graphics processor units. He has led research into mobile graphics and general-purpose computation on graphics processor units in Computer Science Department at Tuskegee University. He serves on the editorial review board for International Journal of Handheld Computing Research (IJHCR).

Shaoen Wu is an assistant professor in the School of Computing at the University of Southern Mississippi and the founder of wireless and mobile computing research program. Before joining USM, he worked as a research staff scientist on networking at ADTRAN Inc. He had been with Bell Labs China for over 3 years as a Member of Technical Staff. Dr. Wu has chaired multiple international conferences and serves in the editorial board of several journals. He has authored about 20 papers in conferences and journals and 3 patents.

Jianliang Xu is an associate professor in the Department of Computer Science, Hong Kong Baptist University. He received the BEng degree in computer science and engineering from Zhejiang University, Hangzhou, China and the PhD degree in computer science from the Hong Kong University of Science and Technology. He was a visiting scholar in the Department of Computer Science and Engineering, Pennsylvania State University, University Park. His research interests include data management, mobile/pervasive computing, wireless sensor networks, and distributed systems. He is a senior member of the IEEE.

Hung-Jen Yang received a BS in Industrial Education from the National Kaohsiung Normal University, an MS in Industrial Technology from the University of North Dakota, and a PhD in Industrial Education and Technology from the Iowa State University in 1984, 1989, and 1991, respectively. He is currently a professor in the Department of Industrial Technology Education and the director of the Center for Instructional and Learning Technology at the National Kaohsiung Normal University, Taiwan. His research interests include computer networks, automation, and technology education.

Ming Yang received his B.S. and M.S. degrees in Electrical Engineering from Tianjin University, China, in 1997 and 2000, respectively, and Ph.D. degree in Computer Science and Engineering from Wright State University, Dayton, Ohio, USA in 2006. He has been with Jacksonville State University as an Assistant Professor in Computer Science since 2006. His research interests include Digital Image/Video Coding, Multimedia Communication & Networking, and Information Security. He is the author/co-author of over twenty publications in leading computer science journals and conference proceedings. He also serves as reviewer of numerous international journals and conferences. He is currently leading a group in Jacksonville State University to conduct research on medical image security and privacy, H.264/AVC video coding, and video streaming over wireless networks.

Ming Yuan Yang received the B.Sc. degree in information engineering form Communication University of China, Beijing, China, in 2000 and his Phd in video coding standards from Loughborough University, UK in 2007. He has worked as a research fellow in video based computer vision related projects at the universities of Loughborough, Central Lancashire and West of Scotland. His main research interests include image and video coding and analysis, video streaming and transmission.

Yong Yao received his B.S. degree from Peking University, and his M.S. and Ph.D. degrees from Cornell University, all in computer science. Dr. Yao's areas of expertise include wireless sensor networks, distributive query processing and optimization, and XML databases. Dr. Yao built one of the first two wireless sensor network database systems, Cougar, with a full-fledge support to declarative query processing in sensor networks. He also made breakthrough contributions on adaptive query processing and power-efficient data retrieval techniques in ad-hoc networks. Dr. Yao has many publications in the form of book chapters, major international conference papers and elite journal papers. His papers have received world-wide recognition, and have been referenced more than 1,200 times by scientists around the world. Dr. Yao is a member of ACM and IEEE. Dr. Yao currently works at IBM Silicon Valley Lab.

Junyang Zhou received a B.Sc. in Applied Mathematics, a M.Sc. in Probability and Mathematic Statistics from Sun Yat-sen University in 2000 and 2003, and a Ph.D. in Computer Science from Hong Kong Baptist University in 2006, respectively. Dr. Zhou current research interests include: Mobile and Location-aware Computing, Ubiquitous/Pervasive Computing, Wireless Communication and Real-Time Networks. He is a member of the ACM and the IEEE.

Index

C

D

K

L

M

N

R

S

T

U

V

W

X

Y

Z